Won-Kee Hong is a Professor of Architectural Engineering at Kyung Hee University, Republic of Korea. He has more than 35 years of professional experience in structural and construction engineering, having worked for Englekirk and Hart, USA; Nihhon Sekkei, Japan; and Samsung Engineering and Construction, Korea. He is the author of *Artificial Neural Network-based Optimized Design of Reinforced Concrete Structures*, also published by CRC Press.

Artificial Neural Network-based Designs of Prestressed Concrete and Composite Structures

Won-Kee Hong

CRC Press is an imprint of the
Taylor & Francis Group, an **informa** business

Cover image: Won-Kee Hong, Shutterstock ©

First edition published 2023
by CRC Press
6000 Broken Sound Parkway NW, Suite 300, Boca Raton, FL 33487-2742

and by CRC Press
4 Park Square, Milton Park, Abingdon, Oxon, OX14 4RN

CRC Press is an imprint of Taylor & Francis Group, LLC

Library of Congress Cataloging-in-Publication Data
Names: Hong, Won-Kee (Professor of architectural engineering), author.
Title: Artificial neural network-based designs of prestressed concrete and
composite structures / Won-Kee Hong.
Description: First edition. | Boca Raton : CRC Press, [2024] |
Includes bibliographical references and index.
Identifiers: LCCN 2023003175 | ISBN 9781032408088 (hbk) |
ISBN 9781032408095 (pbk) | ISBN 9781003354796 (ebk)
Subjects: LCSH: Concrete construction. | Structural design—Data
processing. | Prestressed concrete. | Neural networks (Computer science).
Classification: LCC TA681.5 .H66 2024 |
DDC 624.1/8340285—dc23/eng/20230215
LC record available at https://lccn.loc.gov/2023003175

ISBN: 978-1-032-40808-8 (hbk)
ISBN: 978-1-032-40809-5 (pbk)
ISBN: 978-1-003-35479-6 (ebk)

DOI: 10.1201/9781003354796

Typeset in Sabon
by codeMantra

Contents

Preface xxi

1 Basic principles of prestressed structures based on Eurocode 2 **1**

1.1 Introduction 1
1.1.1 Why prestressing? 1
1.1.2 Prestressing principle 2
1.1.3 Pretensioning and posttensioning 3
1.1.3.1 Pretensioning 3
1.1.3.2 Posttensioning 4
1.1.4 Bonded and unbonded tendon 5
1.1.4.1 Bonded system 5
1.1.4.2 Unbonded system 6
1.1.5 A total length of tendon that can be pulled once 6
1.2 Materials 7
1.2.1 Concrete 7
1.2.1.1 Concrete strength 7
1.2.1.2 Concrete elasticity 8
1.2.1.3 Stress–strain relationship 8
1.2.1.4 Creep and shrinkage 10
1.2.2 Tendon 10
1.2.3 Rebar 12
1.3 Hyperstatic moments 13
1.3.1 General 13
1.3.2 Equivalent load 14
1.3.3 Hyperstatic moments due to secondary prestressing effects 16
1.3.3.1 Why do hyperstatic moments occur? 16
1.3.3.2 Effect of secondary moment caused by prestressing sequence 18

1.3.3.3 *Experimental investigations of effect of hyperstatic forces on a flexural capacity of the posttensioned frame* 21
1.4 *Prestress losses* 26
1.4.1 *Type of prestressing losses* 26
1.4.2 *Assessment of prestressing losses* 27
1.4.3 *Friction losses* 28
1.4.4 *Anchorage draw-in* 30
1.4.5 *Elastic shortening* 31
1.4.6 *Time-dependent losses* 32
1.4.7 *Shrinkage strain of concrete* 32
1.4.8 *Creep of concrete* 32
1.4.9 *Relaxation of the tendon* 33
1.4.10 *Simplified method to evaluate time-dependent losses* 33
1.4.11 *Jacking force* 33
1.5 *Flexural design* 34
1.5.1 *Section analysis* 34
1.5.2 *Flexural design at service limit stage* 35
1.5.2.1 *Stress control* 35
1.5.2.2 *Crack control* 37
1.5.2.3 *Deflection control* 40
1.5.3 *Flexural design at ultimate stage* 43
References 43

2 Holistic design of pretensioned concrete beams based on artificial neural networks **45**

2.1 *Introduction* 45
2.1.1 *Previous studies* 45
2.1.2 *Reverse design scenarios of pretensioned concrete beams* 46
2.1.3 *Significance of the Chapter 2* 46
2.1.4 *Tasks readers can perform after reading this chapter* 47
2.2 *Input and output parameters for forward designs* 48
2.2.1 *Description of input parameters* 48
2.2.2 *Network training* 49
2.2.2.1 *Description of output parameters* 49
2.2.2.2 *Generation of large structural datasets and network training* 51
2.3 *Design scenarios* 60

2.4 *Training methods* 62
2.4.1 TED 62
2.4.2 *Parallel training method (PTM)* 70
2.4.2.1 *Training accuracies and design tables* 70
2.4.2.2 *Design charts corresponding to ρ_p for forward design* 70
2.5 *Reverse designs based on PTM using deep neural networks (DNN)* 86
2.5.1 *Reverse design 1* 86
2.5.1.1 *Formulations of two steps based on back-substitution (BS) applicable to reverse designs* 86
2.5.1.2 *Design accuracies* 95
2.5.2 *Reverse design 2* 96
2.5.2.1 *Formulations of two steps based on back-substitution (BS) applicable to reverse designs* 96
2.5.2.2 *Design accuracies* 98
2.5.2.3 *Design charts* 113
2.5.3 *Reverse design 3* 113
2.5.3.1 *Formulations of two steps based on back-substitution (BS) applicable to reverse designs* 113
2.5.3.2 *Design accuracies by adjusting reverse input parameters to enhance design accuracies* 114
2.5.3.3 *Influence of layer and neurons on design accuracies* 130
2.5.4 *Designs based on CRS using deep neural networks (DNN)* 136
2.5.4.1 *Formulation of ANNs based on back-substitution (BS) applicable to reverse designs* 136
2.5.4.2 *Design accuracies* 139
2.5.4.3 *Design tables and charts* 145
2.5.4.4 *An influence of a number of layers-neurons for Step 1 on design accuracies* 147
2.5.5 *Design based on CRS using SNN* 159
2.5.6 *Comparison of design based on CRS with those based on PTM* 170
2.6 *Conclusions* 178
References 180

3 An optimized design of pretensioned concrete beams using an ANN-based Lagrange algorithm 183

3.1 Introduction and data generation following Eurocode 183
3.1.1 Previous research 183
3.1.2 Significance of the Chapter 3 183
3.1.3 Tasks readers can perform after reading this chapter 184
3.2 Large datasets generated following Eurocode 2 186
3.2.1 Selection of input parameters following Eurocode 186
3.2.1.1 Fourteen inputs parameters describing beam sections and material properties 186
3.2.1.2 Seven inputs parameters describing load and environment conditions 190
3.2.2 Selection of output parameters based on design requirements of Eurocode 191
3.2.2.1 Selection of output parameters at transfer stage (three outputs: Δ_{trans}/L, $\sigma_{c,trans}/f_{cki}$, w_{trans}*) 192*
3.2.2.2 Selection of output parameters at quasi-permanent stage (two outputs: Δ_{lt}/L, Δ_{incr}/L*) 192*
3.2.2.3 Selection of output parameters at a frequent stage (one output: w_{freq}*) 192*
3.2.2.4 Selection of output parameters at a characteristic stage (three outputs: σ_c/f_{ck}, $\sigma_{s,bot}/f_{yk}$, $\sigma_{p,bot}/f_{pk}$*) 192*
3.2.2.5 Selection of output parameters at an ultimate limit stage (ten outputs: M_{Ed}, M_{Rd}, M_{cr}, x/d, $\varepsilon_{s,bot}/\varepsilon_{sy}$, $\varepsilon_{p,bot}/\varepsilon_{py}$, V_{Ed}, $V_{Rd,max}$, $\rho_{sw,sp}$, $\rho_{sw,mid}$*) 193*
3.2.2.6 Selection of output parameters for design efficiency (two outputs) 194
3.2.3 Generation of Large datasets 194
3.2.3.1 Crack width 196
3.2.3.2 Deflections 197
3.2.3.3 Flexural strength 197
3.2.3.4 Shear strength 198
3.3 Derivation of objective functions based on forward ANNs 198
3.3.1 Training ANN 198
3.3.2 Formulation of an optimization for designs 198
3.3.2.1 Derivation of ANN-based objective functions 200

3.3.2.2 *Derivation of ANN-based Lagrange functions* 201
3.3.2.3 *Formulation of KKT conditions based on equality and inequality constraints* 203
3.4 *ANN-based design charts minimizing cost based on single objective function with varying live loads* 207
3.4.1 *A design example to minimize PT beam cost CI_b optimization when $q_L = 400\,kN/m$ ($c_{16} = q_L - 400 = 0$)* 207
3.4.2 *Designs criteria based on crack widths; $w_{SER} \le 0.2\,mm$ for $0\,kN/m \le q_L \le 50\,kN/m$ when CI_b is optimized* 208
3.4.3 *Design criteria based on tendon and concrete stresses ($\rho_{p,bot}/f_{pk} \le 0.75$ between $50\,kN/m \le q_L \le 625\,kN/m$ and $\sigma_c/f_{ck} \le 0.6$ between $200\,kN/m \le q_L \le 625\,kN/m$) when CI_b is optimized* 210
3.4.3.1 *Figure 3.4.2* 210
3.4.3.2 *Figure 3.4.3* 211
3.4.3.3 *Figure 3.4.4* 212
3.4.4 *Design criteria based on deflections for $0\,kN/m \le q_L \le 1{,}125\,kN$ when CI_b is minimized* 216
3.4.5 *Design criteria based on moment demand (M_{Ed}), moment resistance (M_{Rd}), and cracking moment (M_{Cr}) for $0\,kN/m \le q_L \le 1{,}125\,kN$ when CI_b is minimized* 217
3.4.6 *Design criteria based on reinforcement strains and neutral axis depth representing beam ductility for $0\,kN/m \le q_L \le 1{,}125\,kN$ when CI_b is minimized* 217
3.4.7 *Design criteria based on shear stirrup ratios required for $0\,kN/m \le q_L \le 1{,}125\,kN$ when CI_b is minimized* 218
3.4.8 *A Minimized CI_b for $0\,kN/m \le q_L \le 1{,}125\,kN$ verified by large datasets and structural mechanics-based design* 218
3.5 *ANN-based design charts minimizing weight based on individual objective function with varying live loads* 219
3.5.1 *Design criteria based on deflections ($\Delta_{INC} \le 24\,mm$ for $0\,kN/m \le q_L \le 150\,kN/m$) when W is minimized* 219

3.5.2 *Design criteria based on concrete stresses (σ_c/ $f_{ck} \leq 0.6$ for 50 kN/m $\leq q_L \leq$ 1,110 kN/m) and tendon stresses ($\sigma_{p.bot}/f_{pk}$=0.75 for 150 kN/m $\leq q_L \leq$ 1,110 kN/m) when W is minimized* 221
3.5.2.1 *Figure 3.5.2* 221
3.5.2.2 *Figure 3.5.3* 222
3.5.3 *Design criteria based on crack widths ($w_{SER} \leq$ 0.2 mm for 0 kN/m $\leq q_L \leq$ 1,110 kN) obtained when W is minimized* 222
3.5.4 *Design criteria based on moment demand (M_{Ed}), moment resistance (M_{Rd}), and cracking moment (M_{Cr}) for 0 kN/m $\leq q_L \leq$ 1,110 kN) when W is minimized* 224
3.5.5 *Design criteria based on reinforcement strains and neutral axis depth representing beam ductility obtained for 0 kN/m $\leq q_L \leq$ 1,110 kN/m when W is minimized* 224
3.5.6 *Design criteria based on shear stirrup ratios required for 0 kN/m $\leq q_L \leq$ 1,110 kN when W is minimized* 225
3.5.7 *Beam weights (W) obtained by the large datasets bounded on the lower weight obtained by the ANN-based optimized design charts* 227
3.6 *ANN-based design charts minimizing beam depths based on individual objective function with varying live loads* 228
3.6.1 *Design criteria based on deflections ($\Delta_{INC} \leq$ 24 mm for 0 kN/m $\leq q_L \leq$ 425 kN/m) when h is minimized* 228
3.6.2 *Design criteria based on concrete stresses obtained ($\sigma_c/f_{ck} \leq 0.6$ for 125 kN/m $\leq q_L \leq$ 1,110 kN/m) when h is minimized* 230
3.6.2.1 *Figure 3.6.2* 230
3.6.2.2 *Figure 3.6.3* 232
3.6.3 *Design criteria based on crack widths; $w_{SER} \leq$ 0.2 mm for 0 kN/m $\leq q_L \leq$ 1,110 kN when h is minimized* 232
3.6.4 *Design criteria based on moment demand (M_{Ed}), moment resistance (M_{Rd}), and cracking moment (M_{Cr}) for 0 kN/m $\leq q_L \leq$ 1,110 kN when h is minimized* 233

3.6.5 *Design criteria based on reinforcement strains and neutral axis depth representing beam ductility for 0 kN/m ≤ q_L ≤ 1,110 kN when h is minimized* 234
3.6.6 *Design criteria based on shear stirrup ratios for 0 kN/m ≤ q_L ≤ 1,110 kN when h is minimized* 235
3.6.7 *Cost and beam weight obtained for 0 kN/m ≤ q_L ≤ 1,110 kN when h is minimized* 235
3.7 *Formulation of ANN-based design charts for pre-tensioned (bonded) concrete beams* 238
3.8 *Design recommendations* 240
3.9 *Conclusions* 241
References 242

4 Multi-objective optimizations (MOO) of pretensioned concrete beams using an ANN-based Lagrange algorithm 245

4.1 *Introduction* 245
4.1.1 *Previous studies* 245
4.1.2 *Motivations of the optimizations based on MOO designs* 246
4.2 *Contributions of the proposed method to PT beam designs* 246
4.2.1 *Significance of the Chapter 4* 246
4.2.2 *Tasks readers can perform after reading this chapter* 248
4.2.3 *What can be changed with future PT beam designs* 249
4.3 *Large datasets generated following Eurocode* 250
4.3.1 *Selection of input parameters following Eurocode* 250
4.3.1.1 *Fourteen forward inputs parameters describing beam sections and material properties* 250
4.3.1.2 *Seven forward inputs parameters describing load and environment conditions* 251
4.3.2 *Selection of forward output parameters based on design requirements of Eurocode* 253
4.3.3 *Bigdata generation* 253
4.4 *MOO design of a PT beam* 257
4.4.1 *Derivation of objective functions based on forward ANNs* 257
4.4.2 *An optimization scenario* 258

4.4.3 *Derivation of a unified function of objective (UFO) for a PT beam* 258
4.5 *A MOO-based PT beam design* 259
4.5.1 *Formulation of an UFO* 259
4.5.2 *Generation of tradeoff ratios contributed by individual objective function* 260
4.5.3 *Negligible errors* 261
4.5.4 *Pareto frontiers based on designated tradeoff ratios* 261
4.5.5 *Optimized objective functions* 262
4.5.5.1 *Two and three-dimensional Pareto frontier* 262
4.5.5.2 *Beam sections among Designs 1, 2, and 3* 265
4.5.5.3 *Beam sections for Design 4* 266
4.5.5.4 *Verification of Pareto frontier* 266
4.5.6 *Influence of three objective functions on concrete sections* 267
4.5.6.1 *Verification of design parameters (b, $\rho_{s,bot}$, $\rho_{s,top}$, and $\rho_{p,bot}$) corresponding to optimized objective functions* 267
4.5.6.2 *Delaying cracks with weight fractions of $w_h : w_{CI_b} : w_{Weight} = 0.4865 : 0.477 : 0.0364$ based on highest cracking moment ($M_{cr} = 1.25M_{E.ser}$)* 268
4.5.6.3 *Design 2 with the deepest concrete section (h = 1,500 mm) leading to an uncracked design based on cracking moment capacity ($M_{cr} = 1.13M_{E.ser}$) and effective camber moment ($M_p = 1.23M_{E.ser}$)* 268
4.5.6.4 *Cracked sections with Designs 1, 3, and 4 and uncracked sections with Design 2 under service loads* 268
4.5.7 *Verification of optimized design parameters ($\rho_{s,bot}$, $\rho_{s,top}$, and $\rho_{p,bot}$) on an output-side* 269
4.6 *Findings and Recommendations* 269
4.6.1 *Practical applications of MOO for PT designs* 269
4.6.2 *Verification of MOO* 269
4.7 *Formulation of ANN-based Pareto frontier for pre-tensioned (bonded) concrete beams* 270
4.8 *Conclusions* 272
References 273

5 **Reverse design charts for steel-reinforced concrete (SRC) beams based on artificial neural networks** 277

5.1 *Introduction* 277
 5.1.1 *Previous studies* 277
 5.1.2 *Significance of Chapter 5* 279
5.2 *Artificial intelligence-based design scenarios for SRC beams* 280
5.3 *Generation of large datasets* 283
 5.3.1 *Cost index (CI_b) and CO_2 emissions of concrete and metals for of SRC beams* 283
 5.3.2 *Generation of large datasets* 284
  5.3.2.1 *Big data generator* 284
  5.3.2.2 *Ranges of input parameters* 284
  5.3.2.3 *100,000 structural datasets based on a flowchart* 286
  5.3.2.4 *Means, standard deviations, variances, and histograms of randomly selected design input and output parameters* 289
 5.3.3 *ANN-based mapping input to output parameters* 289
5.4 *Networks based on back substitution* 292
 5.4.1 *Direct and back-substitution (BS) method* 292
 5.4.2 *Direct design method and enhanced accuracies of back-substitution (BS) method* 293
5.5 *Reverse scenarios* 293
 5.5.1 *Reverse Scenario 1* 293
  5.5.1.1 *An ANN formulation* 293
  5.5.1.2 *Training in two steps for back-substitution method* 318
  5.5.1.3 *Verifications* 319
 5.5.2 *Reverse Scenario 2* 319
  5.5.2.1 *An ANN formulation* 319
  5.5.2.2 *Training in two steps for back-substitution method* 320
  5.5.2.3 *Verifications* 320
 5.5.3 *Reverse Scenario 3* 321
  5.5.3.1 *An ANN formulation* 321
  5.5.3.2 *Training in two steps for back-substitution method* 321
  5.5.3.3 *Verifications* 322
 5.5.4 *Reverse Scenario 4* 323

5.5.4.1 *An ANN formulation* 323
5.5.4.2 *Training in two steps for back-substitution method* 324
5.5.4.3 *Verifications* 324
5.5.5 *Development of design charts* 325
5.5.5.1 *Design charts for reverse scenario 4* 325
5.5.5.2 *A design moment (ϕM_n)* 326
5.5.5.3 *A tensile rebar strain (ε_{rt})* 326
5.5.5.4 *A safety factor (SF)* 326
5.5.5.5 *Effective depths (d) and tensile rebar ratios (ρ_{rt}) vs. tensile rebar strains (ε_{rt})* 327
5.5.5.6 *Tensile steel strains (ε_{st}) and curvature ductilities (μ_ϕ) vs. tensile rebar strains (ε_{rt})* 328
5.5.5.7 *Immediate and long-term deflections (Δ_{imme} and Δ_{long}) vs. tensile rebar strains (ε_{rt})* 329
5.5.5.8 *A cost index of an SRC beam (CI_b), and CO_2 emissions, and beam weight (BW) vs. tensile rebar strains (ε_{rt})* 330
5.5.6 *Application of design charts* 331
5.6 *Conclusions* 342
References 343

6 An optimization of steel-reinforced concrete beams using an ANN-based Lagrange algorithm **347**

6.1 *Introduction and significance of chapter 6* 347
6.1.1 *Previous studies* 347
6.1.2 *Innovations and significances* 348
6.2 *ANN-based design scenarios for SRC beams* 348
6.3 *Generation of large datasets* 350
6.4 *Formulations of objective functions using ANN-based Lagrange algorithm for SRC beams* 352
6.4.1 *Training ANN on each output parameter based on a parallel training method (PTM)* 352
6.4.2 *Optimized objective functions using ANN-based Lagrange algorithm* 354
6.4.2.1 *Formulation of objective functions based on ANNs* 354
6.4.2.2 *Ten equality and 16 inequality constraints* 360

6.4.3 *Optimization designs of SRC beams based on ANN-based Lagrange forward network* 361
6.4.3.1 *An optimized design for costs of SRC beams (CI_b)* 361
6.4.3.2 *An optimized design for CO_2 emissions of SRC beams* 364
6.4.3.3 *An optimized design for beam weights (W) of SRC beams* 367
6.4.3.4 *Overall designs efficiencies optimizing objective functions of CI_b, CO_2 emission, and W based on ANN-based Lagrange algorithm* 370
6.5 *Conclusion* 375
References 376

7 Multi-objective optimization (MOO) for steel-reinforced concrete beam developed based on ANN-based Lagrange algorithm **381**

7.1 *Introduction and significance of chapter 7* 381
7.1.1 *Introduction* 381
7.1.2 *Research significance* 382
7.2 *ANN-based design scenarios for SRC beams* 383
7.3 *Generation of large datasets* 384
7.4 *MOO using ANN-based LAGRANGE algorithm for SRC beams* 391
7.4.1 *Training ANNs based on PTM* 391
7.4.2 *Optimized objective functions using ANN-based Lagrange algorithm* 393
7.4.2.1 *Derivation of objective functions based on forward neural networks* 393
7.4.2.2 *Ten equality and 16 inequality constraints* 394
7.4.2.3 *Derivation of a unified function of objective (UFO) for a SRC beam* 395
7.4.3 *Results of optimization MOO of SRC beams* 396
7.4.3.1 *Definition of four specific cases on a Pareto frontier based on a three-dimensional Pareto frontier* 396
7.4.3.2 *Four design cases verified by big datasets* 402

7.4.3.3 *Design case 1 only optimizing costs CI_b based on Pareto frontier projected on CI_b –W and CI_b –CO_2 planes* 402
7.4.3.4 *Design case 2 only optimizing CO_2 emissions based on Pareto frontier projected on CO_2 – W and CI_b – CO_2 planes* 402
7.4.3.5 *Design case 3 only optimizing weights (W) based on Pareto frontier projected on CI_b – W and CO_2 – W planes* 403
7.4.3.6 *Design case 4 with evenly weight fractions $w_b : w_{CI_b} : w_{Weight}$ = 1/3 : 1/3 : 1/3 based on Pareto frontier projected on CI_b – W, CO_2 – W, and CI_b – CO_2 planes* 403
7.4.3.7 *Verifications* 406
7.5 *Conclusions* 407
References 407

8 An ANN-based reverse design of reinforced concrete columns encasing H-shaped steel section **411**

8.1 *Introduction* 411
8.1.1 *Previous studies* 411
8.1.2 *Motivations of the optimizations* 411
8.2 *Contributions of the proposed method to SRC column designs* 413
8.2.1 *Research significance based on novelty and innovation* 413
8.2.2 *Innovations readers will learn in this chapter* 414
8.3 *Reverse design-based optimizations* 417
8.3.1 *Generation of large structural datasets* 417
8.3.2 *BS procedure to perform reverse design* 420
8.3.3 *Network training for four reverse outputs (h, b, h_s, and b_s) based on BS-CRS* 424
8.4 *Reverse designs to minimize CI_c, CO_2, W_c* 424
8.4.1 *Reverse design using BS method* 424
8.4.2 *Reverse design-based optimization using back-substitution (BS) method* 425
8.4.3 *Optimized designs of the CI_c and CO_2 emission* 427
8.4.4 *Optimized designs of W_c* 430
8.5 *Conclusion* 432
References 434

9 Design optimizations of concrete columns encasing H-shaped steel sections under a biaxial bending using an ANN-based Lagrange algorithm 437

9.1 Introduction and significance of this study 437
9.1.1 Research background 437
9.1.2 Research innovations and significances 438
9.1.3 Tasks readers can perform after this chapter 438
9.2 SRC column designs using ANNs and Lagrange multipliers 440
9.2.1 Design scenario for SRC columns 440
9.2.2 Objective and constraining functions for Lagrange optimizations 442
9.2.3 Generation of large structural datasets 444
9.2.4 Network training based on PTM 444
9.3 Design optimization using the Lagrange multiplier method 448
9.3.1 Formulation of the Lagrange function 448
9.3.2 Derivation of AI-based objective functions for CI_c, CO_2 emissions, and weight (W_c) and their Lagrange functions to optimize 451
9.3.3 Formulation of active and inactive conditions 456
9.4 Optimization verifications 457
9.4.1 Optimized cost (CI_c) of an SRC column 457
9.4.1.1 Optimized cost (CI_c) verified by structural design software (AutoSRCHCol) 457
9.4.1.2 Column interaction diagram corresponding to an optimized CI_c 458
9.4.1.3 Verification of optimized CI_c based on ANN by large datasets 460
9.4.2 Optimized CO_2 emissions of an SRC column 462
9.4.2.1 Optimized CO_2 emissions verified by structural design software (AutoSRCHCol) 462
9.4.2.2 Column interaction diagram corresponding to an optimized CO_2 emissions 463
9.4.2.3 Verification of an optimized CO_2 emissions based on ANN by large datasets 463
9.4.3 Optimized column weight (W_c) of an SRC column 466

9.4.3.1 *Optimized column weight (W_c) verified by structural design software (AutoSRCHCol) 466*
9.4.3.2 *Column interaction diagram corresponding to an optimized column weight (W_c) 467*
9.4.3.3 *Verification of an optimized column weight (W_c) based on ANN by large datasets 467*
9.5 *Conclusion 470*
References 472

10 A Pareto frontier using an ANN-based multi-objective optimization (MOO) for concrete columns encasing H-shaped steel sustaining multi-biaxial loads 477

10.1 *Introduction and Significance of the Chapter 10 477*
10.1.1 *Research background 477*
10.1.2 *Research objectives and innovations 478*
10.1.3 *Research significance 479*
10.1.4 *Tasks readers can perform after this chapter 479*
10.2 *Design of SRC columns sustaining multiple biaxial loads optimizing UFO 481*
10.2.1 *SRC column section for ANN-based design scenario 481*
10.2.2 *Generation of large datasets 483*
10.2.3 *Training ANNs on each output parameter based on PTM 484*
10.3 *ANN-based column design scenario 486*
10.3.1 *Fourteen forward input parameters and 11 forward output parameters for a design 486*
10.3.2 *ANN-based Lagrange multi-objective optimization (MOO) 488*
10.3.3 *Formulation of weigh matrices for concrete columns encasing H-shaped steel sustaining multi-biaxial loads 490*
10.3.4 *Generalized ANN-nLP subject to n multiple biaxial load pairs based on network module 493*
10.3.4.1 *Generalized ANN-nLP from ANN-1LP 493*
10.3.4.2 *Formulation of generalized ANN-nLP considering two biaxial load pairs by network duplication 496*

10.4 Design of an SRC column encasing H-shaped steel section sustaining three biaxial load pairs 500
10.4.1 Design scenario 500
10.4.1.1 Selection of design parameters and ranges based on structural analysis 500
10.4.1.2 Establishing equality and inequality constraints imposed by design codes 500
10.4.2 Formulation of ANN forward network subjected to three biaxial load pairs 503
10.4.3 Normalized UFO capturing three objective functions 504
10.4.4 ANN-based Lagrange optimization based on UFO 505
10.4.4.1 Five steps for UFO optimization 505
10.4.4.2 Two gradient vectors $\nabla f(x)$, $\nabla c(x)$ defining a contact of functions based on KKT conditions 508
10.4.4.3 Newton–Rapson iteration to solve for constraining stationary points optimizing UFO based on KKT conditions 509
10.4.5 ANN-based Pareto frontier based on Lagrange optimizations 511
10.4.5.1 Verification of Pareto frontier 511
10.4.5.2 A design results optimizing UFO based on tradeoff ratios 513
10.4.6 Discussion of the diversity of the proposed ANN-based Lagrange algorithm 516
10.5 Conclusions 516
References 518

Index 523

Preface

It is difficult to intuitively find optimized solutions, while satisfying all code requirements simultaneously. Engineers facing several code-restricted design requirements commonly make design decisions based on empirical observations. Mathematician Joseph-Louis Lagrange proposed a method of Lagrange multipliers to identify local maxima and minima of objective functions subject to one or more equality constraints. However, this great idea faces a predicament today because the explicit formulations of objective and constraining functions are complex when it is applied to structural engineering areas where design targets called objective functions are optimized while satisfying all constraints imposed by code requirements and engineer's interests.

An application of an optimization to structural designs has been overlooked, being missed in enhancing design balances and related economy. In this book, an artificial neural network (ANN)-based Lagrange method provides a breakthrough in optimized structural designs. A novel way to perform an optimization of structural designs of prestressed concrete and composite structures is introduced in this book. Hong et al. published several papers, highlighting the importance of optimizations for structural designs in general, and prestressed concrete and composite structures in particular. This book titled *Artificial Neural Network-based Designs of Prestressed Concrete and Composite Structures* is to introduce a design including an optimization of prestressed concrete and composite structures to readers. Readers will have opportunities to learn how an ANN-based structural design can be implemented in designing and optimizing designated objectives and targets.

In Chapters 1–4, pretensioned concrete (PT) beams are holistically designed and optimized based on Lagrange function using artificial neural networks. In the proposed method, optimized engineering designs meet various code restrictions imposed by equality and inequality constraints individually or simultaneously. Design charts for any type of PT beams with different design parameters can be constructed. Especially, Chapter 4 aims to assist structural engineers and decision-makers to design PT beams based on a unified function of objective (*UFO*), resulting in optimizing three objective functions, namely, construction and material cost, beam weights, and beam depth ($\boldsymbol{CI_b}$, **W**, and $\boldsymbol{h}$), simultaneously, which may conflict each other.

In Chapters 5–7, steel-reinforced concrete (SRC) beams are holistically designed and optimized using artificial neural networks, in which Chapter 5 is devoted to reverse design charts for SRC beams based on artificial neural networks, whereas Chapter 6 presents an optimization using an ANN-based Lagrange algorithm. Chapter 7 is written for multi-objective optimization (MOO) based on ANN-based Lagrange algorithm.

In Chapters 8–10, reinforced concrete columns encasing H-shaped steel section are holistically designed and optimized using artificial neural networks. Chapter 8 is devoted to a reverse design based on artificial neural networks, whereas Chapter 9 presents an optimization based on an ANN-based Lagrange algorithm. Chapter 10 is prepared for a Pareto frontier using an ANN-based MOO for concrete columns encasing H-shaped steel sustaining multi-biaxial loads. Explicit objective and constraint functions with respect to design variables are replaced by ANN-based functions obtained using the proposed method. An optimization using ANN-based objective functions obtained by being trained on large datasets accompanies design parameters optimizing the multiple objective functions at simultaneously, resulting in designs that can meet various code restrictions at the same time. Optimized results were verified by conventional structural calculation using design parameters obtained by ANNs and brute force approach.

The author would like to appreciate his students, Nguyen Dinh Han, Tien Dat Pham, Thuc-Anh Le, Manh Cuong Nguyen, and Van Tien Nguyen for their contribution to the birth of this book. The author would like to thank for the support of the National Research Foundation of Korea (NRF) grant funded by the Korean government [MSIT 2019R1A2C2004965]. Finally, the author could not have published this book without the unflagging spiritual support of his wife Debbie, his son David, and daughter-in-law Sharon. His family has been with him during the tough time of the preparation of this book. The author reminds himself of Psam 23:6 at the end; goodness and love will follow me all the days of my life, and I will dwell in the house of the Lord forever.

Chapter 1

Basic principles of prestressed structures based on Eurocode 2

1.1 INTRODUCTION

1.1.1 Why prestressing?

Prestressed concrete structures are commonly used in building structures, in which it can produce a longer span, prevent cracks, reduce structural thickness, and save cost compared to reinforced concrete structures. A main function of prestressing is to provide initial compressive stresses for concrete members to counteract tensile stresses induced by imposed loads, making structural members more resistant to loads, shock, and vibration than conventional concrete. This technique allows to construct a lighter and shallower concrete structure without sacrificing strength. Prestressing steel can be located either inside concrete members (internal prestressing) or outside the members (external prestressing). Internal prestressed member is generally classified into two main types:

- Pretensioned concrete: prestressing steel is directly bonded to concrete.
- Posttensioned concrete: prestressing steel can be either bonded or unbonded to concrete.

Prestressed concrete offers an advantage over reinforced concrete in that only a small prestressed concrete member is required for given structural frames and loading conditions. Savings from self-weight loads are notably significant in long-span structures like bridges, stations, and halls, where a self-weight load adds up a large percentage of a total load. A reduction in self-weight load can also be a significant factor in reducing foundation costs, especially in areas of poor soil quality. Moreover, in cases where a total height of a building is limited, a reduction of story height caused by prestressing techniques would allow a construction of additional stories within a limited height. A further benefit of prestressed concrete is that it can prevent cracks with appropriate prestressing, which is important to provide durability and resist impact, vibration, and shock for infrastructures including liquid-retaining structures, piperack frames, and buildings.

DOI: 10.1201/9781003354796-1

In contrast to the advantages of prestressed concrete discussed above, some disadvantages also exist. Prestressed concrete sections are mainly compressed under service load due to prestressing, so any inherent problems such as long-term deflection resulting from creeps can be exacerbated. From a construction point of view, the prestressing process requires complicated tensioning equipment, anchoring devices, and skilled labors, which may not be available in some remoted areas.

1.1.2 Prestressing principle

Primary effects of prestress are caused by tendon force and upward load, so-called equivalent load, due to eccentricities of prestressing tendon with beam centroids at every section as shown in Figure 1.1.1. Upward loads can carry part of applied downward load, such as dead load, and live load, directly to supports. On the other hand, compressive stress not only reduces

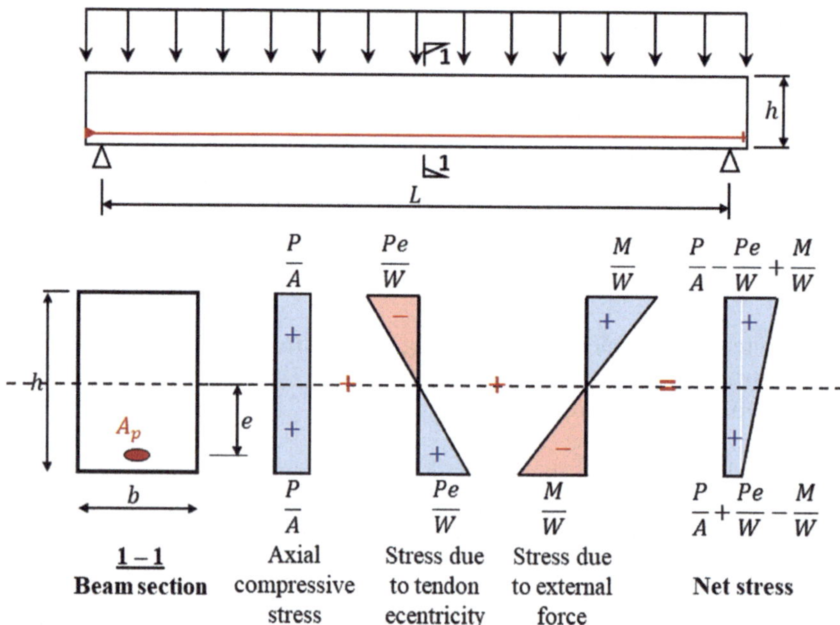

Figure 1.1.1 Principle of restress effects.

tensile stress in concrete but also enhances shear resistance of prestressed beam. Both precompression and upward loads caused by prestressing reduce tensile stresses in concrete as shown in Figure 1.1.1, thus reducing deflections and cracking under service loads. An amount of prestress is designed to prevent all or part of tensile cracking depending on environmental condition and structural type. For example, a structure should be designed with uncracked section under service conditions when environmental condition is highly corrosive.

1.1.3 Pretensioning and posttensioning

Tendons can be stressed either before or after concrete casting, referred to as pretensioning or posttensioning, respectively.

1.1.3.1 Pretensioning

In "pretensioning", strands are stressed against external anchors before casting concrete as shown in Figure 1.1.2, allowing bonded contact between strands and concrete. As the concrete hardens, external anchors are released, allowing prestressing forces to be transferred to concretes only based on a bond mechanism between strands and surrounding concretes. This method is particularly well suited for precast structures, in which concrete members are fabricated remotely and then delivered to the work site once cured. Casting beds, including anchorages at two ends, are often much longer than precast concrete members, allowing multiple members to be prestressed in one pretensioning session as shown in Figure 1.1.2. However, profiles of pretensioned strands are usually straight lines as shown in Figure 1.1.2. External deflecting devices are necessary to bend pretensioned strands as a

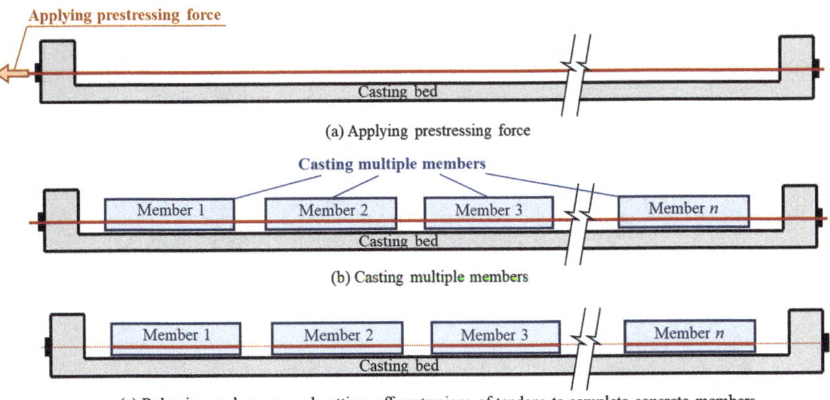

Figure 1.1.2 Pretensioned concrete.

series of straight lines. However, installing deflecting devices slows down production and considerably increases manufacturing costs.

1.1.3.2 Posttensioning

The term "posttensioning" refers to a type of prestressed concretes in which prestressing force is applied by jacking tendons after concrete are cast. Tendons are encased in ducts, or sheaths, which allow them to move freely within ducts or sheaths as shown in Figures 1.1.3 and 1.1.4. Upon reaching

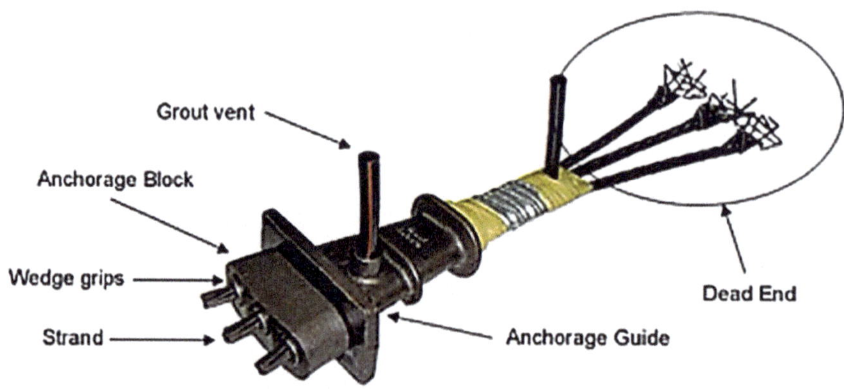

Figure 1.1.3 Bonded posttensioning. (Utracon system [1.1].)

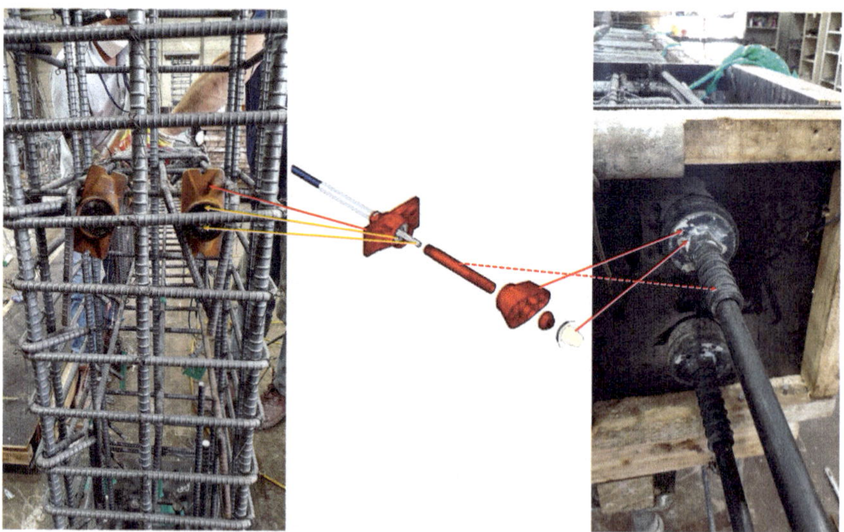

Figure 1.1.4 Unbonded posttensioning. (Utracon system [1.1].)

sufficient strength of concrete, tendons are stressed directly against the concrete, transferring prestressing force to the concrete through anchorages embedded in concrete. Concentrated forces applied through anchorages cause the concrete around anchorages to be under extreme stress. Additional reinforcement is necessary around anchorages for concrete to be kept from splitting. Posttensioning has an advantage over pretensioning in that the tension shape can be selected in a flexible way, such as a parabolic, harped, trapezoidal profile, or any combinations of them. Another advantage of posttensioning over pretensioning is that tensioning can be applied in phases with some tendons or with all. This is useful when loads are applied in phases that are clearly defined such as when designing transfer beams.

1.1.4 Bonded and unbonded tendon

Both bonded and unbonded tendons referred to Figures 1.1.3 and 1.1.4, respectively, may be used for posttensioned systems. Both methods have their own pros and cons. Below are some points to consider.

1.1.4.1 Bonded system

For a bonded system shown in Figure 1.1.3, strands are installed inside galvanized steel or plastic ducts placed at the desired profile in the concrete section, creating a void where the strands will enter. Once strands have been stressed, void around strands is grouted with a high-strength cement, providing ful bonds between tendons and surrounding concretes. Ducts and strands contained within the ducts are referred to as a tendon.

The following are the main features of a bonded system:

- The full strength of the strand can be utilized at the ultimate limit state thanks to strain compatibility of strand and concrete, and hence, there is a lower requirement to use unstressed reinforcement.
- Once ducts have been grouted, anchorages are less critical since a tendon force can be transferred to a concrete section through bonded mechanisms.
- Shear capacity of a concrete section can increase not only by precompression force but also by a bonded tendon itself.
- A tendon that is accidentally damaged results in a localized loss of prestressing force.
- A high friction loss between strands and galvanized steel ducts is observed.
- As hydraulic jacks for multi-strand tendon are too heavy for manual handling, crane time is needed during a stressing operation. In addition, most bonded systems including grouting equipment require cranes to move grouting equipment from one position to another.

1.1.4.2 Unbonded system

In an unbonded system shown in Figure 1.1.4, individual steel strands are encapsulated in a polyurethane sheath and voids between sheaths and strands are filled with a rust-inhibiting grease. Sheaths and grease are installed under controlled factory conditions, whereas a completed tendon undergoes an electronic test ensuring the process has been carried out satisfactorily. Each of these tendons is attached to an anchorage system at each end. Once a concrete section has attained a required strength, tendons are jacked to apply design prestressing force.

The following are the main features of an unbonded system:

- With smaller tendon diameters and a reduced need for cover, it is possible to increase an eccentricity from a neutral axis to a centroid of a tendon, thereby requiring less prestressing force.
- Compared to bonded tendons, friction loss is lower with unbonded tendon because grease inhibits friction.
- Unbonded tendons are flexible and easily bendable in horizontal direction to conform to curved structures.
- Unbonded tendon can be fabricated off-site.
- Unbonded tendons can be replaced (typically with a smaller diameter).
- Prestress along the entire length of an unbonded tendon is lost in broken tendon, and hence, design must be carefully performed to avoid progressive collapse of unbonded system.

1.1.5 A total length of tendon that can be pulled once

A tendon can be jacked in one end or two ends depending on friction losses along a tendon length. As recommended in Technical Report 43, a total length of tendon is restricted as shown in Table 1.1.1.

Longer lengths are achievable, but friction losses should be carefully considered.

Table 1.1.1 Recommended jacking length of tendons (TR43, 1994 [1.2])

Tendon type	*Structural type*	*Single-end stressed*	*Double-end stressed*
Bonded	Beam	<25 m	<50 m
	Slab	<35 m	<65 m
Unbonded	Beam	-	-
	Slab	<45 m	<100 m

1.2 MATERIALS

Various materials are used in prestressed structures, including concrete, reinforcement, and high-strength steel strands.

1.2.1 Concrete

Specifications for concrete should be made in accordance with relevant standards such as ACI 318-19, Eurocode 2, TCVN 5574-2018, etc. Concrete types and grades are chosen according to an expected service life, early strength gain requirements, material availability, and economic considerations. In reinforced concrete, strength and durability are the most important properties; other properties such as strength development, elastic modulus, shrinkage, and creep are secondary importance. In contrast, high modulus of elasticity, low shrinkage, and low creep are desirable properties for prestressed concrete.

1.2.1.1 Concrete strength

Concrete with a strength of 30–50 MPa is most commonly used in prestressed structures. It becomes possible to use a shallower section with a higher strength to reduce stresses, which may be more economical even if unit cost of concrete is high. Material stiffness is not directly proportional to flexural strength, and hence, a shallow slab of concrete with high strength will deflect more than a deeper slab of concrete with lower strength. For prestressed structures, concrete is designed to develop high early strength prior to early stressing, which reduces the risk of shrinkage cracks. It is recommended that concrete must possess a minimum strength at which prestressing can be applied. Concrete strength at transfer of prestress typically reaches 20 MPa or above after 4–7 days. As concrete ages, its strength increases, but the rate of increase is strongly influenced by cure conditions and constituent properties. Eurocode 2 provides Section 3.1.2 as a general guidance for the development rate of concrete strength. In reinforced concrete design, the tensile strength of concrete is not of great significance, as its contribution to bending resistance is ignored. Tensile strength of prestressed concrete, on the other hand, is important, which is used for controlling cracks and deflection. A tensile strength of concrete can be estimated by Eq. (1.2.1) (Table 3.1 in Eurocode 2 (Eurocode 2, 2004) [1.3]).

$$f_{ctm} = 0.3 f_{ck}^{(2/3)} \qquad \text{for } f_{ck} \le 50 \text{ MPa} \tag{1.2.1-1}$$

$$f_{ctm} = 2.12 \ln\left(1 + \left(f_{cm}/10\right)\right) \qquad \text{for } f_{ck} > 50 \text{ MPa} \tag{1.2.1-2}$$

1.2.1.2 Concrete elasticity

Known as Young's modulus, the modulus of elasticity of concrete (E_{cm}) is an indicator of a short-term strain produced by an applied stress in concrete. Concrete's modulus of elasticity (E_{cm}) is a key consideration when estimating deflections, as well as when predicting prestress losses, caused by elastic shortening of the concrete. By using compression tests to determine E_{cm}, the same value is assumed to be true in tension. Eurocode 2 proposes a relationship between E_{cm} and mean compressive strength of concrete (f_{cm}) for general applications as shown in Eq. (1.2.2) (Table 3.1 of Eurocode 2, 2004 [1.3]). The structures should, however, be assessed separately if they are likely to be affected by deviations from these general values.

$$E_{cm} = 22\left(f_{cm}/10\right)^{0.3}\ (\text{GPa})\left(f_{cm} \text{ in MPa}\right) \tag{1.2.2}$$

Time-dependent modulus of elasticity can be estimated by Eq. (3.5) of Eurocode 2.

1.2.1.3 Stress–strain relationship

A parabola–rectangle stress–strain model for a concrete compression follows (Eurocode 2 (Eurocode 2, 2004) [1.3]). The behavior of the concrete in tension is assumed as linear elastic as illustrated in Figure 1.2.1. For the concrete in tension, the stress can be computed based on Eq. (1.2.3).

$$\sigma_{ct} = E_c \varepsilon_{ct} \tag{1.2.3}$$

For a concrete in compression, a stress–strain relationship is composed of two phrases as shown in Eq. (1.2.4).

$$\sigma_c = f_{ck}\left[1-\left(1-\frac{\varepsilon_c}{\varepsilon_{c2}}\right)^n\right] \text{ when } 0 < \varepsilon_c < \varepsilon_{c2} \tag{1.2.4-1}$$

$$\sigma_c = f_{ck} \text{ when } \varepsilon_{c2} < \varepsilon_c < \varepsilon_{cu2} \tag{1.2.4-2}$$

where n is an exponent.

$$n = 2 \text{ for } f_{ck} \le 50 \text{ MPa} \tag{1.2.5-1}$$

$$n = 1.4 + 23.4\left[\left(\frac{90 - f_{ck}}{100}\right)\right]^4 \text{ for } f_{ck} > 50 \text{ MPa} \tag{1.2.5-2}$$

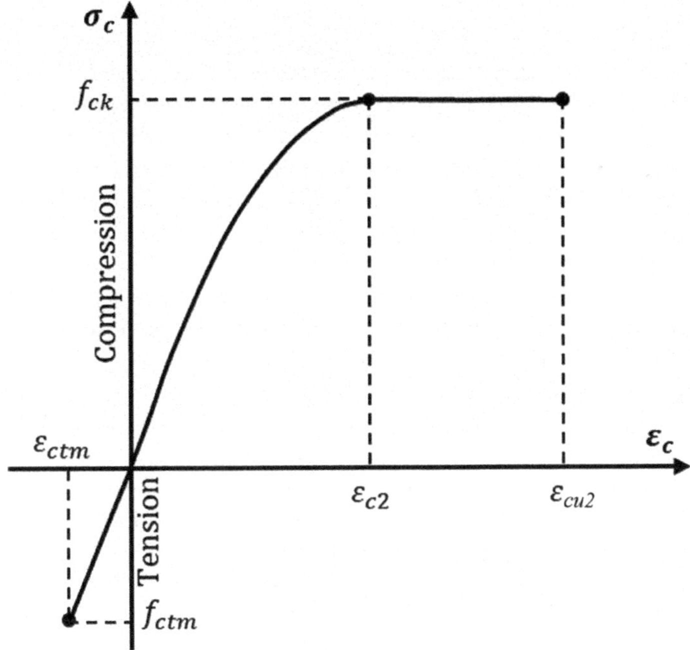

Figure 1.2.1 Stress–strain relationship for concrete in tension and compression. (Nguyen and Hong, 2022 [1.4] and STANDARD, British. Eurocode 2: Design of concrete structures. Part, 2004, 1.1: 230.)

ε_{c2} is the compressive strain reaching peak stress as shown in Eq. (1.2.6).

$$\varepsilon_{c2} = 0.002 \text{ for } f_{ck} \leq 50 \text{ MPa for } f_{ck} \leq 50 \text{ MPa} \tag{1.2.6-1}$$

$$\varepsilon_{c2} = 0.002 + 8.5 \times 10^{-5} \left(f_{ck} - 50\right)^{0.53} \text{ for } f_{ck} > 50 \text{ MPa} \tag{1.2.6-2}$$

ε_{cu2} is the ultimate compressive strain in concrete as shown in Eq. (1.2.7).

$$\varepsilon_{cu2} = 0.0035 \text{ for } f_{ck} \leq 50 \text{ MPa} \tag{1.2.7-1}$$

$$\varepsilon_{cu2} = 0.0026 + 0.035 \left[\left(\frac{90 - f_{ck}}{100}\right)\right]^4 \text{ for } f_{ck} > 50 \text{ MPa} \tag{1.2.7-2}$$

1.2.1.4 Creep and shrinkage

1. **Creep**

   Creep is a significant factor in prestressed structures, which influences both long-term deflections and prestress losses. The basic mechanism of creep in concrete is that of gradual change in a length of concrete under a sustained load. Deformations due to creep occur mostly often in the direction in which a force is applied, such as when a concrete column is subjected to compressions, or when a beam bends. Prestressed concrete members exhibit more creep than reinforced concrete members since they are more compressed due to prestress. Figure 3.1 of Eurocode 2 provides a value for creep coefficients where great accuracy is not required. For more information, including how creep develops over time, see Eurocode's Annex B.2.1.4.2.

   The creep effects in prestressed concrete can be more influential than those of conventional concrete structures due to higher compression stresses. Creeps shall be considered in calculating stress losses and long-term deflections of prestressed members. Empirically, stress losses due to creep can be taken at around 5% and 6% for posttensioned and pretensioned concrete structures [1.5].

   ACI Multiplier Method can be implemented in calculating the effects of creeps and other time-dependent factors, which should be considered in determining long-term deflections [1.6, 1.7].

2. **Shrinkage**

   Concrete members will shrink as leftover water that has not been used to hydrate the cement evaporates. It is dependent on environmental conditions surrounding concrete and external loads applied to the member that determines an amount of shrinkage which occurs. Loss of moisture will be much greater if a concrete is located in a hot dry climate than if it is kept in a moist environment. Section 3.1.4 of Eurocode 2 outlines that shrinkage strains are composed of two components: drying shrinkage strains and autogenous shrinkage strains. Drying shrinkage strains occur gradually as it results from a passage of water through hardened concretes. Autogenous shrinkage strains exhibit itself during a hardening of concretes, which is why shrinkage occurs mainly in the early days following casting. Both drying shrinkage and autogenous shrinkage strains can be calculated in accordance with Section 3.1.4 and Annex B of Eurocode 2.

1.2.2 Tendon

In general, wires have diameters between 3 and 7 mm and a carbon content between 0.70% and 0.85%. Hot-rolled rods are used to make wires, which are heated to 1,000°C and then cooled until they are drawable. The wires are wound on to capstans after being drawn several times in order to boost their tensile strengths. A tendon normally consists of one or multiple seven-wire strands. Six wires are helical wound around a seventh center wire as

Figure 1.2.2 **7-wire strand.**

Table 1.2.1 Specification of commonly used strands

Technical Data

Strand Type		0.5"		0.6"		0.62"		
Code/Specification		BS 5896	ASTM A416	BS 5896	ASTM A416	BS 5896	BS 5896	ASTM A416
		Y1860S7	Grade 270	Y1860S7	Grade 270	Y1770S7	Y1860S7	Grade 270
Nominal diameter	mm	12.9	12.7	15.2	15.2	15.7	15.7	15.7
Ultimate strength, f_{pk}	N/mm²	1860	1860	1860	1860	1770	1860	1860
Cross-sectional area	mm²	100	98.7	139	140	150	150	150
Weight	g/m	781	780	1086	1100	1172	1172	1200
Ultimate load	kN	186	184	259	261	266	279	279
Yield load, $f_{p0.1k}$	kN	164(1)	165.3(2)	228(1)	234.6(2)	234(1)	246(1)	251.4(2)
Modulus of elasticity	N/mm²	~ 195,000						
Relaxation(3) after 1000 hours	%	Max. 2.5						

note:
(1) Yield measured at characteristic value of 0.1% proof load.
(2) Yield measured at 1% extension.
(3) Applicable for relaxation Class 2 according to BS 5896 and SS475: or low relaxation complying with ASTM A416.

shown in Figure 1.2.2. The properties of the various types of strands are summarized in Table 1.2.1.

A typical stress–strain relationship for strands is shown in Figure 1.2.3. The line is linear at low stress levels, and then, it begins to curve at yield point indicating an onset of non-proportional strains even if a yielding point is not clearly defined. According to Eurocode 2, a yielding strength of tendon is taken as a characteristic value of a 0.1% proof load at which 0.1% permanent strain occurs as shown in Figure 1.2.3. In design, the bi-linear stress–strain relationship illustrated in Figure 1.2.4 is used for prestressing steel as shown in Eq. (1.2.8).

$$\sigma_p = E_p\varepsilon_p \text{ when } \varepsilon_p \le \varepsilon_{p0.1k} = f_{p0.1k}/E_p \tag{1.2.8-1}$$

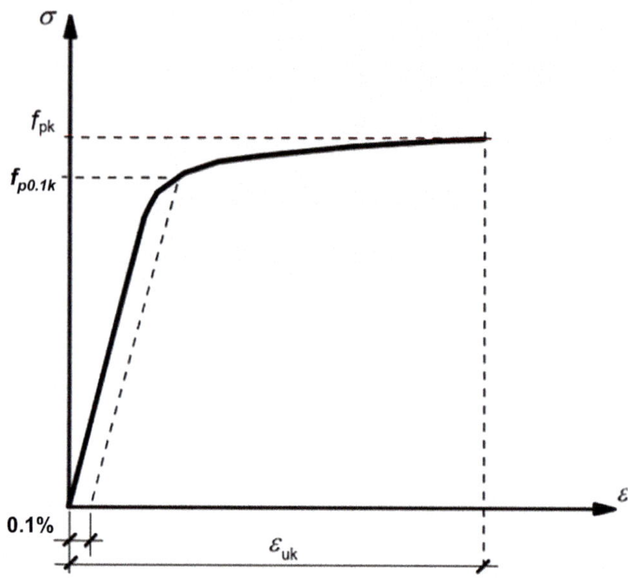

Figure 1.2.3 Stress–strain relationship for typical prestressing strand. (Eurocode 2, 2004 [1.3].)

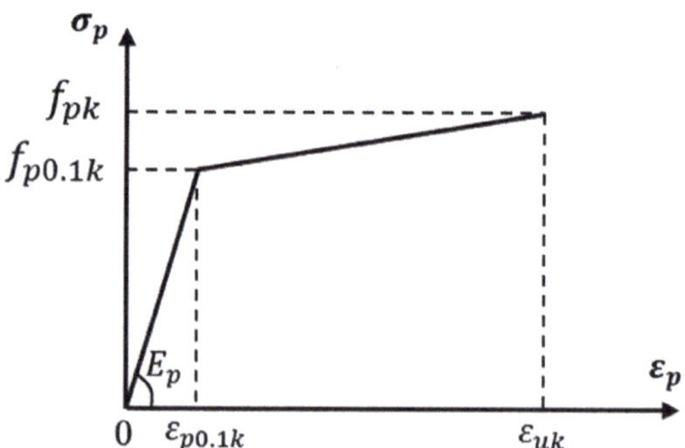

Figure 1.2.4 Bi-linear stress–strain relationship for prestressing strands.

$$\sigma_p = f_{p0.1k} + \frac{f_{pk} - f_{p0.1k}}{\varepsilon_{uk} - \varepsilon_{p0.1k}}\left(\varepsilon_p - \varepsilon_{p0.1k}\right) \text{ when } \varepsilon_{p0.1k} < \varepsilon_p \leq \varepsilon_{uk} \qquad (1.2.8\text{-}2)$$

1.2.3 Rebar

Bi-linear stress–strain relationship of rebars is given in Eq. (1.2.9).

$$\sigma_s = E_s\varepsilon_s \text{ when } \varepsilon_s \leq \varepsilon_{sy} \qquad (1.2.9\text{-}1)$$

$$\sigma_s = f_{sy} + \frac{f_{su} - f_{sy}}{\varepsilon_{su} - \varepsilon_{sy}}(\varepsilon_s - \varepsilon_{sy}) \text{ when } \varepsilon_{sy} < \varepsilon_s \leq \varepsilon_{su} \tag{1.2.9-2}$$

1.3 HYPERSTATIC MOMENTS

1.3.1 General

It is ideal for tendon profiles to produce bending moment diagrams that are similar in shape but have an opposite sign of moments generated by applied loads. In practice, this is not always possible due to various loading conditions and geometric restrictions of structures. As shown in Figure 1.3.1, a variety of shapes of tendon profiles are used to produce equivalent loads similar to external loads. Figure 1.3.1(a) shows an ideal tendon profile for a two-span beam. In the first span where a load is uniformly distributed, the tendon profile is parabolic to generate an upward distributed equivalent

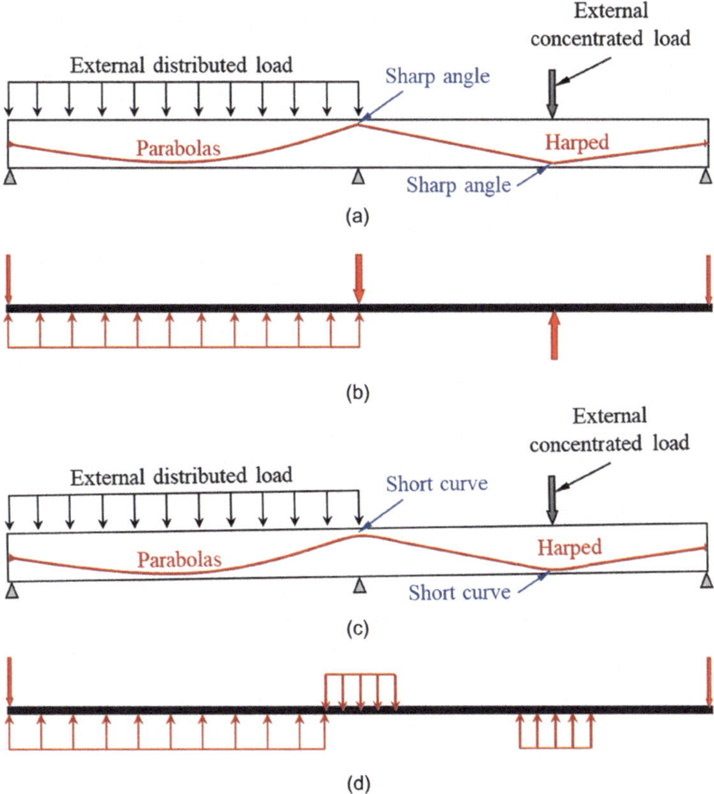

Figure 1.3.1 Tendon profile on a continuous beam. (a) Ideal tendon profile with parabola and sharp angles. (b) Equivalent load of ideal tendon profile with parabola and sharp angles. (c) Tendon profile with short curves. (d) Equivalent load of ideal tendon profile with parabola and short curves.

load. In the second span of Figure 1.3.1(a), a triangular harped shape produces an upward concentrated equivalent load to counteract an externally concentrated load at mid-span. Triangular shapes at support also generate downward equivalent loads, passing directly to the supports to counteract an externally concentrated load at support. Figure 1.3.1(b) shows an equivalent load corresponding to the profile shown in Figure 1.3.1(a). Despite the equivalent loads obtained by the profile shown in Figure 1.3.1(a), there is no way to make an ideal tendon profile because tendons are difficult to be bent at sharp angles. As shown in Figure 1.3.1(d), a parabola with short curves shown in Figure 1.3.1(c) is placed at each corner of the tendon profile to provide equivalent loads.

1.3.2 Equivalent load

Most prestressed structural design software can calculate tendon geometry, equivalent loads, and secondary moments. Nonetheless, understanding equivalent loads generated by different profiles and their combinations can be of great help in designing prestresses. A concentrated upward force is required to maintain the force equilibrium caused by a deflected tendon as shown in Figure 1.3.2.

If a constantly curving tendon is used, a distributed force needs to be applied to maintain a tendon in its intended position. The equivalent load in a curved tendon can be calculated by examining a small but finite section of a tendon in Figure 1.3.3.

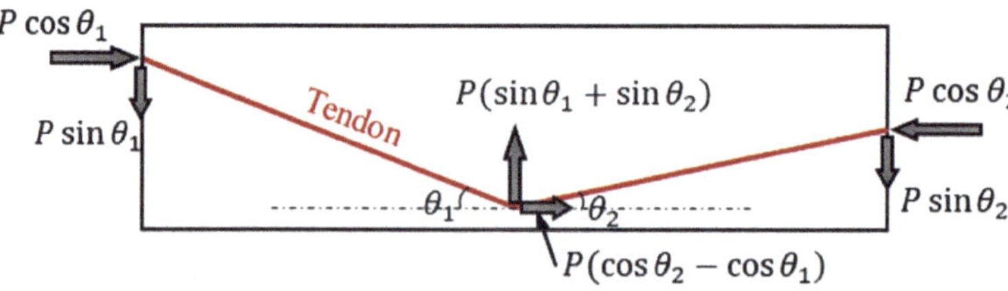

Figure 1.3.2 General harped profile.

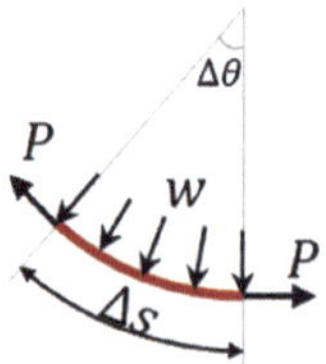

Figure 1.3.3 Small length of tendon.

Neglecting friction forces between tendons and surrounding concretes, forces in a tendon at both ends of an element are equal to a jacking force, P. A uniformly distributed load on a tendon required to maintain it in position is determined based on Eq. (1.3.1).

$$w\Delta s = 2P\sin(\Delta\theta/2) \tag{1.3.1}$$

For small angular changes, $\sin(\Delta\theta/2) = \Delta\theta/2$. If an element is made smaller and smaller, w can be calculated using Eq. (1.3.2), where $r = ds/d\theta$ is a radius of curvature.

$$w = P\frac{d\theta}{ds} \tag{1.3.2-1}$$

$$w = P/r \tag{1.3.2-2}$$

Theoretically, a uniformly distributed load on a tendon is oriented toward the curvature center, as shown in Figure 1.3.3. In practice, most tendon profiles are relatively flat, so forces can be assumed to be vertical at every point. An equation for a parabolic tendon profile is given in Eq. (1.3.3) if x and y coordinates are given with a left-hand origin as shown in Figure 1.3.4.

$$y = 4ax(L - x)/L^2 \tag{1.3.3}$$

For a relatively flat curve, $1/r$ may be approximated by d^2y/dx^2 as shown in Eq. (1.3.4).

$$1/r = d^2y/dx^2 = -8a/L^2 \tag{1.3.4}$$

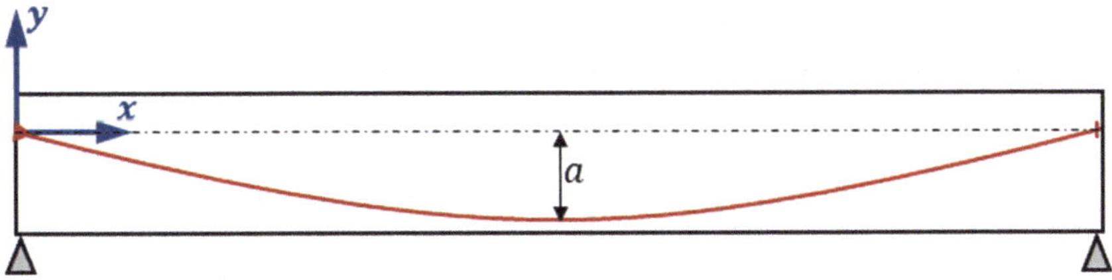

Figure 1.3.4 General parabolic profile.

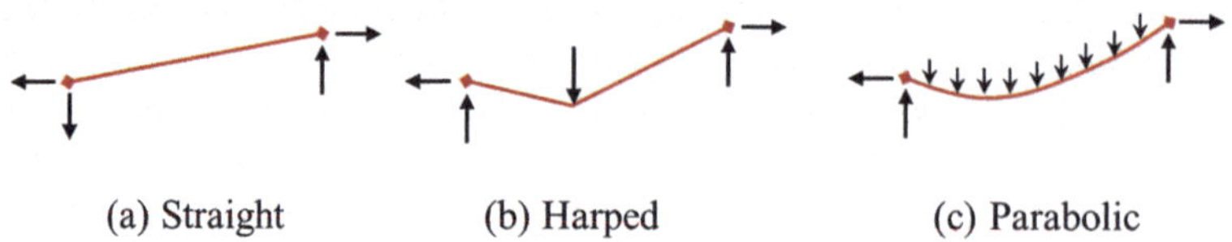

Figure 1.3.5 Basic tendon profiles.

An equivalent load caused by a parabolic profile can be obtained based on Eq. (1.3.5).

$$w = P/r = -8Pa/L^2 \tag{1.3.5}$$

where w is an upwards load.

There are distinct load patterns associated with each profile. Unlike a straight tendon, a harped profile tendon imparts a concentrated load on concrete members, while a parabolic tendon imparts a uniformly distributed load as summarized in Figure 1.3.5.

1.3.3 Hyperstatic moments due to secondary prestressing effects

1.3.3.1 Why do hyperstatic moments occur?

Secondary prestressing effects resulting in hyperstatic moments must be considered when big datasets are generated for an AI-based design of prestressed frames with and without rigid joints (statically indeterminate frames). Primary effect due to prestressing forces causes a beam to bend, shorten, deflect, and rotate as shown in Figure 1.3.8(a), whereas hyperstatic moments due to secondary prestressing effects try to restore shapes deflected by primary effect due to prestressing forces as shown in Figure 1.3.8(b) and (f) when connections of prestressed frames are restrained as rigid. Hyperstatic moments due to secondary prestressing effects can be unwanted, and hence, should always be considered in a design of rigid frames. For frame shown in Figure 1.3.8(b) and (f), hyperstatic beam moments are positive moments throughout the entire beam length, which cause negative effect at mid-span, adding bending moments to those caused by applied vertical loads at mid-span, but reducing bending moments at supports. A combination of axial forces and upward forces generated by tendons in beams are known as 'equivalent' or 'balanced' loads, in which upward forces counterbalance all or a portion of the downward forces due to dead and live loads. Equivalent loads automatically establish both primary and secondary prestressing effects-based hyperstatic moments when prestressing forces are applied to rigid structures. It is only needed to subtract primary effects from total prestressing effects calculated by equivalent loads to obtain hyperstatic moments due to secondary prestressing effects.

Similarly, moments due to secondary prestressing effects (called hyperstatic moments or secondary moments) in columns shown in Figure 1.3.8(b) are identical to column moments caused by equivalent loads shown in Figure 1.3.8(c) because primary moments in columns are zero, as shown in Figure 1.3.8(a). It is noted that frame moments caused by equivalent loads shown in Figure 1.3.8(c) are equal to secondary moments (hyperstatic moments) shown in Figure 1.3.8(b) plus primary moments (zero moments for columns and non-zero moments for beam) shown in Figure 1.3.8(a). In a serviceability design, a separation of primary and secondary hyperstatic effects is not needed since effects of prestressing forces are treated as applied upward loads, counteracting applied downward loads such as dead load and live load. However, effects of primary and secondary hyperstatic should be separated in strength design because secondary hyperstatic effects are considered benefited or not when structural configurations with and without rigid joints are to be maintained, whereas primary prestressing effects are considered as tendon forces always contributing to ultimate section capacity. The total prestressing effects (comprising of both primary effects and hypostatic effects) are regarded as imposed loads in service designs and are consequently included in each load combination. In service designs, stresses and deflections are checked. Thus, prestressing effects (comprising of both primary effects and hyperstatic effects) are regarded as imposed loads in service designs because they cause stresses and deformations in structures. In the strength designs, the classification of primary effects as either sectional strengths or imposed loads depends on the selected design approaches. Let's clarify the design concepts by establishing several definitions:

M1 - Section strength without prestressing moments
Mu1 - Moments induced by factored imposed loads, excluding all prestressing moments
Mpr - Primary moments induced by prestressing effects
Mhp - Hyperstatic moments induced by prestressing effects

Under Design approach 1, primary effects are regarded as sectional strengths (Sectional strength = M1 + Mpr), while the imposed load consists of Mu1 and Mhp (Factored load = Mu1 + Mhp). Design requirements are met when the sectional strength is greater than or equal to the factored load (M1 + Mpr $\geq$ Mu1 + Mhp).

Alternatively, Design approach 2 considers primary effects as imposed load (Factored load = Mu1 + Mhp – Mpr), with sectional strengths equivalent to M1 (Sectional strength = M1). Designs meet the criteria when the sectional strength is greater than or equal to the factored load (M1 $\geq$ Mu1 + Mhp – Mpr).

It is noted that, both design approaches yield the same outcomes. However, Design approach 1 offers a more straightforward calculation of sectional strengths. Consequently, Design approach 1, which treats primary effects as sectional strengths and hyperstatic effects as imposed loads, is conventionally employed in designs.

1.3.3.2 Effect of secondary moment caused by prestressing sequence

Two erection sequences are available in prestressing construction. In Case 1 of Figure 1.3.6, where frames are erected and prestressed level by level in which statically indeterminate frames have weak constraints. Constraining conditions are maintained mostly by hyperstatic moments in beams, while small hyperstatic moments occur with columns because columns are not stiff enough as beams when prestressing forces are applied for Case 1. Beams are mostly responsible for maintaining constraint conditions, producing larger secondary effects of hyperstatic moments in beams, but smaller secondary effects of hyperstatic moments in columns. However, in Case 2 shown in Figure 1.3.7, prestressing forces are applied after an entire erection of the frames, which are statically indeterminate having strong constraints provided by columns. Columns are stiff enough to maintain constraining conditions in Case 2, and hence, constraints are mostly maintained by hyperstatic moments in columns, while smaller secondary effects of hyperstatic moments are acting in beams. Note that the hyperstatic moments in columns are acting opposite against moment due to vertical loads as shown in Figure 1.3.8(b) and (e), and therefore, hyperstatic moments are beneficial to column designs for both Cases 1 and 2.

Secondary moments need to be considered with statically indeterminate frames (refer to Cases 1 and 2 of Figure 1.3.6 and 1.3.7, respectively) to maintain the configuration with redundant reactions at joints. Redundant reactions are caused by redundant constraints when prestressing forces are applied to statically indeterminate frames. These additional moments are called hyperstatic moments which do not exist with statically determinate frames because there are no constraining conditions to maintain at joints, and hence, no redundant reactions exist when prestressing forces are applied to statically determinate frames in which prestressing forces are entirely used to cause camber. In the case of statically determinate frames, primary moments due to prestressing forces ($M=pe$) due to eccentricities of tendons exist, while secondary hyperstatic moments do not exist in which beam ends rotate. However, ends of statically indeterminate beams do maintain their rigid configurations, being required to rotate back when the joints are rigid, because they are constrained and fixed. As shown in statically indeterminate frames of Case 1 of Figure 1.3.6, constraining conditions must be maintained by extra beam moments when prestressing forces are applied to destroy an original beam configuration having redundant constraints. Hyperstatic forces are the forces that prevent camber from occurring. The extra hyperstatic forces must be available in addition to the primary flexural loads ($M=p\times e$) which occur due to eccentricities of tendons. Hyperstatic forces occurring in columns for Cases 1 are also beneficial for a beam design at beam ends even if they are small. However, extra hyperstatic forces acting downwards require extra rebars to be added at mid-section for beam designs for Case 1, whereas hyperstatic moments in beams for Case 2 are small which do not influence a beam design. Hyperstatic forces occur mainly in beams for Case 1, whereas hyperstatic forces occur mainly in columns for

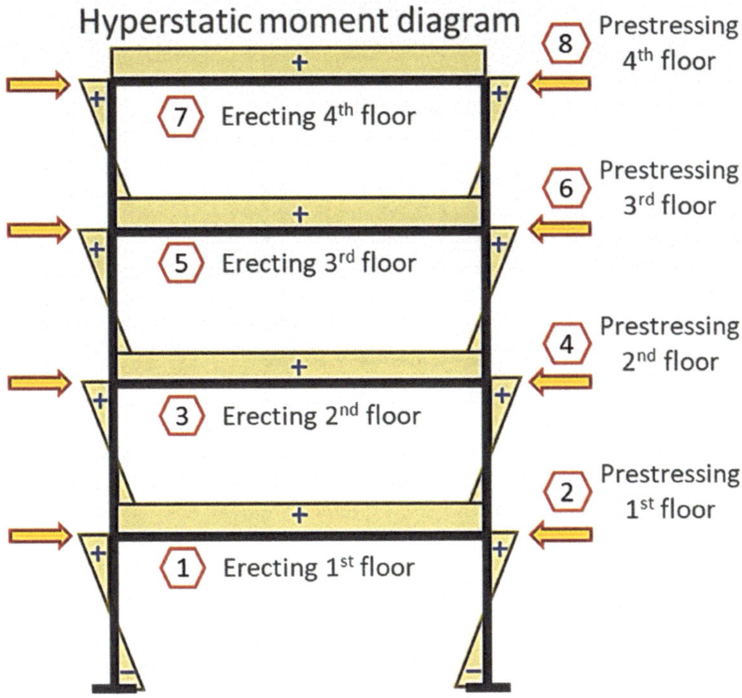

Figure 1.3.6 Secondary hyperstatic moment in posttensioned frame, which is erected and prestressed level by level (PT software like ADAPT (2021) cannot do this); hyperstatic forces prevent camber from occurring. Hyperstatic forces are mainly in beam in Case 1, whereas in Case 2, hyperstatic forces are mainly in column.

Case 2. Hyperstatic forces occurring mainly in columns for Case 2 are beneficial to column designs because the hyperstatic moments in columns are acting opposite against moment due to vertical loads as shown in Figure 1.3.8(b) and (e). Hyperstatic moments are beneficial to column designs for both Cases 1 and 2.

It is noted that PT software like Adapt (2012) cannot design Case 1 shown in Figure 1.3.6 in which beams are posttensioned level by level. They can only calculate secondary effect of hyperstatic moments for Case 2 shown in Figure 1.3.7 where prestressing forces of all floors are applied after an entire erection of frames is completed, capable of designing posttensioned frames accordingly. Software like Adapt (2012) erects entire floors first and applies posttension. When off-site frames are pretensioned level by level, PT software capable of erecting level by level must be used.

Case 3 is for pretensioned beams which do not develop hyperstatic moments. There are no secondary effects occurring with pretensioned beams. PT software like Adapt (2021) cannot consider an effect of

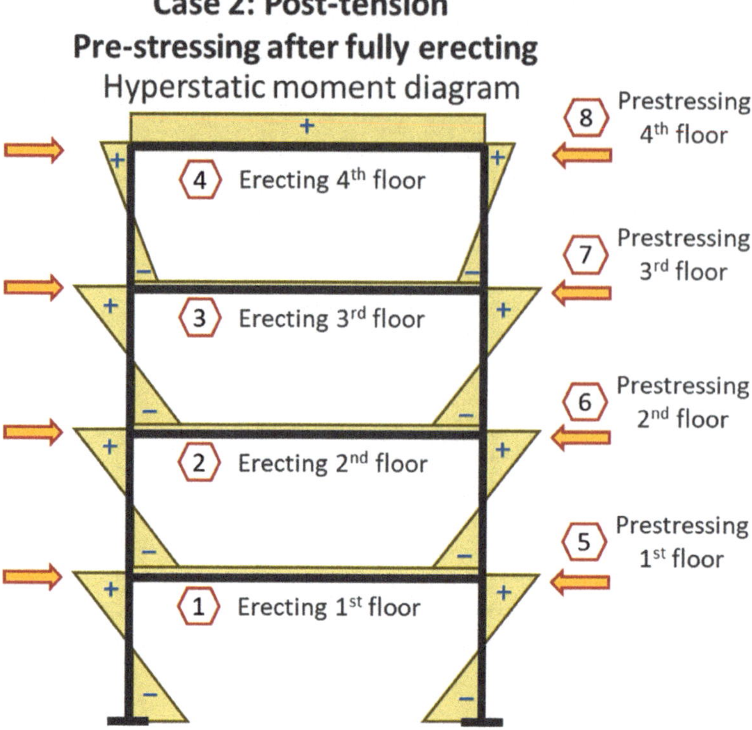

Figure 1.3.7 Secondary hyperstatic moment in posttensioned frame, which is prestressed after an entire erection of frames (PT software like ADAPT (2021) can do this); hyperstatic forces prevent camber from occurring. Hyperstatic forces are mainly in beam in Case 1, whereas in Case 2, hyperstatic forces are mainly in column.

development length, and hence, assuming full bonding lengths for development length design, overestimating bonding stresses between tendon and concrete when full bonding lengths are assumed. Slippage significantly degrades the strength of beams, thus, must be considered, indicating that PT software which cannot consider an effect of development length are incapable of designing pretensioned beams when development length needs to be precisely considered. Note that some PT software only consider global secondary effect of hyperstatic moments for Case 2, and hence, they cannot design pretensioned beams for Case 3. MATLAB-based code developed to generate large structural datasets for artificial neural network-based designs introduced in this book can apply posttensioning after an entire erection of frames for Case 2 shown in Figure 1.3.7. However, code needs to be revised if frames are posttensioned level by level shown in Case 1 shown in Figure 1.3.6. Pretension can be assumed to be applied level by level when all frames are prestressed off-site and erected level by level on-site.

1.3.3.3 Experimental investigations of effect of hyperstatic forces on a flexural capacity of the posttensioned frame

Total effects of prestressing forces consist of primary and secondary hyperstatic effects as shown in Figure 1.3.8. Primary effect due to prestressing forces causes a beam to bend, shorten, deflect, and rotate, whereas hyperstatic moments due to secondary prestressing effects restore shapes

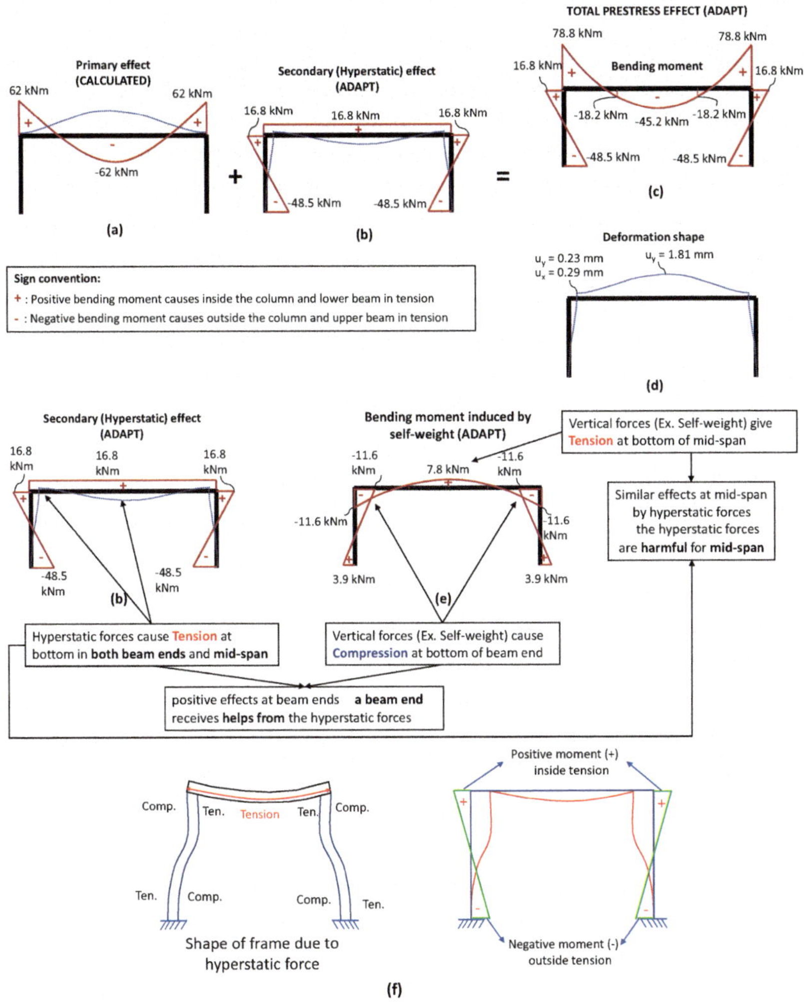

Figure 1.3.8 Primary and secondary hyperstatic effects caused by prestressing forces. (a) Primary effect. (b) Secondary (hyperstatic) effect calculated by Adapt. (c) Total prestressing effect calculated by Adapt. (d) Deformation shape. (e) Bending moment induced by self-weight calculated by Adapt. (f) Shape of frame due to secondary hyperstatic force: column with compression on top exterior face, column with compression on bottom interior face.

deflected by primary effect due to prestressing forces when connections of prestressed frames are restrained. Hyperstatic moments help column resist prestressing forces.

The 1.81 mm upward camber due to prestressing is calculated using Adapt as shown in Figure 1.3.8(d). Hydrostatic forces contribute to increasing a flexural strength of columns of the fixed posttensioned frame, because a bending moment diagram for columns due to the hydrostatic forces shown in Figure 1.3.8(b) and 1.3.8(f) are plotted opposite to those due to vertical self-weigh loads shown in Figure 1.3.8(e). Figure 1.3.8(f) shows that top inside and bottom outside the columns are subject to tensile stresses, whereas top outside and bottom inside the columns are subject to compressions when axial compression caused by prestressing force is applied to beam. However, beams at mid-span and ends of the beam are differently affected by the hydrostatic forces because entire bottom of the beam length shown in Figure 1.3.8(b) is subject to tensions due to the hydrostatic forces. The hydrostatic forces are unwanted and harmful, degrading the bottom of a beam at a mid-span because a mid-span is subject to tensile stresses with the same direction resulted by both hydrostatic forces and vertical loads as shown in Figure 1.3.8(b) and (e), whereas a beam at both ends of the beam receives helps from the hydrostatic forces because tensile stresses caused by hydrostatic forces are opposite to compressive stresses caused by vertical loads at both end. Figure 1.3.8(f) shows a shape of frame due to hyperstatic force, causing the entire bottom of the beam length shown in Figure 1.3.8(b) to be subject to tension.

Using Adapt (2021), bending moment diagrams of the frame due to the camber caused by total prestressing forces are calculated in Figure 1.3.8(c) or (a), whereas bending moment diagrams due to self-weight are calculated in Figure 1.3.8(e) or (b). Figure 1.3.9(c), then, shows bending moment diagram for the frame induced by a test vertical load of 118 kN. Figure 1.3.9(d) shows bending moment diagram for the frame due to a resultant force added from in Figure 1.3.9(a)–(c).

In Figure 1.3.9(d), the loading area of the beam at mid-span is subject to not only bending moment M = 5.1 kN·m due to a test load of 118 kN applied vertically, but also axial compression caused by prestressing force, and hence, it is possible to have compressive stresses in both sides, as indicated in Figure 1.3.9(e) and (f). Crack will not occur in the lower beam section at the applied load of 118 kN because tensile stress in beam of 0.15 MPa shown in Figure 1.3.9(f) is smaller than f_{ct} = 3.92 MPa.

Figure 1.3.9(d) shows that top outside the columns is subject to tensile cracks, whereas bottom inside the columns is subject to no tensile cracks when axial compression caused by prestressing force is applied to beam. The crack patterns are shown in Figure 1.3.9. It is noted that a flexural strength of the columns of the fixed posttensioned frame receives help from hydrostatic forces, contributing to increasing the flexural strength of the columns, because the bending moment diagram for columns due to hydrostatic forces shown in Figure 1.3.8(b) is plotted opposite to those due to vertical loads shown in Figure 1.3.9(b) or (c).

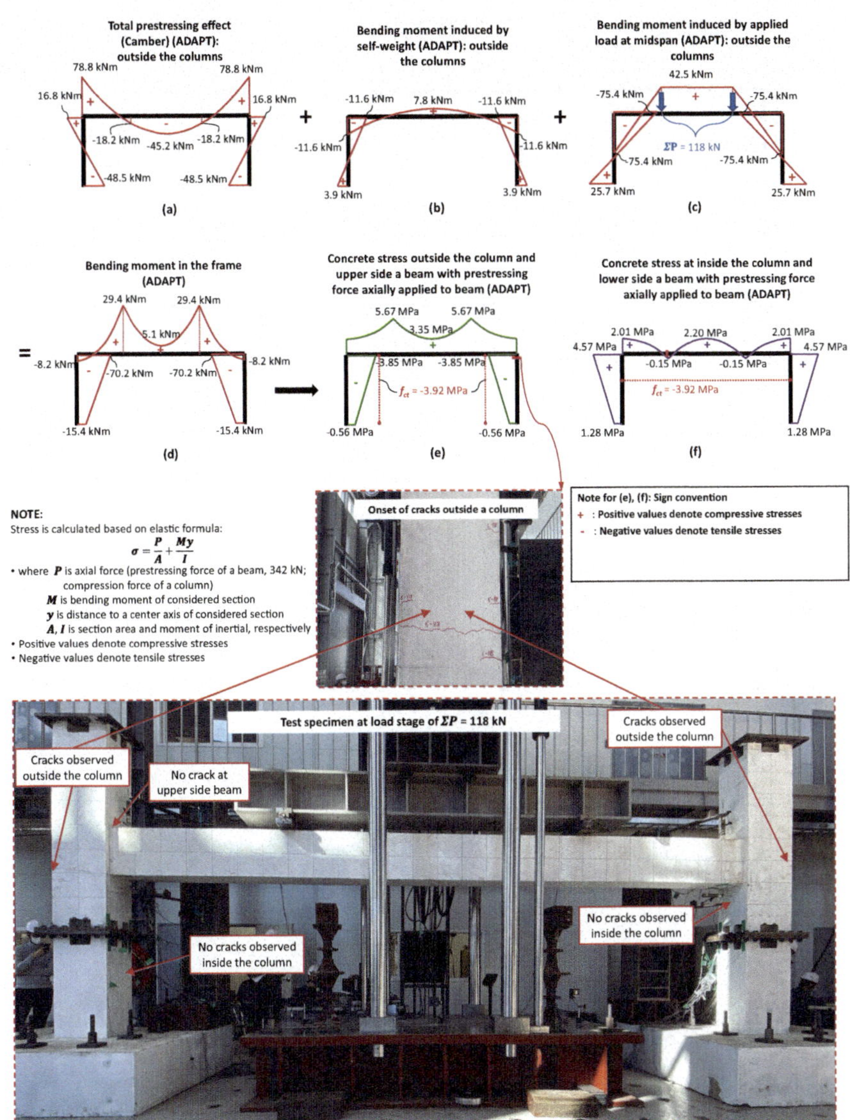

Figure 1.3.9 Onset of column cracks at an applied load of 118 kN [1.8].

Figure 1.3.10 shows propagation of cracks due to a test vertical load of 270 kN. Bending moment diagrams of the frame due to a camber caused by total prestressing forces and due to self-weight are calculated in Figure 1.3.10(a) and (b), respectively, using Adapt (2021). Figure 1.3.10(c) shows bending moment diagram of the frame induced by a vertical load of 270 kN. Figure 1.3.10(d) shows bending moment diagram for the frame due to a resultant force added from Figure 1.3.10(a)–(c). In Figure 1.3.10(d),

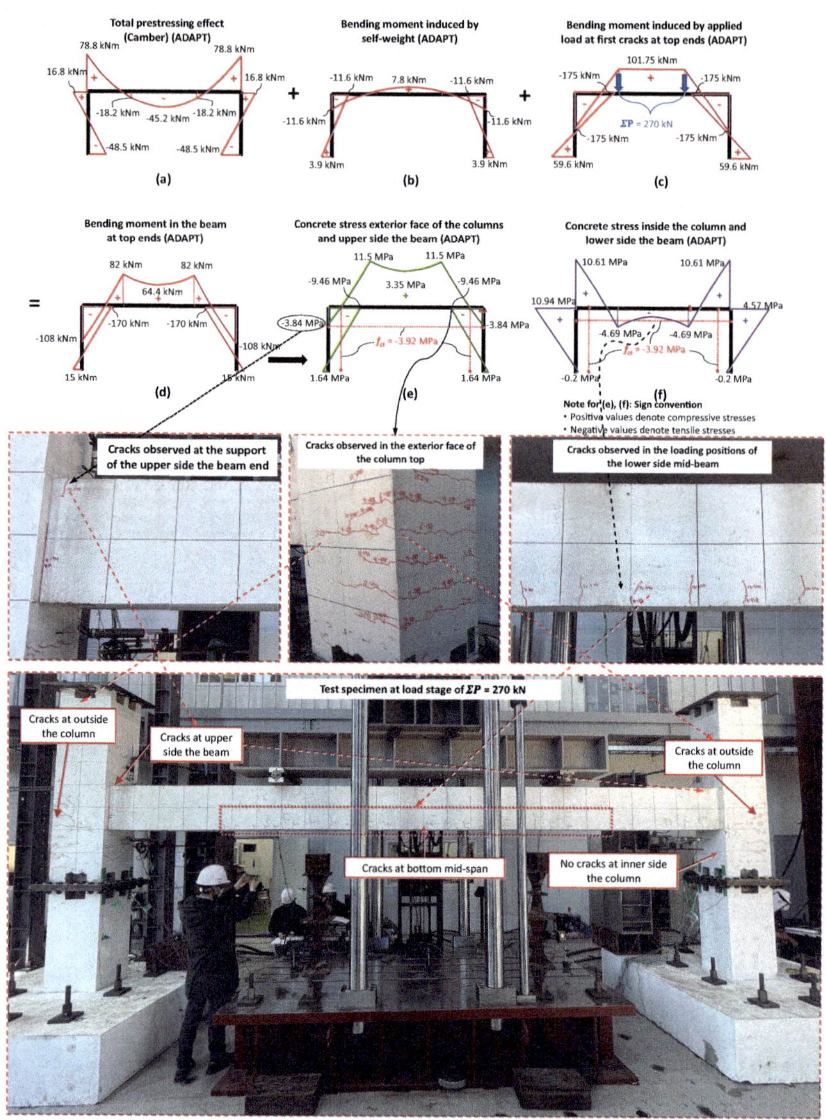

Figure 1.3.10 Crack propagation at an applied load of 270 kN [1.8].

not only bending moment $M = 64.4$ to 82 kN m due to vertical load of 270 kN, but also axial compression caused by prestressing forces are applied Figure 1.3.10(d) shows that top outside the columns is subject to tensile cracks, whereas bottom inside the columns is subject to no tensile cracks when axial compression caused by prestressing force is applied to beam. The crack patterns are shown in Figure 1.3.10.

Cracks of upper beam sections at supports are observed at an applied test load of 270 kN because tensile stress in a beam of 3.84 MPa shown in Figure 1.3.10(e) is in the vicinity of the cracking stress f_{ct} = 3.92 MPa, whereas crack of upper beam sections in the loading area will not occur at the applied load of 270 kN because stress in a beam of 3.35–11.5 MPa is under compressions as shown in Figure 1.3.10(e). However, cracks of lower beam sections in the loading area will occur at an applied load of 270 kN because tensile stress in beam of 4.69 MPa is larger than the cracking stress f_{ct} = 3.92 MPa as shown in Figure 1.3.10(f), whereas cracks of lower beam sections at supports will not occur in which a compression of 10.61 MPa is observed at an applied load of 270 kN. Photos taken during the test verify beam cracks calculated analytically as indicated by arrows shown in Figure 1.3.10. Cracks at an exterior face of the column top are observed at an applied load of 270 kN because a tensile stress of 9.46 MPa shown in Figure 1.3.10(e) is larger than the cracking stress f_{ct} = 3.92 MPa, whereas crack an exterior face of the column bottom will not occur at an applied load of 270 kN because a column stress of 1.64 MPa is under compressions as shown in Figure 1.3.10(e). No cracks at an interior face of the column top are observed at an applied load of 270 kN because this section is under compressive stress of 10.94 MPa as shown in Figure 1.3.10(f). Crack at an interior face of the column bottom will not occur at an applied load of 270 kN because a tensile column stress of 0.2 MPa is less than the cracking stress f_{ct} = 3.92 MPa as shown in Figure 1.3.10(f). Figure 1.3.10(e) and (f) shows that the exterior face of the column bottom and the interior face of the column top are subject to compressions. Photos taken during the test verify column cracks calculated analytically as indicated by arrows shown in Figure 1.3.10. Beams at mid-span and ends are differently affected by the hydrostatic forces because the entire bottom of the beam length is subject to tensions due to the hydrostatic forces at a vertical load of 270 kN as shown in Figure 1.3.10, indicating that the hydrostatic forces are harmful to a beam at mid-span, whereas a beam at both ends is benefited from the hydrostatic forces. It is noted that cracks are exaggerated by pens. 1.3.3.4 Prestressing and hyperstatic moments of high-rise frame vs. low-rise frame.

Figure 1.3.11 shows prestressing moments in three cases, illustrating moments due to prestressing forces in a high-rise frame with multi-spans, high-rise frame with single-span, and Low-rise frame with single-span. Column moments on the first floor with fixed base and columns on second floors which are influenced by columns on the first floors are different from others due to the fixed constraining conditions.

Figure 1.3.12 presents hyperstatic moments in three types of prestressed frames which represent the resisting moments against camber deformations, reducing camber deformations. It is noted that hypestatic moments are caused by the support reactions due to fixed constraints of the first floor, so hyperstatic moments on the first and second floors are much higher than ones on the other floors.

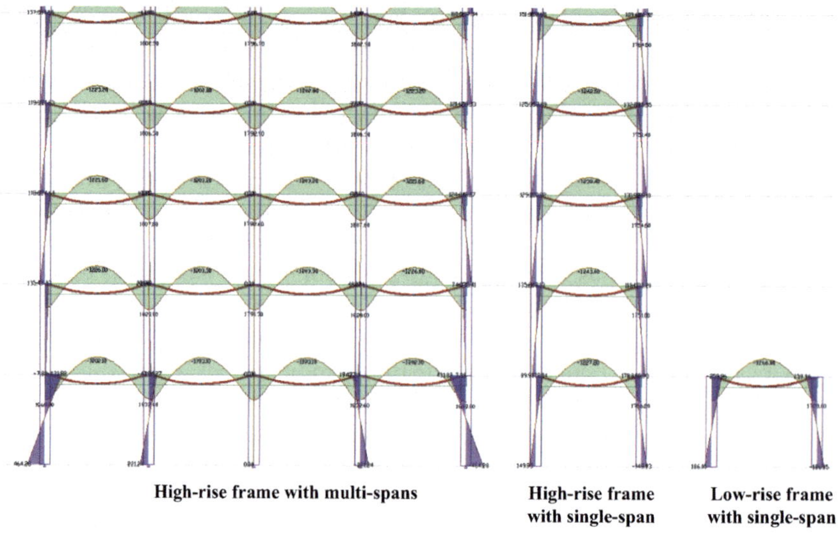

Figure 1.3.11 Prestressing moments in frame.

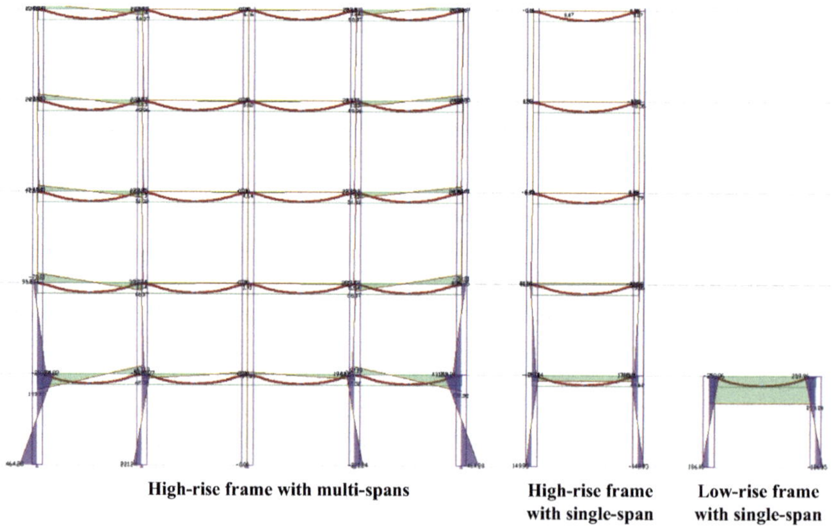

Figure 1.3.12 Hyperstatic moments in frame.

1.4 PRESTRESS LOSSES

1.4.1 Type of prestressing losses

During and after prestressing, some of the applied forces are lost, consequently inducing a compressive force to concrete to a lesser extent than

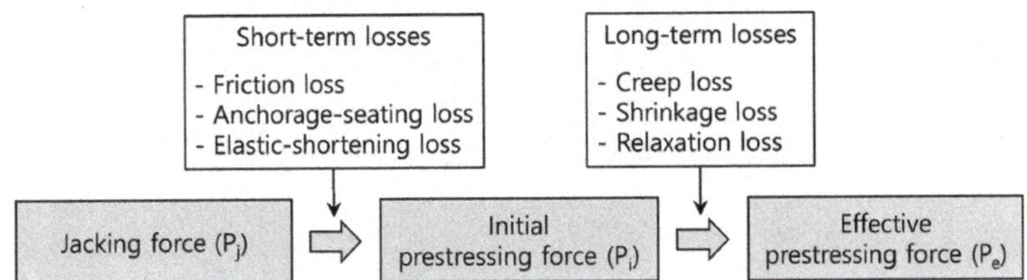

Figure 1.4.1 Type of prestressing losses.

initial jacking. Various types of prestress losses occur which, then, reduce tensile stresses in tendon from time to time. They can be categorized into two groups, short-term and long-term losses as shown in Figure 1.4.1. Short-term losses are caused by frictions, wedge draw-in (refer to Figure 1.4.3(a)), and an elastic shortening of concretes. Anchorage-seating loss is also considered as short-term loss, whereas long-term losses consist of concrete creep, concrete shrinkage, and tendon relaxation.

1.4.2 Assessment of prestressing losses

Prestressing losses, whether larger or smaller than calculated ones, do not substantially affect a design strength of the member, but affect a serviceability design (deflections, camber, and cracking loads) and connections, as mentioned in ACI 318-19 (Standard, 2019 [1.9]). Historically, an arbitrary value has been assumed for the overall loss; however, this estimate is maculate. In service loads, overestimating prestress losses can be nearly as hazardous as underestimating them, since the former can lead to abnormal camber and lateral movement. In preliminary designs, an estimated initial prestressing force f_{pi} occurring in both short- and long-term losses can be assumed around 28% of the tendon strength f_{pu}, whereas an average effective prestressing stress (or force) f_{pe} in tendon is approximately taken between $0.55f_{pk}$ (or f_{pu}) and 0.65 f_{pk} (or f_{pu}). However, prestress losses should be checked in final designs.

Figure 1.4.2 shows prestressing losses where f_{pu} is the tendon strength. We cannot pull the tendon with a full stress f_{pu}, and hence, should pull tendons up to jacking force f_{pj} (refer to Figure 1.4.3(b)) which is smaller value between 80% of a tendon strength f_{pu} (which is prior to any loss at anchorage) and 94% of tensile yield strength of a tendon. We only pull 80% f_{pu} (f_{pj}=0.8 f_{pu}) to avoid tendon yielding. Short-term losses f_{pi} (initial prestressing force) during prestressing are 10% of jacking force f_{pi} (=0.9 f_{pj}, or 0.72 f_{pu}), and hence, initial prestressing force f_{pi} is 0.9 f_{pj}. After f_{pi}, long-term losses of 20% of initial prestressing force f_{pi} are considered to calculate effective prestressing force f_{pe}=0.8 f_{pi} that is permanent tendon forces for design. Finally, total losses are $1-0.8\times0.9\times0.8$=42.4% of the tendon strength f_{pu}, allowing to use 57.6% of a tendon strength f_{pu}

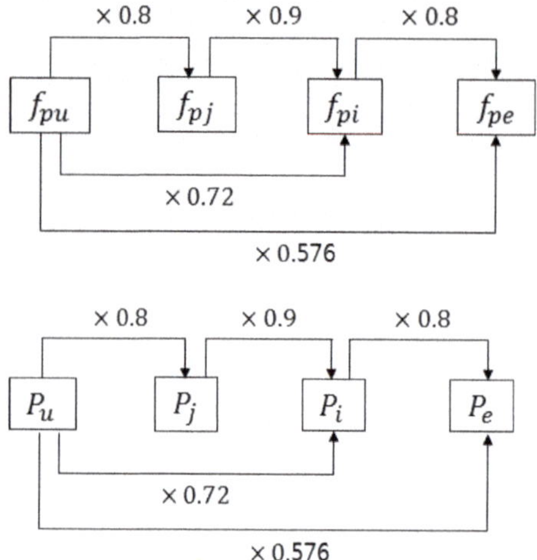

Figure 1.4.2 Assessment of prestressing losses.

for design. Total losses are based on $1-0.8\times0.9\times0.9=35.2\%$ of the tendon strength f_{pu} when long-term losses of 10% of initial prestressing force f_{pi} are considered to calculate effective prestressing force, allowing to use 65% of a tendon strength f_{pu} for design. It is noted that prestressing losses are commonly calculated as shown in Figure 1.4.2. Short-term losses=Jacking forces f_{pj} – initial prestressing force f_{pi}, whereas long-term losses=initial prestressing force f_{pi} – effective prestressing force f_{pe}.

1.4.3 Friction losses

Frictions exist between prestressing tendons and inside the greased plastic extrusion for unbonded tendons, whereas there are frictions from preformed sheath for bonded tendons. There are two basic mechanisms by which friction occurs. One is an intentional contact due to tendon curvature, as when a tendon is draped in the shaped of a parabolic or harped profile. Another is an unintentional deviation of tendons from their intended profile. Losses resulting from intentional tendon contact are referred to as curvature friction, while unintentional contact is referred to as wobble friction. In the first case, loss due to curvature friction is proportional to the sum of all angles of curvature. In the second case, loss due to a wobble friction is correlated with a length of tendons from a live anchor. Both cases result in losses increasing from zero at a live anchor to a maximum at a dead anchor, or at a midpoint of a tendon if it is jacked from both ends. The two losses due to friction and wobble are summarized as shown in Eq. (1.4.1).

$$P(x) = P_{\max}e^{-(\mu\theta+\kappa x)} \tag{1.4.1}$$

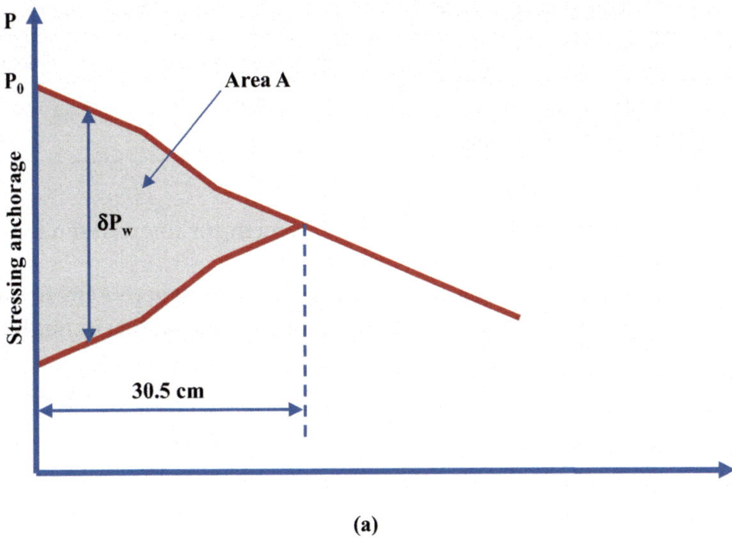

(a)

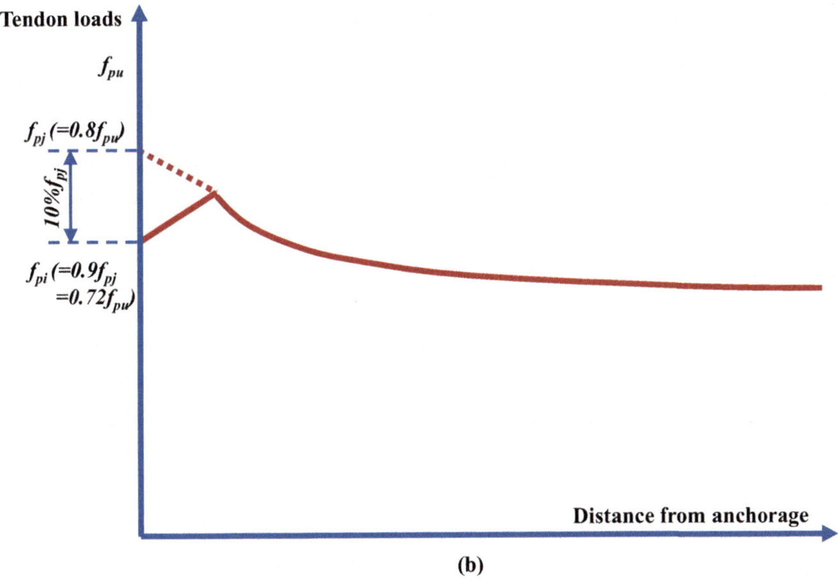

(b)

Figure 1.4.3 Short-term losses. (a) Loss of prestresses due to wedge draw-in (TR43, 1994 [1.2]). (b) Loss of prestresses due to jacking force f_{pj}.

where:

θ is the sum of angular displacements over a distance x (irrespective of direction or sign).

μ is a coefficient of friction between a tendon and its duct.

Table 1.4.1 Typical friction coefficients and wobble factors (TR43, 1994 [1.2])

	Unbonded tendons	*Bonded tendons*
Friction coefficient, μ	0.06	0.20
Wobble coefficient, κ (rad/m)	0.05	0.0085

κ is an unintentional angular displacement for internal tendons (per unit length).
x is a distance along a tendon from the point where a prestressing force is equal to $P_{\max}$ (a force at an active end during tensioning).

This equation is equivalent to Eq. (5.45) in Eurocode 2. Values of μ and κ will depend on prestressing system chosen. The value μ depends on the surface characteristics of tendons and ducts, the elongation of tendons, and a tendon profile. The value κ for unintentional angular displacement depends on a quality of workmanship, a distance between tendon supports, the type of duct or sheath employed, and the degree of vibration used in placing concretes. In an absence of detailed information from prestress system supplier, it is recommended that μ and κ be taken in Table 1.4.1.

1.4.4 Anchorage draw-in

It is necessary to account for the losses arising from wedge draw-in of anchorage devices during anchoring after tensioning. A deformation of an anchorage itself also results in the losses. A small amount of strand movements must occur in anchorages in order for wedges to grip. This inward movement, which is usually a minimum of 6 mm, reduces prestressing forces. The draw-in effect is illustrated in Figure 1.4.3(a).

The force loss is calculated as shown in Eq. (1.4.2):

$$Area\ A = \Delta E_p A_p \tag{1.4.2-1}$$

$$Area\ A = \int_0^{l'} \delta P_w dx \tag{1.4.2-2}$$

where

Δ is the wedge draw-in.
δP_w is the force loss.
E_p is the modulus of elasticity of tendon.
A_p is the area of tendon.
l' is the length of tendon affected by draw-in.

An approximate linear force profile can be obtained if an angle of tendon changes uniformly per unit length. Therefore, if l' is less than a length of tendons; thus, the following equation can be used to calculate l' as shown in Eq. (1.4.3):

$$l' = \sqrt{\Delta E_p A_p / p'} \tag{1.4.3}$$

where
p' is the slope of the force profile.

The force loss of δP_w, then, can be calculated as shown in Eq. (1.4.4):

$$\delta P_w = 2p'(l' - x) \tag{1.4.4}$$

If the wedge draw-in affects the entire length of tendons, the force loss is then shown in Eq. (1.4.5).

$$\delta P_w = \Delta E_p A_p / l + p'l \text{ at live anchorage} \tag{1.4.5-1}$$

$$\delta P_w = \Delta E_p A_p / l - p'l \text{ at dead anchorage} \tag{1.4.5-2}$$

The loss of prestress resulting from wedge draw-in increases as the length of tendons decreases. Short tendon may suffer from a highly detrimental loss of prestresses due to wedge draw-in. Suppose, for example, that a 6 m long tendon that has an elongation of 40 mm will lose 15% of its force due to a wedge draw-in that is 6 mm.

1.4.5 Elastic shortening

Tendons cause concrete to compress when they are stressed, resulting in a slight reduction in the length of a concrete member. When the second tendon is stressed, it may not suffer losses as a result of its own stress, but it may reduce the force in the first tendon. Similarly, the forces in the two previously stressed tendons are reduced when the third tendon is stressed. This shortening is insignificant in most buildings, but it may be significant in structures with high prestressing. The force loss is given by Eq. (1.4.6).

$$\delta P_{el} = E_p A_p \sum \left[\frac{j \cdot \Delta\sigma(t)}{E_c(t)} \right] \tag{1.4.6}$$

where
$\Delta\sigma(t)$ is the variation of stress at a center of gravity of the tendons applied at time t.
j is a coefficient equal to $(n-1)/2n$.
n is the number of identical tendons successively prestressed.

As an approximation for posttensioned structure, j may be taken as 1/2 (Eurocode 2, Clause 5.10.5.1). In the case of pretensioned tendons, the assumption is that a total force is transferred to a member at once, and so j is usually taken to be 1.

1.4.6 Time-dependent losses

The time-dependent losses may be calculated by considering the following two reductions of stress:

- Due to the reduction of strain caused by the deformation of concrete due to creep and shrinkage under permanent loads.
- The reduction of stress in steels due to relaxations under tension.

1.4.7 Shrinkage strain of concrete

Shrinkage strain is reduced by placing reinforcements. Reinforcements on a slab are normally placed with a smaller quantity than on a beam which has a reinforcement cage. This means that shrinkage is smaller in a beam compared to a slab. Considering different shrinkage rates for slabs and beams, however, would not be a practical strategy. Normal practice allows shrinkage rates to be calculated by either ignoring reinforcements or assuming a constant value for an entire floor. Thin members, such as slabs, subjected to low humidity should be handled with special care because shrinkage of more than 0.0004 can occur. Force loss due to shrinkage can be calculated based on Eq. (1.4.7):

$$\delta P_{sh} = \varepsilon_{sh} E_p A_p \tag{1.4.7}$$

where ε_{sh} is the shrinkage strain of concrete.

1.4.8 Creep of concrete

A principal effect of creep on prestressed concrete members is the same as that due to shrinkage. A reduction in prestressing force is also caused by shortening of members with time. It is a complicated process to estimate prestress losses due to creep as there are many factors involved. Force loss due to creep can be calculated based on Eq. (1.4.8).

$$\delta P_c = \varepsilon_c E_p A_p \tag{1.4.8}$$

where

$\varepsilon_c = \sigma_{co}\varphi(t,t_0)/E_{c.}$

σ_c is the stress in the concrete adjacent to tendons.

$\varphi(t,t_0)$ is the creep coefficient.

Table 1.4.2 Relaxation for Class 2 low-relaxation steels (Table B2 of TR43, 1994 [1.2])

Force at transfer as a % of characteristic strength of tendon	1000-hour relaxation	Relaxation factor	Force loss as a % of force at transfer
80%	4.5%	1.5	6.75%
70%	2.5%	1.5	3.75%
60%	1.0%	1.5	1.50%

Notes:

1. Characteristic strength of tendon = $f_{pu} \times A_{ps}$
2. The 1000-hour relaxation values can be replaced with the manufacturer's values if available.

1.4.9 Relaxation of the tendon

Tendon stresses decrease with time due to relaxation of tendons. The relaxation is dependent on an initial tension and a type of strands used. Data for relaxation of Class 2 low-relaxation steels are given in Table 1.4.2.

1.4.10 Simplified method to evaluate time-dependent losses

As mentioned in Section 5.10.6 of Eurocode 2, a simplified method to evaluate time-dependent losses at location under permanent loads is given by Eq. (1.4.9). Readers are referred to Section 5.10.6 of Eurocode 2 for further description of Eq. (1.4.9).

$$\Delta P_{c+S+r} = A_p \Delta\sigma_{p,C+S+r} = A_p \frac{\varepsilon_{cs} E_p + 0{,}8\Delta\sigma_{pr} + \frac{E_p}{E_{cm}}\varphi(t,t_0).\sigma_{c,QP}}{1+\frac{E_p}{E_{cm}}\frac{A_p}{A_c}(1+\frac{A_c}{I_c}Z_{cp}^2)[1+0{,}8\varphi(t,t_0)]} \quad (1.4.9)$$

In Eq. (1.4.9), compression stresses and strains should be interpreted as positive. Local stress values are used for bonded tendons, while mean stress values are used for unbonded tendons.

1.4.11 Jacking force

In accordance with Eurocode 2, a tendon shall not be subjected to a jacking force exceeding the following value as shown in Eq. (1.4.10).

$$P_{\max} = A_p \sigma_{p,\max} \quad (1.4.10)$$

where

A_p is the cross-sectional area of tendons.

$\sigma_{p,\max} = \min\left[0.8 f_{pk}; 0.9 f_{p0.1k}\right]$ is the maximum stress applied to tendons.

In order to qualify for overstressing, maximum prestressing forces Pmax may increase to $0.95 f_{p0.1k}$

1.5 FLEXURAL DESIGN

1.5.1 Section analysis

Designing a prestressed concrete structure requires numerous considerations, primarily determining a distribution of stresses of structural members. Prestressed concrete structures must consider all sections to be critical, and stress distributions must be examined at all stages of loadings because prestress forces induce high stresses throughout entire members. Some basic principles are employed to perform section analysis as follows:

- Plane sections remain plane.
- Interactions between concretes and rebars, between concretes and bonded prestressing tendons are perfectly bonded.
- Material stress–strain curve for both concrete shown in Figure 1.2.1 and steels are followed.
- Concrete tensile strength can be ignored after cracking.

A basic difference between analysis of sections at a serviceability and ultimate limit stage is that concrete strain reaches a ultimate strain of 0.0035 at ultimate limit stage, whereas in service stage, concrete strain varies following service load. A stress–strain of a cross section is shown in Figure 1.5.1. The compression force of concrete, C_c, can be calculated as shown in Eq. (1.5.1), which follows a concrete stress–strain relationship.

$$C_c = \int_0^c \sigma_{c,y} b d_y \tag{1.5.1}$$

Equation (1.5.2) expresses a distance between extreme compressive fiber and a centroid of concrete compressive block.

$$g = c - \int_0^c y\sigma_c b d_y \Big/ \int_0^c \sigma_c b d_y \tag{1.5.2}$$

A contribution of concrete in tension zones is significant in the uncracked regions and should be considered when analyzing sections. Tensile forces in concrete are ignored once a crack appears. Rebar and tendon stresses can be calculated using section compatibility. Tendon force consists of initial jacking forces (T_{p0}) and increasing forces (ΔT_p). Strains in bonded tendons and rebar can be calculated in the same manner using section compatibility. An increase of stress in unbonded tendons should be calculated by considering a deformation of the entire member. Incremental effective prestress in tendons at ultimate limit state can be taken as 100 MPa when detailed calculations are not available (Eurocode 2, 2004). A trial-and-error method

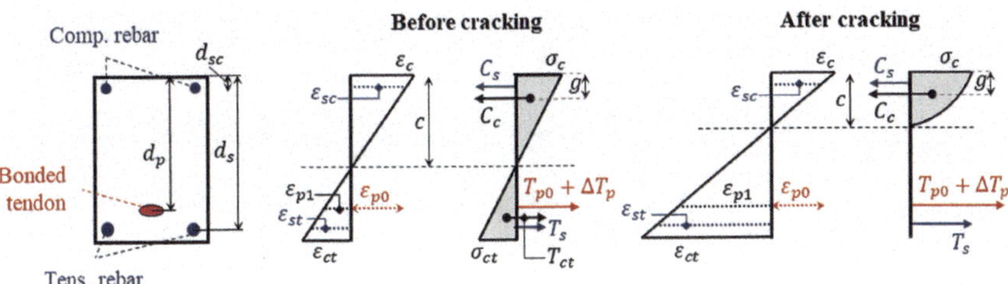

Figure 1.5.1 Stress–strain distribution of a cross section before and after cracking.

is used to determine c, a neutral axis depth, based on the equilibrium condition as shown in Figure 1.5.1.

1.5.2 Flexural design at service limit stage

In prestressed concrete structures, it is not immediately apparent whether the serviceability or the ultimate strength state, or both, are critical until structural analyses are performed. Consequently, serviceability checks of stresses, cracks, and deflections are crucial when designing prestressed concrete structures. For service stage, dead load and total prestressing effects including all short- and long-term losses should be combined with live load, wind, and seismic load to result in maximum stresses, crack width, and deflections. Structural responses including stresses, cracks, and deflections at service stage should be controlled under service criteria specified in design standards such as Eurocode 2 and ACI.

1.5.2.1 Stress control

A limitation of stresses may be necessary for tensile and compressive concrete stresses, and for tensile stresses of rebars and tendons under service load conditions. Limiting the tensile stresses in concrete is an effective method of reducing a risk of cracking. A purpose of limiting compressive concrete stresses is to prevent irreversible strains and longitudinal cracks which is parallel to compressive strain. Tensile stresses of rebars and tendons should be kept below a yield strength with adequate safety margins to prevent uncontrolled cracking. In calculating stresses, cracked section analysis must be considered if a section is expected to crack under service loads. In accordance with Eurocode 2, a cross section is assumed to be cracked if flexural tensile stress exceeds mean value of axial tensile strength of concrete, f_{ctm}. Stress is calculated using section properties associated with either uncracked condition or fully cracked state, depending on situations. Additionally, effects of creep, shrinkage of concrete, and relaxation of tendon as well as temperature variations on structures should be considered if needed.

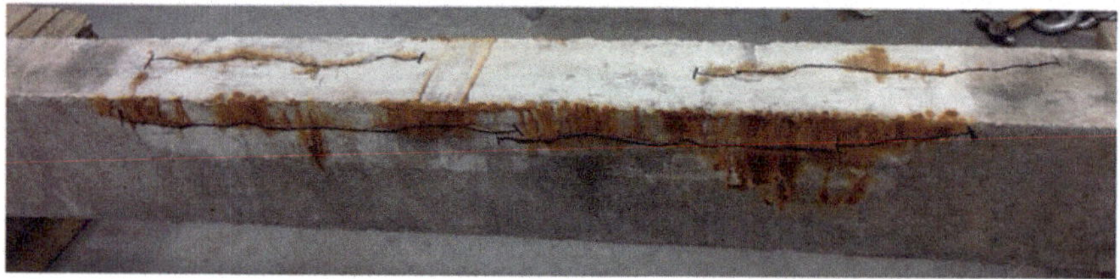

Figure 1.5.2 Corrosion crack pattern for specimen FCU [1.10].

1. Stress control of concrete

   Under service load, concrete suffering from excessive compressive stress may crack longitudinally and creep to a high and unpredictable degree, with seriously detrimental effects on prestress losses. Separations of corroded rebars and concrete cause longitudinal cracks appearing along rebars as shown in Figure 1.5.2. Longitudinal cracks may develop when concrete compressive stress level is exceeded by a critical value under service load. Cracking of this nature can reduce a durability of a structure, and hence, compressive stresses should be limited to prevent longitudinal cracks, microcracks, or high creep levels in concrete, which may impair structural functions.

   The stress limitation at $k_1 \times f_{ck}$ is required for structures in extreme environments, such as corrosions by chlorides from seawater (XS), corrosion induced by chlorides (XD), and freeze/thaw attack (XF) as shown in Table 4.1 of EC2. In these harmful conditions, rebars are strongly affected by corrosions when there are micro cracks in concrete covers, and hence, code limits the concrete compressive stresses. EC2 limits compressive stresses in concrete at $k_1 \times f_{ck}$ when increments in concrete covers or confinement by transverse reinforcements are not sufficiently provided. In the absence of detail measures, concrete compressive stresses may be limited to $0.6f_{ck}$ under final stages when neither enough cover to reinforcements of the compressive zone nor confinements by transverse reinforcement is not available, as mentioned in Eurocode 2. Concrete compressive stresses in transfer state are limited at $0.6f_{ck}$ for all environments even if extra concrete covers or confinement by transverse reinforcements are provided. It may be possible to increase concrete compressive stresses in transfer state up to $0.7f_{ck}(t)$ for pretensioned members, provided that longitudinal cracking is prevented by tests or experience. There is a need to consider non-linear creep if the compressive stress permanently exceeds $0.45f_{ck}(t)$.

   A value of k_1 varies in National Annexes in some countries. Polish allows $k_1=1$, allowing concrete to fully activate compression strengths under the characteristic load levels. However, EC2 conservativity recommend $k_1=0.6$, allowing concrete to activate 60% of compression strengths under the characteristic load levels [1.11].

2. Stress control of tendons and rebars

   Under serviceability conditions, steels could be inelastically deformed due to excessive tensile stresses, which could lead to wide and permanent open cracks. Under the service loads according to Eurocode 2, excessive cracking or deformation is considered avoided if tensile stresses in tendons and rebars are below $0.75f_{pk}$ and $0.8f_{yk}$, respectively.

1.5.2.2 Crack control

The presence of cracks does not necessarily indicate a lack of durability or serviceability. In concrete structures, cracks caused by tension, bending, shearing, or torsion are often unavoidable and may not affect serviceability or durability. Cracks may also result from other causes such as plastic shrinkage, chemical reactions, or expansion of hardened concrete. This chapter does not cover the avoidance and control of such cracks.

1. Crack width calculation

   In terms of determining crack width in practice, it is important to keep in mind that cracking is a highly probabilistic process. Crack widths based on a crack calculation are nominal values for comparing to nominal limit values. These nominal crack widths may not match the crack widths observed on the actual structure. Accordingly, comparing calculated crack widths with nominal crack width limits may only be an approximate method of determining whether the design criteria have been met. Accuracy may not be achieved on this basis. In accordance with Eurocode 2, crack width, w_k, may be calculated from Eq. (1.5.3).

$$w_k = s_{r,\max}\left(\varepsilon_{sm} - \varepsilon_{cm}\right) \tag{1.5.3}$$

   where $s_{r,\max}$ is the maximum crack spacing. ε_{sm} is the mean strain in the reinforcements under a relevant combination of loads, including an effect of imposed deformations considering effects of tension stiffening. Stresses of steel reinforcement are influenced by concrete acting in tension between cracks which is defined as tension stiffening. Stresses in the concrete and reinforcement gradually increase along the force imposed on structures. The tension-stiffening effect disappears when the tensile capacity of the concrete is reached or the bond between the concrete and reinforcement disappears, and hence, only additional tensile strains beyond a state of zero strain of the concrete at the same level are considered. ε_{cm} is the mean strain in the concrete between cracks. $\left(\varepsilon_{sm} - \varepsilon_{cm}\right)$ may be calculated from Eq. (1.5.4).

$$\varepsilon_{sm} - \varepsilon_{cm} = \frac{\sigma_s - \dfrac{k_t f_{ct,eff}}{\rho_{p,eff}}\left(1 + \alpha_e \rho_{p,eff}\right)}{E_s} \geq 0.6\frac{\sigma_s}{E_s} \tag{1.5.4}$$

where σ_s is a stress in tension reinforcements assuming a cracked section. For pretensioned members, σ_s may be replaced by a stress variation $\Delta\sigma_p$ in prestressing tendons from a state of zero strain of the concrete at the same level. α_e is a ratio of elastic modulus of rebar to elastic modulus of concrete, E_s/E_{cm}. $\rho_{p,eff}$ is an effective reinforcement ratio shown in Eq. (1.5.5). k_t is a factor dependent on a duration of a load.

$k_t = 0.6$ for short-term loading

$k_t = 0.4$ for long-term loading

$$\rho_{p,eff} = \left(A_s + \xi_1^2 A_p^{'}\right)/A_{c,eff} \tag{1.5.5}$$

where

A_s is a rebar area under tension. $A_p^{'}$ is an area of pre or post-tensioned tendons within $A_{c,eff}$. $A_{c,eff}$ is an effective area of concrete in tension surrounding reinforcements or prestressing tendons of depth $h_{c,ef}$ where $h_{c,ef}$ is the lesser of $2{,}5(h-d)$, $(h-x)/3$, or $h/2$ shown in Figure 7.1 of Eurocode 2, 2004 [1.3]. $\xi_1 = \sqrt{\xi \cdot \frac{\phi_s}{\phi_p}}$ is an adjusted ratio of bond strength considering different diameters of prestressing and reinforcing steels. ξ is a ratio of bond strength between bonded tendons and ribbed steel in concrete. The values of ξ given in Table 6.2 of Eurocode 2, 2004 [1.3] may be referred. ϕ_s is the largest bar diameter of reinforcing steels. ϕ_p is the equivalent diameter of tendons.

$\phi_p = 1.6\sqrt{A_p}$ for bundles

$\phi_p = 1.75\phi_{wire}$ for single 7 wire strands

$\phi_p = 1.20\phi_{wire}$ for single 3 wire strands

Equation (1.5.6) can be used to calculate a maximum final crack spacing in tension zones where bonded reinforcements are installed at a very close spacing of smaller than $5\left(c+\frac{\phi}{2}\right)$ shown in Figure 7.2 of Eurocode 2, 2004 [1.3].

$$s_{r,\max} = 3.4c + 0.25k_1k_2\ \phi\,/\,\rho_{p,eff} \tag{1.5.6}$$

where ϕ is the bar diameter. c is a cover to longitudinal reinforcements. k_1 is a coefficient which considers bond properties of bonded reinforcements, whereas k_2 is a coefficient considering a distribution of strain.

$k_1 = 0.8$ for high bond bars.

$k_1 = 1.6$ for bars with an effectively plain surface, such as prestressing tendons.

$k_2 = 0.5$ for bending.

$k_2 = 1$ for pure tension.

prestressing steel	ξ		
	pre-tensioned	bonded, post-tensioned	
		≤ C50/60	≥ C70/85
smooth bars and wires	Not applicable	0,3	0,15
strands	0,6	0,5	0,25
indented wires	0,7	0,6	0,3
ribbed bars	0,8	0,7	0,35
Note: For intermediate values between C50/60 and C70/85 interpolation may be used.			

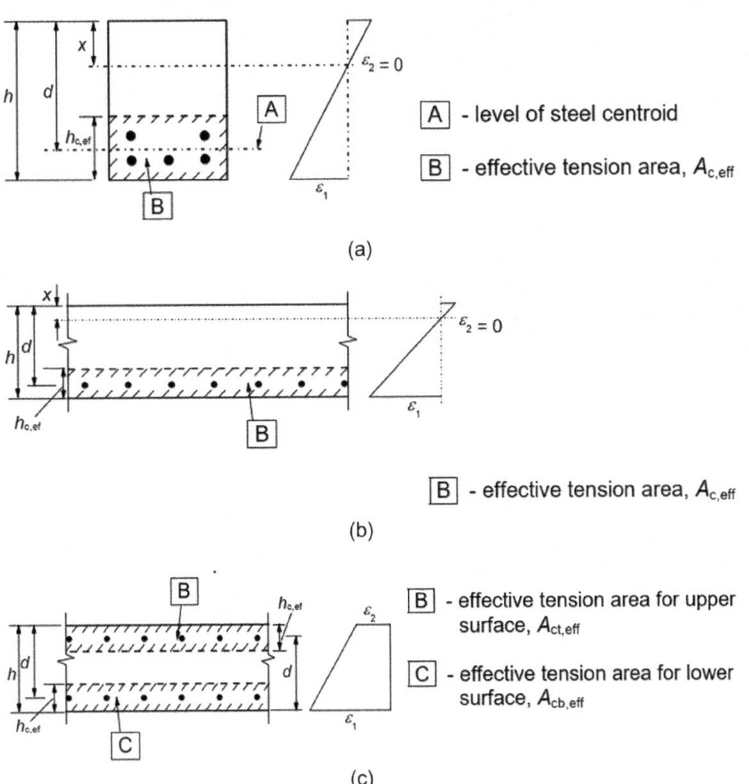

An upper bound of a crack spacing may be determined following Eq. (1.5.7) when bonded reinforcement spacing exceeds $5\left(c+\frac{\phi}{2}\right)$ shown in Figure 1.5.4, or there are no bonded reinforcements in a tension zone.

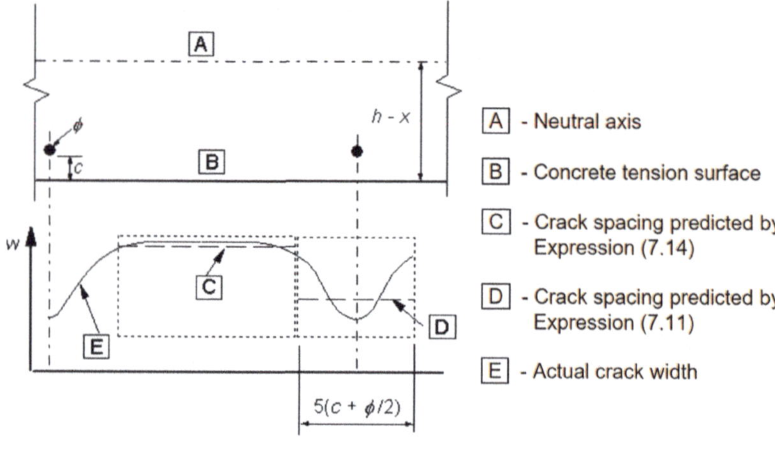

$$s_{r,\max} = 1.3(h - x) \tag{1.5.7}$$

2. Crack width limitation

   An appropriate crack width limitation $w_{\max}$ should be determined based on functions and nature of structures, as well as costs associated with crack control. For concrete inside a building with low air humidity, a crack width should be restricted to 0.2 and 0.4 mm for prestressed members with bonded and unbonded tendons, respectively, unless the client specifies a more restrictive value. The recommended crack width limitation $w_{\max}$ for relevant exposure environments can be found in Table 7.1N of Eurocode 2, 2004 [1.3].

1.5.2.3 Deflection control

In concrete structures, deflections cannot be predicted with a high degree of accuracy due to numerous non-linear variables at play. Concrete is non-linear in both compression and tension, and prestressed members have non-linear load-deflection characteristics, resulting from an abrupt change in stiffness following cracking of concrete sections. A method of deflection calculation outlined in this section should be regarded as giving only estimates of deflections, describing how to approximately calculate deflections. Testing a model of structures using similar materials is the only method that provides reliable results. In contrast to reinforced concrete members, prestressed beams can neutralize all deflections under given loadings through a use of a suitable prestressing. It is generally associated with upward deflections which is known as camber. Cambers from prestresses can neutralize deflections from vertical loads when suitable prestressing is provided, and hence, final deflections in prestressed beams can be zero. Camber deflections in prestressed concrete members should be checked even with no applied loads. Suitable prestressing should be verified by checking beam

Recommended values of w_{max} (mm) (Table 7.1N of Eurocode 2, 2004 [1.3])

Exposure Class	Reinforced members and prestressed members with unbonded tendons	Prestressed members with bonded tendons
	Quasi-permanent load combination	Frequent load combination
X0, XC1	0,4[1]	0,2
XC2, XC3, XC4	0,3	0,2[2]
XD1, XD2, XS1, XS2, XS3		Decompression

Note 1: For X0, XC1 exposure classes, crack width has no influence on durability and this limit is set to guarantee acceptable appearance. In the absence of appearance conditions this limit may be relaxed.
Note 2: For these exposure classes, in addition, decompression should be checked under the quasi-permanent combination of loads.

In the absence of specific requirements (e.g. water-tightness), it may be assumed that limiting the calculated crack widths to the values of w_{max} given in Table 7.1N, under the quasi-permanent combination of loads, will generally be satisfactory for reinforced concrete members in buildings with respect to appearance and durability.
The durability of prestressed members may be more critically affected by cracking. In the absence of more detailed requirements, it may be assumed that limiting the calculated crack widths to the values of w_{max} given in Table 7.1N, under the frequent combination of loads, will generally be satisfactory for prestressed concrete members. The decompression limit requires that all parts of the bonded tendons or duct lie at least 25 mm within concrete in compression.

deflections in all stages. Both cambers at transfer stages and deflections at service stages should not exceed the deflection limits. Controlling deflections with prestress level complicates the process of estimating a size of members based on span/depth ratios, as is the practice for reinforced concrete members. However, PT beam is advantageous over non-prestressed beams because deflections at a final stage "can be" zero.

1. Short-term deflection calculation

   An elastic deflection can be used if prestressed concrete members are assumed to be free from cracks at the considered loads. A more rigorous analysis method must be used when sections are cracked. In general, relationship between curvature φ at a point x along a member and the corresponding deflection y is given by Eq. (1.5.8).

$$\varphi = \frac{d^2y}{dx^2} \tag{1.5.8}$$

   A deflection at any point within a member can be found by integrating Eq. (1.5.8) twice, and this is usually done numerically. Uncracked and cracked curvatures are calculated using Eqs. (1.5.9) and (1.5.10), respectively, where ε_c and c are a concrete compressive strain at extreme fiber and neutral axis depth, respectively, as shown in Figure 1.5.1. Cracked members which is not fully cracked exhibit behavior, intermediating between uncracked and fully cracked conditions, and hence, an adequate prediction of effective curvature φ_{ef} is given by Eq. (1.5.11) as mentioned in Eurocode 2.

$$\varphi_{uncr} = M/E_{cm}I_g \tag{1.5.9}$$

$$\varphi_{cr} = \varepsilon_c/c \tag{1.5.10}$$

$$\varphi_{ef} = \zeta\varphi_{cr} + (1-\zeta)\varphi_{uncr} \tag{1.5.11}$$

ζ is a distribution coefficient allowing tension stiffening at a section and is given by Eq. (1.5.12), where
β is a coefficient considering an influence of the duration of loadings or of repeated loadings on the average strain.
$\beta = 1$ for a single short-term loading.
$\beta = 0.5$ for sustained loadings or many cycles of repeated loadings.
σ_s is a stress in tension reinforcements calculated based on a cracked section. σ_{sr} is a stress in tension reinforcements calculated based on a cracked section under loading conditions causing first cracking.

$$\zeta = 1-\beta\left(\frac{\sigma_{sr}}{\sigma_s}\right)^2 \tag{1.5.12}$$

2. Long-term deflection calculation

Deflections of prestressed concrete members have been determined up to now only on a short-term basis. As the same as reinforced concrete members, deflections of prestressed concrete members will increase with time due to shrinkage and creep movements. A total deformation including creep may be calculated by using an effective modulus of elasticity for concretes according to Eq. (1.5.13) for loads with a duration causing creep.

$$E_{c,eff} = \frac{E_{cm}}{1+\varphi(\infty,t_0)} \tag{1.5.13}$$

where $\varphi(\infty,t_0)$ is a creep coefficient relevant for loads and time intervals. A detail calculation procedure of $\varphi(\infty,t_0)$ can be found in Annex B of Eurocode 2. In accordance with Eurocode 2, shrinkage curvatures φ_{cs} may be assessed using Eq. (1.5.14).

$$\varphi_{cs} = \varepsilon_{cs}\alpha_e S/I \tag{1.5.14}$$

where ε_{cs} is the free shrinkage strain. S is a first moment of area of reinforcements about a centroid of a beam section. I is a second moment of area of a beam section. $\alpha_e = E_s / E_{c,eff}$ is an effective modular ratio. S and I should be calculated for uncracked conditions and fully cracked conditions. A final curvature is assessed by use of Eq. (1.5.11).

3. Deflection control

   A deflection of a member or structure should not adversely affect a function or an appearance of a member or structure. Appropriate deflection limits should be determined considering factors including the nature of structures, finishes, partitions, and fixings as well as their functions. In accordance with Eurocode 2, an appearance and general utility of a structure could be impaired when a calculated sag of a beam, slab, or cantilever subjected to quasi-permanent loads exceeds span/250. Quasi-permanent load levels are calculated from statistical data based on a 50% probability of being exceeded by loads on buildings during service [1.3]. A sag is assessed relative to the supports. Deflections that could damage adjacent parts of a structure should be limited. For a deflection after construction, span/500 is normally an appropriate limit for quasi-permanent loads. Other limits may be considered, depending on the sensitivity of adjacent parts.

1.5.3 Flexural design at ultimate stage

In addition to the Serviceability Limit State analysis and design previously discussed in Section 1.5.2, an ultimate limit state check according to Eurocode 2, Section 6 is required on all prestressed concrete members. Under ultimate limit state, both factored dead and live loads are taken into consideration with secondary prestressing effects. In strength design, primary prestressing effects are commonly regarded as a part of a section strength, whereas secondary prestressing effects can be detrimental to mid-spans of beams. In design, primary effects are commonly treated as sectional strengths and hyperstatic effects as imposed loads. Readers are referred to Section 1.3.3.1 for further explanations. A goal of ultimate strength design is to prevent prestressed concrete members from failing before they reach a design factored load. A fundamental design requirement for a flexural strength is that a resisting design moment capacity M_{Rd} at any section should be greater than or equal to a maximum applied bending moment M_{Ed} for all load combinations.

REFERENCES

[1.1] Utracon System. (2019). Types of Post Tensioning, Utracon Structural Systems Pvt Ltd, 2019. Retrieved from: https://www.utraconindia.com/types-of-post-tensioning.html

[1.2] TR43, C. S. (1994). *Post-tensioned Concrete Floors-Design Handbook* (p. 162). Technical Report 43, Concrete Society.

[1.3] European Committee for Standardization. (2004). *BS EN 1992-1-1:2004. Eurocode 2: Design of Concrete Structures – Part1-1: General Rules and Rules for Buildings*. European Committee for Standardization, Brussels, Belgium.

[1.4] Nguyen, M. C., & Hong, W. K. (2022). Analytical prediction of nonlinear behaviors of beams post-tensioned by unbonded tendons considering shear deformation and tension stiffening effect. *Journal of Asian Architecture and Building Engineering*, *21*(3), 908–929.

[1.5] BYJU'S Exam Prep. (2022, August 3). Losses in Prestressed Concrete: Types, Causes, Loss. Gradeup. https://byjusexamprep.com/losses-in-prestress-i

[1.6] Risa. (2022, December 14). RISA | Automatic Calculation of Long-Term Load Factors in ADAPT-Builder. Risa. https://blog.risa.com/post/automatic-calculation-of-long-term-load-factors-in-adapt-builder

[1.7] ACI Committee 318. (2019). Building code requirements for structural concrete: (ACI 318-19); and commentary (ACI 318R-19). Farmington Hills, MI: American Concrete Institute

[1.8] Hong, W. K., Pham, T. D., Nguyen, D. H., Le, T. A., Nguyen, M. C., & Nguyen, V. T. (2023). Experimental and nonlinear finite element analysis for post-tensioned precast frames with mechanical joints. Journal of Asian Architecture and Building Engineering, 1-24.

[1.9] An ACI Standard. (2019). *Building Code Requirements for Structural Concrete (ACI 318-19)*. American Concrete Institute, Farmington Hills, MI.

[1.10] Elghazy, M. (2018). FRCM composites for strengthening corrosion-damaged structures: experimental and numerical investigations (Doctoral dissertation, Université Laval).

[1.11] Jędrzejczak, M., & Klempka, K. (2017). Limitation of stresses in concrete according to Eurocode 2. *MATEC Web of Conferences*, *117*, 00067. https://doi.org/10.1051/matecconf/201711700067

Chapter 2

Holistic design of pretensioned concrete beams based on artificial neural networks

2.1 INTRODUCTION

2.1.1 Previous studies

Several artificial neural network (ANN)-based studies of prestressed beams have been conducted. Torky and Aburawwash [2.1] published a simple prestressed beam to demonstrate the viability of neural networks over traditional approaches. They proposed a deep learning approach to optimizing prestressed beams. Their data are limited to beam depth, beam width, bending moment, eccentricity, and a number of strands. Sumangala and Jeyasehar [2.2] also studied a damage assessment procedure using an ANN for prestressed concrete beams. They formulated a methodology using results obtained from an experimental study conducted in the laboratory. The measured output from both static and dynamic tests was taken as input to train a neural network based on MATLAB. A quantitative evaluation of the degree of damage was possible by ANN using a natural frequency and stiffness in the postcracking range.

However, these studies are not dedicated to a holistic design of prestressed beams for practical applications. In this book, both forward and reverse designs for prestressed beams are possible, with real-world big datasets of 50,000 generated using the full scale of structural mechanics-based software. Forward and reverse ANNs are trained using their previously suggested training methods shown below. Hong et al. [2.3] proposed training methods based on a feature selection for a reverse design of doubly reinforced concrete beams. They trained large datasets using training entire data (TED), parallel training method (PTM), chained training scheme (CTS), and chained training scheme with revised sequence (CRS) that they developed. Hong et al. also presented artificial intelligence-based novel design charts for doubly reinforced concrete beams [2.4] and novel design methods for reinforced concrete columns [2.5]. Hong and Pham [2.6] performed reverse designs of doubly reinforced concrete beams using Gaussian process regression models enhanced by sequence training/designing techniques based on feature selection algorithms. The training methods

DOI: 10.1201/9781003354796-2

developed in [2.3–2.6] are used to map 15 input parameters to 18 output parameters for a holistic design of pretensioned concrete beams. The large datasets, which contain 33 parameters including 15 input and 18 output parameters, aid the accurate training of ANNs that represent prestressed beams. The proposed design replaces conventional software. This chapter was written based on the previous paper by Hong et al. [2.7].

2.1.2 Reverse design scenarios of pretensioned concrete beams

In this chapter, ANNs are used to holistically design pretensioned concrete (PT) beams. Large input and output data are generated using a mechanics-based software AutoPTbeam to establish reverse design scenarios in which back-substitution (BS) method using ANN-trained reverse-forward networks is proposed to provide reverse designs with 15 input and 18 output parameters for engineers. ANNs for reverse designs of PT beams are formulated based on 15 input structural parameters to investigate the performances of PT beams with pin-pin boundaries. Reverse designs based on neural networks are suggested when preferable control parameters representing design targets are placed on an input-side. The preferable control parameters include four output parameters ($q_{L/250}$, $q_{0.2mm}$, q_{str}, μ_{Δ}) which are reversely preassigned on an input-side for reverse scenarios. All associated design parameters, including crack width, rebar strains at transfer load stage, rebar strains, and displacement ductility ratio at the ultimate load stage, are then calculated on an output-side. For a reverse network of Step 1 of BS method, better design accuracies for a deflection ductility ratios (μ_{Δ}) reversely preassigned on an input-side were observed when deep neural networks were trained by CRS than when training ANNs based on PTM and based on shallow neural networks trained by CRS.

2.1.3 Significance of the Chapter 2

A design of the PT structures can be performed traditionally by numerous computer-aided engineering tools, including Computer-Aided Design (CAD) packages, FEM software, and self-written calculation codes. In this chapter, ANNs are used to design PT structures in both forward and reverse directions based on EC2 [2.8]. A design can be progressed using ANNs trained on big data when big datasets are good enough to show a tendency of PT beam behaviors. Forward and reverse designs using ANNs can be performed based on large datasets, without having primary engineering knowledge to accurately design parameters. Both forward and reverse designs for PT beams are possible using the proposed networks, which is challenging to achieve with classic techniques. Classic design software can be completely replaced by ANN based PT beam design with sufficient training accuracies while exhibiting excellent productivity for both forward and reverse designs. Numerical examples are provided to demonstrate the design steps, constructing design charts under the proposed design scenarios. Design charts can be extended as further as possible to meet the requirements of engineers. The proposed method is advantageous in that it is less dependent on problem types such as column, beam, frame, seismic design, and so on, and instead, the ANN-based method relies on characteristics of large datasets of the considered problem. The applications of proposed method are not limited to designing PT beams but can also be extended to other structures.

2.1.4 Tasks readers can perform after reading this chapter

This chapter demonstrates how to design pretensioned concrete beams based on ANNs. Large amounts of input and output parameters are generated to establish forward and reverse design scenarios. Designs with 15 input and 18 output parameters are proposed to train ANN-based networks for engineers. Tasks readers can perform after reading this chapter include:

1. BS method based on two-step networks to solve reverse design scenarios; in the reverse network of Step 1, reverse output parameters are determined which are, then, used as input parameters in the forward network of Step 2 to determine the rest of the design parameters.
2. Diverse training methods; TED, PTM, and CRS can be used for both the reverse and forward networks with selected training parameters such as a number of hidden layers, neurons, and validation checks. Extra input features can be picked from an output-side when training ANNs based on CRS which is not possible with TED, in which all inputs are mapped to entire outputs simultaneously [2.3].
3. Selection of training parameters; formulating training networks depends on many parameters such as feature scores, volume of datasets, and types of big datasets. For the reverse network in Step 1 of BS method, Deep artificial Neural Networks (DNNs) trained with CRS outperform Shallow artificial Neural Networks (SNNs) trained with CRS and ANNs trained with PTM. When the volume of datasets is insufficient, DNNs provide less accurate design accuracies than those provided by SNNs.
4. Use of ANNs in practical designs; ANNs trained on datasets generated from pretensioned concrete beams produce acceptable design accuracies for use in practical reverse designs. Design parameter tables derived from reverse designs can be extended to plot design charts that inter-connect all design parameters to achieve entire designs in a streak. The reverse design facilitates a rapid identification of design parameters, helping engineers obtain fast decisions with acceptable accuracies.

2.2 INPUT AND OUTPUT PARAMETERS FOR FORWARD DESIGNS

2.2.1 Description of input parameters

ANNs are used to design bonded pretensioned beams with pin-pin ends shown in Figure 2.2.1, in which beam width (mm) b, beam depth (mm) h, and beam length (mm) L of bonded pretensioned beams with pin-pin ends are displayed. In Table 2.2.1, 15 input parameters are selected to generate big datasets using AutoPTbeam for an ANN-based reverse design. Concrete properties include compressive cylinder strength of concrete at 28 days (MPa, f_{ck}), compressive cylinder strength of concrete at transfer stage (MPa, f_{cki}), and long-term deflection factor considering creep effect

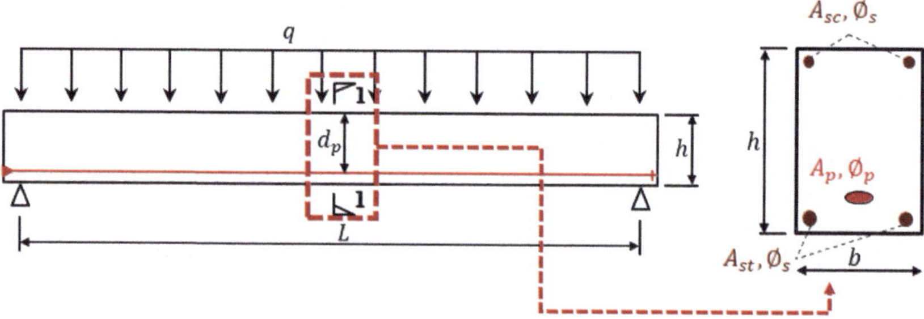

Figure 2.2.1 Bonded pretensioned beams with pin-pin ends for an AI-based design.

Table 2.2.1 Fifteen input parameters to generate big datasets using AutoPTbeam for pretensioned beams

15 Inputs					
No.	*Symbol*	*Definition*	*No*	*Symbol*	*Definition*
1	b	Beam width (mm)	9	ϕ_s	Preferred rebar diameter (mm)
2	h	Beam depth (mm)	10	ρ_p	Tendon ratio
3	f_{ck}	Concrete strength (MPa)	11	f_{py}	Yielding stress of tendon (MPa)
4	f_{ck}	Concrete strength at transfer stage (MPa)	12	ϕ_p	Preferred tendon diameter (mm)
5	ρ_{st}	Bottom rebar ratio	13	$PTloss_{lt}$	Long-term prestress losses
6	ρ_{sc}	Top rebar ratio	14	L	Beam length (mm)
7	f_{sy}	Yielding stress of rebar (MPa)	15	K_c	Creep coefficient
8	f_{sw}	Yielding stress of stirrup (MPa)			

(normally 2.5 ~ 3.5, K_c). Reinforcing bar properties include top, bottom rebar ratios (ρ_{st} and ρ_{sc}), yield stress of rebar (MPa, f_{sy}), yield stress of stirrup (MPa, f_{sw}), and preferred diameter (mm, ϕ_s). Tendon properties include tendon ratios (ρ_p), yield stress (MPa, f_{py}), and preferred diameter (mm, ϕ_p). Long-term prestress loss ($PTLoss_{lt}$) is also one of fifteen input parameters. Tensile strength of tendon f_{pu} is set as a constant value of 1,860 MPa, and hence, f_{pu} is not included in inputs of big datasets.

2.2.2 Network training

2.2.2.1 Description of output parameters

Three load stages are investigated for prestressed designs as shown in Figure 2.2.2. Table 2.2.2 presents 18 output parameters including seven output parameters at the transfer load stage, five output parameters at the service load limit stages, and six outputs at the ultimate load limit stage which are explored in this chapter. Seven outputs of forward designs at a transverse stage are presented in Table 2.2.2(a) where EC2 [2.8] requires concrete compressive stress (σ_c) to be smaller than $0.6f_{cki}$, crack widths to be under a limitation depending on environmental conditions, and top rebar stress (σ_{sc}) to be less than a smaller number between $0.8f_{su}$ and f_{sy}. Cracks are limited based on six classes of environmental exposures by EC2 [2.8], which are classified based on a risk of being attacked by carbonation-induced corrosions, chloride-induced corrosions, freeze thaw, or chemicals.

Five nominal strengths corresponding to maximum applied loads are shown in Table 2.2.2(b), in which service limitations should not be violated as required by EC2 [2.8]. Five design limitations of PT beams are required to meet, by EC2 [2.8], at the three sub-stages such as quasi-permanent, frequent, and characteristic in a service load stage.

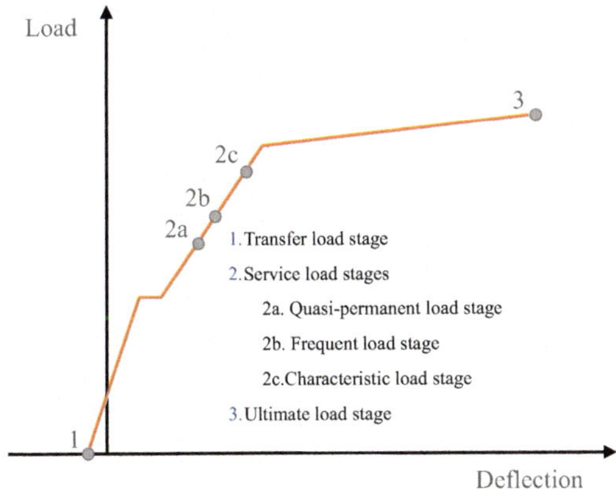

Figure 2.2.2 Three load stages of PT beams.

Table 2.2.2 18 output of forward designs of a PT beam

(a) Seven outputs at transfer load stages		
No.	*Symbol*	*Definition*
1	σ_c/f_{cki}	Top concrete stress per concrete strength at transfer stage
2	ε_{st}	Bottom rebar strain at transfer stage
3	ε_{sc}	Top rebar strain at transfer stage
4	ε_p	Tendon strain at transfer stage
5	σ_{ct}/f_{cki}	Bottom concrete stress per concrete strength at transfer stage
6	c_w	Crack width at transfer stage
7		Defection at transfer stage
(b) Five output parameters at SLL stage (Quasi, frequent, characteristic)		
No.	*Symbol*	*Definition*
8	$q_{L/250}$	Nominal strength when deflection reaches to L/250
9	$q_{0.2mm}$	Nominal strength when crack width reaches to 0.2 mm
10	$q_{0.6f_{ck}}$	Nominal strength when concrete stress reaches to stress limitation ($0.6f_{ck}$)
11	$q_{0.8fsu}$	Nominal strength when rebar stress reaches to stress limitation $\left(\min\left[0.8f_{su};f_{sy}\right]\right)$
12	$q_{0.75fpu}$	Nominal strength when tendon stress reaches to stress limitation $\left(\min\left[0.75f_{pu};f_{py}\right]\right)$
(c) Six output parameters at ULL stage, considering both strength and ductility of member		
No.	*Symbol*	*Definition*
13	q_{str}	Maximum nominal strength that a PT beam
14	ε_{st}	Bottom rebar strain at ultimate stage
15	ε_{sc}	Top rebar strain at ultimate stage
16	ε_p	Tendon strain at ultimate stage
17	ρ_{sw}	Stirrup ratio
18	μ_Δ	Deflection ductility

First, deflections should not exceed a limitation of L/250 at a quasi-permanent load stage. Secondly, crack widths under frequent load combinations should not exceed a limitation (uncrack, 0.1, 0.2, or 0.4 mm). Lastly, concrete compressive stresses (σ_c) under characteristic load combinations should be smaller than $0.6f_{ck}$ to avoid longitudinal cracks. Finally, rebar stresses are limited at a minimum of $0.8f_{su}$ and f_{sy}, whereas tendon stresses (σ_P) should not exceed a smaller number between $0.75f_{pu}$ and f_{py} to avoid unacceptable cracks or deformations at characteristic load combinations.

Six output parameters of forward designs at ultimate stages are presented in Table 2.2.2(c). Maximum nominal strength, rebar strains, tendon strains, and deflection ductility at ultimate stage are calculated by AutoPTbeam, providing a clear insight into structural behaviors under fracture loads by which concrete strain reaches the limitation 0.0035 based on EC2. All 33 input and output parameters used for generating large datasets to train ANNs are presented in Tables 2.2.1 and 2.2.2. ANNs are trained on the large datasets for ANN-based forward (Section 2.4) and reverse designs (Section 2.5) of simply supported bonded pretensioned beams for all stages.

2.2.2.2 Generation of large structural datasets and network training

A list of random variables, their corresponding ranges, and data distributions from the large structural datasets are presented in Table 2.2.3 and Figures 2.2.3–2.2.6. Figure 2.2.7(a) shows the algorithm of AutoPTbeam which was developed for a forward design of bonded pretensioned beams by Nguyen and Hong [2.9]. Flow charts for generating large structural datasets for training ANNs are also presented in Figure 2.2.7(b), whereas large structural datasets with selected non-normalized 15 inputs and 18 outputs generated based on Figure 2.2.7(b) are illustrated.

Regression models are often evaluated using several common metrics, such as Mean Absolute Error (MAE), Mean Absolute Percentage Error, Root Mean Squared Error, and Coefficient of Determination (R^2) [2.10]. The present chapter employs MAE and correlation coefficients (R) calculated according to Eqs. (2.2.1) and (2.2.2) to preliminarily benchmark performances of ANN models.

Reliabilities of ANN models should be concluded based on errors in practical designs, and hence, engineers should always check differences between results provided by ANN predictions and structure mechanic calculations. In the present study, a training method such as CRS training scheme is used to enhance design accuracies based on adjusted numbers of layers and neurons.

$$\boldsymbol{MSE} = \frac{\sum_{i=1}^{n} \boldsymbol{E}_i^2}{\boldsymbol{n}} \tag{2.2.1}$$

$$\boldsymbol{R} = \frac{1}{n-1}\sum_{i=1}^{N}\left(\frac{A_i-\mu_A}{\sigma_A}\right)\left(\frac{P_i-\mu_P}{\sigma_P}\right) \tag{2.2.2}$$

where:

A – Target values in test datasets.
P – Predicted values in test datasets.
E – Errors in test procedures; $E = A - P$.
n – A number of data in test datasets.
μ_A – Mean value of targets in test datasets.
σ_A – Standard deviation of targets in test datasets.
μ_P – Mean value of prediction in test datasets.
μ_P – Standard deviation of prediction in test datasets.

Table 2.2.3 List of random variables and corresponding ranges of the big structural datasets

(a) Big datasets for bonded pretensioned beams; non-normalized 15 inputs

	b (mm)	h (mm)	f_{ck}(MPa)	f_{cki} (MPa)	ρ_{st}	ρ_{sc}	f_{sy} (MPa)	f_{sw}(MPa)
Large data	555.27	837	45	33.43	0.0300	0.0225	447	329
	1,034.07	646	44	26.11	0.0199	0.0214	524	341
	395.70	893	42	20.70	0.0329	0.0106	545	322
	2,970.61	980	35	23.94	0.0148	0.0174	526	486
				...				
Max	3,977.23	1,000	50	40.00	0.0400	0.0400	600	500
Min	203.58	500	30	12.00	0.0012	0.0005	400	300
Mean	1,645.86	749.84	40.04	23.97	0.0208	0.0203	499.46	400.11
Standard deviation	857.44	144.67	6.07	5.92	0.0110	0.0113	57.99	58.27
Variance	735,210.02	20,931.70	36.89	35.03	0.0001	0.0001	3,363.86	3,396.45

	ϕ_s	ρ_p	f_{py}(MPa)	ϕ_p	$PTloss_{lt}$	L(mm)	K_c
i. Large data	29	0.0020	1,693	20.2	0.11	14,224	3.28
	22	0.0012	1,662	19.5	0.16	12,262	2.93
	14	0.0024	1,600	10.7	0.14	9,115	2.94
	25	0.0028	1,615	18.2	0.14	11,109	2.88
				...			
Max	32	0.0032	1,700	21.5	0.20	19,892	3.50
Min	10	0.0006	1,600	9.5	0.10	4,058	2.50
Mean	21	0.0018	1,650.08	15.5	0.15	10,476	3.00
Standard deviation	6.35	0.0006	28.98	3.4	0.029	3,332	0.288
Variance	40.41	4.7675	839.88	11.9	0.0008	11,106,931	0.083

(*Continued*)

Table 2.2.3 (Continued) List of random variables and corresponding ranges of the big structural datasets

(b) Big datasets for bonded pretensioned beams; non-normalized 18 outputs							
(b-1) Non-normalized seven outputs of a transfer stage							
	Transfer stage						
	σ_c/f_{cki}	ε_{st}	ε_{sc}	ε_p	σ_{ct}/f_{cki}	cw	Δ(mm)
ii. Large data	–0.039	–0.00014	0.00002	0.006953	0.151	0.000	–3.74
	–0.025	–0.00010	0.00000	0.007001	0.103	0.000	–1.02
	–0.108	–0.00023	0.00005	0.006685	0.239	0.000	–3.42
	0.000	–0.00036	0.00014	0.006621	0.354	0.018	–7.44
				…			
Max	0.035	–0.00001	0.00105	0.007086	0.797	0.323	20.63
Min	–0.129	–0.00089	–0.00003	0.006107	0.010	0.000	–32.59
Mean	–0.044	–0.00020	0.00005	0.006862	0.204	0.005	–3.57
Standard deviation	0.034	0.0001	6.6e-05	0.00012	0.107	0.013	3.249
Variance	0.001	1.18e-08	4.36e-09	1.49e-08	0.0115	0.00017	10.55

(b-2) Non-normalized five outputs of a service stage					
	iii. Service stage				
	$q_{L/250}$ (kN/m)	$q_{0.2mm}$ (kN/m)	$q_{0.6fck}$ (kN/m)	$q_{0.8fsu}$ (kN/m)	$q_{0.75fpu}$ (kN/m)
iv. Large data	49.17	106.10	95.44	161.74	81.18
	44.50	92.81	111.31	172.16	97.73
	164.26	258.75	179.56	396.24	253.55
	814.41	1,004.05	963.57	1,585.56	1,015.59
			…		
Max	3,219.04	4,194.60	3,589.44	6,130.08	5,236.87
Min	2.45	2.65	8.28	7.11	2.87
Mean	300.80	441.55	426.48	686.60	429.37
Standard deviation	341.78	439.91	387.76	664.14	430.99
Variance	116,819.81	193,525.337	150,362.42	441,092.31	185,759.85

(Continued)

Table 2.2.3 (Continued) List of random variables and corresponding ranges of the big structural datasets

(b-3) Non-normalized six outputs of a ultimate stage						
	Ultimate stage					
	q_{str} (kN/m)	ε_{st}	ε_{sc}	ε_p	ρ_{sw}	μ_Δ
Large data	178.02	0.0099	–0.00239	0.01641	0.0052	1.66
	188.93	0.0150	–0.00158	0.02127	0.0031	1.64
	208.93	0.0011	–0.00121	0.00703	0.0274	1.00
	1,698.61	0.0154	–0.00220	0.02172	0.0034	1.87
			...			
Max	4,622.31	0.0526	0.00105	0.05913	0.0334	10.26
Min	7.80	0.0003	–0.00313	0.00598	0.0009	1.00
Mean	595.59	0.0135	–0.00140	0.01971	0.0065	1.97
Standard deviation	527.73	0.0113	0.00064	0.01137	0.0064	1.0037
Variance	278,503.34	0.000128	4.106e-07	0.000129	4.096e-05	1.007

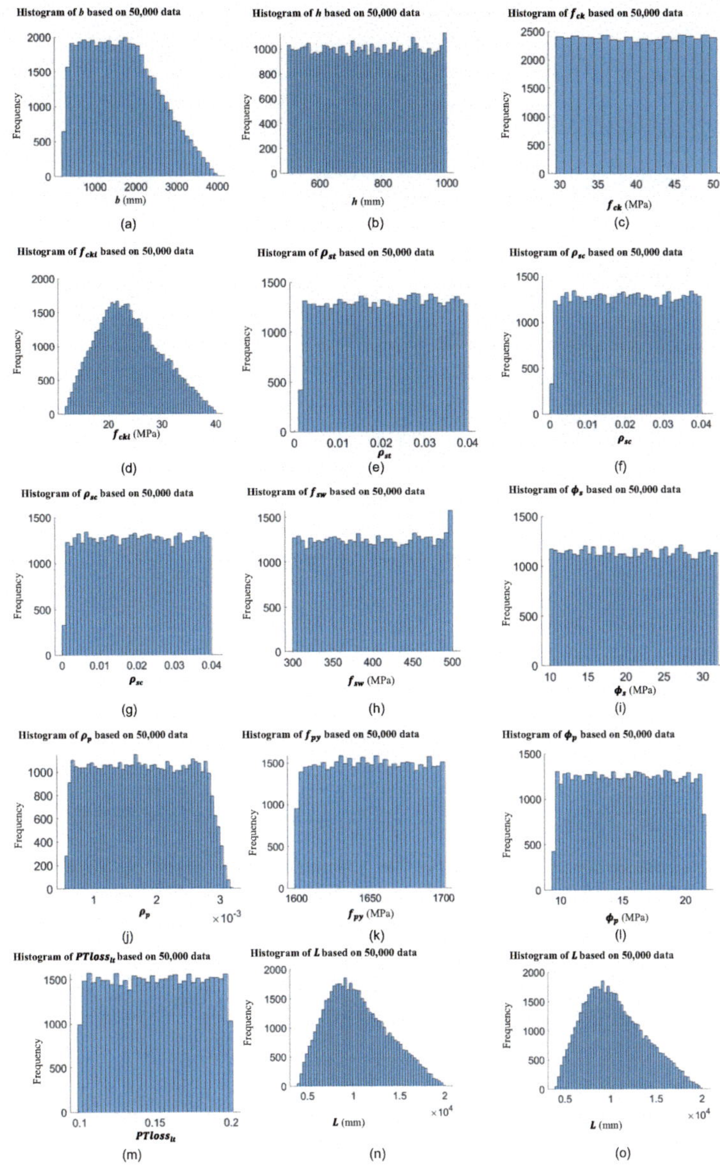

Figure 2.2.3 Histogram of 15 inputs based on 50.000 data. (a) Histogram of b based on 50,000 data. (b) Histogram of h based on 50,000 data. (c) Histogram of f_{ck} based on 50,000 data. (d) Histogram of f_{cki} based on 50,000 data. (e) Histogram of ρ_{st} based on 50,000 data. (f) Histogram of ρ_{sc} based on 50,000 data. (g) Histogram of f_{sy} based on 50,000 data. (h) Histogram of f_{sw} based on 50,000 data. (i) Histogram of ϕ_s based on 50,000 data. (j) Histogram of ρ_p based on 50,000 data. (k) Histogram of f_{py} based on 50,000 data. (l) Histogram of ϕ_p based on 50,000 data. (m) Histogram of $PTloss_{lt}$ based on 50,000 data. (n) Histogram of L based on 50,000 data. (o) Histogram of K_c based on 50,000 data.

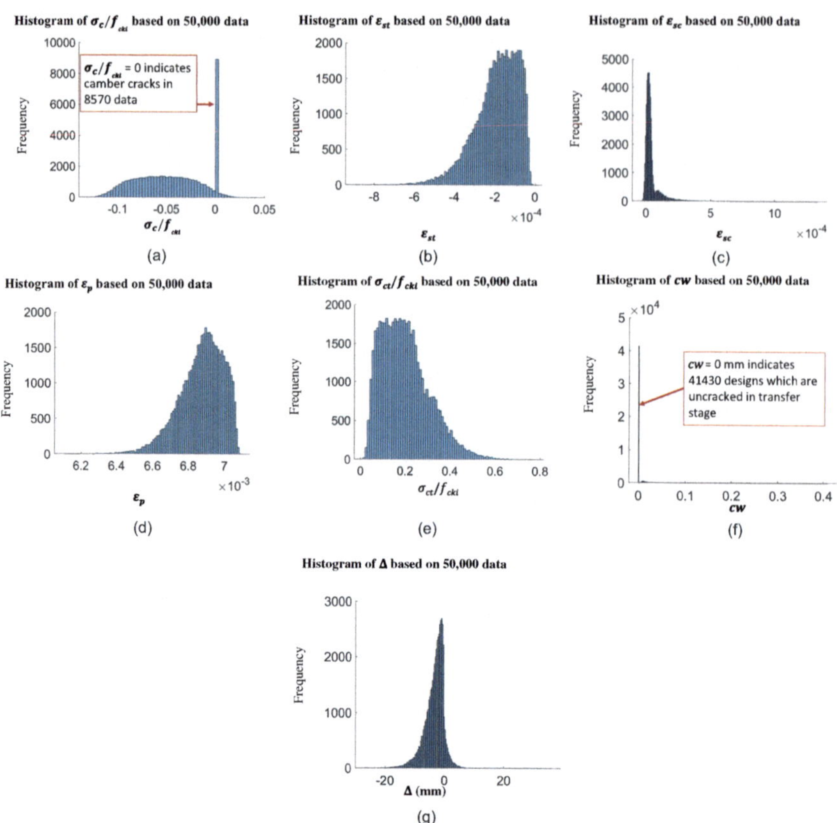

Figure 2.2.4 Histogram of seven outputs of a transfer stage based on 50.000 data. (a) Histogram of σ_c/f_{cki} based on 50,000 data. (b) Histogram of ε_{st} based on 50,000 data. (c) Histogram of ε_{sc} based on 50,000 data. (d) Histogram of ε_p based on 50,000 data. (e) Histogram of σ_{ct}/f_{cki} based on 50,000 data. (f) Histogram of c_w based on 50,000 data. (g) Histogram of Δ based on 50,000 data.

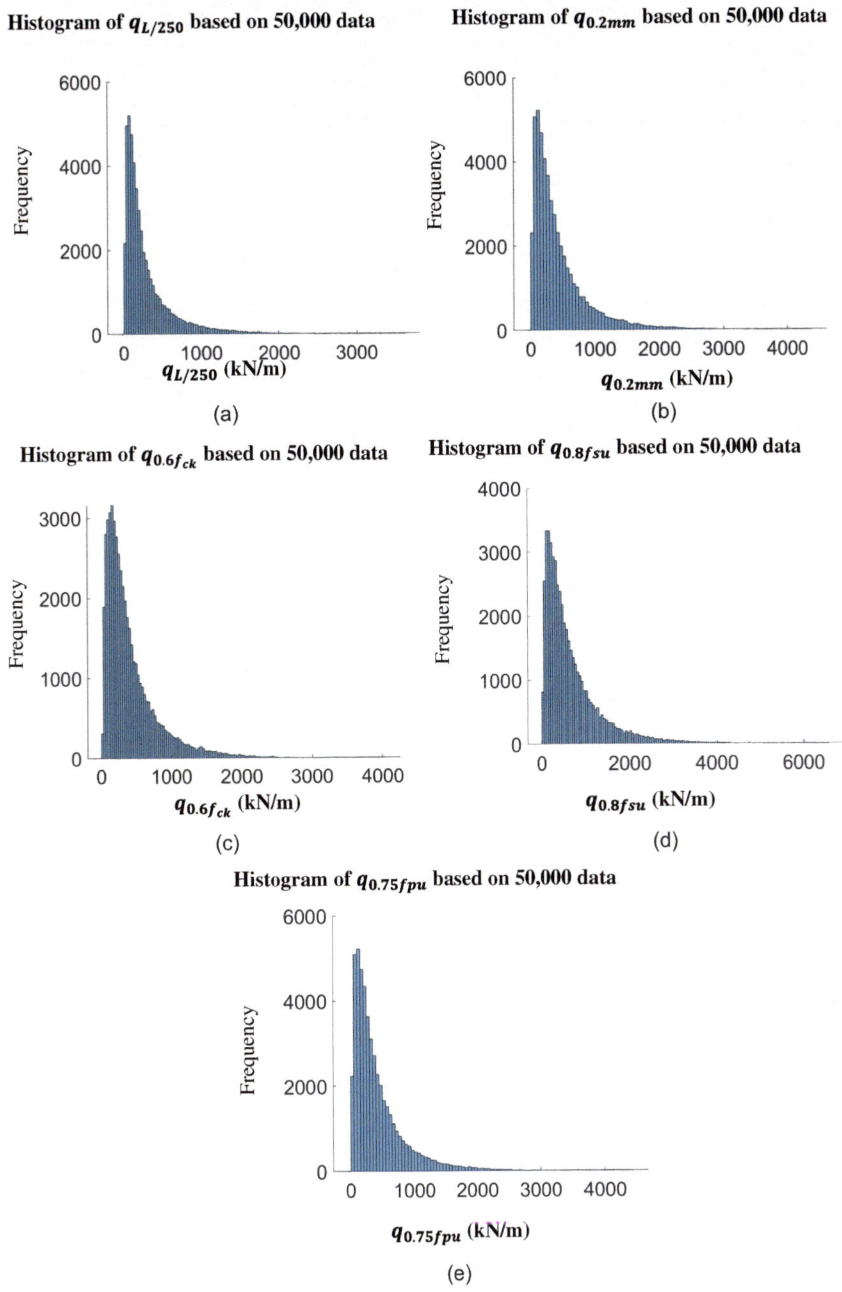

Figure 2.2.5 Histogram of five outputs of a service stage based on 50.000 data. (a) Histogram of $q_{L/250}$ based on 50,000 data. (b) Histogram of $q_{0.2mm}$ based on 50,000 data. (c) Histogram of $q_{0.6f_{ck}}$ based on 50,000 data. (d) Histogram of $q_{0.8fsu}$ based on 50,000 data. (e) Histogram of $q_{0.75fpu}$ based on 50,000 data.

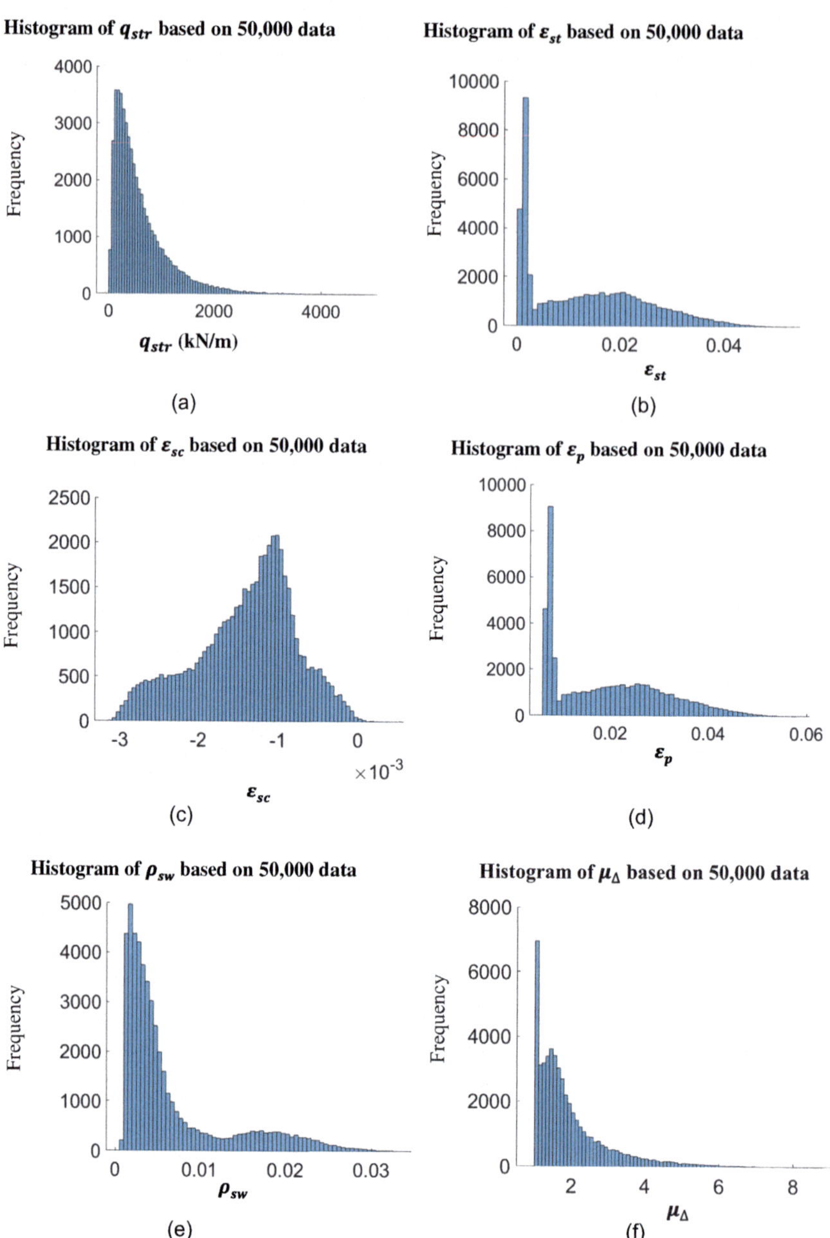

Figure 2.2.6 Histogram of six outputs of an ultimate stage based on 50.000 data. (a) Histogram of q_{str} based on 50,000 data. (b) Histogram of ε_{st} based on 50,000 data. (c) Histogram of ε_{sc} based on 50,000 data. (d) Histogram of ε_{ρ} based on 50,000 data. (e) Histogram of ρsw based on 50,000 data. (f) Histogram of μ_Δ based on 50,000 data.

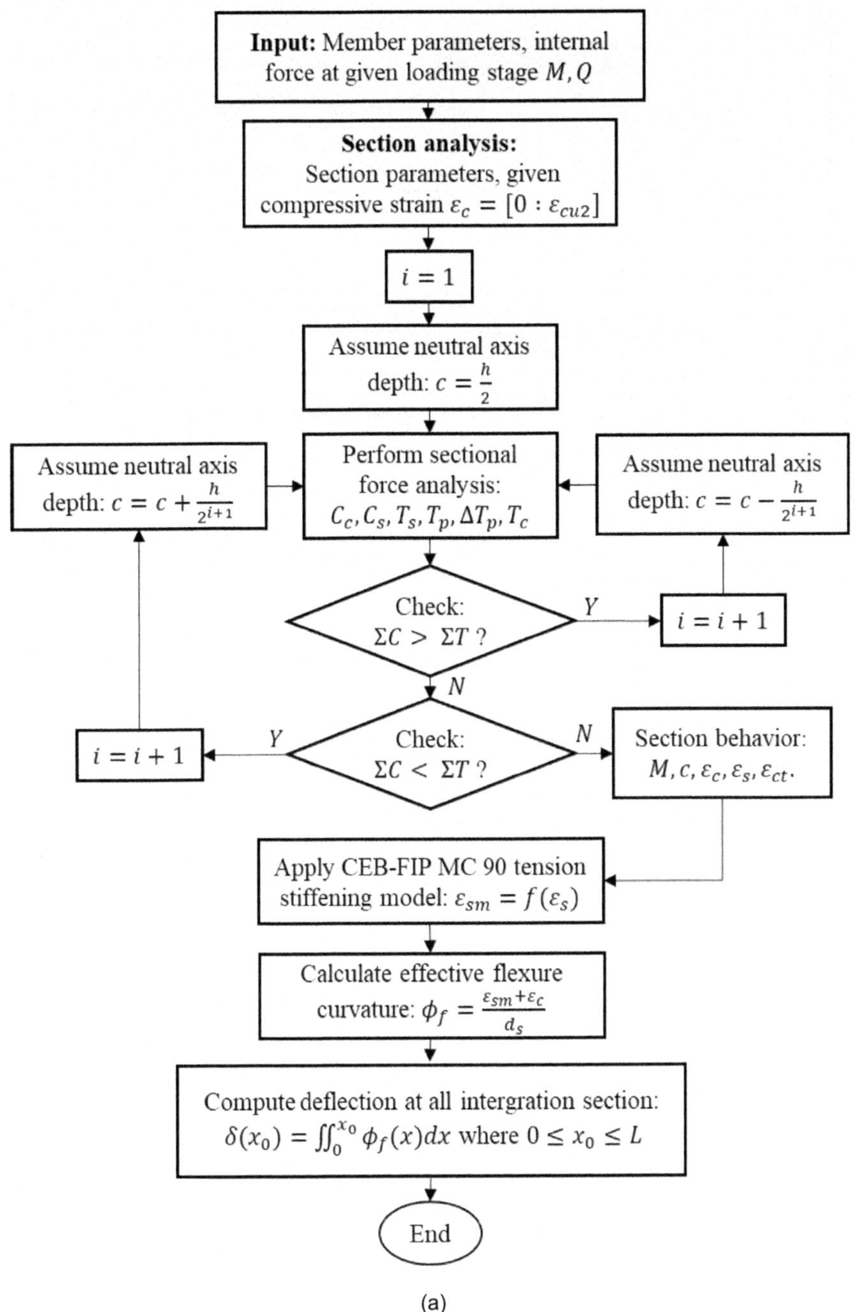

Figure 2.2.7 Generation of large datasets of bonded pretensioned beams for AI-based design. (a) Algorithm for designing bonded pretensioned beams (AutoPTbeam). (b) Generation of large datasets.

(Continued)

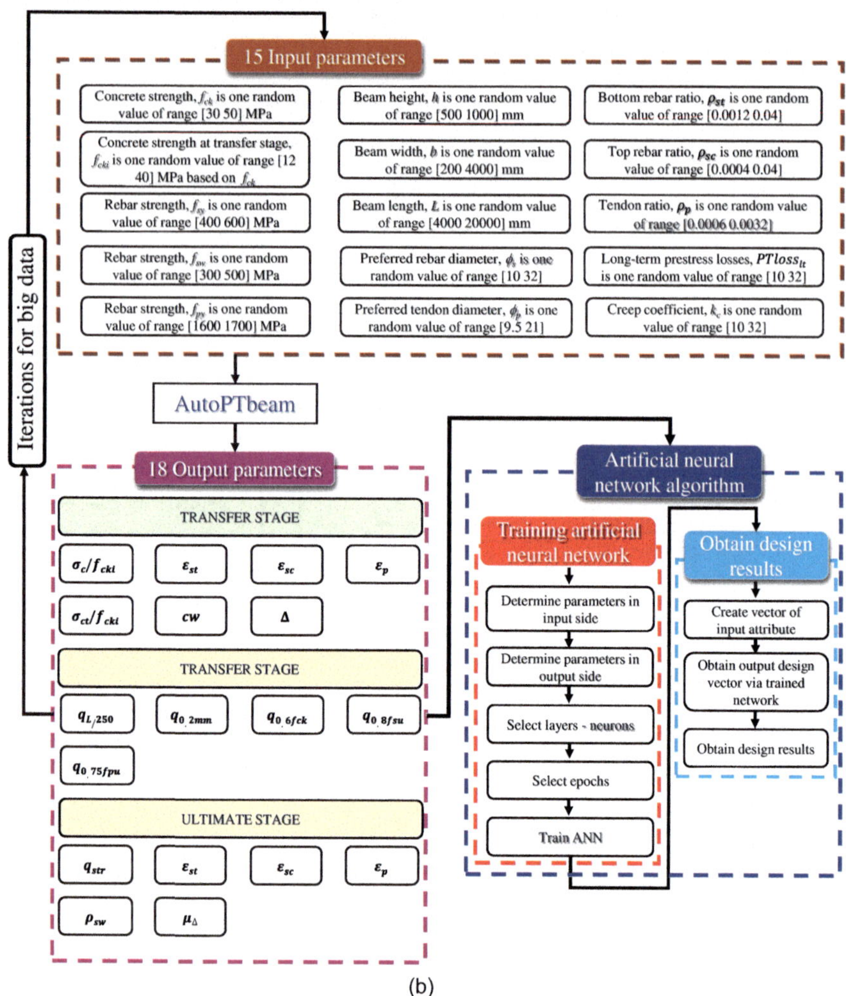

(b)

Figure 2.2.7 (Continued) Generation of large datasets of bonded pretensioned beams for AI-based design. (a) Algorithm for designing bonded pretensioned beams (AutoPTbeam). (b) Generation of large datasets.

2.3 DESIGN SCENARIOS

Design scenarios based on one forward design and three reverse designs are presented in Figure 2.3.1(a) and (b), respectively which investigates one forward design and three reverse designs. An ANN predicts 18 outputs for given 15 inputs.

Figure 2.3.1(b) illustrates three reverse scenarios. In reverse scenario 1, two design parameters (ρ_{st} and ρ_p) shown in green Boxes 5 and 10 of Figure 2.3.1(b-1) are calculated on an output-side when preassigning nominal

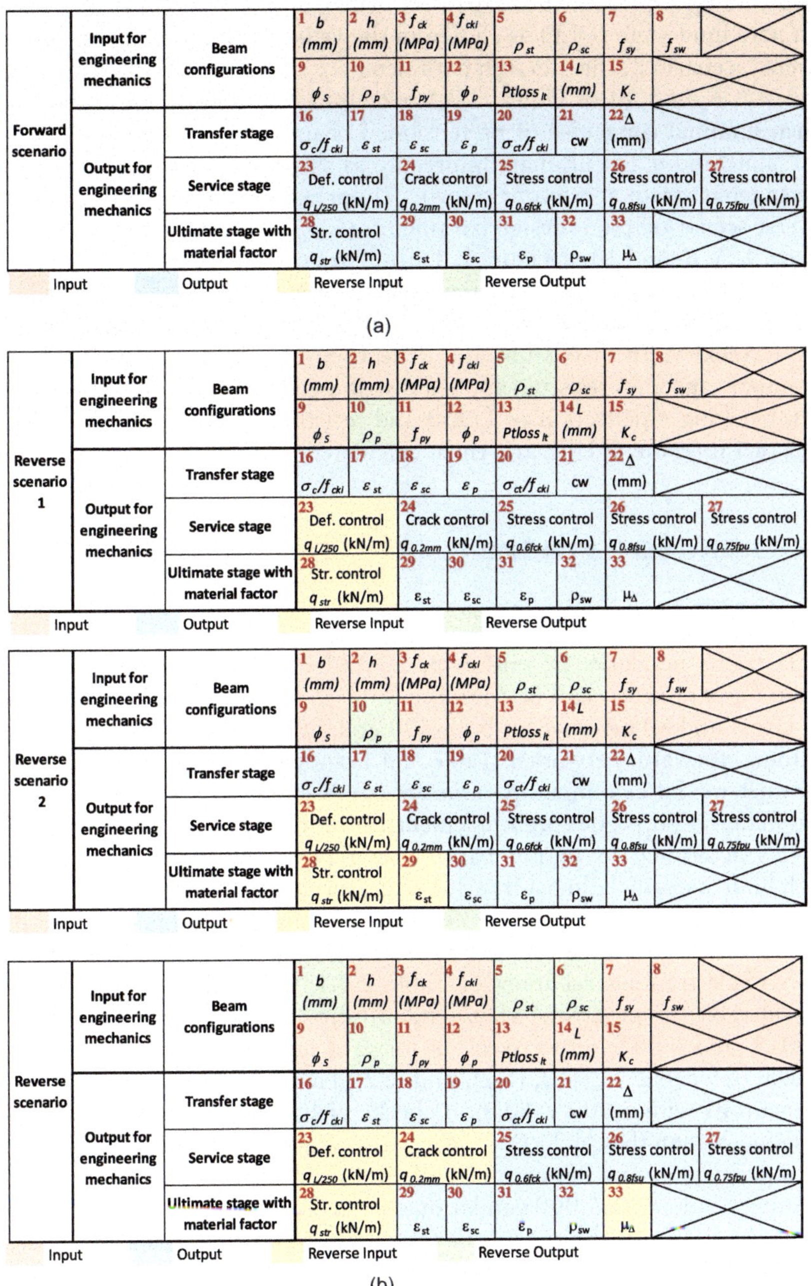

Figure 2.3.1 Design scenarios; one forward and three reverse designs. (a) Forward scenario. (b)-1 reverse scenario 1. (b)-2 reverse scenario 2. (b)-3 reverse scenario 3.

capacities ($q_{L/250}$ reaching L/250 at service load limit state (SLL) and q_{str} at ultimate limit state (ULL) as shown in the yellow cell) on an input-side. In reverse scenario 2, three design parameters (ρ_{st}, ρ_{sc}, and ρ_p) shown in green Boxes 5, 6, and 10 of Figure 2.3.1(b-2) are calculated on an output-side when nominal capacities of pretensioned beams ($q_{L/250}$ reaching L/250 at SLL state and q_{str} at ultimate limit state as shown in the yellow cells) and lower rebar strain at transfer stage ε_{st} are preassigned on an input-side. In reverse scenario 3, four design parameters (b, ρ_{st}, ρ_{sc}, and ρ_p) shown in green Boxes 1, 5, 6, and 10 of Figure 2.3.1(b-3) are calculated on an output-side when nominal capacities of pretensioned beams ($q_{L/250}$, $q_{0.2mm}$, q_{str}, and μ_Δ) are prescribed on an input-side. $q_{0.2mm}$ and μ_Δ represent applied loads reaching a crack width of 0.2 mm and deflection ductility ratios, respectively. Nominal capacities resisting applied distributed load ($q_{L/250}$) of 70 kN/m at SLL reaching a deflection of L/250 and resisting applied distributed load (q_{str}) of 150 kN/m at ULL are, then, calculated.

2.4 TRAINING METHODS

2.4.1 TED

TED trains networks on entire inputs and output data simultaneously. Training parameters can be found in Table 2.4.1 where TED implements 80 layers and 80 neurons for both 500 and 5,000 validation checks when training forward networks. Table 2.4.1 shows the remaining training parameters. Fifteen input parameters representing beam configurations and material properties are implemented in forward designs of PT beams. ANNs determine 18 output parameters representing beam behaviors at each limit state, including a transfer stage, service limit stages, and ultimate limit stage. A total number of 33 parameters including input and output parameters is too large to train networks based on TED as can be seen with weak training accuracy of Table 2.4.1. Tables 2.4.2(a)–(c) present considerable design errors when tendon ratios are preassigned as 0.0007, 0.0015, and 0.002, respectively, in forward design. ANNs trained using TED.b (refer to 2.4.2(a-2), (b-2), and (c-2)) based on 80 hidden layers and 80 neurons with 5,000 validation checks yield a better accuracy to MSE & R-index than that of TED.a (refer to 2.4.2(a-1), (b-1), and (c1)) based on 80 hidden layers and 80 neurons with 500 validation checks, showing that training results with 5,000 validation checks are better than 500 validation checks as shown in Table 2.4.1. Crack width (cw) is regarded as zero when networks predict uncracked section as shown in Table 2.4.2(a), whereas non-zero crack width (cw) is calculated when networks predict cracked section as shown in Tables 2.4.2(b) and (c) where PT beam section is denoted as "Cracked" in σ_{ct}/f_{cki}. The accuracies of trainings based on TED are found unacceptable for designs.

Table 2.4.1 Training summary of forward designs based on TED

No.	Data	Layers	Neurons	*Validation check*	*Required epoch*	*Best epoch for training*	*Stopped epoch*	*Mean square error*	*R at best epoch*	*Comments*
FORWARD PROBLEM: 15 Inputs (b, h, fck, f_{cki}, ρst, ρsc, f_{sy}, f_{sw}, ϕ_S, ρ_p, f_{py}, ϕ_P, $PTloss_{lt}$, L, K_c) – (7+6+5) Outputs: (Transfer 7 Outputs: $\sigma c/fcki$, εst, εsc, ε_p, $\sigma ct/fcki$, cw, Δ) (Service Load 5 Outputs: $q_{L/250}$, $q_{0.2mm}$, $q_{0.6f_{ck}}$, $q_{0.8f_{su}}$, $q_{0.75f_{pu}}$) (Ultimate Load 6 Outputs: $qstr$, εst, εsc, εp, ρsw, $\mu\Delta$)										
15 Inputs (b, h, fck, f_{cki}, ρst, ρsc, f_{sy}, f_{sw}, ϕ_S, ρ_p, f_{py}, ϕ_P, $PTloss_{lt}$, L, K_c) – 18 Outputs – TED										
TED.**a**	50,000	80	80	500	100,000	44,023	44,523	0.001768	0.98778	Neutral
TED.**b**	50,000	80	80	5,000	100,000	31,376	36,376	0.001497	0.9908	Neutral

Table 2.4.2 Design accuracies based on TED.a vs. TED.b

(a) Tendon ratio of 0.0007

(a-1) Tendon ratio of 0.0007 based on TED.a

PARAMETERS		Training model	Network test result	Auto PT-BEAM check	Error
BEAM CONFIGURATION	b (mm)		500	500.00	0.00%
	h (mm)		700	700.00	0.00%
	f_{ck} (MPa)		40	40.00	0.00%
	f_{cki} (MPa)		20	20.00	0.00%
	ρ_{st} (mm^2/mm^2)		0.005	0.0050	0.00%
	ρ_{sc} (mm^2/mm^2)		0.005	0.0050	0.00%
	f_{sy} (MPa)		500	500.00	0.00%
	f_{sw} (MPa)		400	400.00	0.00%
	ϕ_S (mm)		16	16.00	0.00%
	ρ_p (mm^2/mm^2)		**0.0007**	**0.0007**	0.00%
	f_{py} (MPa)		1650	1650.00	0.00%
	ϕ_p (mm)		12.7	12.70	0.00%
	$PTloss_{lt}$		0.15	0.15	0.00%
	L (mm)		10000	10000.00	0.00%
	K_c		3	3.00	0.00%
TRANSFER LOAD STAGE	σ_c/f_{cki}		-0.0312	-0.04	Uncracked
	ε_{st} (mm/mm)		-1.08E-04	-0.00009	16.17%
	ε_{sc} (mm/mm)		4.41E-06	0.00001	-144.89%
	ε_p (mm/mm)		0.00695	0.00701	-0.81%
	σ_{ct}/f_{cki}		0.11787	0.09720	17.54%
	cw (mm)		0.001	0.000	100.00%
	Δ (mm)		-1.161	-0.554	52.26%
SERVICE LOAD COMBINATION	$q_{L/250}$ (kN/m)		30.01	21.50	28.37%
	$q_{0.2mm}$ (kN/m)	TED.a	50.28	24.55	51.19%
	$q_{0.6fck}$ (kN/m)		78.51	46.93	40.23%
	$q_{0.8fpu}$ (kN/m)		94.95	42.98	54.73%
	$q_{0.75fpu}$ (kN/m)		43.76	25.02	42.82%
ULTIMATE LOAD COMBINATION	q_{str} (kN/m)		73.02	46.90	35.77%
	ε_{st} (mm/mm)		0.02493	0.02560	-2.69%
	ε_{sc} (mm/mm)		-0.00065	-0.00087	-34.68%
	ε_p (mm/mm)		0.03088	0.03200	-3.61%
	ρ_{sw} (mm^2/mm^2)		0.0017	0.0013	24.44%
	μ_Δ		2.77	3.01	-8.47%

Unacceptable errors

(Continued)

Table 2.4.2 (Continued) Design accuracies based on TED.a vs. TED.b

(a-2) Tendon ratio of 0.0007 based on TED.b

PARAMETERS		Training model	Network test result	Auto PT-BEAM check	Error
BEAM CONFIGURATION	b (mm)		500	500.00	0.00%
	h (mm)		700	700.00	0.00%
	f_{ck} (MPa)		40	40.00	0.00%
	f_{ci} (MPa)		20	20.00	0.00%
	ρ_{st} (mm^2/mm^2)		0.005	0.0050	0.00%
	ρ_{sc} (mm^2/mm^2)		0.005	0.0050	0.00%
	f_{sy} (MPa)		500	500.00	0.00%
	f_{sw} (MPa)		400	400.00	0.00%
	ϕ_S (mm)		16	16.00	0.00%
	ρ_p (mm^2/mm^2)		**0.0007**	**0.0007**	0.00%
	f_{py} (MPa)		1650	1650.00	0.00%
	ϕ_p (mm)		12.7	12.70	0.00%
	$PTloss_{lt}$		0.15	0.15	0.00%
	L (mm)		10000	10000.00	0.00%
	K_c		3	3.00	0.00%
TRANSFER LOAD STAGE	σ_c/f_{ci}	TED.b	-0.0212	-0.04	Uncracked
	ε_{st} (mm/mm)		-8.61E-05	-0.00009	-5.64%
	ε_{sc} (mm/mm)		2.22E-06	0.00001	-385.42%
	ε_p (mm/mm)		0.00698	0.00701	-0.39%
	σ_{ct}/f_{ci}		0.09030	0.09720	-7.64%
	cw (mm)		0.000	0.000	0.00%
	Δ (mm)		-0.374	-0.554	-48.27%
SERVICE LOAD COMBINATION	$q_{L/250}$ (kN/m)		60.77	21.50	64.63%
	$q_{0.2mm}$ (kN/m)		80.52	24.55	69.52%
	$q_{0.6fck}$ (kN/m)		168.96	46.93	72.22%
	$q_{0.8fsu}$ (kN/m)		158.40	42.98	72.87%
	$q_{0.75fpu}$ (kN/m)		98.34	25.02	74.56%
ULTIMATE LOAD COMBINATION	q_{str} (kN/m)		173.70	46.90	73.00%
	ε_{st} (mm/mm)		0.02664	0.02560	3.90%
	ε_{sc} (mm/mm)		-0.00069	-0.00087	-26.79%
	ε_p (mm/mm)		0.03309	0.03200	3.30%
	ρ_{sw} (mm^2/mm^2)		0.0017	0.0013	25.98%
	μ_Δ		2.71	3.01	-10.86%

Unacceptable errors

(Continued)

Table 2.4.2 (Continued) Design accuracies based on TED.a vs. TED.b

(b) Tendon ratio of 0.0015

(b-1) Tendon ratio of 0.0015 based on TED.a

PARAMETERS		Training model	Network test result	Auto PT-BEAM check	Error
BEAM CONFIGURATION	b (mm)		500	500.00	0.00%
	h (mm)		700	700.00	0.00%
	f_{ck} (MPa)		40	40.00	0.00%
	f_{cki} (MPa)		20	20.00	0.00%
	ρ_{st} (mm²/mm²)		0.005	0.0050	0.00%
	ρ_{sc} (mm²/mm²)		0.005	0.0050	0.00%
	f_{sy} (MPa)		500	500.00	0.00%
	f_{sw} (MPa)		400	400.00	0.00%
	ϕ_S (mm)		16	16.00	0.00%
	ρ_p (mm²/mm²)		**0.0015**	**0.0015**	0.00%
	f_{py} (MPa)		1650	1650.00	0.00%
	ϕ_p (mm)		12.7	12.70	0.00%
	$PTloss_{lt}$		0.15	0.15	0.00%
	L (mm)		10000	10000.00	0.00%
	K_c		3	3.00	0.00%
TRANSFER LOAD STAGE	σ_c/f_{cki}	TED.a	~0	0.00	Cracked
	ε_{st} (mm/mm)		-2.86E-04	-0.00028	3.84%
	ε_{sc} (mm/mm)		0.000055	0.000148	-169.07%
	ε_p (mm/mm)		0.00678	0.00682	-0.61%
	σ_{ct}/f_{cki}		0.30205	0.29216	3.27%
	cw (mm)		0.003	0.021	-600.00%
	Δ (mm)		-7.106	-4.787	32.64%
SERVICE LOAD COMBINATION	$q_{L/250}$ (kN/m)		-11.74	36.25	408.86%
	$q_{0.2mm}$ (kN/m)		27.74	41.65	-50.13%
	$q_{0.6fck}$ (kN/m)		-24.59	57.51	333.87%
	$q_{0.8fsu}$ (kN/m)		65.62	63.13	3.80%
	$q_{0.75fpu}$ (kN/m)		-9.61	42.47	541.80%
ULTIMATE LOAD COMBINATION	q_{str} (kN/m)		38.09	66.40	-74.31%
	ε_{st} (mm/mm)		0.01680	0.01946	-15.84%
	ε_{sc} (mm/mm)		-0.00123	-0.00143	-16.28%
	ε_p (mm/mm)		0.02385	0.02578	-8.07%
	ρ_{sw} (mm²/mm²)		0.0023	0.0013	41.29%
	μ_Δ		2.15	2.70	-25.24%

Unacceptable errors

(Continued)

Table 2.4.2 (Continued) Design accuracies based on TED.a vs. TED.b

(b-2) Tendon ratio of 0.0015 based on TED.b

PARAMETERS		Training model	Network test result	Auto PT-BEAM check	Error
BEAM CONFIGURATION	b (mm)		500	500.00	0.00%
	h (mm)		700	700.00	0.00%
	f_{ck} (MPa)		40	40.00	0.00%
	f_{cki} (MPa)		20	20.00	0.00%
	ρ_{st} (mm^2/mm^2)		0.005	0.0050	0.00%
	ρ_{sc} (mm^2/mm^2)		0.005	0.0050	0.00%
	f_{sy} (MPa)		500	500.00	0.00%
	f_{sw} (MPa)		400	400.00	0.00%
	ϕ_S (mm)		16	16.00	0.00%
	ρ_p (mm^2/mm^2)		**0.0015**	**0.0015**	0.00%
	f_{py} (MPa)		1650	1650.00	0.00%
	ϕ_p (mm)		12.7	12.70	0.00%
	$PTloss_{lt}$		0.15	0.15	0.00%
	L (mm)		10000	10000.00	0.00%
	K_c		3	3.00	0.00%
TRANSFER LOAD STAGE	σ_c/f_{cki}	TED.b	~0	0.00	Cracked
	ε_{st} (mm/mm)		-4.24E-04	-0.00028	34.99%
	ε_{sc} (mm/mm)		0.000116	0.000148	-27.17%
	ε_p (mm/mm)		0.00663	0.00682	-2.78%
	σ_{ct}/f_{cki}		0.42154	0.29216	30.69%
	cw (mm)		0.012	0.021	-75.00%
	Δ (mm)		-9.248	-4.787	48.24%
SERVICE LOAD COMBINATION	$q_{L/250}$ (kN/m)		194.10	36.25	81.32%
	$q_{0.2mm}$ (kN/m)		237.41	41.65	82.46%
	$q_{0.6fck}$ (kN/m)		271.62	57.51	78.83%
	$q_{0.8fsu}$ (kN/m)		343.27	63.13	81.61%
	$q_{0.75fpu}$ (kN/m)		216.26	42.47	80.36%
ULTIMATE LOAD COMBINATION	q_{str} (kN/m)		331.90	66.40	79.99%
	ε_{st} (mm/mm)		0.02105	0.01946	7.55%
	ε_{sc} (mm/mm)		-0.00115	-0.00143	-23.77%
	ε_p (mm/mm)		0.02669	0.02578	3.44%
	ρ_{sw} (mm^2/mm^2)		0.0022	0.0013	38.53%
	μ_Δ		2.67	2.70	-0.88%

Unacceptable errors (arrows point to: ε_{st}, ε_{sc}, σ_{ct}/f_{cki}, cw, Δ of transfer stage; all service load rows; q_{str}, ε_{sc}, ρ_{sw} of ultimate load combination)

(Continued)

Table 2.4.2 (Continued) Design accuracies based on TED.a vs. TED.b

(c) Tendon ratio of 0.002

(c-1) Tendon ratio of 0.002 based on TED.a

PARAMETERS		Training model	Network test result	Auto PT-BEAM check	Error
BEAM CONFIGURATION	b (mm)		500	500.00	0.00%
	h (mm)		700	700.00	0.00%
	f_{ck} (MPa)		40	40.00	0.00%
	f_{cki} (MPa)		20	20.00	0.00%
	ρ_{st} (mm^2/mm^2)		0.005	0.0050	0.00%
	ρ_{sc} (mm^2/mm^2)		0.005	0.0050	0.00%
	f_{sy} (MPa)		500	500.00	0.00%
	f_{sw} (MPa)		400	400.00	0.00%
	ϕ_S (mm)		16	16.00	0.00%
	ρ_p (mm^2/mm^2)		**0.002**	**0.0020**	0.00%
	f_{py} (MPa)		1650	1650.00	0.00%
	ϕ_P (mm)		12.7	12.70	0.00%
	$PTloss_{lt}$		0.15	0.15	0.00%
	L (mm)		10000	10000.00	0.00%
	K_c		3	3.00	0.00%
TRANSFER LOAD STAGE	σ_c/f_{cki}	TED.a	~0	0.00	Cracked
	ε_{st} (mm/mm)		-4.06E-04	-0.00038	5.81%
	ε_{sc} (mm/mm)		0.000183	0.000238	-29.65%
	ε_p (mm/mm)		0.00666	0.00671	-0.76%
	σ_{ct}/f_{cki}		0.41643	0.39383	5.43%
	cw (mm)		0.030	0.034	-13.33%
	Δ (mm)		-9.511	-8.003	15.86%
SERVICE LOAD COMBINATION	$q_{L/250}$ (kN/m)		42.52	44.96	-5.73%
	$q_{0.2mm}$ (kN/m)		37.25	52.25	-40.27%
	$q_{0.6fck}$ (kN/m)		66.27	63.64	3.97%
	$q_{0.8fsu}$ (kN/m)		128.14	75.30	41.24%
	$q_{0.75fpu}$ (kN/m)		88.04	53.11	39.67%
ULTIMATE LOAD COMBINATION	q_{str} (kN/m)		116.96	78.13	33.20%
	ε_{st} (mm/mm)		0.01390	0.01643	-18.20%
	ε_{sc} (mm/mm)		-0.00174	-0.00170	2.63%
	ε_p (mm/mm)		0.02038	0.02272	-11.48%
	ρ_{sw} (mm^2/mm^2)		0.0022	0.0016	27.75%
	μ_Δ		1.90	2.41	-26.86%

Unacceptable errors

(Continued)

Table 2.4.2 (Continued) Design accuracies based on TED.a vs. TED.b

(c-2) Tendon ratio of 0.002 based on TED.b

PARAMETERS		Training model	Network test result	Auto PT-BEAM check	Error
BEAM CONFIGURATION	b (mm)		500	500.00	0.00%
	h (mm)		700	700.00	0.00%
	f_{ck} (MPa)		40	40.00	0.00%
	f_{cki} (MPa)		20	20.00	0.00%
	ρ_{st} (mm^2/mm^2)		0.005	0.0050	0.00%
	ρ_{sc} (mm^2/mm^2)		0.005	0.0050	0.00%
	f_{sy} (MPa)		500	500.00	0.00%
	f_{sw} (MPa)		400	400.00	0.00%
	ϕ_S (mm)		16	16.00	0.00%
	ρ_p (mm^2/mm^2)		**0.002**	**0.0020**	0.00%
	f_{py} (MPa)		1650	1650.00	0.00%
	ϕ_p (mm)		12.7	12.70	0.00%
	$PTloss_{lt}$		0.15	0.15	0.00%
	L (mm)		10000	10000.00	0.00%
	K_c		3	3.00	0.00%
TRANSFER LOAD STAGE	σ_c/f_{cki}	TED.b	~0	0.00	Cracked
	ε_{st} (mm/mm)		-3.94E-04	-0.00038	2.91%
	ε_{sc} (mm/mm)		0.000141	0.000238	-68.22%
	ε_p (mm/mm)		0.00666	0.00671	-0.65%
	σ_{ct}/f_{cki}		0.40057	0.39383	1.68%
	cw (mm)		0.019	0.034	-78.95%
	Δ (mm)		-9.034	-8.003	11.41%
SERVICE LOAD COMBINATION	$q_{L/250}$ (kN/m)		73.50	44.96	38.83%
	$q_{0.2mm}$ (kN/m)		106.91	52.25	51.12%
	$q_{0.6fck}$ (kN/m)		115.71	63.64	45.00%
	$q_{0.8fsu}$ (kN/m)		160.55	75.30	53.10%
	$q_{0.75fpu}$ (kN/m)		108.04	53.11	50.84%
ULTIMATE LOAD COMBINATION	q_{str} (kN/m)		172.45	78.13	54.69%
	ε_{st} (mm/mm)		0.01602	0.01643	-2.59%
	ε_{sc} (mm/mm)		-0.00152	-0.00170	-11.82%
	ε_p (mm/mm)		0.02161	0.02272	-5.12%
	ρ_{sw} (mm^2/mm^2)		0.0023	0.0016	32.27%
	μ_Δ		2.15	2.41	-11.78%

Unacceptable errors

2.4.2 Parallel training method (PTM)

2.4.2.1 *Training accuracies and design tables*

Table 2.4.3 shows a training summary of forward designs based on PTM in which each output parameter of the forward ANNs is mapped by 15 input parameters separately with 500 validation checks. The FW.PTM a, b, and c indicate that ANNs are trained based on 15 layers-15 neurons, 20 layers-20 neurons, and 25 layers-25 neurons, respectively. In Table 2.4.4, for example, an ANN is denoted by FW.PTM.16a which indicates a forward ANN trained by PTM based on 15 layers-15 neurons, while 16a implies output sequential number for the 16th output which is σ_{ct}/f_{cki}. Tables 2.4.3(a)–(c) summarize training accuracies of forward networks with a specified number of layers and neurons based on PTM a, b and, c shown in Table 2.4.3 for 18 output parameters, presenting training accuracies in mapping outputs 16–22, outputs 23–27, and outputs 28–33, respectively. In Tables 2.4.4(a)–(c), three types of designs with tendon ratios set as 0.001, 0.0015, and 0.002 are performed, respectively, where output parameters are obtained in Boxes 16–22, 23–27, and 28–33, respectively, for given input parameters shown in Boxes 1–15. Networks are trained by FW.PTM.a, FW.PTM.b, and FW.PTM.c based on 15 layers-15 neurons, 20 layers-20 neurons, and 25 layers-25 neurons, respectively. Structural mechanics-based software AutoPTbeam is, then, used to verify the design accuracies. It is shown that the errors of –5.66%, –8.96%, –3.47% for an upper rebar strain ε_{st} and –19.57%, –193.33%, –7.94% for concrete strain ε_{sc} are observed at a transfer stage, respectively, based on an ANN trained by FW.PTM.18c when tendon ratios ρ_p of 0.0010, 0.0015, and 0.002 are used as shown in Tables 2.4.4(a-3)–(c-3). As shown in Tables 2.4.2 and 2.4.4, design accuracies based on PTM increase compared with those based on TED when networks are trained and predicted on each output parameter separately; however, the error is still significant based on PTM. Design accuracies are further enhanced by CTS or CRS methods that were developed by Hong [2.11]. Hong [2.12] suggested ANN-based Lagrange optimizations with CTS or CRS methods that demonstrate acceptable design accuracies for use in design applications.

2.4.2.2 *Design charts corresponding to ρ_p for forward design*

Design charts shown in Figure 2.4.1 are established based on the following parameters: preassigned inputs are parameters b = 500 mm, h = 700 mm, f_{ck} = 40 MPa, f_{cki} = 0 MPa, ρ_{st} = 0.005, ρ_{sc} = 0.005, f_{sy} = 500 MPa, f_{sw} = 400 MPa, ϕ_s = 16 mm, f_{py} = 1650 MPa, ϕ_p = 12.7 mm, $PTLoss_{lt}$ = 0.15, L = 10,000 mm, and K_c = 3; output parameter is σ_c/f_{cki} at transfer stage.

Figure 2.4.1 plots design parameters (σ_c/f_{cki}, ε_{st}, ε_{sc}, ε_p, and σ_{ct}/f_{cki}) in Y-axis by varying tendon ratio (ρ_p) which is shown with X-axis. In Figure 2.4.1,

Table 2.4.3 Training summary based on PTM for a forward design

(a) *Training on Output 16–22*												
FORWARD PROBLEM: *15 Inputs* (b, h, f_{ck}, f_{cki}, ρ_{st}, ρ_{sc}, f_{sy}, f_{sw}, ϕ_S, ρ_p, f_{py}, ϕ_P, $PTloss_{lt}$, L, K_c) – (7+6+5) *Outputs:* (*Transfer 7 Outputs:* σ_c/f_{cki}, ε_{st}, ε_{sc}, ε_p, σ_{ct}/f_{cki}, c_w, Δ) (*Service Load 5 Outputs:* $q_{L/250}$, $q_{0.2mm}$, $q_{0.6f_{ck}}$, $q_{0.8f_{su}}$, $q_{0.75f_{pu}}$) (*Ultimate Load 6 Outputs:* q_{str}, ε_{st}, ε_{sc}, ε_p, ρ_{sw}, μ_Δ)												
No.			*Data*	*Layers*	*Neurons*	*Validation check*	*Required epoch*	*Best epoch for training*	*Stopped epoch*	*Mean square error*	*R at best epoch*	*Comments*
15 Inputs (b, h, f_{ck}, f_{cki}, ρ_{st}, ρ_{sc}, f_{sy}, f_{sw}, ϕ_S, ρ_p, f_{py}, ϕ_P, $PTloss_{lt}$, L, K_c) **–1 Output** (*Transfer* σ_c/f_{cki}) - PTM												
FW.PTM.	16	a	50,000	15	15	500	50,000	1,071	1,571	0.004782	0.944072	Insufficient
FW.PTM.	16	b	50,000	20	20	500	50,000	986	1,486	0.005467	0.940664	Insufficient
FW.PTM.	16	c	50,000	25	25	500	50,000	1,322	1,822	0.006158	0.937434	Insufficient
15 Inputs (b, h, f_{ck}, f_{cki}, ρ_{st}, ρ_{sc}, f_{sy}, f_{sw}, ϕ_S, ρ_p, f_{py}, ϕ_P, $PTloss_{lt}$, L, K_c) **–1 Output** (*Transfer* ε_{st}) - PTM												
FW.PTM.	17	a	50,000	15	15	500	50,000	8,397	8,897	3.18E-05	0.998924	Good
FW.PTM.	17	b	50,000	20	20	500	50,000	3,935	4,435	4.21E-05	0.998705	Good
FW.PTM.	17	c	50,000	25	25	500	50,000	12,330	12,830	3.90E-05	0.998705	Good
15 Inputs (b, h, f_{ck}, f_{cki}, ρ_{st}, ρ_{sc}, f_{sy}, f_{sw}, ϕ_S, ρ_p, f_{py}, ϕ_P, $PTloss_{lt}$, L, K_c) **–1 Output** (*Transfer* ε_{sc}) - PTM												
FW.PTM.	18	a	50,000	15	15	500	50,000	3,708	4,208	9.75E-05	0.983229	Neutral
FW.PTM.	18	b	50,000	20	20	500	50,000	6,081	5,581	1.34E-04	0.979253	Neutral
FW.PTM.	18	c	50,000	25	25	500	50,000	4,425	4,925	1.32E-04	0.982499	Neutral

(Continued)

Table 2.4.3 (Continued) Training summary based on PTM for a forward design

(a) Training on Output 16–22

FORWARD PROBLEM:
15 Inputs (b, h, f_{ck}, f_{cki}, ρ_{st}, ρ_{sc}, f_{sy}, f_{sw}, ϕ_S, ρ_p, f_{py}, ϕ_P, $PTloss_{lt}$, L, K_c) – (7+6+5) Outputs:
(Transfer 7 Outputs: σ_c/f_{cki}, ε_{st}, ε_{sc}, ε_p, σ_{ct}/f_{cki}, c_w, Δ)
(Service Load 5 Outputs: $q_{L/250}$, $q_{0.2mm}$, $q_{0.6f_{ck}}$, $q_{0.8f_{su}}$, $q_{0.75f_{pu}}$)
(Ultimate Load 6 Outputs: q_{str}, ε_{st}, ε_{sc}, ε_p, ρ_{sw}, μ_Δ)

No.			*Data*	*Layers*	*Neurons*	*Validation check*	*Required epoch*	*Best epoch for training*	*Stopped epoch*	*Mean square error*	*R at best epoch*	*Comments*
15 Inputs (b, h, f_{ck}, f_{cki}, ρ_{st}, ρ_{sc}, f_{sy}, f_{sw}, ϕ_S, ρ_p, f_{py}, ϕ_P, $PTloss_{lt}$, L, K_c) **–1 Output** (*Transfer* ε_p) - *PTM*												
FW.PTM.	19	a	50,000	15	15	500	50,000	7,784	8,284	2.78E-05	0.999136	Good
FW.PTM.	19	b	50,000	20	20	500	50,000	10,014	10,514	4.16E-05	0.998751	Good
FW.PTM.	19	c	50,000	25	25	500	50,000	15,211	15,711	2.38E-05	0.999307	Good
15 Inputs (b, h, f_{ck}, f_{cki}, ρ_{st}, ρ_{sc}, f_{sy}, f_{sw}, ϕ_S, ρ_p, f_{py}, ϕ_P, $PTloss_{lt}$, L, K_c) **–1 Output** (*Transfer* σ_{ct}/f_{cki}) - *PTM*												
FW.PTM.	20	a	50,000	15	15	500	50,000	5,531	6,031	4.75E-05	0.99888	Good
FW.PTM.	20	b	50,000	20	20	500	50,000	4,262	4,762	5.99E-05	0.998598	Good
FW.PTM.	20	c	50,000	25	25	500	50,000	14,267	14,767	5.57E-05	0.998716	Good
15 Inputs (b, h, f_{ck}, f_{cki}, ρ_{st}, ρ_{sc}, f_{sy}, f_{sw}, ϕ_S, ρ_p, f_{py}, ϕ_P, $PTloss_{lt}$, L, K_c) **–1 Output** (*Transfer cw*) - *PTM*												
FW.PTM.	21	a	50,000	15	15	500	50,000	1,690	2,190	8.41E-05	0.966521	Neutral
FW.PTM.	21	b	50,000	20	20	500	50,000	4,241	4,741	1.06E-04	0.958348	Neutral
FW.PTM.	21	c	50,000	25	25	500	50,000	8,261	8,761	9.61E-05	0.962456	Neutral
15 Inputs (b, h, f_{ck}, f_{cki}, ρ_{st}, ρ_{sc}, f_{sy}, f_{sw}, ϕ_S, ρ_p, f_{py}, ϕ_P, $PTloss_{lt}$, L, K_c) **–1 Output** (*Transfer* Δ) - *PTM*												
FW.PTM.	22	a	50,000	15	15	500	50,000	6,270	6,770	9.37E-06	0.998423	Good
FW.PTM.	22	b	50,000	20	20	500	50,000	3,303	3,803	2.97E-05	0.993811	Good
FW.PTM.	22	c	50,000	25	25	500	50,000	4,559	5,059	5.53E-05	0.992654	Good

(Continued)

Table 2.4.3 (Continued) Training summary based on PTM for a forward design

(b) Training on Output 23–27

FORWARD PROBLEM
15 Inputs (b, h, f_{ck}, f_{cki}, ρ_{st}, ρ_{sc}, f_{sy}, f_{sw}, ϕ_S, ρ_p, f_{py}, ϕ_P, $PTloss_{lt}$, L, K_c) – (7+6+5) Outputs:
(Transfer 7 Outputs: σ_c/f_{cki}, ε_{st}, ε_{sc}, ε_p, σ_{ct}/f_{cki}, c_w, Δ)
(Service Load 5 Outputs: $q_{L/250}$, $q_{0.2mm}$, $q_{0.6f_{ck}}$, $q_{0.8f_{su}}$, $q_{0.75f_{pu}}$)
(Ultimate Load 6 Outputs: q_{str}, ε_{st}, ε_{sc}, ε_p, ρ_{sw}, μ_Δ)

No.	Data	Layers	Neurons	Validation check	Required Epoch	Best Epoch for Training	Stopped Epoch	Mean Square Error	R at Best Epoch	Comments
15 Inputs (b, h, f_{ck}, f_{cki}, ρ_{st}, ρ_{sc}, f_{sy}, f_{sw}, ϕ_S, ρ_p, f_{py}, ϕ_P, $PTloss_{lt}$, L, K_c) **–1 Output** (Service $q_{L/250}$) - PTM										
FW.PTM. 23 a	50,000	15	15	500	50,000	4,892	5,392	2.30E-05	0.998656	Good
FW.PTM. 23 b	50,000	20	20	500	50,000	10,102	10,602	2.61E-06	0.99986	Very good
FW.PTM. 23 c	50,000	25	25	500	50,000	14,343	14,843	2.32E-06	0.999879	Very good
15 Inputs (b, h, f_{ck}, f_{cki}, ρ_{st}, ρ_{sc}, f_{sy}, f_{sw}, ϕ_S, ρ_p, f_{py}, ϕ_P, $PTloss_{lt}$, L, K_c) **–1 Output** (Service $q_{0.2mm}$) - PTM										
FW.PTM. 24 a	50,000	15	15	500	50,000	18,517	19,017	2.37E-07	0.999989	Very good
FW.PTM. 24 b	50,000	20	20	500	50,000	16,097	16,597	2.63E-06	0.999876	Very good
FW.PTM. 24 c	50,000	25	25	500	50,000	17,348	17,848	8.69E-07	0.999958	Very good
15 Inputs (b, h, f_{ck}, f_{cki}, ρ_{st}, ρ_{sc}, f_{sy}, f_{sw}, ϕ_S, ρ_p, f_{py}, ϕ_P, $PTloss_{lt}$, L, K_c) **–1 Output** (Service $q_{0.6f_{ck}}$) - PTM										
FW.PTM. 25 a	50,000	15	15	500	50,000	15,311	15,811	7.29E-07	0.999962	Very good
FW.PTM. 25 b	50,000	20	20	500	50,000	15,522	16,022	7.52E-07	0.999962	Very good
FW.PTM. 25 c	50,000	25	25	500	50,000	18,775	19,275	9.69E-07	0.999954	Very good
15 Inputs (b, h, f_{ck}, f_{cki}, ρ_{st}, ρ_{sc}, f_{sy}, f_{sw}, ϕ_S, ρ_p, f_{py}, ϕ_P, $PTloss_{lt}$, L, K_c) **–1 Output** (Service $q_{0.8f_{su}}$) - PTM										
FW.PTM. 26 a	50,000	15	15	500	50,000	13,024	13,524	3.92E-06	0.999809	Very good
FW.PTM. 26 b	50,000	20	20	500	50,000	5,021	5,521	2.11E-05	0.999004	Good
FW.PTM. 26 c	50,000	25	25	500	50,000	20,844	21,344	7.11E-06	0.999717	Very good
15 Inputs (b, h, f_{ck}, f_{cki}, ρ_{st}, ρ_{sc}, f_{sy}, f_{sw}, ϕ_S, ρ_p, f_{py}, ϕ_P, $PTloss_{lt}$, L, K_c) **–1 Output** (Service $q_{0.75f_{pu}}$) - PTM										
FW.PTM. 27 a	50,000	15	15	500	50,000	4,490	4,990	1.85E-05	0.999126	Good
FW.PTM. 27 b	50,000	20	20	500	50,000	9,634	10,134	5.49E-06	0.999703	Very good
FW.PTM. 27 c	50,000	25	25	500	50,000	6,148	6,648	2.24E-05	0.999177	Good

(Continued)

Table 2.4.3 (Continued) Training summary based on PTM for a forward design

(c) Training on Output 28–33												
FORWARD PROBLEM *15 Inputs* (b, h, f_{ck}, f_{cki}, ρ_{st}, ρ_{sc}, f_{sy}, f_{sw}, ϕ_S, ρ_p, f_{py}, ϕ_P, $PTloss_{lt}$, L, K_c) *– (7+6+5) Outputs:* *(Transfer 7 Outputs:* σ_c/f_{cki}, ε_{st}, ε_{sc}, ε_p, σ_{ct}/f_{cki}, c_w, Δ*)* *(Service Load 5 Outputs:* $q_{L/250}$, $q_{0.2mm}$, $q_{0.6f_{ck}}$, $q_{0.8f_{su}}$, $q_{0.75f_{pu}}$*)* *(Ultimate Load 6 Outputs:* q_{str}, ε_{st}, ε_{sc}, ε_p, ρ_{sw}, μ_Δ*)*												
No.			*Data*	*Layers*	*Neurons*	*Validation check*	*Required Epoch*	*Best Epoch for Training*	*Stopped Epoch*	*Mean Square Error*	*R at Best Epoch*	*Comments*
15 Inputs (b, h, f_{ck}, f_{cki}, ρ_{st}, ρ_{sc}, f_{sy}, f_{sw}, ϕ_S, ρ_p, f_{py}, ϕ_P, $PTloss_{lt}$, L, K_c) **–1 Output** (*Ultimate* q_{str}) *- PTM*												
FW.PTM.	28	a	50,000	15	15	500	50,000	17,103	17,603	1.01E-05	0.99962	Good
FW.PTM.	28	b	50,000	20	20	500	50,000	8,466	8,966	1.36E-05	0.999481	Good
FW.PTM.	28	c	50,000	25	25	500	50,000	8,643	9,143	2.35E-05	0.99908	Good
15 Inputs (b, h, f_{ck}, f_{cki}, ρ_{st}, ρ_{sc}, f_{sy}, f_{sw}, ϕ_S, ρ_p, f_{py}, ϕ_P, $PTloss_{lt}$, L, K_c) **–1 Output** (*Ultimate* ε_{st}) *- PTM*												
FW.PTM.	29	a	50,000	15	15	500	50,000	2,002	2,502	1.19E-04	0.999265	Neutral
FW.PTM.	29	b	50,000	20	20	500	50,000	4,198	4,698	8.08E-05	0.999451	Good
FW.PTM.	29	c	50,000	25	25	500	50,000	11,860	12,360	2.60E-05	0.999857	Good
15 Inputs (b, h, f_{ck}, f_{cki}, ρ_{st}, ρ_{sc}, f_{sy}, f_{sw}, ϕ_S, ρ_p, f_{py}, ϕ_P, $PTloss_{lt}$, L, K_c) **–1 Output** (*Ultimate* ε_{sc}) *- PTM*												
FW.PTM.	30	a	50,000	15	15	500	50,000	5,690	6,190	1.00E-04	0.998833	Neutral
FW.PTM.	30	b	50,000	20	20	500	50,000	8,964	9,464	8.03E-05	0.999406	Good
FW.PTM.	30	c	50,000	25	25	500	50,000	10,401	10,901	4.00E-05	0.999559	Good
15 Inputs (b, h, f_{ck}, f_{cki}, ρ_{st}, ρ_{sc}, f_{sy}, f_{sw}, ϕ_S, ρ_p, f_{py}, ϕ_P, $PTloss_{lt}$, L, K_c) **–1 Output** (*Ultimate* ε_p) *- PTM*												
FW.PTM.	31	a	50,000	15	15	500	50,000	7,733	8,233	2.79E-05	0.99979	Good
FW.PTM.	31	b	50,000	20	20	500	50,000	5,486	5,986	4.92E-05	0.999644	Good
FW.PTM.	31	c	50,000	25	25	500	50,000	8,574	9,074	6.39E-05	0.999697	Good

(Continued)

Table 2.4.3 (Continued) Training summary based on PTM for a forward design

(c) Training on Output 28–33

FORWARD PROBLEM

15 Inputs (b, h, f_{ck}, f_{cki}, ρ_{st}, ρ_{sc}, f_{sy}, f_{sw}, ϕ_S, ρ_p, f_{py}, ϕ_P, $PTloss_{lt}$, L, K_c) – (7+6+5) Outputs:
(Transfer 7 Outputs: σ_c/f_{cki}, ε_{st}, ε_{sc}, ε_p, σ_{ct}/f_{cki}, c_w, Δ)
(Service Load 5 Outputs: $q_{L/250}$, $q_{0.2mm}$, $q_{0.6f_{ck}}$, $q_{0.8f_{su}}$, $q_{0.75f_{pu}}$)
(Ultimate Load 6 Outputs: q_{str}, ε_{st}, ε_{sc}, ε_p, ρ_{sw}, μ_Δ)

No.			*Data*	*Layers*	*Neurons*	*Validation check*	*Required Epoch*	*Best Epoch for Training*	*Stopped Epoch*	*Mean Square Error*	*R at Best Epoch*	*Comments*
15 Inputs (b, h, f_{ck}, f_{cki}, ρ_{st}, ρ_{sc}, f_{sy}, f_{sw}, ϕ_S, ρ_p, f_{py}, ϕ_P, $PTloss_{lt}$, L, K_c) **–1 Output** (*Ultimate ρ_{sw}*) *- PTM*												
FW.PTM.	32	a	50,000	15	15	500	50,000	10,874	11,374	1.56E-05	0.999823	Good
FW.PTM.	32	b	50,000	20	20	500	50,000	12,317	12,817	1.76E-05	0.999865	Good
FW.PTM.	32	c	50,000	25	25	500	50,000	13,555	14,055	1.30E-05	0.999914	Good
15 Inputs (b, h, f_{ck}, f_{cki}, ρ_{st}, ρ_{sc}, f_{sy}, f_{sw}, ϕ_S, ρ_p, f_{py}, ϕ_P, $PTloss_{lt}$, L, K_c) **–1 Output** (*Ultimate μ_Δ*) *- PTM*												
FW.PTM.	33	a	50,000	15	15	500	50,000	6,413	6,913	6.33E-05	0.998257	Good
FW.PTM.	33	b	50,000	20	20	500	50,000	9,294	9,794	4.27E-05	0.998717	Good
FW.PTM.	33	c	50,000	25	25	500	50,000	7,693	8,193	7.33E-05	0.99824	Good

Note:

Three PTMs for each output:

- PTM a: 15 layers, 15 neurons, 500 validation checks.
- PTM b: 20 layers, 20 neurons, 500 validation checks.
- PTM c: 25 layers, 25 neurons, 500 validation checks.

FW.PTM.**16a**

- → Version of training model
- → Output number
- → Training method.
- → Design scenario; FW = Forward ANN

Table 2.4.4 Design accuracies based on PTM

(a) Design accuracies; tendon ratio of 0.001 based on PTM

(a-1) Design accuracies; tendon ratio of 0.001 based on FW.PTM.a

PARAMETERS			Training Model	Network test result	Auto PT-BEAM check	Error
BEAM CONFIGURATION	1	b (mm)		500	500	0.00%
	2	h (mm)		700	700	0.00%
	3	f_{ck} (MPa)		40	40	0.00%
	4	f_{cki} (MPa)		20	20	0.00%
	5	ρ_{st} (mm^2/mm^2)		0.005	0.005	0.00%
	6	ρ_{sc} (mm^2/mm^2)		0.005	0.005	0.00%
	7	f_{sy} (MPa)		500	500	0.00%
	8	f_{sw} (MPa)		400	400	0.00%
	9	ϕ_S (mm)		16	16	0.00%
	10	ρ_p (mm^2/mm^2)		**0.0010**	**0.0010**	**0.00%**
	11	f_{py} (MPa)		1650	1650	0.00%
	12	ϕ_p (mm)		12.7	12.7	0.00%
	13	$PTloss_{lt}$		0.15	0.15	0.00%
	14	L (mm)		10000	10000	0.00%
	15	K_c		3	3	0.00%
TRANSFER LOAD STAGE	16	σ_c/f_{cki}	FW.PTM. **16 a**	-0.0634	-0.07	Uncracked
	17	ε_{st} (mm/mm)	FW.PTM. **17 a**	-0.00014	-0.00015	-4.94%
	18	ε_{sc} (mm/mm)	FW.PTM. **18 a**	0.0000182	0.0000248	-36.44%
	19	ε_p (mm/mm)	FW.PTM. **19 a**	0.00696	0.00695	0.14%
	20	σ_{ct}/f_{cki}	FW.PTM. **20 a**	0.15230	0.15759	-3.47%
	21	cw (mm)	FW.PTM. **21 a**	0.000	0.000	0.00%
	22	Δ (mm)	FW.PTM. **22 a**	-1.752	-2.054	-17.23%
SERVICE LOAD COMBINATION	23	$q_{L/250}$ (kN/m)	FW.PTM. **23 a**	22.41	27.26	-21.65%
	24	$q_{0.2mm}$ (kN/m)	FW.PTM. **24 a**	30.17	31.05	-2.91%
	25	$q_{0.6fck}$ (kN/m)	FW.PTM. **25 a**	49.82	51.39	-3.15%
	26	$q_{0.8fsu}$ (kN/m)	FW.PTM. **26 a**	48.63	50.58	-4.00%
	27	$q_{0.75fpu}$ (kN/m)	FW.PTM. **27 a**	30.97	31.62	-2.09%
ULTIMATE LOAD COMBINATION	28	q_{str} (kN/m)	FW.PTM. **28 a**	56.34	54.32	3.59%
	29	ε_{st} (mm/mm)	FW.PTM. **29 a**	0.02321	0.02309	0.51%
	30	ε_{sc} (mm/mm)	FW.PTM. **30 a**	-0.00111	-0.00110	1.01%
	31	ε_p (mm/mm)	FW.PTM. **31 a**	0.02925	0.02946	-0.72%
	32	ρ_{sw} (mm^2/mm^2)	FW.PTM. **32 a**	0.0013	0.0013	1.80%
	33	μ_Δ	FW.PTM. **33 a**	2.93	2.95	-0.60%

Significant errors

(red box): Given Inputs for ANN (blue box): Inputs for engineering mechanic

(Continued)

Table 2.4.4 (Continued) Design accuracies based on PTM

(a-2) Design accuracies; tendon ratio of 0.001 based on FW.PTM.b

PARAMETERS			Training Model	Network test result	Auto PT-BEAM check	Error
BEAM CONFIGURATION	1	b (mm)		500	500	0.00%
	2	h (mm)		700	700	0.00%
	3	f_{ck} (MPa)		40	40	0.00%
	4	f_{cki} (MPa)		20	20	0.00%
	5	ρ_{st} (mm^2/mm^2)		0.005	0.005	0.00%
	6	ρ_{sc} (mm^2/mm^2)		0.005	0.005	0.00%
	7	f_{sy} (MPa)		500	500	0.00%
	8	f_{sw} (MPa)		400	400	0.00%
	9	ϕ_S (mm)		16	16	0.00%
	10	ρ_p (mm^2/mm^2)		**0.0010**	**0.0010**	**0.00%**
	11	f_{py} (MPa)		1650	1650	0.00%
	12	ϕ_p (mm)		12.7	12.7	0.00%
	13	$PTloss_{lt}$		0.15	0.15	0.00%
	14	L (mm)		10000	10000	0.00%
	15	K_c		3	3	0.00%
TRANSFER LOAD STAGE	16	σ_c/f_{cki}	FW.PTM. **16 b**	-0.0668	-0.07	Uncracked
	17	ε_{st} (mm/mm)	FW.PTM. **17 b**	-0.00014	-0.00015	-4.22%
	18	ε_{sc} (mm/mm)	FW.PTM. **18 b**	0.0000179	0.0000248	-38.75%
	19	ε_p (mm/mm)	FW.PTM. **19 b**	0.00696	0.00695	0.21%
	20	σ_{ct}/f_{cki}	FW.PTM. **20 b**	0.14783	0.15759	-6.61%
	21	cw (mm)	FW.PTM. **21 b**	0.000	0.000	0.00%
	22	Δ (mm)	FW.PTM. **22 b**	-1.909	-2.054	-7.62%
SERVICE LOAD COMBINATION	23	$q_{L/250}$ (kN/m)	FW.PTM. **23 b**	25.80	27.26	-5.68%
	24	$q_{0.2mm}$ (kN/m)	FW.PTM. **24 b**	28.14	31.05	-10.34%
	25	$q_{0.6fck}$ (kN/m)	FW.PTM. **25 b**	50.55	51.39	-1.66%
	26	$q_{0.8fsu}$ (kN/m)	FW.PTM. **26 b**	51.13	50.58	1.08%
	27	$q_{0.75fpu}$ (kN/m)	FW.PTM. **27 b**	28.32	31.62	-11.66%
ULTIMATE LOAD COMBINATION	28	q_{str} (kN/m)	FW.PTM. **28 b**	55.00	54.32	1.23%
	29	ε_{st} (mm/mm)	FW.PTM. **29 b**	0.02322	0.02309	0.54%
	30	ε_{sc} (mm/mm)	FW.PTM. **30 b**	-0.00109	-0.00110	-0.61%
	31	ε_p (mm/mm)	FW.PTM. **31 b**	0.02946	0.02946	0.00%
	32	ρ_{sw} (mm^2/mm^2)	FW.PTM. **32 b**	0.0012	0.0013	-1.47%
	33	μ_Δ	FW.PTM. **33 b**	2.95	2.95	0.25%

Significant errors (rows 18, 24, 27)

Red box: Given Inputs for ANN; Blue box: Inputs for engineering mechanic

(*Continued*)

Table 2.4.4 (Continued) Design accuracies based on PTM

(a-3) Design accuracies; tendon ratio of 0.001 based on FW.PTM.c

PARAMETERS			Training Model	Network test result	Auto PT-BEAM check	Error
BEAM CONFIGURATION	1	b (mm)		500	500	0.00%
	2	h (mm)		700	700	0.00%
	3	f_{ck} (MPa)		40	40	0.00%
	4	f_{cki} (MPa)		20	20	0.00%
	5	ρ_{st} (mm^2/mm^2)		0.005	0.005	0.00%
	6	ρ_{sc} (mm^2/mm^2)		0.005	0.005	0.00%
	7	f_{sy} (MPa)		500	500	0.00%
	8	f_{sw} (MPa)		400	400	0.00%
	9	ϕ_S (mm)		16	16	0.00%
	10	ρ_p (mm^2/mm^2)		**0.0010**	**0.0010**	**0.00%**
	11	f_{py} (MPa)		1650	1650	0.00%
	12	ϕ_p (mm)		12.7	12.7	0.00%
	13	$PTloss_{lt}$		0.15	0.15	0.00%
	14	L (mm)		10000	10000	0.00%
	15	K_c		3	3	0.00%
TRANSFER LOAD STAGE	16	σ_c/f_{cki}	FW.PTM. **16 c**	-0.0676	-0.07	Uncracked
	17	ε_{st} (mm/mm)	FW.PTM. **17 c**	-0.00014	-0.00015	-5.66%
	18	ε_{sc} (mm/mm)	FW.PTM. **18 c**	0.0000207	0.0000248	-19.57%
	19	ε_p (mm/mm)	FW.PTM. **19 c**	0.00696	0.00695	0.11%
	20	σ_{ct}/f_{cki}	FW.PTM. **20 c**	0.14974	0.15759	-5.24%
	21	cw (mm)	FW.PTM. **21 c**	0.000	0.000	0.00%
	22	Δ (mm)	FW.PTM. **22 c**	-1.812	-2.054	-13.39%
SERVICE LOAD COMBINATION	23	$q_{L/250}$ (kN/m)	FW.PTM. **23 c**	27.31	27.26	0.17%
	24	$q_{0.2mm}$ (kN/m)	FW.PTM. **24 c**	29.79	31.05	-4.21%
	25	$q_{0.6fck}$ (kN/m)	FW.PTM. **25 c**	48.10	51.39	-6.82%
	26	$q_{0.8fsu}$ (kN/m)	FW.PTM. **26 c**	50.16	50.58	-0.82%
	27	$q_{0.75fpu}$ (kN/m)	FW.PTM. **27 c**	31.40	31.62	-0.70%
ULTIMATE LOAD COMBINATION	28	q_{str} (kN/m)	FW.PTM. **28 c**	55.50	54.32	2.12%
	29	ε_{st} (mm/mm)	FW.PTM. **29 c**	0.02315	0.02309	0.24%
	30	ε_{sc} (mm/mm)	FW.PTM. **30 c**	-0.00110	-0.00110	0.01%
	31	ε_p (mm/mm)	FW.PTM. **31 c**	0.02935	0.02946	-0.36%
	32	ρ_{sw} (mm^2/mm^2)	FW.PTM. **32 c**	0.0012	0.0013	-1.96%
	33	μ_Δ	FW.PTM. **33 c**	2.95	2.95	0.19%

Significant errors (rows 18 and 22)

[red box]: Given Inputs for ANN [blue box]: Inputs for engineering mechanic

(*Continued*)

Table 2.4.4 (Continued) Design accuracies based on PTM

(b) Design accuracies; tendon ratio of 0.0015 based on PTM

(b-1) Design accuracies; tendon ratio of 0.0015 based on FW.PTM.a

PARAMETERS			Training Model	Network test result	Auto PT-BEAM check	Error
BEAM CONFIGURATION	1	b (mm)		500	500	0.00%
	2	h (mm)		700	700	0.00%
	3	f_{ck} (MPa)		40	40	0.00%
	4	f_{cki} (MPa)		20	20	0.00%
	5	ρ_{st} (mm^2/mm^2)		0.005	0.005	0.00%
	6	ρ_{sc} (mm^2/mm^2)		0.005	0.005	0.00%
	7	f_{sy} (MPa)		500	500	0.00%
	8	f_{sw} (MPa)		400	400	0.00%
	9	ϕ_S (mm)		16	16	0.00%
	10	ρ_p (mm^2/mm^2)		**0.0015**	**0.0015**	**0.00%**
	11	f_{py} (MPa)		1650	1650	0.00%
	12	ϕ_p (mm)		12.7	12.7	0.00%
	13	$PTloss_{lt}$		0.15	0.15	0.00%
	14	L (mm)		10000	10000	0.00%
	15	K_c		3	3	0.00%
TRANSFER LOAD STAGE	16	σ_c/f_{cki}	FW.PTM. **16 a**	~0	0.00	Cracked
	17	ε_{st} (mm/mm)	FW.PTM. **17 a**	-0.00025	-0.00028	-10.65%
	18	ε_{sc} (mm/mm)	FW.PTM. **18 a**	0.0000589	0.0001480	-151.23%
	19	ε_p (mm/mm)	FW.PTM. **19 a**	0.00685	0.00682	0.50%
	20	σ_{ct}/f_{cki}	FW.PTM. **20 a**	0.25612	0.29216	-14.07%
	21	cw (mm)	FW.PTM. **21 a**	0.002	0.021	-950.00%
	22	Δ (mm)	FW.PTM. **22 a**	-4.549	-4.787	-5.22%
SERVICE LOAD COMBINATION	23	$q_{L/250}$ (kN/m)	FW.PTM. **23 a**	28.73	36.25	-26.19%
	24	$q_{0.2mm}$ (kN/m)	FW.PTM. **24 a**	40.76	41.65	-2.18%
	25	$q_{0.6fck}$ (kN/m)	FW.PTM. **25 a**	57.48	57.51	-0.06%
	26	$q_{0.8fsu}$ (kN/m)	FW.PTM. **26 a**	60.43	63.13	-4.47%
	27	$q_{0.75fpu}$ (kN/m)	FW.PTM. **27 a**	43.00	42.47	1.25%
ULTIMATE LOAD COMBINATION	28	q_{str} (kN/m)	FW.PTM. **28 a**	67.99	66.40	2.34%
	29	ε_{st} (mm/mm)	FW.PTM. **29 a**	0.01972	0.01946	1.35%
	30	ε_{sc} (mm/mm)	FW.PTM. **30 a**	-0.00144	-0.00143	0.83%
	31	ε_p (mm/mm)	FW.PTM. **31 a**	0.02558	0.02578	-0.76%
	32	ρ_{sw} (mm^2/mm^2)	FW.PTM. **32 a**	0.0014	0.0013	1.73%
	33	μ_Δ	FW.PTM. **33 a**	2.67	2.70	-1.08%

Significant errors

: Given Inputs for ANN   : Inputs for engineering mechanic

(Continued)

Table 2.4.4 (Continued) Design accuracies based on PTM

(b-2) Design accuracies; tendon ratio of 0.0015 based on FW.PTM.b

PARAMETERS			Training Model	Network test result	Auto PT-BEAM check	Error
BEAM CONFIGURATION	1	b (mm)		500	500	0.00%
	2	h (mm)		700	700	0.00%
	3	f_{ck} (MPa)		40	40	0.00%
	4	f_{cki} (MPa)		20	20	0.00%
	5	ρ_{st} (mm^2/mm^2)		0.005	0.005	0.00%
	6	ρ_{sc} (mm^2/mm^2)		0.005	0.005	0.00%
	7	f_{sy} (MPa)		500	500	0.00%
	8	f_{sw} (MPa)		400	400	0.00%
	9	ϕ_S (mm)		16	16	0.00%
	10	ρ_p (mm^2/mm^2)		**0.0015**	**0.0015**	**0.00%**
	11	f_{py} (MPa)		1650	1650	0.00%
	12	ϕ_p (mm)		12.7	12.7	0.00%
	13	$PTloss_{lt}$		0.15	0.15	0.00%
	14	L (mm)		10000	10000	0.00%
	15	K_c		3	3	0.00%
TRANSFER LOAD STAGE	16	σ_c/f_{cki}	FW.PTM. **16 b**	~0	0.00	Cracked
	17	ε_{st} (mm/mm)	FW.PTM. **17 b**	-0.00025	-0.00028	-11.79%
	18	ε_{sc} (mm/mm)	FW.PTM. **18 b**	0.0000433	0.0001480	-241.60%
	19	ε_p (mm/mm)	FW.PTM. **19 b**	0.00685	0.00682	0.47%
	20	σ_{ct}/f_{cki}	FW.PTM. **20 b**	0.25275	0.29216	-15.59%
	21	cw (mm)	FW.PTM. **21 b**	0.005	0.021	-320.00%
	22	Δ (mm)	FW.PTM. **22 b**	-4.559	-4.787	-4.99%
SERVICE LOAD COMBINATION	23	$q_{L/250}$ (kN/m)	FW.PTM. **23 b**	33.32	36.25	-8.78%
	24	$q_{0.2mm}$ (kN/m)	FW.PTM. **24 b**	38.63	41.65	-7.81%
	25	$q_{0.6fck}$ (kN/m)	FW.PTM. **25 b**	57.61	57.51	0.17%
	26	$q_{0.8fsu}$ (kN/m)	FW.PTM. **26 b**	60.21	63.13	-4.85%
	27	$q_{0.75fpu}$ (kN/m)	FW.PTM. **27 b**	38.48	42.47	-10.36%
ULTIMATE LOAD COMBINATION	28	q_{str} (kN/m)	FW.PTM. **28 b**	66.08	66.40	-0.49%
	29	ε_{st} (mm/mm)	FW.PTM. **29 b**	0.01963	0.01946	0.90%
	30	ε_{sc} (mm/mm)	FW.PTM. **30 b**	-0.00142	-0.00143	-0.27%
	31	ε_p (mm/mm)	FW.PTM. **31 b**	0.02583	0.02578	0.20%
	32	ρ_{sw} (mm^2/mm^2)	FW.PTM. **32 b**	0.0014	0.0013	2.21%
	33	μ_Δ	FW.PTM. **33 b**	2.68	2.70	-0.80%

Significant errors

(Red box): Given Inputs for ANN (Blue box): Inputs for engineering mechanic

(*Continued*)

Table 2.4.4 (Continued) Design accuracies based on PTM

(b-3) Design accuracies; tendon ratio of 0.0015 based on FW.PTM.c

PARAMETERS			Training Model	Network test result	Auto PT-BEAM check	Error
BEAM CONFIGURATION	1	b (mm)		500	500	0.00%
	2	h (mm)		700	700	0.00%
	3	f_{ck} (MPa)		40	40	0.00%
	4	f_{cki} (MPa)		20	20	0.00%
	5	ρ_{st} (mm^2/mm^2)		0.005	0.005	0.00%
	6	ρ_{sc} (mm^2/mm^2)		0.005	0.005	0.00%
	7	f_{sy} (MPa)		500	500	0.00%
	8	f_{sw} (MPa)		400	400	0.00%
	9	ϕ_S (mm)		16	16	0.00%
	10	ρ_p (mm^2/mm^2)		**0.0015**	**0.0015**	**0.00%**
	11	f_{py} (MPa)		1650	1650	0.00%
	12	ϕ_p (mm)		12.7	12.7	0.00%
	13	$PTloss_{lt}$		0.15	0.15	0.00%
	14	L (mm)		10000	10000	0.00%
	15	K_c		3	3	0.00%
TRANSFER LOAD STAGE	16	σ_c/f_{cki}	FW.PTM. **16 c**	~0	0.00	Cracked
	17	ε_{st} (mm/mm)	FW.PTM. **17 c**	-0.00025	-0.00028	-8.96%
	18	ε_{sc} (mm/mm)	FW.PTM. **18 c**	0.0000505	0.0001480	-193.33%
	19	ε_p (mm/mm)	FW.PTM. **19 c**	0.00684	0.00682	0.39%
	20	σ_{ct}/f_{cki}	FW.PTM. **20 c**	0.25811	0.29216	-13.19%
	21	cw (mm)	FW.PTM. **21 c**	0.007	0.021	-200.00%
	22	Δ (mm)	FW.PTM. **22 c**	-4.258	-4.787	-12.41%
SERVICE LOAD COMBINATION	23	$q_{L/250}$ (kN/m)	FW.PTM. **23 c**	36.22	36.25	-0.07%
	24	$q_{0.2mm}$ (kN/m)	FW.PTM. **24 c**	41.25	41.65	-0.96%
	25	$q_{0.6fck}$ (kN/m)	FW.PTM. **25 c**	55.49	57.51	-3.65%
	26	$q_{0.8fsu}$ (kN/m)	FW.PTM. **26 c**	62.77	63.13	-0.57%
	27	$q_{0.75fpu}$ (kN/m)	FW.PTM. **27 c**	42.46	42.47	-0.01%
ULTIMATE LOAD COMBINATION	28	q_{str} (kN/m)	FW.PTM. **28 c**	71.55	66.40	7.19%
	29	ε_{st} (mm/mm)	FW.PTM. **29 c**	0.01952	0.01946	0.35%
	30	ε_{sc} (mm/mm)	FW.PTM. **30 c**	-0.00143	-0.00143	0.12%
	31	ε_p (mm/mm)	FW.PTM. **31 c**	0.02572	0.02578	-0.22%
	32	ρ_{sw} (mm^2/mm^2)	FW.PTM. **32 c**	0.0014	0.0013	2.94%
	33	μ_Δ	FW.PTM. **33 c**	2.66	2.70	-1.22%

Significant errors (rows 18, 20, 21, 22)

Red box: Given Inputs for ANN; Blue box: Inputs for engineering mechanic

(Continued)

Table 2.4.4 (Continued) Design accuracies based on PTM

(c) Design accuracies; tendon ratio of 0.002 based on PTM

(c-1) Design accuracies; tendon ratio of 0.002 based on FW.PTM.a

PARAMETERS			Training Model	Network test result	Auto PT-BEAM check	Error
BEAM CONFIGURATION	1	b (mm)		500	500	0.00%
	2	h (mm)		700	700	0.00%
	3	f_{ck} (MPa)		40	40	0.00%
	4	f_{cki} (MPa)		20	20	0.00%
	5	ρ_{st} (mm^2/mm^2)		0.005	0.005	0.00%
	6	ρ_{sc} (mm^2/mm^2)		0.005	0.005	0.00%
	7	f_{sy} (MPa)		500	500	0.00%
	8	f_{sw} (MPa)		400	400	0.00%
	9	ϕ_S (mm)		16	16	0.00%
	10	ρ_p (mm^2/mm^2)		**0.0020**	**0.0020**	**0.00%**
	11	f_{py} (MPa)		1650	1650	0.00%
	12	ϕ_p (mm)		12.7	12.7	0.00%
	13	$PTloss_{lt}$		0.15	0.15	0.00%
	14	L (mm)		10000	10000	0.00%
	15	K_c		3	3	0.00%
TRANSFER LOAD STAGE	16	σ_c/f_{cki}	FW.PTM. **16 a**	~0	0.00	Cracked
	17	ε_{st} (mm/mm)	FW.PTM. **17 a**	-0.00037	-0.00038	-3.00%
	18	ε_{sc} (mm/mm)	FW.PTM. **18 a**	0.0002171	0.0002378	-9.53%
	19	ε_p (mm/mm)	FW.PTM. **19 a**	0.00673	0.00671	0.31%
	20	σ_{ct}/f_{cki}	FW.PTM. **20 a**	0.37910	0.39383	-3.89%
	21	cw (mm)	FW.PTM. **21 a**	0.029	0.034	-17.24%
	22	Δ (mm)	FW.PTM. **22 a**	-7.797	-8.003	-2.65%
SERVICE LOAD COMBINATION	23	$q_{L/250}$ (kN/m)	FW.PTM. **23 a**	35.28	44.96	-27.45%
	24	$q_{0.2mm}$ (kN/m)	FW.PTM. **24 a**	51.33	52.25	-1.80%
	25	$q_{0.6fck}$ (kN/m)	FW.PTM. **25 a**	64.02	63.64	0.60%
	26	$q_{0.8fsu}$ (kN/m)	FW.PTM. **26 a**	71.85	75.30	-4.81%
	27	$q_{0.75fpu}$ (kN/m)	FW.PTM. **27 a**	53.93	53.11	1.53%
ULTIMATE LOAD COMBINATION	28	q_{str} (kN/m)	FW.PTM. **28 a**	79.59	78.13	1.83%
	29	ε_{st} (mm/mm)	FW.PTM. **29 a**	0.01670	0.01643	1.57%
	30	ε_{sc} (mm/mm)	FW.PTM. **30 a**	-0.00171	-0.00170	0.50%
	31	ε_p (mm/mm)	FW.PTM. **31 a**	0.02259	0.02272	-0.58%
	32	ρ_{sw} (mm^2/mm^2)	FW.PTM. **32 a**	0.0015	0.0016	-2.88%
	33	μ_Δ	FW.PTM. **33 a**	2.42	2.41	0.54%

Significant errors

[red box] : Given Inputs for ANN [blue box] : Inputs for engineering mechanic

(Continued)

Table 2.4.4 (Continued) Design accuracies based on PTM

(c-2) Design accuracies; tendon ratio of 0.002 based on FW.PTM.b

PARAMETERS			Training Model	Network test result	Auto PT-BEAM check	Error
BEAM CONFIGURATION	1	b (mm)		500	500	0.00%
	2	h (mm)		700	700	0.00%
	3	f_{ck} (MPa)		40	40	0.00%
	4	f_{cki} (MPa)		20	20	0.00%
	5	ρ_{st} (mm^2/mm^2)		0.005	0.005	0.00%
	6	ρ_{sc} (mm^2/mm^2)		0.005	0.005	0.00%
	7	f_{sy} (MPa)		500	500	0.00%
	8	f_{sw} (MPa)		400	400	0.00%
	9	ϕ_S (mm)		16	16	0.00%
	10	ρ_p (mm^2/mm^2)		**0.0020**	**0.0020**	**0.00%**
	11	f_{py} (MPa)		1650	1650	0.00%
	12	ϕ_p (mm)		12.7	12.7	0.00%
	13	$PTloss_{lt}$		0.15	0.15	0.00%
	14	L (mm)		10000	10000	0.00%
	15	K_c		3	3	0.00%
TRANSFER LOAD STAGE	16	σ_c/f_{cki}	FW.PTM. **16 b**	~0	0.00	Cracked
	17	ε_{st} (mm/mm)	FW.PTM. **17 b**	-0.00037	-0.00038	-3.42%
	18	ε_{sc} (mm/mm)	FW.PTM. **18 b**	0.0002032	0.0002378	-17.03%
	19	ε_p (mm/mm)	FW.PTM. **19 b**	0.00673	0.00671	0.29%
	20	σ_{ct}/f_{cki}	FW.PTM. **20 b**	0.37898	0.39383	-3.92%
	21	cw (mm)	FW.PTM. **21 b**	0.030	0.034	-13.33%
	22	Δ (mm)	FW.PTM. **22 b**	-7.764	-8.003	-3.09%
SERVICE LOAD COMBINATION	23	$q_{L/250}$ (kN/m)	FW.PTM. **23 b**	41.69	44.96	-7.85%
	24	$q_{0.2mm}$ (kN/m)	FW.PTM. **24 b**	49.57	52.25	-5.42%
	25	$q_{0.6fck}$ (kN/m)	FW.PTM. **25 b**	63.91	63.64	0.43%
	26	$q_{0.8fsu}$ (kN/m)	FW.PTM. **26 b**	67.58	75.30	-11.43%
	27	$q_{0.75fpu}$ (kN/m)	FW.PTM. **27 b**	48.16	53.11	-10.28%
ULTIMATE LOAD COMBINATION	28	q_{str} (kN/m)	FW.PTM. **28 b**	77.20	78.13	-1.21%
	29	ε_{st} (mm/mm)	FW.PTM. **29 b**	0.01659	0.01643	0.96%
	30	ε_{sc} (mm/mm)	FW.PTM. **30 b**	-0.00169	-0.00170	-0.34%
	31	ε_p (mm/mm)	FW.PTM. **31 b**	0.02280	0.02272	0.37%
	32	ρ_{sw} (mm^2/mm^2)	FW.PTM. **32 b**	0.0016	0.0016	-0.53%
	33	μ_Δ	FW.PTM. **33 b**	2.41	2.41	-0.09%

Significant errors (rows 18, 21, 26, 27)

(red box) : Given Inputs for ANN (blue box) : Inputs for engineering mechanic

(Continued)

Table 2.4.4 (Continued) Design accuracies based on PTM

(c-3) Design accuracies; tendon ratio of 0.002 based on FW.PTM.c

PARAMETERS			Training Model	Network test result	Auto PT-BEAM check	Error
BEAM CONFIGURATION	1	b (mm)		500	500	0.00%
	2	h (mm)		700	700	0.00%
	3	f_{ck} (MPa)		40	40	0.00%
	4	f_{cki} (MPa)		20	20	0.00%
	5	ρ_{st} (mm^2/mm^2)		0.005	0.005	0.00%
	6	ρ_{sc} (mm^2/mm^2)		0.005	0.005	0.00%
	7	f_{sy} (MPa)		500	500	0.00%
	8	f_{sw} (MPa)		400	400	0.00%
	9	ϕ_S (mm)		16	16	0.00%
	10	ρ_p (mm^2/mm^2)		**0.0020**	**0.0020**	**0.00%**
	11	f_{py} (MPa)		1650	1650	0.00%
	12	ϕ_p (mm)		12.7	12.7	0.00%
	13	$PTloss_{lt}$		0.15	0.15	0.00%
	14	L (mm)		10000	10000	0.00%
	15	K_c		3	3	0.00%
TRANSFER LOAD STAGE	16	σ_c/f_{cki}	FW.PTM. 16 c	~0	0.00	Cracked
	17	ε_{st} (mm/mm)	FW.PTM. 17 c	-0.00037	-0.00038	-3.47%
	18	ε_{sc} (mm/mm)	FW.PTM. 18 c	0.0002203	0.0002378	-7.94%
	19	ε_p (mm/mm)	FW.PTM. 19 c	0.00673	0.00671	0.25%
	20	σ_{ct}/f_{cki}	FW.PTM. 20 c	0.38204	0.39383	-3.09%
	21	cw (mm)	FW.PTM. 21 c	0.028	0.034	-21.43%
	22	Δ (mm)	FW.PTM. 22 c	-7.230	-8.003	-10.70%
SERVICE LOAD COMBINATION	23	$q_{L/250}$ (kN/m)	FW.PTM. 23 c	45.02	44.96	0.13%
	24	$q_{0.2mm}$ (kN/m)	FW.PTM. 24 c	52.18	52.25	-0.13%
	25	$q_{0.6fck}$ (kN/m)	FW.PTM. 25 c	61.69	63.64	-3.17%
	26	$q_{0.8fsu}$ (kN/m)	FW.PTM. 26 c	74.04	75.30	-1.70%
	27	$q_{0.75fpu}$ (kN/m)	FW.PTM. 27 c	52.17	53.11	-1.81%
ULTIMATE LOAD COMBINATION	28	q_{str} (kN/m)	FW.PTM. 28 c	84.28	78.13	7.29%
	29	ε_{st} (mm/mm)	FW.PTM. 29 c	0.01653	0.01643	0.55%
	30	ε_{sc} (mm/mm)	FW.PTM. 30 c	-0.00170	-0.00170	0.30%
	31	ε_p (mm/mm)	FW.PTM. 31 c	0.02270	0.02272	-0.09%
	32	ρ_{sw} (mm^2/mm^2)	FW.PTM. 32 c	0.0016	0.0016	0.16%
	33	μ_Δ	FW.PTM. 33 c	2.42	2.41	0.34%

Significant errors

: Given Inputs for ANN　　: Inputs for engineering mechanic

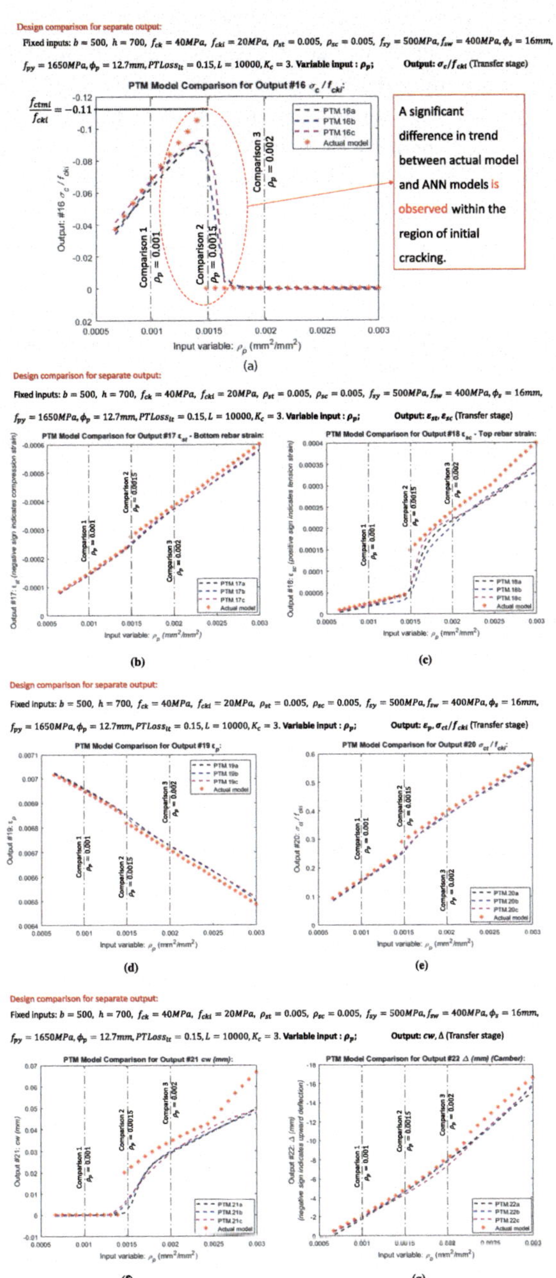

Figure 2.4.1 Design charts corresponding to ρ_p. (a) σ_c/f_{cki} (upper concrete stress per concrete strength at transfer stage) vs. ρ_p. (b) ε_{st} vs. ρ_p. (c) ε_{sc} vs. ρ_p. (d) ε_p vs. ρ_p. (e) σ_{ct}/f_{cki} (lower concrete stress per concrete strength at transfer stage) vs. ρ_p. (f) c_w vs. ρ_p. (g) Δ vs. ρ_p.

σ_c/f_{cki} (output # 16) represents upper concrete stress per concrete strength at the transfer stage in which upper concrete is in tension during the transfer stage when the load is upward. A tensile concrete stress (σ_{ct} output # 20) immediately decreases to 0 when reaching its tensile concrete strength at the transfer stage (f_{cki}), leading to tensile concrete strength which is lost after cracking. A discontinuity is demonstrated in tensile concrete sections due to this cracking, whereas a feed-forward ANN approximates a discontinuity based on real-valued continuous functions. A significant difference between tensile concrete sections and ANNs is observed at the vicinity of initial cracking.

Figure 2.4.1(a)–(e) plotted based on Table 2.4.4 presents design aides for σ_{ct}/f_{cki}, ε_{st}, ε_{sc}, ε_p, and σ_{ct}/f_{cki} (output # 16, 17, 18, 19, and 20, respectively) as a function of tendon ratio ρ_p. Design parameters (ε_{st}, ε_{sc}, ε_p, and σ_{ct}/f_{cki}) are obtained rapidly and accurately for a given tendon ratio (ρ_p). Figure 2.4.1(f), and (g) also shows plots for cw (Output # 21) vs. ρ_p and Δ (Output # 22) vs. ρ_p, respectively.

2.5 REVERSE DESIGNS BASED ON PTM USING DEEP NEURAL NETWORKS (DNN)

2.5.1 Reverse design 1

2.5.1.1 *Formulations of two steps based on back-substitution (BS) applicable to reverse designs*

In Reverse Scenario 1 shown in Figure 2.5.1, design parameters shown in ρ_{st} and ρ_p in green Boxes 5 and 10 of Figure 2.5.1 are calculated on an output-side when preassigning targeted nominal capacities, $q_{L/250}$ and q_{str}, of pretensioned beams shown in the yellow Boxes 23 and 28 of

Reverse scenario 1	Input for engineering mechanics	Beam configurations	1 b (mm)	2 h (mm)	3 f_{ck} (MPa)	4 f_{cki} (MPa)	5 ρ_{st}	6 ρ_{sc}	7 f_{sy}	8 f_{sw}	
			9 ϕ_s	10 ρ_p	11 f_{py}	12 ϕ_p	13 $Ptloss_{lt}$	14 L (mm)	15 K_c		
	Output for engineering mechanics	Transfer stage	16 σ_c/f_{cki}	17 ε_{st}	18 ε_{sc}	19 ε_p	20 σ_{ct}/f_{cki}	21 cw	22 Δ (mm)		
		Service stage	23 Def. control $q_{L/250}$ (kN/m)		24 Crack control $q_{0.2mm}$ (kN/m)		25 Stress control $q_{0.6fck}$ (kN/m)		26 Stress control $q_{0.8fsu}$ (kN/m)		27 Stress control $q_{0.75fpu}$ (kN/m)
		Ultimate stage with material factor	28 Str. control q_{str} (kN/m)		29 ε_{st}	30 ε_{sc}	31 ε_p	32 ρ_{sw}	33 μ_Δ		

Input Output Reverse Input Reverse Output

Figure 2.5.1 Scenario table for reverse scenario 1.

Figure 2.5.1 on an input-side. In Figure 2.5.1 and Table 2.5.2, the targeted nominal load capacities ($q_{L/250}$ and q_{str}) are referred to reverse input parameters, whereas ρ_{st} and ρ_p in green Boxes 5 and 10 are referred to reverse output parameters. Training accuracies of the 5th output parameter (ρ_{st}) based on R1.PTM.5a and the 10th output parameter (ρ_p) based on R1.PTM.10a and R1.PTM.10b are summarized in Table 2.5.1 where R1 indicates a reverse ANN for Design Scenario 1, whereas 5a is mapped based on 20 layers-20 neurons, 10a is mapped based on 20 layers-20 neurons, and 10b is mapped based on 40 layers-40 neurons. In a conventional forward design, targeted reverse parameters such as nominal load capacities ($q_{L/250}$ and q_{str}) of pretensioned beams are calculated on an output-side, whereas design parameters such as lower tensile rebar ratio (ρ_{st}) and tendon ratio (ρ_p) are input parameters. In a conventional forward design, a reverse design is difficult because engineers are unable to prescribe targeted nominal load capacities ($q_{L/250}$ and q_{str}) shown in the yellow Boxes 23 and 28 of Figure 2.5.1 on an input-side to obtain design parameters such as lower tensile rebar ratio (ρ_{st}) and tendon ratio (ρ_p) on an output-side.

The BS method proposed by Hong [2.3] is used, in which two independent networks shown in Table 2.5.2(a) are formulated to perform a reverse design of PT beams. In the reverse network of Step 1, only reverse output parameters ρ_{st}, and ρ_p shown in Boxes 5 and 10 of Table 2.5.2 are precalculated on an output-side for given targeting nominal capacities $q_{L/250}$ and q_{str} preassigned reversely in Boxes 23 and 28 of an input-side. Two reverse output parameters ρ_{st} and ρ_p are, then, used as input parameters for the forward network of Step 2. The targeted nominal capacities ($q_{L/250}$ and q_{str}) of pretensioned beams are referred to reverse input parameters. This is also described in Table 2.5.1 where PTM is implemented in obtaining reverse output parameters ρ_{st}, and ρ_p on an output-side the Step 1, whereas nominal capacities $q_{L/250}$ and q_{str} are reversely preassigned on an input-side the Step 1. In the forward network of Step 2, the 18 outputs including targeted reverse parameters ($q_{L/250}$ and q_{str} shown in Boxes 23 and 28 which are reversely prescribed on an input-side Step 1) are calculated in Boxes 16–33 when inputs including design parameters (ρ_{st} and ρ_p) calculated in Boxes 5 and 10 of an output-side the reverse network of Step 1 are used in Boxes 1–15 of an input-side the forward network of Step 2 as shown in Tables 2.5.2(a)–(c). As shown in the reverse network of Step 1, two ANNs, R1.PTM.5a and R1.PTM.10a, based on 20 layers and 20 neurons (refer to Table 2.5.1) are used to determine ρ_{st}, and ρ_p shown in Boxes 5 and 10, respectively, for a given reversely preassigned nominal strength, $q_{L/250}$ and q_{str}, shown in Boxes 23 and 28. Fifteen input parameters shown in Boxes 1–15 including ρ_{st} and ρ_p precalculated in Boxes 5 and 10 are implemented in determining 18 outputs in the forward network of Step 2 based on FW.PTM.c with 50 layers and 50 neurons.

Table 2.5.1 Training accuracies of reverse scenario 1 based on PTM; reverse network of Step 1

REVERSE NETWORK OF STEP 1-REVERSE SCENARIO 1

15 Inputs b, h, f_{ck}, f_{cki}, ρ_{sc}, f_{sy}, f_{sw}, ϕ_S, f_{py}, ϕ_P, $PTloss_{lt}$, L, K_c, $q_{\frac{L}{250}}$, qstr) – 2 Outputs (ρ_{st}, ρ_p)

No.	Data	Layers	Neurons	Validation check	Required epoch	Best epoch for training	Stopped epoch	Mean square error	R at best epoch	Comments
15 Inputs (b, h, f_{ck}, f_{cki}, ρ_{sc}, f_{sy}, f_{sw}, ϕ_S, f_{py}, ϕ_P, $PTloss_{lt}$, L, K_c, $q_{L/250}$, q_{str}) **–1 Output** (ρ_{st}) - *PTM*										
R1.PTM. **5** **a**	50,000	20	20	500	50,000	3,514	4,014	0.01546	0.916086	Insufficient
15 Inputs (b, h, f_{ck}, f_{cki}, ρ_{sc}, f_{sy}, f_{sw}, ϕ_S, f_{py}, ϕ_P, $PTloss_{lt}$, L, K_c, $q_{L/250}$, q_{str}) **–1 Output** (ρ_p) - *PTM*										
R1.PTM. **10** **a**	50,000	20	20	500	50,000	6,744	7,244	1.22E-02	0.916305	Insufficient
R1.PTM. **10** **b**	50,000	40	40	5000	100,000	6,651	11,651	0.013101	0.91199	Insufficient

Note: R1.PTM.**10a**

→ Version of training model

→ Output number

→ Training method.

→ Design scenario;

R1 = Reverse ANN for Scenario 1

Table 2.5.2 Design accuracies of Reverse scenario I

(a) $q_{L/250}$, (70 kN/m) at a deflection of L/250 (SLL) and q_{str} (150 kN/m) at an ULL state

(a-1) Step 1: Reverse network with reverse inputs ($q_{L/250}$, q_{str}) in Boxes 23 and 28 to calculate reverse outputs (ρ_{st}, ρ_p) in Boxes 5 and 10

PARAMETERS			Training Model	Network test result	Auto PT-BEAM check	Error
BEAM CONFIGURATION	1	b (mm)		500	500	0.00%
	2	h (mm)		700	700	0.00%
	3	f_{ck} (MPa)		40	40	0.00%
	4	f_{cki} (MPa)		20	20	0.00%
	5	ρ_{st} (mm^2/mm^2)	R1.PTM. 5 a	**0.01038**	**0.01038**	**0.00%**
	6	ρ_{sc} (mm^2/mm^2)		0.005	0.005	0.00%
	7	f_{sy} (MPa)		500	500	0.00%
	8	f_{sw} (MPa)		400	400	0.00%
	9	ϕ_S (mm)		16	16	0.00%
	10	ρ_p (mm^2/mm^2)	R1.PTM. 10 a	**0.00109**	**0.00109**	**0.00%**
	11	f_{py} (MPa)		1650	1650	0.00%
	12	ϕ_p (mm)		12.7	12.7	0.00%
	13	$PTloss_{lt}$		0.15	0.15	0.00%
	14	L (mm)		8000	8000	0.00%
	15	K_c		3	3	0.00%
R1 Inputs	23	$q_{L/250}$ (kN/m)		**70.00**	**69.22**	**1.12%**
	28	q_{str} (kN/m)		**150.00**	**149.29**	**0.47%**

: Given Inputs for ANN R1 : Inputs for engineering mechanic

: Reverse Outputs calculated by reverse ANN model is used as inputs to calculate other outputs by forward ANN model .

ρ_{st} and ρ_p are substituted into a forward network of Step 2

(Continued)

Two steps are summarized again as follows:

Step 1 (reverse network); reverse ANNs with reverse inputs ($q_{L/250}$ and q_{str}) calculate reverse outputs (ρ_{st} and ρ_p).

Step 2 (forward network); Using reverse outputs ρ_{st} and ρ_p calculated on an output-side the Step 1, forward ANNs calculate the 18 output parameters, including targeted parameters $q_{L/250}$ and q_{str}, which are already reversely prescribed on an input-side the Step 1. A reverse design based on Steps 1 and 2 can be validated by comparing pre-assigned reverse input parameters, $q_{L/250}$ and q_{str}, in the reverse network of Step 1 by those calculated in the forward network of Step 2, demonstrating 1.12% and 0.47% for $q_{L/250}$ and q_{str}, respectively, as shown in Boxes 23 and 28 of Tables 2.5.2(a-1) and (a-3). All 18 output parameters including $q_{L/250}$ and q_{str} are also obtained in Boxes 16–33 of Table 2.5.2(a-2).

Table 2.5.2 (Continued) Design accuracies of Reverse scenario 1

(a-2) Step 2: Forward network using reverse outputs (from Step 1) as input parameters to calculate 18 output parameters in Boxes 16–33 including $q_{L/250}$ and q_{str}

ρ_{st} and ρ_p are calculated by a reverse network of Step 1

PARAMETERS			Training Model	Network test result	Auto PT-BEAM check	Error
BEAM CONFIGURATION	1	b (mm)		500	500	0.00%
	2	h (mm)		700	700	0.00%
	3	f_{ck} (MPa)		40	40	0.00%
	4	f_{cki} (MPa)		20	20	0.00%
	5	ρ_{st} (mm^2/mm^2)	R1.PTM. 5 a	**0.01038**	**0.01038**	**0.00%**
	6	ρ_{sc} (mm^2/mm^2)		0.005	0.005	0.00%
	7	f_{sy} (MPa)		500	500	0.00%
	8	f_{sw} (MPa)		400	400	0.00%
	9	ϕ_S (mm)		16	16	0.00%
	10	ρ_p (mm^2/mm^2)	R1.PTM. 10 a	**0.00109**	**0.00109**	**0.00%**
	11	f_{py} (MPa)		1650	1650	0.00%
	12	ϕ_p (mm)		12.7	12.7	0.00%
	13	$PTloss_{lt}$		0.15	0.15	0.00%
	14	L (mm)		8000	8000	0.00%
	15	K_c		3	3	0.00%
REMAIN OUTPUTS CALCULATED BY REVERSE-FORWARD STEP	16	σ_c/f_{cki}	FW.PTM. 16 c	-0.0762	-0.07	Uncracked
	17	ε_{st} (mm/mm)	FW.PTM. 17 c	-0.00015	-0.00015	-0.18%
	18	ε_{sc} (mm/mm)	FW.PTM. 18 c	0.0000261	0.0000259	0.64%
	19	ε_p (mm/mm)	FW.PTM. 19 c	0.00695	0.00695	0.02%
	20	σ_{ct}/f_{cki}	FW.PTM. 20 c	0.15598	0.15542	0.36%
	21	cw (mm)	FW.PTM. 21 c	0.000	0.000	0.00%
	22	Δ (mm)	FW.PTM. 22 c	-1.758	-1.941	-10.43%
	24	$q_{0.2mm}$ (kN/m)	FW.PTM. 24 c	83.40	85.08	-2.02%
	25	$q_{0.6fck}$ (kN/m)	FW.PTM. 25 c	103.70	106.01	-2.23%
	26	$q_{0.8fsu}$ (kN/m)	FW.PTM. 26 c	141.79	138.55	2.28%
	27	$q_{0.75fpu}$ (kN/m)	FW.PTM. 27 c	92.50	83.71	9.50%
	29	ε_{st} (mm/mm)	FW.PTM. 29 c	0.01321	0.01324	-0.18%
	30	ε_{sc} (mm/mm)	FW.PTM. 30 c	-0.00198	-0.00199	-0.38%
	31	ε_p (mm/mm)	FW.PTM. 31 c	0.01953	0.01948	0.27%
	32	ρ_{sw} (mm^2/mm^2)	FW.PTM. 32 c	0.0023	0.0023	0.82%
	33	μ_Δ	FW.PTM. 33 c	1.85	1.87	-0.58%

: Reverse-forward inputs : Inputs for engineering mechanic
: Reverse-forward ouputs

(Continued)

Table 2.5.2 (Continued) Design accuracies of Reverse scenario 1

(a-3) Step 3: Summary of design based on two steps of BS

Summary of design based on two steps of BS

PARAMETERS			Training Model	Network test result	Auto PT-BEAM check	Error
BEAM CONFIGURATION	1	b (mm)		500	500	0.00%
	2	h (mm)		700	700	0.00%
	3	f_{ck} (MPa)		40	40	0.00%
	4	f_{cki} (MPa)		20	20	0.00%
	5	ρ_{st} (mm^2/mm^2)	R1.PTM. 5 a	0.01038	0.01038	0.00%
	6	ρ_{sc} (mm^2/mm^2)		0.005	0.005	0.00%
	7	f_{sy} (MPa)		500	500	0.00%
	8	f_{sw} (MPa)		400	400	0.00%
	9	ϕ_S (mm)		16	16	0.00%
	10	ρ_p (mm^2/mm^2)	R1.PTM. 10 a	0.00109	0.00109	0.00%
	11	f_{py} (MPa)		1650	1650	0.00%
	12	ϕ_p (mm)		12.7	12.7	0.00%
	13	$PTloss_{lt}$		0.15	0.15	0.00%
	14	L (mm)		8000	8000	0.00%
	15	K_c		3	3	0.00%
TRANSFER LOAD STAGE	16	σ_c/f_{cki}	FW.PTM. 16 c	-0.0762	-0.07	Uncrack
	17	ε_{st} (mm/mm)	FW.PTM. 17 c	-0.00015	-0.00015	-0.18%
	18	ε_{sc} (mm/mm)	FW.PTM. 18 c	0.0000261	0.0000259	0.64%
	19	ε_p (mm/mm)	FW.PTM. 19 c	0.00695	0.00695	0.02%
	20	σ_{ct}/f_{cki}	FW.PTM. 20 c	0.15598	0.15542	0.36%
	21	cw (mm)	FW.PTM. 21 c	0.000	0.000	0.00%
	22	Δ (mm)	FW.PTM. 22 c	-1.758	-1.941	-10.43%
SERVICE LOAD COMBINATION	23	$q_{L/250}$ (kN/m)		**70.00**	**69.22**	**1.12%**
	24	$q_{0.2mm}$ (kN/m)	FW.PTM. 24 c	83.40	85.08	-2.02%
	25	$q_{0.6fck}$ (kN/m)	FW.PTM. 25 c	103.70	106.01	-2.23%
	26	$q_{0.8fsu}$ (kN/m)	FW.PTM. 26 c	141.79	138.55	2.28%
	27	$q_{0.75fpu}$ (kN/m)	FW.PTM. 27 c	92.50	83.71	9.50%
ULTIMATE LOAD COMBINATION	28	q_{str} (kN/m)		**150.00**	**149.29**	**0.47%**
	29	ε_{st} (mm/mm)	FW.PTM. 29 c	0.01321	0.01324	-0.18%
	30	ε_{sc} (mm/mm)	FW.PTM. 30 c	-0.00198	-0.00199	-0.38%
	31	ε_p (mm/mm)	FW.PTM. 31 c	0.01953	0.01948	0.27%
	32	ρ_{sw} (mm^2/mm^2)	FW.PTM. 32 c	0.0023	0.0023	0.82%
	33	μ_Δ	FW.PTM. 33 c	1.85	1.87	-0.58%

(red box): Given Inputs for ANN R1 (blue box): Inputs for engineering mechanic

(purple box): Reverse Outputs calculated by reverse ANN model is used as inputs to calculate other outputs by forward ANN model .

(green box): Reverse-forward inputs (grey box): Reverse-forward ouputs

(Continued)

Table 2.5.2 (Continued) Design accuracies of Reverse scenario 1

(b) $q_{L/250}$, (100 kN/m) at a deflection of L/250 (SLL) and q_{str} (200 kN/m) at an ULL state

PARAMETERS			Training Model	Network test result	Auto PT-BEAM check	Error
BEAM CONFIGURATION	1	b (mm)		500	500	0.00%
	2	h (mm)		700	700	0.00%
	3	f_{ck} (MPa)		40	40	0.00%
	4	f_{cki} (MPa)		20	20	0.00%
	5	ρ_{st} (mm²/mm²)	R1.PTM. 5 a	0.01853	0.01853	0.00%
	6	ρ_{sc} (mm²/mm²)		0.005	0.005	0.00%
	7	f_{sy} (MPa)		500	500	0.00%
	8	f_{sw} (MPa)		400	400	0.00%
	9	ϕ_s (mm)		16	16	0.00%
	10	ρ_p (mm²/mm²)	R1.PTM. 10 a	0.00189	0.00189	0.00%
	11	f_{py} (MPa)		1650	1650	0.00%
	12	ϕ_p (mm)		12.7	12.7	0.00%
	13	$PTloss_{lt}$		0.15	0.15	0.00%
	14	L (mm)		8000	8000	0.00%
	15	K_c		3	3	0.00%
TRANSFER LOAD STAGE	16	σ_c/f_{cki}	FW.PTM. 16 c	~0	0.00	Cracked
	17	ε_{st} (mm/mm)	FW.PTM. 17 c	-0.00024	-0.00026	-5.38%
	18	ε_{sc} (mm/mm)	FW.PTM. 18 c	0.0000874	0.0001435	-64.22%
	19	ε_p (mm/mm)	FW.PTM. 19 c	0.00685	0.00684	0.20%
	20	σ_{ct}/f_{cki}	FW.PTM. 20 c	0.25869	0.27390	-5.88%
	21	cw (mm)	FW.PTM. 21 c	0.007	0.020	-185.71%
	22	Δ (mm)	FW.PTM. 22 c	-3.548	-3.684	-3.84%
SERVICE LOAD COMBINATION	23	$q_{L/250}$ (kN/m)		**100.00**	**107.20**	**-7.20%**
	24	$q_{0.2mm}$ (kN/m)	FW.PTM. 24 c	154.25	155.76	-0.98%
	25	$q_{0.6fck}$ (kN/m)	FW.PTM. 25 c	129.68	134.30	-3.56%
	26	$q_{0.8fsu}$ (kN/m)	FW.PTM. 26 c	238.36	241.29	-1.23%
	27	$q_{0.75fpu}$ (kN/m)	FW.PTM. 27 c	157.58	148.17	5.97%
ULTIMATE LOAD COMBINATION	28	q_{str} (kN/m)		**200.00**	**228.93**	**-14.46%**
	29	ε_{st} (mm/mm)	FW.PTM. 29 c	0.00460	0.00454	1.27%
	30	ε_{sc} (mm/mm)	FW.PTM. 30 c	-0.00278	-0.00277	0.31%
	31	ε_p (mm/mm)	FW.PTM. 31 c	0.00993	0.01067	-7.44%
	32	ρ_{sw} (mm²/mm²)	FW.PTM. 32 c	0.0052	0.0052	-0.36%
	33	μ_Δ	FW.PTM. 33 c	1.28	1.30	-1.44%

Significant errors

: Given Inputs for ANN R1 : Inputs for engineering mechanic

: Reverse Outputs calculated by reverse ANN model is used as inputs to calculate other outputs by forward ANN model .

: Reverse-forward inputs : Reverse-forward ouputs

(Continued)

Table 2.5.2 (Continued) Design accuracies of Reverse scenario 1

(c) $q_{L/250}$, (100, 120 kN/m) at a deflection of L/250 (SLL) and q_{str} (300, 350 kN/m) at an ULL state,

(c-1) $q_{L/250}$, (100 kN/m) at a deflection of L/250 (SLL) and q_{str} (300 kN/m) at an ULL state

PARAMETERS			Training Model	Network test result	Auto PT-BEAM check	Error
BEAM CONFIGURATION	1	b (mm)		1200	1200	0.00%
	2	h (mm)		900	900	0.00%
	3	f_{ck} (MPa)		40	40	0.00%
	4	f_{cki} (MPa)		20	20	0.00%
	5	ρ_{st} (mm^2/mm^2)	R1.PTM. 5 a	0.02119	0.02119	0.00%
	6	ρ_{sc} (mm^2/mm^2)		0.005	0.005	0.00%
	7	f_{sy} (MPa)		500	500	0.00%
	8	f_{sw} (MPa)		400	400	0.00%
	9	ϕ_s (mm)		16	16	0.00%
	10	ρ_p (mm^2/mm^2)	R1.PTM. 10 a	0.00156	0.00156	0.00%
	11	f_{py} (MPa)		1650	1650	0.00%
	12	ϕ_p (mm)		12.7	12.7	0.00%
	13	$PTloss_{lt}$		0.15	0.15	0.00%
	14	L (mm)		14000	14000	0.00%
	15	K_c		3	3	0.00%
TRANSFER LOAD STAGE	16	σ_c/f_{cki}	FW.PTM. 16 c	-0.0798	-0.07	Uncracked
	17	ε_{st} (mm/mm)	FW.PTM. 17 c	-0.00017	-0.00017	0.77%
	18	ε_{sc} (mm/mm)	FW.PTM. 18 c	0.0000351	0.0000273	22.14%
	19	ε_p (mm/mm)	FW.PTM. 19 c	0.00693	0.00693	0.01%
	20	σ_{ct}/f_{cki}	FW.PTM. 20 c	0.17620	0.17331	1.64%
	21	cw (mm)	FW.PTM. 21 c	0.000	0.000	0.00%
	22	Δ (mm)	FW.PTM. 22 c	-3.002	-3.250	-8.24%
SERVICE LOAD COMBINATION	23	$q_{L/250}$ (kN/m)		**100.00**	**100.37**	**-0.37%**
	24	$q_{0.2mm}$ (kN/m)	FW.PTM. 24 c	198.67	198.93	-0.13%
	25	$q_{0.6fck}$ (kN/m)	FW.PTM. 25 c	166.86	165.96	0.54%
	26	$q_{0.8fsu}$ (kN/m)	FW.PTM. 26 c	329.58	320.94	2.62%
	27	$q_{0.75fpu}$ (kN/m)	FW.PTM. 27 c	185.59	185.15	0.23%
ULTIMATE LOAD COMBINATION	28	q_{str} (kN/m)		**300.00**	**329.35**	**-9.78%**
	29	ε_{st} (mm/mm)	FW.PTM. 29 c	0.00397	0.00411	-3.35%
	30	ε_{sc} (mm/mm)	FW.PTM. 30 c	-0.00297	-0.00298	-0.13%
	31	ε_p (mm/mm)	FW.PTM. 31 c	0.01012	0.01021	-0.91%
	32	ρ_{sw} (mm^2/mm^2)	FW.PTM. 32 c	0.0045	0.0045	0.22%
	33	μ_Δ	FW.PTM. 33 c	1.23	1.24	-0.94%

Significant errors

: Given Inputs for ANN R1 : Inputs for engineering mechanic

: Reverse Outputs calculated by reverse ANN model is used as inputs to calculate other outputs by forward ANN model .

: Reverse-forward inputs : Reverse-forward ouputs

(Continued)

Table 2.5.2 (Continued) Design accuracies of Reverse scenario I

(c-2) $q_{L/250}$, (120 kN/m) at a deflection of L/250 (SLL) and q_{str} (350 kN/m) at an ULL state

PARAMETERS			Training Model	Network test result	Auto PT-BEAM check	Error
BEAM CONFIGURATION	1	b (mm)		1200	1200	0.00%
	2	h (mm)		900	900	0.00%
	3	f_{ck} (MPa)		40	40	0.00%
	4	f_{cki} (MPa)		20	20	0.00%
	5	ρ_{st} (mm^2/mm^2)	R1.PTM. 5 a	0.01971	0.01971	0.00%
	6	ρ_{sc} (mm^2/mm^2)		0.005	0.005	0.00%
	7	f_{sy} (MPa)		500	500	0.00%
	8	f_{sw} (MPa)		400	400	0.00%
	9	ϕ_s (mm)		16	16	0.00%
	10	ρ_p (mm^2/mm^2)	R1.PTM. 10 a	0.00226	0.00226	0.00%
	11	f_{py} (MPa)		1650	1650	0.00%
	12	ϕ_p (mm)		12.7	12.7	0.00%
	13	$PTloss_{lt}$		0.15	0.15	0.00%
	14	L (mm)		14000	14000	0.00%
	15	K_c		3	3	0.00%
TRANSFER LOAD STAGE	16	σ_c/f_{cki}	FW.PTM. 16 c	~0	0.00	Cracked
	17	ε_{st} (mm/mm)	FW.PTM. 17 c	-0.00029	-0.00029	-1.04%
	18	ε_{sc} (mm/mm)	FW.PTM. 18 c	0.0001558	0.0001432	8.08%
	19	ε_p (mm/mm)	FW.PTM. 19 c	0.00681	0.00680	0.07%
	20	σ_{ct}/f_{cki}	FW.PTM. 20 c	0.29777	0.29881	-0.35%
	21	cw (mm)	FW.PTM. 21 c	0.022	0.020	9.09%
	22	Δ (mm)	FW.PTM. 22 c	-7.349	-7.511	-2.20%
SERVICE LOAD COMBINATION	23	$q_{L/250}$ (kN/m)		**120.00**	**119.76**	**0.20%**
	24	$q_{0.2mm}$ (kN/m)	FW.PTM. 24 c	218.65	218.10	0.25%
	25	$q_{0.6fck}$ (kN/m)	FW.PTM. 25 c	173.77	174.10	-0.19%
	26	$q_{0.8fsu}$ (kN/m)	FW.PTM. 26 c	344.02	336.46	2.20%
	27	$q_{0.75fpu}$ (kN/m)	FW.PTM. 27 c	203.62	205.26	-0.80%
ULTIMATE LOAD COMBINATION	28	q_{str} (kN/m)		**350.00**	**338.29**	**3.35%**
	29	ε_{st} (mm/mm)	FW.PTM. 29 c	0.00373	0.00381	-2.10%
	30	ε_{sc} (mm/mm)	FW.PTM. 30 c	-0.00299	-0.00300	-0.16%
	31	ε_p (mm/mm)	FW.PTM. 31 c	0.00988	0.00991	-0.37%
	32	ρ_{sw} (mm^2/mm^2)	FW.PTM. 32 c	0.0047	0.0047	-0.06%
	33	μ_Δ	FW.PTM. 33 c	1.22	1.22	0.07%

: Given Inputs for ANN R1 ; : Inputs for engineering mechanic

: Reverse Outputs calculated by reverse ANN model is used as inputs to calculate other outputs by forward ANN model .

: Reverse-forward inputs ; : Reverse-forward ouputs

2.5.1.2 Design accuracies

In Table 2.5.2 of Reverse Scenario 1, input parameters are preassigned as follows: b=500mm, h=700mm, f_{ck}=40MPa, f_{cki}=20MPa, f_{sy}=500MPa, f_{sw}=400MPa, ϕ_s=16mm (diameter of rebar), f_{py}=1,650MPa, ϕ_p=12.7mm (diameter of tendon), $PTLoss_{lt}$=0.15, L =8,000, and K_c=3, whereas the following reverse parameters are reversely preassigned on an input-side; $q_{L/250}$=70 KN/m (SLL) and q_{str}=150 KN/m (ULL) for Table 2.5.2(a), $q_{L/250}$=100 KN/m (SLL), and q_{str}=200 KN/m (ULL) for Table 2.5.2(b), and $q_{L/250}$=100 KN/m (SLL), and q_{str}=300 KN/m (ULL) for Table 2.5.2(c). In Reverse Scenario 1, the lower tensile rebar ratio of 0.01038 (ρ_{st}) shown in Box 5 of Table 2.5.2(a) and the tendon ratio of 0.00109 (ρ_p) shown in Box 10 of Table 2.5.2(a-1) are calculated by the reverse network of Step 1. These design parameters are calculated for reversely assigned targeted nominal capacities resisting applied distributed load ($q_{L/250}$=70 kN/m) with deflection reaching L/250 at SLL shown in Box 23 of Table 2.5.2(a-1) and nominal capacities resisting applied distributed load (q_{str}=150 kN/m) at ULL shown in Box 28 of Table 2.5.2(a-1). Table 2.5.2 (a-2) is forward network which uses reverse outputs (calculated from Step 1) as input parameters to calculate 18 output parameters including $q_{L/250}$ and q_{str} in Boxes 16–33. Table 2.5.2(a-2) demonstrates acceptable accuracies for most design parameters calculated except for a couple of parameters such as deflection with the error of –10.43% shown in Box 22 of Tables 2.5.2(a-2) and (a-3); however, the absolute error is small as shown with –1.758 mm for ANN-based deflection vs. –1.941 mm for structural mechanics-based software. Similarly, a lower tensile rebar ratio (ρ_{st}) of 0.01853 shown in Box 5 of Table 2.5.2(b) and tendon ratio (ρ_p) of 0.00189 shown in Box 10 of Table 2.5.2(b) are precalculated by the reverse network of Step 1 when considering given targeted nominal capacities resisting applied distributed load ($q_{L/250}$=100 kN/m) with deflection reaching L/250 at SLL shown in Box 23 of Table 2.5.2(b) and nominal capacities resisting applied distributed load (q_{str}=200 kN/m) at ULL shown in Box 28 of Table 2.5.2(b). The 18 output parameters are calculated in Boxes 16–33 of Table 2.5.2(b) based on the forward network of Step 2 trained by FW.PTM.c based on 50 layers-50 neurons.

Table 2.5.2(c-1) presents a design error of 22.14% for concrete strain ε_{sc} shown in Box 18 at transfer stage when nominal strength ($q_{L/250}$) of 100 kN/m at SLL and maximum nominal strength (q_{str}) of 300 kN/m at ULL are reversely preassigned. However, a design error for concrete strain ε_{st} shown in Box 18 at the transfer stage reduces to 8.08% shown in Table 2.5.2(c-2) when nominal strength ($q_{L/250}$) of 100 kN/m at SLL and maximum nominal strength (q_{str}) of 300 kN/m at ULL are adjusted to 120 kN/m and 350 kN/m, respectively. It should be noted that, as shown in Tables 2.5.2(a)–(c), removing input conflicts by increasing nominal strength ($q_{L/250}$) to 120 kN/m at SLL and maximum nominal strength (q_{str}) of 350 kN/m at ULL can reduce design errors. Tables demonstrate acceptable accuracies for most design parameters calculated in

Table 2.5.2 except for a couple of parameters such as deflection with the error of –10.43% shown in Box 22 of Table 2.5.2(a-3); however, the absolute error is small as shown with –1.758 mm for ANN-based deflection vs. –1.941 mm for structural mechanics-based software. Deflection error decreases to –2.20% shown in Box 22 of Table 2.5.2(c-2), when input conflicts are adjusted.

2.5.2 Reverse design 2

2.5.2.1 Formulations of two steps based on back-substitution (BS) applicable to reverse designs

In Reverse Scenario 2 shown in Figure 2.5.2 and Table 2.5.3, design parameters (ρ_{st}, ρ_{sc}, and ρ_p shown in green shown in Boxes 5, 6, and 10) are calculated on an output-side the reverse network of Step 1 when targeted nominal capacities of pretensioned beams ($q_{L/250}$ = 70 KN/m at SLL, q_{str} = 150 KN/m at ULL and ε_{st} at ULL shown in the yellow shown in Boxes 23, 28, and 29) are reversely preassigned on an input-side. The ANNs trained by R2.PTM.5a, R2.PTM.6a, and R2.PTM.10a are implemented in determining ρ_{st}, ρ_{sc}, and ρ_p, respectively. Targeted nominal load capacities ($q_{L/250}$, q_{str}, and ε_{st}) are referred to reverse input parameters, whereas ρ_{st}, ρ_{sc}, and ρ_p are referred to reverse output parameters. R2.PTM.5a, R2.PTM.6a, and R2.PTM.10a map input parameters to the 5th, 6th, and 10th output parameters ρ_{st}, ρ_{sc}, and ρ_p, respectively, shown in Figure 2.5.2 and Table 2.5.3 based on PTM using 20 layers-20 neurons.

Table 2.5.3 summarizes the training accuracies based on PTM, whereas Table 2.5.4 presents the design accuracies. In a conventional forward design, targeted reverse parameters such as nominal load capacities ($q_{L/250}$=70 KN/m at SLL, q_{str}=150 KN/m at ULL) of pretensioned beams and strains (ε_{st}) at ULL are calculated on an output-side, whereas design parameters such as lower tensile rebar ratios (ρ_{st}), upper compressive rebar ratios (ρ_{sc}), and tendon ratios (ρ_p) are input parameters. In Table 2.5.4, three reverse input parameters ($q_{L/250}$, q_{str}, and ε_{st}) are reversely preassigned in Boxes 23, 28, and 29 of the reverse network of Step 1 to precalculate three reverse output parameters ρ_{st}, ρ_{sc}, and ρ_p in Boxes 5, 6, and 10. R2.PTM.5a,

Reverse scenario 2	Input for engineering mechanics	Beam configurations	1 b (mm)	2 h (mm)	3 f_{ck} (MPa)	4 f_{cki} (MPa)	5 ρ_{st}	6 ρ_{sc}	7 f_{sy}	8 f_{sw}	
			9 ϕ_s	10 ρ_p	11 f_{py}	12 ϕ_p	13 $Ptloss_{lt}$	14 L (mm)	15 K_c		
	Output for engineering mechanics	Transfer stage	16 σ_c/f_{cki}	17 ε_{st}	18 ε_{sc}	19 ε_p	20 σ_{ct}/f_{cki}	21 cw	22 Δ (mm)		
		Service stage	23 Def. control $q_{L/250}$ (kN/m)		24 Crack control $q_{0.2mm}$ (kN/m)		25 Stress control $q_{0.6fck}$ (kN/m)		26 Stress control $q_{0.8fsu}$ (kN/m)		27 Stress control $q_{0.75fpu}$ (kN/m)
		Ultimate stage with material factor	28 Str. control q_{str} (kN/m)		29 ε_{st}	30 ε_{sc}	31 ε_p	32 ρ_{sw}	33 μ_Δ		

Input   Output   Reverse Input   Reverse Output

Figure 2.5.2 Scenario table for reverse scenario 2.

Table 2.5.3 Training accuracies of reverse scenario 2 based on PTM; reverse network of Step 1

v. REVERSE NETWORK OF STEP 1-REVERSE SCENARIO 2
15 Inputs (b, h, f_{ck}, f_{cki}, f_{sy}, f_{sw}, ϕ_S, f_{py}, ϕ_P, $PTloss_{lt}$, L, K_c, $q_{L/250}$, q_{str}, ε_{st}) – 3 Outputs (ρ_{st}, ρ_{sc}, ρ_p)

No.			*Data*	*Layers*	*Neurons*	*Validation check*	*Required epoch*	*Best epoch for training*	*Stopped epoch*	*Mean square error*	*R at best epoch*	*Comments*
15 Inputs (b, h, f_{ck}, f_{cki}, f_{sy}, f_{sw}, ϕ_S, f_{py}, ϕ_P, $PTloss_{lt}$, L, K_c, $q_{L/250}$, q_{str}, ε_{st}) **–1 Output** (ρ_{st}) **- PTM**												
R2.PTM.	**5**	**a**	50,000	20	20	500	100,000	33,916	34,416	1.00E-04	0.999404	Good
R2.PTM.	**5**	**b**	50,000	30	30	1000	100,000	69,188	70,188	6.49E-05	0.999461	Good
15 Inputs (b, h, f_{ck}, f_{cki}, f_{sy}, f_{sw}, ϕ_S, f_{py}, ϕ_P, $PTloss_{lt}$, L, K_c, $q_{L/250}$, q_{str}, ε_{st}) **–1 Output** (ρ_{sc}) **- PTM**												
R2.PTM.	**6**	**a**	50,000	20	20	500	100,000	19,791	20,291	1.12E-03	0.993705	Neutral
R2.PTM.	**6**	**b**	50,000	30	30	1,000	100,000	10,408	11,408	0.001698	0.991312	Neutral
15 Inputs (b, h, f_{ck}, f_{cki}, f_{sy}, f_{sw}, ϕ_S, f_{py}, ϕ_P, $PTloss_{lt}$, L, K_c, $q_{L/250}$, q_{str}, ε_{st}) **–1 Output** (ρ_p) **- PTM**												
R2.PTM.	**10**	**a**	50,000	20	20	500	100,000	61,476	61,976	4.10E-04	0.997294	Neutral
R2.PTM.	**10**	**b**	50,000	30	30	1000	100,000	74,177	75,177	0.000714	0.99527	Neutral

Note:R2.PTM.**10a**

- → Version of training model
- → Output number
- → Training method.
- → Design scenario;
  R2 = Reverse ANN for Scenario 2

R2.PTM.6a, and R2.PTM.10a based on 20 layers and 20 neurons are used for the reverse network of Step 1. Fifteen input parameters including design parameters (ρ_{st}, ρ_{sc}, and ρ_p) calculated on an output-side the reverse network of Step 1 of Table 2.5.4 are, then, used in Boxes 1–15 on an input-side the forward network of Step 2. 15 forward ANNs trained by FW.PTM.c based on 40 layers and 40 neurons are implemented in obtaining 18 output parameters in Boxes 16–33 including three reverse input parameters ($q_{L/250}$, q_{str}, and ε_{st}) as shown in Figure 2.5.2 and Table 2.5.4.

2.5.2.2 Design accuracies

2.5.2.2.1 Adjusting reverse input parameters to enhance design accuracies

In Table 2.5.4(a) of Reverse Scenario 2, input parameters are preassigned as follows: b=500 mm, h=700 mm, f_{ck}=40 MPa, f_{cki}=20 MPa, f_{sy}=500 MPa, f_{sw}=400 MPa, ϕ_s=16 mm (diameter of rebar), f_{py}=1,650 MPa, ϕ_p=12.7 mm

Table 2.5.4 Design accuracies of Reverse Scenario 2

(a-1) Design results based on 20 layers-20 neurons for Step 1; $q_{L/250}$ = 70 kN/m reaching a deflection of L/250 at SLL, q_{str} = 150 kN/m at ULL, and lower tensile rebar strain (ε_{st}) of 0.0075 at ULL

(a-1-1) Step 1: Reverse network with reverse inputs ($q_{L/250}$, q_{str}, ε_{st}) to calculate reverse outputs (ρ_{st}, ρ_{sc}, ρ_p)

PARAMETERS			Training Model	Network test result	Auto PT-BEAM check	Error
BEAM CONFIGURATION	1	b (mm)		500.00	500.00	0.00%
	2	h (mm)		700	700	0.00%
	3	f_{ck} (MPa)		40	40	0.00%
	4	f_{cki} (MPa)		20	20	0.00%
	5	ρ_{st} (mm²/mm²)	R2.PTM. 5a	0.01053	0.01053	0.00%
	6	ρ_{sc} (mm²/mm²)	R2.PTM. 6a	0.00093	0.00093	0.00%
	7	f_{sy} (MPa)		500	500	0.00%
	8	f_{sw} (MPa)		400	400	0.00%
	9	ϕ_s (mm)		16	16	0.00%
	10	ρ_p (mm²/mm²)	R2.PTM. 10a	0.00118	0.00118	0.00%
	11	f_{py} (MPa)		1650	1650	0.00%
	12	ϕ_p (mm)		12.7	12.7	0.00%
	13	$PTloss_{lt}$		0.15	0.15	0.00%
	14	L (mm)		8000	8000	0.00%
	15	K_c		3	3	0.00%
R2 INPUTS	23	$q_{L/250}$ (kN/m)		70.00	70.19	-0.27%
	28	q_{str} (kN/m)		150.00	147.90	1.40%
	29	ε_{st} (mm/mm)		0.00750	0.00824	-9.85%

ρ_{st}, ρ_{sc}, and ρ_p are substituted into a forward network of Step 2

: Given Inputs for ANN R3 : Inputs for engineering mechanic

: Reverse Outputs calculated by reverse ANN model is used as inputs to calculate other outputs by forward ANN model.

(Continued)

Table 2.5.4 (Continued) Design accuracies of Reverse Scenario 2

(a-1-2) Step 2; Forward network of Step 2 using 15 input parameters including three reverse output parameters (ρ_{st}, ρ_{sc}, and ρ_p) calculated in Boxes 5, 6, and 10 of Step 1 to calculate 18 output parameters including $q_{L/250}$ and q_{str} in Boxes 16 to 33 and Step 3: Summary of design based on two steps of BS

Summary of design based on two steps of BS →

PARAMETERS			Training Model	Network test result	Auto PT-BEAM check	Error
BEAM CONFIGURATION	1	b (mm)		500.00	500.00	0.00%
	2	h (mm)		700	700	0.00%
	3	f_{ck} (MPa)		40	40	0.00%
	4	f_{cki} (MPa)		20	20	0.00%
	5	ρ_{st} (mm^2/mm^2)	R2.PTM. 5 a	0.01053	0.01053	0.00%
	6	ρ_{sc} (mm^2/mm^2)	R2.PTM. 6 a	0.00093	0.00093	0.00%
	7	f_{sy} (MPa)		500	500	0.00%
	8	f_{sw} (MPa)		400	400	0.00%
	9	ϕ_S (mm)		16	16	0.00%
	10	ρ_p (mm^2/mm^2)	R2.PTM. 10 a	0.00118	0.00118	0.00%
	11	f_{py} (MPa)		1650	1650	0.00%
	12	ϕ_p (mm)		12.7	12.7	0.00%
	13	$PTloss_{lt}$		0.15	0.15	0.00%
	14	L (mm)		8000	8000	0.00%
	15	K_c		3	3	0.00%
TRANSFER LOAD STAGE	16	σ_c/f_{cki}	FW.PTM. 16 c	-0.0850	-0.08	Uncracked
	17	ε_{st} (mm/mm)	FW.PTM. 17 c	-0.00016	-0.00016	0.28%
	18	ε_{sc} (mm/mm)	FW.PTM. 18 c	0.0000342	0.0000320	6.50%
	19	ε_p (mm/mm)	FW.PTM. 19 c	0.00694	0.00693	0.02%
	20	σ_{ct}/f_{cki}	FW.PTM. 20 c	0.17535	0.17273	1.50%
	21	cw (mm)	FW.PTM. 21 c	0.000	0.000	0.00%
	22	Δ (mm)	FW.PTM. 22 c	-2.077	-2.251	-8.41%
SERVICE LOAD COMBINATION	23	$q_{L/250}$ (kN/m)		**70.00**	**70.19**	**-0.27%**
	24	$q_{0.2mm}$ (kN/m)	FW.PTM. 24 c	86.51	88.12	-1.86%
	25	$q_{0.6fck}$ (kN/m)	FW.PTM. 25 c	97.82	100.92	-3.17%
	26	$q_{0.8fsu}$ (kN/m)	FW.PTM. 26 c	143.94	141.90	1.42%
	27	$q_{0.75fpu}$ (kN/m)	FW.PTM. 27 c	97.65	86.63	11.29%
ULTIMATE LOAD COMBINATION	28	q_{str} (kN/m)		**150.00**	**147.90**	**1.40%**
	29	ε_{st} (mm/mm)		**0.00750**	**0.00824**	**-9.85%**
	30	ε_{sc} (mm/mm)	FW.PTM. 30 c	-0.00244	-0.00244	-0.03%
	31	ε_p (mm/mm)	FW.PTM. 31 c	0.01430	0.01442	-0.83%
	32	ρ_{sw} (mm^2/mm^2)	FW.PTM. 32 c	0.0026	0.0026	-0.02%
	33	μ_Δ	FW.PTM. 33 c	1.58	1.56	1.30%

: Given Inputs for ANN R3
: Inputs for engineering mechanic
: Reverse Outputs calculated by reverse ANN model is used as inputs to calculate other outputs by forward ANN model .
: Reverse-forward inputs
: Reverse-forward ouputs

(Continued)

Table 2.5.4 (Continued) Design accuracies of Reverse Scenario 2

(a-2) Design results based on 20 layers-20 neurons for Step 1; $q_{L/250}$= 70 kN/m reaching a deflection of L/250 at SLL (at a deflection of L/250), q_{str} =150 kN/m at ULL

(a-2-1) Lower tensile rebar strain (ε_{st}) of 0.004 with input conflicts causing large errors

PARAMETERS			Training Model	Network test result	Auto PT-BEAM check	Error
BEAM CONFIGURATION	1	b (mm)		500.00	500.00	0.00%
	2	h (mm)		700	700	0.00%
	3	f_{ck} (MPa)		40	40	0.00%
	4	f_{cki} (MPa)		20	20	0.00%
	5	ρ_{st} (mm²/mm²)	R2.PTM. 5 a	0.01255	0.01255	0.00%
	6	ρ_{sc} (mm²/mm²)	R2.PTM. 6 a	-0.00037	-0.00037	0.00%
	7	f_{sy} (MPa)		500	500	0.00%
	8	f_{sw} (MPa)		400	400	0.00%
	9	ϕ_s (mm)		16	16	0.00%
	10	ρ_p (mm²/mm²)	R2.PTM. 10 a	0.00113	0.00113	0.00%
	11	f_{py} (MPa)		1650	1650	0.00%
	12	ϕ_p (mm)		12.7	12.7	0.00%
	13	$PTloss_{lt}$		0.15	0.15	0.00%
	14	L (mm)		8000	8000	0.00%
	15	K_c		3	3	0.00%
TRANSFER LOAD STAGE	16	σ_c/f_{cki}	FW.PTM. 16 c	~0	0.00	Cracked
	17	ε_{st} (mm/mm)	FW.PTM. 17 c	-0.00015	-0.00021	-45.48%
	18	ε_{sc} (mm/mm)	FW.PTM. 18 c	0.0000308	0.0057639	-18620.47%
	19	ε_p (mm/mm)	FW.PTM. 19 c	0.00695	0.00680	2.12%
	20	σ_{ct}/f_{cki}	FW.PTM. 20 c	0.15971	0.64324	-302.75%
	21	cw (mm)	FW.PTM. 21 c	0.000	3.747	Inf
	22	Δ (mm)	FW.PTM. 22 c	-1.823	-17.328	-850.75%
SERVICE LOAD COMBINATION	23	$q_{L/250}$ (kN/m)		70.00	72.65	-3.79%
	24	$q_{0.2mm}$ (kN/m)	FW.PTM. 24 c	95.39	97.12	-1.82%
	25	$q_{0.6fck}$ (kN/m)	FW.PTM. 25 c	99.04	102.86	-3.86%
	26	$q_{0.8fpu}$ (kN/m)	FW.PTM. 26 c	158.25	157.73	0.33%
	27	$q_{0.75fpu}$ (kN/m)	FW.PTM. 27 c	107.18	94.38	11.95%
ULTIMATE LOAD COMBINATION	28	q_{str} (kN/m)		150.00	161.13	-7.42%
	29	ε_{st} (mm/mm)		0.00400	0.00604	-50.96%
	30	ε_{sc} (mm/mm)	FW.PTM. 30 c	-0.00264	-0.00264	0.16%
	31	ε_p (mm/mm)	FW.PTM. 31 c	0.01195	0.01219	-1.99%
	32	ρ_{sw} (mm²/mm²)	FW.PTM. 32 c	0.0031	0.0031	0.37%
	33	μ_Δ	FW.PTM. 33 c	1.39	1.34	3.59%

Big errors due to $\rho_{sc} < 0$

Preassigning $\varepsilon_{st} = 0.004$ gives big errors

: Given Inputs for ANN R3 : Inputs for engineering mechanic

: Reverse Outputs calculated by reverse ANN model is used as inputs to calculate other outputs by forward ANN model .

: Reverse-forward inputs : Reverse-forward ouputs

(Continued)

Table 2.5.4 (Continued) Design accuracies of Reverse Scenario 2

(a-2-2) Lower tensile rebar strain (ε_{st}) of adjusted from 0.004 to 0.0075 at ULL (identical to Table (a-1-1))

PARAMETERS			Training Model	Network test result	Auto PT-BEAM check	Error
BEAM CONFIGURATION	1	b (mm)		500.00	500.00	0.00%
	2	h (mm)		700	700	0.00%
	3	f_{ck} (MPa)		40	40	0.00%
	4	f_{cki} (MPa)		20	20	0.00%
	5	ρ_{st} (mm²/mm²)	R2.PTM. 5 a	0.01053	0.01053	0.00%
	6	ρ_{sc} (mm²/mm²)	R2.PTM. 6 a	0.00093	0.00093	0.00%
	7	f_{sy} (MPa)		500	500	0.00%
	8	f_{sw} (MPa)		400	400	0.00%
	9	ϕ_S (mm)		16	16	0.00%
	10	ρ_p (mm²/mm²)	R2.PTM. 10 a	0.00118	0.00118	0.00%
	11	f_{py} (MPa)		1650	1650	0.00%
	12	ϕ_p (mm)		12.7	12.7	0.00%
	13	$PTloss_{lt}$		0.15	0.15	0.00%
	14	L (mm)		8000	8000	0.00%
	15	K_c		3	3	0.00%
TRANSFER LOAD STAGE	16	σ_c/f_{cki}	FW.PTM. 16 c	-0.0850	-0.08	Uncracked
	17	ε_{st} (mm/mm)	FW.PTM. 17 c	-0.00016	-0.00016	0.28%
	18	ε_{sc} (mm/mm)	FW.PTM. 18 c	0.0000342	0.0000320	6.50%
	19	ε_p (mm/mm)	FW.PTM. 19 c	0.00694	0.00693	0.02%
	20	σ_{ct}/f_{cki}	FW.PTM. 20 c	0.17535	0.17273	1.50%
	21	cw (mm)	FW.PTM. 21 c	0.000	0.000	0.00%
	22	Δ (mm)	FW.PTM. 22 c	-2.077	-2.251	-8.41%
SERVICE LOAD COMBINATION	23	$q_{L/250}$ (kN/m)		**70.00**	**70.19**	**-0.27%**
	24	$q_{0.2mm}$ (kN/m)	FW.PTM. 24 c	86.51	88.12	-1.86%
	25	$q_{0.6fck}$ (kN/m)	FW.PTM. 25 c	97.82	100.92	-3.17%
	26	$q_{0.8fpu}$ (kN/m)	FW.PTM. 26 c	143.94	141.90	1.42%
	27	$q_{0.75fpu}$ (kN/m)	FW.PTM. 27 c	97.65	86.63	11.29%
ULTIMATE LOAD COMBINATION	28	q_{str} (kN/m)		**150.00**	**147.90**	**1.40%**
	29	ε_{st} (mm/mm)		**0.00750**	**0.00824**	**-9.85%**
	30	ε_{sc} (mm/mm)	FW.PTM. 30 c	-0.00244	-0.00244	-0.03%
	31	ε_p (mm/mm)	FW.PTM. 31 c	0.01430	0.01442	-0.83%
	32	ρ_{sw} (mm²/mm²)	FW.PTM. 32 c	0.0026	0.0026	-0.02%
	33	μ_Δ	FW.PTM. 33 c	1.58	1.56	1.30%

Errors are reduced

Adjusting ε_{st} from 0.004 to **0.0075** gives reasonable design accuracy

: Given Inputs for ANN R3   : Inputs for engineering mechanic

: Reverse Outputs calculated by reverse ANN model is used as inputs to calculate other outputs by forward ANN model .

: Reverse-forward inputs   : Reverse-forward ouputs

(Continued)

(diameter of tendon), $PTLoss_{lt}$=0.15, L =8,000, and K_c=3 when the following reverse parameters are preassigned on an input-side: $q_{L/250}$=70 KN/m (SLL), q_{str}=150 KN/m (ULL), and lower tensile rebar strain (ε_{st}) of 0.0075. Table 2.5.4(a-1) presents design results based on the lower tensile rebar strain (ε_{st}) of 0.0075 at ULL while maintaining $q_{L/250}$ as 70 KN/m at SLL reaching a deflection of L/250 and q_{str} as 150 KN/m at ULL.

Table 2.5.4 (Continued) Design accuracies of Reverse Scenario 2

(a-3) Design results based on 30 layers-30 neurons for Step 1; $q_{L/250}$= 70 kN/m reaching a deflection of L/250 at SLL, q_{str} =150 kN/m at ULL,

(a-3-1) Lower tensile rebar strain (ε_{st}) of 0.004 with input conflicts causing large errors

PARAMETERS			Training Model	Network test result	Auto PT-BEAM check	Error
BEAM CONFIGURATION	1	b (mm)		500.00	500.00	0.00%
	2	h (mm)		700	700	0.00%
	3	f_{ck} (MPa)		40	40	0.00%
	4	f_{cki} (MPa)		20	20	0.00%
	5	ρ_{st} (mm²/mm²)	R2.PTM. 5 b	0.01305	0.01305	1.33E-16
	6	ρ_{sc} (mm²/mm²)	R2.PTM. 6 b	-0.00180	-0.00180	-1.20E-16
	7	f_{sy} (MPa)		500	500	0.00%
	8	f_{sw} (MPa)		400	400	0.00%
	9	ϕ_S (mm)		16	16	0.00%
	10	ρ_p (mm²/mm²)	R2.PTM. 10 b	0.00118	0.00118	0.00%
	11	f_{py} (MPa)		1650	1650	0.00%
	12	ϕ_p (mm)		12.7	12.7	0.00%
	13	$PTloss_{lt}$		0.15	0.15	0.00%
	14	L (mm)		8000	8000	0.00%
	15	K_c		3	3	0.00%
TRANSFER LOAD STAGE	16	σ_c/f_{cki}	FW.PTM. 16 c	-0.0834	-0.03	Uncracked
	17	ε_{st} (mm/mm)	FW.PTM. 17 c	-0.00015	-0.00017	-12.38%
	18	ε_{sc} (mm/mm)	FW.PTM. 18 c	0.0000327	0.0001020	-212.39%
	19	ε_p (mm/mm)	FW.PTM. 19 c	0.00694	0.00692	0.31%
	20	σ_{ct}/f_{cki}	FW.PTM. 20 c	0.16739	0.19124	-14.25%
	21	cw (mm)	FW.PTM. 21 c	0.000	-0.592	Inf
	22	Δ (mm)	FW.PTM. 22 c	-1.956	-3.189	-63.00%
SERVICE LOAD COMBINATION	23	$q_{L/250}$ (kN/m)		**70.00**	**74.15**	**-5.93%**
	24	$q_{0.2mm}$ (kN/m)	FW.PTM. 24 c	99.36	100.92	-1.58%
	25	$q_{0.6fck}$ (kN/m)	FW.PTM. 25 c	97.74	102.03	-4.39%
	26	$q_{0.8fsu}$ (kN/m)	FW.PTM. 26 c	162.39	162.97	-0.36%
	27	$q_{0.75fpu}$ (kN/m)	FW.PTM. 27 c	111.80	97.83	12.49%
ULTIMATE LOAD COMBINATION	28	q_{str} (kN/m)		**150.00**	**162.47**	**-8.31%**
	29	ε_{st} (mm/mm)		**0.00400**	**0.00498**	**-24.48%**
	30	ε_{sc} (mm/mm)	FW.PTM. 30 c	-0.00274	-0.00273	0.35%
	31	ε_p (mm/mm)	FW.PTM. 31 c	0.01087	0.01112	-2.32%
	32	ρ_{sw} (mm²/mm²)	FW.PTM. 32 c	0.0034	0.0033	1.07%
	33	μ_Δ	FW.PTM. 33 c	1.30	1.24	4.74%

Model **R2.PTM.b** is used

Big errors due to $\rho_{sc} < 0$

Preassigning $\varepsilon_{st} = 0.004$ gives big errors

: Given Inputs for ANN R3 : Inputs for engineering mechanic

: Reverse Outputs calculated by reverse ANN model is used as inputs to calculate other outputs by forward ANN model .

: Reverse-forward inputs : Reverse-forward ouputs

(Continued)

Table 2.5.4 (Continued) Design accuracies of Reverse Scenario 2

(a-3-2) Lower tensile rebar strain (ε_{st}) adjusted from of 0.004 to 0.0075 at ULL

PARAMETERS			Training Model	Network test result	Auto PT-BEAM check	Error
BEAM CONFIGURATION	1	b (mm)		500.00	500.00	0.00%
	2	h (mm)		700	700	0.00%
	3	f_{ck} (MPa)		40	40	0.00%
	4	f_{cki} (MPa)		20	20	0.00%
	5	ρ_{st} (mm²/mm²)	R2.PTM. 5 b	0.01064	0.01064	0.00%
	6	ρ_{sc} (mm²/mm²)	R2.PTM. 6 b	0.00074	0.00074	0.00%
	7	f_{sy} (MPa)		500	500	0.00%
	8	f_{sw} (MPa)		400	400	0.00%
	9	ϕ_S (mm)		16	16	0.00%
	10	ρ_p (mm²/mm²)	R2.PTM. 10 b	0.00122	0.00122	0.00%
	11	f_{py} (MPa)		1650	1650	0.00%
	12	ϕ_p (mm)		12.7	12.7	0.00%
	13	$PTloss_{lt}$		0.15	0.15	0.00%
	14	L (mm)		8000	8000	0.00%
	15	K_c		3	3	0.00%
TRANSFER LOAD STAGE	16	σ_c/f_{cki}	FW.PTM. 16 c	-0.0871	-0.08	Uncracked
	17	ε_{st} (mm/mm)	FW.PTM. 17 c	-0.00017	-0.00017	0.57%
	18	ε_{sc} (mm/mm)	FW.PTM. 18 c	0.0000362	0.0000336	7.16%
	19	ε_p (mm/mm)	FW.PTM. 19 c	0.00693	0.00693	0.03%
	20	σ_{ct}/f_{cki}	FW.PTM. 20 c	0.18161	0.17889	1.50%
	21	cw (mm)	FW.PTM. 21 c	0.000	0.000	0.00%
	22	Δ (mm)	FW.PTM. 22 c	-2.187	-2.358	-7.86%
SERVICE LOAD COMBINATION	23	$q_{L/250}$ (kN/m)		70.00	71.31	-1.88%
	24	$q_{0.2mm}$ (kN/m)	FW.PTM. 24 c	88.28	89.85	-1.78%
	25	$q_{0.6fck}$ (kN/m)	FW.PTM. 25 c	98.19	101.38	-3.25%
	26	$q_{0.8fsu}$ (kN/m)	FW.PTM. 26 c	145.91	144.14	1.21%
	27	$q_{0.75fpu}$ (kN/m)	FW.PTM. 27 c	99.44	88.29	11.22%
ULTIMATE LOAD COMBINATION	28	q_{str} (kN/m)		**150.00**	**149.62**	**0.26%**
	29	ε_{st} (mm/mm)		**0.00750**	**0.00789**	**-5.16%**
	30	ε_{sc} (mm/mm)	FW.PTM. 30 c	-0.00247	-0.00247	0.02%
	31	ε_p (mm/mm)	FW.PTM. 31 c	0.01392	0.01406	-1.01%
	32	ρ_{sw} (mm²/mm²)	FW.PTM. 32 c	0.0026	0.0026	-0.03%
	33	μ_Δ	FW.PTM. 33 c	1.55	1.53	1.26%

: Given Inputs for ANN R3 : Inputs for engineering mechanic

: Reverse Outputs calculated by reverse ANN model is used as inputs to calculate other outputs by forward ANN model .

: Reverse-forward inputs : Reverse-forward ouputs

Errors are reduced

Adjusting ε_{st} from 0.004 to **0.0075** gives reasonable design accuracy

(Continued)

2.5.2.2.1.1 TABLE 2.5.4(A-1)

The reverse network of Step 1 is used to precalculate three reverse output parameters (ρ_{st}, ρ_{sc}, and ρ_p) shown in Boxes 5, 6, and 10 of an output-side the reverse network of Step 1. Three reverse output parameters (ρ_{st}, ρ_{sc}, and ρ_p) are calculated as 0.01053, 0.00093, and 0.00118 shown in Boxes 5, 6, and 10 of an input-side reverse network of Step 1, respectively, for a given $q_{L/250}$ (70 KN/m), q_{str} (150 KN/m), and ε_{st} (0.0075), as shown in Table 2.5.4(a-1). In the forward network of Step 2, 15 input parameters including reverse output parameters (ρ_{st}, ρ_{sc}, and ρ_p) precalculated in reverse network of Step 1 are implemented in Boxes 1–15 of the forward network of Step 2 to obtain 18 output parameters including three parameters ($q_{L/250}$, q_{str}, and ε_{st})

Table 2.5.4 (Continued) Design accuracies of Reverse Scenario 2

(a-4) Design results based on 20 layers-20 neurons for Step 1; $q_{L/250}$ adjusted from 70 kN/m to 140 kN/m reaching a deflection of L/250 at SLL, q_{str} adjusted from 150 kN/m to 235 kN/m at ULL based on lower tensile rebar strain (ε_{st}) of 0.004 with input conflicts causing large errors

(a-4-1) Design accuracies without adjusting $q_{L/250}$ = 70 kN/m and q_{str} = 150 kN/m

PARAMETERS			Training Model	Network test result	Auto PT-BEAM check	Error
BEAM CONFIGURATION	1	b (mm)		500.00	500.00	0.00%
	2	h (mm)		700	700	0.00%
	3	f_{ck} (MPa)		40	40	0.00%
	4	f_{cki} (MPa)		20	20	0.00%
	5	ρ_{st} (mm²/mm²)	R2.PTM. 5 a	0.01255	0.01255	0.00%
	6	ρ_{sc} (mm²/mm²)	R2.PTM. 6 a	-0.00037	-0.00037	0.00%
	7	f_{sy} (MPa)		500	500	0.00%
	8	f_{sw} (MPa)		400	400	0.00%
	9	ϕ_s (mm)		16	16	0.00%
	10	ρ_p (mm²/mm²)	R2.PTM. 10 a	0.00113	0.00113	0.00%
	11	f_{py} (MPa)		1650	1650	0.00%
	12	ϕ_p (mm)		12.7	12.7	0.00%
	13	$PTloss_{lt}$		0.15	0.15	0.00%
	14	L (mm)		8000	8000	0.00%
	15	K_c		3	3	0.00%
TRANSFER LOAD STAGE	16	σ_c/f_{cki}	FW.PTM. 16 c	~0	0.00	Cracked
	17	ε_{st} (mm/mm)	FW.PTM. 17 c	-0.00015	-0.00021	-45.48%
	18	ε_{sc} (mm/mm)	FW.PTM. 18 c	0.0000308	0.0057639	-18620.47%
	19	ε_p (mm/mm)	FW.PTM. 19 c	0.00695	0.00680	2.12%
	20	σ_{ct}/f_{cki}	FW.PTM. 20 c	0.15971	0.64324	-302.75%
	21	cw (mm)	FW.PTM. 21 c	0.000	3.747	Inf
	22	Δ (mm)	FW.PTM. 22 c	-1.823	-17.328	-850.75%
SERVICE LOAD COMBINATION	23	$q_{L/250}$ (kN/m)		**70.00**	**72.65**	**-3.79%**
	24	$q_{0.2mm}$ (kN/m)	FW.PTM. 24 c	95.39	97.12	-1.82%
	25	$q_{0.6fck}$ (kN/m)	FW.PTM. 25 c	99.04	102.86	-3.86%
	26	$q_{0.8fsu}$ (kN/m)	FW.PTM. 26 c	158.25	157.73	0.33%
	27	$q_{0.75fpu}$ (kN/m)	FW.PTM. 27 c	107.18	94.38	11.95%
ULTIMATE LOAD COMBINATION	28	q_{str} (kN/m)		**150.00**	**161.13**	**-7.42%**
	29	ε_{st} (mm/mm)		**0.00400**	**0.00604**	**-50.96%**
	30	ε_{sc} (mm/mm)	FW.PTM. 30 c	-0.00264	-0.00264	0.16%
	31	ε_p (mm/mm)	FW.PTM. 31 c	0.01195	0.01219	-1.99%
	32	ρ_{sw} (mm²/mm²)	FW.PTM. 32 c	0.0031	0.0031	0.37%
	33	μ_Δ	FW.PTM. 33 c	1.39	1.34	3.59%

Significant errors

: Given Inputs for ANN R3 : Inputs for engineering mechanic

: Reverse Outputs calculated by reverse ANN model is used as inputs to calculate other outputs by forward ANN model .

: Reverse-forward inputs : Reverse-forward ouputs

(Continued)

Table 2.5.4 (Continued) Design accuracies of Reverse Scenario 2

(a-4-2) Improved design accuracies when adjusting to $q_{L/250}$ = 140 kN/m and q_{str} = 235 kN/m

PARAMETERS			Training Model	Network test result	Auto PT-BEAM check	Error
BEAM CONFIGURATION	1	b (mm)		500.00	500.00	0.00%
	2	h (mm)		700	700	0.00%
	3	f_{ck} (MPa)		40	40	0.00%
	4	f_{cki} (MPa)		20	20	0.00%
	5	ρ_{st} (mm²/mm²)	R2.PTM. 5 a	0.01397	0.01397	0.00%
	6	ρ_{sc} (mm²/mm²)	R2.PTM. 6 a	0.00484	0.00484	0.00%
	7	f_{sy} (MPa)		500	500	0.00%
	8	f_{sw} (MPa)		400	400	0.00%
	9	ϕ_S (mm)		16	16	0.00%
	10	ρ_p (mm²/mm²)	R2.PTM. 10 a	0.00346	0.00346	-1.25E-16
	11	f_{py} (MPa)		1650	1650	0.00%
	12	ϕ_p (mm)		12.7	12.7	0.00%
	13	$PTloss_{lt}$		0.15	0.15	0.00%
	14	L (mm)		8000	8000	0.00%
	15	K_c		3	3	0.00%
TRANSFER LOAD STAGE	16	σ_c/f_{cki}	FW.PTM. 16 c	0.0000	0.00	Cracked
	17	ε_{st} (mm/mm)	FW.PTM. 17 c	-0.00053	-0.00054	-3.46%
	18	ε_{sc} (mm/mm)	FW.PTM. 18 c	0.0004049	0.0003729	7.89%
	19	ε_p (mm/mm)	FW.PTM. 19 c	0.00655	0.00654	0.15%
	20	σ_{ct}/f_{cki}	FW.PTM. 20 c	0.52812	0.53297	-0.92%
	21	cw (mm)	FW.PTM. 21 c	0.045	0.060	-33%
	22	Δ (mm)	FW.PTM. 22 c	-9.555	-11.198	-17.19%
SERVICE LOAD COMBINATION	23	$q_{L/250}$ (kN/m)		**140.00**	**137.70**	**1.64%**
	24	$q_{0.2mm}$ (kN/m)	FW.PTM. 24 c	179.28	181.21	-1.08%
	25	$q_{0.6fck}$ (kN/m)	FW.PTM. 25 c	142.24	147.65	-3.80%
	26	$q_{0.8fpu}$ (kN/m)	FW.PTM. 26 c	249.44	256.40	-2.79%
	27	$q_{0.75fpu}$ (kN/m)	FW.PTM. 27 c	177.18	175.95	0.69%
ULTIMATE LOAD COMBINATION	28	q_{str} (kN/m)		**235.00**	**235.46**	**-0.19%**
	29	ε_{st} (mm/mm)		**0.00400**	**0.00422**	**-5.45%**
	30	ε_{sc} (mm/mm)	FW.PTM. 30 c	-0.00282	-0.00280	0.47%
	31	ε_p (mm/mm)	FW.PTM. 31 c	0.01027	0.01035	-0.80%
	32	ρ_{sw} (mm²/mm²)	FW.PTM. 32 c	0.0055	0.0055	-0.40%
	33	μ_Δ	FW.PTM. 33 c	1.28	1.29	-1.12%

Errors are reduced

Other parameters (q_{str} and $q_{L/250}$) have to be adjusted if keeping ε_{st} = 0.004

: Given Inputs for ANN R3
: Inputs for engineering mechanic
: Reverse Outputs calculated by reverse ANN model is used as inputs to calculate other outputs by forward ANN model .
: Reverse-forward inputs
: Reverse-forward ouputs

(Continued)

which were reversely preassigned in Boxes 23, 28, and 29 of reverse network of Step 1. In Table 2.5.4(a-1-2), the largest error is found with 11.29% for $q_{0.75fpu}$ shown in Box 27 that represents nominal strength when tendons reach stresses smaller of 0.75 f_{pu} or f_{py}. Description of $q_{0.75fpu}$ is found in Table 2.2.2(b). The next largest error for the lower tensile rebar strain (ε_{st}) is −9.85%, with a ε_{st} of 0.0075 found in Box 29 based on an ANN compared to 0.00824 based on a structural mechanics-based AutoPTBEAM at ULL. A reverse design based on BS method of Steps 1 and 2 is validated by comparing preassigned reverse input parameters ($q_{L/250}$, q_{str}, and ε_{st}) with those obtained in the forward network of Step 2, demonstrating −0.27%,

Table 2.5.4 (Continued) Design accuracies of Reverse Scenario 2

(b-1) Design results based on 20 layers-20 neurons for Step 1; $q_{L/250}$ = 70 kN/m reaching a deflection of L/250 at SLL, q_{str} = 150 kN/m at ULL based on lower tensile rebar strain of 0.015 and 0.025

(b-1-1) Lower tensile rebar strain of 0.015

PARAMETERS			Training Model	Network test result	Auto PT-BEAM check	Error
BEAM CONFIGURATION	1	b (mm)		500.00	500.00	0.00%
	2	h (mm)		700	700	0.00%
	3	f_{ck} (MPa)		40	40	0.00%
	4	f_{cki} (MPa)		20	20	0.00%
	5	ρ_{st} (mm²/mm²)	R2.PTM. 5 a	0.00988	0.00988	0.00%
	6	ρ_{sc} (mm²/mm²)	R2.PTM. 6 a	0.00666	0.00666	0.00%
	7	f_{sy} (MPa)		500	500	0.00%
	8	f_{sw} (MPa)		400	400	0.00%
	9	ϕ_S (mm)		16	16	0.00%
	10	ρ_p (mm²/mm²)	R2.PTM. 10 a	0.00113	0.00113	0.00%
	11	f_{py} (MPa)		1650	1650	0.00%
	12	ϕ_p (mm)		12.7	12.7	0.00%
	13	$PTloss_{lt}$		0.15	0.15	0.00%
	14	L (mm)		8000	8000	0.00%
	15	K_c		3	3	0.00%
TRANSFER LOAD STAGE	16	σ_c/f_{cki}	FW.PTM. 16 c	-0.0787	-0.07	Uncracked
	17	ε_{st} (mm/mm)	FW.PTM. 17 c	-0.00016	-0.00016	-0.12%
	18	ε_{sc} (mm/mm)	FW.PTM. 18 c	0.0000275	0.0000272	0.95%
	19	ε_p (mm/mm)	FW.PTM. 19 c	0.00694	0.00694	0.02%
	20	σ_{ct}/f_{cki}	FW.PTM. 20 c	0.16513	0.16464	0.30%
	21	cw (mm)	FW.PTM. 21 c	0.000	0.000	0.00%
	22	Δ (mm)	FW.PTM. 22 c	-1.902	-2.086	-9.73%
SERVICE LOAD COMBINATION	23	$q_{L/250}$ (kN/m)		**70.00**	**69.72**	**0.39%**
	24	$q_{0.2mm}$ (kN/m)	FW.PTM. 24 c	82.54	84.09	-1.87%
	25	$q_{0.6fck}$ (kN/m)	FW.PTM. 25 c	105.94	108.01	-1.96%
	26	$q_{0.8fsu}$ (kN/m)	FW.PTM. 26 c	139.36	136.02	2.40%
	27	$q_{0.75fpu}$ (kN/m)	FW.PTM. 27 c	90.83	83.01	8.60%
ULTIMATE LOAD COMBINATION	28	q_{str} (kN/m)		**150.00**	**146.80**	**2.13%**
	29	ε_{st} (mm/mm)		**0.01500**	**0.01561**	**-4.09%**
	30	ε_{sc} (mm/mm)	FW.PTM. 30 c	-0.00177	-0.00177	-0.25%
	31	ε_p (mm/mm)	FW.PTM. 31 c	0.02185	0.02189	-0.16%
	32	ρ_{sw} (mm²/mm²)	FW.PTM. 32 c	0.0022	0.0022	0.92%
	33	μ_Δ	FW.PTM. 33 c	1.99	2.00	-0.74%

ε_{st} greater than ***0.0075*** *gives good accuracy*

[red box] : Given Inputs for ANN R3 [blue box] : Inputs for engineering mechanic

[purple box] : Reverse Outputs calculated by reverse ANN model is used as inputs to calculate other outputs by forward ANN model .

[green box] : Reverse-forward inputs [gray box] : Reverse-forward ouputs

(Continued)

Table 2.5.4 (Continued) Design accuracies of Reverse Scenario 2

(b-1-2) Lower tensile rebar strain of 0.025

		PARAMETERS	Training Model	Network test result	Auto PT-BEAM check	Error
BEAM CONFIGURATION	1	b (mm)		500.00	500.00	0.00%
	2	h (mm)		700	700	0.00%
	3	f_{ck} (MPa)		40	40	0.00%
	4	f_{ckt} (MPa)		20	20	0.00%
	5	ρ_{st} (mm^2/mm^2)	R2.PTM. 5 a	0.01073	0.01073	0.00%
	6	ρ_{sc} (mm^2/mm^2)	R2.PTM. 6 a	0.02005	0.02005	0.00%
	7	f_{sy} (MPa)		500	500	0.00%
	8	f_{sw} (MPa)		400	400	0.00%
	9	ϕ_S (mm)		16	16	0.00%
	10	ρ_p (mm^2/mm^2)	R2.PTM. 10 a	0.00091	0.00091	0.00%
	11	f_{py} (MPa)		1650	1650	0.00%
	12	ϕ_p (mm)		12.7	12.7	0.00%
	13	$PTloss_{lt}$		0.15	0.15	0.00%
	14	L (mm)		8000	8000	0.00%
	15	K_c		3	3	0.00%
TRANSFER LOAD STAGE	16	σ_c/f_{ckt}	FW.PTM. 16 c	-0.0538	-0.05	Uncracked
	17	ε_{st} (mm/mm)	FW.PTM. 17 c	-0.00012	-0.00011	1.37%
	18	ε_{sc} (mm/mm)	FW.PTM. 18 c	0.0000130	0.0000143	-10.12%
	19	ε_p (mm/mm)	FW.PTM. 19 c	0.00698	0.00698	-0.01%
	20	σ_{ct}/f_{ckt}	FW.PTM. 20 c	0.12128	0.12084	0.36%
	21	cw (mm)	FW.PTM. 21 c	0.000	0.000	0.00%
	22	Δ (mm)	FW.PTM. 22 c	-1.171	-1.401	-19.64%
SERVICE LOAD COMBINATION	23	$q_{L/250}$ (kN/m)		**70.00**	**70.09**	**-0.13%**
	24	$q_{0.2mm}$ (kN/m)	FW.PTM. 24 c	81.34	82.65	-1.62%
	25	$q_{0.6fck}$ (kN/m)	FW.PTM. 25 c	127.92	129.24	-1.03%
	26	$q_{0.8fsu}$ (kN/m)	FW.PTM. 26 c	143.44	137.84	3.90%
	27	$q_{0.75fpu}$ (kN/m)	FW.PTM. 27 c	85.52	81.14	5.12%
ULTIMATE LOAD COMBINATION	28	q_{str} (kN/m)		**150.00**	**149.77**	**0.15%**
	29	ε_{st} (mm/mm)		**0.02500**	**0.02518**	**-0.73%**
	30	ε_{sc} (mm/mm)	FW.PTM. 30 c	-0.00091	-0.00091	-0.05%
	31	ε_p (mm/mm)	FW.PTM. 31 c	0.03121	0.03158	-1.18%
	32	ρ_{sw} (mm^2/mm^2)	FW.PTM. 32 c	0.0021	0.0021	0.11%
	33	μ_Δ	FW.PTM. 33 c	2.37	2.38	-0.41%

*ε_{st} greater than **0.0075** gives good accuracy*

Red box: Given Inputs for ANN R3; Blue box: Inputs for engineering mechanic

Purple box: Reverse Outputs calculated by reverse ANN model is used as inputs to calculate other outputs by forward ANN model .

Green box: Reverse-forward inputs; Grey box: Reverse-forward ouputs

(Continued)

Table 2.5.4 (Continued) Design accuracies of Reverse Scenario 2

(b-2) Design results based on 30 layers-30 neurons for Step 1; $q_{L/250}$ = 70 kN/m reaching a deflection of L/250 at SLL, q_{str} = 150 kN/m at ULL based on lower tensile rebar strain of 0.015 and 0.025

(b-2-1) Lower tensile rebar strain of 0.015

Model **R2.PTM.b** is used

PARAMETERS			Training Model	Network test result	Auto PT-BEAM check	Error
BEAM CONFIGURATION	1	b (mm)		500.00	500.00	0.00%
	2	h (mm)		700	700	0.00%
	3	f_{ck} (MPa)		40	40	0.00%
	4	f_{cki} (MPa)		20	20	0.00%
	5	ρ_{st} (mm^2/mm^2)	R2.PTM. 5 b	0.01019	0.01019	0.00%
	6	ρ_{sc} (mm^2/mm^2)	R2.PTM. 6 b	0.00658	0.00658	0.00%
	7	f_{sy} (MPa)		500	500	0.00%
	8	f_{sw} (MPa)		400	400	0.00%
	9	ϕ_s (mm)		16	16	0.00%
	10	ρ_p (mm^2/mm^2)	R2.PTM. 10 b	0.00116	0.00116	0.00%
	11	f_{py} (MPa)		1650	1650	0.00%
	12	ϕ_p (mm)		12.7	12.7	0.00%
	13	$PTloss_{lt}$		0.15	0.15	0.00%
	14	L (mm)		8000	8000	0.00%
	15	K_c		3	3	0.00%
TRANSFER LOAD STAGE	16	σ_c/f_{cki}	FW.PTM. 16 c	-0.0802	-0.08	Uncracked
	17	ε_{st} (mm/mm)	FW.PTM. 17 c	-0.00016	-0.00016	-0.13%
	18	ε_{sc} (mm/mm)	FW.PTM. 18 c	0.0000286	0.0000281	1.70%
	19	ε_p (mm/mm)	FW.PTM. 19 c	0.00694	0.00694	0.02%
	20	σ_{ct}/f_{cki}	FW.PTM. 20 c	0.16915	0.16857	0.34%
	21	cw (mm)	FW.PTM. 21 c	0.000	0.000	0.00%
	22	Δ (mm)	FW.PTM. 22 c	-1.969	-2.153	-9.33%
SERVICE LOAD COMBINATION	23	$q_{L/250}$ (kN/m)		**70.00**	**71.32**	**-1.88%**
	24	$q_{0.2mm}$ (kN/m)	FW.PTM. 24 c	85.18	86.73	-1.82%
	25	$q_{0.6fck}$ (kN/m)	FW.PTM. 25 c	107.12	109.23	-1.97%
	26	$q_{0.8fsu}$ (kN/m)	FW.PTM. 26 c	143.14	139.97	2.21%
	27	$q_{0.75fpu}$ (kN/m)	FW.PTM. 27 c	93.41	85.47	8.50%
ULTIMATE LOAD COMBINATION	28	q_{str} (kN/m)		**150.00**	**151.01**	**-0.67%**
	29	ε_{st} (mm/mm)		**0.01500**	**0.01498**	**0.16%**
	30	ε_{sc} (mm/mm)	FW.PTM. 30 c	-0.00182	-0.00183	-0.38%
	31	ε_p (mm/mm)	FW.PTM. 31 c	0.02122	0.02124	-0.08%
	32	ρ_{sw} (mm^2/mm^2)	FW.PTM. 32 c	0.0023	0.0023	0.89%
	33	μ_Δ	FW.PTM. 33 c	1.95	1.96	-0.59%

*ε_{st} greater than **0.0075** gives good accuracy*

[red box] : Given Inputs for ANN R3 [blue box] : Inputs for engineering mechanic

[purple box] : Reverse Outputs calculated by reverse ANN model is used as inputs to calculate other outputs by forward ANN model .

[green box] : Reverse-forward inputs [gray box] : Reverse-forward ouputs

(Continued)

Table 2.5.4 (Continued) Design accuracies of Reverse Scenario 2

(b-2-2) Lower tensile rebar strain of 0.025

Model **R2.PTM.b** is used

PARAMETERS			Training Model	Network test result	Auto PT-BEAM check	Error
BEAM CONFIGURATION	1	b (mm)		500.00	500.00	0.00%
	2	h (mm)		700	700	0.00%
	3	f_{ck} (MPa)		40	40	0.00%
	4	f_{cki} (MPa)		20	20	0.00%
	5	ρ_{st} (mm^2/mm^2)	R2.PTM. 5 b	0.01079	0.01079	0.00%
	6	ρ_{sc} (mm^2/mm^2)	R2.PTM. 6 b	0.01945	0.01945	0.00%
	7	f_{sy} (MPa)		500	500	0.00%
	8	f_{sw} (MPa)		400	400	0.00%
	9	ϕ_S (mm)		16	16	0.00%
	10	ρ_p (mm^2/mm^2)	R2.PTM. 10 b	0.00091	0.00091	0.00%
	11	f_{py} (MPa)		1650	1650	0.00%
	12	ϕ_p (mm)		12.7	12.7	0.00%
	13	$PTloss_{lt}$		0.15	0.15	0.00%
	14	L (mm)		8000	8000	0.00%
	15	K_c		3	3	0.00%
TRANSFER LOAD STAGE	16	σ_c/f_{cki}	FW.PTM. 16 c	-0.0545	-0.05	Uncracked
	17	ε_{st} (mm/mm)	FW.PTM. 17 c	-0.00012	-0.00011	1.27%
	18	ε_{sc} (mm/mm)	FW.PTM. 18 c	0.0000133	0.0000147	-10.49%
	19	ε_p (mm/mm)	FW.PTM. 19 c	0.00698	0.00698	-0.01%
	20	σ_{ct}/f_{cki}	FW.PTM. 20 c	0.12237	0.12205	0.26%
	21	cw (mm)	FW.PTM. 21 c	0.000	0.000	0.00%
	22	Δ (mm)	FW.PTM. 22 c	-1.190	-1.420	-19.37%
SERVICE LOAD COMBINATION	23	$q_{L/250}$ (kN/m)		**70.00**	**70.31**	**-0.44%**
	24	$q_{0.2mm}$ (kN/m)	FW.PTM. 24 c	81.86	83.20	-1.63%
	25	$q_{0.6fck}$ (kN/m)	FW.PTM. 25 c	127.34	128.58	-0.98%
	26	$q_{0.8fsu}$ (kN/m)	FW.PTM. 26 c	144.18	138.49	3.94%
	27	$q_{0.75fpu}$ (kN/m)	FW.PTM. 27 c	86.11	81.65	5.18%
ULTIMATE LOAD COMBINATION	28	q_{str} (kN/m)		**150.00**	**150.65**	**-0.44%**
	29	ε_{st} (mm/mm)		**0.02500**	**0.02485**	**0.59%**
	30	ε_{sc} (mm/mm)	FW.PTM. 30 c	-0.00094	-0.00094	-0.07%
	31	ε_p (mm/mm)	FW.PTM. 31 c	0.03087	0.03124	-1.21%
	32	ρ_{sw} (mm^2/mm^2)	FW.PTM. 32 c	0.0021	0.0021	0.17%
	33	μ_Δ	FW.PTM. 33 c	2.35	2.35	-0.10%

*ε_{st} greater than **0.0075** gives good accuracy*

: Given Inputs for ANN R3 : Inputs for engineering mechanic

: Reverse Outputs calculated by reverse ANN model is used as inputs to calculate other outputs by forward ANN model .

: Reverse-forward inputs : Reverse-forward ouputs

1.4%, and –9.85% for $q_{L/250}$, q_{str}, and ε_{st}, respectively, as shown in Boxes 23, 28, and 29 of Table 2.5.4(a-1-2). The error of –9.85% for ε_{st} shown in Box 29 of Table 2.5.4(a-1) appears large; however, the absolute strains between preassigned reverse input parameter and one obtained based on an ANN are 0.00750 and 0.00824 which are close enough. In Step 2, ANNs are trained on FW.PTM.c with 50 layers and 50 neurons, as shown in Table 2.5.4(a-1-2).

2.5.2.2.1.2 TABLE 2.5.4(a-2)

Table 2.5.4(a-2) shows design accuracies of Reverse Scenario 2 based on $q_{L/250}$ of 70 kN/m at SLL reaching a deflection of L/250, q_{str} of 150 kN/m at ULL, and lower tensile rebar strain (ε_{st}) of 0.004. Lower tensile rebar ratio (ρ_{st}) of 0.01255, upper compressive rebar ratio (ρ_{sc}) of –0.00037, and tendon ratio (ρ_p) of 0.00113 are precalculated in Boxes 5, 6, and 10 of an output-side the reverse network of Step 1 of Table 2.5.4(a-2-1) when lower tensile rebar strain (ε_{st}) of 0.004 shown in Box 29 is preassigned as a reverse input. The reverse network of Step 1 calculates a negative rebar ratio of –0.00037 for the upper compressive rebar (ρ_{sc}) of –0.00037, resulting in significant errors when preassigned reverse input parameters such as the lower tensile rebar strain (ε_{st}) of 0.004 cause input conflicts that violate structural mechanics. Consequently, significant errors are caused in a design shown in Table 2.5.4(a-2-1) due to the lower tensile rebar strain ε_{st} of 0.004 which is not properly preassigned as reverse input parameters, being unable to yield reasonable compressive rebar ratio (ρ_{sc}) at ULL in Step 1. It is noted that lower tensile rebar strains (ε_{st}) of 0.004 are not found between the lower (0.0075) and lower (0.0075) and upper (0.0295) and bounds at ULL under the design conditions of $q_{L/250}$ of 70 kN/m at SLL reaching a deflection of L/250 and q_{str} of 150 kN/m at ULL as shown in Tables 2.5.4(a-1-1), (a-1–2), respectively, which are graphically shown in Figure 2.5.3. Errors are significantly reduced when the lower tensile rebar strain (ε_{st}) at ULL is adjusted to a data range between 0.0075 and 0.03 as shown in Table 2.5.4(a-2-2). When the lower tensile rebar strain (ε_{st}) of 0.004 increases to 0.0075, some errors are significantly reduced. Lower tensile rebar ratio (ρ_{st}) of 0.01053, compressive rebar ratio (ρ_{sc}) of 0.00093, and tendon ratio (ρ_p) of 0.00118 are precalculated in Boxes 5, 6, and 10 of an output-side the reverse network of Step 1 of Table 2.5.4(a-2-2) when lower tensile rebar strain (ε_{st}) of 0.004 is adjusted to 0.0075 as shown in Box 29. Accordingly, the largest error reaches –9.85% (0.0075 vs. 0.00824) and 11.29% (97.65 kN/m vs. 86.63 kN/m) for ε_{st} and $q_{0.75fpu}$ at ULL, respectively. These design parameters also result in targeted nominal capacities resisting applied distributed loads of $q_{L/250}$ = 70 kN/m shown in Box 23 of Table 2.5.4(a-2-2) with an error equal to –0.27% (70 kN/m vs. 70.19 kN/m) when deflection reaches L/250 at SLL. The error of maximum nominal strength q_{str} = 150 kN/m that a PT beam can resist at ULL shown in Box 28 of Table 2.5.4(a-2-2)

Fixed inputs: $b = 500,\ h = 700,\ f_{ck} = 40MPa,\ f_{cki} = 20MPa,$

$f_{sy} = 500MPa, f_{sw} = 400MPa, \phi_s = 16mm, f_{py} = 1650MPa,$

$\phi_p = 12.7mm, PTLoss_{lt} = 0.15, L = 8000, K_c = 3,$

$\boldsymbol{q_{L/250} = 70\ kN/m\,, q_{str} = 150\ kN/m}$. **Variable input :** $\boldsymbol{\varepsilon_{st}}$ (ULL);

Output: $\boldsymbol{\rho_{sc}}$

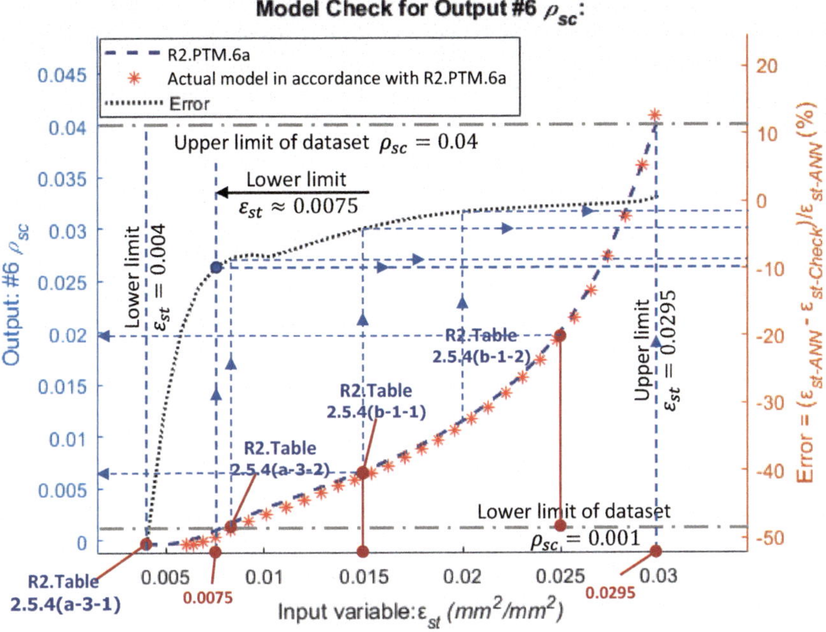

Figure 2.5.3 Upper compressive rebar ratio (ρ_{sc}) vs. lower tensile rebar strain (ε_{st}) at $q_{L/250}$ = 70 kN/m at SLL reaching a deflection of L/250) and q_{str} = 150 kN/m at ULL.

also decrease with an error equal to 1.4% from –7.42% which is based on the lower tensile rebar strain (ε_{st}) of 0.004 shown in Table 2.5.4(a-2-1) as demonstrated by 150 kN/m vs. 147.90 kN/m. In Tables 2.5.4(a-2-1) and (a-2-2), R2.PTM.5a, R2.PTM.6a, and R2.PTM.10a based on 20 layers and 20 neurons are used for the reverse network of Step 1, whereas for the forward network of Step 2, 18 forward ANNs trained by FW.PTM.c based on 40 layers and 40 neurons are implemented in obtaining 18 output parameters in Boxes 16–33 including three reverse input parameters ($q_{L/250}$, q_{str}, and ε_{st}).

2.5.2.2.1.3 TABLE 2.5.4(a-3)

In Table 2.5.4(a-3), R2.PTM.5b, R2.PTM.6b, and R2.PTM.10b based on 30 layers and 30 neurons are used for the reverse network of Step 1, whereas, for the forward network of Step 2, 18 forward ANNs trained by FW.PTM.c based on 40 layers and 40 neurons are implemented in obtaining 18 output parameters in Boxes 16–33 including three reverse input parameters ($q_{L/250}$, q_{str}, and ε_{st}). When the lower tensile rebar strain (ε_{st}) of 0.0075 shown in Box 29 is preassigned as a reverse input, the error for the lower tensile rebar strain (ε_{st}) at ULL shown in Box 29 based on 30 layers-30 neurons shown in Table 2.5.4(a-3-2) decreases to –5.16% (0.0075 vs. 0.00789) from –9.85% (0.0075 vs. 0.00824) based on 20 layers-20 neurons shown in Box 29 of the forward network of Step 2 of Table 2.5.4(a-2-2). Similarly, when the lower tensile rebar strain (ε_{st}) of 0.004 shown in Box 29 is preassigned as a reverse input, the error for the lower tensile rebar strains (ε_{st}) at ULL shown in Box 29 based on 30 layers-30 neurons reduces significantly to –24.48% (0.004 vs. 0.00498 shown in Table 2.5.4(a-3-1) from –50.96% (0.004 vs. 0.00604 shown in Table 2.5.4(a-2-1)) based on 20 layers-20 neurons shown in Box 29 of the forward network of Step 2. It is noted that use of 30 layers-30 neurons enhances design accuracies more significantly than when using 20 layers-20 neurons in BS Step 1. The accuracies become even better when input conflict is removed; the lower tensile rebar strain (ε_{st}) of 0.0075 shown in Table 2.5.4(a-3-2) should be used to replace 0.004 shown in Table 2.5.4(a-3-1).

2.5.2.2.1.4 TABLE 2.5.4(a-4)

In the forward networks of Step 2 of Table 2.5.4(a-4-2), errors with reverse input parameters improperly preassigned on an input-side decrease significantly when adjusting $q_{L/250}$ of 70 to–140 kN/m at SLL and q_{str} of 150 to–235 kN/m at ULL, while maintaining lower tensile rebar strain (ε_{st}) of 0.004. It is noted that design accuracies found from the forward networks of Step 2 shown in Table 2.5.4(a-4-2) with adjusted $q_{L/250}$ of 140 kN/m and q_{str} of 235 kN/m improve from those found from Table 2.5.4(a-4-1) with $q_{L/250}$ of 70 kN/m at SLL and q_{str} of 150 kN/m at ULL. This is because those input conflicts are removed by increasing $q_{L/250}$ (SLL) and q_{str} (ULL) for ε_{st} to reach 0.004.

2.5.2.2.2 *Influence of layer and neurons on design accuracies*

Table 2.5.4(b-1) for the forward network of Step 1 based on 20 layers-20 neurons and Table 2.5.4(b-2) for the forward network of Step 1 based on 30 layers-30 neurons present design accuracies for the lower tensile rebar strain (ε_{st}) of 0.015 and 0.025, respectively, while maintaining $q_{L/250}$ of 70 kN/m at SLL and q_{str} of 150 kN/m at ULL. Because the lower tensile rebar strains (ε_{st}) of 0.015 and 0.025 at ULL are located between the upper (0.0295) and lower (0.0075) bounds of the generated big structural datasets, both design errors are small for practical use, as shown in Table 2.5.4. An ANN trained by 30 layers-30 neurons shown in Table 2.5.4(b-2) used in calculating design parameters (ρ_{st}, ρ_{sc}, and ρ_p in Boxes 5, 6, and 10 of an output-side the reverse network of Step 1) provides design accuracies slightly improved than those obtained by 20 layers-20 neurons shown in Table 2.5.4(b-1).

2.5.2.3 Design charts

A design chart is created by varying the lower tensile rebar strains (ε_{st}) at ULL from 0.004 to 0.03 using Table 2.5.4. Upper compressive rebar ratio (ρ_{sc}) vs. lower tensile rebar (ε_{st}) at ULL are plotted as shown in Figure 2.5.3, where upper rebar ratios ρ_{sc} against lower rebar strains (ε_{st}) at ULL are identified when $q_{L/250}$ of 70 kN/m at SLL and q_{str} of 150 kN/m at ULL are imposed. The input parameters are set at:

$$b = 500,\ h = 700,\ f_{ck} = 40\ \text{MPa},\ f_{cki} = 20\text{MPa},\ f_{sy} = 500\ \text{MPa},$$

$$f_{sw} = 400\ \text{MPa},\ \phi_s = 16\ \text{mm},\ f_{py} = 1{,}650\ \text{MPa},\ \phi_p = 12.7\ \text{mm},$$

$$PTLoss_{lt} = 0.15,\ L = 8{,}000,\ K_c = 3.$$

Table 2.5.4(a-2) is obtained by setting the lower tensile rebar strains (ε_{st}) at 0.004 and 0.0075, respectively, as shown in Table 2.5.4(a-2-1) and (a-2-2) whereas Tables 2.5.4(b-1-1) and (b-1-2) are also obtained by setting the lower tensile rebar strains (ε_{st}) at 0.0015 and 0.0025. An ANN trained by R2.PTM.6a is used to calculate upper compressive rebar ratio ρ_{sc} against lower tensile rebar strains (ε_{st}) between 0.004 and 0.0075. The right ordinate of Figure 2.5.3 shows that errors based on ANNs and structural mechanics-based AutoPTbeam increase rapidly when lower tensile rebar strains (ε_{st}) less than 0.0075 are assigned. When the lower tensile rebar strains (ε_{st}) at ULL are between the upper (0.0295) and lower (0.0075) bounds, design errors are small enough for practical use. Figure 2.5.3 shows the upper and lower bounds of ε_{st} at ULL are 0.0295 and 0.0075, respectively, when $q_{L/250}$ and q_{str} are set at 70 kN/m at SLL and 150 kN/m at ULL, respectively. Table 2.5.4(a-3-1), 2.5.4(a-3-2), 2.5.4(b-1-1), and 2.5.4(b-1-2) is used to construct design points shown in the Figure 2.5.3.

2.5.3 Reverse design 3

2.5.3.1 Formulations of two steps based on back-substitution (BS) applicable to reverse designs

In a Reverse Scenario 3 shown in Figure 2.5.4 and Table 2.5.5, design parameters (b, ρ_{st}, ρ_{sc}, and ρ_p in green Boxes 1, 5, 6, and 10 of Figure 2.5.4) are predicted on an output-side the reverse network of Step 1, targeting

Reverse scenario	Input for engineering mechanics	Beam configurations	1 b (mm)	2 h (mm)	3 f_{ck} (MPa)	4 f_{cki} (MPa)	5 ρ_{st}	6 ρ_{sc}	7 f_{sy}	8 f_{sw}	
			9 ϕ_s	10 ρ_p	11 f_{py}	12 ϕ_p	13 $Ptloss_{lt}$	14 L (mm)	15 K_c		
	Output for engineering mechanics	Transfer stage	16 σ_c/f_{cki}	17 ε_{st}	18 ε_{sc}	19 ε_p	20 σ_{ct}/f_{cki}	21 cw	22 Δ (mm)		
		Service stage	23 Def. control $q_{L/250}$ (kN/m)		24 Crack control $q_{0.2mm}$ (kN/m)		25 Stress control $q_{0.6fck}$ (kN/m)		26 Stress control $q_{0.8fsu}$ (kN/m)		27 Stress control $q_{0.75fpu}$ (kN/m)
		Ultimate stage with material factor	28 Str. control q_{str} (kN/m)		29 ε_{st}	30 ε_{sc}	31 ε_p	32 ρ_{sw}	33 μ_Δ		

Input Output Reverse Input Reverse Output

Figure 2.5.4 Scenario table for reverse scenario 3.

nominal capacities ($q_{L/250}$, $q_{0.2mm}$, q_{str}, and μ_Δ in the yellow in Boxes 23, 24, 28, and 33 of Figure 2.5.4). The training accuracy based on PTM is summarized in Table 2.5.5 where an ANN trained by R3.PTM.a based on 20 layers-20 neurons is used to predict design parameters (b, ρ_{st}, ρ_{sc}, and ρ_p) in Boxes 1, 5, 6, and 10 on an output-side the Step 1 of Table 2.5.6(a). An ANN with FW.PTM.c using 40 layers and 40 neurons is implemented in training the forward network of Step 2 when 18 output parameters which include reverse input parameters ($q_{L/250}$, $q_{0.2mm}$, q_{str}, and μ_Δ which were already prescribed in Boxes 23, 24, 28, and 33 in Step 1) are predicted in Boxes 16–33 on an output-side Step 2 of Table 2.5.6(b). Preassigned and predicted reverse input parameters ($q_{L/250}$, $q_{0.2mm}$, q_{str}, and μ_Δ) can be compared for verifications. Fifteen input design parameters including b, ρ_{st}, ρ_{sc}, and ρ_p precalculated on an output-side the reverse network of Step 1 are used in Boxes 1–15 of the forward network of Step 2. Tables 2.5.6(a-2), (b-2), and (c-2) compare calculated parameters ($q_{L/250}$, $q_{0.2mm}$, q_{str}, and μ_Δ) in Step 2 with preassigned reverse input parameters ($q_{L/250}$, $q_{0.2mm}$, q_{str}, and μ_Δ which were already prescribed in Boxes 23, 24, 28, and 33 in Step 1).

2.5.3.2 Design accuracies by adjusting reverse input parameters to enhance design accuracies

Table 2.5.6 of a Reverse Scenario 3 presents selected design parameters including input parameters; h =700 mm, f_{ck}=40 MPa, f_{cki}=20 MPa, f_{sy}=500 MPa, f_{sw}=400 MPa, ϕ_s=16 mm, (diameter of rebar), f_{py}=1,650 MPa, ϕ_p=12.7 mm (diameter of tendon), $PTLoss_{lt}$=0.15, L = 10,000, and K_c=3 when the following reverse parameters are reversely preassigned on an input-side; $q_{L/250}$ = 80 KN/m (SLL), $q_{0.2\ mm}$=100 KN/m (SLL), q_{str}=150 KN/m (ULL), and deflection ductility ratios (μ_Δ) of 1.5.

Table 2.5.5 Training accuracies of a Reverse Scenario 3 based on PTM; reverse network of Step 1

REVERSE NETWORK OF STEP 1-REVERSE SCENARIO 3 15 Inputs (h, fck, f_{cki}, f_{sy}, f_{sw}, ϕ_S, f_{py}, ϕ_P, $PTloss_{lt}$, L, K_c, $q_{L/250}$, $q_{0.2mm}$, $qstr$, μ_Δ) – 4 Outputs (b, ρst, ρsc, ρ_p)										
No.	*Data*	*Layers*	*Neurons*	*Validation check*	*Required epoch*	*Best epoch for training*	*Stopped epoch*	*Mean square error*	*R at best epoch*	*Comments*
15 Inputs (h, fck, f_{cki}, f_{sy}, f_{sw}, ϕ_S, f_{py}, ϕ_P, $PTloss_{lt}$, L, K_c, $q_{L/250}$, $q_{0.2mm}$, $qstr$, μ_Δ) **–1 Output** (b) - *PTM*										
R3.PTM. 1 a	50,000	20	20	500	50,000	31,407	31,907	0.001083	0.990338	Neural
R3.PTM. 1 b	50,000	30	30	1,000	70,000	67,507	68,507	0.000714	0.994255	Neural
R3.PTM. 1 c	50,000	40	40	5,000	100,000	55,324	60,324	0.000764	0.994578	Neural
15 Inputs (h, fck, f_{cki}, f_{sy}, f_{sw}, ϕ_S, f_{py}, ϕ_P, $PTloss_{lt}$, L, K_c, $q_{L/250}$, $q_{0.2mm}$, $qstr$, μ_Δ) **–1 Output** (ρst) - *PTM*										
R3.PTM. 5 a	50,000	20	20	500	50,000	27,052	27,552	1.67E-03	0.989345	Insufficient
R3.PTM. 5 b	50,000	30	30	1,000	70,000	47,114	48,114	0.001197	0.994218	Neural
R3.PTM. 5 c	50,000	40	40	5,000	100,000	48,774	53,774	9.18E-04	0.995224	Neural
15 Inputs (h, fck, f_{cki}, f_{sy}, f_{sw}, ϕ_S, f_{py}, ϕ_P, $PTloss_{lt}$, L, K_c, $q_{L/250}$, $q_{0.2mm}$, $qstr$, μ_Δ) **–1 Output** (ρsc) - *PTM*										
R3.PTM. 6 a	50,000	20	20	500	50,000	18,532	19,032	1.45E-02	0.911693	Insufficient
R3.PTM. 6 b	50,000	30	30	1,000	70,000	26,751	27,751	0.013178	0.929903	Insufficient
R3.PTM. 6 c	50,000	40	40	5,000	100,000	35,302	40,302	1.29E-02	0.9375	Insufficient
15 Inputs (h, fck, f_{cki}, f_{sy}, f_{sw}, ϕ_S, f_{py}, ϕ_P, $PTloss_{lt}$, L, K_c, $q_{L/250}$, $q_{0.2mm}$, $qstr$, μ_Δ) **–1 Output** (ρ_p) - *PTM*										
R3.PTM. 10 a	50,000	20	20	500	50,000	18,887	19,387	7.82E-03	0.937977	Insufficient
R3.PTM. 10 b	50,000	30	30	1,000	70,000	30,611	31,611	0.004052	0.976151	Neural
R3.PTM. 10 c	50,000	40	40	5,000	100,000	37,338	42,338	3.41E-03	0.9796	Neural

Note:R3.PTM.**10a**

- **a** → Version of training model
- **10** → Output number
- PTM → Training method.
- R3 → Design scenario; R3 = Reverse ANN for Scenario 3

Table 2.5.6 Design accuracies of reverse scenario 3

(a) Design results based on 20 layers-20 neurons Steps 1 and 2; $q_{L/250}$= 80 kN/m reaching a deflection of L/250 at SLL, $q_{0.2\ mm}$= 100 KN/m at SLL, q_{str} =150 kN/m at ULL based on deflection ductility ratio (μ_Δ) of 1.50 at ULL

(a-1) Step 1: Reverse network with reverse inputs ($q_{L/250}$, $q_{0.2mm}$, q_{str}, μ_Δ) to calculate reverse outputs (b, ρ_{st}, ρ_{sc}, ρ_p)

PARAMETERS			Training Model	Network test result	Auto PT-BEAM check	Error
BEAM CONFIGURATION	1	b (mm)	R3.PTM. 1 a	1252.12	1252.12	0.00%
	2	h (mm)		700	700	0.00%
	3	f_{ck} (MPa)		40	40	0.00%
	4	f_{cki} (MPa)		20	20	0.00%
	5	ρ_{st} (mm^2/mm^2)	R3.PTM. 5 a	0.00564	0.00564	0.00%
	6	ρ_{sc} (mm^2/mm^2)	R3.PTM. 6 a	0.00122	0.00122	0.00%
	7	f_{sy} (MPa)		500	500	0.00%
	8	f_{sw} (MPa)		400	400	0.00%
	9	ϕ_S (mm)		16	16	0.00%
	10	ρ_p (mm^2/mm^2)	R3.PTM. 10 a	0.00197	0.00197	0.00%
	11	f_{py} (MPa)		1650	1650	0.00%
	12	ϕ_p (mm)		12.7	12.7	0.00%
	13	$PTloss_{lt}$		0.15	0.15	0.00%
	14	L (mm)		10000	10000	0.00%
	15	K_c		3	3	0.00%
REVERSE 3 INPUTS	23	$q_{L/250}$ (kN/m)		80.00	111.41	-39.26%
	24	$q_{0.2mm}$ (kN/m)		100.00	133.75	-33.75%
	28	q_{str} (kN/m)		150.00	198.24	-32.16%
	33	μ_Δ		1.50	1.95	-30.18%

: Given Inputs for ANN R3 : Inputs for engineering mechanic

: Reverse Outputs calculated by reverse ANN model is used as inputs to calculate other outputs by forward ANN model .

b, ρ_{st}, ρ_{sc}, and ρ_p are substituted into a forward network of Step 2

(*Continued*)

2.5.3.2.1 Table 2.5.6(a)

The reverse network of Step 1 produces a beam width (b) of 1252.12 mm, a lower tensile rebar ratio (ρ_{st}) of 0.00564, a compressive rebar ratio (ρ_{sc}) of 0.00122, and a tendon ratio (ρ_p) of 0.00197 in Boxes 1, 5, 6, and 10 on the output-side of Table 2.5.6(a-1) when a deflection ductility ratio (μ_Δ) of 1.50 is preassigned. In Table 2.5.6(a-2), significant errors are caused in the forward network of Step 2 with the largest errors reaching −41.94% (0.062 mm vs. 0.088 mm) shown in Box 21 of Table 2.5.6(a-2) and −21.38% (−7.055 mm vs. −8.564 mm) shown in Box 22 of Table 2.5.6(a-2) for crack width (c_w) and deflection (camber, Δ) at transfer load limit state, respectively. This is because that reverse input parameter such as deflection ductility ratio (μ_Δ) of 1.50 is incorrectly preassigned out of data range at ULL

Table 2.5.6 (Continued) Design accuracies of reverse scenario 3

(a-2) Step 2: Reverse-forward network using reverse outputs (from Step 1) as input parameters to calculate the rest of output parameters

b, ρ_{st}, ρ_{sc}, and ρ_p are calculated by a reverse network of Step 1

	PARAMETERS		Training Model	Network test result	Auto PT-BEAM check	Error
BEAM CONFIGURATION	1	b (mm)	R3.PTM. 1 a	**1252.12**	**1252.12**	**0.00%**
	2	h (mm)		700	700	0.00%
	3	f_{ck} (MPa)		40	40	0.00%
	4	f_{cki} (MPa)		20	20	0.00%
	5	ρ_{st} (mm^2/mm^2)	R3.PTM. 5 a	**0.00564**	**0.00564**	**0.00%**
	6	ρ_{sc} (mm^2/mm^2)	R3.PTM. 6 a	**0.00122**	**0.00122**	**0.00%**
	7	f_{sy} (MPa)		500	500	0.00%
	8	f_{sw} (MPa)		400	400	0.00%
	9	ϕ_S (mm)		16	16	0.00%
	10	ρ_p (mm^2/mm^2)	R3.PTM. 10 a	**0.00197**	**0.00197**	**0.00%**
	11	f_{py} (MPa)		1650	1650	0.00%
	12	ϕ_p (mm)		12.7	12.7	0.00%
	13	$PTloss_{lt}$		0.15	0.15	0.00%
	14	L (mm)		10000	10000	0.00%
	15	K_c		3	3	0.00%
REMAIN OUTPUTS CALCULATED BY REVERSE-FORWARD STEP	16	σ_c/f_{cki}	FW.PTM. 16 c	~0	0.00	Cracked
	17	ε_{st} (mm/mm)	FW.PTM. 17 c	-0.00038	-0.00040	-6.20%
	18	ε_{sc} (mm/mm)	FW.PTM. 18 c	0.0003789	0.0003995	-5.44%
	19	ε_p (mm/mm)	FW.PTM. 19 c	0.00671	0.00669	0.39%
	20	σ_{ct}/f_{cki}	FW.PTM. 20 c	0.40381	0.42381	-4.95%
	21	cw (mm)	FW.PTM. 21 c	0.062	0.088	-41.94%
	22	Δ (mm)	FW.PTM. 22 c	-7.055	-8.564	-21.38%
	25	$q_{0.6fck}$ (kN/m)	FW.PTM. 25 c	152.11	153.27	-0.77%
	26	$q_{0.8fu}$ (kN/m)	FW.PTM. 26 c	193.21	195.11	-0.98%
	27	$q_{0.75fpu}$ (kN/m)	FW.PTM. 27 c	136.96	135.23	1.26%
	29	ε_{st} (mm/mm)	FW.PTM. 29 c	0.01079	0.01084	-0.46%
	30	ε_{sc} (mm/mm)	FW.PTM. 30 c	-0.00220	-0.00220	-0.10%
	31	ε_p (mm/mm)	FW.PTM. 31 c	0.01690	0.01705	-0.89%
	32	ρ_{sw} (mm^2/mm^2)	FW.PTM. 32 c	0.0017	0.0017	-0.89%

Green box: Reverse-forward inputs; Blue box: Inputs for engineering mechanic; Grey box: Reverse-forward ouputs

(Continued)

for Step 1 of Table 2.5.6(a-1). Errors of crack width (c_w) and deflection (camber, Δ) at transfer load limit state decrease to –8.00% (0.050 mm vs. 0.054 mm) and 15.58% (–10.985 mm vs. –12.696 mm), respectively, when deflection ductility ratio (μ_Δ) of 1.50 is adjusted to 1.95 as shown in Table 2.5.6(b-2).

Table 2.5.6 (Continued) Design accuracies of reverse scenario 3

(a-3) Step 3: Summary of design based on two steps of BS

Summary of design based on two steps of BS

PARAMETERS			Training Model	Network test result	Auto PT-BEAM check	Error
BEAM CONFIGURATION	1	b (mm)	R3.PTM. **1 a**	1252.12	1252.12	0.00%
	2	h (mm)		700	700	0.00%
	3	f_{ck} (MPa)		40	40	0.00%
	4	f_{cki} (MPa)		20	20	0.00%
	5	ρ_{st} (mm^2/mm^2)	R3.PTM. **5 a**	0.00564	0.00564	0.00%
	6	ρ_{sc} (mm^2/mm^2)	R3.PTM. **6 a**	0.00122	0.00122	0.00%
	7	f_{sy} (MPa)		500	500	0.00%
	8	f_{sw} (MPa)		400	400	0.00%
	9	ϕ_S (mm)		16	16	0.00%
	10	ρ_p (mm^2/mm^2)	R3.PTM. **10 a**	0.00197	0.00197	0.00%
	11	f_{py} (MPa)		1650	1650	0.00%
	12	ϕ_p (mm)		12.7	12.7	0.00%
	13	$PTloss_{lt}$		0.15	0.15	0.00%
	14	L (mm)		10000	10000	0.00%
	15	K_c		3	3	0.00%
TRANSFER LOAD STAGE	16	σ_c/f_{cki}	FW.PTM. **16 c**	~0	0.00	Cracked
	17	ε_{st} (mm/mm)	FW.PTM. **17 c**	-0.00038	-0.00040	-6.20%
	18	ε_{sc} (mm/mm)	FW.PTM. **18 c**	0.0003789	0.0003995	-5.44%
	19	ε_p (mm/mm)	FW.PTM. **19 c**	0.00671	0.00669	0.39%
	20	σ_{ct}/f_{cki}	FW.PTM. **20 c**	0.40381	0.42381	-4.95%
	21	cw (mm)	FW.PTM. **21 c**	0.062	0.088	-41.94%
	22	Δ (mm)	FW.PTM. **22 c**	-7.055	-8.564	-21.38%
SERVICE LOAD COMBINATION	23	$q_{L/250}$ (kN/m)		**80.00**	**111.41**	**-39.26%**
	24	$q_{0.2mm}$ (kN/m)		**100.00**	**133.75**	**-33.75%**
	25	$q_{0.6fck}$ (kN/m)	FW.PTM. **25 c**	152.11	153.27	-0.77%
	26	$q_{0.8fsu}$ (kN/m)	FW.PTM. **26 c**	193.21	195.11	-0.98%
	27	$q_{0.75fpu}$ (kN/m)	FW.PTM. **27 c**	136.96	135.23	1.26%
ULTIMATE LOAD COMBINATION	28	q_{str} (kN/m)		**150.00**	**198.24**	**-32.16%**
	29	ε_{st} (mm/mm)	FW.PTM. **29 c**	0.01079	0.01084	-0.46%
	30	ε_{sc} (mm/mm)	FW.PTM. **30 c**	-0.00220	-0.00220	-0.10%
	31	ε_p (mm/mm)	FW.PTM. **31 c**	0.01690	0.01705	-0.89%
	32	ρ_{sw} (mm^2/mm^2)	FW.PTM. **32 c**	0.0017	0.0017	-0.89%
	33	μ_Δ		**1.50**	**1.95**	**-30.18%**

: Given Inputs for ANN R3 : Inputs for engineering mechanic

: Reverse Outputs calculated by reverse ANN model is used as inputs to calculate other outputs by forward ANN model .

: Reverse-forward inputs : Reverse-forward ouputs

(Continued)

Table 2.5.6 (Continued) Design accuracies of reverse scenario 3

(b) Design results based on 20 layers-20 neurons for Steps 1 and 2; $q_{L/250}$= 80 kN/m reaching a deflection of L/250 at SLL, $q_{0.2\ mm}$=100 KN/m at SLL, q_{str}=150 kN/m at ULL based on deflection ductility ratio (μ_Δ) of 1.50 adjusted to 1.95 at ULL

(b-1) Deflection ductility ratio (μ_Δ) of 1.50 (identical to Table 2.5.6(a-3))

PARAMETERS			Training Model	Network test result	Auto PT-BEAM check	Error
BEAM CONFIGURATION	1	b (mm)	R3.PTM. 1 a	1252.12	1252.12	0.00%
	2	h (mm)		700	700	0.00%
	3	f_{ck} (MPa)		40	40	0.00%
	4	f_{cki} (MPa)		20	20	0.00%
	5	ρ_{st} (mm^2/mm^2)	R3.PTM. 5 a	0.00564	0.00564	0.00%
	6	ρ_{sc} (mm^2/mm^2)	R3.PTM. 6 a	0.00122	0.00122	0.00%
	7	f_{sy} (MPa)		500	500	0.00%
	8	f_{sw} (MPa)		400	400	0.00%
	9	ϕ_S (mm)		16	16	0.00%
	10	ρ_p (mm^2/mm^2)	R3.PTM. 10 a	0.00197	0.00197	0.00%
	11	f_{py} (MPa)		1650	1650	0.00%
	12	ϕ_p (mm)		12.7	12.7	0.00%
	13	$PTloss_h$		0.15	0.15	0.00%
	14	L (mm)		10000	10000	0.00%
	15	K_c		3	3	0.00%
TRANSFER LOAD STAGE	16	σ_c/f_{cki}	FW.PTM. 16 c	~0	0.00	Cracked
	17	ε_{st} (mm/mm)	FW.PTM. 17 c	-0.00038	-0.00040	-6.20%
	18	ε_{sc} (mm/mm)	FW.PTM. 18 c	0.0003789	0.0003995	-5.44%
	19	ε_p (mm/mm)	FW.PTM. 19 c	0.00671	0.00669	0.39%
	20	σ_{ct}/f_{cki}	FW.PTM. 20 c	0.40381	0.42381	-4.95%
	21	cw (mm)	FW.PTM. 21 c	0.062	0.088	-41.94%
	22	Δ (mm)	FW.PTM. 22 c	-7.055	-8.564	-21.38%
SERVICE LOAD COMBINATION	23	$q_{L/250}$ (kN/m)		**80.00**	**111.41**	**-39.26%**
	24	$q_{0.2mm}$ (kN/m)		**100.00**	**133.75**	**-33.75%**
	25	$q_{0.6fck}$ (kN/m)	FW.PTM. 25 c	152.11	153.27	-0.77%
	26	$q_{0.8fsu}$ (kN/m)	FW.PTM. 26 c	193.21	195.11	-0.98%
	27	$q_{0.75fpu}$ (kN/m)	FW.PTM. 27 c	136.96	135.23	1.26%
ULTIMATE LOAD COMBINATION	28	q_{str} (kN/m)		**150.00**	**198.24**	**-32.16%**
	29	ε_{st} (mm/mm)	FW.PTM. 29 c	0.01079	0.01084	-0.46%
	30	ε_{sc} (mm/mm)	FW.PTM. 30 c	-0.00220	-0.00220	-0.10%
	31	ε_p (mm/mm)	FW.PTM. 31 c	0.01690	0.01705	-0.89%
	32	ρ_{sw} (mm^2/mm^2)	FW.PTM. 32 c	0.0017	0.0017	-0.89%
	33	μ_Δ		**1.50**	**1.95**	**-30.18%**

Significant errors

Preassigning μ_Δ= 1.5 gives big errors

: Given Inputs for ANN R3 : Inputs for engineering mechanic

: Reverse Outputs calculated by reverse ANN model is used as inputs to calculate other outputs by forward ANN model .

: Reverse-forward inputs : Reverse-forward ouputs

(Continued)

Table 2.5.6 (Continued) Design accuracies of reverse scenario 3

(b-2) Deflection ductility ratio (μ_Δ) adjusted from 1.50 to 1.95

PARAMETERS			Training Model	Network test result	Auto PT-BEAM check	Error
BEAM CONFIGURATION	1	b (mm)	R3.PTM. 1 a	790.90	790.90	0.00%
	2	h (mm)		700	700	0.00%
	3	f_{ck} (MPa)		40	40	0.00%
	4	f_{cki} (MPa)		20	20	0.00%
	5	ρ_{st} (mm²/mm²)	R3.PTM. 5 a	0.00612	0.00612	0.00%
	6	ρ_{sc} (mm²/mm²)	R3.PTM. 6 a	0.00363	0.00363	0.00%
	7	f_{sy} (MPa)		500	500	0.00%
	8	f_{sw} (MPa)		400	400	0.00%
	9	ϕ_S (mm)		16	16	0.00%
	10	ρ_p (mm²/mm²)	R3.PTM. 10 a	0.00249	0.00249	0.00%
	11	f_{py} (MPa)		1650	1650	0.00%
	12	ϕ_p (mm)		12.7	12.7	0.00%
	13	$PTloss_{lt}$		0.15	0.15	0.00%
	14	L (mm)		10000	10000	0.00%
	15	K_c		3	3	0.00%
TRANSFER LOAD STAGE	16	σ_c/f_{cki}	FW.PTM. 16 c	~0	0.00	Cracked
	17	ε_{st} (mm/mm)	FW.PTM. 17 c	-0.00047	-0.00048	-3.49%
	18	ε_{sc} (mm/mm)	FW.PTM. 18 c	0.0003103	0.0003483	-12.22%
	19	ε_p (mm/mm)	FW.PTM. 19 c	0.00663	0.00661	0.34%
	20	σ_{ct}/f_{cki}	FW.PTM. 20 c	0.47159	0.48417	-2.67%
	21	cw (mm)	FW.PTM. 21 c	0.050	0.054	-8.00%
	22	Δ (mm)	FW.PTM. 22 c	-10.985	-12.696	-15.58%
SERVICE LOAD COMBINATION	23	$q_{L/250}$ (kN/m)		**80.00**	**86.15**	**-7.68%**
	24	$q_{0.2mm}$ (kN/m)		**100.00**	**104.70**	**-4.70%**
	25	$q_{0.6fck}$ (kN/m)	FW.PTM. 25 c	109.22	111.49	-2.08%
	26	$q_{0.8fpu}$ (kN/m)	FW.PTM. 26 c	144.17	148.24	-2.82%
	27	$q_{0.75fpu}$ (kN/m)	FW.PTM. 27 c	101.02	105.45	-4.39%
ULTIMATE LOAD COMBINATION	28	q_{str} (kN/m)		**150.00**	**151.02**	**-0.68%**
	29	ε_{st} (mm/mm)	FW.PTM. 29 c	0.01088	0.01104	-1.39%
	30	ε_{sc} (mm/mm)	FW.PTM. 30 c	-0.00219	-0.00219	0.31%
	31	ε_p (mm/mm)	FW.PTM. 31 c	0.01687	0.01725	-2.24%
	32	ρ_{sw} (mm²/mm²)	FW.PTM. 32 c	0.0021	0.0021	-0.67%
	33	μ_Δ		**1.95**	**1.91**	**1.85%**

Red box: Given Inputs for ANN R3; Blue box: Inputs for engineering mechanic

Purple box: Reverse Outputs calculated by reverse ANN model is used as inputs to calculate other outputs by forward ANN model .

Green box: Reverse-forward inputs; Grey box: Reverse-forward ouputs

Errors are reduced

Adjusting μ_Δ from **1.5** to **1.95** gives reasonable design accuracy

(Continued)

2.5.3.2.2 *Table 2.5.6(b)*

Table 2.5.6(a-3) summarizes design results based on 20 layers-20 neurons and 40 layers-40 neurons for Steps 1 and 2, respectively, in which $qL/250$ = 80 kN/m reaching a deflection of L/250 at SLL, q0.2 mm = 100 KN/m at SLL, $qstr$ = 150 kN/m at ULL, and deflection ductility ratios ($\mu\Delta$) of 1.50 at ULL. Targeted nominal capacity of q_{str} =150 kN/m is verified with an error equal to –7.68% ($qL/250$ = 80 kN/m of ANN vs. 86.15 kN/m of AutoPTbeam) shown in Box 23 of Table 2.5.6(b-2) when the deflection reaches L/250 at SLL. An error of –39.26% of $q_{L/250}$ shown in Box 23 of

Table 2.5.6 (Continued) Design accuracies of reverse scenario 3

(c) Design results based on 20 layers-20 neurons for Steps 1 and 2; $q_{L/250}$ = 80 kN/m reaching a deflection of L/250 at SLL, $q_{0.2\ mm}$ = 100 KN/m at SLL, q_{str} = 150 kN/m at ULL based on deflection ductility ratio (μ_Δ) of 3.00 adjusted to 2.75 at ULL

(c-1) Deflection ductility ratio (μ_Δ) of 3.00

PARAMETERS			Training Model	Network test result	Auto PT-BEAM check	Error
BEAM CONFIGURATION	1	b (mm)	R3.PTM. 1 a	476.82	476.82	0.00%
	2	h (mm)		700	700	0.00%
	3	f_{ck} (MPa)		40	40	0.00%
	4	f_{cki} (MPa)		20	20	0.00%
	5	ρ_{s} (mm²/mm²)	R3.PTM. 5 a	0.01165	0.01165	0.00%
	6	ρ_{sc} (mm²/mm²)	R3.PTM. 6 a	0.03446	0.03446	0.00%
	7	f_{sy} (MPa)		500	500	0.00%
	8	f_{sw} (MPa)		400	400	0.00%
	9	ϕ_{S} (mm)		16	16	0.00%
	10	ρ_{p} (mm²/mm²)	R3.PTM. 10 a	0.00268	0.00268	0.00%
	11	f_{py} (MPa)		1650	1650	0.00%
	12	ϕ_{p} (mm)		12.7	12.7	0.00%
	13	$PTloss_{lt}$		0.15	0.15	0.00%
	14	L (mm)		10000	10000	0.00%
	15	K_c		3	3	0.00%
TRANSFER LOAD STAGE	16	σ_c/f_{cki}	FW.PTM. 16 c	~0	0.00	Cracked
	17	ε_{s} (mm/mm)	FW.PTM. 17 c	-0.00039	-0.00039	-1.90%
	18	ε_{sc} (mm/mm)	FW.PTM. 18 c	0.0000795	0.0000817	-2.73%
	19	ε_{p} (mm/mm)	FW.PTM. 19 c	0.00671	0.00670	0.18%
	20	σ_{ct}/f_{cki}	FW.PTM. 20 c	0.38730	0.39206	-1.23%
	21	cw (mm)	FW.PTM. 21 c	0.011	0.010	9.09%
	22	Δ (mm)	FW.PTM. 22 c	-8.078	-8.438	-4.45%
SERVICE LOAD COMBINATION	23	$q_{L/250}$ (kN/m)		**80.00**	**69.17**	**13.53%**
	24	$q_{0.2mm}$ (kN/m)		**100.00**	**89.10**	**10.90%**
	25	$q_{0.6fck}$ (kN/m)	FW.PTM. 25 c	108.55	109.09	-0.50%
	26	$q_{0.8fsu}$ (kN/m)	FW.PTM. 26 c	126.51	130.42	-3.09%
	27	$q_{0.75fpu}$ (kN/m)	FW.PTM. 27 c	84.22	87.54	-3.95%
ULTIMATE LOAD COMBINATION	28	q_{str} (kN/m)		**150.00**	**137.81**	**8.13%**
	29	ε_{s} (mm/mm)	FW.PTM. 29 c	0.02454	0.02448	0.26%
	30	ε_{sc} (mm/mm)	FW.PTM. 30 c	-0.00098	-0.00097	0.43%
	31	ε_{p} (mm/mm)	FW.PTM. 31 c	0.03063	0.03086	-0.74%
	32	ρ_{sw} (mm²/mm²)	FW.PTM. 32 c	0.0027	0.0027	0.48%
	33	μ_Δ		**3.00**	**2.58**	**14.11%**

Significant errors

Preassigning μ_Δ = 3 gives big errors

: Given Inputs for ANN R3 : Inputs for engineering mechanic

: Reverse Outputs calculated by reverse ANN model is used as inputs to calculate other outputs by forward ANN model.

: Reverse-forward inputs : Reverse-forward ouputs

(Continued)

Table 2.5.6(b-1) was observed for the deflection ductility ratio (μ_Δ) of 1.50. An error of −0.68% (150 kN/m of ANN vs. 151.02 kN/m of AutoPTbeam) for nominal capacities of q_{str} = 150 kN/m shown in Box 28 of Table 2.5.6(b-2) is also enhanced from an error of −32.16% shown in Box 28 of Table 2.5.6(b-1) when deflection ductility ratio (μ_Δ) of 1.50 is adjusted to 1.95.

Table 2.5.6 (Continued) Design accuracies of reverse scenario 3

(c-2) Deflection ductility ratio (μ_Δ) of 2.75

		PARAMETERS	Training Model	Network test result	Auto PT-BEAM check	Error
BEAM CONFIGURATION	1	b (mm)	R3.PTM. 1 a	520.87	520.87	0.00%
	2	h (mm)		700	700	0.00%
	3	f_{ck} (MPa)		40	40	0.00%
	4	f_{cki} (MPa)		20	20	0.00%
	5	ρ_{st} (mm^2/mm^2)	R3.PTM. 5 a	0.01062	0.01062	0.00%
	6	ρ_{sc} (mm^2/mm^2)	R3.PTM. 6 a	0.03481	0.03481	0.00%
	7	f_{sy} (MPa)		500	500	0.00%
	8	f_{sw} (MPa)		400	400	0.00%
	9	ϕ_s (mm)		16	16	0.00%
	10	ρ_p (mm^2/mm^2)	R3.PTM. 10 a	0.00271	0.00271	0.00%
	11	f_{py} (MPa)		1650	1650	0.00%
	12	ϕ_p (mm)		12.7	12.7	0.00%
	13	$PTloss_{lt}$		0.15	0.15	0.00%
	14	L (mm)		10000	10000	0.00%
	15	K_c		3	3	0.00%
TRANSFER LOAD STAGE	16	σ_c/f_{cki}	FW.PTM. 16 c	~0	0.00	Cracked
	17	ε_{st} (mm/mm)	FW.PTM. 17 c	-0.00040	-0.00041	-1.90%
	18	ε_{sc} (mm/mm)	FW.PTM. 18 c	0.0000830	0.0000846	-2.00%
	19	ε_p (mm/mm)	FW.PTM. 19 c	0.00670	0.00668	0.18%
	20	σ_{ct}/f_{cki}	FW.PTM. 20 c	0.40005	0.40481	-1.19%
	21	cw (mm)	FW.PTM. 21 c	0.012	0.010	16.67%
	22	Δ (mm)	FW.PTM. 22 c	-8.355	-8.830	-5.69%
SERVICE LOAD COMBINATION	23	$q_{L/250}$ (kN/m)		**80.00**	**75.04**	**6.21%**
	24	$q_{0.2mm}$ (kN/m)		**100.00**	**94.13**	**5.87%**
	25	$q_{0.6fck}$ (kN/m)	FW.PTM. 25 c	117.03	117.71	-0.58%
	26	$q_{0.8fsu}$ (kN/m)	FW.PTM. 26 c	131.63	136.48	-3.68%
	27	$q_{0.75fpu}$ (kN/m)	FW.PTM. 27 c	88.68	92.90	-4.77%
ULTIMATE LOAD COMBINATION	28	q_{str} (kN/m)		**150.00**	**143.94**	**4.04%**
	29	ε_{st} (mm/mm)	FW.PTM. 29 c	0.02516	0.02511	0.18%
	30	ε_{sc} (mm/mm)	FW.PTM. 30 c	-0.00092	-0.00092	0.34%
	31	ε_p (mm/mm)	FW.PTM. 31 c	0.03127	0.03150	-0.75%
	32	ρ_{sw} (mm^2/mm^2)	FW.PTM. 32 c	0.0026	0.0025	0.68%
	33	μ_Δ		**2.75**	**2.71**	**1.41%**

Errors are reduced

Adjusting μ_Δ *from* **3** *to* **2.75** *gives reasonable design accuracy*

: Given Inputs for ANN R3 : Inputs for engineering mechanic

: Reverse Outputs calculated by reverse ANN model is used as inputs to calculate other outputs by forward ANN model .

: Reverse-forward inputs : Reverse-forward ouputs

(*Continued*)

2.5.3.2.3 Table 2.5.6(c)

Table 2.5.6(c) shows a design improvement where the deflection ductility ratio (μ_Δ) of 3.00 is adjusted to 2.75 at ULL. Errors shown in Table 2.5.6(c-1) are significantly larger than those shown in Table 2.5.6(c-2) when the deflection ductility ratio (μ_Δ) of 3.00 is reversely preassigned on an input-side. This indicates there is an input conflict with a deflection ductility ratio (μ_Δ) of 3.00, and hence, an error for the deflection ductility ratio (μ_Δ) shown in Table 2.5.6(c-1) reaches 14.11%. An error shown in Table 2.5.6(c-2)

Table 2.5.6 (Continued) Design accuracies of reverse scenario 3

(d) Influence of neurons on design accuracies based on deflection ductility ratio (μ_Δ) of 2.00 and 2.50 at an ULL

(d-1) Design results based on 20 layers-20 neurons for Steps 1 and 2; $q_{L/250}$ = 80 kN/m reaching a deflection of L/250 at SLL, $q_{0.2\ mm}$ = 100 KN/m at SLL, q_{str} = 150 kN/m at ULL based on and deflection ductility ratio (μ_Δ) of 2.00 and 2.50 at an ULL

(d-1-1) Deflection ductility ratio (μ_Δ) of 2.00

PARAMETERS			Training Model	Network test result	Auto PT-BEAM check	Error
BEAM CONFIGURATION	1	b (mm)	R3.PTM. 1 a	761.19	761.19	0.00%
	2	h (mm)		700	700	0.00%
	3	f_{ck} (MPa)		40	40	0.00%
	4	f_{cki} (MPa)		20	20	0.00%
	5	ρ_{st} (mm²/mm²)	R3.PTM. 5 a	0.00627	0.00627	0.00%
	6	ρ_{sc} (mm²/mm²)	R3.PTM. 6 a	0.00404	0.00404	0.00%
	7	f_{sy} (MPa)		500	500	0.00%
	8	f_{sw} (MPa)		400	400	0.00%
	9	ϕ_s (mm)		16	16	0.00%
	10	ρ_p (mm²/mm²)	R3.PTM. 10 a	0.00254	0.00254	0.00%
	11	f_{py} (MPa)		1650	1650	0.00%
	12	ϕ_p (mm)		12.7	12.7	0.00%
	13	$PTloss_{lt}$		0.15	0.15	0.00%
	14	L (mm)		10000	10000	0.00%
	15	K_c		3	3	0.00%
TRANSFER LOAD STAGE	16	σ_c/f_{cki}	FW.PTM. 16 c	~0	0.00	Cracked
	17	ε_{st} (mm/mm)	FW.PTM. 17 c	-0.00047	-0.00049	-3.34%
	18	ε_{sc} (mm/mm)	FW.PTM. 18 c	0.0003011	0.0003354	-11.38%
	19	ε_p (mm/mm)	FW.PTM. 19 c	0.00663	0.00660	0.33%
	20	σ_{ct}/f_{cki}	FW.PTM. 20 c	0.47352	0.48601	-2.64%
	21	cw (mm)	FW.PTM. 21 c	0.047	0.050	-6.38%
	22	Δ (mm)	FW.PTM. 22 c	-11.247	-12.778	-13.61%
SERVICE LOAD COMBINATION	23	$q_{L/250}$ (kN/m)		**80.00**	**84.52**	**-5.65%**
	24	$q_{0.2mm}$ (kN/m)		**100.00**	**103.03**	**-3.03%**
	25	$q_{0.6fck}$ (kN/m)	FW.PTM. 25 c	106.74	109.08	-2.19%
	26	$q_{0.8fsu}$ (kN/m)	FW.PTM. 26 c	141.52	145.75	-2.99%
	27	$q_{0.75fpu}$ (kN/m)	FW.PTM. 27 c	98.68	103.68	-5.07%
ULTIMATE LOAD COMBINATION	28	q_{str} (kN/m)		**150.00**	**148.79**	**0.81%**
	29	ε_{st} (mm/mm)	FW.PTM. 29 c	0.01101	0.01118	-1.60%
	30	ε_{sc} (mm/mm)	FW.PTM. 30 c	-0.00218	-0.00217	0.41%
	31	ε_p (mm/mm)	FW.PTM. 31 c	0.01699	0.01740	-2.43%
	32	ρ_{sw} (mm²/mm²)	FW.PTM. 32 c	0.0021	0.0021	-0.51%
	33	μ_Δ		**2.00**	**1.92**	**4.17%**

Significant errors

Preassigning $\mu_\Delta = 2$ gives big errors

: Given Inputs for ANN R3
: Inputs for engineering mechanic
: Reverse Outputs calculated by reverse ANN model is used as inputs to calculate other outputs by forward ANN model .
: Reverse-forward inputs
: Reverse-forward ouputs

(Continued)

Table 2.5.6 (Continued) Design accuracies of reverse scenario 3

(d-1-2) Deflection ductility ratio (μ_Δ) of 2.50

PARAMETERS			Training Model	Network test result	Auto PT-BEAM check	Error
BEAM CONFIGURATION	1	b (mm)	R3.PTM. 1 a	575.67	575.67	0.00%
	2	h (mm)		700	700	0.00%
	3	f_{ck} (MPa)		40	40	0.00%
	4	f_{cki} (MPa)		20	20	0.00%
	5	ρ_{st} (mm^2/mm^2)	R3.PTM. 5 a	0.00907	0.00907	0.00%
	6	ρ_{sc} (mm^2/mm^2)	R3.PTM. 6 a	0.02044	0.02044	0.00%
	7	f_{sy} (MPa)		500	500	0.00%
	8	f_{sw} (MPa)		400	400	0.00%
	9	ϕ_S (mm)		16	16	0.00%
	10	ρ_p (mm^2/mm^2)	R3.PTM. 10 a	0.00272	0.00272	0.00%
	11	f_{py} (MPa)		1650	1650	0.00%
	12	ϕ_p (mm)		12.7	12.7	0.00%
	13	$PTloss_{lt}$		0.15	0.15	0.00%
	14	L (mm)		10000	10000	0.00%
	15	K_c		3	3	0.00%
TRANSFER LOAD STAGE	16	σ_c/f_{cki}	FW.PTM. 16 c	~0	0.00	Cracked
	17	ε_{st} (mm/mm)	FW.PTM. 17 c	-0.00043	-0.00044	-2.31%
	18	ε_{sc} (mm/mm)	FW.PTM. 18 c	0.0001226	0.0001306	-6.57%
	19	ε_p (mm/mm)	FW.PTM. 19 c	0.00667	0.00666	0.20%
	20	σ_{ct}/f_{cki}	FW.PTM. 20 c	0.42476	0.43063	-1.38%
	21	cw (mm)	FW.PTM. 21 c	0.016	0.017	-6.25%
	22	Δ (mm)	FW.PTM. 22 c	-9.693	-10.027	-3.45%
SERVICE LOAD COMBINATION	23	$q_{L/250}$ (kN/m)		**80.00**	**76.57**	**4.29%**
	24	$q_{0.2mm}$ (kN/m)		**100.00**	**96.21**	**3.79%**
	25	$q_{0.6fck}$ (kN/m)	FW.PTM. 25 c	108.73	109.52	-0.73%
	26	$q_{0.8fsu}$ (kN/m)	FW.PTM. 26 c	135.33	138.43	-2.29%
	27	$q_{0.75fpu}$ (kN/m)	FW.PTM. 27 c	87.83	95.57	-8.82%
ULTIMATE LOAD COMBINATION	28	q_{str} (kN/m)		**150.00**	**145.42**	**3.05%**
	29	ε_{st} (mm/mm)	FW.PTM. 29 c	0.02108	0.02103	0.25%
	30	ε_{sc} (mm/mm)	FW.PTM. 30 c	-0.00128	-0.00128	-0.26%
	31	ε_p (mm/mm)	FW.PTM. 31 c	0.02723	0.02737	-0.50%
	32	ρ_{sw} (mm^2/mm^2)	FW.PTM. 32 c	0.0024	0.0024	0.86%
	33	μ_Δ		**2.50**	**2.44**	**2.21%**

Errors are reduced

Adjusting μ_Δ from **2** to **2.5** gives reasonable design accuracy

: Given Inputs for ANN R3 : Inputs for engineering mechanic

: Reverse Outputs calculated by reverse ANN model is used as inputs to calculate other outputs by forward ANN model .

: Reverse-forward inputs : Reverse-forward ouputs

(Continued)

is only1.41% when the deflection ductility ratio (μ_Δ) is adjusted to 2.75. Recalculated deflection ductility ratio (μ_Δ) based on AutoPTbeam shown in Table 2.5.6(c-2) is 2.71 close to that (2.75) reversely preassigned on an input-side. Based on the adjusted deflection ductility ratio (μ_Δ) of 2.75, a nominal capacity with an error equal to 6.21% (80 kN/m vs. 75.04 kN/m) is calculated as shown in Box 23 of Table 2.5.6(c-2) which resists applied distributed load of $q_{L/250}$=80 kN/m, while reaching deflection of L/250 at SLL. Error is 13.53% for $q_{L/250}$ when the deflection ductility ratio (μ_Δ) of

Table 2.5.6 (Continued) Design accuracies of reverse scenario 3

(d-2) Design results based on 30 layers-30 neurons for Steps 1 and 2; $q_{L/250}$ = 80 kN/m reaching a deflection of L/250 at SLL, $q_{0.2\,mm}$ = 100 KN/m at SLL, q_{str} = 150 kN/m at ULL based on deflection ductility ratio (μ_Δ) of 2.00 and 2.50 at an ULL

(d-2-1) Deflection ductility ratio (μ_Δ) of 2.00

		PARAMETERS	Training Model	Network test result	Auto PT-BEAM check	Error
BEAM CONFIGURATION	1	b (mm)	R3.PTM. 1 b	744.91	744.91	0.00%
	2	h (mm)		700	700	0.00%
	3	f_{ck} (MPa)		40	40	0.00%
	4	f_{cki} (MPa)		20	20	0.00%
	5	ρ_{st} (mm^2/mm^2)	R3.PTM. 5 b	0.00709	0.00709	0.00%
	6	ρ_{sc} (mm^2/mm^2)	R3.PTM. 6 b	0.00469	0.00469	0.00%
	7	f_{sy} (MPa)		500	500	0.00%
	8	f_{sw} (MPa)		400	400	0.00%
	9	ϕ_S (mm)		16	16	0.00%
	10	ρ_p (mm^2/mm^2)	R3.PTM. 10 b	0.00239	0.00239	0.00%
	11	f_{py} (MPa)		1650	1650	0.00%
	12	ϕ_p (mm)		12.7	12.7	0.00%
	13	$PTloss_{lt}$		0.15	0.15	0.00%
	14	L (mm)		10000	10000	0.00%
	15	K_c		3	3	0.00%
TRANSFER LOAD STAGE	16	σ_c/f_{cki}	FW.PTM. 16 c	~0	0.00	Cracked
	17	ε_{st} (mm/mm)	FW.PTM. 17 c	-0.00042	-0.00044	-3.41%
	18	ε_{sc} (mm/mm)	FW.PTM. 18 c	0.0002569	0.0002863	-11.44%
	19	ε_p (mm/mm)	FW.PTM. 19 c	0.00667	0.00665	0.28%
	20	σ_{ct}/f_{cki}	FW.PTM. 20 c	0.43322	0.44534	-2.80%
	21	cw (mm)	FW.PTM. 21 c	0.039	0.042	-7.69%
	22	Δ (mm)	FW.PTM. 22 c	-9.645	-10.952	-13.55%
SERVICE LOAD COMBINATION	23	$q_{L/250}$ (kN/m)		**80.00**	**80.36**	**-0.45%**
	24	$q_{0.2mm}$ (kN/m)		**100.00**	**100.64**	**-0.64%**
	25	$q_{0.6fck}$ (kN/m)	FW.PTM. 25 c	105.57	107.41	-1.74%
	26	$q_{0.8fpu}$ (kN/m)	FW.PTM. 26 c	141.54	144.97	-2.42%
	27	$q_{0.75fpu}$ (kN/m)	FW.PTM. 27 c	96.89	100.91	-4.15%
ULTIMATE LOAD COMBINATION	28	q_{str} (kN/m)		**150.00**	**149.40**	**0.40%**
	29	ε_{st} (mm/mm)	FW.PTM. 29 c	0.01137	0.01152	-1.28%
	30	ε_{sc} (mm/mm)	FW.PTM. 30 c	-0.00215	-0.00214	0.34%
	31	ε_p (mm/mm)	FW.PTM. 31 c	0.01738	0.01774	-2.07%
	32	ρ_{sw} (mm^2/mm^2)	FW.PTM. 32 c	0.0021	0.0021	-0.38%
	33	μ_Δ		**2.00**	**1.91**	**4.61%**

Model **R3.PTM.b** is used

Significant errors

μ_Δ between 1.75 and 2.75 show good accuracy.

: Given Inputs for ANN R3 : Inputs for engineering mechanic

: Reverse Outputs calculated by reverse ANN model is used as inputs to calculate other outputs by forward ANN model .

: Reverse-forward inputs : Reverse-forward ouputs

(Continued)

3.00 is used as shown in Table 2.5.6(c-1). An error of nominal capacity (q_{str} =150 kN/m) at ULL is also improved to 4.04% (150 kN/m vs. 143.94 kN/m) as shown in Box 28 of Table 2.5.6(c-2) which is compared to an error of 8.13% based on the deflection ductility ratio (μ_Δ) of 3.00 as shown Table 2.5.6(c-1). This indicates that input conflicts are removed when the

Table 2.5.6 (Continued) Design accuracies of reverse scenario 3

(d-2-2) Deflection ductility ratio (μ_Δ) of 2.50

Model **R3.PTM.b** is used

		PARAMETERS	Training Model	Network test result	Auto PT-BEAM check	Error
BEAM CONFIGURATION	1	b (mm)	R3.PTM. **1 b**	565.56	565.56	0.00%
	2	h (mm)		700	700	0.00%
	3	f_{ck} (MPa)		40	40	0.00%
	4	f_{cki} (MPa)		20	20	0.00%
	5	ρ_{st} (mm^2/mm^2)	R3.PTM. **5 b**	0.00915	0.00915	0.00%
	6	ρ_{sc} (mm^2/mm^2)	R3.PTM. **6 b**	0.02260	0.02260	0.00%
	7	f_{sy} (MPa)		500	500	0.00%
	8	f_{sw} (MPa)		400	400	0.00%
	9	ϕ_S (mm)		16	16	0.00%
	10	ρ_p (mm^2/mm^2)	R3.PTM. **10 b**	0.00287	0.00287	0.00%
	11	f_{py} (MPa)		1650	1650	0.00%
	12	ϕ_p (mm)		12.7	12.7	0.00%
	13	$PTloss_{lt}$		0.15	0.15	0.00%
	14	L (mm)		10000	10000	0.00%
	15	K_c		3	3	0.00%
TRANSFER LOAD STAGE	16	σ_c/f_{cki}	FW.PTM. **16 c**	~0	0.00	Cracked
	17	ε_{st} (mm/mm)	FW.PTM. **17 c**	-0.00045	-0.00046	-2.27%
	18	ε_{sc} (mm/mm)	FW.PTM. **18 c**	0.0001220	0.0001296	-6.22%
	19	ε_p (mm/mm)	FW.PTM. **19 c**	0.00665	0.00663	0.21%
	20	σ_{ct}/f_{cki}	FW.PTM. **20 c**	0.44377	0.44942	-1.27%
	21	cw (mm)	FW.PTM. **21 c**	0.016	0.017	-6.25%
	22	Δ (mm)	FW.PTM. **22 c**	-10.028	-10.624	-5.94%
SERVICE LOAD COMBINATION	23	$q_{L/250}$ (kN/m)		**80.00**	**78.95**	**1.31%**
	24	$q_{0.2mm}$ (kN/m)		**100.00**	**98.84**	**1.16%**
	25	$q_{0.6fck}$ (kN/m)	FW.PTM. **25 c**	111.07	112.02	-0.85%
	26	$q_{0.8fsu}$ (kN/m)	FW.PTM. **26 c**	137.48	141.16	-2.68%
	27	$q_{0.75fpu}$ (kN/m)	FW.PTM. **27 c**	89.57	98.14	-9.57%
ULTIMATE LOAD COMBINATION	28	q_{str} (kN/m)		**150.00**	**148.01**	**1.33%**
	29	ε_{st} (mm/mm)	FW.PTM. **29 c**	0.02164	0.02157	0.33%
	30	ε_{sc} (mm/mm)	FW.PTM. **30 c**	-0.00123	-0.00124	-0.12%
	31	ε_p (mm/mm)	FW.PTM. **31 c**	0.02779	0.02791	-0.45%
	32	ρ_{sw} (mm^2/mm^2)	FW.PTM. **32 c**	0.0025	0.0025	1.03%
	33	μ_Δ		**2.50**	**2.49**	**0.59%**

μ_Δ between ***1.75*** *and* ***2.75*** *show good accuracy.*

Legend:
- Red box: Given Inputs for ANN R3
- Blue box: Inputs for engineering mechanic
- Purple box: Reverse Outputs calculated by reverse ANN model is used as inputs to calculate other outputs by forward ANN model .
- Green box: Reverse-forward inputs
- Grey box: Reverse-forward ouputs

(*Continued*)

Table 2.5.6 (Continued) Design accuracies of reverse scenario 3

(d-3) Design results based on 40 layers-40 neurons for Steps 1 and 2; $q_{L/250}$ = 80 kN/m reaching a deflection of L/250 at SLL, $q_{0.2\ mm}$ = 100 KN/m at SLL, q_{str} = 150 kN/m at ULL based on deflection ductility ratio (μ_Δ) of 2.00 and 2.50 at an ULL

(d-3-1) Deflection ductility ratio (μ_Δ) of 2.00

Model **R3.PTM.c** is used

		PARAMETERS	Training Model		Network test result	Auto PT-BEAM check	Error
BEAM CONFIGURATION	1	b (mm)	R3.PTM.	1 c	756.38	756.38	0.00%
	2	h (mm)			700	700	0.00%
	3	f_{ck} (MPa)			40	40	0.00%
	4	f_{cki} (MPa)			20	20	0.00%
	5	ρ_{s} (mm²/mm²)	R3.PTM.	5 c	0.00707	0.00707	0.00%
	6	$\rho_{s'}$ (mm²/mm²)	R3.PTM.	6 c	0.00539	0.00539	0.00%
	7	f_{sy} (MPa)			500	500	0.00%
	8	f_{sw} (MPa)			400	400	0.00%
	9	ϕ_S (mm)			16	16	0.00%
	10	ρ_p (mm²/mm²)	R3.PTM.	10 c	0.00242	0.00242	0.00%
	11	f_{py} (MPa)			1650	1650	0.00%
	12	ϕ_p (mm)			12.7	12.7	0.00%
	13	$PTloss_{lt}$			0.15	0.15	0.00%
	14	L (mm)			10000	10000	0.00%
	15	K_c			3	3	0.00%
TRANSFER LOAD STAGE	16	σ_c/f_{cki}	FW.PTM. 16 c		~0	0.00	Cracked
	17	ε_{s} (mm/mm)	FW.PTM. 17 c		-0.00043	-0.00044	-3.36%
	18	$\varepsilon_{s'}$ (mm/mm)	FW.PTM. 18 c		0.0002439	0.0002711	-11.17%
	19	ε_p (mm/mm)	FW.PTM. 19 c		0.00667	0.00665	0.27%
	20	σ_{ct}/f_{cki}	FW.PTM. 20 c		0.43394	0.44607	-2.80%
	21	cw (mm)	FW.PTM. 21 c		0.036	0.039	-8.33%
	22	Δ (mm)	FW.PTM. 22 c		-9.853	-10.976	-11.40%
SERVICE LOAD COMBINATION	23	$q_{L/250}$ (kN/m)			**80.00**	**82.68**	**-3.35%**
	24	$q_{0.2mm}$ (kN/m)			**100.00**	**103.27**	**-3.27%**
	25	$q_{0.6fck}$ (kN/m)	FW.PTM. 25 c		108.85	110.62	-1.63%
	26	$q_{0.8fsu}$ (kN/m)	FW.PTM. 26 c		144.96	148.43	-2.40%
	27	$q_{0.75fpu}$ (kN/m)	FW.PTM. 27 c		99.11	103.56	-4.49%
ULTIMATE LOAD COMBINATION	28	q_{str} (kN/m)			**150.00**	**153.37**	**-2.25%**
	29	ε_{s} (mm/mm)	FW.PTM. 29 c		0.01208	0.01217	-0.77%
	30	$\varepsilon_{s'}$ (mm/mm)	FW.PTM. 30 c		-0.00209	-0.00208	0.17%
	31	ε_p (mm/mm)	FW.PTM. 31 c		0.01811	0.01840	-1.62%
	32	ρ_{sw} (mm²/mm²)	FW.PTM. 32 c		0.0021	0.0021	-0.24%
	33	μ_Δ			**2.00**	**1.96**	**1.82%**

Significant errors

μ_Δ between 1.75 and 2.75 show good accuracy.

: Given Inputs for ANN R3 : Inputs for engineering mechanic

: Reverse Outputs calculated by reverse ANN model is used as inputs to calculate other outputs by forward ANN model .

: Reverse-forward inputs : Reverse-forward ouputs

(Continued)

deflection ductility ratio (μ_Δ) of 3.00 is adjusted to 2.75 as shown in Tables 2.5.6(c-1) and (c-2). Some errors such as crack width (c_w) shown in Box 21 and deflection at transfer stage (Δ) shown in Box 22 increase when the deflection ductility ratio (μ_Δ) of 3.00 is adjusted to 2.75, however, these errors can be reduced when more neurons are used in training ANNs.

Table 2.5.6 (Continued) Design accuracies of reverse scenario 3

(d-3-2) Deflection ductility ratio (μ_Δ) of 2.50

PARAMETERS			Training Model	Network test result	Auto PT-BEAM check	Error
BEAM CONFIGURATION	1	b (mm)	R3.PTM. 1 c	572.65	572.65	0.00%
	2	h (mm)		700	700	0.00%
	3	f_{ck} (MPa)		40	40	0.00%
	4	f_{ci} (MPa)		20	20	0.00%
	5	ρ_{st} (mm^2/mm^2)	R3.PTM. 5 c	0.00924	0.00924	0.00%
	6	ρ_{sc} (mm^2/mm^2)	R3.PTM. 6 c	0.02299	0.02299	0.00%
	7	f_{sy} (MPa)		500	500	0.00%
	8	f_{sw} (MPa)		400	400	0.00%
	9	ϕ_S (mm)		16	16	0.00%
	10	ρ_p (mm^2/mm^2)	R3.PTM. 10 c	0.00290	0.00290	0.00%
	11	f_{py} (MPa)		1650	1650	0.00%
	12	ϕ_p (mm)		12.7	12.7	0.00%
	13	$PTloss_{lt}$		0.15	0.15	0.00%
	14	L (mm)		10000	10000	0.00%
	15	K_c		3	3	0.00%
TRANSFER LOAD STAGE	16	σ_c/f_{ci}	FW.PTM. 16 c	~0	0.00	Cracked
	17	ε_{st} (mm/mm)	FW.PTM. 17 c	-0.00045	-0.00046	-2.28%
	18	ε_{sc} (mm/mm)	FW.PTM. 18 c	0.0001219	0.0001293	-6.08%
	19	ε_p (mm/mm)	FW.PTM. 19 c	0.00664	0.00663	0.22%
	20	σ_{ct}/f_{ci}	FW.PTM. 20 c	0.44668	0.45229	-1.26%
	21	cw (mm)	FW.PTM. 21 c	0.016	0.017	-6.25%
	22	Δ (mm)	FW.PTM. 22 c	-10.056	-10.715	-6.56%
SERVICE LOAD COMBINATION	23	$q_{L/250}$ (kN/m)		**80.00**	**80.76**	**-0.95%**
	24	$q_{0.2mm}$ (kN/m)		**100.00**	**101.20**	**-1.20%**
	25	$q_{0.6fck}$ (kN/m)	FW.PTM. 25 c	113.46	114.42	-0.85%
	26	$q_{0.8fsu}$ (kN/m)	FW.PTM. 26 c	140.57	144.42	-2.74%
	27	$q_{0.75fpu}$ (kN/m)	FW.PTM. 27 c	91.65	100.45	-9.60%
ULTIMATE LOAD COMBINATION	28	q_{str} (kN/m)		**150.00**	**151.42**	**-0.94%**
	29	ε_{st} (mm/mm)	FW.PTM. 29 c	0.02167	0.02159	0.34%
	30	ε_{sc} (mm/mm)	FW.PTM. 30 c	-0.00123	-0.00123	-0.10%
	31	ε_p (mm/mm)	FW.PTM. 31 c	0.02782	0.02794	-0.45%
	32	ρ_{sw} (mm^2/mm^2)	FW.PTM. 32 c	0.0025	0.0025	1.05%
	33	μ_Δ		**2.50**	**2.48**	**0.65%**

Model **R3.PTM.c** is used

μ_Δ between 1.75 and 2.75 show good accuracy.

: Given Inputs for ANN R3 : Inputs for engineering mechanic

: Reverse Outputs calculated by reverse ANN model is used as inputs to calculate other outputs by forward ANN model .

: Reverse-forward inputs : Reverse-forward ouputs

(Continued)

Table 2.5.6 (Continued) Design accuracies of reverse scenario 3

(e) Influence of neurons on design accuracies based on deflection ductility ratio (μ_Δ) of 3.00 and 2.75 at an ULL state

(e-1) Design results based on 30 layers-30 neurons for Steps 1 and 2; $q_{L/250}$ = 80 kN/m at SLL reaching a deflection of L/250, $q_{0.2\,mm}$ = 100 KN/m at SLL, q_{str} = 150 kN/m at ULL based on deflection ductility ratio (μ_Δ) of 1.50 and 1.75 at an ULL state

(e-1-1) Deflection ductility ratio (μ_Δ) of 1.50

Model **R3.PTM.b** is used

Big errors due to $\rho_{sc} < 0$

Preassigning $\mu_\Delta = 1.5$ gives big errors

		PARAMETERS	Training Model	Network test result	Auto PT-BEAM check	Error
BEAM CONFIGURATION	1	b (mm)	R3.PTM. 1 b	1081.11	1081.11	0.00%
	2	h (mm)		700	700	0.00%
	3	f_{ck} (MPa)		40	40	0.00%
	4	f_{cki} (MPa)		20	20	0.00%
	5	ρ_{st} (mm^2/mm^2)	R3.PTM. 5 b	0.00581	0.00581	0.00%
	6	ρ_{sc} (mm^2/mm^2)	R3.PTM. 6 b	-0.00202	-0.00202	0.00%
	7	f_{sy} (MPa)		500	500	0.00%
	8	f_{sw} (MPa)		400	400	0.00%
	9	ϕ_S (mm)		16	16	0.00%
	10	ρ_p (mm^2/mm^2)	R3.PTM. 10 b	0.00148	0.00148	0.00%
	11	f_{py} (MPa)		1650	1650	0.00%
	12	ϕ_P (mm)		12.7	12.7	0.00%
	13	$PTloss_{lt}$		0.15	0.15	0.00%
	14	L (mm)		10000	10000	0.00%
	15	K_c		3	3	0.00%
TRANSFER LOAD STAGE	16	σ_c/f_{cki}	FW.PTM. 16 c	-0.0579	-0.14	-147.14%
	17	ε_{st} (mm/mm)	FW.PTM. 17 c	-0.00027	-0.00024	13.43%
	18	ε_{sc} (mm/mm)	FW.PTM. 18 c	0.0005232	0.0000322	93.84%
	19	ε_p (mm/mm)	FW.PTM. 19 c	0.00679	0.00686	-1.01%
	20	σ_{ct}/f_{cki}	FW.PTM. 20 c	0.30253	0.24537	18.89%
	21	cw (mm)	FW.PTM. 21 c	0.138	0.581	-321.01%
	22	Δ (mm)	FW.PTM. 22 c	-4.243	-80.497	-1797.20%
SERVICE LOAD COMBINATION	23	$q_{L/250}$ (kN/m)		**80.00**	**76.38**	**4.53%**
	24	$q_{0.2mm}$ (kN/m)		**100.00**	**93.68**	**6.32%**
	25	$q_{0.6fck}$ (kN/m)	FW.PTM. 25 c	113.32	114.29	-0.85%
	26	$q_{0.8fsu}$ (kN/m)	FW.PTM. 26 c	141.41	143.85	-1.73%
	27	$q_{0.75fpu}$ (kN/m)	FW.PTM. 27 c	97.15	94.86	2.36%
ULTIMATE LOAD COMBINATION	28	q_{str} (kN/m)		**150.00**	**143.70**	**4.20%**
	29	ε_{st} (mm/mm)	FW.PTM. 29 c	0.00897	0.00891	0.71%
	30	ε_{sc} (mm/mm)	FW.PTM. 30 c	-0.00238	-0.00238	0.26%
	31	ε_p (mm/mm)	FW.PTM. 31 c	0.01507	0.01510	-0.21%
	32	ρ_{sw} (mm^2/mm^2)	FW.PTM. 32 c	0.0016	0.0016	0.99%
	33	μ_Δ		**1.50**	**1.70**	**-13.21%**

: Given Inputs for ANN R3 : Inputs for engineering mechanic

: Reverse Outputs calculated by reverse ANN model is used as inputs to calculate other outputs by forward ANN model .

: Reverse-forward inputs : Reverse-forward ouputs

(Continued)

Table 2.5.6 (Continued) Design accuracies of reverse scenario 3

(e-1-2) Deflection ductility ratio (μ_Δ) of 1.75

		PARAMETERS	Training Model		Network test result	Auto PT-BEAM check	Error
BEAM CONFIGURATION	1	b (mm)	R3.PTM.	1 b	881.33	881.33	0.00%
	2	h (mm)			700	700	0.00%
	3	f_{ck} (MPa)			40	40	0.00%
	4	f_{cki} (MPa)			20	20	0.00%
	5	ρ_{st} (mm^2/mm^2)	R3.PTM.	5 b	0.00640	0.00640	0.00%
	6	ρ_{sc} (mm^2/mm^2)	R3.PTM.	6 b	0.00085	0.00085	0.00%
	7	f_{sy} (MPa)			500	500	0.00%
	8	f_{sw} (MPa)			400	400	0.00%
	9	ϕ_S (mm)			16	16	0.00%
	10	ρ_p (mm^2/mm^2)	R3.PTM.	10 b	0.00194	0.00194	0.00%
	11	f_{py} (MPa)			1650	1650	0.00%
	12	ϕ_p (mm)			12.7	12.7	0.00%
	13	$PTloss_{lt}$			0.15	0.15	0.00%
	14	L (mm)			10000	10000	0.00%
	15	K_c			3	3	0.00%
TRANSFER LOAD STAGE	16	σ_c/f_{cki}	FW.PTM. 16 c		~0	0.00	Cracked
	17	ε_{st} (mm/mm)	FW.PTM. 17 c		-0.00037	-0.00039	-7.12%
	18	ε_{sc} (mm/mm)	FW.PTM. 18 c		0.0004051	0.0004227	-4.35%
	19	ε_p (mm/mm)	FW.PTM. 19 c		0.00673	0.00670	0.44%
	20	σ_{ct}/f_{cki}	FW.PTM. 20 c		0.39224	0.41688	-6.28%
	21	cw (mm)	FW.PTM. 21 c		0.066	0.103	-56.06%
	22	Δ (mm)	FW.PTM. 22 c		-6.659	-8.042	-20.77%
SERVICE LOAD COMBINATION	23	$q_{L/250}$ (kN/m)			**80.00**	**78.32**	**2.11%**
	24	$q_{0.2mm}$ (kN/m)			**100.00**	**97.31**	**2.69%**
	25	$q_{0.6fck}$ (kN/m)	FW.PTM. 25 c		107.91	109.35	-1.34%
	26	$q_{0.8fsu}$ (kN/m)	FW.PTM. 26 c		141.44	143.77	-1.65%
	27	$q_{0.75fpu}$ (kN/m)	FW.PTM. 27 c		98.41	98.01	0.40%
ULTIMATE LOAD COMBINATION	28	q_{str} (kN/m)			**150.00**	**145.71**	**2.86%**
	29	ε_{st} (mm/mm)	FW.PTM. 29 c		0.00962	0.00965	-0.28%
	30	ε_{sc} (mm/mm)	FW.PTM. 30 c		-0.00231	-0.00231	-0.02%
	31	ε_p (mm/mm)	FW.PTM. 31 c		0.01568	0.01585	-1.04%
	32	ρ_{sw} (mm^2/mm^2)	FW.PTM. 32 c		0.0019	0.0019	-1.18%
	33	μ_Δ			**1.75**	**1.79**	**-2.51%**

Model **R3.PTM.b** is used

Errors are reduced

Adjusting μ_Δ *from 1.5 to* ***1.75*** *improves design accuracy*

: Given Inputs for ANN R3 : Inputs for engineering mechanic

: Reverse Outputs calculated by reverse ANN model is used as inputs to calculate other outputs by forward ANN model .

: Reverse-forward inputs : Reverse-forward ouputs

(Continued)

2.5.3.3 *Influence of layer and neurons on design accuracies*

2.5.3.3.1 Table 2.5.6(d)

Table 2.5.6(d) explores the influence of neurons on design accuracies of Steps 1 and 2 for Reverse Scenario 3 in which $q_{L/250}$ = 80 kN/m reaching a deflection of L/250 at SLL, $q_{0.2\,mm}$ = 100 KN/m at SLL, q_{str} = 150 kN/m at ULL are preassigned. In Tables 2.5.6(d-1)–(d-3), how a number of hidden layers and neurons influence the accuracies of reverse designs are explored. The reverse networks of Step 1 shown in Tables 2.5.6(d-1)–(d-3) use R3.PTM.a, R3.PTM.b, and R3.PTM.c based on 20 layers-20 neurons,

Table 2.5.6 (Continued) Design accuracies of reverse scenario 3

(e-2) Design results based on 40 layers-40 neurons for Steps 1 and 2; $q_{L/250}$ = 80 kN/m at SLL reaching a deflection of L/250, $q_{0.2\,mm}$ = 100 KN/m at SLL, q_{str} = 150 kN/m at ULL based on deflection ductility ratio (μ_Δ) of 1.50 and 1.75 at an ULL state

(e-2-1) Deflection ductility ratio (μ_Δ) of 1.50

PARAMETERS			Training Model	Network test result	Auto PT-BEAM check	Error
BEAM CONFIGURATION	1	b (mm)	R3.PTM. 1 c	1019.45	1019.45	0.00%
	2	h (mm)		700	700	0.00%
	3	f_{ck} (MPa)		40	40	0.00%
	4	f_{cki} (MPa)		20	20	0.00%
	5	ρ_{st} (mm^2/mm^2)	R3.PTM. 5 c	0.00609	0.00609	0.00%
	6	ρ_{sc} (mm^2/mm^2)	R3.PTM. 6 c	-0.00041	-0.00041	0.00%
	7	f_{sy} (MPa)		500	500	0.00%
	8	f_{sw} (MPa)		400	400	0.00%
	9	ϕ_S (mm)		16	16	0.00%
	10	ρ_p (mm^2/mm^2)	R3.PTM. 10 c	0.00150	0.00150	0.00%
	11	f_{py} (MPa)		1650	1650	0.00%
	12	ϕ_p (mm)		12.7	12.7	0.00%
	13	$PTloss_{lt}$		0.15	0.15	0.00%
	14	L (mm)		10000	10000	0.00%
	15	K_c		3	3	0.00%
TRANSFER LOAD STAGE	16	σ_c/f_{cki}	FW.PTM. 16 c	~0	0.00	Cracked
	17	ε_{st} (mm/mm)	FW.PTM. 17 c	-0.00026	-0.00042	-61.37%
	18	ε_{sc} (mm/mm)	FW.PTM. 18 c	0.0003409	0.0064266	-1785.13%
	19	ε_p (mm/mm)	FW.PTM. 19 c	0.00681	0.00658	3.39%
	20	σ_{ct}/f_{cki}	FW.PTM. 20 c	0.27942	0.79920	-186.02%
	21	cw (mm)	FW.PTM. 21 c	0.050	4.313	-8526.00%
	22	Δ (mm)	FW.PTM. 22 c	-4.244	-183.903	-4233.46%
SERVICE LOAD COMBINATION	23	$q_{L/250}$ (kN/m)		**80.00**	**74.22**	**7.22%**
	24	$q_{0.2mm}$ (kN/m)		**100.00**	**91.19**	**8.81%**
	25	$q_{0.6fck}$ (kN/m)	FW.PTM. 25 c	111.63	112.42	-0.71%
	26	$q_{0.8fsu}$ (kN/m)	FW.PTM. 26 c	137.86	140.33	-1.79%
	27	$q_{0.75fpu}$ (kN/m)	FW.PTM. 27 c	93.76	92.20	1.66%
ULTIMATE LOAD COMBINATION	28	q_{str} (kN/m)		**150.00**	**143.15**	**4.57%**
	29	ε_{st} (mm/mm)	FW.PTM. 29 c	0.01034	0.01027	0.67%
	30	ε_{sc} (mm/mm)	FW.PTM. 30 c	-0.00225	-0.00226	-0.33%
	31	ε_p (mm/mm)	FW.PTM. 31 c	0.01651	0.01648	0.18%
	32	ρ_{sw} (mm^2/mm^2)	FW.PTM. 32 c	0.0016	0.0016	0.20%
	33	μ_Δ		**1.50**	**1.87**	**-24.34%**

Model **R3.PTM.c** is used

Big errors due to $\rho_{sc} < 0$

Preassigning $\mu_\Delta = 1.5$ gives big errors

: Given Inputs for ANN R3 ; : Inputs for engineering mechanic

: Reverse Outputs calculated by reverse ANN model is used as inputs to calculate other outputs by forward ANN model .

: Reverse-forward inputs ; : Reverse-forward ouputs

(Continued)

Table 2.5.6 (Continued) Design accuracies of reverse scenario 3

(e-2-2) Deflection ductility ratio (μ_Δ) of 1.75

PARAMETERS			Training Model		Network test result	Auto PT-BEAM check	Error
BEAM CONFIGURATION	1	b (mm)	R3.PTM.	1 c	890.56	890.56	0.00%
	2	h (mm)			700	700	0.00%
	3	f_{ck} (MPa)			40	40	0.00%
	4	f_{cki} (MPa)			20	20	0.00%
	5	ρ_{st} (mm^2/mm^2)	R3.PTM.	5 c	0.00647	0.00647	0.00%
	6	ρ_{sc} (mm^2/mm^2)	R3.PTM.	6 c	0.00135	0.00135	0.00%
	7	f_{sy} (MPa)			500	500	0.00%
	8	f_{sw} (MPa)			400	400	0.00%
	9	ϕ_S (mm)			16	16	0.00%
	10	ρ_p (mm^2/mm^2)	R3.PTM.	10 c	0.00200	0.00200	0.00%
	11	f_{py} (MPa)			1650	1650	0.00%
	12	ϕ_p (mm)			12.7	12.7	0.00%
	13	$PTloss_{lt}$			0.15	0.15	0.00%
	14	L (mm)			10000	10000	0.00%
	15	K_c			3	3	0.00%
TRANSFER LOAD STAGE	16	σ_c/f_{cki}	FW.PTM.	16 c	~0	0.00	Cracked
	17	ε_{st} (mm/mm)	FW.PTM.	17 c	-0.00038	-0.00040	-5.86%
	18	ε_{sc} (mm/mm)	FW.PTM.	18 c	0.0003656	0.0003879	-6.12%
	19	ε_p (mm/mm)	FW.PTM.	19 c	0.00672	0.00669	0.38%
	20	σ_{ct}/f_{cki}	FW.PTM.	20 c	0.39870	0.41838	-4.94%
	21	cw (mm)	FW.PTM.	21 c	0.058	0.083	-43.10%
	22	Δ (mm)	FW.PTM.	22 c	-7.048	-8.475	-20.25%
SERVICE LOAD COMBINATION	23	$q_{L/250}$ (kN/m)			**80.00**	**81.44**	**-1.80%**
	24	$q_{0.2mm}$ (kN/m)			**100.00**	**101.23**	**-1.23%**
	25	$q_{0.6fck}$ (kN/m)	FW.PTM.	25 c	111.43	112.89	-1.31%
	26	$q_{0.8fsu}$ (kN/m)	FW.PTM.	26 c	146.51	148.87	-1.61%
	27	$q_{0.75fpu}$ (kN/m)	FW.PTM.	27 c	101.93	101.91	0.02%
ULTIMATE LOAD COMBINATION	28	q_{str} (kN/m)			**150.00**	**151.30**	**-0.87%**
	29	ε_{st} (mm/mm)	FW.PTM.	29 c	0.00985	0.00989	-0.39%
	30	ε_{sc} (mm/mm)	FW.PTM.	30 c	-0.00229	-0.00229	-0.04%
	31	ε_p (mm/mm)	FW.PTM.	31 c	0.01591	0.01609	-1.17%
	32	ρ_{sw} (mm^2/mm^2)	FW.PTM.	32 c	0.0019	0.0019	-1.19%
	33	μ_Δ			**1.75**	**1.82**	**-3.78%**

(Continued)

30 layers-30 neurons, and 40 layers-40 neurons, respectively, where the deflection ductility ratio (μ_Δ) of 2.00 is adjusted to 2.50. Eighteen output parameters are obtained in Boxes 16–33 by the forward network of Step 2 trained with FW.PTM.c based on 40 layers and 40 neurons. Accuracies of network parameters depend on a number of hidden layers and neurons. ANN using R3.PTM.c based on 40 layers-40 neurons yields the best design accuracies for a deflection ductility ratio (μ_Δ) of 2.50. Input conflicts should also be removed by selecting reversely preassigned input parameters that satisfy structural mechanics.

Table 2.5.6 (Continued) Design accuracies of reverse scenario 3

(e-3) Design results based on 30 layers-30 neurons for Steps 1 and 2; $q_{L/250}$ = 80 kN/m at SLL reaching a deflection of L/250, $q_{0.2\,mm}$ = 100 KN/m at SLL, q_{str} = 150 kN/m at ULL based on deflection ductility ratio (μ_Δ) of 3.00 and 2.75 at an ULL

(e-3-1) Deflection ductility ratio (μ_Δ) of 3.00

Model **R3.PTM.b** is used

PARAMETERS			Training Model		Network test result	Auto PT-BEAM check	Error
BEAM CONFIGURATION	1	b (mm)	R3.PTM.	1 b	448.14	448.14	0.00%
	2	h (mm)			700	700	0.00%
	3	f_{ck} (MPa)			40	40	0.00%
	4	f_{cki} (MPa)			20	20	0.00%
	5	ρ_{st} (mm²/mm²)	R3.PTM.	5 b	0.01200	0.01200	0.00%
	6	ρ_{sc} (mm²/mm²)	R3.PTM.	6 b	0.03527	0.03527	0.00%
	7	f_{sy} (MPa)			500	500	0.00%
	8	f_{sw} (MPa)			400	400	0.00%
	9	ϕ_S (mm)			16	16	0.00%
	10	ρ_p (mm²/mm²)	R3.PTM.	10 b	0.00299	0.00299	0.00%
	11	f_{py} (MPa)			1650	1650	0.00%
	12	ϕ_p (mm)			12.7	12.7	0.00%
	13	$PTloss_k$			0.15	0.15	0.00%
	14	L (mm)			10000	10000	0.00%
	15	K_c			3	3	0.00%
TRANSFER LOAD STAGE	16	σ_c/f_{cki}	FW.PTM. 16 c		~0	0.00	Cracked
	17	ε_{st} (mm/mm)	FW.PTM. 17 c		-0.00043	-0.00044	-2.39%
	18	ε_{sc} (mm/mm)	FW.PTM. 18 c		0.0000914	0.0000911	0.36%
	19	ε_p (mm/mm)	FW.PTM. 19 c		0.00667	0.00665	0.19%
	20	σ_{ct}/f_{cki}	FW.PTM. 20 c		0.42394	0.42956	-1.33%
	21	cw (mm)	FW.PTM. 21 c		0.013	0.011	15.38%
	22	Δ (mm)	FW.PTM. 22 c		-8.696	-9.780	-12.46%
SERVICE LOAD COMBINATION	23	$q_{L/250}$ (kN/m)			**80.00**	**70.49**	**11.89%**
	24	$q_{0.2mm}$ (kN/m)			**100.00**	**91.00**	**9.00%**
	25	$q_{0.6fck}$ (kN/m)	FW.PTM. 25 c		105.99	106.45	-0.43%
	26	$q_{0.8fsu}$ (kN/m)	FW.PTM. 26 c		127.34	131.57	-3.32%
	27	$q_{0.75fpu}$ (kN/m)	FW.PTM. 27 c		85.00	89.32	-5.08%
ULTIMATE LOAD COMBINATION	28	q_{str} (kN/m)			**150.00**	**138.61**	**7.60%**
	29	ε_{st} (mm/mm)	FW.PTM. 29 c		0.02392	0.02385	0.30%
	30	ε_{sc} (mm/mm)	FW.PTM. 30 c		-0.00104	-0.00103	0.61%
	31	ε_p (mm/mm)	FW.PTM. 31 c		0.02998	0.03022	-0.80%
	32	ρ_{sw} (mm²/mm²)	FW.PTM. 32 c		0.0029	0.0028	0.36%
	33	μ_Δ			**3.00**	**2.53**	**15.52%**

Significant errors

Preassigning μ_Δ = 3 gives big errors

: Given Inputs for ANN R3 : Inputs for engineering mechanic

: Reverse Outputs calculated by reverse ANN model is used as inputs to calculate other outputs by forward ANN model .

: Reverse-forward inputs : Reverse-forward ouputs

(Continued)

2.5.3.3.2 Table 2.5.6(e)

Another reverse design is based on ANNs implementing R3.PTM.b based on 30 layers-30 neurons for the reverse network of Step 1 as shown in Tables 2.5.6(e-1) and (e-3) where the deflection ductility ratios (μ_Δ) of 1.50 and 1.75 are adjusted to 3.00 and 2.75, respectively. ANNs implementing

Table 2.5.6 (Continued) Design accuracies of reverse scenario 3

(e-3-2) Deflection ductility ratio (μ_Δ) of 2.75

PARAMETERS			Training Model		Network test result	Auto PT-BEAM check	Error
BEAM CONFIGURATION	1	b (mm)	R3.PTM.	1 b	499.52	499.52	0.00%
	2	h (mm)			700	700	0.00%
	3	f_{ck} (MPa)			40	40	0.00%
	4	f_{cki} (MPa)			20	20	0.00%
	5	ρ_{st} (mm^2/mm^2)	R3.PTM.	5 b	0.01055	0.01055	0.00%
	6	ρ_{sc} (mm^2/mm^2)	R3.PTM.	6 b	0.03603	0.03603	0.00%
	7	f_{sy} (MPa)			500	500	0.00%
	8	f_{sw} (MPa)			400	400	0.00%
	9	ϕ_S (mm)			16	16	0.00%
	10	ρ_p (mm^2/mm^2)	R3.PTM.	10 b	0.00295	0.00295	0.00%
	11	f_{py} (MPa)			1650	1650	0.00%
	12	ϕ_p (mm)			12.7	12.7	0.00%
	13	$PTloss_{lt}$			0.15	0.15	0.00%
	14	L (mm)			10000	10000	0.00%
	15	K_c			3	3	0.00%
TRANSFER LOAD STAGE	16	σ_c/f_{cki}	FW.PTM.	16 c	~0	0.00	Cracked
	17	ε_{st} (mm/mm)	FW.PTM.	17 c	-0.00044	-0.00045	-2.31%
	18	ε_{sc} (mm/mm)	FW.PTM.	18 c	0.0000913	0.0000915	-0.24%
	19	ε_p (mm/mm)	FW.PTM.	19 c	0.00666	0.00664	0.20%
	20	σ_{ct}/f_{cki}	FW.PTM.	20 c	0.43163	0.43686	-1.21%
	21	cw (mm)	FW.PTM.	21 c	0.013	0.011	15.38%
	22	Δ (mm)	FW.PTM.	22 c	-8.836	-9.996	-13.13%
SERVICE LOAD COMBINATION	23	$q_{L/250}$ (kN/m)			**80.00**	**76.28**	**4.65%**
	24	$q_{0.2mm}$ (kN/m)			**100.00**	**95.30**	**4.70%**
	25	$q_{0.6fck}$ (kN/m)	FW.PTM.	25 c	115.60	116.44	-0.72%
	26	$q_{0.8fpu}$ (kN/m)	FW.PTM.	26 c	130.68	136.44	-4.41%
	27	$q_{0.75fpu}$ (kN/m)	FW.PTM.	27 c	88.88	94.09	-5.86%
ULTIMATE LOAD COMBINATION	28	q_{str} (kN/m)			**150.00**	**143.51**	**4.33%**
	29	ε_{st} (mm/mm)	FW.PTM.	29 c	0.02501	0.02497	0.16%
	30	ε_{sc} (mm/mm)	FW.PTM.	30 c	-0.00093	-0.00093	0.47%
	31	ε_p (mm/mm)	FW.PTM.	31 c	0.03109	0.03136	-0.86%
	32	ρ_{sw} (mm^2/mm^2)	FW.PTM.	32 c	0.0027	0.0026	0.68%
	33	μ_Δ			**2.75**	**2.72**	**1.07%**

Model **R3.PTM.b** is used

Errors are reduced

Adjusting μ_Δ from **3** to **2.75** improves design accuracy

Red box: Given Inputs for ANN R3; Blue box: Inputs for engineering mechanic

Purple box: Reverse Outputs calculated by reverse ANN model is used as inputs to calculate other outputs by forward ANN model.

Green box: Reverse-forward inputs; Black box: Reverse-forward ouputs

(*Continued*)

R3.PTM.c based on 40 layers-40 neurons are used in the reverse network of Step 1 as shown in Tables 2.5.6(e-2) and (e-4) where the deflection ductility ratios (μ_Δ) of 1.50 and 1.75 are adjusted to 3.00 and 2.75, respectively. An ANN with FW.PTM.c based on 40 layers and 40 neurons is also implemented in Step 2 to train the forward network when 18 outputs are obtained in Boxes 16–33. Acceptable design accuracy with 0.19% for lower tensile rebar strains (*εst*) is found for ANNs using R3.PTM.c based on 40 layers-40 neurons, with deflection ductility ratio (μ_Δ) of 2.75 which is corresponding to lower tensile rebar strains (ε_{st}) of 0.02479 shown in Box 29 of Table 2.5.6(e-4-2) at ULL. Small errors are demonstrated for the lower tensile rebar strains (ε_{st}) of 0.02479 and 0.02475 with 0.19% obtained based on

Table 2.5.6 (Continued) Design accuracies of reverse scenario 3

(e-4) Design results based on 40 layers-40 neurons for Steps 1 and 2 1 neurons for Step 1; $q_{L/250}$ = 80 kN/m at SLL reaching a deflection of L/250, $q_{0.2\,mm}$ = 100 KN/m at SLL, q_{str} = 150 kN/m at ULL based on deflection ductility ratio (μ_Δ) of 3.00 and 2.75 at an ULL

(e-4-1) Deflection ductility ratio (μ_Δ) of 3.00

Model **R3.PTM.c** is used

PARAMETERS			Training Model	Network test result	Auto PT-BEAM check	Error
BEAM CONFIGURATION	1	b (mm)	R3.PTM. 1 c	445.93	445.93	0.00%
	2	h (mm)		700	700	0.00%
	3	f_{ck} (MPa)		40	40	0.00%
	4	f_{cki} (MPa)		20	20	0.00%
	5	ρ_{st} (mm²/mm²)	R3.PTM. 5 c	0.01225	0.01225	0.00%
	6	ρ_{sc} (mm²/mm²)	R3.PTM. 6 c	0.03710	0.03710	0.00%
	7	f_{sy} (MPa)		500	500	0.00%
	8	f_{sw} (MPa)		400	400	0.00%
	9	ϕ_S (mm)		16	16	0.00%
	10	ρ_p (mm²/mm²)	R3.PTM. 10 c	0.00304	0.00304	0.00%
	11	f_{py} (MPa)		1650	1650	0.00%
	12	ϕ_p (mm)		12.7	12.7	0.00%
	13	$PTloss_{lt}$		0.15	0.15	0.00%
	14	L (mm)		10000	10000	0.00%
	15	K_e		3	3	0.00%
TRANSFER LOAD STAGE	16	σ_c/f_{cki}	FW.PTM. 16 c	~0	0.00	Cracked
	17	ε_{st} (mm/mm)	FW.PTM. 17 c	-0.00043	-0.00044	-2.60%
	18	ε_{sc} (mm/mm)	FW.PTM. 18 c	0.0000899	0.0000889	1.15%
	19	ε_p (mm/mm)	FW.PTM. 19 c	0.00666	0.00665	0.19%
	20	σ_{ct}/f_{cki}	FW.PTM. 20 c	0.42731	0.43317	-1.37%
	21	cw (mm)	FW.PTM. 21 c	0.013	0.011	15.38%
	22	Δ (mm)	FW.PTM. 22 c	-8.639	-9.907	-14.67%
SERVICE LOAD COMBINATION	23	$q_{L/250}$ (kN/m)		**80.00**	**71.65**	**10.44%**
	24	$q_{0.2mm}$ (kN/m)		**100.00**	**92.52**	**7.48%**
	25	$q_{0.6fck}$ (kN/m)	FW.PTM. 25 c	108.45	108.58	-0.12%
	26	$q_{0.8fpu}$ (kN/m)	FW.PTM. 26 c	128.77	133.61	-3.76%
	27	$q_{0.75fpu}$ (kN/m)	FW.PTM. 27 c	86.73	90.74	-4.62%
ULTIMATE LOAD COMBINATION	28	q_{str} (kN/m)		**150.00**	**140.73**	**6.18%**
	29	ε_{st} (mm/mm)	FW.PTM. 29 c	0.02415	0.02409	0.24%
	30	ε_{sc} (mm/mm)	FW.PTM. 30 c	-0.00101	-0.00101	0.69%
	31	ε_p (mm/mm)	FW.PTM. 31 c	0.03018	0.03047	-0.96%
	32	ρ_{sw} (mm²/mm²)	FW.PTM. 32 c	0.0029	0.0029	0.18%
	33	μ_Δ		**3.00**	**2.55**	**15.06%**

Significant errors

Preassigning $\mu_\Delta = 3$ gives big errors

: Given Inputs for ANN R3 : Inputs for engineering mechanic

: Reverse Outputs calculated by reverse ANN model is used as inputs to calculate other outputs by forward ANN model .

: Reverse-forward inputs : Reverse-forward ouputs

(Continued)

ANNs and structural calculations, respectively. Similarly, Box 29 of Table 2.5.6(e-4-1) at ULL shows lower tensile rebar strains (ε_{st}) of 0.02415 with deflection ductility ratio (μ_Δ) of 3.00 which is acceptable design accuracy with 0.24%. However, it is noted that overall significant error is observed with deflection ductility ratio (μ_Δ) of 3.00 due to an input conflict as shown in Table 2.5.6(e-4-1) whereas error is improved as 2.16% with deflection ductility ratio ($\mu\Delta$) of 2.75 as shown in Table 2.5.6(e-4-2).

Table 2.5.6 (Continued) Design accuracies of reverse scenario 3

(e-4-2) Deflection ductility ratio (μ_Δ) of 2.75

PARAMETERS			Training Model	Network test result	Auto PT-BEAM check	Error
BEAM CONFIGURATION	1	b (mm)	R3.PTM. 1 c	502.35	502.35	0.00%
	2	h (mm)		700	700	0.00%
	3	f_{ck} (MPa)		40	40	0.00%
	4	f_{cki} (MPa)		20	20	0.00%
	5	ρ_{st} (mm²/mm²)	R3.PTM. 5 c	0.01069	0.01069	0.00%
	6	ρ_{sc} (mm²/mm²)	R3.PTM. 6 c	0.03585	0.03585	0.00%
	7	f_{sy} (MPa)		500	500	0.00%
	8	f_{sw} (MPa)		400	400	0.00%
	9	ϕ_S (mm)		16	16	0.00%
	10	ρ_p (mm²/mm²)	R3.PTM. 10 c	0.00300	0.00300	0.00%
	11	f_{py} (MPa)		1650	1650	0.00%
	12	ϕ_p (mm)		12.7	12.7	0.00%
	13	$PTloss_{lt}$		0.15	0.15	0.00%
	14	L (mm)		10000	10000	0.00%
	15	K_c		3	3	0.00%
TRANSFER LOAD STAGE	16	σ_c/f_{cki}	FW.PTM. 16 c	~0	0.00	Cracked
	17	ε_{st} (mm/mm)	FW.PTM. 17 c	-0.00044	-0.00045	-2.41%
	18	ε_{sc} (mm/mm)	FW.PTM. 18 c	0.0000932	0.0000933	-0.17%
	19	ε_p (mm/mm)	FW.PTM. 19 c	0.00665	0.00664	0.20%
	20	σ_{ct}/f_{cki}	FW.PTM. 20 c	0.43678	0.44210	-1.22%
	21	cw (mm)	FW.PTM. 21 c	0.014	0.012	14.29%
	22	Δ (mm)	FW.PTM. 22 c	-8.909	-10.170	-14.16%
SERVICE LOAD COMBINATION	23	$q_{L/250}$ (kN/m)		**80.00**	77.71	**2.86%**
	24	$q_{0.2mm}$ (kN/m)		**100.00**	**97.37**	**2.63%**
	25	$q_{0.6fck}$ (kN/m)	FW.PTM. 25 c	116.81	117.71	-0.77%
	26	$q_{0.8fsu}$ (kN/m)	FW.PTM. 26 c	133.47	139.30	-4.37%
	27	$q_{0.75fpu}$ (kN/m)	FW.PTM. 27 c	90.49	96.09	-6.18%
ULTIMATE LOAD COMBINATION	28	q_{str} (kN/m)		**150.00**	**146.44**	**2.37%**
	29	ε_{st} (mm/mm)	FW.PTM. 29 c	0.02479	0.02475	0.19%
	30	ε_{sc} (mm/mm)	FW.PTM. 30 c	-0.00095	-0.00095	0.53%
	31	ε_p (mm/mm)	FW.PTM. 31 c	0.03087	0.03113	-0.84%
	32	ρ_{sw} (mm²/mm²)	FW.PTM. 32 c	0.0027	0.0027	0.66%
	33	μ_Δ		**2.75**	**2.69**	**2.16%**

Model **R3.PTM.c** is used

Errors are reduced

Adjusting μ_Δ from 3 to **2.75** improves design accuracy

: Given Inputs for ANN R3 : Inputs for engineering mechanic

: Reverse Outputs calculated by reverse ANN model is used as inputs to calculate other outputs by forward ANN model .

: Reverse-forward inputs : Reverse-forward ouputs

2.5.4 Designs based on CRS using deep neural networks (DNN)

2.5.4.1 Formulation of ANNs based on back-substitution (BS) applicable to reverse designs

The CRS method is used in training reverse networks [2.3] where a sequence of input feature indices is determined based on the feature scores shown in Table 2.5.7(a) obtained using Neighborhood Component Analysis (NCA) method [2.6]. In a reverse network of Step 1, reverse output parameters (b, ρ_{st}, ρ_{sc}, and ρ_p) are calculated for reversely preassigned input parameters ($q_{L/250}$, $q_{0.2mm}$, q_{str}, and μ_Δ) as shown in Reverse Scenario 3 of Figure 2.5.4. Table 2.5.7 presents training accuracies based on CRS with DNN. CRS method rather than PTM is used to train a reverse ANN in Step 1. Training with CRS method begins with beam width (b) because beam width (b) can

Table 2.5.7 Training sequence of a reverse scenario based on CRS (DNN)

(a) Feature scores based on NCA

	h (mm)	f_{ck} (MPa)	f_{cki} (MPa)	f_{sy}	f_{sw}	ϕ_S	f_{py}	ϕ_P	Ptloss $_{lt}$	L (mm)	K_c	$q_{L/250}$ (kN/m)	$q_{0.2mm}$ (kN/m)	q_{str} (kN/m)	μ_Δ	b (mm)	ρ_{st}	ρ_p	ρ_{sc}
b (mm)	5.77	0.00	0.00	0.00	0.00	0.00	0.00	0.00	0.00	10.92	0.00	4.14	19.05	3.04	0.00		7.34	5.10	0.00
ρ_{st}	5.12	5.04	0.00	5.48	0.00	1.78	0.07	0.00	0.00	0.00	0.00	0.25	0.00	2.23	19.05	0.00		3.79	7.56
ρ_p	30.45	0.02	0.00	0.00	0.03	25.93	37.00	27.67	0.00	0.00	0.00	0.00	0.00	0.00	0.00	0.00	18.70		0.00
ρ_{sc}	5.70	4.00	0.00	4.97	0.00	0.00	0.00	0.00	0.00	0.00	0.00	0.00	0.00	5.18	19.91	0.00	11.36	3.74	

(b) Training accuracies of a reverse Scenario based on CRS (DNN)

REVERSE NETWORK OF STEP 1-REVERSE SCENARIO

15 Inputs (h, fck, f_{cki}, f_{sy}, f_{sw}, ϕ_S, f_{py}, ϕ_P, $PTloss_{lt}$, L, K_c, $q_{L/250}$, $q_{0.2mm}$, $qstr$, μ_Δ) – (4) Outputs (b, ρst, ρsc, ρ_p)

No.			Data	Layers	Neurons	Validation check	Required epoch	Best epoch for training	Stopped epoch	Mean square error	R at best epoch	Comments
15 Inputs (h, f_{ck}, f_{cki}, f_{sy}, f_{sw}, ϕ_S, f_{py}, ϕ_P, $PTloss_{lt}$, L, K_c, $q_{L/250}$, $q_{0.2mm}$, q_{str}, μ_Δ) **–1 Output** (b) - PTM												
R3.PTM.	1	a	50,000	20	20	500	50,000	31,407	31,907	1.1E-03	1.1E-03	0.99034
R3.PTM.	1	b	50,000	30	30	1000	70,000	67,507	68,507	7.1E-04	7.3E-04	0.99425
R3.PTM.	1	c	50,000	40	40	5,000	100,000	55,324	60,324	7.6E-04	7.5E-04	0.99458
16 Inputs (h, f_{ck}, f_{cki}, f_{sy}, f_{sw}, ϕ_S, f_{py}, ϕ_P, $PTloss_{lt}$, L, K_c, $q_{L/250}$, $q_{0.2mm}$, q_{str}, μ_Δ, b) **–1 Output** (ρ_{st}) - CRS												
R3.CRS.	5	a	50,000	20	20	500	50,000	24,914	25,414	1.7.E-04	1.7.E-04	0.99908
R3.CRS.	5	b	50,000	30	30	1,000	100,000	40,601	41,601	9.9.E-05	1.1.E-04	0.99947
R3.CRS.	5	c	50,000	40	40	5,000	100,000	51,497	56,497	7.3.E-05	9.1.E-05	0.99969

(Continued)

Table 2.5.7 (Continued) Training sequence of a reverse scenario based on CRS (DNN)

(b) Training accuracies of a reverse Scenario based on CRS (DNN)												
17 Inputs (h, f_{ck}, f_{cki}, f_{sy}, f_{sw}, ϕ_S, f_{py}, ϕ_P, $PTloss_{lt}$, L, K_c, $q_{L/250}$, $q_{0.2mm}$, q_{str}, μ_Δ, b, ρ_{st}) **–1 Output** (ρ_{sc}) - *CRS*												
R3.CRS.	10	a	50,000	20	20	500	50,000	26,261	26,761	2.3.E-04	2.3.E-04	0.99835
R3.CRS.	10	b	50,000	30	30	1000	100,000	35,490	36,490	1.2.E-04	1.2.E-04	0.99919
R3.CRS.	10	c	50,000	40	40	5,000	100,000	99,850	100,000	3.2.E-05	3.2.E-05	0.99982
18 Inputs (h, f_{ck}, f_{cki}, f_{sy}, f_{sw}, ϕ_S, f_{py}, ϕ_P, $PTloss_{lt}$, L, K_c, $q_{L/250}$, $q_{0.2mm}$, q_{str}, μ_Δ, b, ρ_{st}, ρ_{sc}) **–1 Output** (ρ_p) - *CRS*												
R3.CRS.	6	a	50,000	20	20	500	50,000	5,096	5,596	2.4.E-03	2.2.E-03	0.98852
R3.CRS.	6	b	50,000	30	30	1,000	100,000	7,630	8,630	2.5.E-03	2.2.E-03	0.98743
R3.CRS.	6	c	50,000	40	40	5,000	100,000	7,887	12,887	2.5.E-03	3.0.E-03	0.98848

Note:R3.CRS.**10a**

- Version of training model
- Output number
- Training method.
- Design scenario; R3 = Reverse ANN for Scenario 3

be used to train networks without other output parameters, such as ρ_{st}, ρ_{sc}, and ρ_p, as feature indexes as shown in Table 2.5.7(a). Important feature indexes affecting beam width (b) include h (5.77), L (10.92), $q_{L/250}$ (4.14), $q_{0.2mm}$ (19.05), and q_{str} (3.04) as shown in Table 2.5.7(a), whereas ρ_{st}, ρ_{st}, and ρ_p are not used as feature indexes when training beam width (b) because they belong to the output-side at the same time. Feature scores selected using the NCA method [2.6] are indicated by numbers in Table 2.5.7(a). Beam width (b) is, then, used as a feature index to train another output parameter ρ_{st}, resulting in higher training accuracy for ρ_{st} as shown in Table 2.5.7(b). Similarly, beam width (b) and ρ_{st} are used as feature indexes to train another output parameter ρ_p before training ρ_{sc} which is trained using all output parameters including beam width (b), ρ_{st}, and ρ_p as feature indexes. It is noted that output parameters placed on an output-side at the same time cannot be used to train other output parameters when using TED training method [2.3].

2.5.4.2 Design accuracies

Lower rebar ratio (ρ_{st}) has a feature score of 18.70 on ρ_p, whereas the upper rebar ratio (ρ_{sc}) is influenced by lower tensile rebar ratio (ρ_{st}) and tendon ratio (ρ_p) as much as 11.36 and 3.74, respectively. In the BS method, ANNs with R3.CRS.a are trained based on 20 layers-20 neurons, respectively, for the reverse networks of Step 1 shown in Table 2.5.8(a-1), whereas ANNs with FW.PTM.c based on 40 layers and 40 neurons are implemented in training the forward network of Step 2 shown in Table 2.5.8(a-2) when 18 outputs including reverse input parameters ($q_{L/250}$, $q_{0.2mm}$, q_{str}, and μ_Δ which are already reversely prescribed in Boxes 23, 24, 28, and 33 in Step 1) are calculated in Boxes 16–33 as summarized in Table 2.5.8(a-3). It is noted that an ANN is trained on beam width b, the first parameter to train, based on R3.PTM.1b with 30 layers-30 neurons. In the reverse design of this chapter, $q_{L/250}$=80 kN/m reaching a deflection of L/250 at SLL, $q_{0.2mm}$ = 100 kN/m at SLL, and q_{str} =150 kN/m at ULL are reversely preassigned on an input-side.

2.5.4.2.1 Table 2.5.8(a), Table 2.5.8(b)

Design accuracies based on the reversely preassigned deflection ductility ratio (μ_Δ) of 2.00 are acceptable as shown in Table 2.5.8 (a-1), (a-2), and (a-3). In Table 2.5.8(b-1), input conflicts occur when the deflection ductility ratio (μ_Δ) is 1.5, and hence, deflection ductility ratio (μ_Δ) is adjusted from 1.50 to 1.75 which is within large data ranges as shown in Table 2.5.8(b-2). As a result, design accuracies based on reversely preassigned deflection ductility ratio (μ_Δ) of 1.75 are enhanced to be acceptable as shown in Table 2.5.8(b-2), whereas those based on preassigned deflection ductility ratio (μ_Δ) of 1.50 yield errors that are unacceptable for use in design practice as shown in Table 2.5.8(b-1).

Table 2.5.8 Design accuracies of a reverse scenario based on CRS (DNN)

(a) Preassigning $q_{L/250}$ = 80 kN/m reaching a deflection of L/250 at SLL, $q_{0.2mm}$ = 100 kN/m at SLL, q_{str} = 150 kN/m at ULL based on deflection ductility ratio (μ_Δ) of 2.00 at ULL; 20 layers-20 neurons for Step 1

(a-1) Step 1: Reverse network with reverse inputs ($q_{L/250}$, $q_{0.2mm}$, q_{str}, μ_Δ) to calculate reverse outputs (b, ρ_{st}, ρ_{sc}, ρ_p)

R3.CRS.Table 2			Training Model	Network test result	Auto PT-BEAM check	Error
BEAM CONFIGURATION	1	b (mm)	R3.PTM. 1 b	744.91	744.91	0.00%
	2	h (mm)		700	700	0.00%
	3	f_{ck} (MPa)		40	40	0.00%
	4	f_{cki} (MPa)		20	20	0.00%
	5	ρ_{st} (mm²/mm²)	R3.CRS. 5 a	0.00695	0.00695	0.00%
	6	ρ_{sc} (mm²/mm²)	R3.CRS. 6 a	0.00574	0.00574	0.00%
	7	f_{sy} (MPa)		500	500	0.00%
	8	f_{sw} (MPa)		400	400	0.00%
	9	ϕ_S (mm)		16	16	0.00%
	10	ρ_p (mm²/mm²)	R3.CRS. 10 a	0.00243	0.00243	0.00%
	11	f_{py} (MPa)		1650	1650	0.00%
	12	ϕ_p (mm)		12.7	12.7	0.00%
	13	$PTloss_{lt}$		0.15	0.15	0.00%
	14	L (mm)		10000	10000	0.00%
	15	K_c		3	3	0.00%
REVERSE 3 INPUT	23	$q_{L/250}$ (kN/m)		80.00	81.60	-2.00%
	24	$q_{0.2mm}$ (kN/m)		100.00	101.47	-1.47%
	28	q_{str} (kN/m)		150.00	150.55	-0.37%
	33	μ_Δ		2.00	2.00	-0.11%

b, ρ_{st}, ρ_{sc}, and ρ_p are substituted into a forward network of Step 2

[red box] : Given Inputs for ANN R3 [blue box] : Inputs for engineering mechanic

[purple box] : Reverse Outputs calculated by reverse ANN model is used as inputs to calculate other outputs by forward ANN model .

(Continued)

2.5.4.2.2 Table 2.5.8(c)

In Table 2.5.8(c-1), input conflicts also occur when the deflection ductility ratio (μ_Δ) is 3.00, and hence, deflection ductility ratio (μ_Δ) is adjusted from 3.00 to 2.75 which is within large data ranges as shown in Table 2.5.8(c-2). As a result, design accuracies based on reversely preassigned deflection ductility ratio (μ_Δ) of 3.00 enhanced to be acceptable as shown in Table 2.5.8 (c-2), whereas those based on preassigned deflection ductility ratio (μ_Δ) of 3.00 yield errors that are unacceptable for use in design practice as shown in Table 2.5.8(c-1).

Table 2.5.8 (Continued) Design accuracies of a reverse scenario based on CRS (DNN)

(a-2) Step 2: Reverse-forward network using reverse outputs (from Step 1) as input parameters to calculate the rest of output parameters

b, ρ_{st}, ρ_{sc}, and ρ_p are calculated by a reverse network of Step 1

	R3.CRS.Table 2		Training Model	Network test result	Auto PT-BEAM check	Error
BEAM CONFIGURATION	1	b (mm)	R3.PTM. 1 b	**744.91**	**744.91**	**0.00%**
	2	h (mm)		700	700	0.00%
	3	f_{ck} (MPa)		40	40	0.00%
	4	f_{cki} (MPa)		20	20	0.00%
	5	ρ_{st} (mm²/mm²)	R3.CRS. 5 a	**0.00695**	**0.00695**	**0.00%**
	6	ρ_{sc} (mm²/mm²)	R3.CRS. 6 a	**0.00574**	**0.00574**	**0.00%**
	7	f_{sy} (MPa)		500	500	0.00%
	8	f_{sw} (MPa)		400	400	0.00%
	9	ϕ_S (mm)		16	16	0.00%
	10	ρ_p (mm²/mm²)	R3.CRS. 10 a	**0.00243**	**0.00243**	**0.00%**
	11	f_{py} (MPa)		1650	1650	0.00%
	12	ϕ_P (mm)		12.7	12.7	0.00%
	13	$PTloss_{lt}$		0.15	0.15	0.00%
	14	L (mm)		10000	10000	0.00%
	15	K_c		3	3	0.00%
REMAIN OUTPUTS CALCULATED BY REVERSE-FORWARD STEP	16	σ_c/f_{cki}	FW.PTM. 16 c	~0	0.00	Cracked
	17	ε_{st} (mm/mm)	FW.PTM. 17 c	-0.00043	-0.00044	-3.34%
	18	ε_{sc} (mm/mm)	FW.PTM. 18 c	0.0002382	0.0002644	-10.99%
	19	ε_p (mm/mm)	FW.PTM. 19 c	0.00666	0.00665	0.27%
	20	σ_{ct}/f_{cki}	FW.PTM. 20 c	0.43450	0.44674	-2.82%
	21	cw (mm)	FW.PTM. 21 c	0.035	0.038	-8.57%
	22	Δ (mm)	FW.PTM. 22 c	-9.946	-10.988	-10.48%
	25	$q_{0.6fck}$ (kN/m)	FW.PTM. 25 c	107.55	109.32	-1.65%
	26	$q_{0.8fsu}$ (kN/m)	FW.PTM. 26 c	142.01	145.58	-2.51%
	27	$q_{0.75fpu}$ (kN/m)	FW.PTM. 27 c	97.15	101.80	-4.79%
	29	ε_{st} (mm/mm)	FW.PTM. 29 c	0.01260	0.01265	-0.43%
	30	ε_{sc} (mm/mm)	FW.PTM. 30 c	-0.00204	-0.00204	0.04%
	31	ε_p (mm/mm)	FW.PTM. 31 c	0.01864	0.01888	-1.30%
	32	ρ_{sw} (mm²/mm²)	FW.PTM. 32 c	0.0021	0.0021	-0.19%

Green box: Reverse-forward inputs; Blue box: Inputs for engineering mechanic; Grey box: Reverse-forward ouputs

(Continued)

2.5.4.2.3 *The influence of a number of layers-neurons on design accuracies*

The influence of a number of layers-neurons on design accuracies of ANNs is investigated in Tables 2.5.8 (d-1) to (d-3) for a reverse design of Step 1 using CRS where ANNs are trained based on R3.CRS.a, R3.CRS.b, and R3.CRS.c using 20 layers-20 neurons, 30 layers-30 neurons, and 40

Table 2.5.8 (Continued) Design accuracies of a reverse scenario based on CRS (DNN)

(a-3) Step 3: Summary of design based on two steps of BS

Summary of design based on two steps of BS

R3.CRS.Table 2			Training Model	Network test result	Auto PT-BEAM check	Error
BEAM CONFIGURATION	1	b (mm)	R3.PTM. 1 b	744.91	744.91	0.00%
	2	h (mm)		700	700	0.00%
	3	f_{ck} (MPa)		40	40	0.00%
	4	f_{cki} (MPa)		20	20	0.00%
	5	ρ_{st} (mm^2/mm^2)	R3.CRS. 5 a	0.00695	0.00695	0.00%
	6	ρ_{sc} (mm^2/mm^2)	R3.CRS. 6 a	0.00574	0.00574	0.00%
	7	f_{sy} (MPa)		500	500	0.00%
	8	f_{sw} (MPa)		400	400	0.00%
	9	ϕ_s (mm)		16	16	0.00%
	10	ρ_p (mm^2/mm^2)	R3.CRS. 10 a	0.00243	0.00243	0.00%
	11	f_{py} (MPa)		1650	1650	0.00%
	12	ϕ_p (mm)		12.7	12.7	0.00%
	13	$PTloss_{lt}$		0.15	0.15	0.00%
	14	L (mm)		10000	10000	0.00%
	15	K_c		3	3	0.00%
TRANSFER LOAD STAGE	16	σ_c/f_{cki}	FW.PTM. 16 c	~0	0.00	Cracked
	17	ε_{st} (mm/mm)	FW.PTM. 17 c	-0.00043	-0.00044	-3.34%
	18	ε_{sc} (mm/mm)	FW.PTM. 18 c	0.0002382	0.0002644	-10.99%
	19	ε_p (mm/mm)	FW.PTM. 19 c	0.00666	0.00665	0.27%
	20	σ_{ct}/f_{cki}	FW.PTM. 20 c	0.43450	0.44674	-2.82%
	21	cw (mm)	FW.PTM. 21 c	0.035	0.038	-8.57%
	22	Δ (mm)	FW.PTM. 22 c	-9.946	-10.988	-10.48%
SERVICE LOAD COMBINATION	23	$q_{L/250}$ (kN/m)		**80.00**	**81.60**	**-2.00%**
	24	$q_{0.2mm}$ (kN/m)		**100.00**	**101.47**	**-1.47%**
	25	$q_{0.6fck}$ (kN/m)	FW.PTM. 25 c	107.55	109.32	-1.65%
	26	$q_{0.8fsu}$ (kN/m)	FW.PTM. 26 c	142.01	145.58	-2.51%
	27	$q_{0.75fpu}$ (kN/m)	FW.PTM. 27 c	97.15	101.80	-4.79%
ULTIMATE LOAD COMBINATION	28	q_{str} (kN/m)		**150.00**	**150.55**	**-0.37%**
	29	ε_{st} (mm/mm)	FW.PTM. 29 c	0.01260	0.01265	-0.43%
	30	ε_{sc} (mm/mm)	FW.PTM. 30 c	-0.00204	-0.00204	0.04%
	31	ε_p (mm/mm)	FW.PTM. 31 c	0.01864	0.01888	-1.30%
	32	ρ_{sw} (mm^2/mm^2)	FW.PTM. 32 c	0.0021	0.0021	-0.19%
	33	μ_Δ		**2.00**	**2.00**	**-0.11%**

[red box] : Given Inputs for ANN R3 [blue box] : Inputs for engineering mechanic

[purple box] : Reverse Outputs calculated by reverse ANN model is used as inputs to calculate other outputs by forward ANN model .

[green box] : Reverse-forward inputs [black box] : Reverse-forward ouputs

(*Continued*)

layers-40 neurons, respectively, when deflection ductility ratios (μ_Δ) of 2.00 and 2.50 are reversely preassigned. Design accuracies similar among those of the three networks using R3.CRS.a, R3.CRS.b, and R3.CRS.c in reverse Step 1 are found when forward network FW.PTM.c in Step 2 is used. For the reverse network of Step 1 trained based on 40 layers-40 neurons

Table 2.5.8 (Continued) Design accuracies of a reverse scenario based on CRS (DNN)

(b) Preassigning $q_{L/250}$ = 80 kN/m reaching a deflection of L/250 at SLL, $q_{0.2mm}$ = 100 kN/m at SLL, q_{str} = 150 kN/m at ULL based on deflection ductility ratio (μ_Δ) of 2.00 at ULL; 20 layers-20 neurons for Step 1

(b-1) Deflection ductility ratios (μ_Δ) of 1.50

R3.CRS.Table 1			Training Model	Network test result	Auto PT-BEAM check	Error
BEAM CONFIGURATION	1	b (mm)	R3.PTM. 1 b	1081.11	1081.11	0.00%
	2	h (mm)		700	700	0.00%
	3	f_{ck} (MPa)		40	40	0.00%
	4	f_{cki} (MPa)		20	20	0.00%
	5	ρ_{st} (mm^2/mm^2)	R3.CRS. 5 a	0.00580	0.00580	0.00%
	6	ρ_{sc} (mm^2/mm^2)	R3.CRS. 6 a	-0.00010	-0.00010	0.00%
	7	f_{sy} (MPa)		500	500	0.00%
	8	f_{sw} (MPa)		400	400	0.00%
	9	ϕ_S (mm)		16	16	0.00%
	10	ρ_p (mm^2/mm^2)	R3.CRS. 10 a	0.00167	0.00167	0.00%
	11	f_{py} (MPa)		1650	1650	0.00%
	12	ϕ_p (mm)		12.7	12.7	0.00%
	13	$PTloss_{lt}$		0.15	0.15	0.00%
	14	L (mm)		10000	10000	0.00%
	15	K_c		3	3	0.00%
TRANSFER LOAD STAGE	16	σ_c/f_{cki}	FW.PTM. 16 c	~0	0.00	Cracked
	17	ε_{st} (mm/mm)	FW.PTM. 17 c	-0.00032	-0.00036	-13.88%
	18	ε_{sc} (mm/mm)	FW.PTM. 18 c	0.0004404	0.0005097	-15.73%
	19	ε_p (mm/mm)	FW.PTM. 19 c	0.00677	0.00673	0.67%
	20	σ_{ct}/f_{cki}	FW.PTM. 20 c	0.34835	0.39769	-14.16%
	21	cw (mm)	FW.PTM. 21 c	0.079	0.189	-139.24%
	22	Δ (mm)	FW.PTM. 22 c	-5.237	-8.560	-63.47%
SERVICE LOAD COMBINATION	23	$q_{L/250}$ (kN/m)		**80.00**	**84.15**	**-5.19%**
	24	$q_{0.2mm}$ (kN/m)		**100.00**	**102.75**	**-2.75%**
	25	$q_{0.6fck}$ (kN/m)	FW.PTM. 25 c	122.04	122.86	-0.67%
	26	$q_{0.8fsu}$ (kN/m)	FW.PTM. 26 c	151.97	154.36	-1.57%
	27	$q_{0.75fpu}$ (kN/m)	FW.PTM. 27 c	105.64	103.92	1.63%
ULTIMATE LOAD COMBINATION	28	q_{str} (kN/m)		**150.00**	**156.68**	**-4.45%**
	29	ε_{st} (mm/mm)	FW.PTM. 29 c	0.01031	0.01028	0.34%
	30	ε_{sc} (mm/mm)	FW.PTM. 30 c	-0.00225	-0.00226	-0.26%
	31	ε_p (mm/mm)	FW.PTM. 31 c	0.01646	0.01649	-0.17%
	32	ρ_{sw} (mm^2/mm^2)	FW.PTM. 32 c	0.0016	0.0016	-0.23%
	33	μ_Δ		**1.50**	**1.89**	**-25.81%**

Big errors due to $\rho_{sc} < 0$

Preassigning $\mu_\Delta = 1.5$ gives big errors

: Given Inputs for ANN R3

: Inputs for engineering mechanic

: Reverse Outputs calculated by reverse ANN model is used as inputs to calculate other outputs by forward ANN model .

: Reverse-forward inputs

: Reverse-forward ouputs

(Continued)

Table 2.5.8 (Continued) Design accuracies of a reverse scenario based on CRS (DNN)

(b-2) Deflection ductility ratios (μ_Δ) of 1.75

R3.CRS.Table 1a			Training Model	Network test result	Auto PT-BEAM check	Error
BEAM CONFIGURATION	1	b (mm)	R3.PTM. 1 b	881.33	881.33	0.00%
	2	h (mm)		700	700	0.00%
	3	f_{ck} (MPa)		40	40	0.00%
	4	f_{cki} (MPa)		20	20	0.00%
	5	ρ_{st} (mm^2/mm^2)	R3.CRS. 5 a	0.00629	0.00629	0.00%
	6	ρ_{sc} (mm^2/mm^2)	R3.CRS. 6 a	0.00113	0.00113	0.00%
	7	f_{sy} (MPa)		500	500	0.00%
	8	f_{sw} (MPa)		400	400	0.00%
	9	ϕ_S (mm)		16	16	0.00%
	10	ρ_p (mm^2/mm^2)	R3.CRS. 10 a	0.00208	0.00208	0.00%
	11	f_{py} (MPa)		1650	1650	0.00%
	12	ϕ_p (mm)		12.7	12.7	0.00%
	13	$PTloss_{lt}$		0.15	0.15	0.00%
	14	L (mm)		10000	10000	0.00%
	15	K_c		3	3	0.00%
TRANSFER LOAD STAGE	16	σ_c/f_{cki}	FW.PTM. 16 c	~0	0.00	100.00%
	17	ε_{st} (mm/mm)	FW.PTM. 17 c	-0.00039	-0.00042	-6.18%
	18	ε_{sc} (mm/mm)	FW.PTM. 18 c	0.0003959	0.0004217	-6.51%
	19	ε_p (mm/mm)	FW.PTM. 19 c	0.00670	0.00667	0.39%
	20	σ_{ct}/f_{cki}	FW.PTM. 20 c	0.41856	0.43919	-4.93%
	21	cw (mm)	FW.PTM. 21 c	0.066	0.093	-40.91%
	22	Δ (mm)	FW.PTM. 22 c	-7.617	-9.436	-23.88%
SERVICE LOAD COMBINATION	23	$q_{L/250}$ (kN/m)		**80.00**	**82.21**	-2.77%
	24	$q_{0.2mm}$ (kN/m)		**100.00**	**101.77**	-1.77%
	25	$q_{0.6fck}$ (kN/m)	FW.PTM. 25 c	110.55	112.22	-1.51%
	26	$q_{0.8fsu}$ (kN/m)	FW.PTM. 26 c	145.92	148.48	-1.75%
	27	$q_{0.75fpu}$ (kN/m)	FW.PTM. 27 c	102.22	102.53	-0.30%
ULTIMATE LOAD COMBINATION	28	q_{str} (kN/m)		**150.00**	**150.10**	**-0.07%**
	29	ε_{st} (mm/mm)	FW.PTM. 29 c	0.00954	0.00958	-0.41%
	30	ε_{sc} (mm/mm)	FW.PTM. 30 c	-0.00232	-0.00232	2.75E-06
	31	ε_p (mm/mm)	FW.PTM. 31 c	0.01557	0.01577	-1.29%
	32	ρ_{sw} (mm^2/mm^2)	FW.PTM. 32 c	0.0019	0.0019	-1.27%
	33	μ_Δ		**1.75**	**1.80**	**-2.68%**

: Given Inputs for ANN R3 : Inputs for engineering mechanic

: Reverse Outputs calculated by reverse ANN model is used as inputs to calculate other outputs by forward ANN model .

: Reverse-forward inputs : Reverse-forward ouputs

Errors are reduced

Adjusting μ_Δ from 1.5 to 1.75 gives reasonable design accuracy

(Continued)

shown in Table 2.5.8(d-3-2), the largest error is found with –9.32% (91.98 kN/m based on ANN vs.100.56 kN/m of AutoPTbeam based on structural mechanics) for $q_{0.75fpu}$ representing nominal strength when tendons reach stresses smaller of 0.75 f_{pu} and f_{py}. The next largest error of deflection (camber c_w) at transfer load limit state reaches –7.27% (–9.896 mm based on ANN vs. –10.616 mm of AutoPTbeam based on structural mechanics). Less input conflicts occur with a deflection ductility ratio (μ_Δ) of 2.50 than with 2.00, resulting in good accuracies among all three designs shown in Tables 2.5.8(d-1)–(d-3) which are obtained when deflection ductility ratios (μ_Δ) within a range between 1.75 and 2.75 shown in Figure 2.5.5(a)–(d) are

Table 2.5.8 (Continued) Design accuracies of a reverse scenario based on CRS (DNN)

(c) Preassigning $q_{L/250}$ = 80 kN/m reaching a deflection of L/250 at SLL, $q_{0.2mm}$ = 100 kN/m at SLL, q_{str} = 150 kN/m at ULL based on deflection ductility ratio (μ_Δ) of 2.00 at ULL; 20 layers-20 neurons for Step 1

(c-1) Deflection ductility ratios (μ_Δ) of 3.00

R3.CRS.Table 4			Training Model	Network test result	Auto PT-BEAM check	Error
BEAM CONFIGURATION	1	b (mm)	R3.PTM. 1 b	448.14	448.14	0.00%
	2	h (mm)		700	700	0.00%
	3	f_{ck} (MPa)		40	40	0.00%
	4	f_{cki} (MPa)		20	20	0.00%
	5	ρ_{st} (mm^2/mm^2)	R3.CRS. 5 a	0.01317	0.01317	0.00%
	6	ρ_{sc} (mm^2/mm^2)	R3.CRS. 6 a	0.03002	0.03002	0.00%
	7	f_{sy} (MPa)		500	500	0.00%
	8	f_{sw} (MPa)		400	400	0.00%
	9	ϕ_S (mm)		16	16	0.00%
	10	ρ_p (mm^2/mm^2)	R3.CRS. 10 a	0.00316	0.00316	0.00%
	11	f_{py} (MPa)		1650	1650	0.00%
	12	ϕ_p (mm)		12.7	12.7	0.00%
	13	$PTloss_{lt}$		0.15	0.15	0.00%
	14	L (mm)		10000	10000	0.00%
	15	K_c		3	3	0.00%
TRANSFER LOAD STAGE	16	σ_c/f_{cki}	FW.PTM. 16 c	~0	0.00	Cracked
	17	ε_{st} (mm/mm)	FW.PTM. 17 c	-0.00044	-0.00046	-2.77%
	18	ε_{sc} (mm/mm)	FW.PTM. 18 c	0.0001054	0.0001070	-1.52%
	19	ε_p (mm/mm)	FW.PTM. 19 c	0.00665	0.00664	0.19%
	20	σ_{ct}/f_{cki}	FW.PTM. 20 c	0.43923	0.44503	-1.32%
	21	cw (mm)	FW.PTM. 21 c	0.015	0.014	6.67%
	22	Δ (mm)	FW.PTM. 22 c	-9.115	-10.431	-14.44%
SERVICE LOAD COMBINATION	23	$q_{L/250}$ (kN/m)		**80.00**	**73.20**	**8.50%**
	24	$q_{0.2mm}$ (kN/m)		**100.00**	**97.81**	**2.19%**
	25	$q_{0.6fck}$ (kN/m)	FW.PTM. 25 c	103.47	104.77	-1.26%
	26	$q_{0.8fsu}$ (kN/m)	FW.PTM. 26 c	139.15	141.48	-1.67%
	27	$q_{0.75fpu}$ (kN/m)	FW.PTM. 27 c	89.72	95.56	-6.52%
ULTIMATE LOAD COMBINATION	28	q_{str} (kN/m)		**150.00**	**149.37**	**0.42%**
	29	ε_{st} (mm/mm)	FW.PTM. 29 c	0.02112	0.02103	0.44%
	30	ε_{sc} (mm/mm)	FW.PTM. 30 c	-0.00129	-0.00128	0.27%
	31	ε_p (mm/mm)	FW.PTM. 31 c	0.02721	0.02737	-0.56%
	32	ρ_{sw} (mm^2/mm^2)	FW.PTM. 32 c	0.0031	0.0031	0.35%
	33	μ_Δ		**3.00**	**2.27**	**24.41%**

Significant errors

Preassigning $\mu_\Delta = 3$ gives big errors

: Given Inputs for ANN R3 : Inputs for engineering mechanic

: Reverse Outputs calculated by reverse ANN model is used as inputs to calculate other outputs by forward ANN model .

: Reverse-forward inputs : Reverse-forward ouputs

(*Continued*)

trained. CRS is a good training method to improve design accuracies of reverse networks for use in the first networks of Step 1.

2.5.4.3 Design tables and charts

Tables 2.5.8(a)–(d) are used to construct design charts to determine design parameters of ρ_{sc}, ρ_p, ρ_{st}, and b for given ductility (μ_Δ) between 1.75 and 2.75. It is noted that upper and lower bounds of deflection ductility ratios

Table 2.5.8 (Continued) Design accuracies of a reverse scenario based on CRS (DNN)

(c-2) Deflection ductility ratios (μ_Δ) of 2.75

R3.CRS.Table 4a			Training Model	Network test result	Auto PT-BEAM check	Error
BEAM CONFIGURATION	1	b (mm)	R3.PTM. 1 b	499.52	499.52	0.00%
	2	h (mm)		700	700	0.00%
	3	f_{ck} (MPa)		40	40	0.00%
	4	f_{cki} (MPa)		20	20	0.00%
	5	ρ_{st} (mm^2/mm^2)	R3.CRS. 5 a	0.01106	0.01106	0.00%
	6	ρ_{sc} (mm^2/mm^2)	R3.CRS. 6 a	0.04027	0.04027	1.72E-16
	7	f_{sy} (MPa)		500	500	0.00%
	8	f_{sw} (MPa)		400	400	0.00%
	9	ϕ_S (mm)		16	16	0.00%
	10	ρ_p (mm^2/mm^2)	R3.CRS. 10 a	0.00309	0.00309	-1.41E-16
	11	f_{py} (MPa)		1650	1650	0.00%
	12	ϕ_p (mm)		12.7	12.7	0.00%
	13	$PTloss_{lt}$		0.15	0.15	0.00%
	14	L (mm)		10000	10000	0.00%
	15	K_e		3	3	0.00%
TRANSFER LOAD STAGE	16	σ_c/f_{cki}	FW.PTM. 16 c	0.0003	0.00	100.00%
	17	ε_{st} (mm/mm)	FW.PTM. 17 c	-0.00045	-0.00046	-2.90%
	18	ε_{sc} (mm/mm)	FW.PTM. 18 c	0.0000888	0.0000876	1.40%
	19	ε_p (mm/mm)	FW.PTM. 19 c	0.00664	0.00663	0.19%
	20	σ_{ct}/f_{cki}	FW.PTM. 20 c	0.44305	0.44845	-1.22%
	21	cw (mm)	FW.PTM. 21 c	0.013	0.011	15.38%
	22	Δ (mm)	FW.PTM. 22 c	-8.731	-10.320	-18.20%
SERVICE LOAD COMBINATION	23	$q_{L/250}$ (kN/m)		**80.00**	**80.39**	**-0.48%**
	24	$q_{0.2mm}$ (kN/m)		**100.00**	**100.49**	**-0.49%**
	25	$q_{0.6fck}$ (kN/m)	FW.PTM. 25 c	123.24	123.19	0.04%
	26	$q_{0.8fsu}$ (kN/m)	FW.PTM. 26 c	135.54	143.64	-5.98%
	27	$q_{0.75fpu}$ (kN/m)	FW.PTM. 27 c	94.31	99.03	-5.01%
ULTIMATE LOAD COMBINATION	28	q_{str} (kN/m)		**150.00**	**150.69**	**-0.46%**
	29	ε_{st} (mm/mm)	FW.PTM. 29 c	0.02541	0.02540	0.02%
	30	ε_{sc} (mm/mm)	FW.PTM. 30 c	-0.00089	-0.00089	5.21E-03
	31	ε_p (mm/mm)	FW.PTM. 31 c	0.03137	0.03180	-1.35%
	32	ρ_{sw} (mm^2/mm^2)	FW.PTM. 32 c	0.0028	0.0028	0.32%
	33	μ_Δ		**2.75**	**2.76**	**-0.19%**

: Given Inputs for ANN R3 : Inputs for engineering mechanic

: Reverse Outputs calculated by reverse ANN model is used as inputs to calculate other outputs by forward ANN model .

: Reverse-forward inputs : Reverse-forward ouputs

Errors are reduced

Adjusting μ_Δ *from 3 to* ***2.75*** *gives reasonable design accuracy*

(*Continued*)

(μ_Δ) are governed by ρ_{sc}, ρ_p, ρ_{st}, and b as can be seen in Figure 2.5.5(a)–(d) where design parameters of ρ_{sc}, ρ_p, ρ_{st}, and b are obtained facilely and accurately based on deflection ductility ratios (μ_Δ) selected between 1.75 and 2.75. Figure 2.5.5(a)–(d) is useful for calculating design parameters, ρ_{sc}, ρ_p, ρ_{st}, and b for specified deflection ductility ratios (μ_Δ) between 1.75 and 2.75. Errors increase rapidly outside the training region of big datasets, diverging AutoPTbeam too. Design charts are constructed using design tables shown in Tables 2.5.6(d-1)–(d-3) and Tables 2.5.8(d-1) and (d-2) which present ρ_{sc}, ρ_p, ρ_{st}, and b for deflection ductility ratios (μ_Δ) of 2.0 and 2.5.

Table 2.5.8 (Continued) Design accuracies of a reverse scenario based on CRS (DNN)

(d-1) Preassigning $q_{L/250}$ = 80 kN/m reaching a deflection of L/250 at SLL, $q_{0.2mm}$ = 100 kN/m at SLL, q_{str} = 150 kN/m at ULL based on deflection ductility ratios (μ_Δ) of 2.00, 2.50 at ULL; 20 layers-20 neurons for Step 1

(d-1-1) Deflection ductility ratios (μ_Δ) of 2.00

R3.CRS.Table 2			Training Model	Network test result	Auto PT-BEAM check	Error
BEAM CONFIGURATION	1	b (mm)	R3.PTM. 1 b	744.91	744.91	0.00%
	2	h (mm)		700	700	0.00%
	3	f_{ck} (MPa)		40	40	0.00%
	4	f_{cki} (MPa)		20	20	0.00%
	5	ρ_{st} (mm²/mm²)	R3.CRS. 5 a	0.00695	0.00695	0.00%
	6	ρ_{sc} (mm²/mm²)	R3.CRS. 6 a	0.00574	0.00574	0.00%
	7	f_{sy} (MPa)		500	500	0.00%
	8	f_{sw} (MPa)		400	400	0.00%
	9	ϕ_S (mm)		16	16	0.00%
	10	ρ_p (mm²/mm²)	R3.CRS. 10 a	0.00243	0.00243	0.00%
	11	f_{py} (MPa)		1650	1650	0.00%
	12	ϕ_p (mm)		12.7	12.7	0.00%
	13	$PTloss_h$		0.15	0.15	0.00%
	14	L (mm)		10000	10000	0.00%
	15	K_c		3	3	0.00%
TRANSFER LOAD STAGE	16	σ_c/f_{cki}	FW.PTM. 16 c	~0	0.00	Cracked
	17	ε_{st} (mm/mm)	FW.PTM. 17 c	-0.00043	-0.00044	-3.34%
	18	ε_{sc} (mm/mm)	FW.PTM. 18 c	0.0002382	0.0002644	-10.99%
	19	ε_p (mm/mm)	FW.PTM. 19 c	0.00666	0.00665	0.27%
	20	σ_{ct}/f_{cki}	FW.PTM. 20 c	0.43450	0.44674	-2.82%
	21	cw (mm)	FW.PTM. 21 c	0.035	0.038	-8.57%
	22	Δ (mm)	FW.PTM. 22 c	-9.946	-10.988	-10.48%
SERVICE LOAD COMBINATION	23	$q_{L/250}$ (kN/m)		**80.00**	**81.60**	**-2.00%**
	24	$q_{0.2mm}$ (kN/m)		**100.00**	**101.47**	**-1.47%**
	25	$q_{0.6fck}$ (kN/m)	FW.PTM. 25 c	107.55	109.32	-1.65%
	26	$q_{0.8fsu}$ (kN/m)	FW.PTM. 26 c	142.01	145.58	-2.51%
	27	$q_{0.75fpu}$ (kN/m)	FW.PTM. 27 c	97.15	101.80	-4.79%
ULTIMATE LOAD COMBINATION	28	q_{str} (kN/m)		**150.00**	**150.55**	**-0.37%**
	29	ε_{st} (mm/mm)	FW.PTM. 29 c	0.01260	0.01265	-0.43%
	30	ε_{sc} (mm/mm)	FW.PTM. 30 c	-0.00204	-0.00204	0.04%
	31	ε_p (mm/mm)	FW.PTM. 31 c	0.01864	0.01888	-1.30%
	32	ρ_{sw} (mm²/mm²)	FW.PTM. 32 c	0.0021	0.0021	-0.19%
	33	μ_Δ		**2.00**	**2.00**	**-0.11%**

Significant errors

μ_Δ between 1.75 and 2.75 show good accuracy.

: Given Inputs for ANN R3
: Inputs for engineering mechanic
: Reverse Outputs calculated by reverse ANN model is used as inputs to calculate other outputs by forward ANN model .
: Reverse-forward inputs
: Reverse-forward ouputs

(Continued)

2.5.4.4 An influence of a number of layers-neurons for Step 1 on design accuracies

2.5.4.4.1 Table 2.5.9(a)

An influence of a number of layers-neurons for Step 1 on design accuracies is investigated using DNN trained by CRS as shown in Table 2.5.9. Tables 2.5.9 (a-1) and (a-2) present ANNs for the reverse networks of Step 1

Table 2.5.8 (Continued) Design accuracies of a reverse scenario based on CRS (DNN)

(d-1-2) Deflection ductility ratios (μ_Δ) of 2.50

R3.CRS.Table 3			Training Model	Network test result	Auto PT-BEAM check	Error
BEAM CONFIGURATION	1	b (mm)	R3.PTM. 1 b	565.56	565.56	0.00%
	2	h (mm)		700	700	0.00%
	3	f_{ck} (MPa)		40	40	0.00%
	4	f_{cki} (MPa)		20	20	0.00%
	5	ρ_{st} (mm^2/mm^2)	R3.CRS. 5 a	0.00924	0.00924	0.00%
	6	ρ_{sc} (mm^2/mm^2)	R3.CRS. 6 a	0.02476	0.02476	0.00%
	7	f_{sy} (MPa)		500	500	0.00%
	8	f_{sw} (MPa)		400	400	0.00%
	9	ϕ_s (mm)		16	16	0.00%
	10	ρ_p (mm^2/mm^2)	R3.CRS. 10 a	0.00294	0.00294	0.00%
	11	f_{py} (MPa)		1650	1650	0.00%
	12	ϕ_p (mm)		12.7	12.7	0.00%
	13	$PTloss_{lt}$		0.15	0.15	0.00%
	14	L (mm)		10000	10000	0.00%
	15	K_c		3	3	0.00%
TRANSFER LOAD STAGE	16	σ_c/f_{cki}	FW.PTM. 16 c	~0	0.00	Cracked
	17	ε_{st} (mm/mm)	FW.PTM. 17 c	-0.00046	-0.00047	-2.24%
	18	ε_{sc} (mm/mm)	FW.PTM. 18 c	0.0001180	0.0001245	-5.51%
	19	ε_p (mm/mm)	FW.PTM. 19 c	0.00664	0.00662	0.22%
	20	σ_{ct}/f_{cki}	FW.PTM. 20 c	0.45049	0.45580	-1.18%
	21	cw (mm)	FW.PTM. 21 c	0.016	0.016	0.00%
	22	Δ (mm)	FW.PTM. 22 c	-9.986	-10.790	-8.05%
SERVICE LOAD COMBINATION	23	$q_{L/250}$ (kN/m)		**80.00**	**81.04**	**-1.30%**
	24	$q_{0.2mm}$ (kN/m)		**100.00**	**101.08**	**-1.08%**
	25	$q_{0.6fck}$ (kN/m)	FW.PTM. 25 c	114.46	115.50	-0.91%
	26	$q_{0.8fpu}$ (kN/m)	FW.PTM. 26 c	139.85	143.94	-2.92%
	27	$q_{0.75fpu}$ (kN/m)	FW.PTM. 27 c	91.56	100.33	-9.58%
ULTIMATE LOAD COMBINATION	28	q_{str} (kN/m)		**150.00**	**150.81**	**-0.54%**
	29	ε_{st} (mm/mm)	FW.PTM. 29 c	0.02235	0.02226	0.40%
	30	ε_{sc} (mm/mm)	FW.PTM. 30 c	-0.00117	-0.00117	0.04%
	31	ε_p (mm/mm)	FW.PTM. 31 c	0.02849	0.02862	-0.43%
	32	ρ_{sw} (mm^2/mm^2)	FW.PTM. 32 c	0.0025	0.0025	1.11%
	33	μ_Δ		**2.50**	**2.53**	**-1.32%**

μ_Δ between 1.75 and 2.75 show good accuracy.

: Given Inputs for ANN R3 ; : Inputs for engineering mechanic

: Reverse Outputs calculated by reverse ANN model is used as inputs to calculate other outputs by forward ANN model .

: Reverse-forward inputs ; : Reverse-forward ouputs

(*Continued*)

trained based on R3.CRS.b and R3.CRS.c using 30 layers-30 neurons and 40 layers-40 neurons, respectively, when the deflection ductility ratios (μ_Δ) of 2.75 and 1.50 at ULL are reversely preassigned based on $q_{L/250}$ = 80 kN/m reaching a deflection of L/250 at SLL and $q_{0.2mm}$ = 100 kN/m at SLL, q_{str} = 150 kN/m at ULL.

Table 2.5.8 (Continued) Design accuracies of a reverse scenario based on CRS (DNN)

(d-2) Preassigning $q_{L/250}$ = 80 kN/m reaching a deflection of L/250 at SLL, $q_{0.2mm}$ = 100 kN/m at SLL, q_{str} = 150 kN/m at ULL based on deflection ductility ratios (μ_Δ) of 2.00, 2.50 at ULL; 30 layers-30 neurons for Step 1

(d-2-1) Deflection ductility ratios (μ_Δ) of 2.00

R3.CRS.Table 2			Training Model	Network test result	Auto PT-BEAM	Error
BEAM CONFIGURATION	1	b (mm)	R3.PTM. 1 b	744.91	744.91	0.00%
	2	h (mm)		700	700	0.00%
	3	f_{ck} (MPa)		40	40	0.00%
	4	f_{cki} (MPa)		20	20	0.00%
	5	ρ_{st} (mm²/mm²)	R3.CRS. 5 b	0.00681	0.00681	0.00%
	6	ρ_{sc} (mm²/mm²)	R3.CRS. 6 b	0.00560	0.00560	0.00%
	7	f_{sy} (MPa)		500	500	0.00%
	8	f_{sw} (MPa)		400	400	0.00%
	9	ϕ_S (mm)		16	16	0.00%
	10	ρ_p (mm²/mm²)	R3.CRS. 10 b	0.00246	0.00246	1.76E-16
	11	f_{py} (MPa)		1650	1650	0.00%
	12	ϕ_p (mm)		12.7	12.7	0.00%
	13	$PTloss_{lt}$		0.15	0.15	0.00%
	14	L (mm)		10000	10000	0.00%
	15	K_c		3	3	0.00%
TRANSFER LOAD STAGE	16	σ_c/f_{cki}	FW.PTM. 16 c	0.0005	0.00	Cracked
	17	ε_{st} (mm/mm)	FW.PTM. 17 c	-0.00044	-0.00045	-3.33%
	18	ε_{sc} (mm/mm)	FW.PTM. 18 c	0.0002450	0.0002717	-10.92%
	19	ε_p (mm/mm)	FW.PTM. 19 c	0.00666	0.00664	0.28%
	20	σ_{ct}/f_{cki}	FW.PTM. 20 c	0.44144	0.45387	-2.82%
	21	cw (mm)	FW.PTM. 21 c	0.036	0.039	-8.33%
	22	Δ (mm)	FW.PTM. 22 c	-10.233	-11.295	-10.38%
SERVICE LOAD COMBINATION	23	$q_{L/250}$ (kN/m)		**80.00**	**81.97**	**-2.46%**
	24	$q_{0.2mm}$ (kN/m)		**100.00**	**101.58**	**-1.58%**
	25	$q_{0.6fck}$ (kN/m)	FW.PTM. 25 c	107.34	109.21	-1.74%
	26	$q_{0.8fsu}$ (kN/m)	FW.PTM. 26 c	141.55	145.28	-2.64%
	27	$q_{0.75fpu}$ (kN/m)	FW.PTM. 27 c	97.14	101.98	-4.99%
ULTIMATE LOAD COMBINATION	28	q_{str} (kN/m)		**150.00**	**150.00**	**0.00%**
	29	ε_{st} (mm/mm)	FW.PTM. 29 c	0.01249	0.01256	-0.50%
	30	ε_{sc} (mm/mm)	FW.PTM. 30 c	-0.00205	-0.00205	0.07%
	31	ε_p (mm/mm)	FW.PTM. 31 c	0.01854	0.01879	-1.38%
	32	ρ_{sw} (mm²/mm²)	FW.PTM. 32 c	0.0021	0.0021	-0.21%
	33	μ_Δ		**2.00**	**2.00**	**-0.14%**

Model **R3.CRS.b** is used

Significant errors

*μ_Δ between **1.75** and **2.75** show good accuracy.*

: Given Inputs for ANN R3 : Inputs for engineering mechanic

: Reverse Outputs calculated by reverse ANN model is used as inputs to calculate other outputs by forward ANN model.

: Reverse-forward inputs : Reverse-forward ouputs

(Continued)

ANNs for the forward network of Step 2 trained based on FW.PTM.c using 40 layers and 40 neurons are used to calculate 18 output parameters in Boxes 16–33 of Tables 2.5.9(a-1) and (a-2). Reverse input parameters ($q_{L/250}$, $q_{0.2mm}$, q_{str}, and μ_Δ) are already reversely prescribed in Boxes 23, 24, 28, and 33 in Step 1, and hence, calculated reverse input parameters can be used to verify the networks used in BS. However, in Tables 2.5.9(a-1-2)

Table 2.5.8 (Continued) Design accuracies of a reverse scenario based on CRS (DNN)

(d-2-2) Deflection ductility ratios (μ_Δ) of 2.50

R3.CRS.Table 3			Training Model	Network test result	Auto PT-BEAM	Error
BEAM CONFIGURATION	1	b (mm)	R3.PTM. 1 b	565.56	565.56	0.00%
	2	h (mm)		700	700	0.00%
	3	f_{ck} (MPa)		40	40	0.00%
	4	f_{ci} (MPa)		20	20	0.00%
	5	ρ_{st} (mm^2/mm^2)	R3.CRS. 5 b	0.00920	0.00920	0.00%
	6	ρ_{sc} (mm^2/mm^2)	R3.CRS. 6 b	0.02442	0.02442	1.42E-16
	7	f_{sy} (MPa)		500	500	0.00%
	8	f_{sw} (MPa)		400	400	0.00%
	9	ϕ_S (mm)		16	16	0.00%
	10	ρ_p (mm^2/mm^2)	R3.CRS. 10 b	0.00295	0.00295	0.00%
	11	f_{py} (MPa)		1650	1650	0.00%
	12	ϕ_p (mm)		12.7	12.7	0.00%
	13	$PTloss_{lt}$		0.15	0.15	0.00%
	14	L (mm)		10000	10000	0.00%
	15	K_c		3	3	0.00%
TRANSFER LOAD STAGE	16	σ_c/f_{ck}	FW.PTM. 16 c	0.0004	0.00	Cracked
	17	ε_{st} (mm/mm)	FW.PTM. 17 c	-0.00046	-0.00047	-2.27%
	18	ε_{sc} (mm/mm)	FW.PTM. 18 c	0.0001196	0.0001264	-5.69%
	19	ε_p (mm/mm)	FW.PTM. 19 c	0.00664	0.00662	0.22%
	20	σ_{ct}/f_{ck}	FW.PTM. 20 c	0.45266	0.45806	-1.19%
	21	cw (mm)	FW.PTM. 21 c	0.016	0.017	-6.25%
	22	Δ (mm)	FW.PTM. 22 c	-10.062	-10.875	-8.08%
SERVICE LOAD COMBINATION	23	$q_{L/250}$ (kN/m)		**80.00**	**81.10**	**-1.38%**
	24	$q_{0.2mm}$ (kN/m)		**100.00**	**101.17**	**-1.17%**
	25	$q_{0.6fck}$ (kN/m)	FW.PTM. 25 c	114.10	115.17	-0.94%
	26	$q_{0.8fsu}$ (kN/m)	FW.PTM. 26 c	139.86	143.97	-2.94%
	27	$q_{0.75fpu}$ (kN/m)	FW.PTM. 27 c	91.53	100.43	-9.73%
ULTIMATE LOAD COMBINATION	28	q_{str} (kN/m)		**150.00**	**150.81**	**-0.54%**
	29	ε_{st} (mm/mm)	FW.PTM. 29 c	0.02220	0.02212	0.39%
	30	ε_{sc} (mm/mm)	FW.PTM. 30 c	-0.00119	-0.00119	0.02%
	31	ε_p (mm/mm)	FW.PTM. 31 c	0.02835	0.02847	-0.43%
	32	ρ_{sw} (mm^2/mm^2)	FW.PTM. 32 c	0.0025	0.0025	1.11%
	33	μ_Δ		**2.50**	**2.53**	**-1.00%**

Model **R3.CRS.b** is used

*μ_Δ between **1.75** and **2.75** show good accuracy.*

: Given Inputs for ANN R3 : Inputs for engineering mechanic

: Reverse Outputs calculated by reverse ANN model is used as inputs to calculate other outputs by forward ANN model.

: Reverse-forward inputs : Reverse-forward ouputs

(Continued)

and (a-2-2), design accuracies are not acceptable because the preassigned deflection ductility ratio (μ_Δ) of 1.5 is not within the range between 1.75 and 2.75, whereas design accuracies are acceptable when the preassigned deflection ductility ratio (μ_Δ) of 2.5 is within the range between 1.75 and 2.75 as shown in Tables 2.5.9(b-1-1) and (b-2-1).

Table 2.5.8 (Continued) Design accuracies of a reverse scenario based on CRS (DNN)

(d-3) Preassigning $q_{L/250}$ = 80 kN/m reaching a deflection of L/250 at SLL, $q_{0.2mm}$ = 100 kN/m at SLL, q_{str} = 150 kN/m at ULL based on deflection ductility ratios (μ_Δ) of 2.00, 2.50 at ULL; 40 layers-40 neurons for Step 1

(d-3-1) Deflection ductility ratios (μ_Δ) of 2.00

R3.CRS.Table 2			Training Model	Network test result	Auto PT-BEAM	Error
BEAM CONFIGURATION	1	b (mm)	R3.PTM. 1 c	756.38	756.38	0.00%
	2	h (mm)		700	700	0.00%
	3	f_{ck} (MPa)		40	40	0.00%
	4	f_{cki} (MPa)		20	20	0.00%
	5	ρ_{st} (mm²/mm²)	R3.CRS. 5 c	0.00703	0.00703	0.00%
	6	ρ_{sc} (mm²/mm²)	R3.CRS. 6 c	0.00570	0.00570	0.00%
	7	f_{sy} (MPa)		500	500	0.00%
	8	f_{sw} (MPa)		400	400	0.00%
	9	ϕ_S (mm)		16	16	0.00%
	10	ρ_p (mm²/mm²)	R3.CRS. 10 c	0.00236	0.00236	0.00%
	11	f_{py} (MPa)		1650	1650	0.00%
	12	ϕ_p (mm)		12.7	12.7	0.00%
	13	$PTloss_{lt}$		0.15	0.15	0.00%
	14	L (mm)		10000	10000	0.00%
	15	K_c		3	3	0.00%
TRANSFER LOAD STAGE	16	σ_c/f_{cki}	FW.PTM. 16 c	0.0005	0.00	Cracked
	17	ε_{st} (mm/mm)	FW.PTM. 17 c	-0.00041	-0.00043	-3.36%
	18	ε_{sc} (mm/mm)	FW.PTM. 18 c	0.0002302	0.0002558	-11.11%
	19	ε_p (mm/mm)	FW.PTM. 19 c	0.00668	0.00666	0.26%
	20	σ_{ct}/f_{cki}	FW.PTM. 20 c	0.42137	0.43324	-2.82%
	21	cw (mm)	FW.PTM. 21 c	0.033	0.037	-12.12%
	22	Δ (mm)	FW.PTM. 22 c	-9.328	-10.293	-10.35%
SERVICE LOAD COMBINATION	23	$q_{L/250}$ (kN/m)		**80.00**	**81.14**	**-1.43%**
	24	$q_{0.2mm}$ (kN/m)		**100.00**	**101.12**	**-1.12%**
	25	$q_{0.6fck}$ (kN/m)	FW.PTM. 25 c	108.35	109.95	-1.48%
	26	$q_{0.8fsu}$ (kN/m)	FW.PTM. 26 c	142.63	145.90	-2.29%
	27	$q_{0.75fpu}$ (kN/m)	FW.PTM. 27 c	97.38	101.44	-4.17%
ULTIMATE LOAD COMBINATION	28	q_{str} (kN/m)		**150.00**	**151.16**	**-0.77%**
	29	ε_{st} (mm/mm)	FW.PTM. 29 c	0.01277	0.01281	-0.31%
	30	ε_{sc} (mm/mm)	FW.PTM. 30 c	-0.00203	-0.00203	8.50E-05
	31	ε_p (mm/mm)	FW.PTM. 31 c	0.01883	0.01904	-1.13%
	32	ρ_{sw} (mm²/mm²)	FW.PTM. 32 c	0.0021	0.0021	-0.23%
	33	μ_Δ		**2.00**	**2.01**	**-0.49%**

Model **R3.CRS.c** is used

Significant errors

μ_Δ between 1.75 and 2.75 show good accuracy.

: Given Inputs for ANN R3   : Inputs for engineering mechanic

: Reverse Outputs calculated by reverse ANN model is used as inputs to calculate other outputs by forward ANN model.

: Reverse-forward inputs   : Reverse-forward ouputs

(Continued)

2.5.4.4.2 Table 2.5.9(b)

Tables 2.5.9(b-1) and (b-2) present ANNs for the reverse networks of Step 1 trained based on R3.CRS.b and R3.CRS.c using 30 layers-30 neurons and 40 layers-40 neurons, respectively, when deflection ductility ratios (μ_Δ) of 2.50 and 3.00 at ULL are reversely preassigned based on $q_{L/250}$ = 80 kN/m reaching a deflection of L/250 at SLL and $q_{0.2mm}$ = 100 kN/m at SLL and q_{str} = 150 kN/m at ULL. ANNs for the forward network of Step 2 trained

Table 2.5.8 (Continued) Design accuracies of a reverse scenario based on CRS (DNN)

(d-3-2) Deflection ductility ratios (μ_Δ) of 2.50

R3.CRS.Table 3			Training Model	Network test result	Auto PT-BEAM	Error
BEAM CONFIGURATION	1	b (mm)	R3.PTM. 1 c	572.65	572.65	0.00%
	2	h (mm)		700	700	0.00%
	3	f_{ck} (MPa)		40	40	0.00%
	4	f_{ci} (MPa)		20	20	0.00%
	5	ρ_{st} (mm^2/mm^2)	R3.CRS. 5 c	0.00922	0.00922	0.00%
	6	ρ_{sc} (mm^2/mm^2)	R3.CRS. 6 c	0.02462	0.02462	0.00%
	7	f_y (MPa)		500	500	0.00%
	8	f_{yw} (MPa)		400	400	0.00%
	9	ϕ_S (mm)		16	16	0.00%
	10	ρ_p (mm^2/mm^2)	R3.CRS. 10 c	0.00290	0.00290	0.00%
	11	f_{py} (MPa)		1650	1650	0.00%
	12	ϕ_p (mm)		12.7	12.7	0.00%
	13	$PTloss_{lt}$		0.15	0.15	0.00%
	14	L (mm)		10000	10000	0.00%
	15	K_c		3	3	0.00%
TRANSFER LOAD STAGE	16	σ_c/f_{ci}	FW.PTM. 16 c	0.0004	0.00	Cracked
	17	ε_{st} (mm/mm)	FW.PTM. 17 c	-0.00045	-0.00046	-2.19%
	18	ε_{sc} (mm/mm)	FW.PTM. 18 c	0.0001167	0.0001231	-5.48%
	19	ε_p (mm/mm)	FW.PTM. 19 c	0.00664	0.00663	0.21%
	20	σ_{ct}/f_{ci}	FW.PTM. 20 c	0.44533	0.45057	-1.18%
	21	cw (mm)	FW.PTM. 21 c	0.016	0.016	0.00%
	22	Δ (mm)	FW.PTM. 22 c	-9.896	-10.616	-7.27%
SERVICE LOAD COMBINATION	23	$q_{L/250}$ (kN/m)		**80.00**	**81.21**	**-1.51%**
	24	$q_{0.2mm}$ (kN/m)		**100.00**	**101.30**	**-1.30%**
	25	$q_{0.6fck}$ (kN/m)	FW.PTM. 25 c	115.33	116.27	-0.81%
	26	$q_{0.8fsu}$ (kN/m)	FW.PTM. 26 c	140.51	144.48	-2.83%
	27	$q_{0.75fpu}$ (kN/m)	FW.PTM. 27 c	91.98	100.56	-9.32%
ULTIMATE LOAD COMBINATION	28	q_{str} (kN/m)		**150.00**	**151.47**	**-0.98%**
	29	ε_{st} (mm/mm)	FW.PTM. 29 c	0.02241	0.02232	0.40%
	30	ε_{sc} (mm/mm)	FW.PTM. 30 c	-0.00117	-0.00117	0.04%
	31	ε_p (mm/mm)	FW.PTM. 31 c	0.02855	0.02868	-0.44%
	32	ρ_{sw} (mm^2/mm^2)	FW.PTM. 32 c	0.0025	0.0025	1.09%
	33	μ_Δ		**2.50**	**2.53**	**-1.39%**

Model **R3.CRS.c** is used

μ_Δ between ***1.75*** *and* ***2.75*** *show good accuracy.*

: Given Inputs for ANN R3 : Inputs for engineering mechanic

: Reverse Outputs calculated by reverse ANN model is used as inputs to calculate other outputs by forward ANN model.

: Reverse-forward inputs : Reverse-forward ouputs

based on FW.PTM.c using 40 layers and 40 neurons are used to calculate 18 output parameters in Boxes 16–33 of Tables 2.5.9(b-1) and (b-2). Reverse input parameters ($q_{L/250}$, $q_{0.2mm}$, q_{str}, and μ_Δ) are already reversely prescribed in Boxes 23, 24, 28, and 33 in Step 1, and hence, calculated reverse input parameters can be used to verify the networks used in BS. However, in Tables 2.5.9(b-1-2) and (b-2-2), design accuracies based on the preassigned

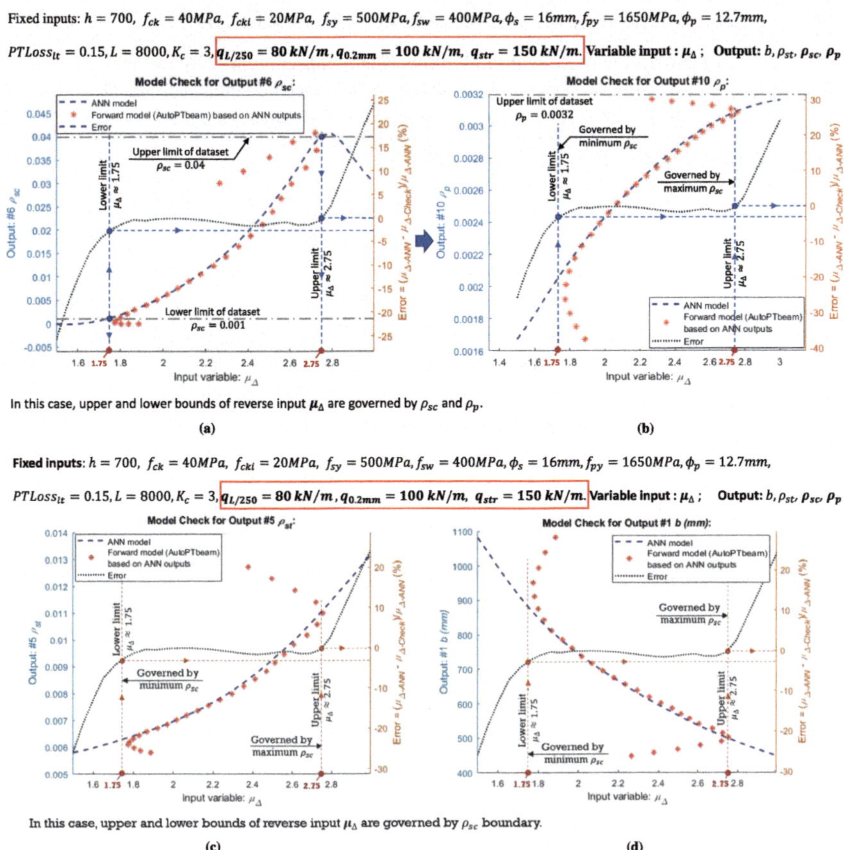

Figure 2.5.5 Reciprocal design charts, errors indicated. (a) ρ_{sc} vs. deflection ductility ratios (μ_Δ). (b) ρ_p vs. deflection ductility ratios (μ_Δ). (c) ρ_{st} vs. deflection ductility ratios (μ_Δ). (d) b vs. deflection ductility ratios (μ_Δ).

deflection ductility ratio (μ_Δ) of 3.00 are not improved as much as those with deflection ductility ratio (μ_Δ) of 2.50, whereas design accuracies based on the preassigned deflection ductility ratio (μ_Δ) of 2.50 are enhanced because the preassigned deflection ductility ratio (μ_Δ) of 2.5 is within the range between 1.75 and 2.75 as shown in Tables 2.5.9(b-1-1) and (b-2-1).

2.5.4.4.3 Table 2.5.9(c)

The ductility ratio (μ_Δ) of 3.00 placed outside the range between 1.75 and 2.75 is shown in Figure 2.5.5. Tables 2.5.9(c-1) and (c-2) present ANNs for the reverse networks of Step 1 trained based on R3.CRS.c and R3.CRS.b using 40 layers-40 neurons and 30 layers-30 neurons, respectively, when

Table 2.5.9 Design accuracies of reverse scenario based on CRS (DNN)

(a-1) Preassigning $q_{L/250}$ = 80 kN/m reaching a deflection of L/250 at SLL, $q_{0.2mm}$ = 100 kN/m at SLL, q_{str} = 150 kN/m at ULL based on deflection ductility ratios (μ_Δ) of 2.75, 1.50 at ULL; 30 layers-30 neurons for Step 1

(a-1-1) Deflection ductility ratios (μ_Δ) of 2.75

R3.CRS.Table 4a			Training Model	Network test result	Auto PT-BEAM check	Error
BEAM CONFIGURATION	1	b (mm)	R3.PTM. 1 b	499.52	499.52	0.00%
	2	h (mm)		700	700	0.00%
	3	f_{ck} (MPa)		40	40	0.00%
	4	f_{cki} (MPa)		20	20	0.00%
	5	ρ_{st} (mm^2/mm^2)	R3.CRS. 5 b	0.01101	0.01101	0.00%
	6	ρ_{sc} (mm^2/mm^2)	R3.CRS. 6 b	0.03880	0.03880	0.00%
	7	f_{y} (MPa)		500	500	0.00%
	8	f_{sw} (MPa)		400	400	0.00%
	9	ϕ_S (mm)		16	16	0.00%
	10	ρ_p (mm^2/mm^2)	R3.CRS. 10 b	0.00312	0.00312	0.00%
	11	f_{py} (MPa)		1650	1650	0.00%
	12	ϕ_p (mm)		12.7	12.7	0.00%
	13	$PTloss_{lt}$		0.15	0.15	0.00%
	14	L (mm)		10000	10000	0.00%
	15	K_c		3	3	0.00%
TRANSFER LOAD STAGE	16	σ_c/f_{cki}	FW.PTM. 16 c	0.0003	0.00	Cracked
	17	ε_{st} (mm/mm)	FW.PTM. 17 c	-0.00045	-0.00047	-2.89%
	18	ε_{sc} (mm/mm)	FW.PTM. 18 c	0.0000923	0.0000913	1.06%
	19	ε_p (mm/mm)	FW.PTM. 19 c	0.00664	0.00662	0.20%
	20	σ_{ct}/f_{cki}	FW.PTM. 20 c	0.44805	0.45351	-1.22%
	21	cw (mm)	FW.PTM. 21 c	0.014	0.011	21.43%
	22	Δ (mm)	FW.PTM. 22 c	-8.862	-10.500	-18.48%
SERVICE LOAD COMBINATION	23	$q_{L/250}$ (kN/m)		**80.00**	**80.52**	**-0.65%**
	24	$q_{0.2mm}$ (kN/m)		**100.00**	**100.86**	**-0.86%**
	25	$q_{0.6fck}$ (kN/m)	FW.PTM. 25 c	121.54	121.95	-0.34%
	26	$q_{0.8fpu}$ (kN/m)	FW.PTM. 26 c	136.55	143.72	-5.26%
	27	$q_{0.75fpu}$ (kN/m)	FW.PTM. 27 c	94.00	99.42	-5.76%
ULTIMATE LOAD COMBINATION	28	q_{str} (kN/m)		**150.00**	**151.00**	**-0.67%**
	29	ε_{st} (mm/mm)	FW.PTM. 29 c	0.02506	0.02504	0.09%
	30	ε_{sc} (mm/mm)	FW.PTM. 30 c	-0.00093	-0.00092	0.61%
	31	ε_p (mm/mm)	FW.PTM. 31 c	0.03107	0.03143	-1.16%
	32	ρ_{sw} (mm^2/mm^2)	FW.PTM. 32 c	0.0028	0.0028	0.42%
	33	μ_Δ		**2.75**	**2.72**	**1.22%**

Model **R3.CRS.b** is used

Big errors due to $\rho_{sc} < 0$

Preassigning μ_Δ = 2.75 gives big errors

: Given Inputs for ANN R3 : Inputs for engineering mechanic

: Reverse Outputs calculated by reverse ANN model is used as inputs to calculate other outputs by forward ANN model .

: Reverse-forward inputs : Reverse-forward ouputs

(Continued)

Table 2.5.9 (Continued) Design accuracies of reverse scenario based on CRS (DNN)

(a-1–2) Deflection ductility ratios (μ_Δ) of 1.50

R3.CRS.Table 1			Training Model	Network test result	Auto PT-BEAM check	Error
BEAM CONFIGURATION	1	b (mm)	R3.PTM. 1 b	1081.11	1081.11	0.00%
	2	h (mm)		700	700	0.00%
	3	f_{ck} (MPa)		40	40	0.00%
	4	f_{cki} (MPa)		20	20	0.00%
	5	ρ_{st} (mm²/mm²)	R3.CRS. 5 b	0.00614	0.00614	0.00%
	6	ρ_{sc} (mm²/mm²)	R3.CRS. 6 b	-0.00138	-0.00138	0.00%
	7	f_{sy} (MPa)		500	500	0.00%
	8	f_{sw} (MPa)		400	400	0.00%
	9	ϕ_S (mm)		16	16	0.00%
	10	ρ_p (mm²/mm²)	R3.CRS. 10 b	0.00165	0.00165	0.00%
	11	f_{py} (MPa)		1650	1650	0.00%
	12	ϕ_p (mm)		12.7	12.7	0.00%
	13	$PTloss_{lt}$		0.15	0.15	0.00%
	14	L (mm)		10000	10000	0.00%
	15	K_c		3	3	0.00%
TRANSFER LOAD STAGE	16	σ_c/f_{cki}	FW.PTM. 16 c	-0.0014	-0.84	Uncracked
	17	ε_{st} (mm/mm)	FW.PTM. 17 c	-0.00031	0.00040	227.52%
	18	ε_{sc} (mm/mm)	FW.PTM. 18 c	0.0005928	-0.0039231	761.76%
	19	ε_p (mm/mm)	FW.PTM. 19 c	0.00677	0.00756	-11.74%
	20	σ_{ct}/f_{cki}	FW.PTM. 20 c	0.35489	-0.64704	282.32%
	21	cw (mm)	FW.PTM. 21 c	0.144	229619.668	Inf
	22	Δ (mm)	FW.PTM. 22 c	-5.057	-8.572	-69.50%
SERVICE LOAD COMBINATION	23	$q_{L/250}$ (kN/m)		**80.00**	**83.49**	**-4.37%**
	24	$q_{0.2mm}$ (kN/m)		**100.00**	**103.53**	**-3.53%**
	25	$q_{0.6fck}$ (kN/m)	FW.PTM. 25 c	120.06	121.15	-0.91%
	26	$q_{0.8fsu}$ (kN/m)	FW.PTM. 26 c	154.67	156.69	-1.31%
	27	$q_{0.75fpu}$ (kN/m)	FW.PTM. 27 c	106.91	104.51	2.24%
ULTIMATE LOAD COMBINATION	28	q_{str} (kN/m)		**150.00**	**156.65**	**-4.43%**
	29	ε_{st} (mm/mm)	FW.PTM. 29 c	0.00871	0.00870	0.10%
	30	ε_{sc} (mm/mm)	FW.PTM. 30 c	-0.00241	-0.00240	0.34%
	31	ε_p (mm/mm)	FW.PTM. 31 c	0.01479	0.01488	-0.66%
	32	ρ_{sw} (mm²/mm²)	FW.PTM. 32 c	0.0017	0.0017	-0.31%
	33	μ_Δ		**1.50**	**1.66**	**-11.00%**

Model R3.CRS.b is used

Significant errors

Preassigning μ_Δ= *1.5* gives big errors

: Given Inputs for ANN R3 : Inputs for engineering mechanic

: Reverse Outputs calculated by reverse ANN model is used as inputs to calculate other outputs by forward ANN model .

: Reverse-forward inputs : Reverse-forward ouputs

(Continued)

the deflection ductility ratios (μ_Δ) of 1.75 at ULL for Table 2.5.9(c-1) and 2.00 at ULL and for Table 2.5.9(c-2) are reversely preassigned based on $q_{L/250}$ = 80 kN/m reaching a deflection of L/250 at SLL and $q_{0.2mm}$ = 100 kN/m at SLL, q_{str} = 150 kN/m at ULL. ANNs for the forward network of Step 2 trained based on FW.PTM.c using 40 layers and 40 neurons are implemented in calculating 18 output parameters in Boxes 16–33 as shown in Tables 2.5.9(c-1) and (c-2). Reverse input parameters ($q_{L/250}$, $q_{0.2mm}$, q_{str}, and μ_Δ) are already reversely prescribed in Boxes 23, 24, 28, and 33 in Step 1, and hence, calculated reverse input parameters can be used to verify the networks used in BS. However, as shown in Table 2.5.9(c-1), design

Table 2.5.9 (Continued) Design accuracies of reverse scenario based on CRS (DNN)

(a-2) Preassigning $q_{L/250}$ = 80 kN/m reaching a deflection of L/250 at SLL, $q_{0.2mm}$ = 100 kN/m at SLL, q_{str} = 150 kN/m at ULL based on deflection ductility ratios (μ_Δ) of 2.75, 1.50 at ULL; 40 layers-40 neurons for Step 1

(a-2-1) Deflection ductility ratios (μ_Δ) of 2.75

Model **R3.CRS.c** is used

	R3.CRS.Table 4a		Training Model		Network test result	Auto PT-BEAM check	Error
BEAM CONFIGURATION	1	b (mm)	R3.PTM.	1 c	502.35	502.35	0.00%
	2	h (mm)			700	700	0.00%
	3	f'_{ck} (MPa)			40	40	0.00%
	4	f'_{ci} (MPa)			20	20	0.00%
	5	ρ_{st} (mm²/mm²)	R3.CRS.	5 c	0.01092	0.01092	0.00%
	6	ρ_{sc} (mm²/mm²)	R3.CRS.	6 c	0.04007	0.04007	0.00%
	7	f_{sy} (MPa)			500	500	0.00%
	8	f_{sw} (MPa)			400	400	0.00%
	9	ϕ_s (mm)			16	16	0.00%
	10	ρ_p (mm²/mm²)	R3.CRS.	10 c	0.00310	0.00310	0.00%
	11	f_{py} (MPa)			1650	1650	0.00%
	12	ϕ_p (mm)			12.7	12.7	0.00%
	13	$PTloss_{lt}$			0.15	0.15	0.00%
	14	L (mm)			10000	10000	0.00%
	15	K_c			3	3	0.00%
TRANSFER LOAD STAGE	16	σ_c/f'_{ci}	FW.PTM.	16 c	0.0003	0.00	Cracked
	17	ε_{st} (mm/mm)	FW.PTM.	17 c	-0.00045	-0.00047	-2.94%
	18	ε_{sc} (mm/mm)	FW.PTM.	18 c	0.0000900	0.0000888	1.32%
	19	ε_p (mm/mm)	FW.PTM.	19 c	0.00664	0.00663	0.20%
	20	σ_{ct}/f'_{ci}	FW.PTM.	20 c	0.44692	0.45226	-1.19%
	21	cw (mm)	FW.PTM.	21 c	0.013	0.011	15.38%
	22	Δ (mm)	FW.PTM.	22 c	-8.790	-10.440	-18.77%
SERVICE LOAD COMBINATION	23	$q_{L/250}$ (kN/m)			**80.00**	**80.95**	**-1.18%**
	24	$q_{0.2mm}$ (kN/m)			**100.00**	**100.96**	**-0.96%**
	25	$q_{0.6fck}$ (kN/m)	FW.PTM.	25 c	123.53	123.59	-0.05%
	26	$q_{0.8fsu}$ (kN/m)	FW.PTM.	26 c	135.91	144.12	-6.04%
	27	$q_{0.75fpu}$ (kN/m)	FW.PTM.	27 c	94.55	99.55	-5.29%
ULTIMATE LOAD COMBINATION	28	q_{str} (kN/m)			**150.00**	**151.05**	**-0.70%**
	29	ε_{st} (mm/mm)	FW.PTM.	29 c	0.02540	0.02539	0.02%
	30	ε_{sc} (mm/mm)	FW.PTM.	30 c	-0.00089	-0.00089	0.52%
	31	ε_p (mm/mm)	FW.PTM.	31 c	0.03137	0.03179	-1.32%
	32	ρ_{sw} (mm²/mm²)	FW.PTM.	32 c	0.0028	0.0027	0.37%
	33	μ_Δ			**2.75**	**2.76**	**-0.54%**

Big errors due to $\rho_{sc} < 0$

: Given Inputs for ANN R3 : Inputs for engineering mechanic

: Reverse Outputs calculated by reverse ANN model is used as inputs to calculate other outputs by forward ANN model .

: Reverse-forward inputs : Reverse-forward ouputs

(Continued)

accuracies based on the ductility ratio (μ_Δ) of 1.75 are not enhanced as much as those with the ductility ratio (μ_Δ) of 2.00, whereas deign accuracies based on the preassigned deflection ductility ratio (μ_Δ) of 2.00 are improved because the ductility ratio (μ_Δ) of 2.00 is within the range between 1.75

Table 2.5.9 (Continued) Design accuracies of reverse scenario based on CRS (DNN)

(a-2-2) Deflection ductility ratios (μ_Δ) of 1.50

R3.CRS.Table 1			Training Model	Network test result	Auto PT-BEAM check	Error
BEAM CONFIGURATION	1	b (mm)	R3.PTM. 1 c	1019.45	1019.45	0.00%
	2	h (mm)		700	700	0.00%
	3	f_{ck} (MPa)		40	40	0.00%
	4	f_{cki} (MPa)		20	20	0.00%
	5	ρ_{st} (mm^2/mm^2)	R3.CRS. 5 c	0.00639	0.00639	0.00%
	6	ρ_{sc} (mm^2/mm^2)	R3.CRS. 6 c	-0.00071	-0.00071	0.00%
	7	f_{sy} (MPa)		500	500	0.00%
	8	f_{sw} (MPa)		400	400	0.00%
	9	ϕ_s (mm)		16	16	0.00%
	10	ρ_p (mm^2/mm^2)	R3.CRS. 10 c	0.00170	0.00170	0.00%
	11	f_{py} (MPa)		1650	1650	0.00%
	12	ϕ_p (mm)		12.7	12.7	0.00%
	13	$PTloss_{lt}$		0.15	0.15	0.00%
	14	L (mm)		10000	10000	0.00%
	15	K_c		3	3	0.00%
TRANSFER LOAD STAGE	16	σ_c/f_{cki}	FW.PTM. 16 c	-0.0006	0.00	Cracked
	17	ε_{st} (mm/mm)	FW.PTM. 17 c	-0.00032	-0.00030	7.26%
	18	ε_{sc} (mm/mm)	FW.PTM. 18 c	0.0005244	-0.0004491	185.64%
	19	ε_p (mm/mm)	FW.PTM. 19 c	0.00676	0.00680	-0.56%
	20	σ_{ct}/f_{cki}	FW.PTM. 20 c	0.35818	0.27887	22.14%
	21	cw (mm)	FW.PTM. 21 c	0.107	-0.111	203.74%
	22	Δ (mm)	FW.PTM. 22 c	-5.311	-64.014	-1105.27%
SERVICE LOAD COMBINATION	23	$q_{L/250}$ (kN/m)		**80.00**	**81.75**	**-2.19%**
	24	$q_{0.2mm}$ (kN/m)		**100.00**	**101.93**	**-1.93%**
	25	$q_{0.6fck}$ (kN/m)	FW.PTM. 25 c	116.78	117.90	-0.96%
	26	$q_{0.8fsu}$ (kN/m)	FW.PTM. 26 c	151.90	153.88	-1.30%
	27	$q_{0.75fpu}$ (kN/m)	FW.PTM. 27 c	104.79	102.74	1.95%
ULTIMATE LOAD COMBINATION	28	q_{str} (kN/m)		**150.00**	**154.76**	**-3.17%**
	29	ε_{st} (mm/mm)	FW.PTM. 29 c	0.00890	0.00890	-0.03%
	30	ε_{sc} (mm/mm)	FW.PTM. 30 c	-0.00238	-0.00238	0.20%
	31	ε_p (mm/mm)	FW.PTM. 31 c	0.01498	0.01509	-0.76%
	32	ρ_{sw} (mm^2/mm^2)	FW.PTM. 32 c	0.0018	0.0018	-0.78%
	33	μ_Δ		**1.50**	**1.70**	**-13.24%**

Model **R3.CRS.c** is used

Big errors due to $\rho_{sc} < 0$

: Given Inputs for ANN R3 : Inputs for engineering mechanic

: Reverse Outputs calculated by reverse ANN model is used as inputs to calculate other outputs by forward ANN model .

: Reverse-forward inputs : Reverse-forward ouputs

(*Continued*)

and 2.75 as shown in Table 2.5.9(c-2). It is noted that ductility ratio (μ_Δ) of 1.75 is placed on the boundary of the data region as shown in Table 2.5.9. Deign accuracies of Tables 2.5.9(c-1) and (c-2) sufficient for design applications are obtained for engineering applications except for lower tensile rebar strains (ε_{st}) and Δ shown in Boxes 18 and 22, respectively.

Table 2.5.9 (Continued) Design accuracies of reverse scenario based on CRS (DNN)

(b-1) Preassigning $q_{L/250}$ = 80 kN/m reaching a deflection of L/250 at SLL, $q_{0.2mm}$ = 100 kN/m at SLL, q_{str} = 150 kN/m at ULL based on deflection ductility ratios (μ_Δ) of 2.50, 3.00 at ULL; 30 layers-30 neurons for Step 1

(b-1-1) Deflection ductility ratios (μ_Δ) of 2.50

Model **R3.CRS.b** is used

R3.CRS.Table 3			Training Model	Network test result	Auto PT-BEAM check	Error
BEAM CONFIGURATION	1	b (mm)	R3.PTM. 1 b	565.56	565.56	0.00%
	2	h (mm)		700	700	0.00%
	3	f_{ck} (MPa)		40	40	0.00%
	4	f_{cki} (MPa)		20	20	0.00%
	5	ρ_{st} (mm^2/mm^2)	R3.CRS. 5 b	0.00920	0.00920	0.00%
	6	ρ_{sc} (mm^2/mm^2)	R3.CRS. 6 b	0.02442	0.02442	1.42E-16
	7	f_{sy} (MPa)		500	500	0.00%
	8	f_{sw} (MPa)		400	400	0.00%
	9	ϕ_s (mm)		16	16	0.00%
	10	ρ_p (mm^2/mm^2)	R3.CRS. 10 b	0.00295	0.00295	0.00%
	11	f_{py} (MPa)		1650	1650	0.00%
	12	ϕ_p (mm)		12.7	12.7	0.00%
	13	$PTloss_{lt}$		0.15	0.15	0.00%
	14	L (mm)		10000	10000	0.00%
	15	K_c		3	3	0.00%
TRANSFER LOAD STAGE	16	σ_c/f_{cki}	FW.PTM. 16 c	0.0004	0.00	Cracked
	17	ε_{st} (mm/mm)	FW.PTM. 17 c	-0.00046	-0.00047	-2.27%
	18	ε_{sc} (mm/mm)	FW.PTM. 18 c	0.0001196	0.0001264	-5.69%
	19	ε_p (mm/mm)	FW.PTM. 19 c	0.00664	0.00662	0.22%
	20	σ_{ct}/f_{cki}	FW.PTM. 20 c	0.45266	0.45806	-1.19%
	21	cw (mm)	FW.PTM. 21 c	0.016	0.017	-6.25%
	22	Δ (mm)	FW.PTM. 22 c	-10.062	-10.875	-8.08%
SERVICE LOAD COMBINATION	23	$q_{L/250}$ (kN/m)		**80.00**	**81.10**	**-1.38%**
	24	$q_{0.2mm}$ (kN/m)		**100.00**	**101.17**	**-1.17%**
	25	$q_{0.6fck}$ (kN/m)	FW.PTM. 25 c	114.10	115.17	-0.94%
	26	$q_{0.8fsu}$ (kN/m)	FW.PTM. 26 c	139.86	143.97	-2.94%
	27	$q_{0.75fpu}$ (kN/m)	FW.PTM. 27 c	91.53	100.43	-9.73%
ULTIMATE LOAD COMBINATION	28	q_{str} (kN/m)		**150.00**	**150.81**	**-0.54%**
	29	ε_{st} (mm/mm)	FW.PTM. 29 c	0.02220	0.02212	0.39%
	30	ε_{sc} (mm/mm)	FW.PTM. 30 c	-0.00119	-0.00119	0.02%
	31	ε_p (mm/mm)	FW.PTM. 31 c	0.02835	0.02847	-0.43%
	32	ρ_{sw} (mm^2/mm^2)	FW.PTM. 32 c	0.0025	0.0025	1.11%
	33	μ_Δ		**2.50**	**2.53**	**-1.00%**

: Given Inputs for ANN R3 ; : Inputs for engineering mechanic

: Reverse Outputs calculated by reverse ANN model is used as inputs to calculate other outputs by forward ANN model .

: Reverse-forward inputs ; : Reverse-forward ouputs

(Continued)

Table 2.5.9 (Continued) Design accuracies of reverse scenario based on CRS (DNN)

(b-1-2) Deflection ductility ratios (μ_Δ) of 3.00

Model **R3.CRS.b** is used

R3.CRS.Table 4			Training Model		Network test result	Auto PT-BEAM check	Error
BEAM CONFIGURATION	1	b (mm)	R3.PTM.	1 b	448.14	448.14	0.00%
	2	h (mm)			700	700	0.00%
	3	f_{ck} (MPa)			40	40	0.00%
	4	f_{cki} (MPa)			20	20	0.00%
	5	ρ_{st} (mm^2/mm^2)	R3.CRS.	5 b	0.01302	0.01302	-1.33E-16
	6	ρ_{sc} (mm^2/mm^2)	R3.CRS.	6 b	0.03704	0.03704	1.87E-16
	7	f_{sy} (MPa)			500	500	0.00%
	8	f_{sw} (MPa)			400	400	0.00%
	9	ϕ_S (mm)			16	16	0.00%
	10	ρ_p (mm^2/mm^2)	R3.CRS.	10 b	0.00324	0.00324	0.00%
	11	f_{py} (MPa)			1650	1650	0.00%
	12	ϕ_p (mm)			12.7	12.7	0.00%
	13	$PTloss_{lt}$			0.15	0.15	0.00%
	14	L (mm)			10000	10000	0.00%
	15	K_c			3	3	0.00%
TRANSFER LOAD STAGE	16	σ_c/f_{cki}	FW.PTM.	16 c	0.0003	0.00	Cracked
	17	ε_{st} (mm/mm)	FW.PTM.	17 c	-0.00045	-0.00047	-3.24%
	18	ε_{sc} (mm/mm)	FW.PTM.	18 c	0.0000957	0.0000942	1.51%
	19	ε_p (mm/mm)	FW.PTM.	19 c	0.00664	0.00663	0.19%
	20	σ_{ct}/f_{cki}	FW.PTM.	20 c	0.44528	0.45171	-1.44%
	21	cw (mm)	FW.PTM.	21 c	0.014	0.012	14.29%
	22	Δ (mm)	FW.PTM.	22 c	-8.804	-10.520	-19.49%
SERVICE LOAD COMBINATION	23	$q_{L/250}$ (kN/m)			**80.00**	**76.00**	**5.00%**
	24	$q_{0.2mm}$ (kN/m)			**100.00**	**99.43**	**0.57%**
	25	$q_{0.6fck}$ (kN/m)	FW.PTM.	25 c	111.93	112.16	-0.20%
	26	$q_{0.8fsu}$ (kN/m)	FW.PTM.	26 c	138.60	143.13	-3.27%
	27	$q_{0.75fpu}$ (kN/m)	FW.PTM.	27 c	92.30	97.25	-5.36%
ULTIMATE LOAD COMBINATION	28	q_{str} (kN/m)			**150.00**	**150.64**	**-0.43%**
	29	ε_{st} (mm/mm)	FW.PTM.	29 c	0.02333	0.02325	0.33%
	30	ε_{sc} (mm/mm)	FW.PTM.	30 c	-0.00109	-0.00108	0.79%
	31	ε_p (mm/mm)	FW.PTM.	31 c	0.02934	0.02962	-0.96%
	32	ρ_{sw} (mm^2/mm^2)	FW.PTM.	32 c	0.0031	0.0031	-0.03%
	33	μ_Δ			**3.00**	**2.46**	**17.85%**

Significant errors

: Given Inputs for ANN R3 : Inputs for engineering mechanic

: Reverse Outputs calculated by reverse ANN model is used as inputs to calculate other outputs by forward ANN model .

: Reverse-forward inputs : Reverse-forward ouputs

(Continued)

2.5.5 Design based on CRS using SNN

Table 2.5.10 presents training accuracies of reverse outputs (b, ρ_{st}, ρ_{sc}, and ρ_p) obtained by the reverse networks of Step 1 in which b is trained by PTMa, PTMb, and PTMc whereas ρst, ρsc, and ρp are trained by CRSa - CRSi as shown in table 2.5.10. DNN based on three types of deep layers and SNN based on six types of shallow layers are implemented. The sequence of the training deep networks for reverse outputs ρst, ρsc, and ρp using CRS is shown in Table 2.5.10. The first reverse output parameter, beam width b, is obtained in Box 1 of Table 2.5.11(a) based on R3.PTM.d. The rest of the reverse output parameters, ρ_{st}, ρ_{sc}, and ρ_p, are also obtained based on R3.CRS.d in Boxes 5, 6, and 10 of Table 2.5.11(a), respectively.

Table 2.5.9 (Continued) Design accuracies of reverse scenario based on CRS (DNN)

(b-2) Preassigning $q_{L/250}$ = 80 kN/m reaching a deflection of L/250 at SLL, $q_{0.2mm}$ = 100 kN/m at SLL, q_{str} = 150 kN/m at ULL based on deflection ductility ratios (μ_Δ) of 2.50, 3.00 at ULL; 40 layers-40 neurons for Step 1

(b-2-1) Deflection ductility ratios (μ_Δ) of 2.50

Model **R3.CRS.c** is used

R3.CRS.Table 3			Training Model	Network test result	Auto PT-BEAM check	Error
BEAM CONFIGURATION	1	b (mm)	R3.PTM. 1 c	572.65	572.65	0.00%
	2	h (mm)		700	700	0.00%
	3	f_{ck} (MPa)		40	40	0.00%
	4	f_{cki} (MPa)		20	20	0.00%
	5	ρ_{st} (mm²/mm²)	R3.CRS. 5 c	0.00922	0.00922	0.00%
	6	ρ_{sc} (mm²/mm²)	R3.CRS. 6 c	0.02462	0.02462	0.00%
	7	f_{sy} (MPa)		500	500	0.00%
	8	f_{sw} (MPa)		400	400	0.00%
	9	ϕ_S (mm)		16	16	0.00%
	10	ρ_p (mm²/mm²)	R3.CRS. 10 c	0.00290	0.00290	0.00%
	11	f_{py} (MPa)		1650	1650	0.00%
	12	ϕ_p (mm)		12.7	12.7	0.00%
	13	$PTloss_{lt}$		0.15	0.15	0.00%
	14	L (mm)		10000	10000	0.00%
	15	K_c		3	3	0.00%
TRANSFER LOAD STAGE	16	σ_c/f_{cki}	FW.PTM. 16 c	0.0004	0.00	Cracked
	17	ε_{st} (mm/mm)	FW.PTM. 17 c	-0.00045	-0.00046	-2.19%
	18	ε_{sc} (mm/mm)	FW.PTM. 18 c	0.0001167	0.0001231	-5.48%
	19	ε_p (mm/mm)	FW.PTM. 19 c	0.00664	0.00663	0.21%
	20	σ_{ct}/f_{cki}	FW.PTM. 20 c	0.44533	0.45057	-1.18%
	21	cw (mm)	FW.PTM. 21 c	0.016	0.016	0.00%
	22	Δ (mm)	FW.PTM. 22 c	-9.896	-10.616	-7.27%
SERVICE LOAD COMBINATION	23	$q_{L/250}$ (kN/m)		**80.00**	**81.21**	**-1.51%**
	24	$q_{0.2mm}$ (kN/m)		**100.00**	**101.30**	**-1.30%**
	25	$q_{0.6fck}$ (kN/m)	FW.PTM. 25 c	115.33	116.27	-0.81%
	26	$q_{0.8fsu}$ (kN/m)	FW.PTM. 26 c	140.51	144.48	-2.83%
	27	$q_{0.75fpu}$ (kN/m)	FW.PTM. 27 c	91.98	100.56	-9.32%
ULTIMATE LOAD COMBINATION	28	q_{str} (kN/m)		**150.00**	**151.47**	**-0.98%**
	29	ε_{st} (mm/mm)	FW.PTM. 29 c	0.02241	0.02232	0.40%
	30	ε_{sc} (mm/mm)	FW.PTM. 30 c	-0.00117	-0.00117	0.04%
	31	ε_p (mm/mm)	FW.PTM. 31 c	0.02855	0.02868	-0.44%
	32	ρ_{sw} (mm²/mm²)	FW.PTM. 32 c	0.0025	0.0025	1.09%
	33	μ_Δ		**2.50**	**2.53**	**-1.39%**

Legend:
- (red box) : Given Inputs for ANN R3
- (blue box) : Inputs for engineering mechanic
- (purple box) : Reverse Outputs calculated by reverse ANN model is used as inputs to calculate other outputs by forward ANN model .
- (green box) : Reverse-forward inputs
- (gray box) : Reverse-forward ouputs

(*Continued*)

Table 2.5.9 (Continued) Design accuracies of reverse scenario based on CRS (DNN)

(b-2-2) Deflection ductility ratios (μ_Δ) of 3.00

Model **R3.CRS.c** is used

R3.CRS.Table 4			Training Model	Network test result	Auto PT-BEAM check	Error
BEAM CONFIGURATION	1	b (mm)	R3.PTM. 1 c	445.93	445.93	0.00%
	2	h (mm)		700	700	0.00%
	3	f_{ck} (MPa)		40	40	0.00%
	4	f_{cki} (MPa)		20	20	0.00%
	5	ρ_{st} (mm^2/mm^2)	R3.CRS. 5 c	0.01286	0.01286	0.00%
	6	ρ_{sc} (mm^2/mm^2)	R3.CRS. 6 c	0.03272	0.03272	0.00%
	7	f_{sy} (MPa)		500	500	0.00%
	8	f_{sw} (MPa)		400	400	0.00%
	9	ϕ_S (mm)		16	16	0.00%
	10	ρ_p (mm^2/mm^2)	R3.CRS. 10 c	0.00327	0.00327	0.00%
	11	f_{py} (MPa)		1650	1650	0.00%
	12	ϕ_p (mm)		12.7	12.7	0.00%
	13	$PTloss_{lt}$		0.15	0.15	0.00%
	14	L (mm)		10000	10000	0.00%
	15	K_c		3	3	0.00%
TRANSFER LOAD STAGE	16	σ_c/f_{cki}	FW.PTM. 16 c	0.0003	0.00	Cracked
	17	ε_{st} (mm/mm)	FW.PTM. 17 c	-0.00046	-0.00047	-3.19%
	18	ε_{sc} (mm/mm)	FW.PTM. 18 c	0.0001045	0.0001051	-0.60%
	19	ε_p (mm/mm)	FW.PTM. 19 c	0.00663	0.00662	0.19%
	20	σ_{ct}/f_{cki}	FW.PTM. 20 c	0.45341	0.45950	-1.34%
	21	cw (mm)	FW.PTM. 21 c	0.015	0.013	13.33%
	22	Δ (mm)	FW.PTM. 22 c	-9.123	-10.846	-18.88%
SERVICE LOAD COMBINATION	23	$q_{L/250}$ (kN/m)		**80.00**	**74.89**	**6.39%**
	24	$q_{0.2mm}$ (kN/m)		**100.00**	**98.66**	**1.34%**
	25	$q_{0.6fck}$ (kN/m)	FW.PTM. 25 c	106.27	107.51	-1.17%
	26	$q_{0.8fsu}$ (kN/m)	FW.PTM. 26 c	138.72	141.79	-2.21%
	27	$q_{0.75fpu}$ (kN/m)	FW.PTM. 27 c	90.46	96.54	-6.72%
ULTIMATE LOAD COMBINATION	28	q_{ur} (kN/m)		**150.00**	**149.30**	**0.47%**
	29	ε_{st} (mm/mm)	FW.PTM. 29 c	0.02206	0.02196	0.47%
	30	ε_{sc} (mm/mm)	FW.PTM. 30 c	-0.00121	-0.00120	0.51%
	31	ε_p (mm/mm)	FW.PTM. 31 c	0.02813	0.02831	-0.65%
	32	ρ_{sw} (mm^2/mm^2)	FW.PTM. 32 c	0.0031	0.0031	0.29%
	33	μ_Δ		**3.00**	**2.36**	**21.40%**

Significant errors

: Given Inputs for ANN R3 : Inputs for engineering mechanic

: Reverse Outputs calculated by reverse ANN model is used as inputs to calculate other outputs by forward ANN model .

: Reverse-forward inputs : Reverse-forward ouputs

(Continued)

R3.CRS.a to R3.CRS.i using three types of deep and six types of shallow layers shown in Table 2.5.10 are tested to determine R3.CRS.d which produces better design accuracy, but its training accuracy is weaker than that of R3.CRS.c. R3.CRS.d produced the best design accuracies for ρ_{st}, and ρ_{sc}. In Table 2.5.10, deep ANNs trained by R3.PTM.a, R3.PTM.b, and R3.PTM.c are used to map 15 input parameters to the first reverse output parameter, beam width b, based on 20, 30, and 40 hidden layers with 20, 30, and 40 neurons. Shallow ANNs trained by R3.PTM.d to R3.PTM.i based on one and two hidden layers with 20, 30, and 40 neurons are also used to

Table 2.5.9 (Continued) Design accuracies of reverse scenario based on CRS (DNN)

(c) Preassigning $q_{L/250}$ = 80 kN/m reaching a deflection of L/250 at SLL, $q_{0.2mm}$ = 100 kN/m at SLL, q_{str} = 150 kN/m at ULL based on deflection ductility ratios (μ_Δ) of 1.75, 2.00 at ULL; 40 layers-40 neurons and 30 layers-30 neurons for Step 1

(c-1) Deflection ductility ratios (μ_Δ) of 1.75

R3.CRS.Table 1a			Training Model	Network test result	Auto PT-BEAM check	Error
BEAM CONFIGURATION	1	b (mm)	R3.PTM. 1 c	890.56	890.56	0.00%
	2	h (mm)		700	700	0.00%
	3	f_{ck} (MPa)		40	40	0.00%
	4	f_{cki} (MPa)		20	20	0.00%
	5	ρ_{st} (mm²/mm²)	R3.CRS. 5 c	0.00649	0.00649	0.00%
	6	ρ_{sc} (mm²/mm²)	R3.CRS. 6 c	0.00095	0.00095	0.00%
	7	f_{sy} (MPa)		500	500	0.00%
	8	f_{sw} (MPa)		400	400	0.00%
	9	ϕ_S (mm)		16	16	0.00%
	10	ρ_p (mm²/mm²)	R3.CRS. 10 c	0.00200	0.00200	0.00%
	11	f_{py} (MPa)		1650	1650	0.00%
	12	ϕ_P (mm)		12.7	12.7	0.00%
	13	$PTloss_{lt}$		0.15	0.15	0.00%
	14	L (mm)		10000	10000	0.00%
	15	K_c		3	3	0.00%
TRANSFER LOAD STAGE	16	σ_c/f_{cki}	FW.PTM. 16 c	0.0005	0.00	Cracked
	17	ε_{st} (mm/mm)	FW.PTM. 17 c	-0.00038	-0.00040	-6.72%
	18	ε_{sc} (mm/mm)	FW.PTM. 18 c	0.0004018	0.0004238	-5.46%
	19	ε_p (mm/mm)	FW.PTM. 19 c	0.00671	0.00668	0.42%
	20	σ_{ct}/f_{cki}	FW.PTM. 20 c	0.40338	0.42626	-5.67%
	21	cw (mm)	FW.PTM. 21 c	0.066	0.099	-50.00%
	22	Δ (mm)	FW.PTM. 22 c	-7.055	-8.671	-22.90%
SERVICE LOAD COMBINATION	23	$q_{L/250}$ (kN/m)		**80.00**	**81.27**	**-1.58%**
	24	$q_{0.2mm}$ (kN/m)		**100.00**	**101.23**	**-1.23%**
	25	$q_{0.6fck}$ (kN/m)	FW.PTM. 25 c	110.73	112.25	-1.37%
	26	$q_{0.8fpu}$ (kN/m)	FW.PTM. 26 c	146.58	148.90	-1.58%
	27	$q_{0.75fpu}$ (kN/m)	FW.PTM. 27 c	102.08	101.90	0.18%
ULTIMATE LOAD COMBINATION	28	q_{str} (kN/m)		**150.00**	**150.70**	**-0.47%**
	29	ε_{st} (mm/mm)	FW.PTM. 29 c	0.00941	0.00944	-0.37%
	30	ε_{sc} (mm/mm)	FW.PTM. 30 c	-0.00233	-0.00233	1.06E-04
	31	ε_p (mm/mm)	FW.PTM. 31 c	0.01545	0.01564	-1.21%
	32	ρ_{sw} (mm²/mm²)	FW.PTM. 32 c	0.0019	0.0019	-1.28%
	33	μ_Δ		**1.75**	**1.78**	**-1.47%**

Model **R3.CRS.c** is used

Significant errors

μ_Δ of 1.75 is on the boundary of the big datasets

: Given Inputs for ANN R3 : Inputs for engineering mechanic

: Reverse Outputs calculated by reverse ANN model is used as inputs to calculate other outputs by forward ANN model.

: Reverse-forward inputs : Reverse-forward ouputs

(Continued)

map 15 input parameters to beam width b. Similarly, input parameters are mapped to the rest of the reverse output parameters, ρ_{st}, ρ_{sc}, and ρ_p, by deep ANNs trained by R3.CRS.a, R3.CRS.b, and R3.CRS.c based on 20, 30, and 40 hidden layers –20, 30, and 40 neurons, respectively, whereas shallow ANNs trained by R3.CRS.d to R3.CRS.i based on one and two hidden layers –20, 30, and 40 neurons. It is noted that, for example, R3.CRS.6e of reverse network of Step 1 of in Table 2.5.10 maps 17 input parameters (h, fck, f_{cki}, f_{sy}, f_{sw}, ϕ_S, f_{py}, ϕ_P, $PTloss_{lt}$, L, K_c, $q_{L/250}$, $q_{0.2mm}$, $qstr$, μ_Δ, b, ρst) to ρ_{sc} shown in Box 6 of Table 2.5.11 using 1 hidden layer-30 neurons,

Table 2.5.9 (Continued) Design accuracies of reverse scenario based on CRS (DNN)

(c-2) Deflection ductility ratios (μ_Δ) of 2.00

R3.CRS.Table 2			Training Model		Network test result	Auto PT-BEAM check	Error
BEAM CONFIGURATION	1	b (mm)	R3.PTM.	1 b	744.91	744.91	0.00%
	2	h (mm)			700	700	0.00%
	3	f_{ck} (MPa)			40	40	0.00%
	4	f_{cki} (MPa)			20	20	0.00%
	5	ρ_{st} (mm²/mm²)	R3.CRS.	5 b	0.00681	0.00681	0.00%
	6	ρ_{sc} (mm²/mm²)	R3.CRS.	6 b	0.00560	0.00560	0.00%
	7	f_{sy} (MPa)			500	500	0.00%
	8	f_{sw} (MPa)			400	400	0.00%
	9	ϕ_S (mm)			16	16	0.00%
	10	ρ_p (mm²/mm²)	R3.CRS.	10 b	0.00246	0.00246	1.76E-16
	11	f_{py} (MPa)			1650	1650	0.00%
	12	ϕ_p (mm)			12.7	12.7	0.00%
	13	$PTloss_{lt}$			0.15	0.15	0.00%
	14	L (mm)			10000	10000	0.00%
	15	K_c			3	3	0.00%
TRANSFER LOAD STAGE	16	σ_c/f_{cki}	FW.PTM. 16 c		0.0005	0.00	Cracked
	17	ε_{st} (mm/mm)	FW.PTM. 17 c		-0.00044	-0.00045	-3.33%
	18	ε_{sc} (mm/mm)	FW.PTM. 18 c		0.0002450	0.0002717	-10.92%
	19	ε_p (mm/mm)	FW.PTM. 19 c		0.00666	0.00664	0.28%
	20	σ_{ct}/f_{cki}	FW.PTM. 20 c		0.44144	0.45387	-2.82%
	21	cw (mm)	FW.PTM. 21 c		0.036	0.039	-8.33%
	22	Δ (mm)	FW.PTM. 22 c		-10.233	-11.295	-10.38%
SERVICE LOAD COMBINATION	23	$q_{L/250}$ (kN/m)			**80.00**	**81.97**	**-2.46%**
	24	$q_{0.2mm}$ (kN/m)			**100.00**	**101.58**	**-1.58%**
	25	$q_{0.6fck}$ (kN/m)	FW.PTM. 25 c		107.34	109.21	-1.74%
	26	$q_{0.8fsu}$ (kN/m)	FW.PTM. 26 c		141.55	145.28	-2.64%
	27	$q_{0.75fpu}$ (kN/m)	FW.PTM. 27 c		97.14	101.98	-4.99%
ULTIMATE LOAD COMBINATION	28	q_{str} (kN/m)			**150.00**	**150.00**	**0.00%**
	29	ε_{st} (mm/mm)	FW.PTM. 29 c		0.01249	0.01256	-0.50%
	30	ε_{sc} (mm/mm)	FW.PTM. 30 c		-0.00205	-0.00205	0.07%
	31	ε_p (mm/mm)	FW.PTM. 31 c		0.01854	0.01879	-1.38%
	32	ρ_{sw} (mm²/mm²)	FW.PTM. 32 c		0.0021	0.0021	-0.21%
	33	μ_Δ			**2.00**	**2.00**	**-0.14%**

Model **R3.CRS.b** is used

Significant errors

μ_Δ between 1.75 and 2.75 show good accuracy.

: Given Inputs for ANN R3 : Inputs for engineering mechanic

: Reverse Outputs calculated by reverse ANN model is used as inputs to calculate other outputs by forward ANN model .

: Reverse-forward inputs : Reverse-forward ouputs

whereas ANNs with FW.PTM.c are implemented in determining 18 output parameters of the forward network of Step 2 shown in Boxes 16–33. A dataset representing 15% of the total large datasets that are not used to train networks is used to verify training accuracies even though users can provide a different dataset ratio for verifying training accuracies. The influence of a number of hidden layers and neurons implemented in reverse networks of Step 1 on design accuracies obtained in forward networks of Step 2 is investigated in Tables 2.5.11(a)–(f). Table 2.5.11(a), (b), and (c) uses one layer with 20, 30, and 40 neurons, respectively whereas Table 2.5.11(d), (e), and (f) uses two layer with 20, 30, and 40 neurons, respectively. Design accuracies of beam width b similar to those obtained by deep ANNs trained by R3.PTM.a, R3.PTM.b, and R3.PTM.c are obtained when using shallow ANNs trained by R3.CRS.d to R3.CRS.i,

Table 2.5.10 Training accuracies of a reverse scenario based on deep and shallow ANNs; training accuracies obtained by mapping *b* based on PTM and ρ_{st}, ρ_{sc}, and ρ_p based on CRS

REVERSE NETWORK OF STEP 1-REVERSE SCENARIO 3 **15 Inputs** (h, f_{ck}, f_{cki}, f_{sy}, f_{sw}, ϕ_S, f_{py}, ϕ_P, $PTloss_{lt}$, L, K_c, $q_{L/250}$, $q_{0.2mm}$, q_{str}, μ_Δ) – (4) **Outputs** (b, ρst, ρsc, ρ_p)												
No.			**Data**	**Layers**	**Neurons**	**Validation check**	**Required epoch**	**Best epoch for training**	**Stopped epoch**	**Mean square error**	**R at best epoch**	*Comments*
15 Inputs (h, fck, f_{cki}, f_{sy}, f_{sw}, ϕ_S, f_{py}, ϕ_P, $PTloss_{lt}$, L, K_c, $q_{L/250}$, $q_{0.2mm}$, $qstr$, μ_Δ) –1 Output (b) - **PTM**												
R3.PTM.	**1**	**a**	50,000	20	20	500	50,000	31,407	31,907	1.1E-03	1.1E-03	0.99034
R3.PTM.	**1**	**b**	50,000	30	30	1,000	70,000	67,507	68,507	7.1E-04	7.3E-04	0.99425
R3.PTM.	**1**	**c**	50,000	40	40	5,000	100,000	55,324	60,324	7.6E-04	7.5E-04	0.99458
R3.PTM.	**1**	**d**	50,000	1	20	500	100,000	23,555	24,055	3.2E-03	3.3E-03	0.96889
R3.PTM.	**1**	**e**	50,000	1	30	500	100,000	27,598	28,098	2.8E-03	2.8E-03	0.97271
R3.PTM.	**1**	**f**	50,000	1	40	500	100,000	34,205	34,705	2.8E-03	2.7E-03	0.9747
R3.PTM.	**1**	**g**	50,000	2	20	500	100,000	55,209	55,709	1.2E-03	1.2E-03	0.98866
R3.PTM.	**1**	**h**	50,000	2	30	500	100,000	23,228	23,728	1.2E-03	1.1E-03	0.98954
R3.PTM.	**1**	**i**	50,000	2	40	500	100,000	31,943	32,443	1.1E-03	1.2E-03	0.99055
16 Inputs (h, f_{ck}, f_{cki}, f_{sy}, f_{sw}, ϕ_S, f_{py}, ϕ_P, $PTloss_{lt}$, L, K_c, $q_{L/250}$, $q_{0.2mm}$, q_{str}, μ_Δ, b) –1 Output (ρ_{st}) - **CRS**												
R3.CRS.	**5**	**a**	50,000	20	20	500	50,000	24,914	25,414	1.7.E-04	1.7.E-04	0.99908
R3.CRS.	**5**	**b**	50,000	30	30	1,000	100,000	40,601	41,601	9.9.E-05	1.1.E-04	0.99947
R3.CRS.	**5**	**c**	50,000	40	40	5,000	100,000	51,497	56,497	7.3.E-05	9.1.E-05	0.99969

(Continued)

Table 2.5.10 (Continued) Training accuracies of a reverse scenario based on deep and shallow ANNs; training accuracies obtained by mapping *b* based on PTM and ρ_{st}, ρ_{sc}, and ρ_p based on CRS

No.			Data	Layers	Neurons	Validation check	Required epoch	Best epoch for training	Stopped epoch	Mean square error	R at best epoch	Comments
REVERSE NETWORK OF STEP 1-REVERSE SCENARIO 3												
15 Inputs (h, f_{ck}, f_{cki}, f_{sy}, f_{sw}, ϕ_S, f_{py}, ϕ_P, $PTloss_{lt}$, L, K_c, $q_{L/250}$, $q_{0.2mm}$, q_{str}, μ_Δ) – (4) Outputs (b, ρst, ρsc, ρ_p)												
R3.CRS.	**5**	**d**	50,000	1	20	500	100,000	50,946	51,446	4.1.E-04	3.9.E-04	0.99756
R3.CRS.	**5**	**e**	50,000	1	30	500	100,000	36,977	37,477	3.3.E-04	3.3.E-04	0.99809
R3.CRS.	**5**	**f**	50,000	1	40	500	100,000	26,860	27,360	2.9.E-04	3.0.E-04	0.99824
R3.CRS.	**5**	**g**	50,000	2	20	500	100,000	29,700	30,200	2.2.E-04	2.2.E-04	0.99875
R3.CRS.	**5**	**h**	50,000	2	30	500	100,000	14,328	14,828	1.9.E-04	1.9.E-04	0.99891
R3.CRS.	**5**	**i**	50,000	2	40	500	100,000	6,637	7,137	2.6.E-04	2.7.E-04	0.99855
17 Inputs (h, f_{ck}, f_{cki}, f_{sy}, f_{sw}, ϕ_S, f_{py}, ϕ_P, $PTloss_{lt}$, L, K_c, $q_{L/250}$, $q_{0.2mm}$, q_{str}, μ_Δ, b, ρ_{st}) –1 Output (ρ_{sc}) - **CRS**												
R3.CRS.	**6**	**a**	50,000	20	20	500	50,000	26,261	26,761	2.3.E-04	2.3.E-04	0.99835
R3.CRS.	**6**	**b**	50,000	30	30	1,000	100,000	35,490	36,490	1.2.E-04	1.2.E-04	0.99919
R3.CRS.	**6**	**c**	50,000	40	40	5,000	100,000	99,850	100,000	3.2.E-05	3.2.E-05	0.99982
R3.CRS.	**6**	**d**	50,000	1	20	500	100,000	43,054	43,554	3.1.E-04	3.3.E-04	0.99776
R3.CRS.	**6**	**e**	50,000	1	30	500	100,000	28,699	29,199	2.7.E-04	2.8.E-04	0.99806
R3.CRS.	**6**	**f**	50,000	1	40	500	100,000	54,119	54,619	1.4.E-04	1.5.E-04	0.99902

(Continued)

Table 2.5.10 (Continued) Training accuracies of a reverse scenario based on deep and shallow ANNs; training accuracies obtained by mapping b based on PTM and ρ_{st}, ρ_{sc}, and ρ_p based on CRS

No.			Data	Layers	Neurons	Validation check	Required epoch	Best epoch for training	Stopped epoch	Mean square error	R at best epoch	Comments
REVERSE NETWORK OF STEP 1-REVERSE SCENARIO 3												
15 Inputs **(**h, f_{ck}, f_{cki}, f_{sy}, f_{sw}, ϕ_S, f_{py}, ϕ_P, $PTloss_{lt}$, L, K_c, $q_{L/250}$, $q_{0.2mm}$, q_{str}, μ_Δ) – (4) *Outputs* (b, ρst, ρsc, ρ_p)												
R3.CRS.	**6**	**g**	50,000	2	20	500	100,000	42,064	42,564	1.4.E-04	1.5.E-04	0.99901
R3.CRS.	**6**	**h**	50,000	2	30	500	100,000	21,752	22,252	1.1.E-04	1.3.E-04	0.99921
R3.CRS.	**6**	**i**	50,000	2	40	500	100,000	42,061	42,561	4.1.E-05	4.8.E-05	0.99972
18 Inputs (h, f_{ck}, f_{cki}, f_{sy}, f_{sw}, ϕ_S, f_{py}, ϕ_P, $PTloss_{lt}$, L, K_c, $q_{L/250}$, $q_{0.2mm}$, q_{str}, μ_Δ, b, ρ_{st}, ρ_{sc}) –1 Output (ρ_p) - **CRS**												
R3.CRS.	**10**	**a**	50,000	20	20	500	50,000	5,096	5,596	2.4.E-03	2.2.E-03	0.98852
R3.CRS.	**10**	**b**	50,000	30	30	1,000	100,000	7,630	8,630	2.5.E-03	2.2.E-03	0.98743
R3.CRS.	**10**	**c**	50,000	40	40	5,000	100,000	7,887	12,887	2.5.E-03	3.0.E-03	0.98848
R3.CRS.	**10**	**d**	50,000	1	20	500	50,000	14,455	14,955	3.1.E-03	3.0.E-03	0.982
R3.CRS.	**10**	**e**	50,000	1	30	500	100,000	58,525	59,025	1.3.E-03	1.5.E-03	0.99171
R3.CRS.	**10**	**f**	50,000	1	40	500	100,000	23,233	23,733	1.2.E-03	1.4.E-03	0.99306
R3.CRS.	**10**	**g**	50,000	2	20	500	50,000	2,527	3,027	2.7.E-03	2.8.E-03	0.98471
R3.CRS.	**10**	**h**	50,000	2	30	500	100,000	12,373	12,873	1.1.E-03	1.2.E-03	0.99442
R3.CRS.	**10**	**i**	50,000	2	40	500	100,000	3,959	4,459	2.5.E-03	2.5.E-03	0.99007

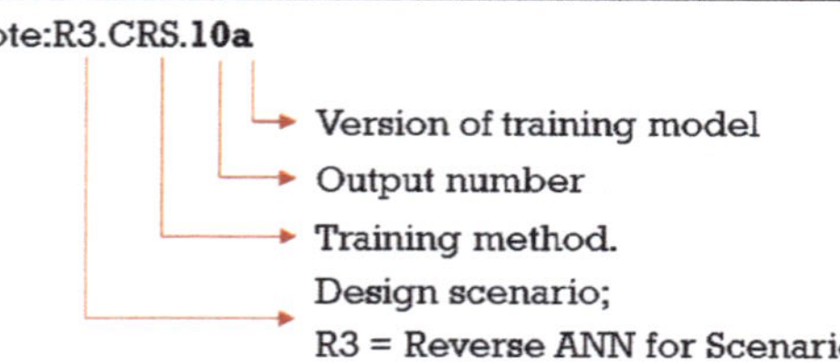

Table 2.5.11 Influence of a number of layers and neurons on the design accuracies of Reverse Scenario 3

(a) Design accuracies obtained based on 1 layer and 20 neurons

(a-1) Deflection ductility ratios (μ_Δ) of 2.00

		PARAMETERS	Training Model	Network test result	Auto PT-BEAM check	Error
BEAM CONFIGURATION	1	b (mm)	R3.PTM. 1 d	751.74	751.74	0.00%
	2	h (mm)		700	700	0.00%
	3	f_{ck} (MPa)		40	40	0.00%
	4	f_{cki} (MPa)		20	20	0.00%
	5	ρ_{st} (mm²/mm²)	R3.CRS. 5 d	0.00635	0.00635	0.00%
	6	ρ_{sc} (mm²/mm²)	R3.CRS. 6 d	0.00432	0.00432	0.00%
	7	f_{sy} (MPa)		500	500	0.00%
	8	f_{sw} (MPa)		400	400	0.00%
	9	ϕ_S (mm)		16	16	0.00%
	10	ρ_p (mm²/mm²)	R3.CRS. 10 d	0.00253	0.00253	0.00%
	11	f_{py} (MPa)		1650	1650	0.00%
	12	ϕ_p (mm)		12.7	12.7	0.00%
	13	$PTloss_{lt}$		0.15	0.15	0.00%
	14	L (mm)		10000	10000	0.00%
	15	K_e		3	3	0.00%
TRANSFER LOAD STAGE	16	σ_c/f_{cki}	FW.PTM. 16 c	0.0005	0.00	Cracked
	17	ε_{st} (mm/mm)	FW.PTM. 17 c	-0.00047	-0.00048	-3.25%
	18	ε_{sc} (mm/mm)	FW.PTM. 18 c	0.0002908	0.0003207	-10.29%
	19	ε_p (mm/mm)	FW.PTM. 19 c	0.00663	0.00661	0.32%
	20	σ_{ct}/f_{cki}	FW.PTM. 20 c	0.46918	0.48161	-2.65%
	21	cw (mm)	FW.PTM. 21 c	0.045	0.046	-2.22%
	22	Δ (mm)	FW.PTM. 22 c	-11.151	-12.570	-12.73%
SERVICE LOAD COMBINATION	23	$q_{L/250}$ (kN/m)		**80.00**	**83.68**	**-4.60%**
	24	$q_{0.2mm}$ (kN/m)		**100.00**	**102.16**	**-2.16%**
	25	$q_{0.6fck}$ (kN/m)	FW.PTM. 25 c	106.03	108.32	-2.16%
	26	$q_{0.8fsu}$ (kN/m)	FW.PTM. 26 c	140.51	144.70	-2.99%
	27	$q_{0.75fpu}$ (kN/m)	FW.PTM. 27 c	97.69	102.77	-5.20%
ULTIMATE LOAD COMBINATION	28	q_{str} (kN/m)		**150.00**	**148.01**	**1.32%**
	29	ε_{st} (mm/mm)	FW.PTM. 29 c	0.01124	0.01140	-1.43%
	30	ε_{sc} (mm/mm)	FW.PTM. 30 c	-0.00216	-0.00215	0.36%
	31	ε_p (mm/mm)	FW.PTM. 31 c	0.01723	0.01762	-2.28%
	32	ρ_{sw} (mm²/mm²)	FW.PTM. 32 c	0.0021	0.0021	-0.45%
	33	μ_Δ		**2.00**	**1.93**	**3.58%**

Model **R3.CRS.d** is used

Significant errors

μ_Δ between 1.75 and 2.75 show good accuracy.

: Given Inputs for ANN R3 : Inputs for engineering mechanic

: Reverse Outputs calculated by reverse ANN model is used as inputs to calculate other outputs by forward ANN model .

: Reverse-forward inputs : Reverse-forward ouputs

(Continued)

implying that ANNs trained using the CRS method in the reverse network of Step 1 can be used to accurately determine reverse outputs (b, ρ_{st}, ρ_{sc}, and ρ_p) with both deep and shallow layers when enough neurons are used. However, significant errors shown in Table 2.5.11 are caused when preassigning a deflection ductility ratio (μ_Δ) out of the ranges between 1.7 and 2.75. The reverse input parameters should be preassigned within the ranges between 1.7 and 2.75 to enhance design accuracies. In Table 2.5.11, design accuracies are compared with those calculated using structural calculations. Tables 2.5.11(a)–(f) where a deflection ductility ratio (μ_Δ) of 2.00 is reversely preassigned on an input-side show acceptable design accuracies.

Table 2.5.11 (Continued) Influence of a number of layers and neurons on the design accuracies of Reverse Scenario 3

(a-2) Deflection ductility ratios (μ_Δ) of 2.50

Model **R3.CRS.d** is used

		PARAMETERS	Training Model		Network test result	Auto PT-BEAM check	Error
BEAM CONFIGURATION	1	b (mm)	R3.PTM.	1 d	665.27	665.27	0.00%
	2	h (mm)			700	700	0.00%
	3	f_{ck} (MPa)			40	40	0.00%
	4	f_{cki} (MPa)			20	20	0.00%
	5	ρ_{st} (mm^2/mm^2)	R3.CRS.	5 d	0.00834	0.00834	0.00%
	6	ρ_{sc} (mm^2/mm^2)	R3.CRS.	6 d	0.01942	0.01942	0.00%
	7	f_{sy} (MPa)			500	500	0.00%
	8	f_{sw} (MPa)			400	400	0.00%
	9	ϕ_S (mm)			16	16	0.00%
	10	ρ_p (mm^2/mm^2)	R3.CRS.	10 d	0.00252	0.00252	0.00%
	11	f_{py} (MPa)			1650	1650	0.00%
	12	ϕ_p (mm)			12.7	12.7	0.00%
	13	$PTloss_{lt}$			0.15	0.15	0.00%
	14	L (mm)			10000	10000	0.00%
	15	K_c			3	3	0.00%
TRANSFER LOAD STAGE	16	σ_c/f_{cki}	FW.PTM. 16 c		0.0005	0.00	Cracked
	17	ε_{st} (mm/mm)	FW.PTM. 17 c		-0.00040	-0.00041	-2.34%
	18	ε_{sc} (mm/mm)	FW.PTM. 18 c		0.0001181	0.0001261	-6.76%
	19	ε_p (mm/mm)	FW.PTM. 19 c		0.00669	0.00668	0.19%
	20	σ_{ct}/f_{cki}	FW.PTM. 20 c		0.40447	0.40993	-1.35%
	21	cw (mm)	FW.PTM. 21 c		0.016	0.016	0.00%
	22	Δ (mm)	FW.PTM. 22 c		-9.103	-9.090	0.14%
SERVICE LOAD COMBINATION	23	$q_{L/250}$ (kN/m)			**80.00**	**82.39**	**-2.98%**
	24	$q_{0.2mm}$ (kN/m)			**100.00**	**102.03**	**-2.03%**
	25	$q_{0.6fck}$ (kN/m)	FW.PTM. 25 c		120.27	120.54	-0.22%
	26	$q_{0.8fsu}$ (kN/m)	FW.PTM. 26 c		144.63	147.33	-1.86%
	27	$q_{0.75fpu}$ (kN/m)	FW.PTM. 27 c		94.91	101.72	-7.17%
ULTIMATE LOAD COMBINATION	28	q_{ur} (kN/m)			**150.00**	**154.90**	**-3.27%**
	29	ε_{st} (mm/mm)	FW.PTM. 29 c		0.02182	0.02176	0.28%
	30	ε_{sc} (mm/mm)	FW.PTM. 30 c		-0.00122	-0.00122	-0.23%
	31	ε_p (mm/mm)	FW.PTM. 31 c		0.02796	0.02811	-0.54%
	32	ρ_{sw} (mm^2/mm^2)	FW.PTM. 32 c		0.0022	0.0022	0.59%
	33	μ_Δ			**2.50**	**2.53**	**-1.18%**

μ_Δ between 1.75 and 2.75 show good accuracy.

: Given Inputs for ANN R3 : Inputs for engineering mechanic

: Reverse Outputs calculated by reverse ANN model is used as inputs to calculate other outputs by forward ANN model .

: Reverse-forward inputs : Reverse-forward ouputs

(*Continued*)

Error ranges −10.77% to −11.68% and −9.46% to −16.62% for upper rebar strains (ε_{sc}) and deflection (camber, c_w) are observed, respectively, at transfer load limit state by a reverse design scenario based on shallow ANNs trained by R3.PTM.d, R3.CRS.e, R3.CRS.f using 1 layer-20 neurons as shown in Tables 2.5.11(a)–(f), respectively, even when using the deflection

Table 2.5.11 (Continued) Influence of a number of layers and neurons on the design accuracies of Reverse Scenario 3

(b) Design accuracies obtained based on 1 layer and 30 neurons

(b-1) Deflection ductility ratios (μ_Δ) of 2.00

		PARAMETERS	Training Model	Network test result	Auto PT-BEAM check	Error
BEAM CONFIGURATION	1	b (mm)	R3.PTM. 1 e	729.79	729.79	0.00%
	2	h (mm)		700	700	0.00%
	3	f_{ck} (MPa)		40	40	0.00%
	4	f_{cit} (MPa)		20	20	0.00%
	5	ρ_{st} (mm²/mm²)	R3.CRS. 5 e	0.00668	0.00668	0.00%
	6	ρ_{sc} (mm²/mm²)	R3.CRS. 6 e	0.00407	0.00407	0.00%
	7	f_{sy} (MPa)		500	500	0.00%
	8	f_{sw} (MPa)		400	400	0.00%
	9	ϕ_S (mm)		16	16	0.00%
	10	ρ_p (mm²/mm²)	R3.CRS. 10 e	0.00257	0.00257	0.00%
	11	f_{py} (MPa)		1650	1650	0.00%
	12	ϕ_p (mm)		12.7	12.7	0.00%
	13	$PTloss_{lt}$		0.15	0.15	0.00%
	14	L (mm)		10000	10000	0.00%
	15	K_c		3	3	0.00%
TRANSFER LOAD STAGE	16	σ_c/f_{cit}	FW.PTM. 16 c	0.0005	0.00	Cracked
	17	ε_{st} (mm/mm)	FW.PTM. 17 c	-0.00047	-0.00049	-3.30%
	18	ε_{sc} (mm/mm)	FW.PTM. 18 c	0.0003017	0.0003342	-10.77%
	19	ε_p (mm/mm)	FW.PTM. 19 c	0.00663	0.00660	0.33%
	20	σ_{ct}/f_{cit}	FW.PTM. 20 c	0.47383	0.48598	-2.57%
	21	cw (mm)	FW.PTM. 21 c	0.047	0.050	-6.38%
	22	Δ (mm)	FW.PTM. 22 c	-11.308	-12.818	-13.36%
SERVICE LOAD COMBINATION	23	$q_{L/250}$ (kN/m)		**80.00**	**82.54**	**-3.17%**
	24	$q_{0.2mm}$ (kN/m)		**100.00**	**101.75**	**-1.75%**
	25	$q_{0.6fck}$ (kN/m)	FW.PTM. 25 c	103.72	106.16	-2.35%
	26	$q_{0.8fsu}$ (kN/m)	FW.PTM. 26 c	140.10	144.44	-3.10%
	27	$q_{0.75fpu}$ (kN/m)	FW.PTM. 27 c	96.91	102.19	-5.45%
ULTIMATE LOAD COMBINATION	28	q_{str} (kN/m)		**150.00**	**147.28**	**1.81%**
	29	ε_{st} (mm/mm)	FW.PTM. 29 c	0.01045	0.01057	-1.16%
	30	ε_{sc} (mm/mm)	FW.PTM. 30 c	-0.00223	-0.00223	0.17%
	31	ε_p (mm/mm)	FW.PTM. 31 c	0.01640	0.01678	-2.35%
	32	ρ_{sw} (mm²/mm²)	FW.PTM. 32 c	0.0022	0.0022	-0.59%
	33	μ_Δ		**2.00**	**1.87**	**6.61%**

Model **R3.CRS.e** is used

Significant errors

μ_Δ *between 1.75 and 2.75 show good accuracy.*

: Given Inputs for ANN R3 : Inputs for engineering mechanic

: Reverse Outputs calculated by reverse ANN model is used as inputs to calculate other outputs by forward ANN model .

: Reverse-forward inputs : Reverse-forward ouputs

(Continued)

ductility ratios (μ_Δ) of 2.00. However, it is noted that these errors are small absolute errors of both upper rebar strains (ε_{sc}) and deflection (camber, c_w) at transfer load limit state. Design accuracies calculated based on the deflection ductility ratios (μ_Δ) of 2.00 shown in Table 2.5.11(e-1) are enhanced when the deflection ductility ratios (μ_Δ) of 2.00 is adjusted to 2.50 to reduce input conflicts as shown in Table 2.5.11(e-2). Training accuracies obtained with R3.CRS.g, R3.CRS.h, R3.CRS.i using 2 layer-20 neurons improved as shown in Table 2.5.10.

2.5.6 Comparison of design based on CRS with those based on PTM

Slightly higher design accuracies based on DNNs trained by R3.CRS.b (30 layers-30 neurons for the reverse network of Step 1) are observed in Table 2.5.9(c-2) compared with both those obtained in Table 2.5.6(d-3-1) based on ANNs trained by R3.PTM.c (40 layers-40 neurons, due to over-fitting) and Table 2.5.11(e-1) based on SNNs trained by R3.CRS.h (2 layers-30

Table 2.5.11 (Continued) Influence of a number of layers and neurons on the design accuracies of Reverse Scenario 3

(b-2) Deflection ductility ratios (μ_Δ) of 2.50

Model **R3.CRS.e** is used

PARAMETERS			Training Model	Network test result	Auto PT-BEAM check	Error
BEAM CONFIGURATION	1	b (mm)	R3.PTM. 1 e	640.78	640.78	0.00%
	2	h (mm)		700	700	0.00%
	3	f_{ck} (MPa)		40	40	0.00%
	4	f_{cki} (MPa)		20	20	0.00%
	5	ρ_{st} (mm^2/mm^2)	R3.CRS. 5 e	0.00880	0.00880	0.00%
	6	ρ_{sc} (mm^2/mm^2)	R3.CRS. 6 e	0.02137	0.02137	0.00%
	7	f_{sy} (MPa)		500	500	0.00%
	8	f_{sw} (MPa)		400	400	0.00%
	9	ϕ_S (mm)		16	16	0.00%
	10	ρ_p (mm^2/mm^2)	R3.CRS. 10 e	0.00257	0.00257	0.00%
	11	f_{py} (MPa)		1650	1650	0.00%
	12	ϕ_p (mm)		12.7	12.7	0.00%
	13	$PTloss_{lt}$		0.15	0.15	0.00%
	14	L (mm)		10000	10000	0.00%
	15	K_c		3	3	0.00%
TRANSFER LOAD STAGE	16	σ_c/f_{cki}	FW.PTM. 16 c	0.0005	0.00	Cracked
	17	ε_{st} (mm/mm)	FW.PTM. 17 c	-0.00040	-0.00041	-2.17%
	18	ε_{sc} (mm/mm)	FW.PTM. 18 c	0.0001120	0.0001190	-6.22%
	19	ε_p (mm/mm)	FW.PTM. 19 c	0.00669	0.00668	0.18%
	20	σ_{ct}/f_{cki}	FW.PTM. 20 c	0.40524	0.41019	-1.22%
	21	cw (mm)	FW.PTM. 21 c	0.015	0.015	0.00%
	22	Δ (mm)	FW.PTM. 22 c	-9.067	-9.098	-0.35%
SERVICE LOAD COMBINATION	23	$q_{L/250}$ (kN/m)		**80.00**	**81.81**	**-2.26%**
	24	$q_{0.2mm}$ (kN/m)		**100.00**	**101.91**	**-1.91%**
	25	$q_{0.6fck}$ (kN/m)	FW.PTM. 25 c	120.14	120.44	-0.25%
	26	$q_{0.8fsu}$ (kN/m)	FW.PTM. 26 c	144.57	147.33	-1.91%
	27	$q_{0.75fpu}$ (kN/m)	FW.PTM. 27 c	94.43	101.38	-7.36%
ULTIMATE LOAD COMBINATION	28	q_{str} (kN/m)		**150.00**	**155.02**	**-3.35%**
	29	ε_{st} (mm/mm)	FW.PTM. 29 c	0.02225	0.02217	0.35%
	30	ε_{sc} (mm/mm)	FW.PTM. 30 c	-0.00118	-0.00118	-0.13%
	31	ε_p (mm/mm)	FW.PTM. 31 c	0.02838	0.02852	-0.50%
	32	ρ_{sw} (mm^2/mm^2)	FW.PTM. 32 c	0.0023	0.0023	0.76%
	33	μ_Δ		**2.50**	**2.53**	**-1.39%**

μ_Δ between 1.75 and 2.75 show good accuracy.

: Given Inputs for ANN R3 ; : Inputs for engineering mechanic

: Reverse Outputs calculated by reverse ANN model is used as inputs to calculate other outputs by forward ANN model .

: Reverse-forward inputs ; : Reverse-forward ouputs

(Continued)

Table 2.5.11 (Continued) Influence of a number of layers and neurons on the design accuracies of Reverse Scenario 3

(c) Design accuracies obtained based on 1 layer and 40 neurons

(c-1) Deflection ductility ratios (μ_Δ) of 2.00

		PARAMETERS	Training Model	Network test result	Auto PT-BEAM check	Error
BEAM CONFIGURATION	1	b (mm)	R3.PTM. 1 f	814.60	814.60	0.00%
	2	h (mm)		700	700	0.00%
	3	f_{ck} (MPa)		40	40	0.00%
	4	f_{cki} (MPa)		20	20	0.00%
	5	ρ_{st} (mm²/mm²)	R3.CRS. 5 f	0.00647	0.00647	0.00%
	6	ρ_{sc} (mm²/mm²)	R3.CRS. 6 f	0.00308	0.00308	0.00%
	7	f_{sy} (MPa)		500	500	0.00%
	8	f_{sw} (MPa)		400	400	0.00%
	9	ϕ_S (mm)		16	16	0.00%
	10	ρ_p (mm²/mm²)	R3.CRS. 10 f	0.00225	0.00225	0.00%
	11	f_{py} (MPa)		1650	1650	0.00%
	12	ϕ_p (mm)		12.7	12.7	0.00%
	13	$PTloss_{lt}$		0.15	0.15	0.00%
	14	L (mm)		10000	10000	0.00%
	15	K_c		3	3	0.00%
TRANSFER LOAD STAGE	16	σ_c/f_{cki}	FW.PTM. 16 c	0.0005	0.00	Cracked
	17	ε_{st} (mm/mm)	FW.PTM. 17 c	-0.00041	-0.00043	-3.87%
	18	ε_{sc} (mm/mm)	FW.PTM. 18 c	0.0002942	0.0003286	-11.68%
	19	ε_p (mm/mm)	FW.PTM. 19 c	0.00668	0.00666	0.30%
	20	σ_{ct}/f_{cki}	FW.PTM. 20 c	0.42860	0.44163	-3.04%
	21	cw (mm)	FW.PTM. 21 c	0.047	0.053	-12.77%
	22	Δ (mm)	FW.PTM. 22 c	-8.836	-10.304	-16.62%
SERVICE LOAD COMBINATION	23	$q_{L/250}$ (kN/m)		**80.00**	**82.11**	**-2.63%**
	24	$q_{0.2mm}$ (kN/m)		**100.00**	**101.56**	**-1.56%**
	25	$q_{0.6fck}$ (kN/m)	FW.PTM. 25 c	108.86	110.63	-1.62%
	26	$q_{0.8fsu}$ (kN/m)	FW.PTM. 26 c	143.53	146.63	-2.16%
	27	$q_{0.75fpu}$ (kN/m)	FW.PTM. 27 c	99.82	102.18	-2.36%
ULTIMATE LOAD COMBINATION	28	q_{str} (kN/m)		**150.00**	**149.99**	**0.00%**
	29	ε_{st} (mm/mm)	FW.PTM. 29 c	0.01081	0.01092	-1.08%
	30	ε_{sc} (mm/mm)	FW.PTM. 30 c	-0.00220	-0.00220	0.20%
	31	ε_p (mm/mm)	FW.PTM. 31 c	0.01684	0.01714	-1.78%
	32	ρ_{sw} (mm²/mm²)	FW.PTM. 32 c	0.0020	0.0020	-0.85%
	33	μ_Δ		**2.00**	**1.91**	**4.42%**

Model **R3.CRS.f** is used

Significant errors

*μ_Δ between **1.75** and **2.75** show good accuracy.*

[red box] : Given Inputs for ANN R3 [blue box] : Inputs for engineering mechanic

[purple box] : Reverse Outputs calculated by reverse ANN model is used as inputs to calculate other outputs by forward ANN model .

[green box] : Reverse-forward inputs [black box] : Reverse-forward ouputs

(*Continued*)

Table 2.5.11 (Continued) Influence of a number of layers and neurons on the design accuracies of Reverse Scenario 3

(c-2) Deflection ductility ratios (μ_Δ) of 2.50

Model **R3.CRS.f** is used

PARAMETERS			Training Model		Network test result	Auto PT-BEAM check	Error
BEAM CONFIGURATION	1	b (mm)	R3.PTM.	1 f	668.56	668.56	0.00%
	2	h (mm)			700	700	0.00%
	3	f_{ck} (MPa)			40	40	0.00%
	4	f_{cki} (MPa)			20	20	0.00%
	5	ρ_{st} (mm^2/mm^2)	R3.CRS.	5 f	0.00843	0.00843	0.00%
	6	ρ_{sc} (mm^2/mm^2)	R3.CRS.	6 f	0.01919	0.01919	0.00%
	7	f_{sy} (MPa)			500	500	0.00%
	8	f_{sw} (MPa)			400	400	0.00%
	9	ϕ_S (mm)			16	16	0.00%
	10	ρ_p (mm^2/mm^2)	R3.CRS.	10 f	0.00247	0.00247	0.00%
	11	f_{py} (MPa)			1650	1650	0.00%
	12	ϕ_P (mm)			12.7	12.7	0.00%
	13	$PTloss_{lt}$			0.15	0.15	0.00%
	14	L (mm)			10000	10000	0.00%
	15	K_c			3	3	0.00%
TRANSFER LOAD STAGE	16	σ_c/f_{cki}	FW.PTM.	16 c	0.0005	0.00	Cracked
	17	ε_{st} (mm/mm)	FW.PTM.	17 c	-0.00039	-0.00040	-2.36%
	18	ε_{sc} (mm/mm)	FW.PTM.	18 c	0.0001154	0.0001231	-6.68%
	19	ε_p (mm/mm)	FW.PTM.	19 c	0.00670	0.00669	0.18%
	20	σ_{ct}/f_{cki}	FW.PTM.	20 c	0.39524	0.40044	-1.32%
	21	cw (mm)	FW.PTM.	21 c	0.016	0.016	0.00%
	22	Δ (mm)	FW.PTM.	22 c	-8.779	-8.716	0.71%
SERVICE LOAD COMBINATION	23	$q_{L/250}$ (kN/m)			**80.00**	**81.54**	**-1.93%**
	24	$q_{0.2mm}$ (kN/m)			**100.00**	**101.25**	**-1.25%**
	25	$q_{0.6fck}$ (kN/m)	FW.PTM.	25 c	120.06	120.22	-0.14%
	26	$q_{0.8fsu}$ (kN/m)	FW.PTM.	26 c	144.40	146.82	-1.67%
	27	$q_{0.75fpu}$ (kN/m)	FW.PTM.	27 c	94.51	100.90	-6.77%
ULTIMATE LOAD COMBINATION	28	q_{str} (kN/m)			**150.00**	**154.54**	**-3.02%**
	29	ε_{st} (mm/mm)	FW.PTM.	29 c	0.02181	0.02175	0.28%
	30	ε_{sc} (mm/mm)	FW.PTM.	30 c	-0.00122	-0.00122	-0.23%
	31	ε_p (mm/mm)	FW.PTM.	31 c	0.02795	0.02810	-0.56%
	32	ρ_{sw} (mm^2/mm^2)	FW.PTM.	32 c	0.0022	0.0022	0.52%
	33	μ_Δ			**2.50**	**2.52**	**-0.87%**

μ_Δ between 1.75 and 2.75 show good accuracy.

: Given Inputs for ANN R3 ; : Inputs for engineering mechanic

: Reverse Outputs calculated by reverse ANN model is used as inputs to calculate other outputs by forward ANN model .

: Reverse-forward inputs ; : Reverse-forward ouputs

(Continued)

Table 2.5.11 (Continued) Influence of a number of layers and neurons on the design accuracies of Reverse Scenario 3

(d) Design accuracies obtained based on 2 layer and 20 neurons

(d-1) Deflection ductility ratios (μ_Δ) of 2.00

PARAMETERS			Training Model	Network test result	Auto PT-BEAM check	Error
BEAM CONFIGURATION	1	b (mm)	R3.PTM. 1 g	741.93	741.93	0.00%
	2	h (mm)		700	700	0.00%
	3	f_{ck} (MPa)		40	40	0.00%
	4	f_{cki} (MPa)		20	20	0.00%
	5	ρ_{st} (mm^2/mm^2)	R3.CRS. 5 g	0.00691	0.00691	0.00%
	6	ρ_{sc} (mm^2/mm^2)	R3.CRS. 6 g	0.00617	0.00617	0.00%
	7	f_{sy} (MPa)		500	500	0.00%
	8	f_{sw} (MPa)		400	400	0.00%
	9	ϕ_S (mm)		16	16	0.00%
	10	ρ_p (mm^2/mm^2)	R3.CRS. 10 g	0.00244	0.00244	0.00%
	11	f_{py} (MPa)		1650	1650	0.00%
	12	ϕ_p (mm)		12.7	12.7	0.00%
	13	$PTloss_{lt}$		0.15	0.15	0.00%
	14	L (mm)		10000	10000	0.00%
	15	K_c		3	3	0.00%
TRANSFER LOAD STAGE	16	σ_c/f_{cki}	FW.PTM. 16 c	0.0005	0.00	Cracked
	17	ε_{st} (mm/mm)	FW.PTM. 17 c	-0.00043	-0.00044	-3.34%
	18	ε_{sc} (mm/mm)	FW.PTM. 18 c	0.0002313	0.0002562	-10.80%
	19	ε_p (mm/mm)	FW.PTM. 19 c	0.00666	0.00665	0.27%
	20	σ_{ct}/f_{cki}	FW.PTM. 20 c	0.43440	0.44667	-2.82%
	21	cw (mm)	FW.PTM. 21 c	0.033	0.037	-12.12%
	22	Δ (mm)	FW.PTM. 22 c	-10.022	-10.970	-9.46%
SERVICE LOAD COMBINATION	23	$q_{L/250}$ (kN/m)		**80.00**	**81.69**	**-2.11%**
	24	$q_{0.2mm}$ (kN/m)		**100.00**	**101.32**	**-1.32%**
	25	$q_{0.6fck}$ (kN/m)	FW.PTM. 25 c	107.88	109.61	-1.61%
	26	$q_{0.8fpu}$ (kN/m)	FW.PTM. 26 c	141.60	145.24	-2.57%
	27	$q_{0.75fpu}$ (kN/m)	FW.PTM. 27 c	96.81	101.68	-5.03%
ULTIMATE LOAD COMBINATION	28	q_{str} (kN/m)		**150.00**	**150.35**	**-0.23%**
	29	ε_{st} (mm/mm)	FW.PTM. 29 c	0.01309	0.01311	-0.15%
	30	ε_{sc} (mm/mm)	FW.PTM. 30 c	-0.00200	-0.00200	-0.06%
	31	ε_p (mm/mm)	FW.PTM. 31 c	0.01915	0.01935	-1.06%
	32	ρ_{sw} (mm^2/mm^2)	FW.PTM. 32 c	0.0021	0.0021	-0.13%
	33	μ_Δ		**2.00**	**2.03**	**-1.74%**

Model **R3.CRS.g** is used

Significant errors

*μ_Δ between **1.75** and **2.75** show good accuracy.*

: Given Inputs for ANN R3

: Inputs for engineering mechanic

: Reverse Outputs calculated by reverse ANN model is used as inputs to calculate other outputs by forward ANN model .

: Reverse-forward inputs

: Reverse-forward ouputs

(Continued)

Table 2.5.11 (Continued) Influence of a number of layers and neurons on the design accuracies of Reverse Scenario 3

(d-2) Deflection ductility ratios (μ_Δ) of 2.50

Model **R3.CRS.g** is used

		PARAMETERS	Training Model		Network test result	Auto PT-BEAM check	Error
BEAM CONFIGURATION	1	b (mm)	R3.PTM.	1 g	578.41	578.41	0.00%
	2	h (mm)			700	700	0.00%
	3	f_{ck} (MPa)			40	40	0.00%
	4	f_{cki} (MPa)			20	20	0.00%
	5	ρ_{st} (mm^2/mm^2)	R3.CRS.	5 g	0.00927	0.00927	0.00%
	6	ρ_{sc} (mm^2/mm^2)	R3.CRS.	6 g	0.02400	0.02400	-1.45E-16
	7	f_{sy} (MPa)			500	500	0.00%
	8	f_{sw} (MPa)			400	400	0.00%
	9	ϕ_S (mm)			16	16	0.00%
	10	ρ_p (mm^2/mm^2)	R3.CRS.	10 g	0.00283	0.00283	1.53E-16
	11	f_{py} (MPa)			1650	1650	0.00%
	12	ϕ_p (mm)			12.7	12.7	0.00%
	13	$PTloss_{lt}$			0.15	0.15	0.00%
	14	L (mm)			10000	10000	0.00%
	15	K_c			3	3	0.00%
TRANSFER LOAD STAGE	16	σ_c/f_{cki}	FW.PTM.	16 c	0.0004	0.00	Cracked
	17	ε_{st} (mm/mm)	FW.PTM.	17 c	-0.00044	-0.00045	-2.15%
	18	ε_{sc} (mm/mm)	FW.PTM.	18 c	0.0001154	0.0001218	-5.55%
	19	ε_p (mm/mm)	FW.PTM.	19 c	0.00665	0.00664	0.21%
	20	σ_{ct}/f_{cki}	FW.PTM.	20 c	0.43614	0.44133	-1.19%
	21	cw (mm)	FW.PTM.	21 c	0.016	0.016	0.00%
	22	Δ (mm)	FW.PTM.	22 c	-9.744	-10.319	-5.90%
SERVICE LOAD COMBINATION	23	$q_{L/250}$ (kN/m)			**80.00**	**80.59**	**-0.74%**
	24	$q_{0.2mm}$ (kN/m)			**100.00**	**100.81**	**-0.81%**
	25	$q_{0.6fck}$ (kN/m)	FW.PTM.	25 c	115.22	116.00	-0.68%
	26	$q_{0.8fsu}$ (kN/m)	FW.PTM.	26 c	140.63	144.34	-2.64%
	27	$q_{0.75fpu}$ (kN/m)	FW.PTM.	27 c	91.82	100.05	-8.97%
ULTIMATE LOAD COMBINATION	28	q_{str} (kN/m)			**150.00**	**151.45**	**-0.97%**
	29	ε_{st} (mm/mm)	FW.PTM.	29 c	0.02228	0.02220	0.39%
	30	ε_{sc} (mm/mm)	FW.PTM.	30 c	-0.00118	-0.00118	-3.13E-05
	31	ε_p (mm/mm)	FW.PTM.	31 c	0.02843	0.02855	-0.45%
	32	ρ_{sw} (mm^2/mm^2)	FW.PTM.	32 c	0.0025	0.0024	1.04%
	33	μ_Δ			**2.50**	**2.52**	**-0.86%**

μ_Δ between 1.75 and 2.75 show good accuracy.

: Given Inputs for ANN R3 : Inputs for engineering mechanic

: Reverse Outputs calculated by reverse ANN model is used as inputs to calculate other outputs by forward ANN model .

: Reverse-forward inputs : Reverse-forward ouputs

(Continued)

Table 2.5.11 (Continued) Influence of a number of layers and neurons on the design accuracies of Reverse Scenario 3

(e) Design accuracies obtained based on 2 layer and 30 neurons

(e-1) Deflection ductility ratios (μ_Δ) of 2.00

PARAMETERS			Training Model	Network test result	Auto PT-BEAM check	Error
BEAM CONFIGURATION	1	b (mm)	R3.PTM. 1 h	769.82	769.82	0.00%
	2	h (mm)		700	700	0.00%
	3	f_{ck} (MPa)		40	40	0.00%
	4	f_{cit} (MPa)		20	20	0.00%
	5	ρ_{st} (mm^2/mm^2)	R3.CRS. 5 h	0.00688	0.00688	0.00%
	6	ρ_{sc} (mm^2/mm^2)	R3.CRS. 6 h	0.00471	0.00471	0.00%
	7	f_{sy} (MPa)		500	500	0.00%
	8	f_{sw} (MPa)		400	400	0.00%
	9	ϕ_S (mm)		16	16	0.00%
	10	ρ_p (mm^2/mm^2)	R3.CRS. 10 h	0.00234	0.00234	0.00%
	11	f_{py} (MPa)		1650	1650	0.00%
	12	ϕ_p (mm)		12.7	12.7	0.00%
	13	$PTloss_{lt}$		0.15	0.15	0.00%
	14	L (mm)		10000	10000	0.00%
	15	K_c		3	3	0.00%
TRANSFER LOAD STAGE	16	σ_c/f_{cit}	FW.PTM. 16 c	0.0005	0.00	Cracked
	17	ε_{st} (mm/mm)	FW.PTM. 17 c	-0.00042	-0.00043	-3.44%
	18	ε_{sc} (mm/mm)	FW.PTM. 18 c	0.0002509	0.0002802	-11.68%
	19	ε_p (mm/mm)	FW.PTM. 19 c	0.00668	0.00666	0.27%
	20	σ_{ct}/f_{cit}	FW.PTM. 20 c	0.42602	0.43814	-2.84%
	21	cw (mm)	FW.PTM. 21 c	0.038	0.041	-7.89%
	22	Δ (mm)	FW.PTM. 22 c	-9.292	-10.485	-12.84%
SERVICE LOAD COMBINATION	23	$q_{L/250}$ (kN/m)		**80.00**	**81.31**	**-1.64%**
	24	$q_{0.2mm}$ (kN/m)		**100.00**	**101.22**	**-1.22%**
	25	$q_{0.6fck}$ (kN/m)	FW.PTM. 25 c	107.89	109.60	-1.58%
	26	$q_{0.8fsu}$ (kN/m)	FW.PTM. 26 c	142.71	145.97	-2.28%
	27	$q_{0.75fpu}$ (kN/m)	FW.PTM. 27 c	98.02	101.62	-3.68%
ULTIMATE LOAD COMBINATION	28	q_{ult} (kN/m)		**150.00**	**150.60**	**-0.40%**
	29	ε_{st} (mm/mm)	FW.PTM. 29 c	0.01187	0.01197	-0.87%
	30	ε_{sc} (mm/mm)	FW.PTM. 30 c	-0.00211	-0.00210	0.21%
	31	ε_p (mm/mm)	FW.PTM. 31 c	0.01791	0.01820	-1.61%
	32	ρ_{sw} (mm^2/mm^2)	FW.PTM. 32 c	0.0021	0.0021	-0.42%
	33	μ_Δ		**2.00**	**1.96**	**1.92%**

Model **R3.CRS.h** is used

Significant errors

μ_Δ between 1.75 and 2.75 show good accuracy.

: Given Inputs for ANN R3 : Inputs for engineering mechanic

: Reverse Outputs calculated by reverse ANN model is used as inputs to calculate other outputs by forward ANN model .

: Reverse-forward inputs : Reverse-forward ouputs

(Continued)

Table 2.5.11 (Continued) Influence of a number of layers and neurons on the design accuracies of Reverse Scenario 3

(e-2) Deflection ductility ratios (μ_Δ) of 2.50

PARAMETERS			Training Model	Network test result	Auto PT-BEAM check	Error
BEAM CONFIGURATION	1	b (mm)	R3.PTM. 1 h	588.26	588.26	0.00%
	2	h (mm)		700	700	0.00%
	3	f_{ck} (MPa)		40	40	0.00%
	4	f_{cki} (MPa)		20	20	0.00%
	5	ρ_{st} (mm²/mm²)	R3.CRS. 5 h	0.00929	0.00929	0.00%
	6	ρ_{sc} (mm²/mm²)	R3.CRS. 6 h	0.02388	0.02388	0.00%
	7	f_{sy} (MPa)		500	500	0.00%
	8	f_{sw} (MPa)		400	400	0.00%
	9	ϕ_s (mm)		16	16	0.00%
	10	ρ_p (mm²/mm²)	R3.CRS. 10 h	0.00277	0.00277	0.00%
	11	f_{py} (MPa)		1650	1650	0.00%
	12	ϕ_p (mm)		12.7	12.7	0.00%
	13	$PTloss_{lt}$		0.15	0.15	0.00%
	14	L (mm)		10000	10000	0.00%
	15	K_c		3	3	0.00%
TRANSFER LOAD STAGE	16	σ_c/f_{cki}	FW.PTM. 16 c	0.0004	0.00	Cracked
	17	ε_{st} (mm/mm)	FW.PTM. 17 c	-0.00043	-0.00044	-2.10%
	18	ε_{sc} (mm/mm)	FW.PTM. 18 c	0.0001129	0.0001191	-5.49%
	19	ε_p (mm/mm)	FW.PTM. 19 c	0.00666	0.00665	0.20%
	20	σ_{ct}/f_{cki}	FW.PTM. 20 c	0.42761	0.43266	-1.18%
	21	cw (mm)	FW.PTM. 21 c	0.016	0.015	6.25%
	22	Δ (mm)	FW.PTM. 22 c	-9.554	-10.029	-4.97%
SERVICE LOAD COMBINATION	23	$q_{L/250}$ (kN/m)		**80.00**	**80.73**	**-0.91%**
	24	$q_{0.2mm}$ (kN/m)		**100.00**	**101.04**	**-1.04%**
	25	$q_{0.6fck}$ (kN/m)	FW.PTM. 25 c	116.48	117.12	-0.55%
	26	$q_{0.8fsu}$ (kN/m)	FW.PTM. 26 c	141.60	145.10	-2.47%
	27	$q_{0.75fpu}$ (kN/m)	FW.PTM. 27 c	92.39	100.28	-8.53%
ULTIMATE LOAD COMBINATION	28	q_{str} (kN/m)		**150.00**	**152.40**	**-1.60%**
	29	ε_{st} (mm/mm)	FW.PTM. 29 c	0.02239	0.02230	0.39%
	30	ε_{sc} (mm/mm)	FW.PTM. 30 c	-0.00117	-0.00117	0.00%
	31	ε_p (mm/mm)	FW.PTM. 31 c	0.02853	0.02866	-0.46%
	32	ρ_{sw} (mm²/mm²)	FW.PTM. 32 c	0.0024	0.0024	1.00%
	33	μ_Δ		**2.50**	**2.52**	**-0.95%**

Model **R3.CRS.h** is used

*μ_Δ between **1.75** and **2.75** show good accuracy.*

: Given Inputs for ANN R3 : Inputs for engineering mechanic

: Reverse Outputs calculated by reverse ANN model is used as inputs to calculate other outputs by forward ANN model .

: Reverse-forward inputs : Reverse-forward ouputs

(Continued)

Table 2.5.11 (Continued) Influence of a number of layers and neurons on the design accuracies of Reverse Scenario 3

(f) Design accuracies obtained based on 2 layer and 40 neurons

(f-1) Deflection ductility ratios (μ_Δ) of 2.00

PARAMETERS			Training Model	Network test result	Auto PT-BEAM check	Error
BEAM CONFIGURATION	1	b (mm)	R3.PTM. 1 i	775.13	775.13	0.00%
	2	h (mm)		700	700	0.00%
	3	f_{ck} (MPa)		40	40	0.00%
	4	f_{cki} (MPa)		20	20	0.00%
	5	ρ_{st} (mm²/mm²)	R3.CRS. 5 i	0.00673	0.00673	0.00%
	6	ρ_{sc} (mm²/mm²)	R3.CRS. 6 i	0.00509	0.00509	0.00%
	7	f_{sy} (MPa)		500	500	0.00%
	8	f_{sw} (MPa)		400	400	0.00%
	9	ϕ_S (mm)		16	16	0.00%
	10	ρ_p (mm²/mm²)	R3.CRS. 10 i	0.00234	0.00234	0.00%
	11	f_{py} (MPa)		1650	1650	0.00%
	12	ϕ_P (mm)		12.7	12.7	0.00%
	13	$PTloss_{lt}$		0.15	0.15	0.00%
	14	L (mm)		10000	10000	0.00%
	15	K_c		3	3	0.00%
TRANSFER LOAD STAGE	16	σ_c/f_{cki}	FW.PTM. 16 c	0.0005	0.00	Cracked
	17	ε_{st} (mm/mm)	FW.PTM. 17 c	-0.00042	-0.00043	-3.41%
	18	ε_{sc} (mm/mm)	FW.PTM. 18 c	0.0002431	0.0002712	-11.56%
	19	ε_p (mm/mm)	FW.PTM. 19 c	0.00668	0.00666	0.27%
	20	σ_{ct}/f_{cki}	FW.PTM. 20 c	0.42572	0.43786	-2.85%
	21	cw (mm)	FW.PTM. 21 c	0.036	0.039	-8.33%
	22	Δ (mm)	FW.PTM. 22 c	-9.358	-10.577	-13.02%
SERVICE LOAD COMBINATION	23	$q_{L/250}$ (kN/m)		**80.00**	**81.87**	**-2.34%**
	24	$q_{0.2mm}$ (kN/m)		**100.00**	**101.36**	**-1.36%**
	25	$q_{0.6fck}$ (kN/m)	FW.PTM. 25 c	108.95	110.60	-1.52%
	26	$q_{0.8fsu}$ (kN/m)	FW.PTM. 26 c	142.57	145.93	-2.36%
	27	$q_{0.75fpu}$ (kN/m)	FW.PTM. 27 c	98.09	101.83	-3.82%
ULTIMATE LOAD COMBINATION	28	q_{str} (kN/m)		**150.00**	**150.70**	**-0.47%**
	29	ε_{st} (mm/mm)	FW.PTM. 29 c	0.01249	0.01255	-0.47%
	30	ε_{sc} (mm/mm)	FW.PTM. 30 c	-0.00205	-0.00205	0.07%
	31	ε_p (mm/mm)	FW.PTM. 31 c	0.01856	0.01879	-1.23%
	32	ρ_{sc} (mm²/mm²)	FW.PTM. 32 c	0.0020	0.0020	-0.36%
	33	μ_Δ		**2.00**	**2.01**	**-0.49%**

Model **R3.CRS.i** is used

Significant errors

μ_Δ between 1.75 and 2.75 show good accuracy.

: Given Inputs for ANN R3   : Inputs for engineering mechanic

: Reverse Outputs calculated by reverse ANN model is used as inputs to calculate other outputs by forward ANN model .

: Reverse-forward inputs   : Reverse-forward ouputs

(*Continued*)

Table 2.5.11 (Continued) Influence of a number of layers and neurons on the design accuracies of Reverse Scenario 3

(f-2) Deflection ductility ratios (μ_Δ) of 2.50

Model **R3.CRS.i** is used

PARAMETERS			Training Model	Network test result	Auto PT-BEAM check	Error
BEAM CONFIGURATION	1	b (mm)	R3.PTM 1 i	623.18	623.18	0.00%
	2	h (mm)		700	700	0.00%
	3	f_{ck} (MPa)		40	40	0.00%
	4	f_{cki} (MPa)		20	20	0.00%
	5	ρ_{st} (mm²/mm²)	R3.CRS. 5 i	0.00890	0.00890	0.00%
	6	ρ_{sc} (mm²/mm²)	R3.CRS. 6 i	0.02160	0.02160	0.00%
	7	f_{sy} (MPa)		500	500	0.00%
	8	f_{sw} (MPa)		400	400	0.00%
	9	ϕ_S (mm)		16	16	0.00%
	10	ρ_p (mm²/mm²)	R3.CRS. 10 i	0.00262	0.00262	0.00%
	11	f_{py} (MPa)		1650	1650	0.00%
	12	ϕ_p (mm)		12.7	12.7	0.00%
	13	$PTloss_{lt}$		0.15	0.15	0.00%
	14	L (mm)		10000	10000	0.00%
	15	K_c		3	3	0.00%
TRANSFER LOAD STAGE	16	σ_c/f_{cki}	FW.PTM. 16 c	0.0005	0.00	Cracked
	17	ε_{st} (mm/mm)	FW.PTM. 17 c	-0.00041	-0.00042	-2.17%
	18	ε_{sc} (mm/mm)	FW.PTM. 18 c	0.0001139	0.0001209	-6.17%
	19	ε_p (mm/mm)	FW.PTM. 19 c	0.00668	0.00667	0.19%
	20	σ_{ct}/f_{cki}	FW.PTM. 20 c	0.41197	0.41708	-1.24%
	21	cw (mm)	FW.PTM. 21 c	0.016	0.016	0.00%
	22	Δ (mm)	FW.PTM. 22 c	-9.266	-9.401	-1.46%
SERVICE LOAD COMBINATION	23	$q_{L/250}$ (kN/m)		**80.00**	**80.97**	**-1.21%**
	24	$q_{0.2mm}$ (kN/m)		**100.00**	**101.05**	**-1.05%**
	25	$q_{0.6fck}$ (kN/m)	FW.PTM. 25 c	117.96	118.32	-0.31%
	26	$q_{0.8fsu}$ (kN/m)	FW.PTM. 26 c	142.85	145.81	-2.08%
	27	$q_{0.75fpu}$ (kN/m)	FW.PTM. 27 c	93.19	100.47	-7.82%
ULTIMATE LOAD COMBINATION	28	q_{str} (kN/m)		**150.00**	**153.32**	**-2.21%**
	29	ε_{st} (mm/mm)	FW.PTM. 29 c	0.02210	0.02202	0.34%
	30	ε_{sc} (mm/mm)	FW.PTM. 30 c	-0.00119	-0.00119	-0.13%
	31	ε_p (mm/mm)	FW.PTM. 31 c	0.02824	0.02838	-0.49%
	32	ρ_{sw} (mm²/mm²)	FW.PTM. 32 c	0.0023	0.0023	0.83%
	33	μ_Δ		**2.50**	**2.52**	**-0.64%**

μ_Δ between 1.75 and 2.75 show good accuracy.

: Given Inputs for ANN R3 : Inputs for engineering mechanic

: Reverse Outputs calculated by reverse ANN model is used as inputs to calculate other outputs by forward ANN model .

: Reverse-forward inputs : Reverse-forward ouputs

neurons, due to over-fitting) when the deflection ductility ratios (μ_Δ) of 2.00 is reversely preassigned on an input-side.

2.6 CONCLUSIONS

Reverse designs with 15 input and 18 output parameters are performed using ANN-trained reverse-forward networks for engineering applications.

Both forward and reverse designs of pretensioned concrete beams are performed in this chapter based on ANNs trained on big datasets. ANN-based PT beam design with sufficient training accuracy can replace classic design software while exhibiting excellent productivity for both forward and reverse designs. Here are some findings of the design examples.

1. Reverse designs are performed based on two-step networks. In the reverse network of Step 1, reverse output parameters are obtained which are, then, back-substituted into forward networks of Step 2 as input parameters to determine the rest of the design parameters.
2. TED, PTM, and CRS can be used for both reverse and forward networks with selected training parameters such as a number of hidden layers, neurons, and validation checks. Extra input features selected from an output-side can be used to train other output parameters as when training ANNs based on CRS that is not possible with TED with which all inputs are mapped to entire outputs simultaneously. Books [2.11] and [2.13] are referred for further information of training methods such as CRS and TED.
3. A selection of training networks depends on many parameters such as feature scores, volume of datasets, and types of big datasets. SNN are recommended when volume of datasets is insufficient, whereas DNN are preferred when sufficient datasets are available. For the reverse network in Step 1, DNNs trained with CRS outperform both ANNs trained with PTM and SNNs trained with CRS.
4. Evidently, ANN models are recommended to expand coverages of big data for interpolations instead of extrapolations when being required by design cases. It is noted that adding more data is not difficult since data used in the proposed method are generated by structural mechanics-based software which is not collected from costly experiments or investigations.
5. The classic design software is often coded in closed forms, dealing with specific design scenarios, for example, calculating rebar areas satisfying strength and service demands. Engineers must either revise programs to re-generate big datasets when new requirements are added, for instance, designing a beam satisfying strength and service demands. However, designing pretensioned concrete beams which meet the specific ductility is challenging in a conventional way. On the other hand, the proposed ANN-based designs can flexibly relocate parameters on an input- or output-sides, allowing users to conveniently control preferred parameters as design targets such as strengths, ductility, and deflections. ANN-based PT beam designs with sufficient accuracies are robust alternatives to conventional designs, providing engineers with a holistic method for all design scenarios.
6. The present designs evaluate ANN models trained based on MAE and correlation coefficients (R). Benchmarking results of all training are presented in Tables 2.4.1, 2.4.3, 2.5.1, 2.5.3, 2.5.5, 2.5.7, and 2.5.10.

However, it is noted that reliabilities of ANN models shall only be concluded based on errors in practical designs. Engineers must always check the differences between results provided by ANN predictions and structural calculations. Training parameters such as a number of layers and neurons should be determined well to avoid overfitting. Training methods such as PTM and Training methods such as PTM and CRS should be also selected carefully to enhance design accuracies.

7. Reverse designs of pretensioned concrete beams performed using ANNs trained on datasets generated from the European Standard (EC2) produce acceptable design accuracies for use in practical designs. Design tables derived from reverse designs can be extended to construct design charts that inter-connect all design parameters to achieve holistic and robust designs in a streak. The reverse design facilitates demonstrated in this chapter a rapid identification of design parameters, helping engineers with fast decisions based on acceptable accuracies.

REFERENCES

[2.1] Torky, A. A., & Aburawwash, A. A. (2018). A deep learning approach to automated structural engineering of prestressed members. *International Journal of Structural and Civil Engineering Research*, 7, 347–352. https://doi.org/10.18178/ijscer.7.4.347-352

[2.2] Sumangala, K., & Antony Jeyasehar, C. (2011). A new procedure for damage assessment of prestressed concrete beams using artificial neural network. *Advances in Artificial Neural Systems*, *2011*, 14. https://doi.org/10.1155/2011/786535

[2.3] Hong, W. K., Pham, T. D., & Nguyen, V. T. (2022). Feature selection based reverse design of doubly reinforced concrete beams. *Journal of Asian Architecture and Building Engineering*, *21*(4), 1472–1496. https://doi.org/10.1080/13467581.2021.1928510

[2.4] Hong, W. K., Nguyen, V. T., & Nguyen, M. C. (2022). Artificial intelligence-based noble design charts for doubly reinforced concrete beams. *Journal of Asian Architecture and Building Engineering*, *21*(4), 1497–1519.https://doi.org/10.1080/13467581.2021.1928511

[2.5] Hong, W.-K., Nguyen, M. C., & Pham, T. D. (2023). Optimized interaction P-M diagram for rectangular reinforced concrete column based on artificial neural networks. *Journal of Asian Architecture and Building Engineering*, *22*(1), 201–225. https://doi.org/10.1080/13467581.2021.2018697

[2.6] Hong, W. K., & Pham, T. D. (2022). Reverse designs of doubly reinforced concrete beams using Gaussian process regression models enhanced by sequence training/designing technique based on feature selection algorithms. *Journal of Asian Architecture and Building Engineering*, *21*(6), 2345–2370. https://doi.org/10.1080/13467581.2021.1971999

[2.7] Hong, W. K., Nguyen, M. C., Pham, T. D., & Le, T. A. (2022). Holistic design of pre-tensioned concrete beams based on artificial intelligence. *Journal of Asian Architecture and Building Engineering*, 1–32. https://doi.org/10.1080/13467581.2022.2097909

[2.8] CEN (European Committee for Standardization). (2004). Eurocode 2 : Part 1-1: General rules and rules for buildings. In *Eurocode 2: Vol. BS En 1992*.

[2.9] Nguyen, M. C., & Hong, W. K. (2022). Analytical prediction of nonlinear behaviors of beams post-tensioned by unbonded tendons considering shear deformation and tension stiffening effect. *Journal of Asian Architecture and Building Engineering*, *21*(3), 908–929. https://doi.org/10.1080/13467581.2021.1908299

[2.10] Naser, M. Z., and A. Alavi. (2020). Insights into performance fitness and error metrics for machine learning. *arXiv preprint arXiv:2006.00887*.

[2.11] Hong, W.K. (2021). *Artificial Intelligence-Based Design of Reinforced Concrete Structures*. Daega, Seoul.

[2.12] W. K., & Nguyen, M. C. (2021). AI-based Lagrange optimization for designing reinforced concrete columns. *Journal of Asian Architecture and Building Engineering*, *21*(6), 2330–2344. https://doi.org/10.1080/13467581.2021.1971998

[2.13] Hong, W.K. (2019). *Hybrid Composite Precast Systems: Numerical Investigation to Construction*. Woodhead Publishing, Elsevier, Cambridge, MA.

Chapter 3

An optimized design of pretensioned concrete beams using an ANN-based Lagrange algorithm

3.1 INTRODUCTION AND DATA GENERATION FOLLOWING EUROCODE

3.1.1 Previous research

A structural optimization is an attractive area, where numerous studies can be found. An early study of [3.1] introduced a software based on a simple algorithm capable of calculating optimal prestressing forces and beam widths to minimize costs or weights of simply supported prestressed beams. In 2015, a hybrid glowworm swarm algorithm was applied to optimize costs and CO_2 emission of precast-prestressed beams in road bridges [3.2]. Many papers, such as [3.3–3.7], proposed to find optimal designs of prestressed beams using evolutionary algorithms, reporting cost reductions of up to 30% compared with conventional designs. However, only a few research applied ANNs in optimizing prestressed beams. One study is [3.8], where an example of minimizing total costs of a prestressed bridge system based on a neural dynamic-based iteration method was demonstrated. In each iteration, gradients of cost function were calculated using structural mechanics-based software, after which these gradients were used to update design parameters for successive iterations based on a neural dynamic learning rule. In the present chapter, ANN-based objective functions, constraining functions, and parameters are derived. Optimized design parameters calculated using Lagrange multipliers and Karush–Kuhn–Tucker (KKT) conditions satisfy code requirements. The proposed design replaces conventional software. This chapter was written based on the previous paper by Hong et al. [3.9].

3.1.2 Significance of the Chapter 3

In this chapter, pretensioned concrete beams (PT beams) are holistically optimized based on Lagrange function using ANNs and ANN-based Lagrange method. In ANN-based Lagrange method, optimized engineering designs meet various code restrictions imposed by equality and

DOI: 10.1201/9781003354796-3

inequality constraints simultaneously. ANNs are trained on 150,000 large datasets with 15 inputs and 18 outputs generated using a structural mechanics-based software AutoPTBEAM following Eurocode. ANN-trained networks are, then, used both to derive ANN-based objective and constraining functions to optimize a Lagrange function. ANN-based functions are generalized to calculate Jacobian and Hessian matrices to solve the KKT equations. Objective functions derived explicitly are replaced by the ANN-based functions after which Lagrange functions are optimized at stationary points representing optimized design parameters of PT beams with pin-pin boundaries. A cost of PT beams under various design requirements is optimized by the proposed method. A design optimizing a PT beam cost satisfies all code requirements, whereas empirical observations of engineers are not used. Diverse design requirements are satisfied without empirical observations of engineers. Design charts for cost-optimized holistic designs for 12 m-span simply supported PT beams under different loading conditions are provided at the end of optimization. Design charts for any type of PT beams with different design parameters similar to those introduced in this chapter can be constructed by repeating procedures introduced in this chapter. The ANN-based Lagrange method is verified to provide minimum costs similar to those obtained by 133,563 conventional designs.

3.1.3 Tasks readers can perform after reading this chapter

Design codes are strictly imposed by 16 equality and 18 inequality conditions. Designs minimizing a PT beam cost satisfy all code requirements while empirical observations of engineers are not used. The readers can perform the following tasks after studying this chapter.

1. Stationary points of Lagrange functions representing design parameters for PT beams with pin-pin boundaries are identified by Newton–Raphson iteration.
2. Cost indexes (CI_b) obtained by the large datasets are bounded on the lower cost obtained based on the ANN-based optimized design charts.
3. Readers can construct design charts to minimize a cost of PT beams under various design requirements based on any design code. In this chapter, diverse ANN-based design charts shown below are provided for holistic designs of PT beams, calculating 21 forward outputs ($\Delta_{trans}/L, \sigma_{c,trans}/f_{cki}, w_{trans}, \Delta_{lt}/L, \Delta_{incr}/L, w_{freq}, \sigma_c/f_{ck}, \sigma_{s,bot}/f_{yk}, \sigma_{p,bot}/f_{pk}, M_{Ed}, M_{Rd}, M_{cr}, x/d, \varepsilon_{s,bot}/\varepsilon_{sy}, \varepsilon_{p,bot}/\varepsilon_{py}, V_{Ed}, V_{Rdmax}, \rho_{sw,sp}, \rho_{sw,mid}, CI_b, Weight$) in arbitrary sequences from 21 forward inputs ($L, b, h, f_{ck}, f_{cki}, \rho_{s,bot}, \rho_{s,top}, f_{yk}, f_{sw}, \phi_p, \rho_{p,bot}, f_{p0.1k}, \sigma_{pm0}, Loss_{lt}, \psi_1, \psi_2, t_0, T, RH, q_D, q_L$), while optimizing design targets. Readers can design section sizes

(b and h), tendon ratios ($\rho_{p,bot}$), and rebar ratios ($\rho_{s,bot}$ and $\rho_{s,top}$) corresponding to applied loads. Readers can also adjust inequalities and equalities to modify design charts tailored for a particular design condition including a span length (L), a material availability (f_{ck}, f_{cki}, f_{yk}, $f_{p0.1k}$, and f_{sw}), prestressing techniques (σ_{pm0}, $Loss_{lt}$, and ϕ_p), combination factors (ψ_1, ψ_2), environmental exposures (t_0, T, and RH), and loadings (q_D and q_L).

The following design charts can be constructed.

1. Design chart based on crack widths ($w_{SER} \leq 0.2$ mm and $w_{Tran} \leq 0.2$ mm) when CI_b is optimized (refer to Figure 3.4.1 when CI_b is minimized; Figure 3.5.4 when W is minimized; Figure 3.6.4 when beam depth h is minimized).
2. Design chart based on stresses; $\sigma_{s,bot}/f_{yk} \leq 0.8$, $\sigma_{p,bot}/f_{pk} \leq 0.75$ and $\sigma_c/f_{ck} \leq 0.6$ when CI_b is optimized (refer to Figures 3.4.2, 3.4.3 when CI_b is minimized; Figure 3.5.2 when W is minimized; Figure 3.6.2 when beam depth h is minimized).
3. Design chart based on deflections ($\Delta_{trans} \leq \frac{L}{250}$, $\Delta_{lt} \leq \frac{L}{250}$, and $\Delta_{incr} \leq \frac{L}{500}$) when CI_b is minimized (refer to Figure 3.4.5 when CI_b is minimized; Figure 3.5.1 when W is minimized; Figure 3.6.1 when beam depth h is minimized).
4. Design chart based on moment demand (M_{Ed}), moment resistance (M_{Rd}), and cracking moment (M_{Cr}) when CI_b is minimized (refer to Figure 3.4.6 when CI_b is minimized; Figure 3.5.5 when W is minimized; Figure 3.6.5 when beam depth h is minimized).
5. Design chart based on tendon strains ($\varepsilon_{p,bot}$), reinforcement strains ($\varepsilon_{s,bot}/\varepsilon_{sy}$), and neutral axis depth ($x/d \leq 0.448$) representing beam ductility when CI_b is minimized (refer to Figure 3.4.7 when CI_b is minimized; Figure 3.5.6 when W is minimized; Figure 3.6.6 when beam depth h is minimized).
6. Design chart based on shear stirrup ratios ($\rho_{sw,sp}$ and $\rho_{sw,mid}$) when CI_b is minimized (refer to Figure 3.4.8 when CI_b is minimized; Figure 3.5.7 when W is minimized; Figure 3.6.7 when beam depth h is minimized).
7. Design chart based on beam cost (CI_b) when CI_b is minimized (refer to Figures 3.4.4 and 3.4.9).
8. Design chart based on beam weight (W) when W is minimized (refer to Figures 3.5.3 and 3.5.9).
9. Design chart based on beam cost (CI_b) when h is minimized (refer to Figure 3.6.8).
10. Design chart based on beam weight (Wt.) when h is minimized (refer to Figure 3.6.9).

3.2 LARGE DATASETS GENERATED FOLLOWING EUROCODE 2

3.2.1 Selection of input parameters following Eurocode

3.2.1.1 Fourteen inputs parameters describing beam sections and material properties

In this chapter, simply supported pretensioned (PT) beams are designed based on 21 input parameters which are listed in Table 3.2.1 including beam sections, material properties, load criteria, environmental conditions, etc. Large datasets are generated to investigate behaviors of PT beams. In Figure 3.2.1(a), seven main parameters, including beam length ($\boldsymbol{L}$), section dimensions ($b \times h$), rebar ratio ($\rho_{s,bot}$, $\rho_{s,top}$), and strand diameter (ϕ_p) and its ratio ($\rho_{p,bot}$), describe a PT beam section. It is noted that ratios of used

Table 3.2.1 Nomenclatures of input and output parameters

21 input parameters		*21 output parameters*	
L	Beam length (mm)	Δ_{trans}/L	Displacement at transfer stage (mm)
b	Beam width (mm)	$\sigma_{c,trans}/f_{cki}$	Ratio of concrete compressive stress at transfer stage
h	Beam height (mm)	w_{trans}	Crack width at transfer stage (mm)
f_{ck}	Characteristic compressive cylinder strength of concrete at 28 days (MPa)	Δ_{lt}/L	Long-term deflection (mm)
f_{cki}	Characteristic compressive cylinder strength of concrete at transfer (MPa)	Δ_{incr}/L	Incremental deflection (mm)
$\rho_{s,bot}$	Bottom rebar ratio	w_{freq}	Crack width at frequent service stage (mm)
$\rho_{s,top}$	Top rebar ratio	σ_c/f_{ck}	Ratio of concrete compressive stress at characteristic service stage
f_{yk}	Characteristic yield strength of longitudinal rebar (MPa)	$\sigma_{s,bot}/f_{yk}$	Ratio of bottom rebar stress at characteristic service stage
f_{sw}	Characteristic yield strength of stirrup (MPa)	$\sigma_{p,bot}/f_{pk}$	Ratio of bottom strand stress at characteristic service stage
ϕ_p	Strand diameter (mm)	M_{Ed}	Design value of the applied internal bending moment (kN·m)
$\rho_{p,bot}$	Bottom strand ratio	M_{Rd}	Moment resistance (kN·m)
$f_{p0.1k}$	Characteristic yield strength of prestressing steel (MPa)	M_{cr}	Cracking moment (kN·m)
σ_{pm0}	Stress in the strand immediately after tensioning or transfer (MPa)	x/d	Ratio of neutral axis depth

(*Continued*)

Table 3.2.1 (Continued) Nomenclatures of input and output parameters

21 input parameters		*21 output parameters*	
$Loss_{lt}$	Long-term loss of stress in strand (%)	$\varepsilon_{s,bot}/\varepsilon_{sy}$	Strain of bottom rebar in terms of yielding strain
ψ_1	Factor for frequent value of a variable action	$\varepsilon_{p,bot}/\varepsilon_{py}$	Strain of bottom strand in terms of yielding strain
ψ_2	Factor for quasi-permanent value of a variable action	V_{Ed}	Design value of the applied shear force (kN)
t_0	The age of concrete at the time of loading (days)	V_{Rdmax}	Maximum shear resistance limited by crushing of the compression struts (kN)
T	Ambient temperature (°C)	$\rho_{sw,sp}$	Stirrup ratio at supports
RH	Relative humidity of ambient environment (%)	$\rho_{sw,mid}$	Stirrup ratio at mid-span
q_D	Distributed dead load (kN/m)	CI_b	Cost index of the beam (KRW/m)
q_L	Distributed live load (kN/m)	$Weight$	Beam weight per length (kN/m)

material area to gross concrete section $\left(e.g., \rho_{s,bot} = {A_{s,bot}}/{b \times h}\right)$ are represented by $\rho_{s,bot}$, $\rho_{s,top}$, and $\rho_{p,bot}$ for bottom rebars, top rebars, and bottom tendons, respectively. An effective depth of rebar (d_s) and strand (d_p) based on section dimensions, concrete cover, and detailing requirements following EC 2 – Section 8 [3.10] are generated in large datasets which are used during design optimizations. Concrete cover to the surface of rebar (c_s) and size of aggregate (d_g), etc. are fixed as 50 and 20 mm, respectively, and hence, they are not generated in large datasets. A structural mechanics-based software, AutoPTBEAM, developed based on algorithm shown in Figure 3.2.1(b) was used to generate large datasets for bonded PT beams. It is noted that the algorithm shown in Figure 3.2.1(b) is developed for both bonded and unbonded beams. AutoPTBEAM uses parabola-rectangle stress–strain relationship provided by EC2 [3.10] shown in Figure 3.2.2 to model a concrete behavior. Material properties, such as axial tensile strength (f_{ctm}) and tangent modulus of elasticity (E_{cm}), are computed from f_{ck} which are based on compressive cylinder strength at 28 days (f_{ck}) following Section 3 – EC 2 [3.10]. f_{cki} is the characteristic compressive cylinder strength of concrete at transfer stage. A stress-strain relationship of rebars shown in Figure 3.2.3(a) is represented by an elasto-plastic model with an ideally plastic branch, using two parameters, yield strength of rebars (f_{yk}) and elastic modulus (E_s). For prestressing strand, an elasto-plastic material with a linear strain hardening is defined by yield strength ($f_{p0.1k}$), tensile strength (f_{pk}), and elastic modulus (E_p) shown in Figure 3.2.3(b).

Elasticity of modulus of rebars (E_s) and strands (E_p) are fixed as 200,000 MPa, whereas tensile strength of strands (f_{pk}) is 1,860 MPa, and hence, these parameters are fixed in large datasets. Important parameters

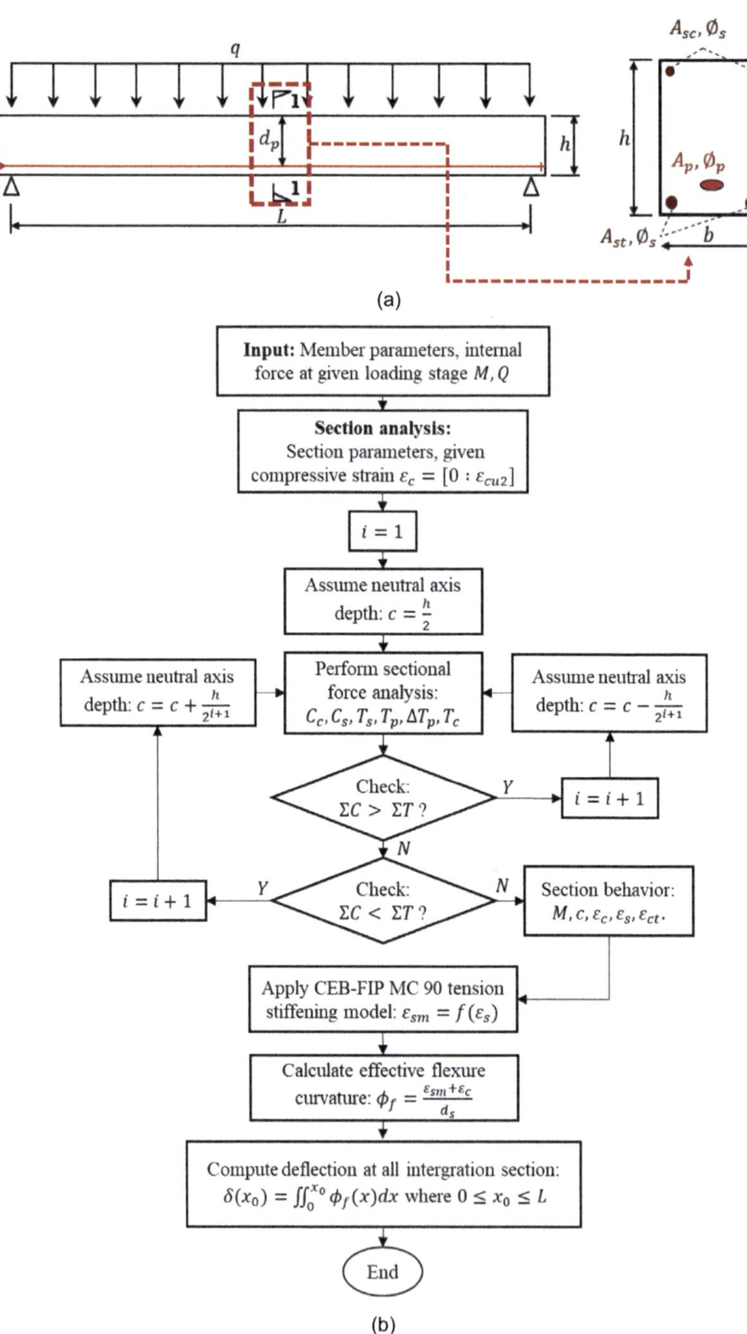

Figure 3.2.1 Beam configurations and design algorithm of AutoPTBEAM [3.11]. (a) Simply supported PT beam configuration. (b) Algorithm for designing bonded pretensioned beams (AutoPTBEAM) [3.11].

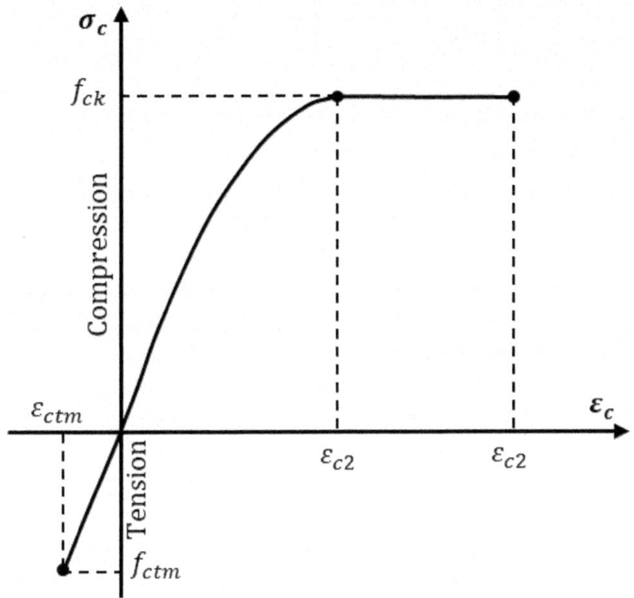

Figure 3.2.2 Behavior of the concrete in tension and compression [3.11].

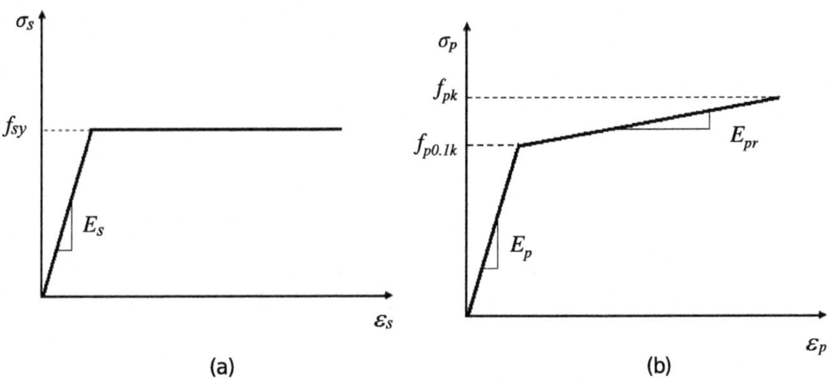

Figure 3.2.3 Stress–strain relationships of rebars and strands. (a) Stress–strain relationships of rebar. (b) Stress–strain relationships of prestressing strands.

describing prestressing of strand material include stresses ($\boldsymbol{\sigma}_{pm0}$) which occurs in the strand immediately after tensioning or transfer (MPa) and percentage of stress losses ($\boldsymbol{Loss}_{lt}$) due to long-term effects including creep, shrinkage, and stress relaxation. Stresses of strands immediately after tensioning should not exceed the smaller of $0.75\boldsymbol{f}_{pk}$ and $0.85\boldsymbol{f}_{p0.1k}$ according to Per EC 2 – Section 5.10.3 [3.10]. In summary, beam section and material properties are described by 14 inputs (L, b, h, f_{ck}, f_{cki}, $\rho_{s,bot}$, $\rho_{s,top}$, f_{yk}, f_{sw}, ϕ_p, $\sigma_{p,bot}$, $f_{p0.1k}$, ρ_{pm0}, $Loss_{lt}$).

3.2.1.2 Seven inputs parameters describing load and environment conditions

Loads can be broadly classified as vertical load, horizontal load, and longitudinal load where, in general, vertical loads are divided into self-weight, dead load, live load, and impact load, whereas horizontal loads consist of wind load and seismic load, which can be ignored for PT beams when the boundary conditions are simply supported. Tractive and braking forces are considered in special cases of designing bridges, gantry girders, and longitudinal load. In this chapter, ANN networks are derived only based on self-weight, dead load, and live load. Self-weight is computed by beam sections. Dead q_D and live loads q_L are considered as distributed loads in this chapter. Other types of loads can be added further based on interests of engineers. Load combinations comprise quasi-permanent combination, frequent combination, characteristic combination, and ultimate load combination in accordance with Eurocode 0 [3.12]. A load combination at transfer stage is conducted to evaluate structural performance after prestressing beams. Table 3.2.2 presents load factors for each type of load combination. Table 3.2.3 [3.12] illustrates the coefficients ψ_1 and ψ_2, which are used in load combinations for assessing the reversible (recoverable) serviceability limit criteria and determining the long-term effects. The reversible (recoverable) serviceability criteria, such as crack widths, are checked under frequent load levels denoting a 1% probability of being exceeded. On the other hand, the calculation of long-term effects involves quasi-permanent load magnitudes, which indicates loads applied with 50% probability of being exceeded for a duration of the design life. Values of ψ_1 and ψ_2 are based on probability of exceedance of considered action. It is noted that ψ_0 is only used when there are more than one variable loads, for example, imposing wind loads and live loads in one combination. An effect of creep on a long-term deflection is calculated based on an age of concrete at the time of loading (t_0), ambient temperature (T), and relative humidity of ambient environment (RH), being referred to Annex B

Table 3.2.2 Load combination following Eurocode 0 [3.12]

Load comb.	*Self-weight*	*Dead load*	*Live load*	*Prestressing*	*Hyperstatic*
Transfer comb.	1	0	0	1 [a]	0
Quasi-permanent comb.	1	1	ψ_2	1 [b]	0
Frequent comb.	1	1	ψ_1	1 [b]	0
Characteristic comb.	1	1	1	1 [b]	0
Ultimate load comb.	1.35	1.35	1.35	0	1

[a] Prestressing effect at transfer stage (without long-term losses).
[b] Prestressing effect at service stage (with long-term losses).

Table 3.2.3 Recommended values for ψ factors for building [3.12]

Action	ψ_0	ψ_1	ψ_2
Imposed loads in buildings, category			
Category A: domestic, residential areas	0.7	0.5	0.3
Category B: office areas	0.7	0.5	0.3
Category C: congregation areas	0.7	0.7	0.6
Category D: shopping areas	0.7	0.7	0.6
Category E: storage areas	1	0.9	0.8
Category F: traffic area, vehicle weight ≤ 30kN	0.7	0.7	0.6
Category G: traffic area, 30kN < vehicle weight ≤ 160kN	0.7	0.5	0.3
Category H: roofs	0	0	0
Snow loads on buildings			
Finland, Iceland, Norway, Sweden	0.7	0.5	0.2
Remainder of CEN Member States, for sites located at altitude H > 1,000 m a.s.l.	0.7	0.5	0.2
Remainder of CEN Member States, for sites located at altitude H ≤ 1,000 m a.s.l.	0.5	0.2	0
Wind loads on buildings	0.6	0.2	0
Temperature (non-fire) in buildings	0.6	0.5	0

of EC 2 [3.10] for further information. In summary, load and environment aspects are described by seven inputs parameters (ψ_1, ψ_2, t_0, T, RH, q_D, q_L).

3.2.2 Selection of output parameters based on design requirements of Eurocode

Structural analysis of PT beam yields a wide variety of output parameters including internal forces, stress, strain, crack width, deflection, etc. for every load combination. However, not all parameters are restrained by the same design requirements at all design limit states as imposed by EC 2 [3.10]. For example, crack width should be carefully checked at frequent load combination but can be ignored at characteristic load combination because frequent load levels are calculated based on 1% probability of being exceeded, resulting in a return period of one week for frequent traffic loads in bridges. Frequent combinations are generally used for checking reversible (recoverable) limit states. Permanent damages, such as excessive cracks, are not allowed in structures under frequent loads. Characteristic loads are higher than frequent loads, and hence, a return period of characteristic wind loads is 50 years, leading to a very small probability of being exceeded by characteristic loads. Characteristic combinations are used for checking irreversible (irrecoverable) limit states, where certain permanent damages such as excessive cracks are allowed. In this chapter, output parameters for ANNs are selected based on design requirements provided by EC 2 [3.10].

3.2.2.1 Selection of output parameters at transfer stage (three outputs: Δ_{trans}/L, $\sigma_{c,trans}/f_{cki}$, w_{trans})

In transfer stage, excessive camber (Δ_{trans}/L), crack width (w_{trans}), and normalized concrete stress in compression zone ($\sigma_{c,trans}/f_{cki}$) caused by prestressing force are avoided by checking serviceability of PT beam. Normalized concrete compressive stress at transfer stage ($\sigma_{c,trans}/f_{cki}$) should be limited to k to prevent longitudinal cracking in accordance with EC 2 – Section 5.10.2.2(5) [3.10]. The value of k of 0.6 is normally used as shown in Eq. (5.42) – EC 2 [3.10].

3.2.2.2 Selection of output parameters at quasi-permanent stage (two outputs: Δ_{lt}/L, Δ_{incr}/L)

Long-term effects and the appearance of PT structures are normally checked by quasi-permanent combinations. Long-term deflection (Δ_{lt}) under quasi-permanent load combinations should not exceed $L/250$ to ensure an appearance of prestressed bonded beams and a safety of general utility of PT structures (Section 7.4.1 (4) – EC 2 [3.10]), whereas incremental deflection (Δ_{incr}) under quasi-permanent load combinations should not exceed $L/550$ not to damage adjacent parts of PT structures (7.4.1 (5) – EC 2 [3.10]). Quasi-permanent loads' levels are calculated from statistical data based on a 50% probability of being exceeded by loads on buildings during service [3.10].

3.2.2.3 Selection of output parameters at a frequent stage (one output: w_{freq})

Frequent combinations are generally used for checking reversible (recoverable) limit states. Cracking shall be limited to not impair the working serviceability or durability of PT structures in which crack widths (w_{freq}) are limited based on various structural types, exposure class, and protection. Table 3.2.4 (Table 7.1N – EC 2 [9]) recommends values of crack widths (w_{freq}). Frequent loads' levels are calculated from statistical data based on a 1% probability of being exceeded by loads (with a return period of one week) on buildings during service [3.10].

3.2.2.4 Selection of output parameters at a characteristic stage (three outputs: σ_c/f_{ck}, $\sigma_{s,bot}/f_{yk}$, $\sigma_{p,bot}/f_{pk}$)

Characteristic combinations are normally used for checking irreversible (irrecoverable) limit states, where certain permanent damages such as excessive cracks are allowed. Normalized compressive stress of concrete

Table 3.2.4 Recommended limitations for crack widths (Table 7.1N – Eurocode 2 [3.10])

Exposure class	*Reinforced members and prestressed members with unbonded tendons*	*Prestressed members with bonded tendons*
	Quasi-permanent load combination	Frequent load combination
X0, XC1	0.4 mm	0.2 mm
XC2, XC3, XC4	0.3 mm	0.2 mm
XD1, XD2, XS1, XS2, XS3		All parts of the bonded tendons or duct lie at least 25 mm within concrete in compression.

(σ_c/f_{ck}) should not exceed k_1 to avoid longitudinal cracks under characteristic stage in accordance with Section 7.2 (2) – EC 2 [9]. Normalized stress in rebars ($\sigma_{s,bot}/f_{yk}$) and strands ($\sigma_{p,bot}/f_{pk}$) under characteristic load combinations should not also exceed k_2 and k_3, respectively, to avoid unacceptable cracking or deformation. National Annex provides values of k_1, k_2, and k_3, for each country. EC 2 [9] recommends values of 0.6, 0.8, and 0.75 for k_1, k_2, and k_3, respectively. Characteristic load levels are calculated from statistical data based on a 0% probability of being exceeded by loads on buildings during service [3.10].

3.2.2.5 Selection of output parameters at an ultimate limit stage (ten outputs: M_{Ed}, M_{Rd}, M_{cr}, x/d, $\varepsilon_{s,bot}/\varepsilon_{sy}$, $\varepsilon_{p,bot}/\varepsilon_{py}$, V_{Ed}, $V_{Rd,max}$, $\rho_{sw,sp}$, $\rho_{sw,mid}$)

Three parameters for a moment resistance (M_{Rd}, a flexural capacity of the section), an applied factored internal bending moment (M_{Ed}), and a cracking moment (M_{cr}) under an ultimate limit stage are calculated as output parameters to verify a flexural capacity of the section ($M_{Rd} > M_{Ed}$), avoiding brittle failure after an onset of flexural cracking ($M_{Rd} > M_{cr}$). Eurocode suggests controlling ductility of a beam by a normalized neutral axis depth (x/d) for ductility requirement. According to Section 5.5 – EC 2 [3.10], for instance, a limitation of $x/d \leq 0.448$ should be verified when concrete strength is smaller than or equal to 50 MPa. It is noted that there is no moment redistribution of a simply supported PT beam which is statically determinate structure. A ductility of simply supported PT beams is evaluated by two other output parameters representing normalized bottom rebar strains ($\varepsilon_{s,bot}/\varepsilon_{sy}$) and normalized tendon strains ($\varepsilon_{p,bot}/\varepsilon_{py}$). Both shear resistance sustained by stirrups and by concrete sections ($V_{Rd,max}$, limited by crushing of compression struts) should be also verified against design shear forces (V_{Ed}). According to Eqs. 6.8, 9.4, and 9.5N of EC 2 [3.10]), stirrup ratios at supports ($\rho_{sw,sp}$) and mid-span ($\rho_{sw,mid}$) can be determined by equating shear resistance provided by stirrup to design shear forces.

3.2.2.6 *Selection of output parameters for design efficiency (two outputs)*

The accuracies and efficiencies of the optimized design are verified by two designs for cost index (CI_b) and beam weight (***Weight***). A cost index of a simply supported PT beam is computed based on the construction material prices in Korea (2021).

3.2.3 Generation of Large datasets

As discussed in Sections 3.2.1 and 3.2.2, 21 input parameters (L, b, h, f_{ck}, f_{cki}, $\rho_{s,bot}$, $\rho_{s,top}$, f_{yk}, f_{sw}, ϕ_p, $\rho_{p,bot}$, $f_{p0.1k}$, σ_{pm0}, $Loss_{lt}$, ψ_1, ψ_2, t_0, T, RH, q_D, q_L) are implemented in AutoPTBEAM to compute 21 corresponding output parameters (Δ_{trans}/L, $\sigma_{c,trans}/f_{cki}$, w_{trans}, Δ_{lt}/L, Δ_{incr}/L, w_{freq}, σ_c/f_{ck}, $\sigma_{p,bot}/f_{yk}$, $\sigma_{p,bot}/f_{pk}$, M_{Ed}, M_{Rd}, M_{cr}, x/d, $\varepsilon_{s,bot}/\varepsilon_{sy}$, $\varepsilon_{p,bot}/\varepsilon_{py}$, V_{Ed}, V_{Rdmax}, $\rho_{sw,sp}$, $\rho_{sw,mid}$, CI_b, ***Weight***), generating large datasets for ANN-based design. AutoPTBEAM is used to randomly generate a total of 200,000 datasets, some of which are listed in Table 3.2.5(a). Forward inputs are used to obtain forward outputs based on Eurocode. Formulae provided by Eurocode are established based on diverse experiments, resulting in reliable design results. For instance, calculations of crack widths are verified by [3.13] based on experiments [3.14–3.23] in which errors between calculated and tested results have a mean value of 0.0022 mm with a standard deviation of 0.0769 mm, illustrating an accuracy and conservativeness provided by Eurocode. Big datasets are generated using design standard verified by tests. It is challenging for conventional designs to achieve design goals that are possible with optimized design of pretensioned concrete beams based on the ANN-based Lagrange method. Randomly generated data are normalized using a min–max normalization technique, which is one of the most common methods to normalize datasets as shown in Table 3.2.5(b). The minima and maxima of datasets of all input features are transformed into –1 and 1 while other datasets are transformed into a decimal between –1 and 1 accordingly.

Table 3.2.5 Large datasets of a simply supported PT beam generated by AutoPTBEAM

(a) Non-normalized data of simply supported PT beams

			200,000 DATASETS GENERATED BY AutoPT											
			Data 1	Data 2	Data 3	Data 4	Data 5	Data 6	Data 7	Data 8	...	MIN	MAX	MEAN
21 INPUT PARAMETERS FOR STRUCTURAL SOFTWARE (AutoSPT)	1	L (mm)	28170.3	41990.7	30821.5	12258.7	21051.1	16745.8	44484.6	36342.6	...	5020.4	62380.8	26276.5
	2	b (mm)	6970.1	1103.0	4770.4	491.7	4279.6	2032.1	5358.8	5563.0	...	205.7	9977.3	3304.3
	3	h (mm)	1948.8	1840.4	1847.7	579.7	1110.5	1095.7	2030.4	2245.3	...	500.0	2500.0	1500.7
	4	f_{ck} (MPa)	44	42	32	32	57	56	42	39	...	30	60	44.96
	5	f_{cki} (MPa)	15	18	22	29	55	39	31	17	...	15	60	30.01
	6	$\rho_{s.bot}$	0.02472	0.03719	0.02124	0.03022	0.03737	0.02597	0.00682	0.02569	...	0.00000	0.04000	0.01997
	7	$\rho_{s.top}$	0.00681	0.00833	0.02033	0.02512	0.01716	0.03959	0.00040	0.02829	...	0.00000	0.04000	0.01995
	8	f_{yk} (MPa)	598	424	556	512	408	559	494	437	...	400	600	500.0
	9	f_{sw} (MPa)	567	459	442	417	389	497	567	548	...	300	600	449.9
	10	ϕ_p (mm)	15.7	17.8	9.5	9.5	12.7	17.8	14.3	15.7	...	9.5	17.8	13.69
	11	$\rho_{p.bot}$	0.00354	0.00161	0.00367	0.00227	0.00195	0.00204	0.00162	0.00260	...	0.00070	0.00370	0.00220
	12	$f_{p,0.1k}$ (MPa)	1663.67	1663.35	1563.29	1560.55	1635.45	1686.85	1666.10	1673.06	...	1500.00	1700.00	1600.15
	13	σ_{pm0} (MPa)	1309.55	1262.09	1252.19	1032.40	1207.20	1070.95	1134.74	1331.68	...	1000.00	1395.00	1176.22
	14	$Loss_{lt}$	0.12	0.15	0.14	0.12	0.14	0.19	0.11	0.19	...	0.10	0.20	0.15
	15	ψ_1	0.2	0.5	0.2	0.9	0.7	0.9	0.5	0.9	...	0.2	1	0.615
	16	ψ_2	0	0.3	0	0.8	0.6	0.8	0.2	0.8	...	0	1	0.415
	17	t_0 (days)	21	16	22	8	7	27	18	20	...	7	28	17.49
	18	T (°C)	11.3	-11.6	-5.8	26.2	-10.3	6.0	-21.5	-8.6	...	-40.0	40.0	0.0
	19	RH (%)	88	98	91	43	78	49	74	56	...	40	100	70
	20	q_D (kN/m)	1450.42	4.43	413.17	51.61	63.32	226.38	56.42	258.29	...	-304.86	8476.64	203.87
	21	q_L (kN/m)	54.28	118.80	379.17	34.74	676.29	292.25	9.79	228.58	...	-244.78	8409.69	368.24
21 OUTPUT PARAMETERS FROM STRUCTURAL SOFTWARE (AutoSPT)	22	Δ_{tranz}/L	-0.0004	0.0023	-0.0002	0.0001	0.0004	0.0000	0.0030	0.0003	...	-0.0012	0.0030	0.0005
	23	$\sigma_{c\text{-}tranz}/f_{cki}$	0.225	0.360	0.268	0.100	0.091	0.099	0.501	0.174	...	0.025	1.000	0.199
	24	w_{tranz} (mm)	0.000	0.044	0.000	0.000	0.000	0.000	0.234	0.000	...	0.000	0.400	0.022
	25	Δ_{lt}/L	0.0065	0.0067	0.0053	0.0067	0.0067	0.0067	0.0067	0.0059	...	-0.0030	0.0067	0.0051
	26	Δ_{incr}/L	0.0030	0.0030	0.0030	0.0030	0.0030	0.0030	0.0030	0.0030	...	-0.0025	0.0030	0.0026
	27	w_{freq} (mm)	0.169	0.181	0.110	0.400	0.176	0.315	0.375	0.102	...	0.000	0.400	0.215
	28	σ_c/f_{ck}	0.730	0.982	0.778	1.000	0.733	0.646	0.646	0.455	...	0.031	1.000	0.667
	29	$\sigma_{s\text{-}bot}/f_{yk}$	0.317	0.771	0.421	0.854	0.663	0.601	0.698	0.277	...	-0.003	1.000	0.594
	30	$\sigma_{p\text{-}bot}/f_{pk}$	0.707	0.705	0.685	0.663	0.661	0.608	0.716	0.632	...	0.430	1.000	0.672
	31	M_{Ed} (kNm)	182943.5	38345.0	120253.7	1755.9	47551.8	20130.6	83664.1	131937.5	...	36.4	930788.0	75623.6
	32	M_{Rd} (kNm)	311624.1	39449.7	201284.4	1822.1	61790.7	28817.9	100167.1	310217.9	...	105.9	1058907.3	97501.3
	33	M_{cr} (kNm)	81962.7	6503.9	48307.3	259.2	10523.9	4792.3	37350.1	69760.6	...	52.4	212731.4	23154.4
	34	x/d	0.601	0.642	0.346	0.432	0.415	0.253	0.236	0.195	...	0.032	0.782	0.294
	35	$\varepsilon_{s\text{-}bot}/\varepsilon_{sy}$	0.78	0.92	2.38	1.80	2.07	3.28	4.58	6.63	...	0.33	43.17	5.24
	36	$\varepsilon_{p\text{-}bot}/\varepsilon_{py}$	0.92	0.79	1.42	1.03	1.02	1.41	1.88	2.19	...	0.51	12.30	2.09
	37	V_{Ed} (kN)	25976.8	3652.7	15606.5	572.9	9035.5	4808.5	7523.0	14521.5	...	22.1	127323.8	9603.4
	38	V_{Rdmax} (kN)	58596.4	6148.4	44790.5	954.2	26716.9	14160.0	67409.5	72081.9	...	149.5	212893.6	32646.1
	39	$\rho_{sw\text{-}sp}$	0.0030	0.0045	0.0024	0.0041	0.0037	0.0029	0.0009	0.0013	...	0.0007	0.0245	0.0016
	40	$\rho_{sw\text{-}mid}$	0.0009	0.0024	0.0010	0.0019	0.0018	0.0015	0.0009	0.0009	...	0.0001	0.0025	0.0003
	41	CI_b (KRW/m)	3328083.9	450094.7	2548307.0	88440.3	1235203.4	834589.6	747397.7	3535355.9	...	5337.7	10554206.5	1324078.2
	42	$Weight$ (kN/m)	339.6	50.7	220.4	7.1	118.8	55.7	272.0	312.3	...	2.6	623.4	142.2

(Continued)

Table 3.2.5 (Continued) Large datasets of a simply supported PT beam generated by AutoPTBEAM

(b) Normalized data of simply supported PT beams based on min–max

			200,000 DATASETS NORMALIZED BY MAPMINMAX [-1;1]											
			Data 1	Data 2	Data 3	Data 4	Data 5	Data 6	Data 7	Data 8	...	MIN	MAX	MEAN
21 INPUT PARAMETERS FOR STRUCTURAL SOFTWARE (*AutoSPT*)	1	L (mm)	-0.1928	0.2891	-0.1004	-0.7476	-0.4411	-0.5912	0.3760	0.0921	...	-1.000	1.000	-0.259
	2	b (mm)	0.3845	-0.8163	-0.0657	-0.9415	-0.1662	-0.6262	0.0547	0.0965	...	-1.000	1.000	-0.366
	3	h (mm)	0.4488	0.3404	0.3477	-0.9203	-0.3895	-0.4043	0.5304	0.7453	...	-1.000	1.000	0.001
	4	f_{ck} (MPa)	-0.0667	-0.2000	-0.8667	-0.8667	0.8000	0.7333	-0.2000	-0.4000	...	-1.000	1.000	-0.002
	5	f_{cki} (MPa)	-1.0000	-0.8667	-0.6889	-0.3778	0.7778	0.0667	-0.2889	-0.9111	...	-1.000	1.000	-0.333
	6	$\rho_{s,bot}$	0.2358	0.8597	0.0618	0.5110	0.8684	0.2984	-0.6592	0.2845	...	-1.000	1.000	-0.001
	7	$\rho_{s,top}$	-0.6597	-0.5835	0.0165	0.2561	-0.1422	0.9794	-0.9801	0.4147	...	-1.000	1.000	-0.002
	8	f_{yk} (MPa)	0.9800	-0.7600	0.5600	0.1200	-0.9200	0.5900	-0.0600	-0.6300	...	-1.000	1.000	0.000
	9	f_{sw} (MPa)	0.7800	0.0600	-0.0533	-0.2200	-0.4067	0.3133	0.7800	0.6533	...	-1.000	1.000	-0.001
	10	ϕ_p (mm)	0.4940	1.0000	-1.0000	-1.0000	-0.2289	1.0000	0.1566	0.4940	...	-1.000	1.000	0.009
	11	$\rho_{p,bot}$	0.8908	-0.3937	0.9819	0.0439	-0.1696	-0.1084	-0.3871	0.2674	...	-1.000	1.000	-0.001
	12	$f_{p,0.1k}$ (MPa)	0.6368	0.6335	-0.3671	-0.3945	0.3545	0.8686	0.6610	0.7306	...	-1.000	1.000	0.002
	13	σ_{pm0} (MPa)	0.5673	0.3271	0.2769	-0.8360	0.0491	-0.6408	-0.3178	0.6794	...	-1.000	1.000	-0.108
	14	$Loss_{lt}$	-0.6320	0.0366	-0.2424	-0.6712	-0.1435	0.8866	-0.8178	0.8723	...	-1.000	1.000	0.001
	15	ψ_1	-1.0000	-0.2500	-1.0000	0.7500	0.2500	0.7500	-0.2500	0.7500	...	-1.000	1.000	0.037
	16	ψ_2	-1.0000	-0.4000	-1.0000	0.6000	0.2000	0.6000	-0.6000	0.6000	...	-1.000	1.000	-0.170
	17	t_0 (days)	0.3333	-0.1429	0.4286	-0.9048	-1.0000	0.9048	0.0476	0.2381	...	-1.000	1.000	-0.001
	18	T (°C)	0.2836	-0.2902	-0.1448	0.6544	-0.2582	0.1504	-0.5366	-0.2149	...	-1.000	1.000	0.001
	19	RH (%)	0.6130	0.9488	0.7021	-0.9057	0.2716	-0.6877	0.1303	-0.4647	...	-1.000	1.000	0.000
	20	q_D (kN/m)	-0.6002	-0.9296	-0.8365	-0.9188	-0.9161	-0.8790	-0.9177	-0.8717	...	-1.000	1.000	-0.884
	21	q_L (kN/m)	-0.9309	-0.9160	-0.8558	-0.9354	-0.7871	-0.8759	-0.9412	-0.8906	...	-1.000	1.000	-0.858
21 OUTPUT PARAMETERS FROM STRUCTURAL SOFTWARE (*AutoSPT*)	22	Δ_{trans}/L	-0.6459	0.6549	-0.5397	-0.3729	-0.2724	-0.4274	1.0000	-0.2992	...	-1.000	1.000	-0.209
	23	$\sigma_{c\text{-}trans}/f_{cki}$	-0.5894	-0.3136	-0.5014	-0.8454	-0.8646	-0.8491	-0.0235	-0.6939	...	-1.000	1.000	-0.642
	24	w_{trans} (mm)	-1.0000	-0.7807	-1.0000	-1.0000	-1.0000	-1.0000	0.1678	-1.0000	...	-1.000	1.000	-0.888
	25	Δ_{lt}/L	0.9655	1.0000	0.7082	1.0000	1.0000	1.0000	1.0000	0.8329	...	-1.000	1.000	0.666
	26	Δ_{incr}/L	1.0000	1.0000	1.0000	1.0000	1.0000	1.0000	1.0000	1.0000	...	-1.000	1.000	0.853
	27	w_{freq} (mm)	-0.1568	-0.0939	-0.4498	1.0000	-0.1191	0.5731	0.8751	-0.4905	...	-1.000	1.000	0.074
	28	σ_c/f_{ck}	0.4436	0.9633	0.5420	1.0000	0.4493	0.2702	0.2687	-0.1236	...	-1.000	1.000	0.312
	29	$\sigma_{s\text{-}bot}/f_{yk}$	-0.3625	0.5429	-0.1534	0.7085	0.3273	0.2037	0.3984	-0.4409	...	-1.000	1.000	0.190
	30	$\sigma_{p\text{-}bot}/f_{pk}$	-0.0300	-0.0372	-0.1050	-0.1852	-0.1921	-0.3773	0.0029	-0.2919	...	-1.000	1.000	-0.151
	31	M_{Ed} (kN·m)	-0.6070	-0.9177	-0.7417	-0.9963	-0.8979	-0.9568	-0.8203	-0.7166	...	-1.000	1.000	-0.838
	32	M_{Rd} (kN·m)	-0.4116	-0.9257	-0.6200	-0.9968	-0.8835	-0.9458	-0.8110	-0.4142	...	-1.000	1.000	-0.816
	33	M_{cr} (kN·m)	-0.2297	-0.9393	-0.5462	-0.9981	-0.9015	-0.9554	-0.6493	-0.3445	...	-1.000	1.000	-0.783
	34	x/d	0.5164	0.6254	-0.1627	0.0665	0.0220	-0.4124	-0.4557	-0.5673	...	-1.000	1.000	-0.303
	35	$\varepsilon_{s\text{-}bot}/\varepsilon_{sy}$	-0.9791	-0.9724	-0.9044	-0.9315	-0.9188	-0.8622	-0.8017	-0.7058	...	-1.000	1.000	-0.771
	36	$\varepsilon_{p\text{-}bot}/\varepsilon_{py}$	-0.9298	-0.9511	-0.8446	-0.9105	-0.9134	-0.8460	-0.7671	-0.7144	...	-1.000	1.000	-0.731
	37	V_{Ed} (kN)	-0.5922	-0.9430	-0.7552	-0.9913	-0.8584	-0.9248	-0.8822	-0.7722	...	-1.000	1.000	-0.849
	38	V_{Rdmax} (kN)	-0.4505	-0.9436	-0.5803	-0.9924	-0.7502	-0.8683	-0.3677	-0.3238	...	-1.000	1.000	-0.695
	39	$\rho_{sw\text{-}sp}$	-0.8780	-0.8011	-0.9095	-0.8204	-0.8398	-0.8849	-0.9902	-0.9717	...	-1.000	1.000	-0.866
	40	$\rho_{sw\text{-}mid}$	-0.9827	-0.8575	-0.9753	-0.8986	-0.9085	-0.9318	-0.9845	-0.9847	...	-1.000	1.000	-0.923
	41	CI_b (KRW/m)	-0.3700	-0.9157	-0.5179	-0.9842	-0.7668	-0.8428	-0.8593	-0.3307	...	-1.000	1.000	-0.750
	42	*Weight* (kN/m)	0.0856	-0.8449	-0.2985	-0.9855	-0.6256	-0.8291	-0.1321	-0.0024	...	-1.000	1.000	-0.550

3.2.3.1 Crack width

Crack widths are calculated using Eq. (3.2.1) based on Section 7.3.4 of EC 2 [3.10].

$$W_k = S_{r,max}\left(\varepsilon_{sm} - \varepsilon_{cm}\right) \tag{3.2.1}$$

where

W_k – Crack width.

$S_{r,max}$ – Maximum crack spacing determined based on Section 7.3.4(3) of EC 2 [3.10].

ε_{sm} – Mean strain in a reinforcement under a relevant combination of loads determined based on Section 7.3.4(2) of EC 2 [3.10].

ε_{cm} – Mean strain in concrete between cracks determined based on Section 7.3.4(2) of EC 2 [3.10].

3.2.3.2 Deflections

A deflection is calculated using AutoPTBEAM based on section curvatures including effects of cracks and creeps based on Section 7.4.3 of EC 2 [3.10]. Effects of cracks are estimated using Eq. (3.2.2).

$$\alpha = \zeta \alpha_{\mathrm{II}} + (1-\zeta)\alpha_{\mathrm{I}} \tag{3.2.2}$$

where

α – Sections curvatures for calculating deflections.

α_{I} – Sections curvatures based on uncracked sections.

α_{II} – Sections curvatures based on cracked sections.

ζ – A distribution coefficient based on Section 7.4.3(3) of EC 2 [3.10].

Effective modulus of elasticity for concrete reflecting effects of creep is calculated using Eq. (3.2.3) based on secant modulus of elasticity of concrete and creep coefficient relevant to the load and time interval.

$$E_{c,eff} = \frac{E_{cm}}{1+\varphi(\infty, t_0)} \tag{3.2.3}$$

where

$E_{c,eff}$ – Effective modulus of elasticity for concrete.

E_{cm} – Secant modulus of elasticity of concrete.

$\varphi(\infty, t_0)$ – Creep coefficient relevant to the load and time interval calculated based on Section 3.1.3 of EC 2 [3.10].

3.2.3.3 Flexural strength

Design flexural strengths of PT beams are determined based on section analyses and partial safety factors of 1.5, 1.15, and 1.15 which are applied to concrete, rebars, and prestressing strands, respectively, according to Section 2.4.2.4 of EC 2 [3.10].

3.2.3.4 *Shear strength*

Shear strength of PT beams is determined using Eq. (3.2.4) according to Section 6.2.3 of EC 2 [3.10].

$$V_{Rd} = \min\left(V_{Rd,s} \text{ and } V_{Rd,max}\right) \tag{3.2.4}$$

where

V_{Rd} – Design shear capacity of beams.

$V_{Rd,s}$ – Design shear capacity which can be sustained by stirrups calculated based on Section 6.2.3(3) of EC 2 [3.10].

$V_{Rd,max}$ – Maximum design shear capacity which can be sustained by a concrete section, limited by crushing of compression struts. $V_{Rd,max}$ is calculated based on Section 6.2.3(3) of EC 2 [3.10].

3.3 DERIVATION OF OBJECTIVE FUNCTIONS BASED ON FORWARD ANNs

3.3.1 Training ANN

Large datasets generated by structural software (AutoPTBEAM) are used to train forward neural networks which are then used to derive ANN-based functions which can be used in derivative-based Lagrange optimization [3.24–3.26] because ANN-based functions are differentiable. For ANN-based forward designs, the relationships between the 21 input variables ($\boldsymbol{L}$, $\boldsymbol{b}, \boldsymbol{h}, \boldsymbol{f_{ck}}, \boldsymbol{f_{cki}}, \boldsymbol{\rho_{s,bot}}, \boldsymbol{\rho_{s,top}}, \boldsymbol{f_{yk}}, \boldsymbol{f_{sw}}, \boldsymbol{\phi_p}, \boldsymbol{\rho_{p,bot}}, \boldsymbol{f_{p0.1k}}, \boldsymbol{\sigma_{pm0}}, \boldsymbol{Loss_{lt}}, \boldsymbol{\psi_1}, \boldsymbol{\psi_2}, \boldsymbol{t_0}, \boldsymbol{T}, \boldsymbol{RH}$, $\boldsymbol{q_D}, \boldsymbol{q_L}$) and 21 output variables ($\boldsymbol{\Delta_{trans}/L}, \boldsymbol{\sigma_{c,trans}/f_{cki}}, \boldsymbol{w_{trans}}, \boldsymbol{\Delta_{lt}/L}, \boldsymbol{\Delta_{incr}/L}, \boldsymbol{w_{freq}}$, $\boldsymbol{\sigma_c/f_{ck}}, \boldsymbol{\sigma_{s,bot}/f_{yk}}, \boldsymbol{\sigma_{p,bot}/f_{pk}}, \boldsymbol{M_{Ed}}, \boldsymbol{M_{Rd}}, \boldsymbol{M_{cr}}, \boldsymbol{x/d}, \boldsymbol{\varepsilon_{s,bot}/\varepsilon_{sy}}, \boldsymbol{\varepsilon_{p,bot}/\varepsilon_{py}}, \boldsymbol{V_{Ed}}, \boldsymbol{V_{Rdmax}}$, $\boldsymbol{\rho_{sw,sp}}, \boldsymbol{\rho_{sw,mid}}, \boldsymbol{CI_b}, \boldsymbol{Weight}$) are approximated based on an ANN by mapping the input parameters to the output parameters using PTM method [3.27]. Training information is summarized in Table 3.3.1. Weight and bias matrices are obtained as a result of training ANNs.

3.3.2 Formulation of an optimization for designs

An ANN-based Lagrange function for objective function of cost index $\boldsymbol{CI_b}$ is minimized for a PT beam based on ANN-based equality and inequality constraints. Algorithm for ANN-based forward neural networks shown in Figure 3.3.1 minimizes an objective function, minimizing designs of a PT beam, where six steps are presented to solve for KKT conditions based on inequality functions. KKT conditions are developed in [3.28] and [3.29] when inequality constraints exist for Lagrange optimizations. Inequality constraints are considered as either active or inactive conditions in which activated inequality constraints are transformed to equality constraints, whereas inactivated inequality constraints are ignored before solving for candidate solutions of Lagrange functions.

Table 3.3.1 Training results based on PTM

Training table: Forward - PTM 21 Inputs ($L, b, h, f_{ck}, f_{cki}, \rho_{s\text{-}bot}, \rho_{s\text{-}top}, f_{yk}, f_{sw}, D_p, \rho_{p\text{-}bot}, f_{p,0.1k}, \sigma_{pm0}, Loss_{lt}, \psi_1, \psi_2, t_0, T, RH, q_D, q_L$) - 21 Outputs ($\Delta_{trans}, \sigma_{c\text{-}trans}/f_{cki}, w_{trans}, \Delta_{lt}, \Delta_{incr}, w_{freq}, \sigma_c/f_{ck}, \sigma_{s\text{-}bot}/f_{yk}, \sigma_{p\text{-}bot}/f_{pk}, M_{Ed}, M_{Rd}, M_{mr}, x/d,$ $\varepsilon_{s\text{-}bot}/\varepsilon_{sy}, \varepsilon_{p\text{-}bot}/\varepsilon_{py}, V_{Ed}, V_{Rdmax}, \rho_{sw\text{-}sp}, \rho_{sw\text{-}mid}, CI_b, Weight$)											
No.	**Output**	**Data**	**Layers**	**Neurons**	**Validation check**	**Suggested Epoch**	**Best Epoch**	**Stopped Epoch**	**MSE Tr.perf**	**MSE T.perf**	**R at Best Epoch**
1	Δ_{trans}	150,000	5	80	1,000	100,000	9,220	10,220	7.5.E-05	1.8.E-04	0.9990
2	$\sigma_{c\text{-}trans}/f_{cki}$	150,000	5	80	1,000	100,000	24,141	25,141	8.2.E-05	2.3.E-04	0.9990
3	w_{trans}	150,000	10	50	1,000	100,000	6,050	7,050	1.1.E-04	1.1.E-03	0.9979
4	Δ_{lt}	150,000	10	80	1,000	100,000	27,481	28,481	1.1.E-04	5.9.E-04	0.9985
5	Δ_{incr}	150,000	10	80	1,000	100,000	14,209	15,209	3.6.E-04	2.0.E-03	0.9972
6	w_{freq}	150,000	5	50	1,000	100,000	24,979	25,979	4.4.E-04	1.2.E-03	0.9973
7	σ_c/f_{ck}	150,000	5	80	1,000	100,000	39,893	40,893	1.8.E-04	4.1.E-04	0.9995
8	$\sigma_{s\text{-}bot}/f_{yk}$	150,000	10	80	1,000	100,000	33,738	34,738	3.6.E-04	9.0.E-04	0.9987
9	$\sigma_{p\text{-}bot}/f_{pk}$	150,000	10	80	1,000	100,000	60,376	61,376	2.1.E-04	4.8.E-04	0.9978
10	M_{Ed}	150,000	5	50	1,000	100,000	68,109	68,118	3.0.E-09	4.6.E-09	1.0000
11	M_{Rd}	150,000	2	50	1,000	100,000	67,642	68,642	1.1.E-06	1.2.E-06	1.0000
12	M_{mr}	150,000	5	50	1,000	100,000	18,692	19,692	2.1.E-05	2.2.E-05	0.9998
13	x/d	150,000	2	50	1,000	100,000	27,025	28,025	3.1.E-05	4.2.E-05	0.9999
14	$\varepsilon_{s\text{-}bot}/\varepsilon_{sy}$	150,000	10	50	1,000	100,000	61,405	62,405	7.4.E-06	8.6.E-06	0.9999
15	$\varepsilon_{p\text{-}bot}/\varepsilon_{py}$	150,000	10	50	1,000	100,000	99,947	100,000	1.7.E-05	2.1.E-05	0.9998
16	V_{Ed}	150,000	5	50	1,000	100,000	67,927	68,814	2.2.E-09	7.2.E-09	1.0000
17	V_{Rdmax}	150,000	10	50	1,000	100,000	99,997	100,000	6.7.E-07	8.0.E-07	1.0000
18	$\rho_{sw\text{-}sp}$	150,000	10	50	1,000	100,000	34,022	35,022	1.1.E-05	5.2.E-05	0.9997
19	$\rho_{sw\text{-}mid}$	150,000	10	80	1,000	100,000	25,258	26,258	4.5.E-05	1.4.E-04	0.9935
20	CI_b	150,000	2	80	1,000	100,000	91,623	91,656	4.7.E-09	5.4.E-09	1.0000
21	$Weight$	150,000	5	50	1,000	100,000	76,571	76,590	1.2.E-09	1.3.E-09	1.0000

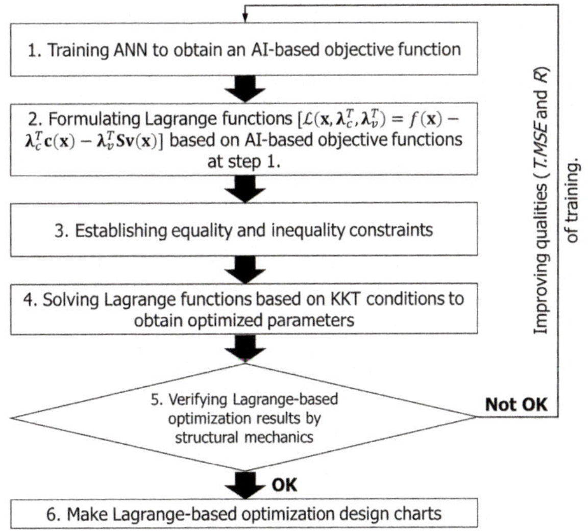

Figure 3.3.1 Algorithm for neural networks minimizing designs of a PT beam.

3.3.2.1 Derivation of ANN-based objective functions

An ANN-based generalizable function is derived for objective functions as shown in Eq. (3.3.1) [3.26, 3.30, and 3.31].

$$f(\mathbf{x}) = g^D\left(f_{lin}^N\left(\mathbf{W}^N f_t^{N-1}\left(\mathbf{W}^{N-1}\dots f_t^1\left(\mathbf{W}^1 g^N(\mathbf{x})+\mathbf{b}^1\right)\dots+\mathbf{b}^{N-1}\right)+\boldsymbol{b}^N\right)\right) \quad (3.3.1)$$

where $\mathbf{x}$ is the input parameters; N is a number of layers including hidden layers and output layer; $\mathbf{W}^n$ is a weight matrix connecting neural values between layer $n-1$ and layer n; $\mathbf{b}^n$ is a bias matrix assigned to neural values of layer n; and g^N, g^D are a normalization and de-normalization functions, respectively [3.26].

Non-linear relationships of networks are activated by activation functions f_t^n at layer n, whereas a linear activation function f_{lin}^n is used for an output layer, allowing output parameters not to be squashed. An ANN-based objective function is adopted for a Lagrange optimization. Either forward or reverse network is used to optimize a Lagrange function depending on design natures. Reverse network-based Lagrange optimization may be selected when input conflicts can be obviously avoided. Otherwise, input conflicts can be removed using forward neural networks. In Table 3.3.1 based on a combination of three types of hidden layers (2, 5, and 10) with two types of neurons (50 and 80), PTM maps 21 input variables ($\boldsymbol{L}, \boldsymbol{b}, \boldsymbol{h}, \boldsymbol{f_{ck}}, \boldsymbol{f_{cki}}, \boldsymbol{\rho_{s,bot}}, \boldsymbol{\rho_{s,top}}, \boldsymbol{f_{yk}}, \boldsymbol{f_{sw}}, \boldsymbol{\phi_p}, \boldsymbol{\rho_{p,bot}}, \boldsymbol{f_{p0.1k}}, \boldsymbol{\sigma_{pm0}}, \boldsymbol{Loss_{lt}}, \boldsymbol{\psi_1}, \boldsymbol{\psi_2}, \boldsymbol{t_0}, \boldsymbol{T}, \boldsymbol{RH}, \boldsymbol{q_D}, \boldsymbol{q_L}$) to 21 output variables ($\boldsymbol{\Delta_{trans}/L}, \boldsymbol{\sigma_{c,trans}/f_{cki}}, \boldsymbol{w_{trans}}, \boldsymbol{\Delta_{lt}/L}, \boldsymbol{\Delta_{incr}/L}, \boldsymbol{w_{freq}}, \boldsymbol{\sigma_c/f_{ck}}, \boldsymbol{\sigma_{s,bot}/f_{yk}}, \boldsymbol{\sigma_{p,bot}/f_{pk}}, \boldsymbol{M_{Ed}}, \boldsymbol{M_{Rd}}, \boldsymbol{M_{cr}}, \boldsymbol{x/d}, \boldsymbol{\varepsilon_{s,bot}/\varepsilon_{sy}}, \boldsymbol{\varepsilon_{p,bot}/\varepsilon_{py}}, \boldsymbol{V_{Ed}}, \boldsymbol{V_{Rdmax}}, \boldsymbol{\rho_{sw,sp}}, \boldsymbol{\rho_{sw,mid}}, \boldsymbol{CI_b}, \boldsymbol{Weight}$). In Table 3.3.1, PTM maps the entire 21 input parameters to each of single output parameter [3.27], presenting the best training accuracies. Cost index (CI_b) of a PT beam is optimized with 80 neurons (refer to Table 3.3.1) based on an ANN-based objective function shown in Eq. (3.3.2) with two hidden layers and one output layer for 21 input variables ($\boldsymbol{L}, \boldsymbol{b}, \boldsymbol{h}, \boldsymbol{f_{ck}}, \boldsymbol{f_{cki}}, \boldsymbol{\rho_{s,bot}}, \boldsymbol{\rho_{s,top}}, \boldsymbol{f_{yk}}, \boldsymbol{f_{sw}}, \boldsymbol{\phi_p}, \boldsymbol{\rho_{p,bot}}, \boldsymbol{f_{p0.1k}}, \boldsymbol{\sigma_{pm0}}, \boldsymbol{Loss_{lt}}, \boldsymbol{\psi_1}, \boldsymbol{\psi_2}, \boldsymbol{t_0}, \boldsymbol{T}, \boldsymbol{RH}, \boldsymbol{q_D}, \boldsymbol{q_L}$). Figure 3.3.2 illustrates an ANN representing one shown in Eq. (3.3.2).

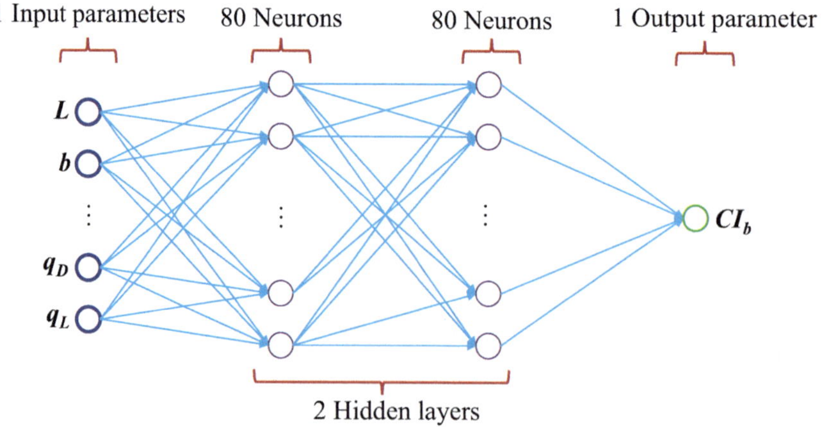

Figure 3.3.2

$$\underbrace{CI_b}_{[1\times1]} = g_{CI_b}^{D}\left(f_l^3\left(\underbrace{\mathbf{W}_{CI_b}^{3}}_{[1\times80]} f_t^2\left(\underbrace{\mathbf{W}_{CI_b}^{2}}_{[80\times80]} f_t^1\left(\underbrace{\mathbf{W}_{CI_b}^{1}}_{[80\times21]}\underbrace{g_{CI_b}^{N}(\mathbf{x})}_{[21\times1]} + \underbrace{\mathbf{b}_{CI_b}^{1}}_{[80\times1]}\right) + \underbrace{\mathbf{b}_{CI_b}^{2}}_{[80\times1]}\right) + \underbrace{\mathbf{b}_{CI_b}^{3}}_{[1\times1]}\right)\right) \tag{3.3.2}$$

3.3.2.2 Derivation of ANN-based Lagrange functions

Equation (3.3.3) represents a Lagrange function $\mathcal{L}$ of beam cost (CI_b) which is derived as a function of input variables $\mathbf{x} = [x_1, x_2,\ldots, x_n]^T$ based on Lagrange multipliers, $\lambda_c = [\lambda_1, \lambda_2,\ldots, \lambda_m]^T$ and $\lambda_v = [\lambda_1, \lambda_2,\ldots, \lambda_l]^T$, applied to equality $\mathbf{c}(\mathbf{x})$ and inequality $\mathbf{v}(\mathbf{x})$ constraints, respectively. Equation (3.3.3) is optimized to minimize beam cost (CI_b).

$$\mathcal{L}(\mathbf{x}, \lambda_c^T, \lambda_v^T) = f(\mathbf{x}) - \lambda_c^T\mathbf{c}(\mathbf{x}) - \lambda_v^T\mathbf{S}\mathbf{v}(\mathbf{x}) = CI_b - \lambda_c^T\mathbf{c}(\mathbf{x}) - \lambda_v^T\mathbf{S}\mathbf{v}(\mathbf{x}) \tag{3.3.3}$$

19 inequality functions constraining optimization are presented as shown in Table 3.3.2(b), whereas 16 simple equalities shown in Table 3.3.2(a) including $\boldsymbol{L}, \boldsymbol{f_{ck}}, \boldsymbol{f_{cki}}, \boldsymbol{f_{yk}}, \boldsymbol{f_{sw}}, \boldsymbol{\phi_p}, \boldsymbol{f_{p0.1k}}, \boldsymbol{\sigma_{pm0}}, \boldsymbol{Loss_{lt}}, \boldsymbol{\psi_1}, \boldsymbol{\psi_2}, \boldsymbol{t_0}, \boldsymbol{T}, \boldsymbol{RH}, \boldsymbol{q_D}, \boldsymbol{q_L}$ are substituted directly into networks [3.26].

Table 3.3.2 Equality and inequality conditions

(a) Equality constraints	
EQUALITY CONSTRAINTS	
$c_1 = L - 12{,}000 = 0$	Beam span fixed at 12 m
$c_2 = f_{ck} - 40 = 0$	Characteristic compressive cylinder strength of concrete at 28 days (MPa)
$c_3 = f_{cki} - 20 = 0$	Characteristic compressive cylinder strength of concrete at transfer (MPa)
$c_4 = f_{yk} - 500 = 0$	Characteristic yield strength of longitudinal rebar (MPa)
$c_5 = f_{sw} - 400 = 0$	Characteristic yield strength of stirrup (MPa)
$c_6 = \phi_p - 15.2 = 0$	Strand diameter (mm)
$c_7 = f_{p0.1k} - 1{,}640 = 0$	Characteristic yield strength of prestressing steel (MPa)
$c_8 = \sigma_{pm0} - 1{,}350 = 0$	Stress of the strand immediately after tensioning or transfer (MPa)
$c_9 = Loss_{lt} - 0.15 = 0$	Long-term loss of stress of strand (%)
$c_{10} = \psi_1 - 0.5 = 0$	Factor for frequent value of a variable action
$c_{11} = \psi_2 - 0.3 = 0$	Factor for quasi-permanent value of a variable action
$c_{12} = t_0 - 28 = 0$	Age of concrete at the time of loading (days)
$c_{13} = T - 20 = 0$	Ambient temperature (°C)
$c_{14} = RH - 70 = 0$	Relative humidity of ambient environment (%)
$c_{15} = q_D - 50 = 0$	Distributed dead load (kN/m)
$c_{16} = q_L - m = 0$ (*)	Distributed live load (kN/m) (*) q_L varies

(Continued)

Table 3.3.2 (Continued) Equality and inequality conditions

(b) Inequality constraints	
INEQUALITY CONSTRAINTS	
$v_1 = -\Delta_{trans}/L + 1/250 \geq 0$	Upward deflection of beams at transfer, Δ_{trans}, should not exceed L/250
$v_2 = -\sigma_{c,trans}/f_{cki} + 0.6 \geq 0$	Compressive concrete stress, σ_c, should not exceed $0.6f_{cki}$ under transfer combination
$v_3 = -w_{trans} + 0.2 \geq 0$	Crack width under transfer stage, w_{trans}, should not exceed 0.2 mm
$v_4 = -\Delta_{lt}/L + 1/250 \geq 0$	Long-term deflection of beams, Δ_{lt}, should not exceed L/250 under quasi-permanent combination
$v_5 = -\Delta_{incr}/L + 1/500 \geq 0$	Incremental (short-term) deflection of beams, Δ_{incr}, should not exceed L/500 under quasi-permanent combination
$v_6 = -w_{freq} + 0.2 \geq 0$	Crack width under frequent combination, w_{freq}, should not exceed 0.2 mm
$v_7 = -\sigma_c/f_{ck} + 0.6 \geq 0$	Compressive concrete stress, σ_c, should not exceed $0.6f_{ck}$ under characteristic combination
$v_8 = -\sigma_{s,bot}/f_{yk} + 0.8 \geq 0$	Tensile stress of rebar, σ_s, should not exceed $0.8f_{yk}$ under characteristic combination
$v_9 = -\sigma_{p,bot}/f_{pk} + 0.75 \geq 0$	Tensile stress of tendon, σ_p, should not exceed $0.75f_{pk}$ under characteristic combination
$v_{10} = M_{Rd} - M_{Ed} \geq 0$	Moment resistance, M_{Rd}, should be greater than or equal to factored bending moment, M_{Ed}, under ultimate load combination
$v_{11} = M_{Rd} - M_{cr} \geq 0$	Moment resistance, M_{Rd}, should be greater than or equal to cracking moment, M_{cr}, under ultimate load combination
$v_{12} = -x/d + 0.448 \geq 0$	To control section ductility, normalized neutral axis depth, x/d, should be smaller than or equal to 0.448
$v_{13} = V_{Rdmax} - V_{Ed} \geq 0$	Maximum shear resistance limited by crushing of the compression struts, V_{Rdmax}, should be greater than or equal to factored shear force, V_{Ed}
$v_{14} = -h + 1,500 \geq 0$	Architectural requirements
$v_{15} = b - 0.4h \geq 0$	
$v_{16} = -b + 1,500 \geq 0$	
$v_{17} = -\rho_{s,bot} + 0.04 \geq 0$	Rebar ratio, ρ_s, should be limited up to 4%
$v_{18} = -\rho_{s,top} + 0.04 \geq 0$	
$v_{19} = -\rho_{p,bot} + 0.0037 \geq 0$	Tendon ratio, $\rho_{p,bot}$, is limited up to 0.37%

KKT conditions are, then, solved by Newton–Raphson iteration as shown in Eq. (3.3.4). A set of partial differential equations with respect to 19 input parameters called a Jacobi of Lagrange functions $\nabla\mathcal{L}(\mathbf{x}, \lambda_c, \lambda_v)$ derived based on gradient vectors are linearized by Hessian matrix. Stationary points of Lagrange function $\mathcal{L}(\mathbf{x}, \lambda_c, \lambda_v)$ are, then, found using Newton–Raphson numerical iteration with respect to a function of input variables $\mathbf{x}$, Lagrange multipliers of equality constraints $\boldsymbol{\lambda}_c$, and Lagrange multipliers of inequality constraints λ_v. Constrained optimization problems are converted to unconstrained problems by Lagrange multipliers λ_c and λ_v for both equality and inequality constraints, respectively. Newton–Raphson iteration indicated in Eq. (3.3.4) is used to update input variable $\mathbf{x}$ and Lagrange multipliers λ. Newton–Raphson approximation is repeated until stationary points of Lagrange function $\mathcal{L}(\mathbf{x}, \lambda_c, \lambda_v)$ converge. Readers are referred to [3.26] for detailed explanation for this iteration.

$$\begin{bmatrix} \mathbf{x}^{(k+1)} \\ \lambda_c^{(k+1)} \\ \lambda_v^{(k+1)} \end{bmatrix} = \begin{bmatrix} \mathbf{x}^{(k)} \\ \lambda_c^{(k)} \\ \lambda_v^{(k)} \end{bmatrix} - \left[\boldsymbol{H}_{\mathcal{L}}\left(\mathbf{x}^{(k)}, \lambda_c^{(k)}, \lambda_v^{(k)}\right)\right]^{-1} \nabla\mathcal{L}\left(\mathbf{x}^{(k)}, \lambda_c^{(k)}, \lambda_v^{(k)}\right) \quad (3.3.4)$$

where $\left[\boldsymbol{H}_{\mathcal{L}}\left(\mathbf{x}^{(k)}, \lambda_c^{(k)}, \lambda_v^{(k)}\right)\right]$ shown in Eqs. (3.3.4)–(3.3.6) is a Hessian matrix of Lagrange functions, whereas $\nabla\mathcal{L}\left(\mathbf{x}^{(k)}, \lambda_c^{(k)}, \lambda_v^{(k)}\right)$ shown in Eqs. (3.3.4)–(3.3.6) is a Jacobi of Lagrange functions $\nabla\mathcal{L}(\mathbf{x}, \lambda_c, \lambda_v)$, which is a first derivation of Lagrange functions calculated to find gradient vectors of both objective functions and equality $\mathbf{c}(\mathbf{x})$, active inequality $\mathbf{v}(\mathbf{x})$ equations constraining objective functions.

$$\nabla\mathcal{L}(\mathbf{x}, \lambda_c, \lambda_v) = \begin{bmatrix} \nabla f(\mathbf{x}) - \boldsymbol{J}_c(\mathbf{x})^T \lambda_c - \boldsymbol{J}_v(\mathbf{x})^T \mathrm{S}\lambda_v \\ -\mathbf{c}(\mathbf{x}) \\ -Sv(\mathbf{x}) \end{bmatrix} \quad (3.3.5)$$

$$\boldsymbol{J}_c(\mathbf{x}) = \begin{bmatrix} \nabla c_1(\mathbf{x}) \\ \nabla c_2(\mathbf{x}) \\ \vdots \\ \nabla c_m(\mathbf{x}) \end{bmatrix} \quad \text{and} \quad \boldsymbol{J}_v(\mathbf{x}) = \begin{bmatrix} \nabla v_1(\mathbf{x}) \\ \nabla v_2(\mathbf{x}) \\ \vdots \\ \nabla v_l(\mathbf{x}) \end{bmatrix} \quad (3.3.6)$$

where $\boldsymbol{J}_c(\mathbf{x})$ and $\boldsymbol{J}_v(\mathbf{x})$ are Jacobian matrices constraining functions $\mathbf{c}$ and $\mathbf{v}$ at $\mathbf{x}$, respectively.

3.3.2.3 Formulation of KKT conditions based on equality and inequality constraints

A Lagrange function $\mathcal{L}$ is derived as a function of input variables $\mathbf{x} = [x_1, x_2, \ldots, x_n]^T$ in terms of Lagrange multipliers which are based on equality constrains $\lambda_c = [\lambda_1, \lambda_2, \ldots, \lambda_m]^T$ shown in Eq. (3.3.7) and inequality constrains $\lambda_v = [\lambda_1, \lambda_2, \ldots, \lambda_l]^T$ shown in Eq. (3.3.8). Lagrange multipliers of equality constraints $\lambda_c = [\lambda_{c1}, \lambda_{c2}, \ldots, \lambda_{c16}]^T$ and inequality constraints $\lambda_v = [\lambda_{v1}, \lambda_{v2}, \ldots, \lambda_{v19}]^T$ are presented in Table 3.3.2. Multivariate objective function $f(\mathbf{x})$ is subjected to equality $\mathbf{c}(\mathbf{x})=\left[c_1(\mathbf{x}),\, c_2(\mathbf{x}),\, \ldots,\, c_m(\mathbf{x})\right]^T=0$ and inequality constraints $\mathbf{v}(\mathbf{x})=\left[v_1(\mathbf{x}),\, v_2(\mathbf{x}),\, \ldots,\, v_l(\mathbf{x})\right]^T \geq 0$. Lagrange multiplier method (LMM) finds stationary points of maxima or minima of a Lagrange functions [3.26].

Equality constrains : $c_i(\mathrm{x})=0, \quad i=1,\ldots,16$

$$\boldsymbol{c}(\boldsymbol{x})=\left[c_1(\boldsymbol{x}),\, c_2(\boldsymbol{x}),\, \ldots,\, c_{16}(\boldsymbol{x})\right]^T \tag{3.3.7}$$

Inequality constrains : $v_j(\mathrm{x})\geq 0, \quad j=1,\ldots,19$

$$\boldsymbol{v}(\boldsymbol{x})=\left[\boldsymbol{v}_1(\boldsymbol{x}),\boldsymbol{v}_2(\boldsymbol{x}),\ldots,\, \boldsymbol{v}_{19}(\boldsymbol{x})\right]^T \tag{3.3.8}$$

Equation (3.3.9) is inequality matrix S which identifies active inequality constraints (v_i) for KKT conditions.

$$\boldsymbol{S}=\begin{bmatrix} s_1 & 0 & \cdots & 0 \\ 0 & s_2 & \cdots & 0 \\ \vdots & \vdots & \ddots & \vdots \\ 0 & 0 & \cdots & s_{19} \end{bmatrix}_{[19\times 19]} \tag{3.3.9}$$

1. **Formulation of inequality matrix [3.26]**
   Inequality constraints are assumed as either active or inactive conditions for KKT equations as explained in [3.26]. Lagrange optimization is formulated with only equality constraints by treating active inequality constraints to equality constraints, whereas ignoring inactive inequality. Non-zero s_i of inequality matrix S identifies active inequality constraints as shown in Eqs. (3.3.9) and (3.3.10) where active inequality constraints (v_i) with non-zero s_i and inactive inequality constraints (v_i) with zero s_i are shown. Inequality constraints (v_i) are activated when $s_i=1$ while inequality constraints (v_i) are not activated when $s_i=0$. As shown in Eq. (3.3.10), an inequality is activated to equality if a non-zero-diagonal matrix of an inequality S is specified, whereas an inactivated inequality is ignored if a zero-diagonal matrix of an inequality S is specified. Calculated stationary points must be found within the range defined by inactivated inequality constraints if diagonal matrix is deactivated by setting $\mathrm{S}=0$. Code requirements are imposed by inequality constraints. As shown in Table 3.3.2(b), EC2 [3.10] controls concrete damages by imposing $v_2=-\sigma_{c,trans}/f_{cki}+0.6\geq 0$ during prestressing states, indicating compressive concrete stress σ_c should not exceed $0.6f_{cki}$ under transfer combination. A concrete stress $\sigma_{c,trans}$ at transfer state is set to a 60% of a concrete strength f_{cki} at a time of prestressing when inequality constraint (v_2) is activated with $s_2=1$, leading to Eq. (3.3.10).

$$\begin{matrix} 1^{st} & 2^{nd} & 3^{rd} & 10^{th} & 1 & 1^{th} \end{matrix}$$

$$\boldsymbol{S} = \begin{bmatrix} s_1 = 0 & 0 & 0 & \cdots & 0 & 0 \\ 0 & s_2 = 1 & 0 & \cdots & 0 & 0 \\ 0 & 0 & 0 & \cdots & 0 & 0 \\ \vdots & \vdots & \vdots & \ddots & \vdots & \vdots \\ 0 & 0 & 0 & \cdots & s_{18} = 0 & 0 \\ 0 & 0 & 0 & \cdots & 0 & s_{19} = 0 \end{bmatrix}_{19\times 19} \begin{matrix} 1^{st} \\ 2^{nd} \\ 3^{rd} \\ \cdots \\ 18^{th} \\ 19^{th} \end{matrix} \tag{3.3.10}$$

Lagrange function with $s_2 = 1$ is obtained in Eq. (3.3.11) by substituting Eq. (3.3.10) into Eq. (3.3.3).

$$\mathcal{L}_{CI_b}(\mathbf{x}, \boldsymbol{\lambda}_c, \boldsymbol{\lambda}_v) = CI_b - \boldsymbol{\lambda}_c^T \mathbf{c}(\mathbf{x}) - \boldsymbol{\lambda}_v^T \times \begin{bmatrix} s_1 = 0 & 0 & \cdots & 0 \\ 0 & s_2 = 1 & \cdots & 0 \\ \vdots & \vdots & \ddots & \vdots \\ 0 & 0 & \cdots & s_{19} = 0 \end{bmatrix}_{[19\times 19]} \mathbf{v}(\mathbf{x}) \tag{3.3.11}$$

where $\boldsymbol{\lambda}_c$ and $\boldsymbol{\lambda}_v$ are Lagrange multipliers with respect to $c_i(\mathbf{x})$ and $v_j(\mathbf{x})$, respectively.

Inequality $v_{17} = -\rho_{s,bot} + 0.04 \geq 0$ of Table 3.3.2(b) controls a maximum limit of tensile rebar ratio. Inequality matrix **S** becomes Eq. (3.3.12) where inequality constraint (v_{17}) is activated when $s_{17} = 1$.

$$\begin{matrix} 1^{st} & 2^{nd} & 3^{rd} & \cdots & 9^{th} & 10^{th} & 11^{th} \end{matrix}$$

$$\boldsymbol{S} = \begin{bmatrix} s_1 = 0 & 0 & 0 & \cdots & 0 & 0 & 0 \\ 0 & s_2 = 0 & 0 & \cdots & 0 & 0 & 0 \\ 0 & 0 & s_3 = 0 & \cdots & 0 & 0 & 0 \\ \vdots & \vdots & \vdots & \ddots & \vdots & \vdots & \vdots \\ 0 & 0 & 0 & \cdots & s_{17} = 1 & 0 & 0 \\ 0 & 0 & 0 & \cdots & 0 & s_{18} = 0 & 0 \\ 0 & 0 & 0 & \cdots & 0 & 0 & s_{19} = 0 \end{bmatrix}_{19\times 19} \begin{matrix} 1^{st} \\ 2^{nd} \\ 3^{rd} \\ \text{M} \\ 17^{th} \\ 18^{th} \\ 19^{th} \end{matrix} \tag{3.3.12}$$

Lagrange function with $s_{17} = 1$ is obtained in Eq. (3.3.13) by substituting Eq. (3.3.12) into Eq. (3.3.3).

$$\mathbf{L}_{CI_b}\left(\mathbf{x},\boldsymbol{\lambda}_c,\boldsymbol{\lambda}_v\right)=CI_b-\boldsymbol{\lambda}_c^T\mathbf{c}\left(\mathbf{x}\right)-\boldsymbol{\lambda}_v^T \times \begin{bmatrix} s_1=0 & 0 & \cdots & \cdots & 0 & 0 \\ 0 & s_2=0 & \cdots & \cdots & 0 & 0 \\ 0 & 0 & \ddots & \cdots & 0 & 0 \\ \vdots & \vdots & \vdots & s_{17}=1 & \vdots & \vdots \\ 0 & 0 & \cdots & \cdots & s_{18}=0 & 0 \\ 0 & 0 & \cdots & \cdots & 0 & s_{19}=0 \end{bmatrix}_{[19\times19]} \mathbf{v}\left(\mathbf{x}\right) \tag{3.3.13}$$

where $\boldsymbol{\lambda}_c$ and $\boldsymbol{\lambda}_v$ are Lagrange multipliers with respect to $c_i\left(\mathbf{x}\right)$ and $v_j\left(\mathbf{x}\right)$, respectively.

2. **Optimization based on KKT conditions with all equality and inequality constraints**

   For the formulation of KKT equations, an inequality constraint is considered inactive or slacked which is ignored; otherwise, they are considered active (or tight, binding). Lagrange multipliers should be greater than zero for equality and active inequality constraints otherwise they should be zeros which are ignored when inequality constraints are inactive during a generation of Lagrange function under KKT conditions. Eq. (3.3.14) shows none of the inequalities which is active with s_i equal to 0. Lagrange functions with multipliers based on all combinations of equality and inequality constraints should be explored for an optimality under KKT conditions.

$$\mathbf{S}=\begin{bmatrix} 0 & 0 & \cdots & 0 \\ 0 & 0 & \cdots & 0 \\ \vdots & \vdots & \ddots & \vdots \\ 0 & 0 & \cdots & 0 \end{bmatrix}_{[19\times19]} \tag{3.3.14}$$

Lagrange function with all s_i equal to 0 is obtained in Eq. (3.3.15) by substituting Eq. (3.3.14) into Eq. (3.3.3).

$$\mathcal{L}_{CI_b}\left(\mathbf{x},\boldsymbol{\lambda}_c,\boldsymbol{\lambda}_v\right)=CI_b-\boldsymbol{\lambda}_c^T\mathbf{c}\left(\mathbf{x}\right)-\boldsymbol{\lambda}_v^T\begin{bmatrix} 0 & 0 & \cdots & 0 \\ 0 & 0 & \cdots & 0 \\ \vdots & \vdots & \ddots & \vdots \\ 0 & 0 & \cdots & 0 \end{bmatrix}_{[19\times19]} \mathbf{v}\left(\mathbf{x}\right) \tag{3.3.15}$$

where $\boldsymbol{\lambda}_c$ and $\boldsymbol{\lambda}_v$ are Lagrange multipliers with respect to $c_i\left(\mathbf{x}\right)$ and $v_j\left(\mathbf{x}\right)$, respectively.

Simple equality and inequality constraints shown in Table 3.3.2 are preassigned as input parameters to reduce complexity of an equality constraint vector (C) and an inequality constraint vector (V). Simple equality and inequality constraint are substituted into 21 input variables (L, b, h, f_{ck}, f_{cki}, $\rho_{s,bot}$, $\rho_{s,top}$, f_{yk}, f_{sw}, ϕ_p, $\rho_{p,bot}$, $f_{p0.1k}$, σ_{pm0}, $Loss_{lt}$, ψ_1, ψ_2, t_0, T, RH, q_D, q_L) for a first iteration based on Newton–Raphson method. A selection of good initial trial parameters is very important to converge Newton–Raphson iteration as rapidly and accurately as possible. Output parameters such as compressive concrete strain $\varepsilon_{rc\_0.003}$ are not engaged with any of objective functions, equality, and inequality constraints. These parameters are obtained as a result of a Lagrange optimization, not appearing in design scenario. Material properties, geometry such as length of the beams, loadings are defined by equality constraints, whereas inequality constraints impose requirements as specified by codes, architectures, or government. For example, a beam span of 12 m is defined by equality $c_1 = L - 12{,}000 = 0$, whereas a beam depth is imposed to be smaller than or equal to 1,500 mm by inequality $v_{14} = -h + 1{,}500 \geq 0$ constraint. Table 3.3.2 presents equalities and inequalities constraining an optimization of a simply supported PT beam. Sixteen equalities including span lengths (c_1), material properties $(c_2, c_3, c_4, c_5$, and $c_7)$, prestressing parameters $(c_6, c_8, c_9$, and $c_{12})$, factors for load combinations $(c_{10}$ and $c_{11})$, environmental conditions $(c_{13}$ and $c_{14})$, loadings $(c_{15}$ and $c_{16})$ are presented in Table 3.3.2(a), whereas Table 3.3.2(b) presents 19 inequalities imposed by the EC2 and architectures, including limitations at a transfer stage $(v_1, v_2$, and $v_3)$, restrictions at service stages (from v_4 to v_9), demands at an ultimate stage $(v_{10}, v_{11}$ and $v_{13})$, and a maximum normalized neutral axis depth (v_{12}). Inequality v_{12} imposes a normalized neutral axis depth x/d to be smaller than or equal to 0.448 to control a section ductility. It is noted that the smaller x/d, the higher neutral axis, resulting in higher rebar strains to become more ductile. In addition, six inequalities $v_{14} - v_{19}$ impose architectural requirements and upper boundary of rebar ratios.

3.4 ANN-BASED DESIGN CHARTS MINIMIZING COST BASED ON SINGLE OBJECTIVE FUNCTION WITH VARYING LIVE LOADS

3.4.1 A design example to minimize PT beam cost CI_b optimization when $q_L = 400$ kN/m $(c_{16} = q_L - 400 = 0)$

As shown in Table 3.4.1 prepared based on an ANN-based Lagrange algorithm, cost (CI_b) for a PT beam is minimized under a live load $q_L = 400$ kN/m, yielding design parameters (b, h, $\rho_{s,bot}$, $\rho_{s,top}$, and $\rho_{p,bot}$), while all 16 equalities and 19 inequalities given in Table 3.3.2 are satisfied. Based on the proposed optimization algorithm, quantities of rebars and tendons are saved to minimize CI_b when the upper boundary of beam height ($h = 1{,}500$ mm)

Table 3.4.1 Design minimizing beam cost with $\boldsymbol{q_L}$ = 400 kN/m ($c_{16} = \boldsymbol{q_L} - 400 = 0$)

Design parameters obtained by ANN-based Hong-Lagrange optimization

CI_b OPTIMIZATION Design table based on PTM								
TRAINING INPUTS			TRAINING OUTPUTS					
	Parameter	Value		Parameter	AI results	Check (AutoPT)	Difference
1	L (mm)	12000	22	Δ_{trans} (L/250=48 mm)	-7.08	-6.27	-0.804
2	b (mm)	1018	23	$\sigma_{c\text{-}trans}/f_{cki}$ (≤ 0.6)	0.354	0.357	-0.003
3	h (mm)	1500	24	w_{trans} (mm) (≤ 0.2 mm)	0.060	0.062	-0.002
4	f_{ck} (MPa)	40	25	Δ_{lt} (mm) (L/250=48 mm)	0.02	-0.48	0.501
5	f_{cki} (MPa)	20	26	Δ_{incr} (mm) (L/500=24 mm)	6.66	6.68	-0.024
6	$\rho_{s,bot}$	0.0030	27	w_{freq} (mm) (≤ 0.2 mm)	0.008	0.000	0.008
7	$\rho_{s,top}$	0.0030	28	σ_c/f_{ck} (≤ 0.6)	0.600	0.600	0.000
8	f_{yk} (MPa)	500	29	$\sigma_{s\text{-}bot}/f_{yk}$ (≤ 0.8)	0.549	0.560	-0.010
9	f_{sw} (MPa)	400	30	$\sigma_{p\text{-}bot}/f_{pk}$ (≤ 0.75)	0.750	0.761	-0.011
10	ϕ_p (mm)	15.2	31	M_{Ed} (kN·m)	8795.8	8787.2	8.63
11	$\rho_{p,bot}$	0.0026	32	M_{Rd} (kN·m) (≥ M_{Ed})	10490.6	10493.2	-2.59
12	$f_{p,0.1k}$ (MPa)	1640	33	M_{cr} (kN·m)	5863.9	5795.3	68.60
13	σ_{pm0} (MPa)	1350	34	x/d (≤ 0.448)	0.1897	0.1895	0.0002
14	$Loss_{lt}$	0.15	35	$\varepsilon_{s\text{-}bot}/\varepsilon_{sy}$	6.00	5.99	0.011
15	ψ_1	0.5	36	$\varepsilon_{p\text{-}bot}/\varepsilon_{py}$	2.47	2.47	0.006
16	ψ_2	0.3	37	V_{Ed} (kN)	2929.33	2929.05	0.28
17	t_0 (days)	28	38	V_{Rdmax} (kN) (≥ V_{Ed})	9729.202	9744.639	-15.44
18	T (°C)	20	39	$\rho_{sw\text{-}sp}$	0.00250	0.00254	-0.00004
19	RH (%)	70	40	$\rho_{sw\text{-}mid}$	0.00127	0.00126	0.00001
20	q_D (kN/m)	50.0	41	CI_b (KRW/m)	131531	131189	341.35
21	q_L (kN/m)	400.0	42	$Weight$ (kN/m)	38.17	38.18	-0.005

Noticeable errors but do not affect design

Note:

- **ANN network:**
  - 21 inputs for ANN forward network.
  - 21 outputs for ANN forward network.
- **Verification by AutoPT:**
  21 forward input (from Boxes 1 to 21) are implemented in *AutoPT* to calculate 21 forward outputs (from Boxes 22 to 42 to verify).

which is automatically selected maximizes moment arms of rebars and tendons forces. Use of concrete volume is also maximized to minimize CI_b, while use of rebars and tendons is minimized. Almost all output parameters except for $\boldsymbol{\Delta}_{lt}$ and $\boldsymbol{w}_{freq}$ are accurately obtained by ANNs. Noticeable errors in $\boldsymbol{\Delta}_{lt}$ and $\boldsymbol{w}_{freq}$ do not affect overall designs because they are small compared with limitations required by EC2 [3.10]. An absolute error of 0.501 mm of $\boldsymbol{\Delta}_{lt}$ is equal to 1.04% of a long-term deflection limit ($L/250$ = 48 mm according to EC2 [3.10]), whereas an absolute error 0.008 mm of $\boldsymbol{w}_{freq}$ is equal to 4% of a crack width limit (0.2 mm according to EC2 [3.10]) as shown in Table 3.4.1.

3.4.2 Designs criteria based on crack widths; $w_{SER} \leq$ 0.2 mm for 0 kN/m $\leq q_L \leq$ 50 kN/m when CI_b is optimized

A design table shown in Table 3.4.1 can be used to construct ANN-based design charts with varying live loads. In Section 3.1.3, ANN-based design charts of PT beam are presented based on live loads (q_L) between

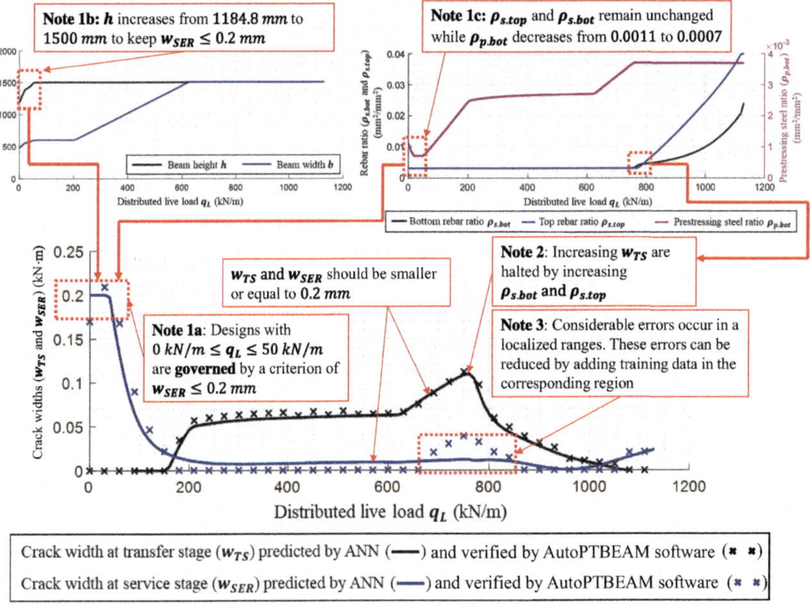

Figure 3.4.1 Designs criteria based on crack widths; $w_{SER} \leq 0.2$ mm for 0 kN/m $\leq q_L \leq$ 50 kN/m when CI_b is optimized.

0 and 1,125 kN/m which can be used to determine design parameters including section dimensions (**b** and **h**), rebar ratios ($\rho_{s.bot}$ and $\rho_{s.top}$), and tendon ratios ($\rho_{p.bot}$) when CI_b and W_c are minimized. Good accuracies are demonstrated by ANN-based design charts compared with structural mechanics-based software AutoPT.

Relationships among design parameters including section dimensions (**b** and **h**), rebar ratios ($\rho_{s.bot}$ and $\rho_{s.top}$), tendon ratios ($\rho_{p.bot}$), and crack widths (w_{TS} and w_{SER}) are presented in Figure 3.4.1 in which Note 1a describes that crack widths w_{SER} should be less than or equal to 0.2 mm when 0 kN/m $\leq q_L \leq$ 50 kN/m while CI_b is optimized. w_{SER} does not exceed 0.2 mm when beam depth h increases from 1,184.8 to 1,500 mm for live loads (q_L) which increases from 0 to 50 $\frac{\text{kN}}{\text{m}}$ as stated by Note 1b. Note 1c of Figure 3.4.1 also states that $\rho_{p.bot}$ decreases from 0.0011 to 0.0007 when $\rho_{s.bot}$ and $\rho_{s.top}$ remain unchanged between 0 kN/m $\leq q_L \leq$ 50 kN/m when increasing h. Material quantities indicated by section dimensions (b and h), tendon ratios ($\rho_{p.bot}$), and rebar ratios ($\rho_{s.bot}$ and $\rho_{s.top}$) are automatically calculated to minimize CI_b, while satisfying 19 inequalities shown in Table 3.3.2(b). A sudden reduction of crack widths (w_{TS}) is observed when rebar ratios ($\rho_{s.bot}$ and $\rho_{s.top}$) increase from q_L = 750 kN/m according to Note 2. Note 3 discusses noticeable design errors that are found with crack width at a service stage (w_{SER}) between 700 kN/m $\leq q_L \leq$ 800 kN/m, even if these errors can be reduced by adding training data in the corresponding ranges.

3.4.3 Design criteria based on tendon and concrete stresses ($\rho_{p,bot}/f_{pk} \leq 0.75$ between $50\,\text{kN/m} \leq q_L \leq 625\,\text{kN/m}$ and $\sigma_c/f_{ck} \leq 0.6$ between $200\,\text{kN/m} \leq q_L \leq 625\,\text{kN/m}$) when CI_b is optimized

3.4.3.1 Figure 3.4.2

Relationships among section dimensions (b and h), rebar ratios ($\rho_{s.bot}$ and $\rho_{s.top}$), tendon ratios ($\rho_{p.bot}$), and stresses (σ_c, $\sigma_{s,bot}$, and $\sigma_{p.bot}$) are shown in Figure 3.4.2 where stress, $\sigma_{p.bot}/f_{pk}$, at tendon should be less than or equal to 0.75 when 50 kN/m $\leq q_L \leq$ 200 kN/m according to Note 1a. Note 1b also states $\sigma_{p.bot}/f_{pk} \leq 0.75$ is kept when bottom prestressing steel ratio $\rho_{p.bot}$ increases rapidly from 0.0007 to 0.0024 for live loads (q_L) which increases from 50 kN/m to 200 kN/m. It is noted that tension forces are distributed to a larger steel area when an area of rebars and tendons in tension, $\left(\rho_{s.bot} + \rho_{p.bot}\right) \times b \times h$ including bottom rebars ($\rho_{s.bot}$) and tendons ($\rho_{p.bot}$), increases by adding tendons ($\rho_{p.bot}$), and hence, stress at bottom rebars ($\sigma_{s.bot}$) and tendons ($\sigma_{p.bot}$) do not increase even when q_L increases. Note 2a also states that $\sigma_c/f_{ck} \leq 0.6$ and $\sigma_{p.bot}/f_{pk} \leq 0.75$ are kept when beam depth

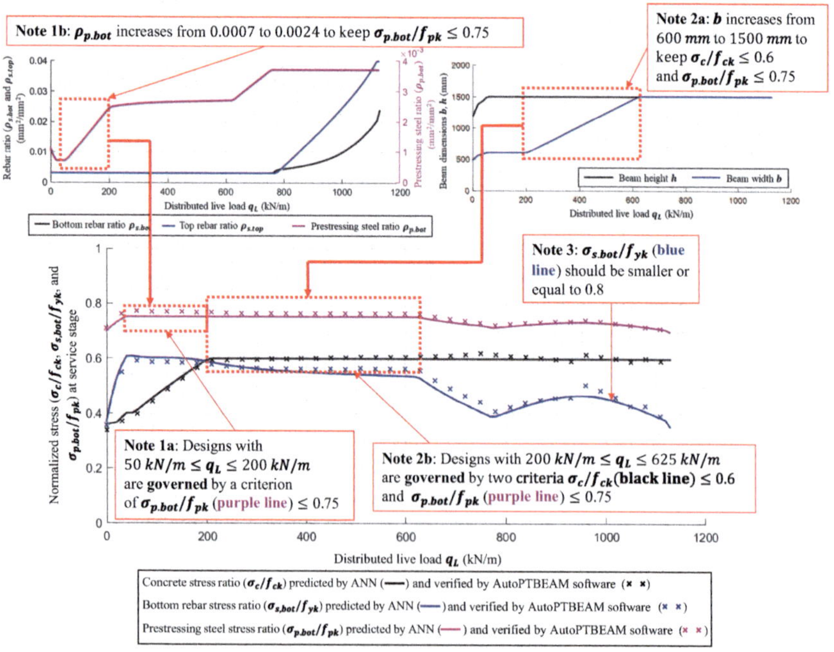

Figure 3.4.2 Design criteria based on stresses; $\sigma_{p.bot}/f_{pk} \leq 0.75$ for 50 kN/m $\leq q_L \leq$ 625 kN/m and $\sigma_c/f_{ck} \leq 0.6$ for 200 kN/m $\leq q_L \leq$ 625 kN/m when CI_b is optimized.

h increases from 600 to 1,500 mm for live loads (q_L) which increases from 200 to 625 kN/m, enlarging concrete area under compression, leading to increased moment arms of rebars and tendons forces. As shown in Figure 3.4.2, concrete stresses (σ_c) and tendon stresses ($\sigma_{p.bot}$) reach, simultaneously, their limits of 0.6 f_{ck} and 0.75 f_{pk}, respectively, in designs with 200 kN/m $\leq q_L \leq$ 625 kN/m. When 200 kN/m $\leq q_L \leq$ 625 kN/m, Note 2b describes that concrete compression stresses (σ_c) are equal to 0.6 f_{ck}, whereas tendon stresses ($\sigma_{p.bot}$) reach their limits 0.75 f_{pk}. Rebar stresses ($\sigma_{s.bot}$) at all loads (q_L) are less than 0.8 as stated by Note 3, meeting a criterion of $\sigma_{s.bot}/f_{yk} \leq 0.8$.

3.4.3.2 Figure 3.4.3

In Figure 3.4.3, designs for live loads between 625 kN/m $\leq q_L \leq$ 1,125 kN/m are governed by a criterion of $\frac{\sigma_c}{f_{ck}} \leq 0.6$ because concrete stresses (σ_c) in the live load range between 625 kN/m $\leq q_L \leq$ 1,125 kN/m reach $0.6f_{ck}$ as stated in Note 1a of Figure 3.4.3 which shows how stresses $\left(\sigma_c, \sigma_{s,bot}, \text{and } \sigma_{p.bot}\right)$ are developed. Note 1a states that concrete stresses, $\frac{\sigma_c}{f_{ck}}$, should be less than

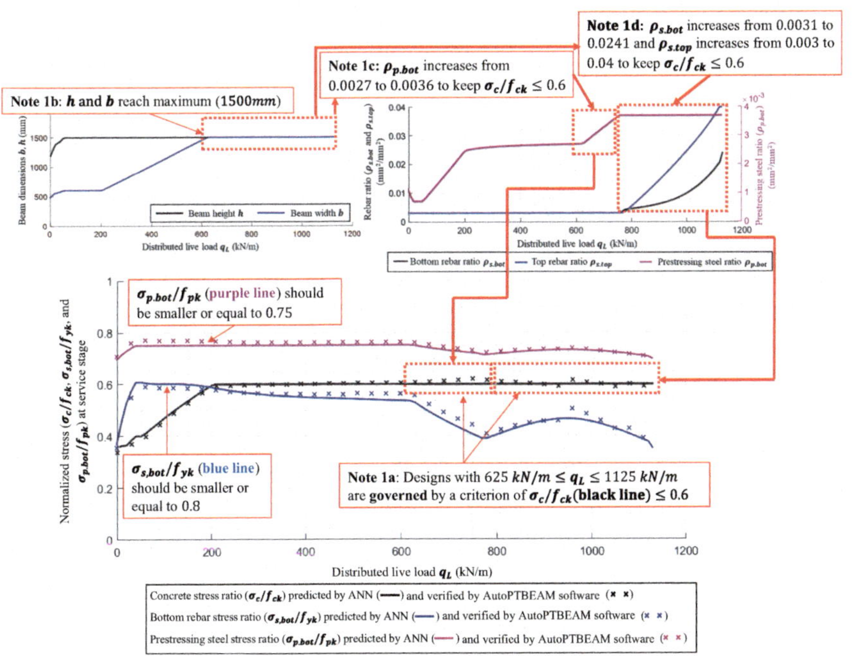

Figure 3.4.3 Design criteria based on concrete stresses; $\sigma_c/f_{ck} \leq 0.6$ for 625 kN/m $\leq q_L \leq$ 1,125 kN/m when CI_b is optimized.

or equal to 0.6 when 625 kN/m $\le q_L \le$ 1,125 kN/m. As stated in Note 1b, section dimensions (*b* and *h*) cannot exceed 1,500 mm due to an inequality constraint, $v_{14} = -h + 1{,}500 \ge 0$, shown in Table 3.3.2(b) even when q_L increase from 625 to 1,125 kN/m, and hence, ANN increases tendon ratios ($\rho_{p.bot}$) from 0.0027 to 0.0036 as an alternative when live loads (q_L) increase from 625 to 750 kN/m as indicated in Note 1c. When live loads (q_L) increase from 750 to 1,125 kN/m, an ANN increases bottom and top rebar ratios ($\rho_{s.bot}$ and $\rho_{s.top}$) from 0.0031 to 0.0241 and from 0.003 to 0.04, respectively, as shown in Note 1d. It is noted that an ANN uses higher tendon ratios ($\rho_{p.bot}$) when live loads (q_L) increase from 625 to 750 kN/m after which an ANN uses higher rebar ratios ($\rho_{s.bot}$ and $\rho_{s.top}$) to keep $\sigma_c/f_{ck} \le 0.6$ when live loads (q_L) increase from 625 to 1,125 kN/m. Compressive stresses in concrete (σ_c) decrease when adding top compression rebars ($\rho_{s.top}$) because compressive stresses in concrete are redistributed to top compression rebars ($\rho_{s.top}$). Compressive strains and stresses in concrete (σ_c) can be lowered by decreasing rebar stresses ($\rho_{s.bot}$). Rebar stresses ($\rho_{s.bot}$) can be efficiently reduced by adding both higher tendon ratios ($\rho_{p.bot}$) and higher bottom rebar ratios ($\rho_{s.bot}$) as can be seen by strain compatibilities. In summary, $\frac{\sigma_c}{f_{ck}} \le 0.6$ can be maintained by adding bottom tendons ($\rho_{p.bot}$), bottom rebars ($\rho_{s.bot}$), and adding top rebars ($\rho_{s.top}$) when live loads (q_L) increase, demonstrating that an ANN-based optimization design automatically add top rebars ($\rho_{s.top}$), bottom rebars ($\rho_{s.bot}$), and bottom tendons ($\rho_{p.bot}$) to keep $\sigma_c/f_{ck} \le 0.6$ while minimizing $\boldsymbol{CI_b}$.

3.4.3.3 Figure 3.4.4

It is noted that stresses of top rebars in compression are not shown in Legends for Figure 3.4.3 because there is no design limitation related to stresses in compression top rebars. As shown in Figure 3.4.4, design table for live load q_L = 900 kN/m is made based on the design charts shown in Figures 3.4.1–3.4.3 and 3.4.5–3.4.10 which are constructed by iterating live loads q_L between 0 and 1,125 kN/m preassigned as an equality. Optimized section sizes (*b* = 1,500 mm and *h* = 1,500 mm shown in Boxes 2 and 3, respectively), a tendon ratio ($\rho_{p.bot} = 0.0037$ shown in Box 11), and rebar ratios ($\rho_{s.bot} = 0.0062$ shown in Box 6 and $\rho_{s.top} = 0.0143$ shown in Box 7) are selected against an imposed live load q_L = 900 kN/m (one point in design chart) shown in Box 21. The design table optimizes cost CI_b of a PT beam based on the ANN-based design charts. A tendon ratio was constrained to be ≤0.0037 based on data ranges of big datasets, in which the upper boundary of tendon ratio ($\rho_{p.bot} = 0.0037$) was found. As a design result, cost is minimized as $\boldsymbol{CI_b}$ = 3,966,694 KRW/m, while all 19 inequalities shown in Table 3.3.2(b) are satisfied, especially meeting a criterion of $v_7 = -\sigma_c/f_{ck} + 0.6 \ge 0$ when 625 kN/m $\le q_L \le$ 1,125 kN/m. Input parameter

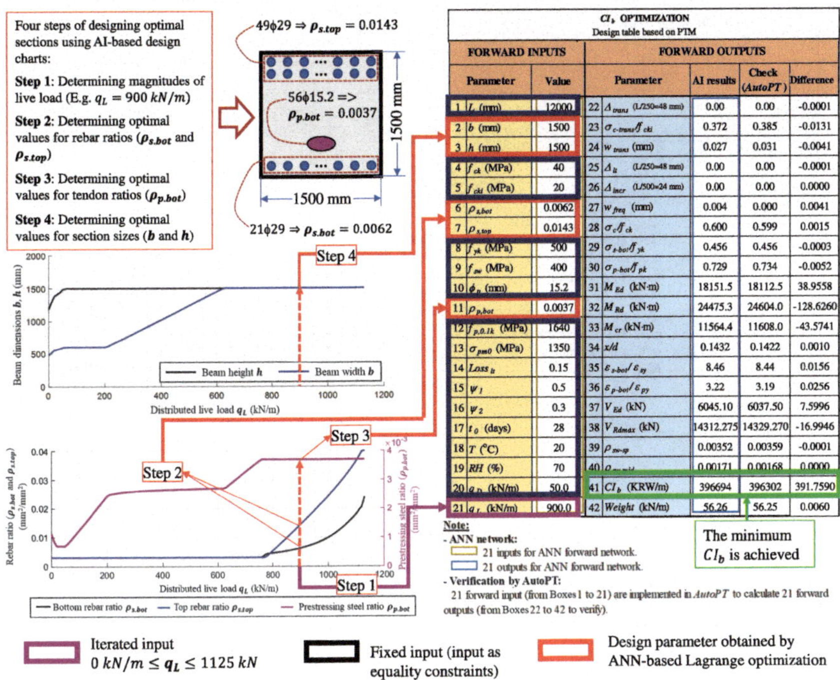

CI_b OPTIMIZATION
Design table based on PTM

FORWARD INPUTS			FORWARD OUTPUTS				
	Parameter	Value		Parameter	AI results	Check (AutoPT)	Difference
1	L (mm)	12000	22	Δ_{trans} (L/250=48 mm)	0.00	0.00	-0.0001
2	b (mm)	1500	23	$\sigma_{c\text{-}trans}/f_{cki}$	0.372	0.385	-0.0131
3	h (mm)	1500	24	w_{trans} (mm)	0.027	0.031	-0.0041
4	f_{ck} (MPa)	40	25	Δ_{lt} (L/250=48 mm)	0.00	0.00	-0.0001
5	f_{cki} (MPa)	20	26	Δ_{incr} (L/500=24 mm)	0.00	0.00	0.0000
6	$\rho_{s,bot}$	0.0062	27	w_{freq} (mm)	0.004	0.000	0.0041
7	$\rho_{s,top}$	0.0143	28	σ_c/f_{ck}	0.600	0.599	0.0015
8	f_{yk} (MPa)	500	29	$\sigma_{s\text{-}bot}/f_{yk}$	0.456	0.456	-0.0003
9	f_{sw} (MPa)	400	30	$\sigma_{p\text{-}bot}/f_{pk}$	0.729	0.734	-0.0052
10	ϕ_p (mm)	15.2	31	M_{Ed} (kN·m)	18151.5	18112.5	38.9558
11	$\rho_{p,bot}$	0.0037	32	M_{Rd} (kN·m)	24475.3	24604.0	-128.6260
12	$f_{p,0.1k}$ (MPa)	1640	33	M_{cr} (kN·m)	11564.4	11608.0	-43.5741
13	σ_{pm0} (MPa)	1350	34	x/d	0.1432	0.1422	0.0010
14	$Loss_{lt}$	0.15	35	$\varepsilon_{s\text{-}bot}/\varepsilon_{sy}$	8.46	8.44	0.0156
15	ψ_1	0.5	36	$\varepsilon_{p\text{-}bot}/\varepsilon_{py}$	3.22	3.19	0.0256
16	ψ_2	0.3	37	V_{Ed} (kN)	6045.10	6037.50	7.5996
17	t_0 (days)	28	38	V_{Rdmax} (kN)	14312.275	14329.270	-16.9946
18	T (°C)	20	39	$\rho_{sw\text{-}sp}$	0.00352	0.00359	-0.0001
19	RH (%)	70	40	$\rho_{sw\text{-}mid}$	0.00171	0.00168	0.0000
20	q_D (kN/m)	50.0	41	CI_b (KRW/m)	396694	396302	391.7590
21	q_L (kN/m)	900.0	42	$Weight$ (kN/m)	56.26	56.25	0.0060

Note:
- ANN network:
21 inputs for ANN forward network.
21 outputs for ANN forward network.
- Verification by AutoPT:
21 forward input (from Boxes 1 to 21) are implemented in *AutoPT* to calculate 21 forward outputs (from Boxes 22 to 42 to verify).

Figure 3.4.4 A design example of CI_b optimization when $q_L = 900$ kN/m.

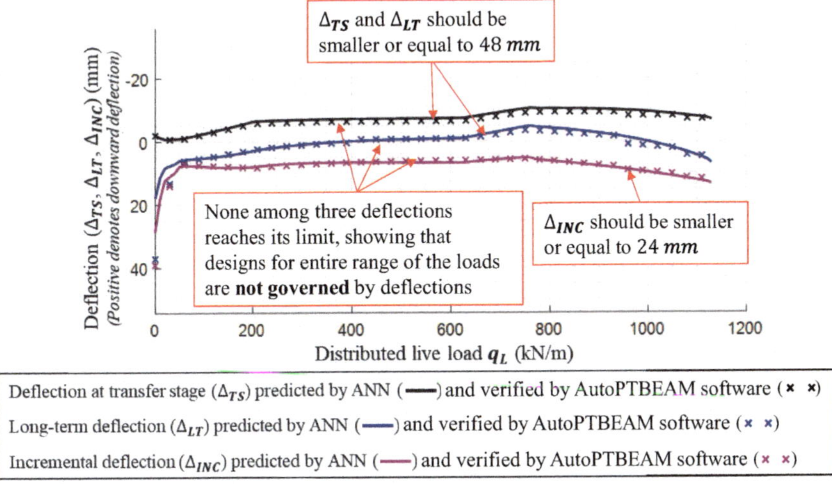

Figure 3.4.5 Design criteria based on deflections for 0 kN/m ≤ q_L ≤ 1,125 kN when CI_b is minimized.

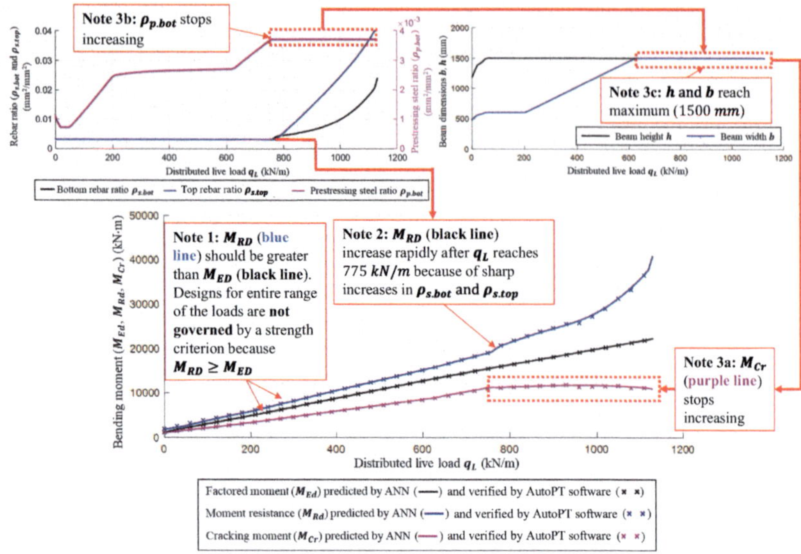

Figure 3.4.6 Design criteria based on moment demand (M_{Ed}), moment resistance (M_{Rd}), and cracking moment (M_{Cr}) for 0 kN/m $\leq q_L \leq$ 1,125 kN when CI_b is minimized.

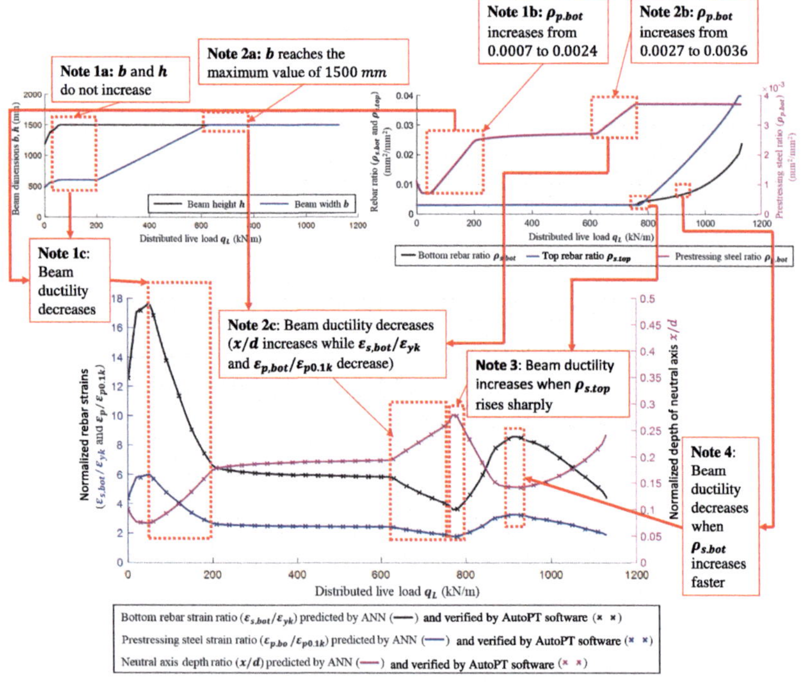

Figure 3.4.7 Design criteria based on reinforcement strains and neutral axis depth representing beam ductility for 0 kN/m $\leq q_L \leq$ 1,125 kN when CI_b is minimized.

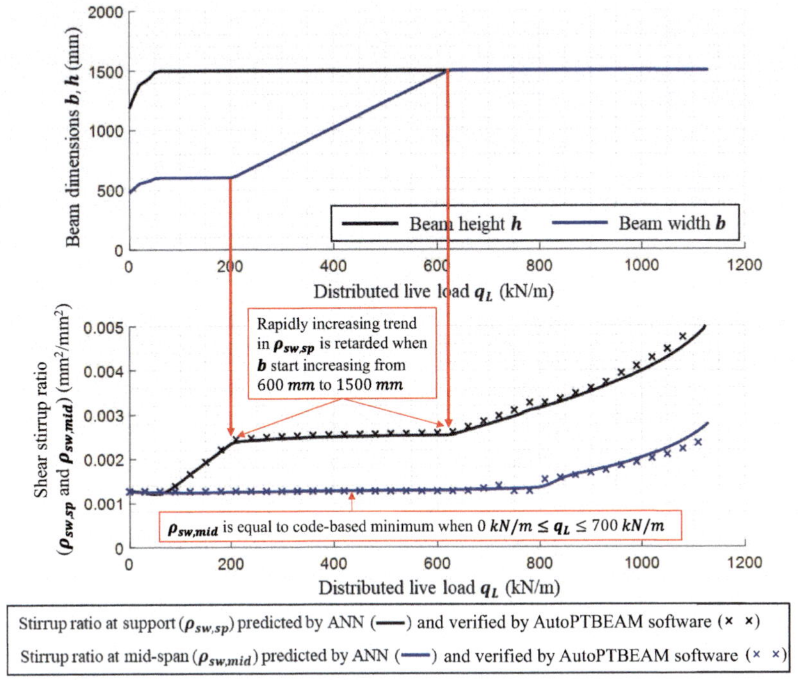

Figure 3.4.8 Design criteria based on shear stirrup ratios required for 0 kN/m $\leq q_L \leq$ 1,125 kN when CI_b is minimized.

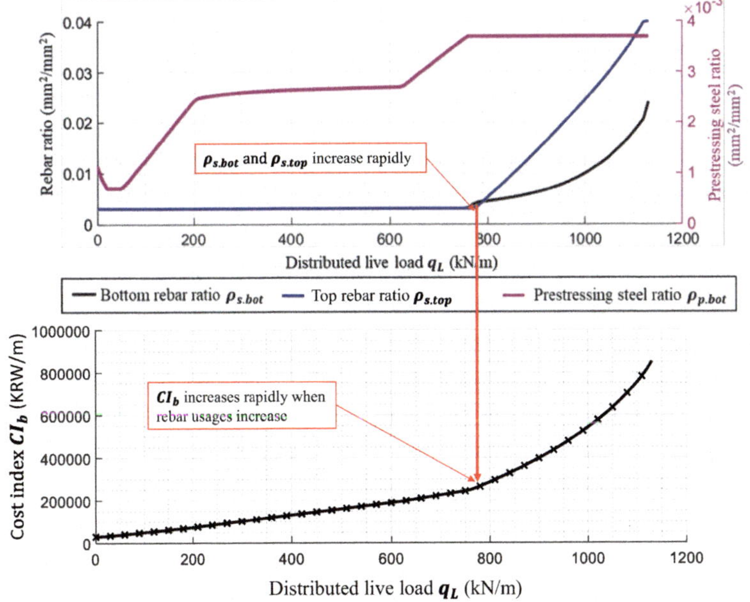

Figure 3.4.9 Minimized CI_b for 0 kN/m $\leq q_L \leq$ 1,125 kN.

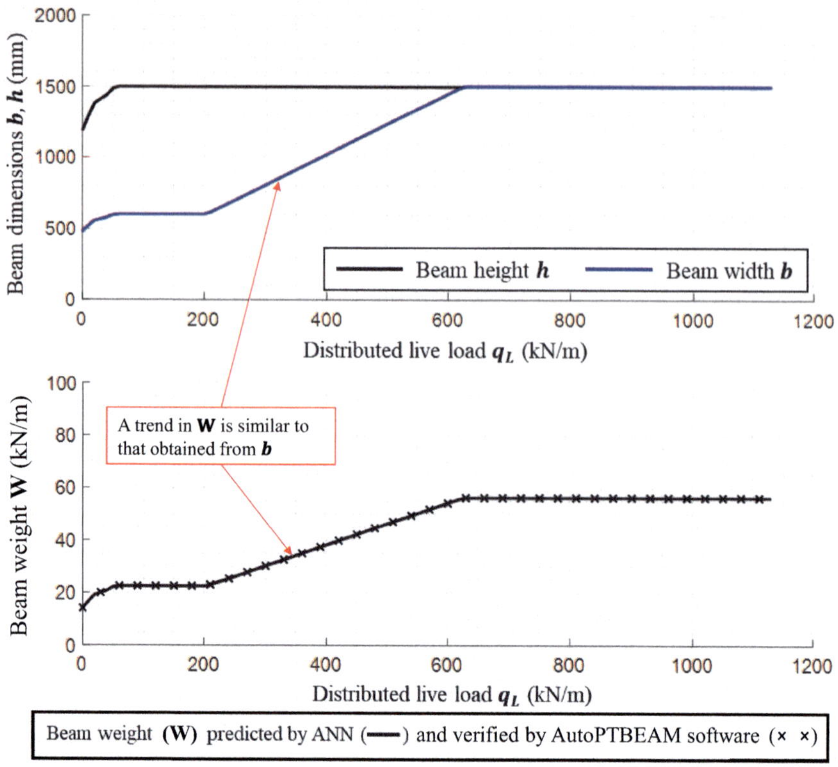

Figure 3.4.10 Beam weight obtained for 0 kN/m ≤ q_L ≤ 1,125 kN when CI_b is minimized.

such as a depth of beam is set at 1,500 mm by equality constraints in the design example, demonstrating that a diverse selection of parameters, such as rebar ratios, tendon ratios, concrete volumes, is possible when the beam is designed for large moment. However, optimized beam depth of a rectangular beam section can be identified by constraining using an inequality.

3.4.4 Design criteria based on deflections for 0 kN/m ≤ q_L ≤ 1,125 kN when CI_b is minimized

As shown in Table 3.3.2(b), deflections in beam at service stages (Δ_{TS}, Δ_{LT}, and Δ_{INC}) are constrained where long-term deflections (Δ_{LT}) and short-term deflections at transfer stages (Δ_{TS}) are limited to $L/250$ = 48 mm, whereas incremental deflections (Δ_{INC}) should be equal to or less than $L/500$ = 24 mm. However, no deflections (Δ_{TS}, Δ_{LT}, and Δ_{INC}) reach a corresponding limitation, indicating that designs are not governed by deflections (Δ_{TS}, Δ_{LT}, and Δ_{INC}) in the present charts according to Figure 3.4.5 which presents deflections obtained for 0 kN/m ≤ q_L ≤ 1,125 kN when CI_b is minimized.

3.4.5 Design criteria based on moment demand (M_{Ed}), moment resistance (M_{Rd}), and cracking moment (M_{Cr}) for 0 kN/m $\leq q_L \leq$ 1,125 kN when CI_b is minimized

Design criteria based on factored moment demand (M_{Ed}), moment resistance (M_{Rd}), and cracking moment (M_{Cr}) are illustrated in Figure 3.4.6 for 0 kN/m $\leq q_L \leq$ 1,125 kN when CI_b is minimized. Designs are not governed by a strength criterion because factored demands (M_{Ed}) are always smaller than resistances (M_{Rd}), $M_{Rd} \geq M_{Ed}$, according to Note 1. As shown in Note 2, beam resistances (M_{Rd}) increase rapidly from q_L = 775 kN/m where rebar ratios ($\rho_{s.bot}$ and $\rho_{s.top}$) start increasing rapidly as shown in Note 3b. A constant cracking moment (M_{Cr}) in designs is observed for the entire range of the loads between 750 kN/m $\leq q_L \leq$ 1,125 kN/m as shown in Note 3a when tendon ratio ($\rho_{p.bot}$) reaches the upper boundary of 0.0037 from q_L = 750 kN/m as shown in Note 3b. Section sizes (h and b) also reach an upper boundary of 1,500 mm from q_L = 750 kN/m as shown in Note 3c. Cracking moments (M_{Cr}) mainly depend on section sizes (h and b) and tendon ratios ($\rho_{p.bot}$) according to relationships described in Notes 3a, 3b, and 3c.

3.4.6 Design criteria based on reinforcement strains and neutral axis depth representing beam ductility for 0 kN/m $\leq q_L \leq$ 1,125 kN when CI_b is minimized

Relationships among section dimensions (b and h), rebar ratios ($\rho_{s.bot}$ and $\rho_{s.top}$), tendon ratios ($\rho_{p.bot}$), bottom rebar strains ($\varepsilon_{s.bot}$), tendon strains ($\varepsilon_{p.bot}$), and depths of neutral axis (x) are illustrated in Figure 3.4.7. It is observed that beam ductility described by strains ($\varepsilon_{s.bot}$ and $\varepsilon_{p.bot}$) and depths of neutral axis (x) are dependent on the tendon ratios ($\rho_{p.bot}$) when live loads (q_L) range between 0 kN/m to 775 kN/m. Normalized tendon strains ($\varepsilon_{p.bot}$) and normalized rebar strains ($\varepsilon_{s.bot}$) increase when tendon ratios ($\rho_{p.bot}$) decrease from 0.0011 to 0.0007 shown for 0 kN/m $\leq q_L \leq$ 50 kN/m of Figure 3.4.7, decreasing depth of normalized neutral axis (x) which increases a beam ductility. However, opposite trends are found in designs of the next region of Figure 3.4.7 where beams become much brittle when tendon ratios ($\varepsilon_{p.bot}$) increase from 0.0007 to 0.0024 for 50 kN/m $\leq q_L \leq$ 200 kN/m as shown in Note 1c. Reductions of beam ductility (Note 1c and Note 2c) are significantly displayed when section dimensions (b and h) do not increase (Note 1a and Note 2a), while tendon ratios ($\rho_{p.bot}$) increase rapidly from 0.0007 to 0.0024 (Note 1b) and from 0.0027 to 0.0036 (Note 2b) for 50 kN/m $\leq q_L \leq$ 200 kN/m and 600 kN/m $\leq q_L \leq$ 775 kN/m, respectively. It is also noted that, for live loads (q_L) between 50 kN/m $\leq q_L \leq$ 200 kN/m and 600 kN/m $\leq q_L \leq$ 775 kN/m, beam x/d increases while $\varepsilon_{s.bot}/\varepsilon_{yk}$ and

$\varepsilon_p/\varepsilon_{po.1k}$ decrease, reducing beam ductility as shown in Note 1c and Note 2c. Beam ductility is governed by rebar ratios ($\rho_{s.bot}$ and $\rho_{s.top}$) as shown in Notes 3 and 4 when the upper boundaries of section sizes (b and h) and tendon ratios ($\rho_{p.bot}$) are reached. Beam ductility increases when $\rho_{s.top}$ rises rapidly as shown in Note 3, whereas beam ductility decreases when $\rho_{s.bot}$ increase faster as shown in Note 4. As shown in Note 3, ductility increases when compressive rebars ($\rho_{s.top}$) increase faster than tensile rebars ($\rho_{s.bot}$) for q_L which increases from 775 to 950 kN/m. After that, beam ductility decreases for q_L which increases 950 kN/m $\leq q_L \leq$ 1,125 kN/m because tensile rebars ($\rho_{s.bot}$) increase faster than compressive rebars ($\rho_{s.top}$) as shown in Note 4.

3.4.7 Design criteria based on shear stirrup ratios required for 0 kN/m $\leq q_L \leq$ 1,125 kN when CI_b is minimized

In Figure 3.4.8, stirrup ratios at beam supports ($\rho_{sw,sp}$) and mid-spans ($\rho_{sw,mid}$) are optimized by ANNs when minimizing beam cost (CI_b). Design parameters including a beam size (b and h) are also obtained. It is noted that shear resistances of beams are contributed by concrete sections and stirrups, and hence, rapidly increasing rate in stirrup ratios at support $\rho_{sw,sp}$ is retarded at supports ($\rho_{sw,sp}$) when b increases from 600 to 1,500 mm for designs with 200 kN/m $\leq q_L \leq$ 625 kN. Small stirrup ratio of 0.0013 is provided at mid-spans ($\rho_{sw,mid}$) when live loads (q_L) range between 0 and 750 kN/m.

3.4.8 A Minimized CI_b for 0 kN/m $\leq q_L \leq$ 1,125 kN verified by large datasets and structural mechanics-based design

Cost indexes (CI_b) and beam weights (W) are optimized in Figures 3.4.9 and 3.4.10, respectively. In Figure 3.4.9, a rate of cost increase changes at q_L = 775 kN/m where rebar ratios ($\rho_{s.bot}$ and $\rho_{s.top}$) start increasing rapidly, indicating that beam prices strongly depend on rebar usages. Figure 3.4.10 also displays beam weights (W) equal to beam volumes ($L \times b \times h$) increases where beam size equal to ($b \times h$) increases, indicating beam weights (W) are directly depending on beam volumes ($L \times b \times h$). The minimized cost obtained by 133,563 conventional designs are used to verify the minimized cost obtained by ANN-based Lagrange method as shown in Figure 3.4.11. It can be observed that cost indexes (CI_b) obtained by the large datasets are bounded on the lower costs calculated using the proposed design charts.

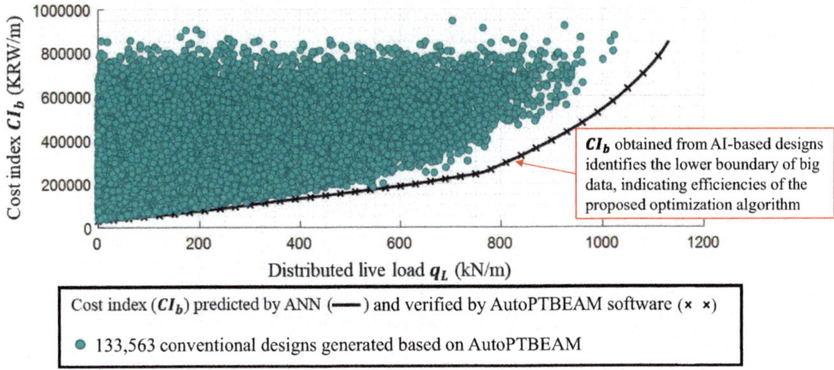

Figure 3.4.11 Cost indexes (CI_b) obtained by the large datasets bounded on the lower cost obtained by the ANN-based optimized design charts.

3.5 ANN-BASED DESIGN CHARTS MINIMIZING WEIGHT BASED ON INDIVIDUAL OBJECTIVE FUNCTION WITH VARYING LIVE LOADS

3.5.1 Design criteria based on deflections ($\Delta_{INC} \leq 24$ mm for 0 kN/m $\leq q_L \leq$ 150 kN/m) when *W* is minimized

In Table 3.5.1, design parameters (b, h, $\rho_{s,bot}$, $\rho_{s,top}$, and $\rho_{p,bot}$) minimizing weight (W) of a PT beam subjected to $q_L = 400$ kN/m are obtained based on ANN-based Lagrange optimization when all 16 equalities and 18 inequalities given in Table 3.3.2 are satisfied. ANN-based Lagrange algorithm minimizes the beam weight (W) by maximizing the beam height (h = 1500 mm), rebar ratios ($\rho_{s.bot} = 0.0292$ and $\rho_{s.top} = 0.04$), and tendon ratio $\rho_{p.bot} = 0.0037$, resulting in the smallest beam width (b) of 608 mm as shown in Table 3.5.1. ANNs select large amounts of rebars and tendons while using less concrete to minimize beam weight (W). Design parameters calculated based on ANNs well match those obtained from structural mechanics-based software AutoPTBEAM. Figure 3.5.1 describes deflections in beam at service stages (Δ_{TS}, Δ_{LT}, and Δ_{INC}). Note 1a indicates incremental deflections (Δ_{INC}) between 0 kN/m $\leq q_L \leq$ 150 kN/m reach L/500=24 mm. ANN autonomously increases h from 959.55 to 1,169.92 mm, b from 383.82 to 467.97 mm (Note 1b), whereas ANN increases $\rho_{s.bot}$ from 0.003 to 0.025, and $\rho_{s.top}$ from 0.0268 to 0.04 (Note 1c) to maintain $\Delta_{INC} \leq 24$ mm between $q_L = 0$ and 150 kN/m.

Table 3.5.1 Design minimized weight **W** with $c_{16} = q_L - 400 = 0$

***Weight* OPTIMIZATION**
Design table based on PTM

TRAINING INPUTS			TRAINING OUTPUTS				
	Parameter	**Value**		**Parameter**	**AI results**	**Check (*AutoPT*)**	**Difference**
1	L (mm)	12000	22	Δ_{trans} (L/250=48 mm)	-6.01	-5.04	-0.976
2	b (mm)	608	23	$\sigma_{c\text{-}trans}/f_{cki}$ (≤ 0.6)	0.224	0.220	0.004
3	h (mm)	1500	24	w_{trans} (mm) (≤ 0.2 mm)	0.000	0.000	0.000
4	f_{ck} (MPa)	40	25	Δ_{lt} (mm) (L/250=48 mm)	10.96	10.18	0.787
5	f_{cki} (MPa)	20	26	Δ_{incr} (mm) (L/500=24 mm)	16.55	15.21	1.340
6	$\rho_{s,bot}$	0.0292	27	w_{freq} (mm) (≤ 0.2 mm)	0.036	0.031	0.005
7	$\rho_{s,top}$	0.0400	28	σ_c/f_{ck} (≤ 0.6)	0.600	0.590	0.010
8	f_{yk} (MPa)	500	29	$\sigma_{s\text{-}bot}/f_{yk}$ (≤ 0.8)	0.289	0.308	-0.018
9	f_{sw} (MPa)	400	30	$\sigma_{p\text{-}bot}/f_{pk}$ (≤ 0.75)	0.674	0.676	-0.002
10	ϕ_p (mm)	15.2	31	M_{Ed} (kN·m)	8546.9	8510.7	36.15
11	$\rho_{p,bot}$	0.0037	32	M_{Rd} (kN·m) (≥ M_{Ed})	17632.2	17716.1	-83.90
12	$f_{p,0.1k}$ (MPa)	1640	33	M_{cr} (kN·m)	3972.3	3996.2	-23.92
13	σ_{pm0} (MPa)	1350	34	x/d (≤ 0.448)	0.2896	0.2928	-0.0032
14	$Loss_{lt}$	0.15	35	$\varepsilon_{s\text{-}bot}/\varepsilon_{sy}$	3.34	3.38	-0.043
15	ψ_1	0.5	36	$\varepsilon_{p\text{-}bot}/\varepsilon_{py}$	1.55	1.52	0.032
16	ψ_2	0.3	37	V_{Ed} (kN)	2839.08	2836.90	2.18
17	t_0 (days)	28	38	V_{Rdmax} (kN) (≥ V_{Ed})	4916.101	4747.416	168.69
18	T (°C)	20	39	$\rho_{sw\text{-}sp}$	0.00514	0.00548	-0.00034
19	RH (%)	70	40	$\rho_{sw\text{-}mid}$	0.00275	0.00256	0.00019
20	q_D (kN/m)	50.0	41	CI_b (KRW/m)	366935	366060	875.49
21	q_L (kN/m)	400.0	42	*Weight* (kN/m)	22.83	22.82	0.017

Design parameters obtained by ANN-based Hong-Lagrange optimization

Good accuracies are obtained

Note:
- **ANN network:**
  - 21 inputs for ANN forward network.
  - 21 outputs for ANN forward network.
- **Verification by AutoPT:**
  21 forward input (from Boxes 1 to 21) are implemented in *AutoPT* to calculate 21 forward outputs (from Boxes 22 to 42 to verify).

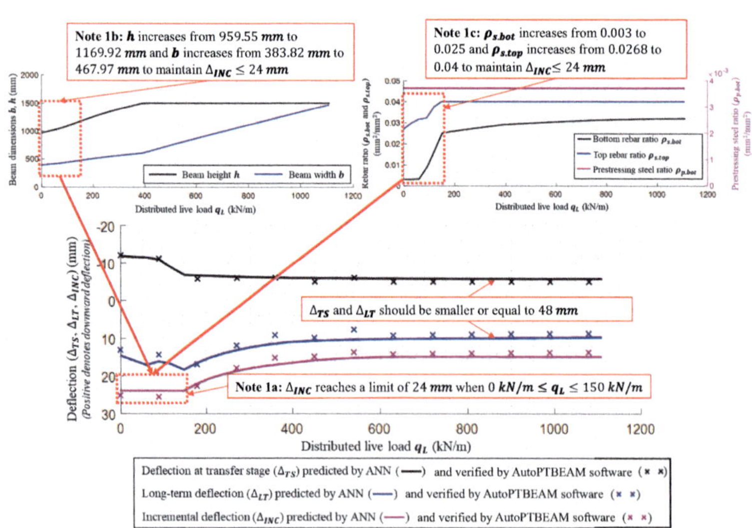

Figure 3.5.1 Design criteria based on deflections; $\Delta_{INC} \leq 24$ mm for 0 kN/m $\leq q_L \leq$ 150 kN/m when W is minimized.

3.5.2 Design criteria based on concrete stresses ($\sigma_c/f_{ck} \leq 0.6$ for 50 kN/m $\leq q_L \leq$ 1,110 kN/m) and tendon stresses ($\sigma_{p.bot}/f_{pk} = 0.75$ for 150 kN/m $\leq q_L \leq$ 1,110 kN/m) when W is minimized

3.5.2.1 *Figure 3.5.2*

Figure 3.5.2 shows relationships among section dimensions (b and h), rebar ratios ($\rho_{s,bot}$ and $\rho_{s,top}$), tendon ratios ($\rho_{p,bot}$), and stresses (σ_c, $\sigma_{s,bot}$, and $\sigma_{p.bot}$). Note 1 in Figure 3.5.2 states that the upper boundary of tendon ratio $\rho_{p.bot} = 0.0037$ (with upper boundary prestressing stresses of $\sigma_{p.bot}/f_{pk} = 0.75$ between 150 kN/m $\leq q_L \leq$ 1,110 kN/m) is selected for the entire range of the loads, providing in smallest beam sections. Note 2 in Figure 3.5.2 shows that designs to minimize **W** between 50 kN/m $\leq q_L \leq$ 1,110 kN/m are governed by a criterion for compressive stresses in concrete (σ_c) of $\sigma_c/f_{ck} = 0.6$ (black line), where a design chart can be divided into three segments.

Between 50 kN/m $\leq q_L \leq$ 150 kN/m, Note 3a states that ANN increases $\rho_{s,top}$ from 0.0322 to 0.04 and $\rho_{s,bot}$ from 0.0030 to 0.025, whereas Note 3b states that h from 1,031.20 to 1,169.92 mm and b from 412.48

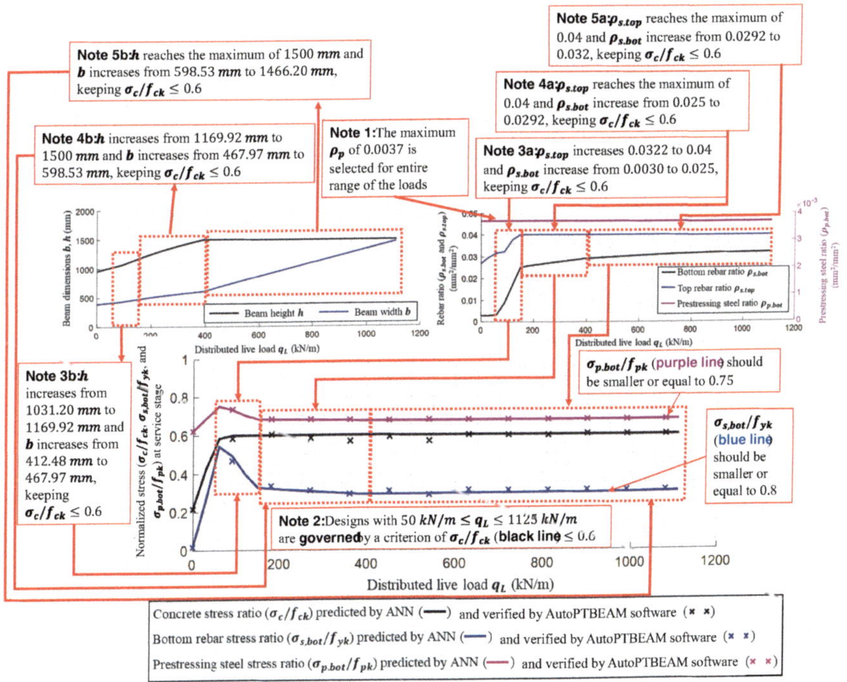

Figure 3.5.2 Design criteria based on concrete stresses; $\sigma_c/f_{ck} \leq 0.6$ for 50 kN/m $\leq q_L \leq$ 1,110 kN/m when W is minimized.

to 467.97 mm, which drastically reducing tensile stresses in rebar ($\sigma_{s,bot}$) to maintain $\sigma_c/f_{ck} \leq 0.6$.

Between 150 kN/m $\leq q_L \leq$ 400 kN/m, Note 4b states h increases from 1169.92 to 1,500 mm and b increases from 467.97 to 598.53 mm, whereas Note states that 4a $\rho_{s,bot}$ increase from 0.025 to 0.0292 while $\rho_{s,top}$ reaches the upper boundary of 0.04, leading to a slow increase in rebar tensile stresses $\sigma_{s,bot}$ (readers follow Note 4b and 5a Note for this verification) but σ_c/f_{ck} does not still exceed 0.6.

Between 150 kN/m $\leq q_L \leq$ 1,125 kN/m, σ_c/f_{ck} is kept under 0.6 by increasing b from 598.53 to 1,466.20 mm because h cannot increase above 1,500 mm (Note 5b) and by increasing $\rho_{s,bot}$ from 0.0292 to 0.032 because $\rho_{s,top}$ cannot increase above 0.04 (Note 5a).

In summary, section dimensions (b and h) and rebar ratios ($\rho_{s,bot}$ and $\rho_{s,top}$) increase to keep $\sigma_c/f_{ck} \leq 0.6$ when q_L increases from 50 to 1,110 kN/m. It is noted that ANN automatically adjusts design parameters to keep $\sigma_c/f_{ck} \leq 0.6$ while minimizing weight (**W**). Figures 3.5.1 and 3.5.2 also indicate that designs with 50 kN/m $\leq q_L \leq$ 150 kN/m are governed by two criteria which are $\Delta_{INC} \leq 24$ mm as shown in Note 1c of Figure 3.5.1 and $\sigma_c/f_{ck} \leq 0.6$ as shown in Notes 3a and 3b of Figure 3.5.2.

3.5.2.2 Figure 3.5.3

In Figure 3.5.3 corresponding to an imposed live load $q_L = 780$ kN/m shown in Box 21, a beam weight **W** of PT beam is minimized using the design charts obtained by ANN-based Lagrange optimization, where optimized section sizes (b = 1,075 mm and h = 1,500 mm shown in Boxes 2 and 3, respectively), a tendon ratio ($\rho_{p.bot} = 0.0037$ shown in Box 11), and rebar ratios ($\rho_{s.bot} = 0.0313$ and $\rho_{s.top} = 0.04$ shown in Boxes 6 and 7, respectively) are calculated. The weight W is minimized as 40.32 kN/m shown in Box 42 while all 18 inequalities in Table 3.3.2(b) are satisfied. Especially, a criterion of $v_7 = -\sigma_c/f_{ck} + 0.6 \geq 0$ governs designs when 50 kN/m $\leq q_L \leq$ 1,110 kN/m.

3.5.3 Design criteria based on crack widths ($w_{SER} \leq 0.2$ mm for 0 kN/m $\leq q_L \leq$ 1,110 kN) obtained when W is minimized

Figure 3.5.4 demonstrates relationships among rebar ratios ($\rho_{s,bot}$ and $\rho_{s,top}$), tendon ratios ($\rho_{p,bot}$), and crack widths (w_{TS} and w_{SER}), where a noticeable reduction in crack widths at service stage (w_{SER}) is observed when $q_L = 100$ kN/m because $\rho_{s,bot}$ starts increasing faster. Crack widths at both transfer stages (w_{TS}) and service stage (w_{SER}) do not reach a 0.2 mm limit, indicating that designs are not governed by crack width limits.

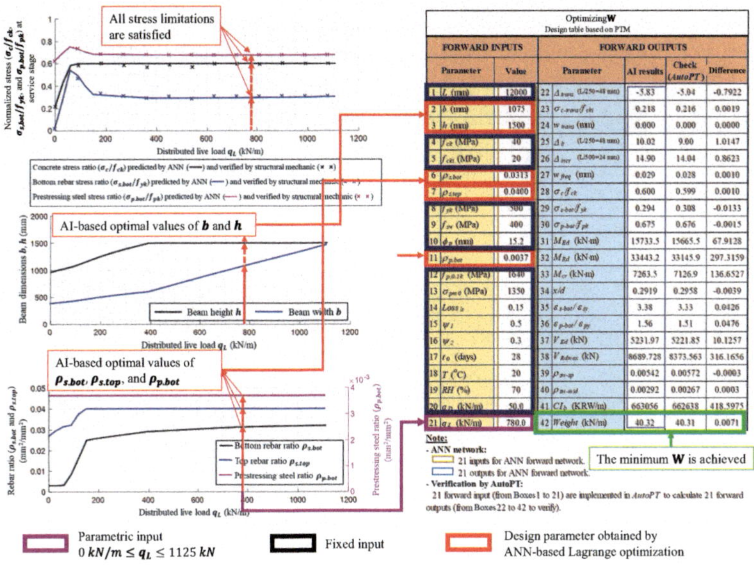

FORWARD INPUTS			FORWARD OUTPUTS				
	Parameter	Value		Parameter	AI results	Check (*AutoPT*)	Difference
1	L (mm)	12000	22	Δ_{trans} (L/250=48 mm)	-5.83	-5.04	-0.7922
2	b (mm)	1075	23	$\sigma_{c\text{-}trans}/f_{cki}$	0.218	0.216	0.0019
3	h (mm)	1500	24	w_{trans} (mm)	0.000	0.000	0.0000
4	f_{ck} (MPa)	40	25	Δ_{lt} (L/250=48 mm)	10.02	9.00	1.0147
5	f_{cki} (MPa)	20	26	Δ_{incr} (L/500=24 mm)	14.90	14.04	0.8623
6	$\rho_{s.bot}$	0.0313	27	w_{freq} (mm)	0.029	0.028	0.0010
7	$\rho_{s.top}$	0.0400	28	σ_c/f_{ck}	0.600	0.599	0.0010
8	f_{yk} (MPa)	500	29	$\sigma_{s\text{-}bot}/f_{yk}$	0.294	0.308	-0.0133
9	f_{yw} (MPa)	400	30	$\sigma_{p\text{-}bot}/f_{pk}$	0.675	0.676	-0.0015
10	ϕ_p (mm)	15.2	31	M_{Ed} (kN·m)	15733.5	15665.5	67.9128
11	$\rho_{p.bot}$	0.0037	32	M_{Rd} (kN·m)	33443.2	33145.9	297.3159
12	$f_{p0.1k}$ (MPa)	1640	33	M_{cr} (kN·m)	7263.5	7126.9	136.6527
13	σ_{pm0} (MPa)	1350	34	x/d	0.2919	0.2958	-0.0039
14	$Loss_{lt}$	0.15	35	$\varepsilon_{s\text{-}bot}/\varepsilon_{sy}$	3.38	3.33	0.0426
15	ψ_1	0.5	36	$\varepsilon_{p\text{-}bot}/\varepsilon_{py}$	1.56	1.51	0.0476
16	ψ_2	0.3	37	V_{Ed} (kN)	5231.97	5221.85	10.1257
17	t_0 (days)	28	38	V_{Rdmax} (kN)	8689.728	8373.563	316.1656
18	T (°C)	20	39	$\rho_{pt\text{-}up}$	0.00542	0.00572	-0.0003
19	RH (%)	70	40	$\rho_{pt\text{-}mid}$	0.00292	0.00267	0.0003
20	q_D (kN/m)	50.0	41	CI_b (KRW/m)	663056	662638	418.5975
21	q_L (kN/m)	780.0	42	$Weight$ (kN/m)	40.32	40.31	0.0071

Figure 3.5.3 A design example of W optimization when $q_L = 780$ kN/m.

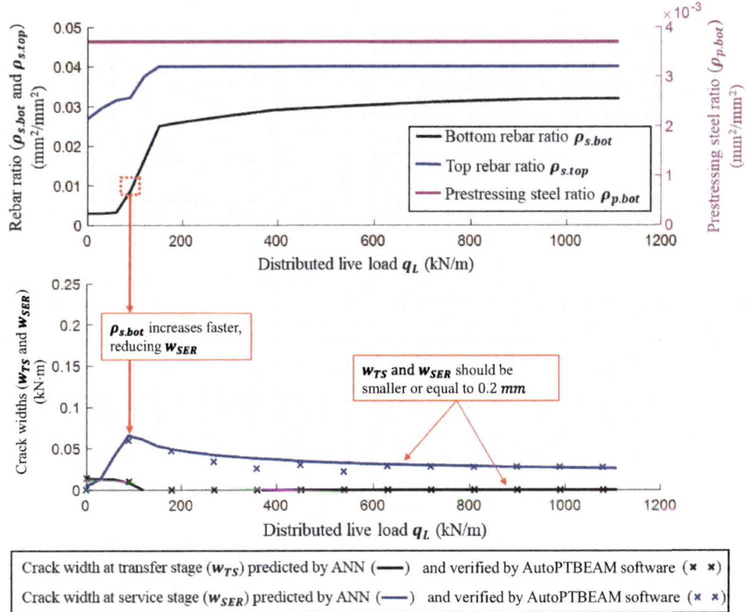

Figure 3.5.4 Design criteria based on crack widths; $w_{SER} \le 0.2$ mm for 0 kN/m $\le q_L \le$ 1,110 kN obtained when W is minimized.

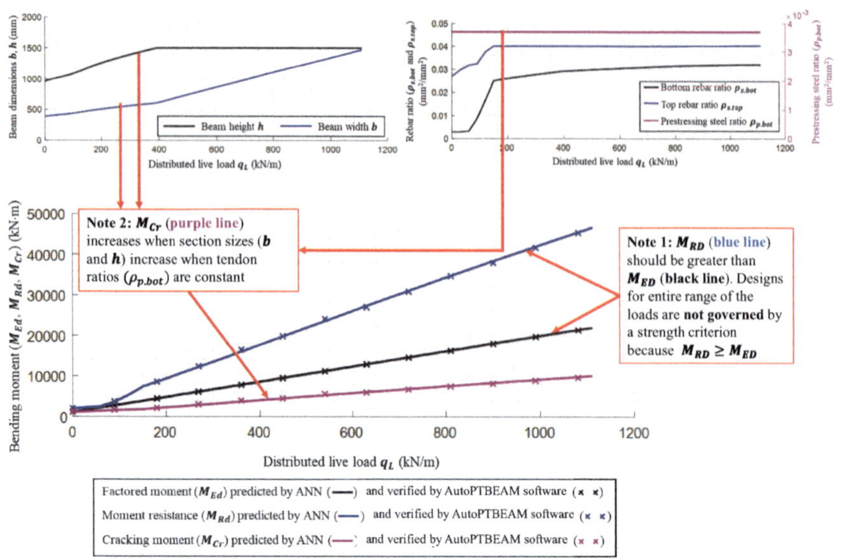

Figure 3.5.5 Design criteria based on moment demand (M_{Ed}), moment resistance (M_{Rd}), and cracking moment (M_{Cr}) for 0 kN/m ≤ q_L ≤ 1,110 kN/m when W is minimized.

3.5.4 Design criteria based on moment demand (M_{Ed}), moment resistance (M_{Rd}), and cracking moment (M_{Cr}) for 0 kN/m ≤ q_L ≤ 1,110 kN) when W is minimized

Figure 3.5.5 illustrates moment demand (M_{Ed}), moment resistance (M_{Rd}), and cracking moment (M_{Cr}) of a PT beam, where Note 1 indicates that demands (M_{Ed}) are always smaller than resistances (M_{Rd}), indicating that designs are not governed by a strength criterion because $M_{Rd} \geq M_{ED}$. In addition, Note 2 of Figure 3.5.5 indicates that beam section dimensions (*b* and *h*) should increase when cracking moments (M_{Cr}) increase for 0 kN/m ≤ q_L ≤ 1,110 kN/m. It is noted that tendon ratios (*ρp,bot*) are constant whereas beam height (*h*) becomes also constant from 370kN/m.

3.5.5 Design criteria based on reinforcement strains and neutral axis depth representing beam ductility obtained for 0 kN/m ≤ q_L ≤ 1,110 kN/m when W is minimized

Figure 3.5.6 explains relationships among rebar ratios ($\rho_{s,bot}$ and $\rho_{s,top}$), tendon ratios ($\rho_{p,bot}$), bottom rebar strains ($\varepsilon_{s,bot}$), tendon strains ($\varepsilon_{p,bot}$), and depths of neutral axis (x). It can be observed that beam ductility indicated

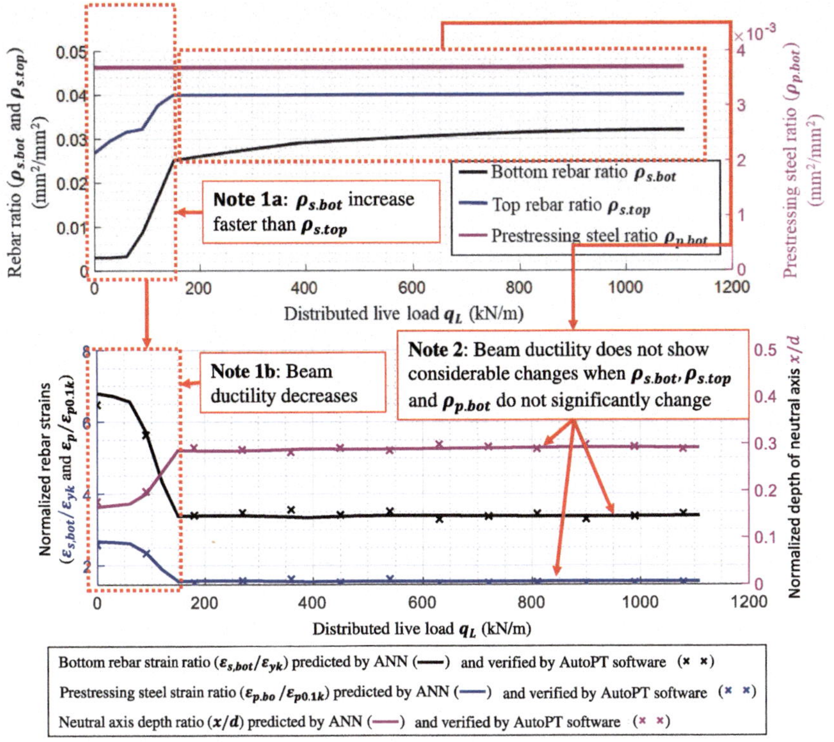

Figure 3.5.6 Design criteria based on reinforcement strains and neutral axis depth representing beam ductility obtained for 0 kN/m ≤ q_L ≤ 1,110 kN/m when W is minimized.

by strains ($\boldsymbol{\varepsilon}_{s,bot}$ and $\boldsymbol{\varepsilon}_{p,bot}$) and depths of neutral axis ($\boldsymbol{x}$) depends on rebar ratios ($\boldsymbol{\rho}_{s,bot}$ and $\boldsymbol{\rho}_{s,top}$) because a constant tendon ratio ($\boldsymbol{\rho}_{p,bot}$) is used for the entire load range. Notes 1a and 1b of Figure 3.5.6 show that beam ductility decreases with 0 kN/m ≤ q_L ≤ 150 kN/m because tensile rebars ($\boldsymbol{\rho}_{s,bot}$) increase faster than compressive rebars ($\boldsymbol{\rho}_{s,top}$). After that, beam ductility becomes stable when q_L increases from 150 to 1,110 kN/m because there is no significant changes in rebar ratios ($\boldsymbol{\rho}_{s,bot}$ and $\boldsymbol{\rho}_{s,top}$) and tendon ratios ($\boldsymbol{\rho}_{p,bot}$), as indicated by Note 2.

3.5.6 Design criteria based on shear stirrup ratios required for 0 kN/m ≤ q_L ≤ 1,110 kN when W is minimized

Figure 3.5.7 shows required stirrup ratios at beam supports ($\boldsymbol{\rho}_{sw,sp}$) and mid-spans ($\boldsymbol{\rho}_{sw,mid}$) calculated by an ANN-based Lagrange optimization when section sizes ($\boldsymbol{b}$ and $\boldsymbol{h}$) are determined to minimize beam weights (**W**) as

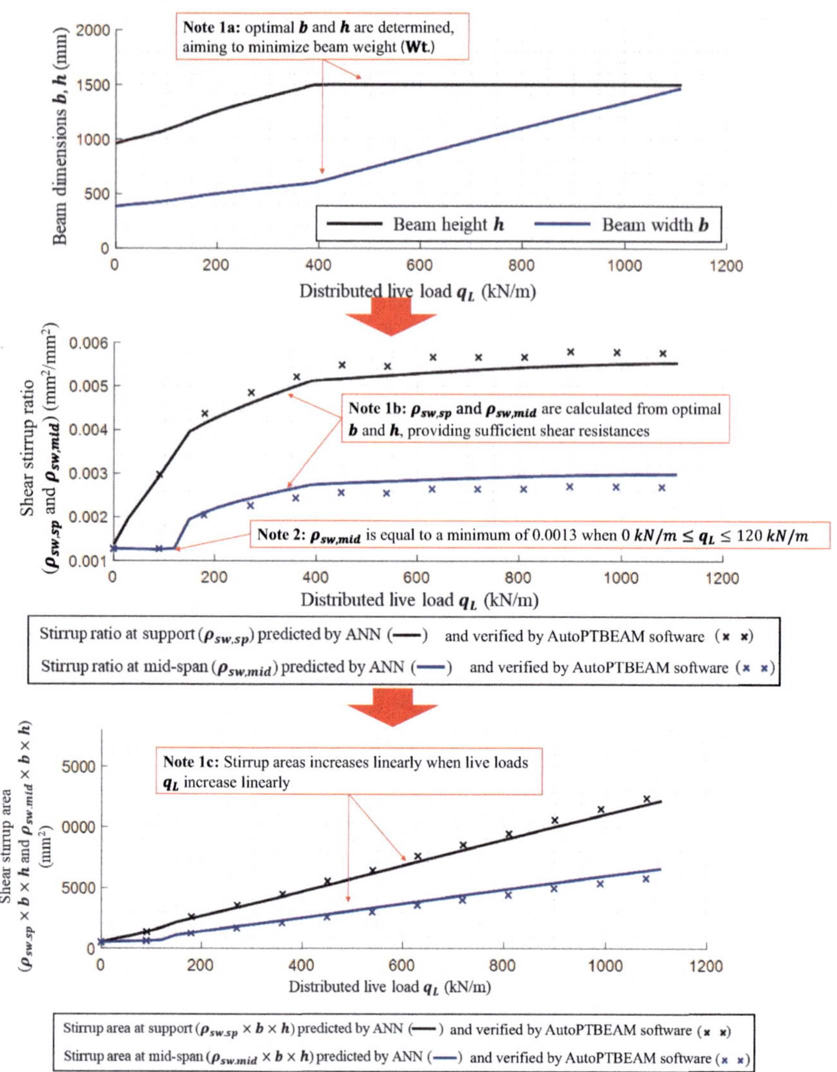

Figure 3.5.7 Design criteria based on shear stirrup ratios required for 0 kN/m $\leq q_L \leq$ 1,110 kN when W is minimized.

discussed in Notes 1a and 1b of Figure 3.5.7. Note 1c indicates that calculated stirrup ratios ($\rho_{sw,sp}$ and $\rho_{sw,mid}$) are linearly increasing when live loads q_L increase. Besides, a minimum stirrup ratio of 0.0013 is calculated at mid-spans ($\rho_{sw,mid}$) when live loads (q_L) range between 0 and 120 kN/m as stated by Note 2.

3.5.7 Beam weights (W) obtained by the large datasets bounded on the lower weight obtained by the ANN-based optimized design charts

Figures 3.5.8 and 3.5.9 illustrate cost indexes (CI_b) and beam weights (**W**) when beam weights (**W**) are minimized, where monotonically increasing tendency is observed for both cost indexes (CI_b) and beam weights (**W**). In Figure 3.5.9, beam height ***h*** reaches maximum of 1500mm earlier than beam width ***b*** which increases gradually to maximum of 1500. Figure 3.5.10 verifies the

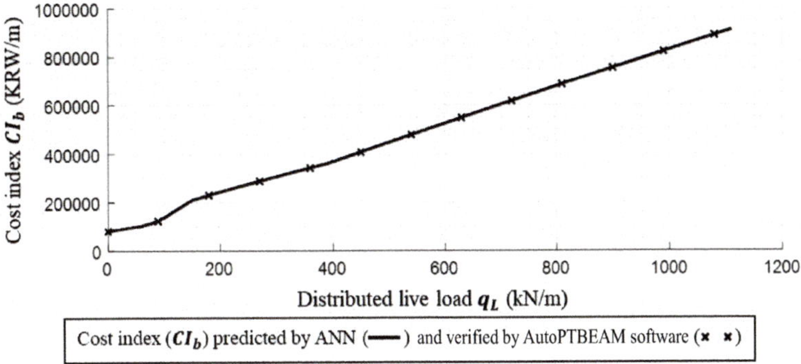

Figure 3.5.8 Cost indexes for 0 kN/m ≤ q_L ≤ 1,110 kN when W is minimized.

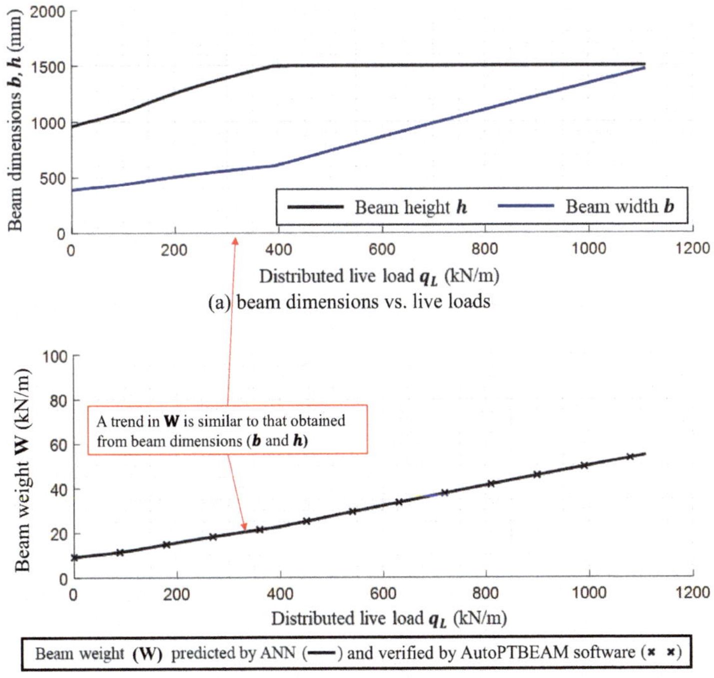

Figure 3.5.9 Minimized W for 0 kN/m ≤ q_L ≤ 1,110 kN.

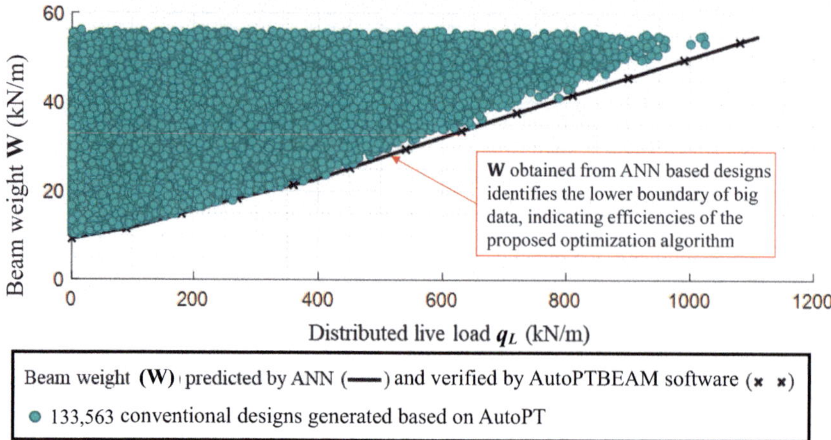

Figure 3.5.10 Beam weights (*W*) obtained by the large datasets bounded on the lower weight obtained by the ANN-based optimized design charts.

provided designs with minimum beam weights (**W**) obtained by 133,563 conventional designs. It can be observed that beam weights (**W**) obtained by the large datasets are bounded on the lower weight obtained by the costs calculated using design charts based on the ANN-based Lagrange optimization (refer to Figure 3.5.9(b) for ANN-based Lagrange optimization).

3.6 ANN-BASED DESIGN CHARTS MINIMIZING BEAM DEPTHS BASED ON INDIVIDUAL OBJECTIVE FUNCTION WITH VARYING LIVE LOADS

3.6.1 Design criteria based on deflections ($\Delta_{INC} \leq 24$ mm for 0 kN/m $\leq q_L \leq$ 425 kN/m) when *h* is minimized

Beams are sometimes required to have minimum depths due to architectural reasons. Table 3.6.1 minimizes beam depth h for a PT beam under a live load $q_L = 400$ kN/m shown in Box 21 based on an ANN-based Lagrange optimization, yielding design parameters (b, h, $\rho_{s,bot}$, $\rho_{s,top}$, and $\rho_{p,bot}$), while all 16 equalities and 18 inequalities given in Table 3.3.2 are satisfied. The upper boundary of the beam width (1,500 mm), rebar ratios ($\rho_{s.bot} = 0.0283$ and $\rho_{s.top} = 0.04$), and tendon ratio $\rho_{p.bot} = 0.0037$ are obtained to minimize beam depth (h). High tendon ratios will reduce deflections and increase moment resistance (M_{Rd}), and hence, beam depths decrease. ANN-based design parameters well match those obtained from structural mechanics-based software AutoPTBEAM. Table 3.6.1 limits rebar ratio to an upper boundary of 0.04 and tendon ratio to an upper boundary of 0.0037

Table 3.6.1 Design minimized h with $c_{16} = q_L - 400 = 0$

***h* OPTIMIZATION** Design table based on PTM							
TRAINING INPUTS			**TRAINING OUTPUTS**				
Parameter		**Value**	**Parameter**		**AI results**	**Check (*AutoPT*)**	**Difference**
1	L (mm)	12000	22	Δ_{trans} (L/250=48 mm)	-6.54	-6.78	0.246
2	b (mm)	1500	23	$\sigma_{c\text{-}trans}/f_{cki}$ (≤ 0.6)	0.222	0.221	0.000
3	h (mm)	966	24	w_{trans} (mm) (≤ 0.2 mm)	0.000	0.000	0.000
4	f_{ck} (MPa)	40	25	Δ_{lt} (mm) (L/250=48 mm)	17.22	17.08	0.148
5	f_{cki} (MPa)	20	26	Δ_{incr} (mm) (L/500=24 mm)	24.00	23.86	0.140
6	$\rho_{s,bot}$	0.0283	27	w_{freq} (mm) (≤ 0.2 mm)	0.035	0.036	0.000
7	$\rho_{s,top}$	0.0400	28	σ_c/f_{ck} (≤ 0.6)	0.600	0.600	0.000
8	f_{yk} (MPa)	500	29	$\sigma_{s\text{-}bot}/f_{yk}$ (≤ 0.8)	0.316	0.320	-0.005
9	f_{sw} (MPa)	400	30	$\sigma_{p\text{-}bot}/f_{pk}$ (≤ 0.75)	0.681	0.680	0.001
10	ϕ_p (mm)	15.2	31	M_{Ed} (kN·m)	8765.1	8752.2	12.90
11	$\rho_{p,bot}$	0.0037	32	M_{Rd} (kN·m) (≥ M_{Ed})	17853.6	17649.6	204.02
12	$f_{p,0.1k}$ (MPa)	1640	33	M_{cr} (kN·m)	4120.9	4068.5	52.35
13	σ_{pm0} (MPa)	1350	34	x/d (≤ 0.448)	0.2912	0.2904	0.0008
14	$Loss_{lt}$	0.15	35	$\varepsilon_{s\text{-}bot}/\varepsilon_{sy}$	3.40	3.42	-0.019
15	ψ_1	0.5	36	$\varepsilon_{p\text{-}bot}/\varepsilon_{py}$	1.54	1.53	0.006
16	ψ_2	0.3	37	V_{Ed} (kN)	2916.91	2917.39	-0.49
17	t_0 (days)	28	38	V_{Rdmax} (kN) (≥ V_{Ed})	7725.662	7523.908	201.75
18	T (°C)	20	39	$\rho_{sw\text{-}sp}$	0.00344	0.00356	-0.00012
19	RH (%)	70	40	$\rho_{sw\text{-}mid}$	0.00132	0.00126	0.00005
20	q_D (kN/m)	50.0	41	CI_b (KRW/m)	575452	575391	60.81
21	q_L (kN/m)	400.0	42	$Weight$ (kN/m)	36.22	36.23	-0.011

Design parameters obtained by ANN-based Hong-Lagrange optimization

Good accuracies are obtained

Note:

- **ANN network:**
  - 21 inputs for ANN forward network.
  - 21 outputs for ANN forward network.
- **Verification by AutoPT:**
  21 forward input (from Boxes 1 to 21) are implemented in *AutoPT* to calculate 21 forward outputs (from Boxes 22 to 42 to verify).

according to ranges of big data. An upper boundary of tendon ratio of 0.0037 can be extended when higher tendon ratio is desired. Figure 3.6.1 demonstrates deflections of a PT beam at service stages (Δ_{TS}, Δ_{LT}, and Δ_{INC}), Note 1a states that PT beam designs with 0 kN/m $\leq q_L \leq$ 425 kN/m are governed by a criterion of $\Delta_{INC} \leq 24$ mm, and hence, ANN increases h from 602.08 to 985.03 mm as shown in Note 1b. ANN also increases $\rho_{s,bot}$ from 0.003 to 0.03 and $\rho_{s,top}$ from 0.0172 to 0.04 as shown in Note 1c to maintain $\Delta_{INC} \leq 24$ mm when q_L increases from 0 to 425 kN/m. It is noted that the upper boundaries of ρ_p of 0.0037 and b of 1,500 mm are calculated for the entire load range as indicated in Note 2, so that h does not need to increase significantly to keep $\Delta_{INC} \leq 24$ mm for 0 kN/m $\leq q_L \leq$ 425 kN/m as indicated in Note 1b. In Table 3.3.2(b), b is constrained to be less than 1,500 mm by $v_{16} = -b + 1500 \geq 0$. It is noted that Figure 3.6.1 describes design criteria based on deflections in beam at service stages (Δ_{TS}, Δ_{LT}, and Δ_{INC}) for 0 kN/m $\leq q_L \leq$ 425 kN/m when h is minimized. In Table 3.3.2, upward

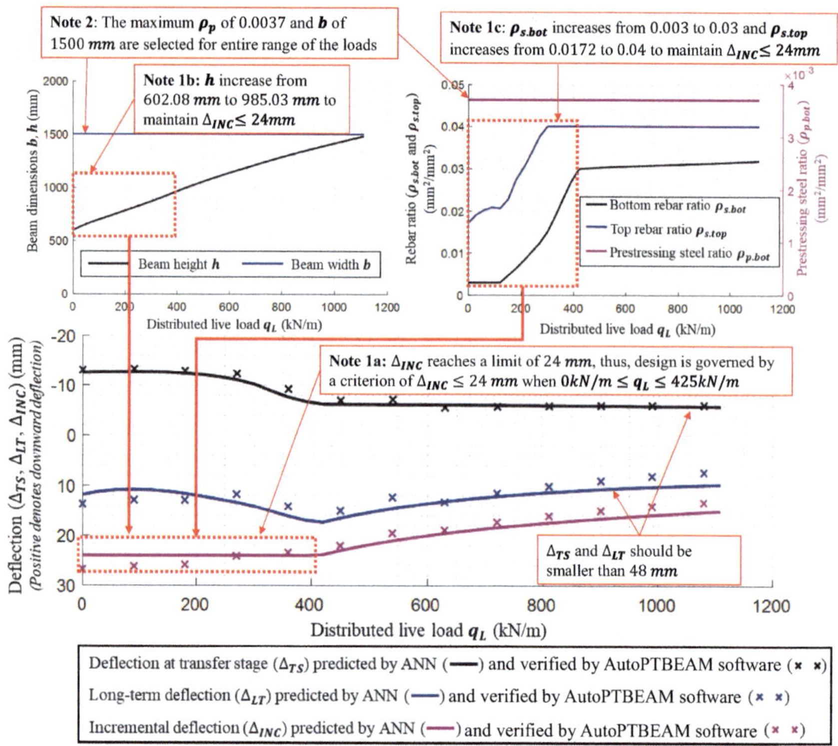

Figure 3.6.1 Design criteria based on deflections; $\Delta_{INC} \leq 24$ mm for 0 kN/m $\leq q_L \leq$ 425 kN/m when h is minimized.

deflection of beams at transfer Δ_{trans} and long-term deflection of beams Δ_{lt} should not exceed L/250=48 mm, whereas incremental deflection of beams Δ_{incr} should not exceed L/500 under quasi-permanent combination.

3.6.2 Design criteria based on concrete stresses obtained ($\sigma_c/f_{ck} \leq 0.6$ for 125 kN/m $\leq q_L \leq$ 1,110 kN/m) when h is minimized

3.6.2.1 Figure 3.6.2

Figure 3.6.2 demonstrates relationships among section dimensions (b and h), rebar ratios ($\rho_{s.bot}$ and $\rho_{s.top}$), tendon ratios ($\rho_{p.bot}$), and stresses (σ_c, $\sigma_{s,bot}$, and $\sigma_{p.bot}$) where the upper boundary of ρ_p of 0.0037 and b of 1,500 mm is calculated for the entire range of the loads as shown in Note 1, and hence, beam depth h does not need to increase significantly to keep $\sigma_c/f_{ck} \leq 0.6$ when q_L increases from 125 to 1,100 kN/m. Compressive stresses in concrete (σ_c) stated in Note 2 indicate that optimized designs are governed by

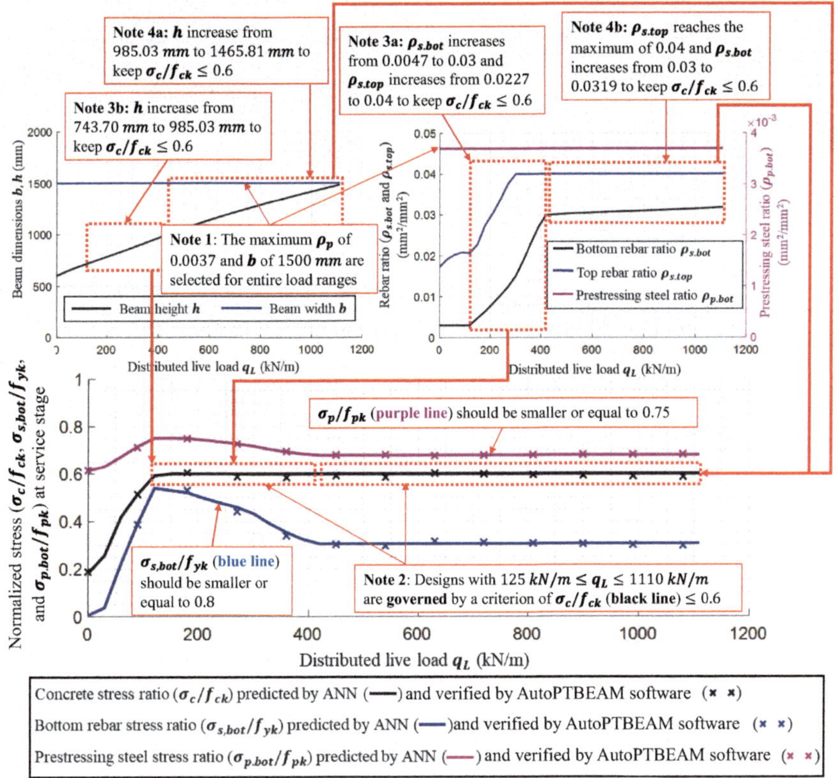

Figure 3.6.2 Design criteria based on concrete stresses obtained; $\sigma_c/f_{ck} \leq 0.6$ for 125 kN/m $\leq q_L \leq$ 1,110 kN/m when h is minimized.

a criterion of $\sigma_c/f_{ck} \leq 0.6$ for 125 kN/m $\leq q_L \leq$ 1,110 kN/m, in which design chart is divided into two parts.

In the first part, ANN increases $\rho_{s.top}$ from 0.00227 to 0.04, $\rho_{s.bot}$ from 0.0047 to 0.03 as shown in Note 3a, h from 743.70 to 985.03 mm as shown in Note 3b, significantly reducing tensile stresses in rebar ($\sigma_{s.bot}$) and maintaining $\sigma_c/f_{ck} \leq 0.6$ for 125 kN/m $\leq q_L \leq$ 425 kN/m.

In the second part, h increases from 985.03 to 1,465.81 mm as shown in Note 4a and $\rho_{s.bot}$ increase slightly from 0.03 to 0.0319, while $\rho_{s.top}$ reaches the upper boundary of 0.04 as shown in Note 4b for 425 kN/m $\leq q_L \leq$ 1,110 kN/m, leading to a constant rebar tensile stresses ($\sigma_{s.bot}$) but σ_c/f_{ck} still does not exceed 0.6 as shown in Note 1.

In summary, ANN automatically increases rebar ratios ($\rho_{s.bot}$ and $\rho_{s.top}$) and selects the upper boundary of ρ_p (0.0037) and b (1,500 mm) to keep $\sigma_c/f_{ck} \leq 0.6$ while h increases. Figures 3.5.1 and 3.5.2 also indicate that designs with 125 kN/m $\leq q_L \leq$ 425 kN/m are governed by the two criteria which are $\Delta_{INC} \leq 24$ mm and $\sigma_c/f_{ck} \leq 0.6$.

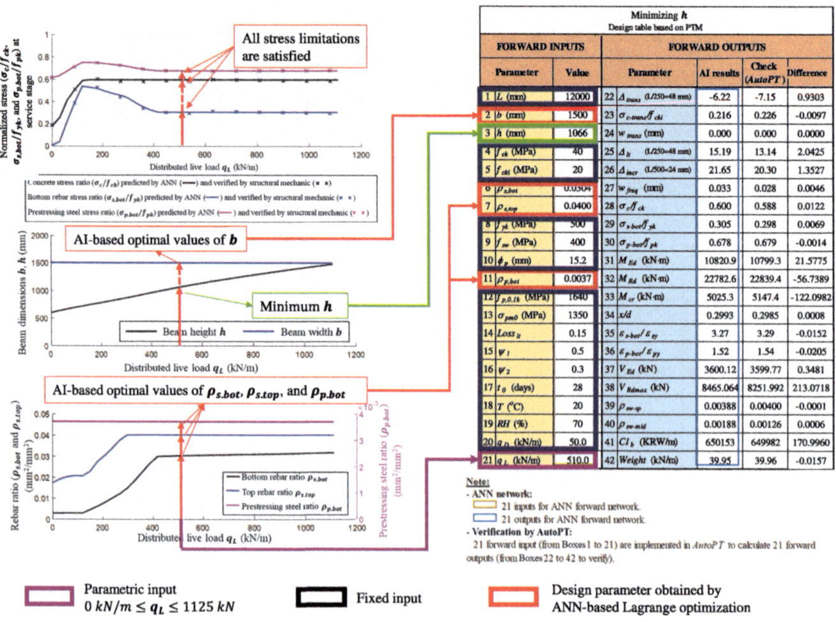

Minimizing h
Design table based on PTM

FORWARD INPUTS			FORWARD OUTPUTS				
	Parameter	Value		Parameter	AI results	Check (*AutoPT*)	Difference
1	L (mm)	12000	22	Δ_{trans} (L/250=48 mm)	-6.22	-7.15	0.9303
2	b (mm)	1500	23	$\sigma_{c\text{-}trans}/f_{cki}$	0.216	0.226	-0.0097
3	h (mm)	1066	24	w_{trans} (mm)	0.000	0.000	0.0000
4	f_{ck} (MPa)	40	25	Δ_{lt} (L/250=48 mm)	15.19	13.14	2.0425
5	f_{cki} (MPa)	20	26	Δ_{incr} (L/500=24 mm)	21.65	20.30	1.3527
6	$\rho_{s,bot}$	0.0304	27	w_{freq} (mm)	0.033	0.028	0.0046
7	$\rho_{s,top}$	0.0400	28	σ_c/f_{ck}	0.600	0.588	0.0122
8	f_{yk} (MPa)	500	29	$\sigma_{s\text{-}bot}/f_{yk}$	0.305	0.298	0.0069
9	f_{sw} (MPa)	400	30	$\sigma_{p\text{-}bot}/f_{pk}$	0.678	0.679	-0.0014
10	ϕ_p (mm)	15.2	31	M_{Ed} (kN·m)	10820.9	10799.3	21.5775
11	$\rho_{p,bot}$	0.0037	32	M_{Rd} (kN·m)	22782.6	22839.4	-56.7389
12	$f_{p,0.1k}$ (MPa)	1640	33	M_{cr} (kN·m)	5025.3	5147.4	-122.0982
13	σ_{pm0} (MPa)	1350	34	x/d	0.2993	0.2985	0.0008
14	$Loss_{lt}$	0.15	35	$\varepsilon_{s\text{-}bot}/\varepsilon_{sy}$	3.27	3.29	-0.0152
15	ψ_1	0.5	36	$\varepsilon_{p\text{-}bot}/\varepsilon_{py}$	1.52	1.54	-0.0205
16	ψ_2	0.3	37	V_{Ed} (kN)	3600.12	3599.77	0.3481
17	t_0 (days)	28	38	V_{Rdmax} (kN)	8465.064	8251.992	213.0718
18	T (°C)	20	39	$\rho_{sw\text{-}sp}$	0.00388	0.00400	-0.0001
19	RH (%)	70	40	$\rho_{sw\text{-}mid}$	0.00188	0.00126	0.0006
20	q_D (kN/m)	50.0	41	CI_b (KRW/m)	650153	649982	170.9960
21	q_L (kN/m)	510.0	42	$Weight$ (kN/m)	39.95	39.96	-0.0157

Note:
- ANN network:
  21 inputs for ANN forward network.
  21 outputs for ANN forward network.
- Verification by AutoPT:
  21 forward input (from Boxes 1 to 21) are implemented in *AutoPT* to calculate 21 forward outputs (from Boxes 22 to 42 to verify).

Figure 3.6.3 A design example of minimization of beam depth (h) when q_L = 510 kN/m.

3.6.2.2 *Figure 3.6.3*

In Figure 3.6.3 corresponding to an imposed live load $q_L = 510$ kN/m shown in Box 21, a PT beam is optimized using the design charts obtained by ANN-based Lagrange optimization, where the upper boundary of $b = 1{,}500$ mm shown in Box 2, the upper boundary of $\rho_{p.bot} = 0.0037$ shown in Box 11, and rebar ratios ($\rho_{s.bot} = 0.0304$ and $\rho_{s.top} = 0.04$ shown in Boxes 6 and 7, respectively) are identified to minimize beam depth h. Shallow beam depth $h = 1{,}066$ mm is obtained in Box 3, while all 18 inequalities in Table 3.3.2(b) are satisfied. Especially, a criterion of $v_7 = -\sigma_c/f_{ck} + 0.6 \geq 0$ which governs designs when 125 kN/m $\leq q_L \leq$ 1,110 kN/m.

3.6.3 Design criteria based on crack widths; $w_{SER} \leq 0.2$ mm for 0 kN/m $\leq q_L \leq$ 1,110 kN when h is minimized

Figure 3.6.4 illustrates relationships among rebar ratios ($\rho_{s.bot}$ and $\rho_{s.top}$), tendon ratios ($\rho_{p.bot}$), and crack widths (w_{TS} and w_{SER}). Crack widths at both transfer stages (w_{TS}) and service stage (w_{SER}) do not reach a limitation of 0.2 mm, indicating that designs are not governed by crack criterion as indicated in Note 1. It is noted that, camber cracks at transfer stage (w_{TS}) are caused by stresses generated by prestressed tendons. Note 2 stated that a noticeable

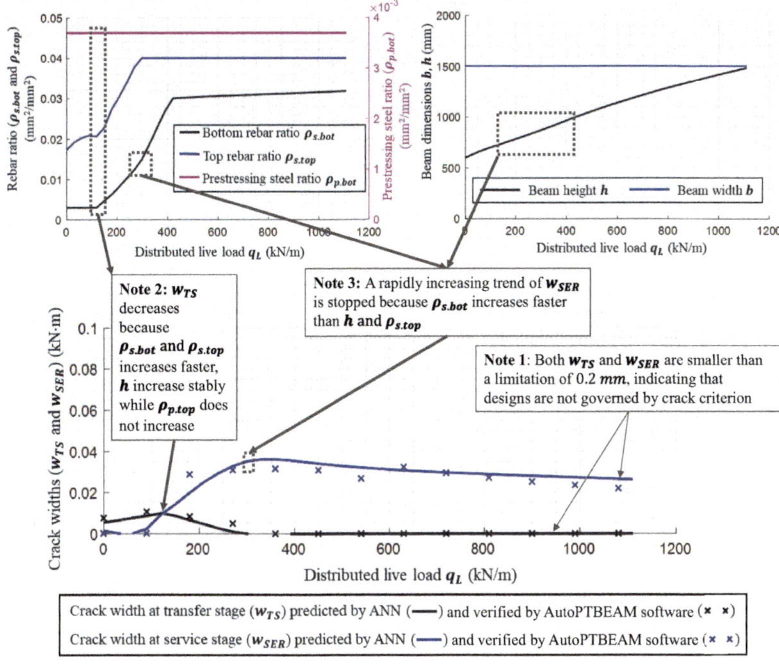

Figure 3.6.4 Design criteria based on crack widths; $w_{SER} \leq 0.2$ mm for 0 kN/m $\leq q_L \leq$ 1,110 kN when h is minimized.

reduction of crack widths at transfer stage (w_{TS}) are observed when q_L equal to 125 kN/m because $\rho_{s.bot}$ and $\rho_{s.top}$ increase faster, whereas h increases slowly while $\rho_{p.top}$ is constant. Note 3 points out w_{SER} stops increasing for q_L = 300 kN/m because $\rho_{s.bot}$ increases faster than and $\rho_{s.top}$.

3.6.4 Design criteria based on moment demand (M_{Ed}), moment resistance (M_{Rd}), and cracking moment (M_{Cr}) for 0 kN/m $\leq q_L \leq$ 1,110 kN when h is minimized

Figure 3.6.5 illustrates design criteria based on moment demand (M_{Ed}), moment resistance (M_{Rd}), and cracking moment (M_{Cr}) for 0 kN/m $\leq q_L \leq 1{,}110$ kN when h is minimized, where Note 1 shows that demands (M_{Ed}) are always smaller than resistances (M_{Rd}), indicating that designs are not governed by a strength criterion because $M_{RD} \geq M_{ED}$. Note 2 states that beam resistances (M_{Rd}) swiftly rise at q_L = 300 kN/m because $\rho_{s.bot}$, h and $\rho_{s.top}$ increases rapidly. Figure 3.6.5 illustrates cracking moments (M_{Cr}) which increase slowly for 0 kN/m $\leq q_L \leq 1{,}110$ kN when beam depth (h) increases, indicating that cracking moments (M_{Cr}) increase as beam

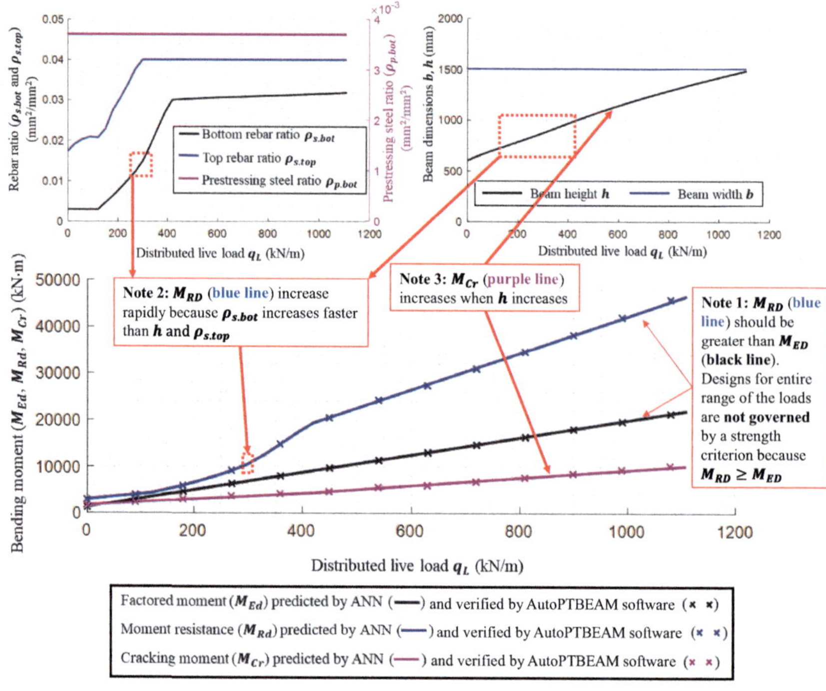

Figure 3.6.5 Design criteria based on moment demand (M_{Ed}), moment resistance (M_{Rd}), and cracking moment (M_{Cr}) for 0 kN/m $\leq q_L \leq$ 1,110 kN when h is minimized.

depths (h) increase when tendon ratios ($\rho_{p.bot}$) and beam width (b) do not change.

3.6.5 Design criteria based on reinforcement strains and neutral axis depth representing beam ductility for 0 kN/m $\leq q_L \leq$ 1,110 kN when h is minimized

Figure 3.6.6 explains relationships among rebar ratios ($\rho_{s.bot}$ and $\rho_{s.top}$), tendon ratios ($\rho_{p.bot}$), bottom rebar strains ($\varepsilon_{s.bot}$), tendon strains ($\varepsilon_{p.bot}$), and depths of neutral axis (x). It can be observed that beam ductility described by strains ($\varepsilon_{s.bot}$ and $\varepsilon_{p.bot}$) and depths of neutral axis (x) depends on rebar ratios ($\rho_{s.bot}$ and $\rho_{s.top}$) when tendon ratio ($\rho_{p.bot}$) does not change for the entire live loads (q_L) range. Note 1 shows that beam ductility decreases in designs with 50 kN/m $\leq q_L \leq$ 425 kN/m because tensile rebars ($\rho_{s.bot}$) increase faster than compressive rebars ($\rho_{s.top}$), after which beam ductility becomes constant when q_L increase from 425 to 1,110 kN/m because there is no significant changes in rebar ratios ($\rho_{s.bot}$ and $\rho_{s.top}$) and tendon ratios ($\rho_{p.bot}$), as stated in Note 2.

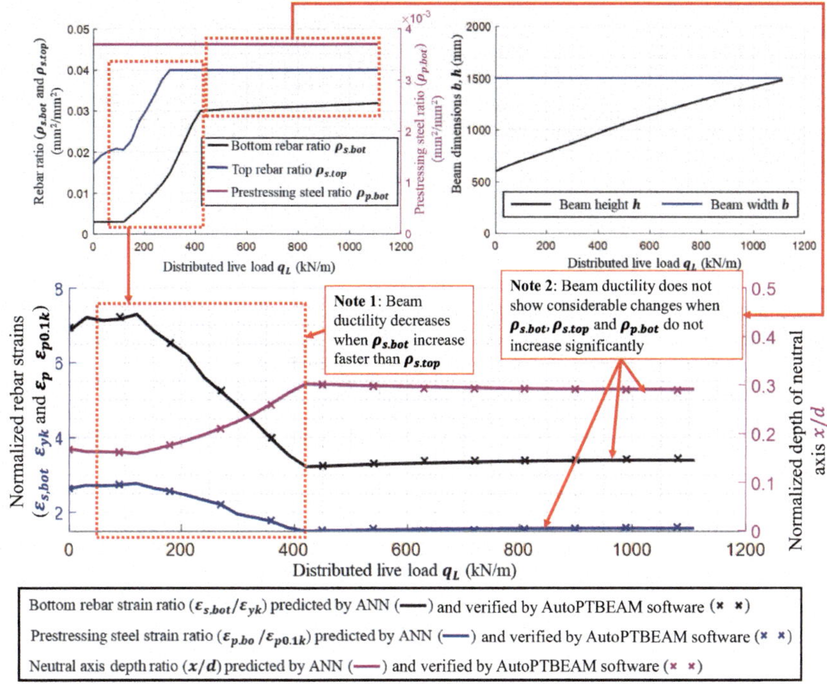

Figure 3.6.6 Design criteria based on reinforcement strains and neutral axis depth representing beam ductility for 0 kN/m $\leq q_L \leq$ 1,110 kN when h is minimized.

3.6.6 Design criteria based on shear stirrup ratios for 0 kN/m $\leq q_L \leq$ 1,110 kN when h is minimized

Figure 3.6.7 shows required ratios of stirrup at beam supports ($\rho_{sw,sp}$) and mid-spans ($\rho_{sw,mid}$) calculated by an ANN-based Lagrange optimization when design parameters including a beam size (b and h) are determined to minimize beam height (h). Note 1 states noticeable design errors of stirrup ratios in mid-spans ($\rho_{sw,mid}$) for 400 kN/m $\leq q_L \leq$ 600 kN/m. These errors can be reduced by adding training data in the corresponding ranges. Note 2 states that a minimum ratio of 0.0013 is selected based on ANNs and software AutoPTBEAM for stirrups at mid-spans ($\rho_{sw,mid}$) with live loads (q_L) range between 0 and 120 kN/m.

3.6.7 Cost and beam weight obtained for 0 kN/m $\leq q_L \leq$ 1,110 kN when h is minimized

Figures 3.6.8 and 3.6.9 illustrate cost indexes (CI_b) and beam weights (W) when beam depths (h) are minimized. Figure 3.6.8 indicates that CI_b

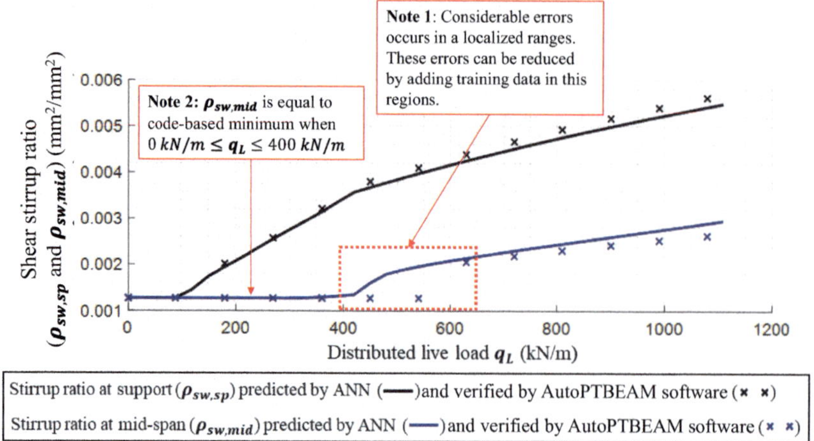

Figure 3.6.7 Design criteria based on shear stirrup ratios for 0 kN/m ≤ q_L ≤1,110 kN when h is minimized.

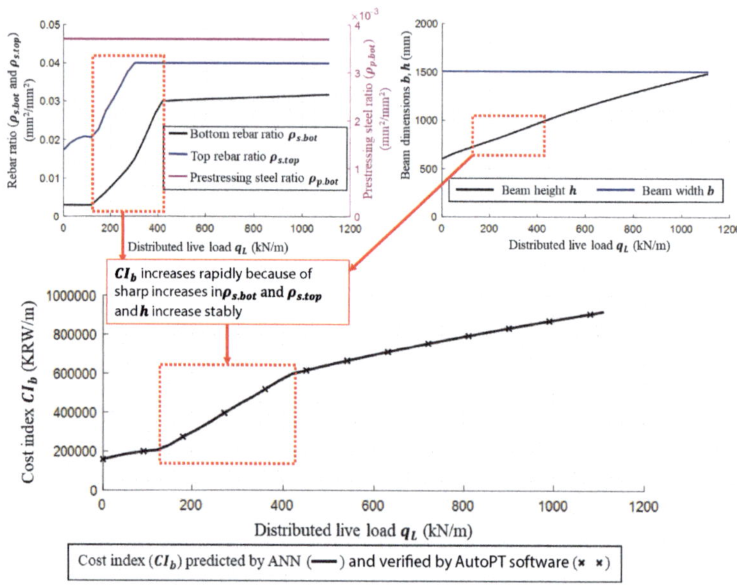

Figure 3.6.8 Cost indexes obtained for 0 kN/m ≤ q_L ≤1,110 kN when h is minimized.

increases rapidly for 125 kN/m ≤ q_L ≤ 425 kN because of sudden increases in $\rho_{s.bot}$ and $\rho_{s.top}$ while b becomes constant. Figure 3.6.9 displays a trend for beam weights (**W**) similar to that of cost indexes (CI_b), indicating that beam weights (**W**) are directly calculated from beam volumes ($L \times b \times h$). Figure 3.6.10 verifies the designs with minimized beam depth obtained

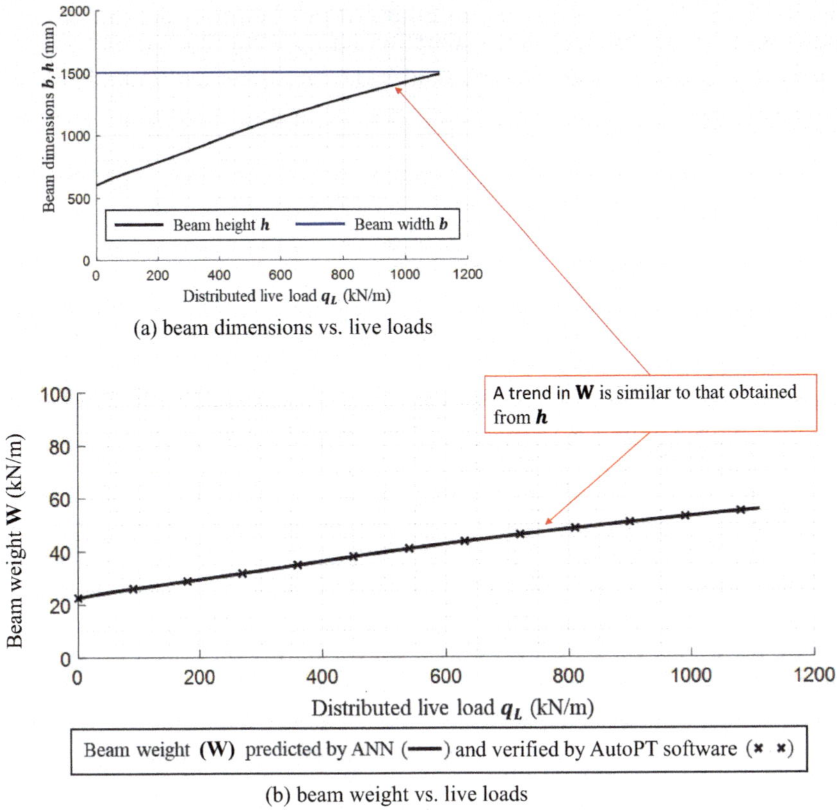

Beam dimensions b, h (mm)

Beam height h — Beam width b

Distributed live load q_L (kN/m)

(a) beam dimensions vs. live loads

A trend in **W** is similar to that obtained from h

Beam weight **W** (kN/m)

Distributed live load q_L (kN/m)

Beam weight (W) predicted by ANN (——) and verified by AutoPT software (✖ ✖)

(b) beam weight vs. live loads

Figure 3.6.9 Beam weight obtained for 0 kN/m $\leq q_L \leq$ 1,110 kN when h is minimized.

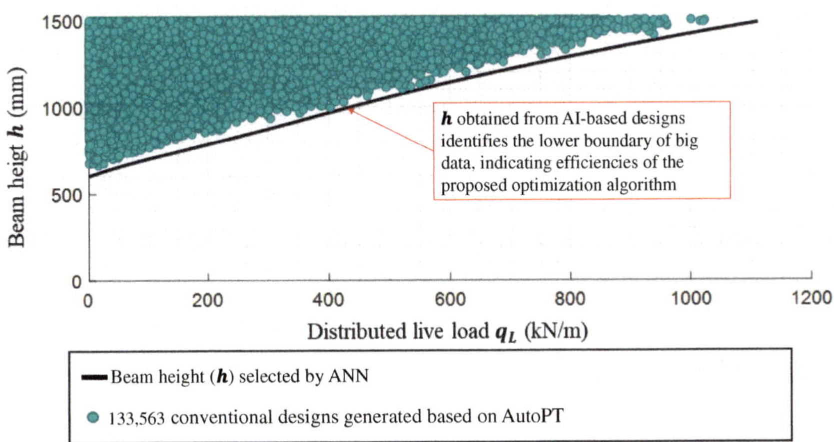

Figure 3.6.10 Beam depth (h) obtained by the large datasets bounded on the lower depth obtained by the ANN-based optimized design charts refer to Figure 3.6.9(b) for ANN-based Lagrange optimization).

by 133,563 conventional designs. It can be observed that beam depths (*h*) obtained by the large datasets are bounded on the lower depth obtained using design charts based on the ANN-based Lagrange optimization, indicating that any shallower section calculated by ANNs is not found from the large datasets.

3.7 FORMULATION OF ANN-BASED DESIGN CHARTS FOR PRE-TENSIONED (BONDED) CONCRETE BEAMS

Section 3.7 reviews how to formulate ANN-based design charts for pre-tensioned (bonded) concrete beams. Figure 3.7.1 depicts an algorithm for constructing ANN-based design charts for pre-tensioned (bonded) concrete beams. **Steps 1.1** to **1.4** outline the process of training Artificial Neural Networks (ANNs), which are assigning ranges of input parameters, generating data, and performing training ANN. **Steps 1.1** to **1.4** develop an ANN-based generalizable objective functions which are used in **Step 1.5**. An initial live load of m = 0 kN/m is employed in **Step 2.1**, which is incorporated into the equalities of **Step 2.2**, establishing equality constraints multiplied with Lagrange multiplier of equality shown in **Step 2.3**. The design requirements imposed by inequalities shown in **Step 3.1** are subsequently multiplied with Lagrange multiplier of inequality shown in **Step 3.2**. The ANN-based objective functions shown in **Step 1.5** are combined with the equality constraints of **Step 2.3** and the inequality constraints of **Step 3.2** to formulate an ANN-based Lagrange function shown in **Step 4**. The optimized parameters are determined in **Step 5** by solving for stationary points of the ANN-based Lagrange function. The design is verified in **Step 6**, using the AutoPT software which is based on structural mechanics using optimized parameters determined by ANNs. The procedure moves to **Step 6.1** if the design accuracy is found insufficient, where a number of large datasets is increased, and training parameters are revised to minimize errors. A design point is added to the design chart in **Step 7** when a sufficient accuracy is attained. The process continues by incrementing the live load denoted as m by 25 kN/m (this increment can be any depending on the design cases) in **Step 8**. **Step 9** verifies if the newly increased live load falls within the range of the large datasets. The algorithm terminates when the magnitude of the live load exceeds the limits of the large datasets. Otherwise, the procedure returns to **Step 2.2** to revise the live load in Equality **C16**, advancing to determine a new design point.

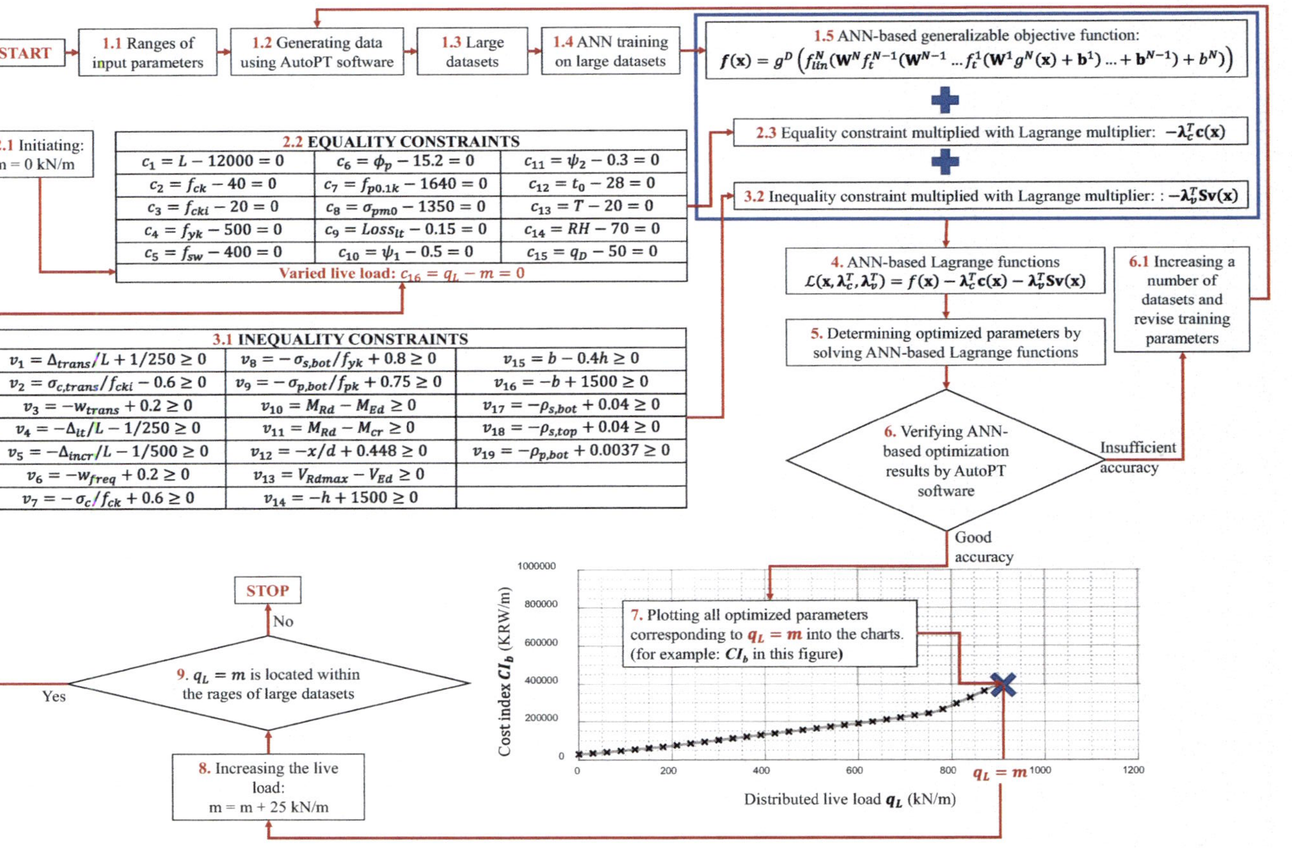

Figure 3.7.1 An algorithm for constructing ANN-based design charts for pre-tensioned (bonded) concrete beams.

3.8 DESIGN RECOMMENDATIONS

Engineers can design PT beams by following the design charts provided in this chapter. ANN-based design charts shown in (1)–(10) for holistic designs of PT beams are provided, calculating 21 forward outputs (Δ_{trans}/L, $\sigma_{c,trans}/f_{cki}$, w_{trans}, Δ_{lt}/L, Δ_{incr}/L, w_{freq}, σ_c/f_{ck}, $\sigma_{s,bot}/f_{yk}$, $\sigma_{p,bot}/f_{pk}$, M_{Ed}, M_{Rd}, M_{cr}, x/d, $\varepsilon_{s,bot}/\varepsilon_{sy}$, $\varepsilon_{p,bot}/\varepsilon_{py}$, V_{Ed}, V_{Rdmax}, $\rho_{sw,sp}$, $\rho_{sw,mid}$, CI_b, ***Weight***), in arbitrary sequences, from 21 forward inputs (L, b, h, f_{ck}, f_{cki}, $\rho_{s,bot}$, $\rho_{s,top}$, f_{yk}, f_{sw}, ϕ_p, $\rho_{p.bot}$, $f_{p0.1k}$, σ_{pm0}, $Loss_{lt}$, ψ_1, ψ_2, t_0, T, RH, q_D, q_L). when optimizing design targets or objective functions.

1. Design chart based on crack widths ($w_{SER} \leq 0.2$ mm and $w_{Tran} \leq 0.2$ mm) when CI_b is optimized (refer to Figure 3.4.1 when CI_b is minimized; Figure 3.5.4 when W is minimized; Figure 3.6.4 when beam depth h is minimized).
2. Design chart based on stresses; $\sigma_{s,bot}/f_{yk} \leq 0.8$, $\sigma_{p,bot}/f_{pk} \leq 0.75$ and $\sigma_c/f_{ck} \leq 0.6$ when CI_b is optimized (refer to Figures 3.4.2 and 3.4.3 when CI_b is minimized; Figure 3.5.2 when W is minimized; Figure 3.6.2 when beam depth h is minimized).
3. Design chart based on deflections ($\Delta_{trans} \leq \frac{L}{250}$, $\Delta_{lt} \leq \frac{L}{250}$, and $\Delta_{incr} \leq \frac{L}{500}$) when CI_b is minimized (refer to Figure 3.4.5 when CI_b is minimized; Figure 3.5.1 when W is minimized; Figure 3.6.1 when beam depth h is minimized).
4. Design chart based on moment demand (M_{Ed}), moment resistance (M_{Rd}), and cracking moment (M_{Cr}) when CI_b is minimized (refer to Figure 3.4.6 when CI_b is minimized; Figure 3.5.5 when W is minimized; Figure 3.6.5 when beam depth h is minimized).
5. Design chart based on tendon strains ($\varepsilon_{p,bot}$), reinforcement strains ($\varepsilon_{s,bot}/\varepsilon_{sy}$), and neutral axis depth ($x/d \leq 0.448$) representing beam ductility when CI_b is minimized (refer to Figure 3.4.7 when CI_b is minimized; Figure 3.5.6 when W is minimized; Figure 3.6.6 when beam depth h is minimized).
6. Design chart based on shear stirrup ratios ($\rho_{sw,sp}$ and $\rho_{sw,mid}$) when CI_b is minimized (refer to Figure 3.4.8 when CI_b is minimized; Figure 3.5.7 when W is minimized; Figure 3.6.7 when beam depth h is minimized).
7. Design chart based on beam cost (CI_b) when CI_b is minimized (refer to Figures 3.4.4 and 3.4.9).
8. Design chart based on beam weight (W) when W is minimized (refer to Figures 3.5.3 and 3.5.9).
9. Design chart based on beam cost (CI_b) when h is minimized (refer to Figure 3.6.8).
10. Design chart based on beam weight (Wt.) when h is minimized (refer to Figure 3.6.9).

Thirteen criteria corresponding to 13 inequalities from v_1 to v_{13} required by EC2 [9] shown in Table 3.3.2(b) are considered in the design charts. ANN-based optimal designs can be governed by one or multiple criteria when CI_b is minimized, for example.

3.9 CONCLUSIONS

ANN-based designs established in [3.32] and [3.33] are extended to holistically optimize PT beam designs. ANNs are trained on large datasets generated with 15 input and 18 output parameters using mechanics-based software AutoPTBEAM following Eurocode. It is now possible for engineers to make more rational engineering decisions, resulting in designs that can meet various code restrictions simultaneously based on the proposed optimization techniques. Designs minimizing a PT beam cost satisfy all code requirements strictly imposed by 16 equalities and 19 inequalities conditions while empirical observations of engineers are not used. A cost of PT beams minimized by the proposed method is validated by 133,563 data points filtered from two million datasets. Major conclusions drawn from this chapter are also provided as follows:

1. Optimized parameters of simply supported PT beams are stationary points of Lagrange functions identified using Newton–Raphson iteration.
2. Big datasets of 133,563 conventional designs verify minimum cost (CI_b) obtained by ANN-based optimized design charts, showing an efficiency of the proposed method.
3. The lowest beam cost is obtained for a selected live load $q_L = 900$ kN/m, while all requirements in terms of 19 inequalities are met by codes and architectures. Design parameters including, h, $\rho_{s,bot}$, $\rho_{s,top}$, and $\rho_{p,bot}$ are also optimized by ANN-based Lagrange functions.
4. Readers can construct design charts to minimize the cost of PT beams, beam weight, and beam depth under various design requirements based on any design code. Design charts can determine optimal section sizes (b and h), tendon ratios ($\rho_{p,bot}$), and rebar ratios ($\rho_{s,bot}$ and $\rho_{s,top}$) for specifically applied loads. Engineers can also adjust inequalities and equalities to modify design charts tailored for a particular design condition including a span length (L), a material availability (f_{ck}, f_{cki}, f_{yk}, $f_{p0.1k}$, and f_{sw}), prestressing techniques (σ_{pm0}, $Loss_{lt}$, and ϕ_p), combination factors (ψ_1, ψ_2), environmental exposures (t_0, T, and RH), and loadings (q_D and q_L).
5. Noticeable design conclusions based on recommendations for the PT design are described below.
   1. Designs with 0 kN/m $\leq q_L \leq$ 50 kN/m are governed by a crack width limit at service stage ($w_{SER} \leq 0.2$ mm defined by inequality

v_6 of Table 3.3.2(b)) while parameters corresponding to other 12 criteria (inequalities from v_1 to v_5 and from v_7 to v_{13}) do not reach their limits.

2. Designs with 50 kN/m $\leq q_L \leq$ 200 kN/m are governed by a criterion of $\sigma_{p,bot}/f_{pk} \leq 0.75$ (inequality v_9 of Table 3.3.2(b)) while parameters corresponding to other 12 criteria (inequalities from v_1 to v_8 and from v_{10} to v_{13}) do not reach their limits.
3. Designs with 200 kN/m $\leq q_L \leq$ 625 kN/m are governed by two criteria of $\sigma_c/f_{ck} \leq 0.6$ (inequality v_7) and $\sigma_{p,bot}/f_{pk} \leq 0.75$ (inequality v_9 of Table 3.3.2(b)) while parameters corresponding to other 11 criteria (inequalities from v_1 to v_6, v_8, and from v_{10} to v_{13}) do not reach their limits.
4. Designs with 625 kN/m $\leq q_L \leq$ 1,125 kN/m are governed by a criterion of $\sigma_c/f_{ck} \leq 0.6$ (inequality v_7 of Table 3.3.2(b)) while parameters corresponding to other 12 criteria (inequalities from v_1 to v_6 and from v_8 to v_{13}) do not reach their limits.

REFERENCES

[3.1] Erbatur, F., Zaid, R. Al, & Dahman, N. A. (1992). Optimization and sensitivity of prestressed concrete beams. *Computers and Structures*, *45*(5–6), 881–886. https://doi.org/10.1016/0045-7949(92)90046-3

[3.2] Yepes, V., Martí, J. V., & García-Segura, T. (2015). Cost and CO_2 emission optimization of precast-prestressed concrete U-beam road bridges by a hybrid glowworm swarm algorithm. *Automation in Construction*, *49*(PA), 123–134. https://doi.org/10.1016/j.autcon.2014.10.013

[3.3] Alqedra, M., Arafa, M., & Ismail, M. (2011). Optimum cost of prestressed and reinforced concrete beams using genetic algorithms. *Journal of Artificial Intelligence*, 4(1), http://hdl.handle.net/20.500.12358/24893

[3.4] Türkeli, E., & Öztürk, H. T. (2017). Optimum design of partially prestressed concrete beams using genetic algorithms. *Structural Engineering and Mechanics*, *64*(5), 579–589. https://doi.org/10.12989/sem.2017.64.5.579

[3.5] AydIn, Z., & Ayvaz, Y. (2010). Optimum topology and shape design of prestressed concrete bridge girders using a genetic algorithm. *Structural and Multidisciplinary Optimization*, *41*(1), 151–162. https://doi.org/10.1007/s00158-009-0404-2

[3.6] Aydın, Z., & Ayvaz, Y. (2013). Overall cost optimization of prestressed concrete bridge using genetic algorithm. *KSCE Journal of Civil Engineering*, *17*(4), 769–776. https://doi.org/10.1007/s12205-013-0355-4

[3.7] Quaranta, G., Fiore, A., & Marano, G. C. (2014). Optimum design of prestressed concrete beams using constrained differential evolution algorithm. *Structural and Multidisciplinary Optimization*, *49*(3), 441–453. https://doi.org/10.1007/s00158-013-0979-5

[3.8] Sirca Jr, G. F., Adeli, H., & Asce, F. (2005). Cost optimization of prestressed concrete bridges. *Journal of Structural Engineering*, *131*(3), 380–388. https://doi.org/10.1061/(ASCE)0733-9445(2005)131

[3.9] Hong, W.K, Nguyen, M.C., Pham, T. D. (2022). An autonomous pre-tensioned beam design based on an AI-based Hong-Lagrange method. *Journal of Asian Architecture and Building Engineering* (under review).

[3.10] CEN (European Committee for Standardization). (2004). Eurocode 2 : Part 1-1: General rules and rules for buildings. In *Eurocode 2: Vol. BS En 1992*.

[3.11] Nguyen, M. C., & Hong, W. K. (2022). Analytical prediction of nonlinear behaviors of beams post-tensioned by unbonded tendons considering shear deformation and tension stiffening effect. *Journal of Asian Architecture and Building Engineering*, *21*(3), 908–929. https://doi.org/10.1080/13467581.2021.1908299

[3.12] EC-Standards. (1990). *EN 1990:2002E - Basis of structural design - EC0*. 1–87.

[3.13] European Concrete Platform (2008). Commentary Eurocode 2.

[3.14] Beeby, A. W. (2001). Calculation of Crack Width, PrEN 1992-1 (Final draft) Chapter 7.3.4.

[3.15] Beeby, A. W., Favre, R., Koprna, M. & Jaccoud, J. P. (1985). Cracking and Deformations, CEB Manual.

[3.16] Elighausen, R., Mallée, R., & Rehm, G. (1976). Rissverhalten von Stahlbetonkörpern bei Zugbeanspruchung, Untersuchungsbericht.

[3.17] Falkner, M. (1969). Zur Frage der Rissbildung durch Eigen – und Zwangspannungen in Folge Temperatur im Stahlbetonbau, Deutscher Ausschuss Für Stahlbeton Heft 208.

[3.18] Hartl, G. (1977). Die Arbeitslinie eingebetteter St¨ale bei Erst – und Kurzzeitbelastung, Dissertation.

[3.19] Jaccoud, J. P., & Charif, H. (1986). Armature Minimale pour le Contrôle de la Fissuration, Publication No 114.

[3.20] Krips, M. (1985). Rissenbreitenbeschränkung im Stahlbeton und Spannbeton, TH Damstadt.

[3.21] Rehm, G., & Rüsch, H. (1963). Versuche mit Betonformstählen Teil I, Deutscher Ausschuss für Stahlbeton Heft 140.

[3.22] Rehm, G., & Rüsch, H. (1963). Versuche mit Betonformstählen Teil II, Deutscher Ausschuss für Stahlbeton Heft 160.

[3.23] Rehm, G., & Rüsch, H. (1964). Versuche mit Betonformstählen Teil III, Deutscher Ausschuss für Stahlbeton Heft 165.

[3.24] Hong, W. K. (2023). *Artificial Neural Network-based Optimized Design of Reinforced Concrete Structures*. Taylor & Francis, Boca Raton, FL.

[3.25] Hong, W. K., Nguyen, V. T., & Nguyen, M. C. (2022). Optimizing reinforced concrete beams cost based on AI-based Lagrange functions. *Journal of Asian Architecture and Building Engineering*, *21*(6), 2426–2443. https://doi.org/10.1080/13467581.2021.2007105

[3.26] Hong, W. K., & Nguyen, M. C. (2022). AI-based Lagrange optimization for designing reinforced concrete columns. *Journal of Asian Architecture and Building Engineering*, 21(6), 2330–2344. https://doi.org/10.1080/13467581.2021.1971998

[3.27] Hong, W. K., Pham, T. D., & Nguyen, V. T. (2022). Feature selection based reverse design of doubly reinforced concrete beams. *Journal of Asian Architecture and Building Engineering*, *21*(4), 1472–1496. https://doi.org/10.1080/13467581.2021.1928510

[3.28] Kuhn, H. W., & Tucker, A. W. (2014). Nonlinear programming. In Giorgi G & Kjeldsen T. H (eds) *Traces and Emergence of Nonlinear Programming* (pp. 247–258). Basel: Birkhäuser.

[3.29] Kuhn, H. W. & Tucker, A. W. (1951). "Nonlinear programming. In *Proceedings of 2nd Berkeley Symposium* (pp. 481–492). Berkeley: University of California Press.

[3.30] Krenker, A., Bešter, J., & Kos, A. (2011). Introduction to the artificial neural networks. In Suzuki K (ed), *Artificial Neural Networks: Methodological Advances and Biomedical Applications* (pp. 1–18). InTech, London.

[3.31] Villarrubia, G., De Paz, J. F., Chamoso, P. & De la Prieta, F. (2018). Artificial neural networks used in optimization problems. *Neurocomputing*, *272*, 10–16. https://doi.org/10.1016/j.neucom.2017.04.075.

[3.32] Hong, W.K. (2019). *Hybrid Composite Precast Systems: Numerical Investigation to Construction*. Woodhead Publishing, Elsevier, Cambridge, MA.

[3.33] Hong, W.K. (2021). *Artificial intelligence-based design of reinforced concrete structures*. Daega, Seoul.

Chapter 4

Multi-objective optimizations (MOO) of pretensioned concrete beams using an ANN-based Lagrange algorithm

4.1 INTRODUCTION

4.1.1 Previous studies

Numerous studies including [4.1–4.5] attempted to optimize PT members using non-derivative methods, resulting in a maximum cost reduction of 30% compared with conventional designs. However, these studies only dealt with a single objective function, while solving multiple objectives optimizations (MOO) requires a weighted sum method suggested by [4.6] to combine all objective functions based on varying weighted fractions necessary to construct Pareto frontiers. Barakat et al. [4.7] presented a reliability-based optimization of PT beams using an ε-constraint method, where overall costs are minimized, and reliability indexes are maximized. In 2017, García-Segura et al. [4.8] introduced a sustainable design of PT bridges, minimizing costs, adverse environmental impacts, and social consequences. Optimized designs were achieved by with harmony search algorithms applying metamodel-assisted MOO based on ANNs integrated. The same algorithms were also used in [4.9] to consider three objective functions namely cost, safety, and corrosion initiation time in designing road bridges. Artificial intelligent techniques were also applied to designs of various types of structures, and these studies include the strength estimation of the tilted angle connectors in [4.10], the assessment of the effect of additives on concrete in [4.11] and [4.12], the behavior prediction of the channel shear connectors embedded in concrete [4.13], the cohesion investigations of the sandy soil reinforced by fibers in [4.14], and in concrete [4.13], the cohesion investigations of the tie sections of concrete deep beams in [4.15].

There are numerous approaches to deal with MOO problems, many of which are conducted based on Pareto optimality. The non-dominated sorting genetic algorithm developed in [4.16], where non-dominated sorting techniques are applied to evolutionary programming to find the Pareto-optimal front of MOO problems. The multi-objective Grey Wolf optimizer introduced in [4.17] employs a concept of Pareto optimality in a metaheuristic method established in [4.18]. The present chapter introduces a specific

DOI: 10.1201/9781003354796-4

example of optimizing multiple objective functions for PT beams which have 21 input and 21 output parameters. The proposed design replaces conventional software. This chapter was written based on the previous paper by Hong et al. [4.19].

4.1.2 Motivations of the optimizations based on MOO designs

Not many studies that optimize three objective functions (h, CI_b, and W) at the same time are available for a PT beam, whereas in Pareto frontiers presented in this chapter, three objective functions are optimized using ANN-based objective functions and the weighted sum method, while code requirements imposed by equalities and inequalities are satisfied by KKT conditions. Designs are optimized based on "one run, one optimization" for a PT beam. In the innovative ANN-based Lagrange method, three objective functions (h, CI_b, and W) are minimized simultaneously for PT beams designed based on European standards. In the design, 21 output parameters (Δ_{trans}/L, $\sigma_{c,trans}/f_{cki}$, w_{trans}, Δ_{lt}/L, Δ_{incr}/L, w_{freq}, σ_c/f_{ck}, $\sigma_{s,bot}/f_{yk}$, $\sigma_{p,bot}/f_{pk}$, M_{Ed}, M_{Rd}, M_{cr}, x/d, $\varepsilon_{s,bot}/\varepsilon_{sy}$, $\varepsilon_{p,bot}/\varepsilon_{py}$, V_{Ed}, V_{Rdmax}, $\rho_{sw,sp}$, $\rho_{sw,mid}$, CI_b, W) are identified for preassigned 21 input variables (L, b, h, f_{ck}, f_{cki}, $\rho_{s,bot}$, $\rho_{s,top}$, f_{yk}, f_{sw}, ϕ_p, $\rho_{p,bot}$, $f_{p0.1k}$, σ_{pm0}, $Loss_{lt}$, ψ_1, ψ_2, t_0, T, RH, q_D, q_L). Table 4.1.1 presents definitions of input and output parameters. How a unified objective function (UFO) is formulated to integrate three objective functions (h, CI_b, and W) into one objective function for simultaneous optimizations of the three objective functions (h, CI_b, and W) is demonstrated. A Pareto frontier is constructed based on tradeoff ratios (or weighted fractions) controlled based on design interests in which a contribution made by each of the three objective functions (h, CI_b, and W) to a total achieved optimization is identified and verified by the designs based on structural mechanics-based software AutoPTBEAM. Optimized designs for PT beams are performed against loads and design requirements where both equality and in equality constraints are imposed based on local design requirements and engineers' interests. Facile, robust but accurate simultaneous optimizations of the three objective functions (h, CI_b, and W) are possible while being verified. Pareto frontiers are verified by the lower boundary of the big datasets obtained using structural software for $q_L = 400$ kN/m.

4.2 CONTRIBUTIONS OF THE PROPOSED METHOD TO PT BEAM DESIGNS

4.2.1 Significance of the Chapter 4

A pretensioned (PT) concrete structure with considerably reduced member sizes, cracks, and deflections is one of the most popular structural frames for buildings, bridges, towers, vessels, etc. Material consumption decreases

Table 4.1.1 Nomenclatures of input and output parameters

21 input parameters		*21 output parameters*	
L	Beam length (mm)	Δ_{trans}/L	Displacement at transfer stage (mm)
b	Beam width (mm)	$\sigma_{c,trans}/f_{cki}$	Ratio of concrete compressive stress at transfer stage
h	Beam height (mm)	w_{trans}	Crack width at transfer stage (mm)
f_{ck}	Characteristic compressive cylinder strength of concrete at 28 days (MPa)	Δ_{lt}/L	Long-term deflection (mm)
f_{cki}	Characteristic compressive cylinder strength of concrete at transfer (MPa)	Δ_{incr}/L	Incremental deflection (mm)
$\rho_{s,bot}$	Bottom rebar ratio	w_{freq}	Crack width at frequent service stage (mm)
$\rho_{s,top}$	Top rebar ratio	σ_c/f_{ck}	Ratio of concrete compressive stress at characteristic service stage
f_{yk}	Characteristic yield strength of longitudinal rebar (MPa)	$\sigma_{s,bot}/f_{yk}$	Ratio of bottom rebar stress at characteristic service stage
f_{sw}	Characteristic yield strength of stirrup (MPa)	$\sigma_{p,bot}/f_{pk}SW$	Ratio of bottom strand stress at characteristic service stage
ϕ_p	Strand diameter (mm)	M_{Ed}	Design value of the applied internal bending moment (kN·m)
$\rho_{p,bot}$	Bottom strand ratio	M_{Rd}	Moment resistance (kN·m)
$f_{p0.1k}$	Characteristic yield strength of prestressing steel (MPa)	M_{cr}	Cracking moment (kN·m)
σ_{pm0}	Stress in the strand immediately after tensioning or transfer (MPa)	x/d	Ratio of neutral axis depth
$Loss_{lt}$	Long-term loss of stress in strand (%)	$\varepsilon_{s,bot}/\varepsilon_{sy}$	Strain of bottom rebar in terms of yielding strain
ψ_1	Factor for frequent value of a variable action	$\varepsilon_{p,bot}/\varepsilon_{py}$	Strain of bottom strand in terms of yielding strain
ψ_2	Factor for quasi-permanent value of a variable action	V_{Ed}	Design value of the applied shear force (kN)
t_0	The age of concrete at the time of loading (days)	V_{Rdmax}	Maximum shear resistance limited by crushing of the compression struts (kN)
T	Ambient temperature (°C)	$\rho_{sw,sp}$	Stirrup ratio at supports
RH	Relative humidity of ambient environment (%)	$\rho_{sw,mid}$	Stirrup ratio at mid-span
q_D	Distributed dead load (kN/m)	CI_b	Cost index of the beam (KRW/m)
q_L	Distributed live load (kN/m)	$Weight$	Beam weight per length (kN/m)

compared with unprestressed reinforced concrete (RC) structures when introducing prestressing forces. Rectangular RC beams are optimized in [4.20] using a computational LMM, whereas calculations of PT elements are much more complex than unprestressed RC structures due to prestressing forces, and hence, special attention should be paid to the calculations of deflections, cambers, and flexibility [4.21]. Common concerning features in PT designs including costs, member weights, and section depths directly affect economy, construction plans, and structure aesthetics. However, research optimizing these three objective functions simultaneously in PT designs is not easy to find.

This chapter aims to assist structural engineers and decision-makers to design PT beams based on a unified function of objective (UFO). Simply supported pretensioned concrete beams (a PT beam) is optimized using ANN-based Lagrange method, resulting in simultaneously optimizing three objective functions, such as construction and material cost, beam weights, and beam depth (CI_b, **W**, and h) which may conflict each other. This chapter provides five steps to optimize PT beams based on a MOO governed by contribution fractions made by each objective function. Examples of optimized designs are demonstrated, which can be used as guides for engineers and decision-makers throughout a preliminary design stage. Competitive and efficient designs optimized based on multi-objective functions with "one run, one optimization" for PT beams are eventually provided. Holistic design charts which is simple, fast to use but accurate to simultaneously optimize the three objective functions (h, CI_b, and W) are now available for a design of PT beams. This chapter performs MOO for a PT beam based on novel method in which objective functions are not limited to the three objective functions, whereas multiple objective functions of any type are simultaneously optimized. The ANN-based method introduced in this chapter is a steppingstone for the next PT designs that can be optimized based on multi-objective functions simultaneously using an ANN-based Lagrange optimization. Other published works in this area that performed a MOO design for PT beams based on ANNs were seldom found.

4.2.2 Tasks readers can perform after reading this chapter

Conflicting design targets for PT beams are simultaneously optimized for real-life design solutions of PT beams. The MOO application will be applied to frames including RC and PT frames for tall buildings. Readers can perform the following tasks after reading this chapter:

1. Readers can provide an ANN-based design optimizing multi-objective functions for PT beams, simultaneously, that can be used by structural engineers and decision-makers. A UFO of simply supported pretensioned concrete beams (a PT beam) is optimized using ANN-based Lagrange optimization, meeting various code restrictions

simultaneously. They can also use the proposed technique to make more rational engineering decisions.

2. Readers can derive a UFO to integrate three objective functions (***h***, ***CI***$_b$, and **W**) as one objective function. A contribution of individual objective function to a total achieved optimization can be controlled by adjusting tradeoff ratios based on design interests.
3. The proposed *UFO* removes dominant influence of one parameter on others by normalizing objective functions with their maxima and minima obtained from ANN-based Lagrange optimizations with an individual objective function.
4. Readers are able to optimize PT beams and find 21 output variables ($\Delta_{trans}/L, \sigma_{c,trans}/f_{cki}, w_{trans}, \Delta_{lt}/L, \Delta_{incr}/L, w_{freq}, \sigma_c/f_{ck}, \sigma_{s,bot}/f_{yk}, \sigma_{p,bot}/f_{pk}, M_{Ed}, M_{Rd}, M_{cr}, x/d, \varepsilon_{s,bot}/\varepsilon_{sy}, \varepsilon_{p,bot}/\varepsilon_{py}, V_{Ed}, V_{Rdmax}, \rho_{sw,sp}, \rho_{sw,mid}, CI_b, Weight$) for preassigned 21 input variables ($L, b, h, f_{ck}, f_{cki}, \rho_{s,bot}, \rho_{s,top}, f_{yk}, f_{sw}, \phi_p, \rho_{p,bot}, f_{p0.1k}, \sigma_{pm0}, Loss_{lt}, \psi_1, \psi_2, t_0, T, RH, q_D, q_L$), while minimizing three objective functions (***h***, ***CI***$_b$, and **W**) simultaneously.
5. A Pareto frontier constructed based on combinations of weight fractions optimizes the three objective functions (***h***, ***CI***$_b$, and **W**), simultaneously, verifying the Pareto frontier well with the lower boundary of the designs randomly generated based on structural software the AutoPTBEAM for $q_L = 400$ kN/m. Designs which are optimized based on the three objective functions (***h***, ***CI***$_b$, and **W**) simultaneously can also be verified by readers.
6. The proposed method can be used as a guide for engineers and decision-makers throughout the preliminary design stage. Readers minimize the three objective functions (***h***, ***CI***$_b$, and **W**), such as costs, weights, and beam depth, which are conflicting, helping engineers and readers perform such a conflicting design optimization to provide real-life design solutions. This chapter eventually provides competitive and efficient designs optimized based on multi-objective functions to deal with such conflicts.
7. Readers can follow the procedures provided in this chapter to establish their own charts to impose their own design requirements.

4.2.3 What can be changed with future PT beam designs

The optimized designs performed in this chapter are simple, fast but accurate, simultaneously optimizing the three objective functions (***h***, ***CI***$_b$, and **W**). For the first time, the present study attempting to solve MOO problems using the ANN-based Lagrange method is now available for PT beam designs. A Pareto-optimal front based on predetermined contributions of individual objective function offers a convenient and effective designing tool for engineers. RC beams [4.22], RC columns [4.23], and steel-reinforced concrete (SRC) columns [4.24] are optimized using ANN-based Lagrange method, showing relatively good accuracy and efficiencies, whereas ANN-based optimized design of doubly reinforced rectangular concrete beams is based on multi-objective functions in [4.25]. ANN-based Lagrange optimization for RC circular columns having multiobjective functions is also performed based on multi-objective functions in [4.53]. The future PT beam designs may be performed based on MOO in which PT beams are designed to simultaneously optimize multi-objective functions including $\boldsymbol{h}$, $\boldsymbol{CI_b}$, and **W** constrained by quality and inequality functions using the ANN-based Lagrange method.

4.3 LARGE DATASETS GENERATED FOLLOWING EUROCODE

4.3.1 Selection of input parameters following Eurocode

4.3.1.1 Fourteen forward inputs parameters describing beam sections and material properties

In Table 4.1.1 based on Eurocode (EC2 [4.27] and EC0 [4.28]), large structural datasets of simply supported pretensioned (PT) beams with bonded tendon are generated based on 21 input and 21 output parameters using a structural mechanics-based software AutoPTBEAM developed in [4.25]. A structural mechanics-based software AutoPTBEAM that designs bonded and unbonded PT beams based on Eurocode (EC2 [4.27] and EC0 [4.28]) is developed in 2022 [4.26]. Figure 4.3.1 illustrates seven parameters ($\boldsymbol{L}$, $\boldsymbol{b}, \boldsymbol{h}$, $\boldsymbol{\rho_{s,bot}}$, $\boldsymbol{\rho_{s,top}}$, $\boldsymbol{\phi_p}$, $\boldsymbol{\rho_{p,bot}}$) representing beam configurations. Material properties shown in Figures 4.3.2, 4.3.3(a), and 4.3.3(b), where constitutive relationships of concrete, rebars, and tendon are presented, respectively, are described by five parameters ($\boldsymbol{f_{ck}}, \boldsymbol{f_{cki}}, \boldsymbol{f_{yk}}, \boldsymbol{f_{sw}}, \boldsymbol{f_{p0.1k}}$). Prestressing forces are calculated based on two inputs $\boldsymbol{\sigma_{pm0}}$ indicating stresses of the strands occurring immediately after tensioning and $\boldsymbol{Loss_{lt}}$ indicating losses of stresses of strands due to long-term effects including creeps, shrinkages, and stress relaxations.

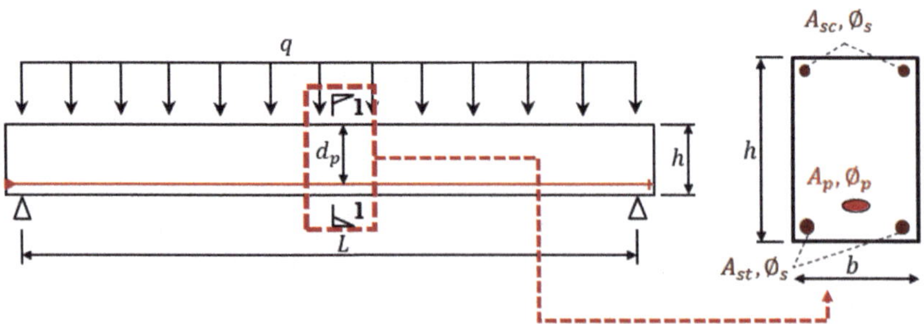

Figure 4.3.1 PT beam configuration.

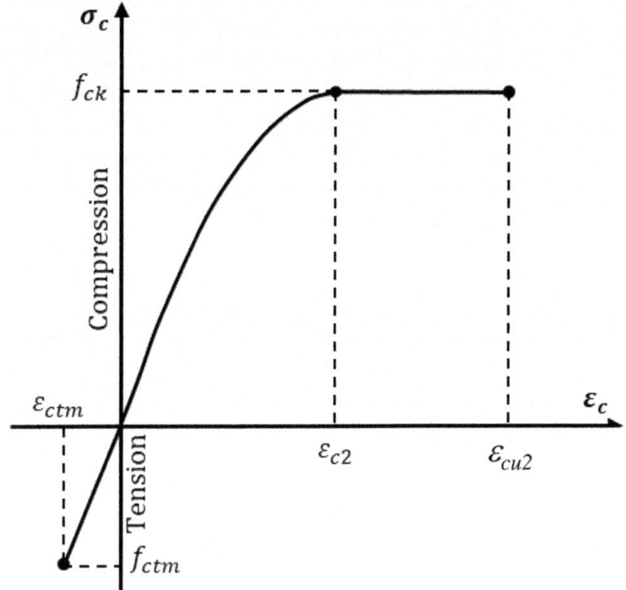

Figure 4.3.2 Behavior of the concrete in tension and compression [4.27].

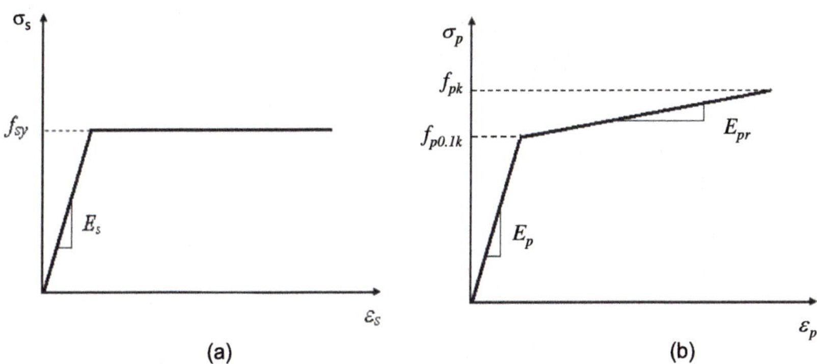

Figure 4.3.3 Stress–strain relationships of rebar and strand. (a) Stress–strain relationships of rebar. (b) Stress–strain relationships of prestressing strands.

4.3.1.2 Seven forward inputs parameters describing load and environment conditions

For a simply supported beam, input parameters ψ_1, ψ_2, t_0, T, RH, q_D, and q_L are used to describe load and environmental conditions, whereas q_D and q_L are used to describe uniformly distributed dead and live loads, respectively. Lateral load on building such as wind loads are neglected in this design example because beams are simply supported. Five load combinations based on Eurocode 0 (EC0) [4.28] are shown in Table 4.3.1, whereas combination factors ψ_1 and ψ_2 recommended for different building types [4.28] are shown in Table 4.3.2. The load combinations are included on an input-side, which reflects loading properties in different building types. Input parameters also include an age of concrete at the time of loading ($\boldsymbol{t_0}$), ambient temperature ($\boldsymbol{T}$), and relative humidity of ambient environment ($\boldsymbol{RH}$) which are used for estimating effect of creeps on a long-term deflection.

Table 4.3.1 Load combination following Eurocode 0 [4.28]

Load comb.	*Self-weight*	*Dead load*	*Live load*	*Prestressing*	*Hyperstatic*
Transfer comb.	1	0	0	1[a]	0
Quasi-permanent comb.	1	1	ψ_2	1[b]	0
Frequent comb.	1	1	ψ_1	1[b]	0
Characteristic comb.	1	1	1	1[b]	0
Ultimate load comb.	1.35	1.35	1.35	0	1

[a] Prestressing effect at transfer stage (without long-term losses).
[b] Prestressing effect at service stage (with long-term losses).

Table 4.3.2 Recommended values for ψ factors for different building types [4.28]

Action	ψ_0	ψ_1	ψ_2
Imposed loads in buildings, category			
Category A: domestic, residential areas	0.7	0.5	0.3
Category B: office areas	0.7	0.5	0.3
Category C: congregation areas	0.7	0.7	0.6
Category D: shopping areas	0.7	0.7	0.6
Category E: storage areas	1	0.9	0.8
Category F: traffic area, vehicle weight ≤ 30kN	0.7	0.7	0.6
Category G: traffic area, 30kN < vehicle weight ≤ 160kN	0.7	0.5	0.3
Category H: roofs	0	0	0
Snow loads on buildings			
Finland, Iceland, Norway, Sweden	0.7	0.5	0.2
Remainder of CEN Member States, for sites located at altitude H > 1,000 m a.s.l.	0.7	0.5	0.2
Remainder of CEN Member States, for sites located at altitude H ≤ 1,000 m a.s.l.	0.5	0.2	0
Wind loads on buildings	0.6	0.2	0
Temperature (non-fire) in buildings	0.6	0.5	0

4.3.2 Selection of forward output parameters based on design requirements of Eurocode

Design adequacies and efficiencies of 21 output parameters are explored in the design example. Deformations and concrete stresses in beam right after releasing tendon anchors are presented by the three outputs at transfer stages (Δ_{trans}/L, $\sigma_{c,trans}/f_{cki}$, w_{trans}). Deformations, cracks, and stresses at the service stage are calculated based on quasi-permanent, frequent, and characteristic load levels, respectively [4.29]. Serviceability of beam is then ensured under service loads by checking long-term deflection (Δ_{lt}) and incremental deflection (Δ_{incr}) at quasi-permanent load levels. Table 5.1N – Eurocode 2 [4.27] shown in Table 4.3.3 recommends limitations for crack widths (w_{freq}). Crack widths are computed with frequent load combinations for bonded tendons and quasi-permanent load combinations for unbonded tendons. Unacceptable cracks and deformations under service loads should be avoided by checking stresses in concrete (σ_c), rebars ($\sigma_{s,bot}$), and tendons $\left(\sigma_{p,bot}\right)$ at characteristic load combinations. Ten strength and ductility criteria M_{Ed}, M_{Rd}, M_{cr}, x/d, $\varepsilon_{s,bot}/\varepsilon_{sy}$, $\varepsilon_{p,bot}/\varepsilon_{py}$, V_{Ed}, $V_{Rd,max}$, $\rho_{sw,sp}$, and $\rho_{sw,mid}$ are also evaluated for ultimate stage load combinations. To conclude designs, 21 output parameters consists of 19 output parameters which must meet code-based design requirements imposed by inequalities v_1 to v_{19} and two more objective output parameters which are cost index (CI_b) and beam weight (W) calculated on an output-side to evaluate design efficiencies. Inequalities shown in Table 4.3.4(b) mostly apply to output parameters, whereas input parameters are constrained by equalities shown in Table 4.3.4(a).

4.3.3 Bigdata generation

Non-normalized and normalized 200,000 datasets generated by AutoPT BEAM are shown in Tables 4.3.5(a) and 4.3.5(b), respectively, where 21 input parameters (L, b, h, f_{ck}, f_{cki}, $\rho_{s,bot}$, $\rho_{s,top}$, f_{yk}, f_{sw}, ϕ_p, $\rho_{p,bot}$, $f_{p0.1k}$, σ_{pm0}, $Loss_{lt}$, ψ_1, ψ_2, t_0, T, RH, q_D, q_L) are randomly selected, resulting in 21

Table 4.3.3 Recommended limitations for crack widths (Table 5.1N – Eurocode 2 [4.27])

Exposure class	*Reinforced members and prestressed members with unbonded tendons*	*Prestressed members with bonded tendons*
	Quasi-permanent load combination	Frequent load combination
X0, XC1	0.4 mm	0.2 mm
XC2, XC3, XC4	0.3 mm	0.2 mm
XD1, XD2, XS1, XS2, XS3		All parts of the bonded tendons or duct lie at least 25 mm within concrete in compression.

Table 4.3.4 Equality and inequality conditions

(a) Equality constraints	
$c_1 = L - 12,000 = 0$	beam span fixed at 12 m
$c_2 = f_{ck} - 40 = 0$	Characteristic compressive cylinder strength of concrete at 28 days (MPa)
$c_3 = f_{cki} - 20 = 0$	Characteristic compressive cylinder strength of concrete at transfer (MPa)
$c_4 = f_{yk} - 500 = 0$	Characteristic yield strength of longitudinal rebar (MPa)
$c_5 = f_{sw} - 400 = 0$	Characteristic yield strength of stirrup (MPa)
$c_6 = \phi_p - 15.2 = 0$	Strand diameter (mm)
$c_7 = f_{p0.1k} - 1,640 = 0$	Characteristic yield strength of prestressing steel (MPa)
$c_8 = \sigma_{pm0} - 1,350 = 0$	Stress of the strand immediately after tensioning or transfer (MPa)
$c_9 = Loss_{lt} - 0.15 = 0$	Long-term loss of stress of strand (%)
$c_{10} = \psi_1 - 0.5 = 0$	Factor for frequent value of a variable action
$c_{11} = \psi_2 - 0.3 = 0$	Factor for quasi-permanent value of a variable action
$c_{12} = t_0 - 28 = 0$	Age of concrete at the time of loading (days)
$c_{13} = T - 20 = 0$	Ambient temperature (°C)
$c_{14} = RH - 70 = 0$	Relative humidity of ambient environment (%)
$c_{15} = q_D - 50 = 0$	Distributed dead load (kN/m)
$c_{16} = q_L - 400 = 0$	Distributed live load (kN/m)
(b) Inequality constraints	
$v_1 = -\Delta_{trans}/L + 1/250 \geq 0$	Upward deflection of beams at transfer, Δ_{trans}, should not exceed L/250
$v_2 = -\sigma_{c,trans}/f_{cki} + 0.6 \geq 0$	Compressive concrete stress, σ_c, should not exceed $0.6f_{cki}$ under transfer combination
$v_3 = -w_{trans} + 0.2 \geq 0$	Crack width under transfer stage, w_{trans}, should not exceed 0.2 mm
$v_4 = -\Delta_{lt}/L - 1/250 \geq 0$	Long-term deflection of beams, Δ_{lt}, should not exceed *L*/250 under quasi-permanent combination
$v_5 = -\Delta_{incr}/L - 1/500 \geq 0$	Incremental deflection of beams, Δ_{incr}, should not exceed *L*/500 under quasi-permanent combination
$v_6 = -w_{freq} + 0.2 \geq 0$	Crack width under frequent combination, w_{freq}, should not exceed 0.2 mm
$v_7 = -\sigma_c/f_{ck} + 0.6 \geq 0$	Compressive concrete stress, σ_c, should not exceed $0.6f_{ck}$ under characteristic combination
$v_8 = -\sigma_{s,bot}/f_{yk} + 0.8 \geq 0$	Tensile stress of rebar, σ_s, should not exceed $0.8f_{yk}$ under characteristic combination
$v_9 = -\sigma_{p,bot}/f_{pk} + 0.75 \geq 0$	Tensile stress of tendon, σ_p, should not exceed $0.75f_{pk}$ under characteristic combination
$v_{10} = M_{Rd} - M_{Ed} \geq 0$	Moment resistance, M_{Rd}, should be greater than or equal to factored bending moment, M_{Ed}, under ultimate load combination
$v_{11} = M_{Rd} - M_{cr} \geq 0$	Moment resistance, M_{Rd}, should be greater than or equal to cracking moment, M_{cr}, under ultimate load combination

(Continued)

Table 4.3.4 (Continued) Equality and inequality conditions

$v_{12} = -x/d + 0.448 \geq 0$	To control section ductility, normalized neutral axis depth, x/d should be smaller than or equal to 0.448
$v_{13} = V_{Rdmax} - V_{Ed} \geq 0$	Maximum shear resistance limited by crushing of the compression struts, V_{Rdmax}, should be greater than or equal to factored shear force, V_{Ed}
$v_{14} = -h + 1{,}500 \geq 0$	Architectural requirements
$v_{15} = b - 0.4h \geq 0$	
$v_{16} = -b + 1{,}500 \geq 0$	
$v_{17} = -\rho_{s,bot} + 0.04 \geq 0$	Rebar ratio, ρ_s, should be limited up to 4%
$v_{18} = -\rho_{s,top} + 0.04 \geq 0$	
$v_{19} = -\rho_{p,bot} + 0.0037 \geq 0$	Tendon ratio, $\rho_{p,bot}$, is limited up to 0.37%

Table 4.3.5 Large datasets of a simply supported PT beam generated by AutoPT

(a) Non-normalized

			200,000 DATASETS GENERATED BY AutoPT											
			Data 1	Data 2	Data 3	Data 4	Data 5	Data 6	Data 7	Data 8	...	MIN	MAX	MEAN
21 INPUT PARAMETERS FOR STRUCTURAL SOFTWARE (AutoSPT)	1	L (mm)	28170.3	41990.7	30821.5	12258.7	21051.1	16745.8	44484.6	36342.6	...	5020.4	62380.8	26276.5
	2	b (mm)	6970.1	1103.0	4770.4	491.7	4279.6	2032.1	5358.8	5563.0	...	205.7	9977.3	3304.3
	3	h (mm)	1948.8	1840.4	1847.7	579.7	1110.5	1095.7	2030.4	2245.3	...	500.0	2500.0	1500.7
	4	f_{ck} (MPa)	44	42	32	32	57	56	42	39	...	30	60	44.96
	5	f_{cki} (MPa)	15	18	22	29	55	39	31	17	...	15	60	30.01
	6	$\rho_{s,bot}$	0.02472	0.03719	0.02124	0.03022	0.03737	0.02597	0.00682	0.02569	...	0.00000	0.04000	0.01997
	7	$\rho_{s,top}$	0.00681	0.00833	0.02033	0.02512	0.01716	0.03959	0.00040	0.02829	...	0.00000	0.04000	0.01995
	8	f_{yk} (MPa)	598	424	556	512	408	559	494	437	...	400	600	500.0
	9	f_{sw} (MPa)	567	459	442	417	389	497	567	548	...	300	600	449.9
	10	ϕ_p (mm)	15.7	17.8	9.5	9.5	12.7	17.8	14.3	15.7	...	9.5	17.8	13.69
	11	$\rho_{p,bot}$	0.00354	0.00161	0.00367	0.00227	0.00195	0.00204	0.00162	0.00260	...	0.00070	0.00370	0.00220
	12	$f_{p,0.1k}$ (MPa)	1663.67	1663.35	1563.29	1560.55	1635.45	1686.85	1666.10	1673.06	...	1500.00	1700.00	1600.15
	13	σ_{pm0} (MPa)	1309.55	1262.09	1252.19	1032.40	1207.20	1070.95	1134.74	1331.68	...	1000.00	1395.00	1176.22
	14	$Loss_{lt}$	0.12	0.15	0.14	0.12	0.14	0.19	0.11	0.19	...	0.10	0.20	0.15
	15	ψ_1	0.2	0.5	0.2	0.9	0.7	0.9	0.5	0.9	...	0.2	1	0.615
	16	ψ_2	0	0.3	0	0.8	0.6	0.8	0.2	0.8	...	0	1	0.415
	17	t_0 (days)	21	16	22	8	7	27	18	20	...	7	28	17.49
	18	T (°C)	11.3	-11.6	-5.8	26.2	-10.3	6.0	-21.5	-8.6	...	-40.0	40.0	0.0
	19	RH (%)	88	98	91	43	78	49	74	56	...	40	100	70
	20	q_D (kN/m)	1450.42	4.43	413.17	51.61	63.32	226.38	56.42	258.29	...	-304.86	8476.64	203.87
	21	q_L (kN/m)	54.28	118.80	379.17	34.74	676.29	292.25	9.79	228.58	...	-244.78	8409.69	368.24
21 OUTPUT PARAMETERS FROM STRUCTURAL SOFTWARE (AutoSPT)	22	Δ_{trans}/L	-0.0004	0.0023	-0.0002	0.0001	0.0004	0.0000	0.0030	0.0003	...	-0.0012	0.0030	0.0005
	23	$\sigma_{c\text{-}trans}/f_{cki}$	0.225	0.360	0.268	0.100	0.091	0.099	0.501	0.174	...	0.025	1.000	0.199
	24	w_{trans} (mm)	0.000	0.044	0.000	0.000	0.000	0.000	0.234	0.000	...	0.000	0.400	0.022
	25	Δ_{lt}/L	0.0065	0.0067	0.0053	0.0067	0.0067	0.0067	0.0067	0.0059	...	-0.0030	0.0067	0.0051
	26	Δ_{incr}/L	0.0030	0.0030	0.0030	0.0030	0.0030	0.0030	0.0030	0.0030	...	-0.0025	0.0030	0.0026
	27	w_{freq} (mm)	0.169	0.181	0.110	0.400	0.176	0.315	0.375	0.102	...	0.000	0.400	0.215
	28	σ_c/f_{ck}	0.730	0.982	0.778	1.000	0.733	0.646	0.646	0.455	...	0.031	1.000	0.667
	29	$\sigma_{s\text{-}bot}/f_{yk}$	0.317	0.771	0.421	0.854	0.663	0.601	0.698	0.277	...	-0.003	1.000	0.594
	30	$\sigma_{p\text{-}bot}/f_{pk}$	0.707	0.705	0.685	0.663	0.661	0.608	0.716	0.632	...	0.430	1.000	0.672
	31	M_{Ed} (kNm)	182943.5	38345.0	120253.7	1755.9	47551.8	20130.6	83664.1	131937.5	...	36.4	930788.0	75623.6
	32	M_{Rd} (kNm)	311624.1	39449.7	201284.4	1822.1	61790.7	28817.9	100167.1	310217.9	...	105.9	1058907.3	97501.3
	33	M_{cr} (kNm)	81962.7	6503.9	48307.3	259.2	10523.9	4792.3	37350.1	69760.6	...	52.4	212731.4	23154.4
	34	x/d	0.601	0.642	0.346	0.432	0.415	0.253	0.236	0.195	...	0.032	0.782	0.294
	35	$\varepsilon_{s\text{-}bot}/\varepsilon_{sy}$	0.78	0.92	2.38	1.80	2.07	3.28	4.58	6.63	...	0.33	43.17	5.24
	36	$\varepsilon_{p\text{-}bot}/\varepsilon_{py}$	0.92	0.79	1.42	1.03	1.02	1.41	1.88	2.19	...	0.51	12.30	2.09
	37	V_{Ed} (kN)	25976.8	3652.7	15606.5	572.9	9035.5	4808.5	7523.0	14521.5	...	22.1	127323.8	9603.4
	38	V_{Rdmax} (kN)	58596.4	6148.4	44790.5	954.2	26716.9	14160.0	67409.5	72081.9	...	149.5	212893.6	32646.1
	39	$\rho_{sv\text{-}sp}$	0.0030	0.0045	0.0024	0.0041	0.0037	0.0029	0.0009	0.0013	...	0.0007	0.0245	0.0016
	40	$\rho_{sv\text{-}mid}$	0.0009	0.0024	0.0010	0.0019	0.0018	0.0015	0.0009	0.0009	...	0.0001	0.0025	0.0003
	41	CI_b (KRW/m)	3328083.9	450094.7	2548307.0	88440.3	1235203.4	834589.6	747397.7	3535355.9	...	5337.7	10554206.5	1324078.2
	42	$Weight$ (kN/m)	339.6	50.7	220.4	7.1	118.8	55.7	272.0	312.3	...	2.6	623.4	142.2

(Continued)

Table 4.3.5 (Continued) Large datasets of a simply supported PT beam generated by AutoPT

(b) Normalized based on min–max

			200,000 DATASETS NORMALIZED BY MAPMINMAX [-1;1]											
			Data 1	Data 2	Data 3	Data 4	Data 5	Data 6	Data 7	Data 8	...	MIN	MAX	MEAN
21 INPUT PARAMETERS FOR STRUCTURAL SOFTWARE (*AutoSPT*)	1	L (mm)	-0.1928	0.2891	-0.1004	-0.7476	-0.4411	-0.5912	0.3760	0.0921	...	-1.000	1.000	-0.259
	2	b (mm)	0.3845	-0.8163	-0.0657	-0.9415	-0.1662	-0.6262	0.0547	0.0965	...	-1.000	1.000	-0.366
	3	h (mm)	0.4488	0.3404	0.3477	-0.9203	-0.3895	-0.4043	0.5304	0.7453	...	-1.000	1.000	0.001
	4	f_{ck} (MPa)	-0.0667	-0.2000	-0.8667	-0.8667	0.8000	0.7333	-0.2000	-0.4000	...	-1.000	1.000	-0.002
	5	f_{cki} (MPa)	-1.0000	-0.8667	-0.6889	-0.3778	0.7778	0.0667	-0.2889	-0.9111	...	-1.000	1.000	-0.333
	6	$\rho_{s,bot}$	0.2358	0.8597	0.0618	0.5110	0.8684	0.2984	-0.6592	0.2845	...	-1.000	1.000	-0.001
	7	$\rho_{s,top}$	-0.6597	-0.5835	0.0165	0.2561	-0.1422	0.9794	-0.9801	0.4147	...	-1.000	1.000	-0.002
	8	f_{yk} (MPa)	0.9800	-0.7600	0.5600	0.1200	-0.9200	0.5900	-0.0600	-0.6300	...	-1.000	1.000	0.000
	9	f_{sw} (MPa)	0.7800	0.0600	-0.0533	-0.2200	-0.4067	0.3133	0.7800	0.6533	...	-1.000	1.000	-0.001
	10	ϕ_p (mm)	0.4940	1.0000	-1.0000	-1.0000	-0.2289	1.0000	0.1566	0.4940	...	-1.000	1.000	0.009
	11	$\rho_{p,bot}$	0.8908	-0.3937	0.9819	0.0439	-0.1696	-0.1084	-0.3871	0.2674	...	-1.000	1.000	-0.001
	12	$f_{p,0.1k}$ (MPa)	0.6368	0.6335	-0.3671	-0.3945	0.3545	0.8686	0.6610	0.7306	...	-1.000	1.000	0.002
	13	σ_{pm0} (MPa)	0.5673	0.3271	0.2769	-0.8360	0.0491	-0.6408	-0.3178	0.6794	...	-1.000	1.000	-0.108
	14	$Loss_{lt}$	-0.6320	0.0366	-0.2424	-0.6712	-0.1435	0.8866	-0.8178	0.8723	...	-1.000	1.000	0.001
	15	ψ_1	-1.0000	-0.2500	-1.0000	0.7500	0.2500	0.7500	-0.2500	0.7500	...	-1.000	1.000	0.037
	16	ψ_2	-1.0000	-0.4000	-1.0000	0.6000	0.2000	0.6000	-0.6000	0.6000	...	-1.000	1.000	-0.170
	17	t_0 (days)	0.3333	-0.1429	0.4286	-0.9048	-1.0000	0.9048	0.0476	0.2381	...	-1.000	1.000	-0.001
	18	T (°C)	0.2836	-0.2902	-0.1448	0.6544	-0.2582	0.1504	-0.5366	-0.2149	...	-1.000	1.000	0.001
	19	RH (%)	0.6130	0.9488	0.7021	-0.9057	0.2716	-0.6877	0.1303	-0.4647	...	-1.000	1.000	0.000
	20	q_D (kN/m)	-0.6002	-0.9296	-0.8365	-0.9188	-0.9161	-0.8790	-0.9177	-0.8717	...	-1.000	1.000	-0.884
	21	q_L (kN/m)	-0.9309	-0.9160	-0.8558	-0.9354	-0.7871	-0.8759	-0.9412	-0.8906	...	-1.000	1.000	-0.858
21 OUTPUT PARAMETERS FROM STRUCTURAL SOFTWARE (*AutoSPT*)	22	Δ_{trans}/L	-0.6459	0.6549	-0.5397	-0.3729	-0.2724	-0.4274	1.0000	-0.2992	...	-1.000	1.000	-0.209
	23	$\sigma_{c\text{-}trans}/f_{cki}$	-0.5894	-0.3136	-0.5014	-0.8454	-0.8646	-0.8491	-0.0235	-0.6939	...	-1.000	1.000	-0.642
	24	w_{trans} (mm)	-1.0000	-0.7807	-1.0000	-1.0000	-1.0000	-1.0000	0.1678	-1.0000	...	-1.000	1.000	-0.888
	25	Δ_{lt}/L	0.9655	1.0000	0.7082	1.0000	1.0000	1.0000	1.0000	0.8329	...	-1.000	1.000	0.666
	26	Δ_{incr}/L	1.0000	1.0000	1.0000	1.0000	1.0000	1.0000	1.0000	1.0000	...	-1.000	1.000	0.853
	27	w_{freq} (mm)	-0.1568	-0.0939	-0.4498	1.0000	-0.1191	0.5731	0.8751	-0.4905	...	-1.000	1.000	0.074
	28	σ_c/f_{ck}	0.4436	0.9633	0.5420	1.0000	0.4493	0.2702	0.2687	-0.1236	...	-1.000	1.000	0.312
	29	$\sigma_{s\text{-}bot}/f_{yk}$	-0.3625	0.5429	-0.1534	0.7085	0.3273	0.2037	0.3984	-0.4409	...	-1.000	1.000	0.190
	30	$\sigma_{p\text{-}bot}/f_{pk}$	-0.0300	-0.0372	-0.1050	-0.1852	-0.1921	-0.3773	0.0029	-0.2919	...	-1.000	1.000	-0.151
	31	M_{Ed} (kN·m)	-0.6070	-0.9177	-0.7417	-0.9963	-0.8979	-0.9568	-0.8203	-0.7166	...	-1.000	1.000	-0.838
	32	M_{Rd} (kN·m)	-0.4116	-0.9257	-0.6200	-0.9968	-0.8835	-0.9458	-0.8110	-0.4142	...	-1.000	1.000	-0.816
	33	M_{cr} (kN·m)	-0.2297	-0.9393	-0.5462	-0.9981	-0.9015	-0.9554	-0.6493	-0.3445	...	-1.000	1.000	-0.783
	34	x/d	0.5164	0.6254	-0.1627	0.0665	0.0220	-0.4124	-0.4557	-0.5673	...	-1.000	1.000	-0.303
	35	$\varepsilon_{s\text{-}bot}/\varepsilon_{sy}$	-0.9791	-0.9724	-0.9044	-0.9315	-0.9188	-0.8622	-0.8017	-0.7058	...	-1.000	1.000	-0.771
	36	$\varepsilon_{p\text{-}bot}/\varepsilon_{py}$	-0.9298	-0.9511	-0.8446	-0.9105	-0.9134	-0.8460	-0.7671	-0.7144	...	-1.000	1.000	-0.731
	37	V_{Ed} (kN)	-0.5922	-0.9430	-0.7552	-0.9913	-0.8584	-0.9248	-0.8822	-0.7722	...	-1.000	1.000	-0.849
	38	V_{Rdmax} (kN)	-0.4505	-0.9436	-0.5803	-0.9924	-0.7502	-0.8683	-0.3677	-0.3238	...	-1.000	1.000	-0.695
	39	$\rho_{sw\text{-}sp}$	-0.8780	-0.8011	-0.9095	-0.8204	-0.8398	-0.8849	-0.9902	-0.9717	...	-1.000	1.000	-0.866
	40	$\rho_{sw\text{-}mid}$	-0.9827	-0.8575	-0.9753	-0.8986	-0.9085	-0.9318	-0.9845	-0.9847	...	-1.000	1.000	-0.923
	41	CI_b (KRW/m)	-0.3700	-0.9157	-0.5179	-0.9842	-0.7668	-0.8428	-0.8593	-0.3307	...	-1.000	1.000	-0.750
	42	$Weight$ (kN/m)	0.0856	-0.8449	-0.2985	-0.9855	-0.6256	-0.8291	-0.1321	-0.0024	...	-1.000	1.000	-0.550

corresponding output parameters (Δ_{trans}/L, $\sigma_{c,trans}/f_{cki}$, w_{trans}, Δ_{lt}/L, Δ_{incr}/L, w_{freq}, σ_c/f_{ck}, $\sigma_{s,bot}/f_{yk}$, $\sigma_{p,bot}/f_{pk}$, M_{Ed}, M_{Rd}, M_{cr}, x/d, $\varepsilon_{s,bot}/\varepsilon_{sy}$, $\varepsilon_{p,bot}/\varepsilon_{py}$, V_{Ed}, V_{Rdmax}, $\rho_{sw,sp}$, $\rho_{sw,mid}$, CI_b, $Weight$) for designs. AutoPT BEAM also calculates parameters related to beam sections, such as effective depth of rebar (d_s), strand (d_p), and concrete cover to the surface of rebars (c_s) based on detailing rules required by EC2 [4.27]. The present example uses big datasets generated based on equations shown in European code EC2 [4.27] which are developed based on extensive experimental observations. They are widely accepted by engineers. For instance, crack widths estimated based on Eurocode are verified by the experimental investigations seen from

[4.30–4.40] in which a mean value of 0.0022 mm and a standard deviation of 0.0769 mm are reported for errors between calculated and tested results, showing the accuracy of Eurocode.

4.4 MOO DESIGN OF A PT BEAM

4.4.1 Derivation of objective functions based on forward ANNs

In this section, objective and constraining functions are generalized using weight and bias matrices obtained by training forward ANNs. During training, 21 input parameters ($\boldsymbol{L}, \boldsymbol{b}, \boldsymbol{h}, f_{ck}, f_{cki}, \rho_{s,bot}, \rho_{s,top}, f_{yk}, f_{sw}, \phi_p, \rho_{p,bot}, f_{p0.1k}, \sigma_{pm0}, \boldsymbol{Loss}_{lt}, \psi_1, \psi_2, t_0, \boldsymbol{T}, \boldsymbol{RH}, q_D, q_L$) are mapped to 21 output parameters ($\Delta_{trans}/L, \sigma_{c,trans}/f_{cki}, w_{trans}, \Delta_{lt}/L, \Delta_{incr}/L, w_{freq}, \sigma_c/f_{ck}, \sigma_{s,bot}/f_{yk}\ \sigma_{p,bot}/f_{pk}, M_{Ed}, M_{Rd}, M_{cr}, x/d, \varepsilon_{s,bot}/\varepsilon_{sy}, \varepsilon_{p,bot}/\varepsilon_{py}, V_{Ed}, V_{Rdmax}, \rho_{sw,sp}, \rho_{sw,mid}, CI_b$, ***Weight***) using PTM method [4.41], [4.54] as shown in Table 4.4.1. Explicit objective functions are replaced by ANN-based objective functions, avoiding

Table 4.4.1 Training results based on PTM

Training table: Forward - PTM

21 Inputs ($L, b, h, f_{ck}, f_{cki}, \rho_{s\text{-}bot}, \rho_{s\text{-}top}, f_{yk}, f_{sw}, D_p, \rho_{p\text{-}bot}, f_{p,0.1k}, \sigma_{pm0}, Loss_{lt}, \psi_1, \psi_2, t_0, T, RH, q_D, q_L$) -
21 Outputs ($\Delta_{trans}, \sigma_{c\text{-}trans}/f_{cki}, w_{trans}, \Delta_{lt}, \Delta_{incr}, w_{freq}, \sigma_c/f_{ck}, \sigma_{s\text{-}bot}/f_{yk}, \sigma_{p\text{-}bot}/f_{pk}, M_{Ed}, M_{Rd}, M_{mr}, x/d, \varepsilon_{s\text{-}bot}/\varepsilon_{sy}, \varepsilon_{p\text{-}bot}/\varepsilon_{py}, V_{Ed}, V_{Rdmax}, \rho_{sw\text{-}sp}, \rho_{sw\text{-}mid}, CI_b$, *Weight*)

No.	Output	Data	Layers	Neurons	Validation check	Suggested Epoch	Best Epoch	Stopped Epoch	MSE Tr.perf	MSE T.perf	R at Best Epoch
1	Δ_{trans}	150,000	5	80	1,000	100,000	9,220	10,220	7.5.E-05	1.8.E-04	0.9990
2	$\sigma_{c\text{-}trans}/f_{cki}$	150,000	5	80	1,000	100,000	24,141	25,141	8.2.E-05	2.3.E-04	0.9990
3	w_{trans}	150,000	10	50	1,000	100,000	6,050	7,050	1.1.E-04	1.1.E-03	0.9979
4	Δ_{lt}	150,000	10	80	1,000	100,000	27,481	28,481	1.1.E-04	5.9.E-04	0.9985
5	Δ_{incr}	150,000	10	80	1,000	100,000	14,209	15,209	3.6.E-04	2.0.E-03	0.9972
6	w_{freq}	150,000	5	50	1,000	100,000	24,979	25,979	4.4.E-04	1.2.E-03	0.9973
7	σ_c/f_{ck}	150,000	5	80	1,000	100,000	39,893	40,893	1.8.E-04	4.1.E-04	0.9995
8	$\sigma_{s\text{-}bot}/f_{yk}$	150,000	10	80	1,000	100,000	33,738	34,738	3.6.E-04	9.0.E-04	0.9987
9	$\sigma_{p\text{-}bot}/f_{pk}$	150,000	10	80	1,000	100,000	60,376	61,376	2.1.E-04	4.8.E-04	0.9978
10	M_{Ed}	150,000	5	50	1,000	100,000	68,109	68,118	3.0.E-09	4.6.E-09	1.0000
11	M_{Rd}	150,000	2	50	1,000	100,000	67,642	68,642	1.1.E-06	1.2.E-06	1.0000
12	M_{mr}	150,000	5	50	1,000	100,000	18,692	19,692	2.1.E-05	2.2.E-05	0.9998
13	x/d	150,000	2	50	1,000	100,000	27,025	28,025	3.1.E-05	4.2.E-05	0.9999
14	$\varepsilon_{s\text{-}bot}/\varepsilon_{sy}$	150,000	10	50	1,000	100,000	61,405	62,405	7.4.E-06	8.6.E-06	0.9999
15	$\varepsilon_{p\text{-}bot}/\varepsilon_{py}$	150,000	10	50	1,000	100,000	99,947	100,000	1.7.E-05	2.1.E-05	0.9998
16	V_{Ed}	150,000	5	50	1,000	100,000	67,927	68,814	2.2.E-09	7.2.E-09	1.0000
17	V_{Rdmax}	150,000	10	50	1,000	100,000	99,997	100,000	6.7.E-07	8.0.E-07	1.0000
18	$\rho_{sw\text{-}sp}$	150,000	10	50	1,000	100,000	34,022	35,022	1.1.E-05	5.2.E-05	0.9997
19	$\rho_{sw\text{-}mid}$	150,000	10	80	1,000	100,000	25,258	26,258	4.5.E-05	1.4.E-04	0.9935
20	CI_b	150,000	2	80	1,000	100,000	91,623	91,656	4.7.E-09	5.4.E-09	1.0000
21	*Weight*	150,000	5	50	1,000	100,000	76,571	76,590	1.2.E-09	1.3.E-09	1.0000

mathematical difficulties met by conventional optimizations based on Lagrange multiplier.

4.4.2 An optimization scenario

In an example of MOO, overall construction and material cost (*CI*), beam weight (**W**), and beam height (*h*) are minimized based on 16 inputs constrained by 16 equalities for fixed beam length (*L*), material properties ($f_{ck}, f_{cki}, f_{yk}, f_{sw}, f_{p0.1k}$), tendon parameters ($\sigma_{pm0}$, $Loss_{lt}, \phi_p$), and applied loads ($\psi_1, \psi_2, t_0, T, RH, q_D, q_L$) as shown in Table 4.3.4(a). In Table 4.3.4(b), inequalities v_1 to v_{13} are also established to impose 13 code-based design requirements to optimize design parameters (b, $\rho_{s,bot}$, $\rho_{s,top}$, $\rho_{p,bot}$) within ranges prescribed in six inequalities from v_{14} to v_{19} of Table 4.3.4(b).

4.4.3 Derivation of a unified function of objective (UFO) for a PT beam

A *UFO* for PT beam shown in Eqs. (4.4.1) and (4.4.2) is defined by an algorithm based on *weighted sum method* [4.6], [4.55]. In Eq. (4.4.1), three objective functions are integrated into one UFO combined with equality and inequality constraints shown in Table 4.3.4 after which Lagrange function of an *UFO* shown in Eqs. (4.4.3)–(4.4.5) is optimized simultaneously. Optimized design parameters are determined by substituting a Lagrange function into a MATLAB built-in optimization toolbox [4.42–4.47]. A flowchart is presented in Figure 4.4.1 where ANN-based Lagrange

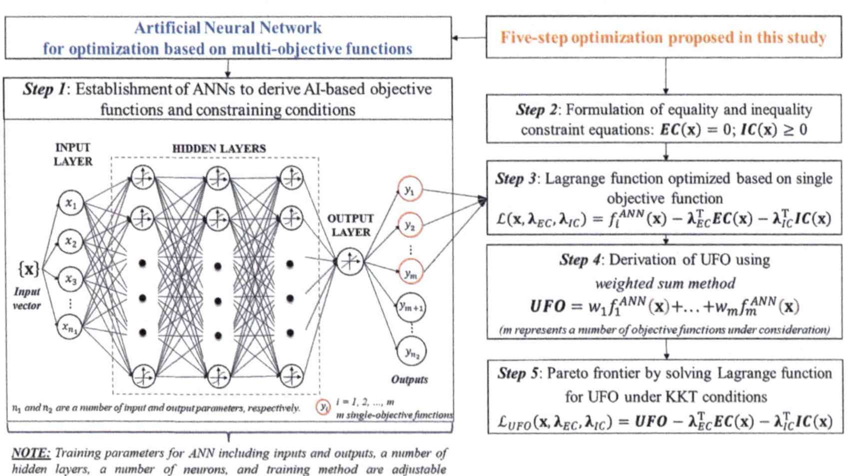

Figure 4.4.1 ANN-based Lagrange optimization algorithm of five steps based on UFO [4.25].

optimization algorithm of five steps based on UFO is shown to solve MOO problems with an example of RC columns [4.48].

$$UFO = \sum_{i=1}^{m} w_i f_i^{ANN}(\mathrm{x}) = w_1 f_1^{ANN}(\mathrm{x}) + \ldots + w_m f_m^{ANN}(\mathrm{x})$$

$$\text{(dimensionless) [4.49] and [4.50]} \quad (4.4.1)$$

$$w_i \in [0,1]; \sum_{i=1}^{m} w_i = 1 \quad \text{(dimensionless)} \quad (4.4.2)$$

$$\mathrm{L}_{UFO}(\mathrm{x}, \lambda_c, \lambda_v) = UFO - \lambda_c^{\mathrm{T}} EC(\mathrm{x}) - \lambda_v^{\mathrm{T}} IC(\mathrm{x}) \quad \text{(dimensionless)} \quad (4.4.3)$$

$$EC(\mathrm{x}) = \left[EC_1(\mathrm{x}), EC_2(\mathrm{x}), \ldots, EC_{m_1}(\mathrm{x})\right]^T; m_1$$

$$\text{is a number of equality constraints} \quad (4.4.4)$$

$$IC(\mathrm{x}) = \left[IC_1(\mathrm{x}), IC_2(\mathrm{x}), \ldots, IC_{m_2}(\mathrm{x})\right]^T; m_2$$

$$\text{is a number of inequality constraints} \quad (4.4.5)$$

4.5 A MOO-BASED PT BEAM DESIGN

4.5.1 Formulation of an UFO

Equality and inequality constraining an optimization of a simply supported PT beam are presented in Table 4.3.4; Table 4.3.4(a) lists 16 equalities including span lengths (c_1), material properties (c_2, c_3,c_4,c_5, and c_7), prestressing parameters $(c_6, c_8, c_9, \text{ and } c_{12})$, factors for load combinations $(c_{10} \text{ and } c_{11})$, environmental conditions $(c_{13} \text{ and } c_{14})$, loadings $(c_{15} \text{ and } c_{16})$, whereas Table 4.3.4(b) lists 19 inequalities imposed by the EC2 [4.27] and architectures, including limitations of a transfer stage $(v_1, v_2, \text{ and } v_3)$, restrictions of service stages (from v_4 to v_9), demands at an ultimate stage $(v_{10}, v_{11} \text{ and } v_{13})$, and a maximum neutral axis depth ratio (v_{12}). Architectural requirements and maximum rebar ratios are also imposed by six inequalities v_{14} to v_{19} in addition to the equality and inequality constraints. An *UFO* formulated in Eq. (4.5.1) by combining three objective functions (***h***, ***CI****$_b$*, and **W**) allows engineers to optimize multiple objective functions simultaneously with adjustable tradeoff ratios contributed by each of the three objective functions which can be controlled based on design interests. Objective functions are normalized with their optimized maxima and minima of an individual objective function to remove dominant influence of one parameter on other parameters in an *UFO*.

$$UFO = w_h \frac{h - h_{min}}{h_{max} - h_{min}} + w_{CI_b} \frac{CI_b - CI_{b\,min}}{CI_{b\,max} - CI_{b\,min}} + w_{Wt.} \frac{W - W_{min}}{W_{max} - W_{min}} \quad (4.5.1)$$

(dimensionless)

where

w_h, w_{CI_b}, and $w_{Wt.}$ - Weight fractions (tradeoff ratios) contributed by individual objective function, $\boldsymbol{h}$, $\boldsymbol{CI_b}$, and $\mathbf{W}$, allowing users to adjust among objectives

$$w_h + w_{CI_b} + w_W = 1$$

$\boldsymbol{h_{min}}$, $\boldsymbol{CI_{bmin}}$, and $\mathbf{W_{min}}$ - The minima of $\boldsymbol{h}$, $\boldsymbol{CI_b}$, and $\mathbf{W}$ determined by ANN-based Lagrange optimizations. For example, the following minima are obtained when equalities and inequalities shown in Table 4.3.4 are imposed with $\boldsymbol{q_L}$ = 400 kN/m.

$\boldsymbol{h_{min}}$ = 966 mm, $\boldsymbol{CI_{bmin}}$ = 131,531 KRW/m, and $\mathbf{W_{min}}$ = 22.83 kN/m

It is noted that $\boldsymbol{h_{min}}$, $\boldsymbol{CI_{bmin}}$, and $\mathbf{W_{min}}$ are taken from the three different beam designs

$\boldsymbol{h_{max}}$, $\boldsymbol{CI_{bmax}}$, and $\mathbf{W_{max}}$ - The maxima of $\boldsymbol{h}$, $\boldsymbol{CI_b}$, and $\mathbf{W}$ determined by ANN-based Lagrange optimizations. For example, the following maxima are obtained when equalities and inequalities shown in Table 4.3.4 are imposed with $\boldsymbol{q_L}$ = 400 kN/m.

$\boldsymbol{h_{max}}$ = 1,500 mm, $\boldsymbol{CI_{bmax}}$ = 1,015,180 KRW/m, and $\mathbf{W_{max}}$ = 56.26 kN/m

It is noted that $\boldsymbol{h_{max}}$, $\boldsymbol{CI_{bmax}}$, and $\mathbf{W_{max}}$ are taken from the three different beam designs

4.5.2 Generation of tradeoff ratios contributed by individual objective function

Design parameters ($\boldsymbol{b}$, $\rho_{s,bot}$, $\rho_{s,top}$, and $\rho_{p,bot}$) of a PT beam are presented in Table 4.5.1 obtained based on ANN-based Lagrange optimization which equally minimizes $\boldsymbol{h}$, $\boldsymbol{CI_b}$, and W when tradeoff ratio of ($w_h : w_{CI_b} : w_{Weight} = 1/3 : 1/3 : 1/3$) are implemented, while all 16 equalities and 19 inequalities given in Table 4.3.4 are satisfied. The proposed optimization automatically minimizes $\boldsymbol{h}$, $\boldsymbol{CI_b}$, and W, simultaneously, based on $UFO = \frac{1}{3}\frac{\boldsymbol{h} - 966}{1500 - 966} + \frac{1}{3}\frac{\boldsymbol{CI_b} - 131531}{1015180 - 131531} + \frac{1}{3}\frac{\text{Wt.} - 22.83}{56.26 - 22.83}$.

Table 4.5.1 MOO with $c_{16} = q_L - 400 = 0$ based on $w_h : w_{CI_b} : w_{Weight} = 1/3 : 1/3 : 1/3$ at $q_L = 400$ kN

MULTIPLE-OBJECTIVE OPTIMIZATION
$w_h : w_{CI_b} : w_{weight} = 1/3 : 1/3 : 1/3$

TRAINING INPUTS			TRAINING OUTPUTS				
	Parameter	**Value**		**Parameter**	**AI results**	**Check (AutoPT)**	**Difference**
1	L (mm)	12000	22	Δ_{trans} (L/250=48 mm)	-10.48	-9.24	-1.233
2	b (mm)	1500	23	$\sigma_{c\text{-}trans}/f_{cki}$ (≤ 0.6)	0.287	0.285	0.002
3	h (mm)	1002	24	w_{trans} (mm) (≤ 0.2 mm)	0.004	0.000	0.004
4	f_{ck} (MPa)	40	25	Δ_{lt} (mm) (L/250=48 mm)	7.02	9.79	-2.762
5	f[illegible] (MPa)	20	26	Δ_{incr} (mm) (L/500=24 mm)	17.62	19.03	-1.410
6	$\rho_{s,bot}$	0.0123	27	w_{freq} (mm) (≤ 0.2 mm)	0.016	0.032	-0.017
7	$\rho_{s,top}$	0.0267	28	σ_c/f_{ck} (≤ 0.6)	0.600	0.619	-0.019
8	f_{yk} (MPa)	500	29	$\sigma_{s\text{-}bot}/f_{yk}$ (≤ 0.8)	0.425	0.459	-0.034
9	f_{sw} (MPa)	400	30	$\sigma_{p\text{-}bot}/f_{pk}$ (≤ 0.75)	0.718	0.718	-0.001
10	ϕ_p (mm)	15.2	31	M_{Ed} (kN·m)	8772.8	8776.2	-3.41
11	$\rho_{p,bot}$	0.0037	32	M_{Rd} (kN·m) (≥ M_{Ed})	13074.8	12797.6	277.18
12	$f_{p,0.1k}$ (MPa)	1640	33	M_{cr} (kN·m)	4897.0	4689.7	207.27
13	σ_{pm0} (MPa)	1350	34	x/d (≤ 0.448)	0.1886	0.1883	0.0003
14	$Loss_{lt}$	0.15	35	$\varepsilon_{s\text{-}bot}/\varepsilon_{sy}$	5.96	6.04	-0.077
15	ψ_1	0.5	36	$\varepsilon_{p\text{-}bot}/\varepsilon_{py}$	2.30	2.30	0.004
16	ψ_2	0.3	37	V_{Ed} (kN)	2924.77	2925.41	-0.64
17	t_0 (days)	28	38	V_{Rdmax} (kN) (≥ V_{Ed})	9213.550	9192.681	20.87
18	T (°C)	20	39	$\rho_{sw\text{-}sp}$	0.00282	0.00292	-0.00010
19	RH (%)	70	40	$\rho_{sw\text{-}mid}$	0.00128	0.00126	0.00001
20	q_D (kN/m)	50.0	41	CI_b (KRW/m)	393664	393564	100.06
21	q_L (kN/m)	400.0	42	$Weight$ (kN/m)	37.56	37.57	-0.012

Design parameters obtained by ANN-based Hong-Lagrange optimization

Noticeable errors but do not affect design.

Note:
- **ANN network:**
  - 21 inputs for ANN forward network.
  - 21 outputs for ANN forward network.
- **Verification by AutoPT:**
  21 forward input (from Boxes 1 to 21) are implemented in *AutoPT* to calculate 21 forward outputs (from Boxes 22 to 42 to verify).

4.5.3 Negligible errors

An ANN provides acceptable calculations for almost all output parameters except for Δ_{lt} and w_{freq} for $q_L = 400$ kN/m. Both output parameters of Δ_{lt} and w_{freq} calculated by an ANN and AutoPT BEAM with errors of absolute errors of –2.762 mm for Δ_{lt} and –0.017 mm for w_{freq} satisfy design limitations, not affecting design results. An absolute error –2.762 mm in Δ_{lt} is equivalent to 5.75% of a long-term deflection limit (48 mm according to EC2 [4.27]), whereas an absolute error –0.017 mm in w_{freq} is equivalent to 8.5% of a crack width limit (0.2 mm according to EC2 [4.27]).

4.5.4 Pareto frontiers based on designated tradeoff ratios

The three objective functions are simultaneously optimized based on combinations of 100 randomly generated tradeoff ratios representing

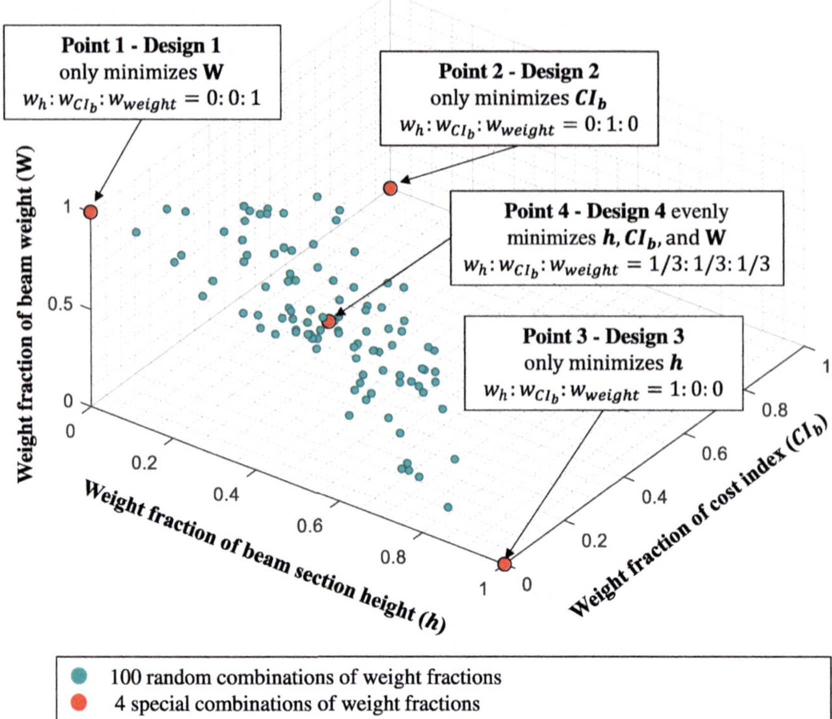

Figure 4.5.1 Weight fractions generated to obtain Pareto frontier.

contributions made by the three individual objective functions. Design parameters ($\boldsymbol{b}$, $\rho_{s,bot}$, $\rho_{s,top}$, and $\rho_{p,bot}$) are obtained on an output-side while all 16 equalities and 19 inequalities given in Table 4.3.4 are satisfied.

In Figure 4.5.1, Pareto frontiers are obtained based on 100 randomly generated and four specially generated combinations of weight fractions in which the sum of the three weight fractions in one combination should be equivalent to 1. Multiple combinations of weight fractions may yield identical designs when a discrete Pareto frontier is calculated. Four special combinations include the three individual objective optimizations represented by Pareto points 1, 2, and 3, respectively. Pareto point 4 equally minimizes among $\boldsymbol{h}, \boldsymbol{CI_b}$, and $\mathbf{W}$.

4.5.5 Optimized objective functions

4.5.5.1 Two and three-dimensional Pareto frontier

Three special combinations of the three objective optimizations are represented by Pareto points 1, 2, and 3 at which weight $\mathbf{W}$, cost $\boldsymbol{CI_b}$, and $\boldsymbol{h}$ are minimized, respectively, separately, whereas all three objective functions (weight **W**, cost CI_b, and h) are equally optimized in Design 4 represented by Pareto point 4.

Figure 4.5.2(a) shows a three-dimensional Pareto frontier for the three objective functions (weight **W**, cost CI_b, and h) along the three-dimensional axes. Figure 4.5.2(a) shows design parameters for the four beam designs

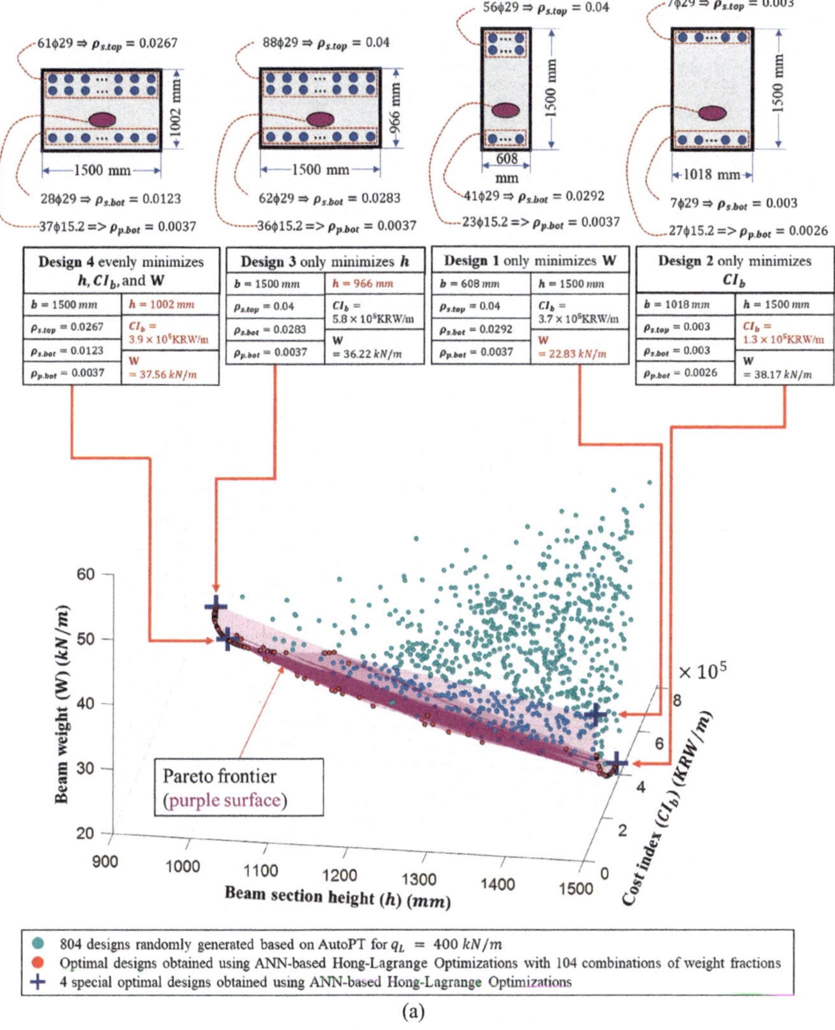

(a)

Figure 4.5.2 Comparison between Pareto-efficient designs and structural datasets at $c_{16} = q_L - 400 = 0$. (a) A dimensional Pareto frontier for the three objective functions (h, CI_b, and W). (b) Design points projected to a $CI_b - W$ plane. (c) Design points projected to a $h - W$ plane. (d) Design points projected to a $h - CI_b$ surface.

(Continued)

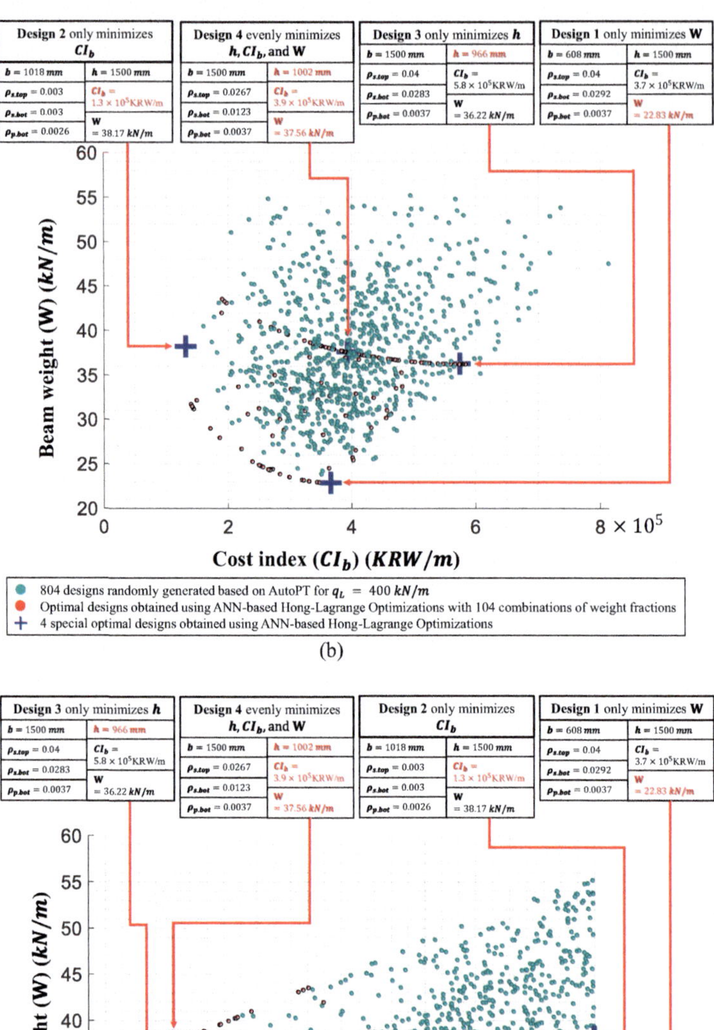

Figure 4.5.2 (Continued) Comparison between Pareto-efficient designs and structural datasets at $c_{I6} = q_L - 400 = 0$. (a) A dimensional Pareto frontier for the three objective functions (h, CI_b, and W). (b) Design points projected to a $CI_b - W$ plane. (c) Design points projected to a $h - W$ plane. (d) Design points projected to a $h - CI_b$ surface.

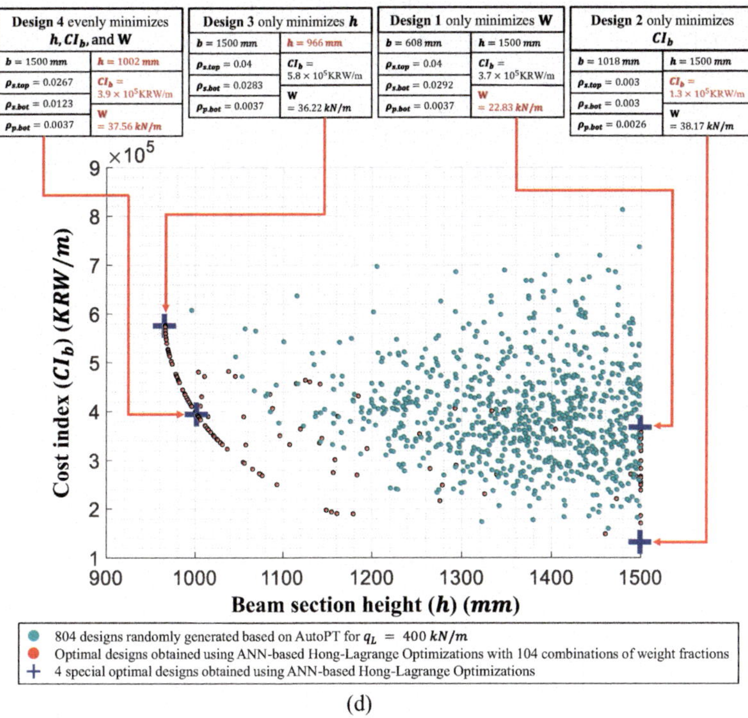

Figure 4.5.2 (Continued) Comparison between Pareto-efficient designs and structural datasets at $c_{16} = q_L - 400 = 0$. (a) A dimensional Pareto frontier for the three objective functions (h, CI_b, and W). (b) Design points projected to a $CI_b - W$ plane. (c) Design points projected to a $h - W$ plane. (d) Design points projected to a $h - CI_b$ surface.

based on the four special tradeoffs' combinations of the three objective functions. Figure 4.5.2(a) illustrates beam designs in which beams sizes and rebar ratios change when they are optimized under Designs 1, 2, 3, and 4. Figure 4.5.2(b) shows design points projected to a $\boldsymbol{CI_b}$ – **W** plane, whereas Figure 4.5.2(c) shows design points projected to a $\boldsymbol{h}$ – **W** plane. Figure 4.5.2(d) also shows design points projected to a $\boldsymbol{h} - \boldsymbol{CI_b}$ plane. In Figure 4.5.2(b)–(d), designs are projected to $\boldsymbol{CI_b}$ – **W**, $\boldsymbol{h}$ – **W**, and $\boldsymbol{h} - \boldsymbol{CI_b}$ planes based on tradeoffs among the three objective functions. It is noted that all four PT beam sections will resist the same load $q_L = 400$ kN/m.

4.5.5.2 Beam sections among Designs 1, 2, and 3

When minimizing cost index ($\boldsymbol{CI_b}$) for Design 2 and beam weight (**W**) for Design 1, maximum beam depth ($\boldsymbol{h}$ = 1,500 mm) which maximizes moment arms for rebars and tendons to minimize cost index ($\boldsymbol{CI_b}$) and beam weight (**W**) is obtained as shown in Figure 4.5.2(a). As shown in Figure 4.5.2(a) and (b), a design minimizing $\boldsymbol{CI_b}$ for Design 2 selects a large beam width

($\boldsymbol{b}$ = 1,018 mm) with small tendon and rebar ratios ($\rho_{p.bot} = 0.0026$, $\rho_{s.top} = 0.003$, and $\rho_{s.bot} = 0.003$) to reduce costs, however, a design minimizing **W** for Design 1 uses a narrow beam width ($\boldsymbol{b}$ = 608 mm) with high tendon and rebar ratios $\left(\rho_{p.bot} = 0.0037,\ \rho_{s.top} = 0.04, \text{ and } \rho_{s.bot} = 0.0292\right)$ to decrease beam weights. It is noted that a beam that costs lesser is heavier, whereas a light beam is expensive. A wide beam width ($\boldsymbol{b} = 1{,}500$ mm), large ratios of tendons and rebars $\left(\rho_{p.bot} = 0.0037,\ \rho_{s.top} = 0.04, \text{ and } \rho_{s.bot} = 0.0283\right)$ are selected when minimizing a beam depth ($\boldsymbol{h} = 966$ mm) for Design 3, producing cost index which is 4.46 times higher than minimized $\boldsymbol{CI_b}$ for Design 2. A beam weight is also 1.59 times heavier than minimized **W** for Design 1.

4.5.5.3 Beam sections for Design 4

A design point for Design 4 with the three equal weight fractions $w_h : w_{CI_b} : w_{Weight} = 1/3 : 1/3 : 1/3$ is located on a center of big datasets as shown in Figure 4.5.2(b) where a $\boldsymbol{CI_b} - \mathbf{W}$ plane is projected. This indicates median values of $\boldsymbol{CI_b} = 393{,}664$ KRW/m and $\mathbf{W} = 37.56$ kN / m over those provided by big datasets (cyan points). A beam height ($\boldsymbol{h}$) of 1,002 mm which is closer to a minimized $\boldsymbol{h_{min}} = 966$ mm obtained for Design 3 is obtained because $\boldsymbol{h} = 1{,}002$ mm is the most favorable value for minimizing $\text{UFO} = \frac{1}{3}\frac{\boldsymbol{h} - 966}{1{,}500 - 966} + \frac{1}{3}\frac{\boldsymbol{CI_b} - 131{,}531}{1{,}015{,}180 - 131{,}531} + \frac{1}{3}\frac{\mathbf{W} - 22.83}{56.26 - 22.83}$. It is noted that Pareto frontier for other load magnitudes can be produced by repeating the same optimization procedures shown in Table 4.5.1 without retraining ANNs.

4.5.5.4 Verification of Pareto frontier

Probable designs obtained by optimizing respective objective function are defined by average values of the ten best designs among the big datasets. Effective savings based on optimized designs of this example are proved compared with probable designs obtained from the big datasets. A minimized weight provided in Design 1 is lighter than a probable weight of 25.476 kN/m calculated from the ten lightest designs of the big datasets which is 10.4% heavier than that obtained by Design 1, whereas a minimized cost given by Design 2 is cheaper than a probable cost of 1.86E5 KRW/m calculated from the ten cheapest designs of the datasets which is 30.1% higher than that obtained by Design 2. A minimized beam depth by Design 3 is shallower than a probable beam depth of 1,063.2 mm obtained from ten smallest designs of the big datasets which is 9.1% deeper than that obtained by Design 3.

4.5.6 Influence of three objective functions on concrete sections

4.5.6.1 *Verification of design parameters (b, $\rho_{s,bot}$, $\rho_{s,top}$, and $\rho_{p,bot}$) corresponding to optimized objective functions*

Design parameters ($\boldsymbol{b}$, $\rho_{s,bot}$, $\rho_{s,top}$, and $\rho_{p,bot}$) corresponding to optimized objective functions are also obtained and verified on an output-side based on 104 combinations of weight fractions shown in Figure 4.5.1 while all 16 equalities and 19 inequalities given in Table 4.3.4 are satisfied for $q_L = 400$ kN/m. All 104 Pareto frontiers plotted in Figure 4.5.1 yield design parameters, including $\boldsymbol{b}$, $\rho_{s,bot}$, $\rho_{s,top}$, and $\rho_{p,bot}$ obtained during optimizing *UFO*. Cracking moment capacities ($\boldsymbol{M_{cr}}$), effective camber moments ($\boldsymbol{M_p}$) caused by prestressing, and total rebar areas ($\boldsymbol{A_r}$) obtained by four special optimal designs (Designs 1–4) are displayed in Figure 4.5.3 which also presents design parameters ($\boldsymbol{b}$, $\rho_{s,bot}$, $\rho_{s,top}$, and $\rho_{p,bot}$) on an output-side corresponding to optimized objective functions. It is noted that $\boldsymbol{M_{cr}}$ and $\boldsymbol{M_p}$ are normalized by a moment ($\boldsymbol{M_{E.ser}}$) induced by service loads. In Figure 4.5.3,

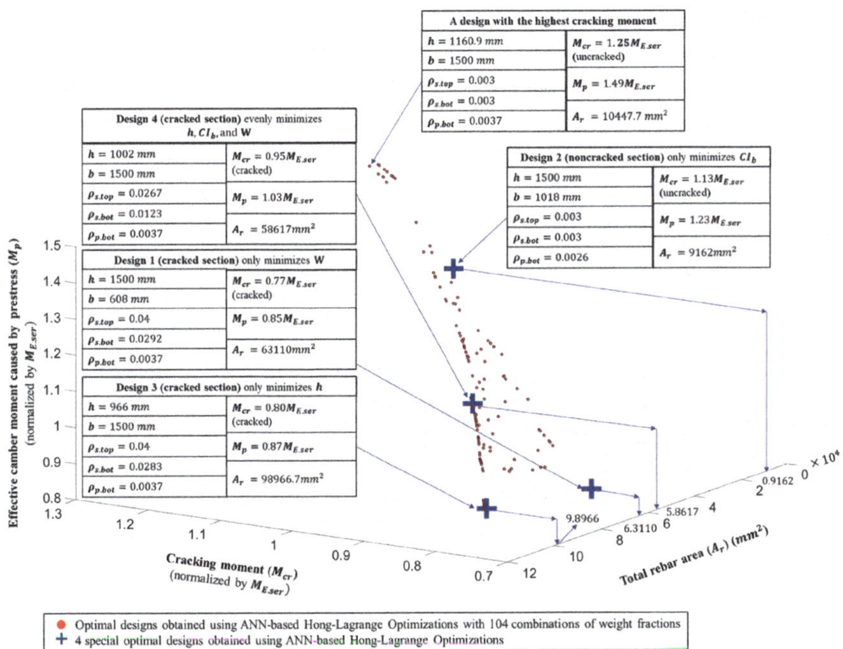

Note: $M_{E.ser}$ is moment induced by service loads excluding prestressing moments. $M_{E.ser}$ is calculated based on a combination of $Selfweight + Dead\ load + 0.5Live\ load$, reflecting 1% probability of being exceeded

Figure 4.5.3 Three dimensional relationships among cracking moments (M_{cr}), effective camber moments (M_p) caused by prestresses, and total rebar areas (A_r).

design parameters including $\boldsymbol{b}$, $\rho_{s,bot}$, $\rho_{s,top}$, and $\rho_{p,bot}$ obtained during optimizing UFO are observed. The $\boldsymbol{M}_{cr}$, $\boldsymbol{M}_{p}$, $\boldsymbol{A}_{r}$, $\boldsymbol{b}$, $\rho_{s,bot}$, $\rho_{s,top}$, and $\rho_{p,bot}$ parameters are not objective functions of the present optimization, thus they are not maximized nor minimized; however, they represent values corresponding to optimized objective functions which are design targets.

4.5.6.2 Delaying cracks with weight fractions of $w_h : w_{CI_b} : w_{Weight} = 0.4865 : 0.477 : 0.0364$ based on highest cracking moment ($M_{cr} = 1.25M_{E.ser}$)

Figure 4.5.3 shows cracking moment capacities ($\boldsymbol{M}_{cr}$) and effective camber moments ($\boldsymbol{M}_{p}$) in terms of $\boldsymbol{M}_{E.ser}$ which is calculated based on a combination of *Selfweight* + *Dead load* + 0.5*Live load*, reflecting a 1% probability of being exceeded during service stages. Effects of prestressing magnitudes on beams are demonstrated by effective camber moments ($\boldsymbol{M}_{p}$) caused by prestressing while crack resistances contributed by concrete sections, prestressing tendons, and rebars are represented by cracking moment capacities ($\boldsymbol{M}_{cr}$). High cracking resistances (cracking moments) are found in designs with high camber moments, whereas cracks take place earlier when beam depths are minimized as indicated in Figure 4.5.3. A design with the highest cracking moment (cracking moment capacities) obtained with $1.49\boldsymbol{M}_{E.ser}$ which delays cracks is found when three weight fractions are $w_h : w_{CI_b} : w_{Weight} = 0.4865 : 0.477 : 0.0364$, where a maximum tendon ratio of 0.0037 is used.

4.5.6.3 Design 2 with the deepest concrete section (h = 1,500 mm) leading to an uncracked design based on cracking moment capacity ($M_{cr} = 1.13M_{E.ser}$) and effective camber moment ($M_p = 1.23M_{E.ser}$)

As shown with relationships among cracking moments ($\boldsymbol{M}_{cr}$), effective camber moments ($\boldsymbol{M}_{p}$) caused by prestresses, and total rebar areas ($\boldsymbol{A}_{r}$) of Figure 4.5.3, Design 2 has the cracking moment capacity ($\boldsymbol{M}_{cr} = 1.13\boldsymbol{M}_{E.ser}$) and effective camber moment ($\boldsymbol{M}_{p} = 1.23\boldsymbol{M}_{E.ser}$) when Design 2 has the deepest concrete section ($\boldsymbol{h}$ = 1,500 mm × $\boldsymbol{b}$ = 1,018 mm), leading to an uncracked design when $\boldsymbol{M}_{cr} > \boldsymbol{M}_{E.ser}$ even though a small tendon ratio ($\rho_{p.bot} = 0.0026$) and rebar area ($\boldsymbol{A}_{r}$ = 9,162 mm^2) are provided.

4.5.6.4 Cracked sections with Designs 1, 3, and 4 and uncracked sections with Design 2 under service loads

In contrast, large rebar areas and tendon ratios are calculated when concrete depth $\boldsymbol{h}$ are minimized of 966 and 1,002 mm for Designs 3 and 4, respectively. Design 3 demands large rebar area of 98,966.7 mm^2 and a tendon ratio of 0.0037 when $\boldsymbol{h}$ is minimized to 966 mm, whereas Design 4 also requires large rebar area of 58,617 mm^2 and a tendon ratio of 0.0037 when $\boldsymbol{h}$ is calculated as 1,002 mm. It is noted that Design 4 requires lesser concrete volume than that of Design 3 due to the deep depth (1,002 mm) of the section. Large tendon ratio ($\rho_{p.bot} = 0.0037$) and rebar area ($A_r = 631{,}10\ mm^2$) are also provided for Design 1, where smallest b of 608 mm is calculated for achieving the lightest weight. Minimized concrete sections calculated by Designs 1, 3, and 4 yield cracked sections under service loads, requiring large tendon and rebar areas, whereas a concrete section calculated by Designs 2 is uncracked section under service loads, requiring maximized concrete sections with lesser tendon and rebar areas than those of Designs 1, 3, and 4.

4.5.7 Verification of optimized design parameters ($\rho_{s,bot}$, $\rho_{s,top}$, and $\rho_{p,bot}$) on an output-side

Figure 4.5.2 (b), (c) and (d) verifies Pareto-based efficient designs in terms of the three objective functions with the structural datasets, demonstrating that the Pareto frontier for the three objective functions is well compared with the lower boundary of the 804 cyan-colored designs randomly generated based on AutoPT for $q_L = 400$ kN/m. The lower boundary of the 804 cyan dots randomly generated based on AutoPTBEAM for $q_L = 400$ kN/m are well predicted by a Pareto frontier denoted by a red curve based on 104 combinations of weight fractions obtained using ANN-based Lagrange method. Figure 4.5.2(a), (b), and (d) shows that a Pareto frontier obtained by ANNs is the lowest ($1.3 \times 10^{5\ KRW/m}$), which is witnessed by random designs obtained by the large datasets.

4.6 FINDINGS AND RECOMMENDATIONS

4.6.1 Practical applications of MOO for PT designs

An ANN-based design optimizing multi-objective functions simultaneously for PT beams is presented for structural engineers and decision-makers. More rational engineering decisions can be obtained by ANN-based Lagrange optimization techniques, resulting in designs that can meet various code restrictions simultaneously. A *UFO* of the three ANN-based objective functions (h, CIb, and W) for simply supported PT beams is optimized simultaneously while meeting various code restrictions imposed by constraining functions. The entire MOO-based design is performed using software which is developed based on MATLAB toolboxes [4.42–4.47]. Optimizing conflicting design targets for PT beams is now possible to provide real-life design solutions. A *UFO* of the three ANN-based objective functions (h, CI_b, and W) for simply supported PT beams is optimized simultaneously while meeting various code restrictions imposed by constraining functions.

4.6.2 Verification of MOO

The MOO designs were verified by structural software AutoPTBEAM with $q_L = 400$ kN/m. Lagrange multipliers and the weighted sum method are implemented in calculating Pareto frontiers optimizing three ANN-based objective functions, while code requirements given as equalities and inequalities are satisfied by KKT conditions [4.51,4.52]. Engineers can also establish both equality and inequality constraints to reflect their own design requirements. Suitability of a Pareto frontier optimizing the three objective functions (***h***, $\boldsymbol{CI_b}$, and W) simultaneously is demonstrated, verifying MOO design by the lower boundary of the big datasets generated using structural software AutoPTBEAM for $\boldsymbol{q_L}$ = 400 kN/m. Merits of the proposed method are also verified, offering a design guide for engineers and decision-makers. Engineers can follow the procedures provided in this chapter to establish their own MOO designs for their PT beams. ANN-based MOO processes will have to perform large computations for trainings and optimizations, requiring computer hardware support. PT frames for tall buildings based on MOO application are in progress.

4.7 FORMULATION OF ANN-BASED PARETO FRONTIER FOR PRE-TENSIONED (BONDED) CONCRETE BEAMS

Section 4.7 reviews how to formulate an ANN-based Pareto frontier for pre-tensioned (bonded) concrete beams. Figure 4.7.1 presents an algorithm for constructing a Pareto frontier for pre-tensioned (bonded) concrete beams. In **Steps 1.1** to **1.4**, large datasets are generated to train Artificial Neural Networks (ANNs) to obtain three individual objective functions of $\boldsymbol{CI_b}$, **W**, and ***h***, presented in Steps **1.5.1**, **1.5.2**, and **1.5.3**, respectively. Equalities and inequalities listed in **Step 2.1** are transformed into constraints in **Steps 2.2** and **2.3**. Each objective function from **Steps 1.5.1**, **1.5.2**, and **1.5.3** is combined separately with the constraints, as shown in **Steps 2.2** and **2.3**, resulting in three individual ANN-based Lagrange functions corresponding to $\boldsymbol{CI_b}$, **W**, and ***h***, as shown in **Steps 3.1**, **3.2**, and **3.3**, respectively. Subsequently, three separate ANN-based objective functions are optimized to determine the maxima and minima of $\boldsymbol{CI_b}$, **W**, and ***h***, as shown in **Steps 4.1**, **4.2**, and **4.3**, respectively. In **Step 5.1**, the algorithm begins with an initial value of i = 1, which is then increased to generates a combination of weight fractions from a weight fraction list shown in **Step 5.2**. A unified objective function (UFO) is then formulated in **Step 6**, by combining the combination of weight fractions obtained from **Step 5.2**, the maxima and minima of $\boldsymbol{CI_b}$, **W**, **h** derived in **Steps 4.1**, **4.2**, **4.3**, and the three ANN-based objective functions obtained by training shown in **Step 1.4**. An ANN-based Lagrange function of the UFO is finally presented in **Step 7** by intergrading the unified objective function (UFO) derived from **Step 6** and the constraints presented in **Steps 2.2** and **2.3**. Optimized parameters corresponding to the ith combination of weight fractions are determined by solving for stationary points of the ANN-based Lagrange function of the UFO. The design is verified in **Step 9** using the AutoPT software which relies on structural mechanics principles using optimized parameters determined by ANNs. The procedure moves to **Step 9.1** if

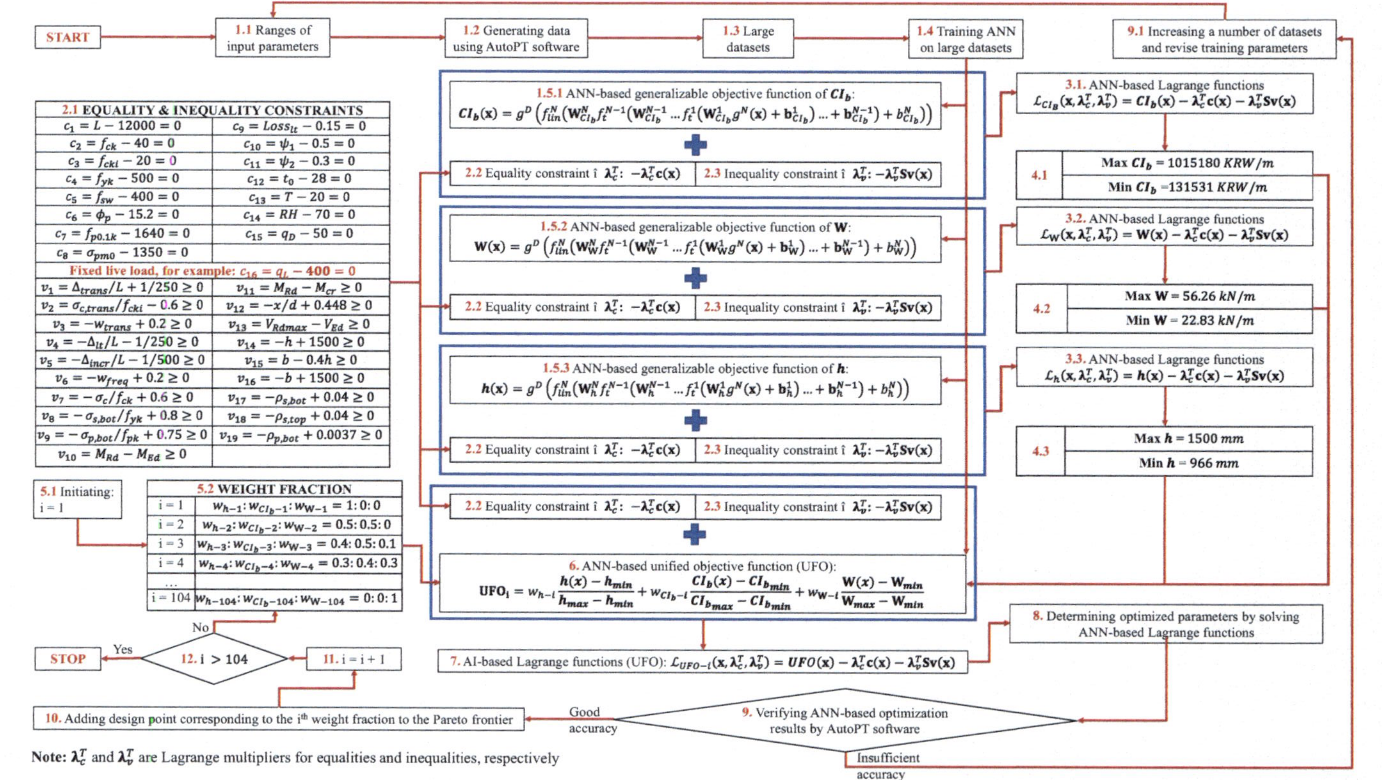

Figure 4.7.1 An algorithm for constructing an ANN-based Pareto frontier for pre-tensioned (bonded) concrete beams.

the design accuracy is found insufficient, where a number of large datasets is increased, and training parameters are revised to minimize errors. A procedure advances to **Step 10** when the accuracies are verified by **Step 9**, adding the next design point of the i+1th combination of weight fractions to the Pareto frontier. The algorithm terminates when the i reaches 104, indicating that all 104 combinations of weight fraction list generated in **Step 5.2** have been investigated. Otherwise, the algorithm increments i by 1, progressing to the next design point.

4.8 CONCLUSIONS

In the ANN-based novel way proposed in this chapter, a solution to MOO for pretensioned concrete (PT) beams is presented, where three conflicting objective functions, namely, cost including construction and material, beam weights, and beam depth, are optimized simultaneously. ANN-based algorithms find optimized design parameters within ranges prescribed by six inequalities shown in $v_{14} - v_{19}$ of Table 4.3.4(b) against applied loads. Twenty-one input parameters including tendon parameters in which five input parameters are constrained by six inequalities $v_{14} - v_{19}$ of Table 4.3.4(b), whereas 16 input parameters are constrained by equalities shown in $c_1 - c_{16}$ of Table 4.3.4(a). Five steps to design PT beams for MOO are proposed in which a Pareto frontier is constructed based on combinations of weight fractions with specified ratios. Reductions of 9.1%, 30.1%, and 10.4% in beam depths, costs, and weights are observed, respectively, compared with probable designs obtained from big data where average values of the ten best designs among the big datasets corresponding to each objective function are defined as the three probable designs. MOO for PT beams in a novel way which is simple, robust but accurate to optimize the three objective functions (***h***, ***CI$_b$***, and W) is available for the first time for a design of PT beams. The following conclusions describing novelty of the proposed methods are drawn to assist engineers to optimize PT beam designs. The large datasets are bounded on the lower designs obtained by a Pareto frontier (104 combinations of weight fractions) calculated based on the ANN-based Lagrange optimization. Other published works in this area that performed a MOO design for PT beams based on ANNs were seldom found.

1. MOO design of simply supported PT beams is provided in which 21 output variables (Δ_{trans}/L, $\sigma_{c,trans}/f_{cki}$, w_{trans}, Δ_{lt}/L, Δ_{incr}/L, w_{freq}, σ_c/f_{ck}, $\sigma_{s,bot}/f_{yk}$, $\sigma_{p,bot}/f_{pk}$, M_{Ed}, M_{Rd}, M_{cr}, x/d, $\varepsilon_{s,bot}/\varepsilon_{sy}$, $\varepsilon_{p,bot}/\varepsilon_{py}$, V_{Ed}, V_{Rdmax}, $\rho_{sw,sp}$, $\rho_{sw,mid}$, CI_b, W) are obtained for given 21 input variables (L, b, h, f_{ck}, f_{cki}, $\rho_{s,bot}$, $\rho_{s,top}$, f_{yk}, f_{sw}, ϕ_p, $\rho_{p,bot}$, $f_{p0.1k}$, σ_{pm0}, $Loss_{lt}$, ψ_1, ψ_2, t_0, T, RH, q_D, q_L) while minimizing three objective functions (design targets: h, CI_b, and W) simultaneously.
2. Design examples demonstrate how an *UFO* is formulated to integrate three objective functions (***h***, ***CI$_b$***, and W) into one objective function. Structural engineers and decision-makers can determine a contribution

of individual objective function to a total achieved optimization by adjusting tradeoff ratios controlled based on design interests.

3. In *UFO*, dominant influence of one parameter on others is removed by normalizing individual objective functions with their maxima and minima obtained using ANN-based optimizations. A number of objective functions to be simultaneously optimized can be any.
4. A Pareto frontier based on combinations of weight fractions optimizing the three objective functions ($\boldsymbol{h}$, $\boldsymbol{CI_b}$, and **W**) simultaneously is well verified with the lower boundary of the designs randomly generated based on structural software the AutoPTBEAM for $q_L = 400$ kN/m. Reductions of 9.1%, 30.1%, and 10.4% in beam depths, costs, and weights, respectively, are observed compared with probable designs obtained from big datasets.
5. In Figure 4.5.3, Designs 1, 3, and 4 with small concrete volume and large tendon, rebar areas yield crack sections under service loads, while Design 2 with large concrete volume and small tendon, rebar areas provide uncrack sections under service loads.
6. The ANN-based Lagrange method that can optimize multi-objective functions simultaneously is a steppingstone for the next PT designs. Robustness and flexibility for MOO designs of PT beams over the current designs are offered using ANN-based optimization techniques, meeting various code restrictions simultaneously.

REFERENCES

[4.1] Alqedra, M., Arafa, M., & Ismail, M. (2011). Optimum cost of prestressed and reinforced concrete beams using genetic algorithms. *Journal of Artificial Intelligence*, *4*(1), 76–88. http://hdl.handle.net/20.500.12358/24893

[4.2] Türkeli, E., & Öztürk, H. T. (2017). Optimum design of partially prestressed concrete beams using Genetic Algorithms. *Structural Engineering and Mechanics*, *64*(5), 579–589. https://doi.org/10.12989/sem.2017.64.5.579

[4.3] AydIn, Z., & Ayvaz, Y. (2010). Optimum topology and shape design of prestressed concrete bridge girders using a genetic algorithm. *Structural and Multidisciplinary Optimization*, *41*(1), 151–162. https://doi.org/10.1007/s00158-009-0404-2

[4.4] Aydın, Z., & Ayvaz, Y. (2013). Overall cost optimization of prestressed concrete bridge using genetic algorithm. *KSCE Journal of Civil Engineering*, *17*(4), 769–776. https://doi.org/10.1007/s12205-013-0355-4

[4.5] Quaranta, G., Fiore, A., & Marano, G. C. (2014). Optimum design of prestressed concrete beams using constrained differential evolution algorithm. *Structural and Multidisciplinary Optimization*, *49*(3), 441–453. https://doi.org/10.1007/s00158-013-0979-5

[4.6] Afshari, H., Hare, W., & Tesfamariam, S. (2019). Constrained multi-objective optimization algorithms: Review and comparison with application in reinforced concrete structures. *Applied Soft Computing Journal*, *83*, 105631. https://doi.org/10.1016/j.asoc.2019.105631

[4.7] Barakat, S., Bani-Hani, K., & Taha, M. Q. (2004). Multi-objective reliability-based optimization of prestressed concrete beams. *Structural Safety*, *26*(3), 311–342. https://doi.org/10.1016/j.strusafe.2003.09.001

[4.8] García-Segura, T., Penadés-Plà, V., & Yepes, V. (2018). Sustainable bridge design by metamodel-assisted multi-objective optimization and decision-making under uncertainty. *Journal of Cleaner Production*, *202*, 904–915. https://doi.org/10.1016/j.jclepro.2018.08.177

[4.9] García-Segura, T., Yepes, V., & Frangopol, D. M. (2017). Multi-objective design of post-tensioned concrete road bridges using artificial neural networks. *Structural and Multidisciplinary Optimization*, *56*(1), 139–150. https://doi.org/10.1007/s00158-017-1653-0

[4.10] Shariati, M., Mafipour, M. S., Mehrabi, P., Shariati, A., Toghroli, A., Trung, N. T., & Salih, M. N. A. (2021). A novel approach to predict shear strength of tilted angle connectors using artificial intelligence techniques. *Engineering with Computers*, *37*(3), 2089–2109. https://doi.org/10.1007/s00366-019-00930-x

[4.11] Shariati, M., Armaghani, D. J., Khandelwal, M., Zhou, J., & Khorami, M. (2021). Assessment of longstanding effects of fly ash and silica fume on the compressive strength of concrete using extreme learning machine and artificial neural network. *Journal of Advanced Engineering and Computation*, *5*(1), 50–74. http://dx.doi.org/10.25073/jaec.202151.308

[4.12] Shariati, M., Mafipour, M. S., Mehrabi, P., Ahmadi, M., Wakil, K., Trung, N. T., & Toghroli, A. (2020). Prediction of concrete strength in presence of furnace slag and fly ash using Hybrid ANN-GA (Artificial Neural Network-Genetic Algorithm). *Smart Structures and Systems, an International Journal*, *25*(2), 183–195.

[4.13] Shariati, M., Mafipour, M. S., Mehrabi, P., Bahadori, A., Zandi, Y., Salih, M. N. A., Nguyen, H., Dou, J., Song, X., & Poi-Ngian, S. (2019). Application of a hybrid artificial neural network-particle swarm optimization (ANN-PSO) model in behavior prediction of channel shear connectors embedded in normal and high-strength concrete. *Applied Sciences*, *9*(24). https://doi.org/10.3390/app9245534

[4.14] Armaghani, D. J., Mirzaei, F., Shariati, M., Trung, N. T., Shariati, M., & Trnavac, D. (2020). Hybrid ANN-based techniques in predicting cohesion of sandy-soil combined with fiber. *Geomechanics and Engineering*, *20*(3), 191–205. https://doi.org/10.12989/gae.2020.20.3.191

[4.15] Mohammadhassani, M., Nezamabadi-Pour, H., Suhatril, M., & Shariati, M. (2013). Identification of a suitable ANN architecture in predicting strain in tie section of concrete deep beams. *Structural Engineering and Mechanics: An International Journal*, *46*(6), 853–868. https://doi.org/10.12989/sem.2013.46.6.853

[4.16] Deb, K., Pratap, A., Agarwal, S., & Meyarivan, T. (2002). A fast and elitist multiobjective genetic algorithm: NSGA-II. *IEEE Transactions on Evolutionary Computation*, *6*(2), 182–197. https://doi.org/10.1109/4235.996017

[4.17] Mirjalili, S., Saremi, S., Mirjalili, S. M., & Coelho, L. D. S. (2016). Multi-objective grey wolf optimizer: A novel algorithm for multi-criterion optimization. *Expert Systems with Applications*, *47*, 106–119. https://doi.org/10.1016/j.eswa.2015.10.039

[4.18] Mirjalili, S., Mirjalili, S. M., & Lewis, A. (2014). Grey wolf optimizer. *Advances in Engineering Software*, *69*, 46–61. https://doi.org/10.1016/j.advengsoft.2013.12.007

[4.19] Hong, W. K., Nguyen, M. C., Pham, T. D. (2022). Pre-tensioned concrete beams optimized with a unified function of objective (UFO) using ANN-based Hong-Lagrange method. *Journal of Asian Architecture and Building Engineering* (under review).

[4.20] Shariat, M., Shariati, M., Madadi, A., & Wakil, K. (2018). Computational Lagrangian multiplier method by using for optimization and sensitivity analysis of rectangular reinforced concrete beams. *Steel and Composite Structures*, *29*(2), 243–256. https://doi.org/10.12989/scs.2018.29.2.243

[4.21] Wilby, C. B. (2013). *Structural Concrete: Materials; Mix Design; Plain, Reinforced and Prestressed Concrete; Design Tables*. Oxfordshire: Elsevier.

[4.22] Hong, W. K., Nguyen, V. T., & Nguyen, M. C. (2022). Optimizing reinforced concrete beams cost based on AI-based Lagrange functions. *Journal of Asian Architecture and Building Engineering*, *21*(6), 2426–2443. https://doi.org/10.1080/13467581.2021.2007105

[4.23] Hong, W. K., & Nguyen, M. C. (2022). AI-based Lagrange optimization for designing reinforced concrete columns. *Journal of Asian Architecture and Building Engineering*, *21*(6), 2330–2344. https://doi.org/10.1080/13467581.2021.1971998

[4.24] Hong, W.-K., Nguyen, V. T., Nguyen, D. H., & Nguyen, M. C. (2022). An AI-based Lagrange optimization for a design for concrete columns encasing H-shaped steel sections under a biaxial bending. *Journal of Asian Architecture and Building Engineering*, 22(2). https://doi.org/10.1080/13467581.2022.2060985

[4.25] Hong, W. K., & Le, T. A. (2023). ANN-based optimized design of doubly reinforced rectangular concrete beams based on multi-objective functions. *Journal of Asian Architecture and Building Engineering*, 1413–1429. https://doi.org/10.1080/13467581.2022.2085720

[4.26] Nguyen, M. C., & Hong, W. K., (2022). Analytical prediction of nonlinear behaviors of beams post-tensioned by unbonded tendons considering shear deformation and tension stiffening effect. *Journal of Asian Architecture and Building Engineering*, *21*(3), 908–929. https://doi.org/10.1080/13467581.2021.1908299

[4.27] CEN (European Committee for Standardization). (2004). Eurocode 2 : Part 1-1: General rules and rules for buildings. In E*urocode 2: Vol. BS En 1992*.

[4.28] EC-Standards. (1990). EN 1990:2002E - Basis of structural design - EC0. 1–87.

[4.29] Group, J. N., Matters, S., Nordic, T., Bre, T., Project, C., & Pg, G. (2002). Discussion on the Rules for Combination of Actions in En 1990 "Basis of Structural Design". October 2004, 1–9.

[4.30] European Concrete Platform. (2008). Commentary Eurocode 2.

[4.31] Beeby, A. W. (2001). Calculation of Crack Width, PrEN 1992-1 (Final draft) Chapter 7.3.4.

[4.32] Beeby, A. W., Favre, R., Koprna, M., & Jaccoud, J. P. (1985). Cracking and Deformations, CEB Manual.

[4.33] Elighausen, R., Mallée, R., & Rehm, G. (1976). Rissverhalten von Stahlbetonkörpern bei Zugbeanspruchung, Untersuchungsbericht.

[4.34] Falkner, M. (1969). Zur Frage der Rissbildung durch Eigen – und Zwangspannungen in Folge Temperatur im Stahlbetonbau, Deutscher Ausschuss Für Stahlbeton Heft 208.

[4.35] Hartl, G. (1977). Die Arbeitslinie eingebetteter St¨ale bei Erst – und Kurzzeitbelastung, Dissertation.

[4.36] Jaccoud, J. P., & Charif, H. (1986). Armature Minimale pour le Contrôle de la Fissuration, Publication N° 114.
[4.37] Krips, M. (1985). Rissenbreitenbeschränkung im Stahlbeton und Spannbeton, TH Damstadt.
[4.38] Rehm, G., & Rüsch, H. (1963). Versuche mit Betonformstählen Teil I, Deutscher Ausschuss für Stahlbeton Heft 140.
[4.39] Rehm, G. & Rüsch, H. (1963). Versuche mit Betonformstählen Teil II, Deutscher Ausschuss für Stahlbeton Heft 160.
[4.40] Rehm, G., & Rüsch, H. (1964). Versuche mit Betonformstählen Teil III, Deutscher Ausschuss für Stahlbeton Heft 165.
[4.41] Hong, W. K., Pham, T. D., & Nguyen, V. T. (2022). Feature selection based reverse design of doubly reinforced concrete beams. *Journal of Asian Architecture and Building Engineering*, *21*(4), 1472–1496. https://doi.org/10.1080/13467581.2021.1928510
[4.42] MathWorks. (2022a). Deep Learning Toolbox: User's Guide (R2022a). Retrieved July 26, 20122 from: https://www.mathworks.com/help/pdf_doc/deeplearning/nnet_ug.pdf
[4.43] MathWorks. (2022a). Parallel Computing Toolbox: Documentation (R2022a). Retrieved July 26, 2022 from: https://uk.mathworks.com/help/parallel-computing/
[4.44] MathWorks. (2022a). Statistics and Machine Learning Toolbox: Documentation (R2022a). Retrieved July 26, 2022, from: https://uk.mathworks.com/help/stats/
[4.45] MathWorks. (2022a). Global Optimization: User's Guide (R2022a). Retrieved July 26, 20122 from: https://www.mathworks.com/help/pdf_doc/gads/gads.pdf
[4.46] MathWorks. (2022a). Optimization Toolbox: Documentation (R2022a). Retrieved July 26, 2022, from: https://uk.mathworks.com/help/optim/
[4.47] MathWorks. (2022). *MATLAB R2022a*. Natick, MA: MathWorks.
[4.48] Hong, W. K., Le, T. A., Nguyen, M. C., & Pham, T. D. (2022). ANN-based Lagrange optimization for RC circular columns having multi-objective functions. *Journal of Asian Architecture and Building Engineering*. https://doi.org/10.1080/13467581.2022.2064864
[4.49] Basheer, I. A., & Hajmeer, M. (2000). Artificial neural networks: Fundamentals, computing, design, and application. *Journal of Microbiological Methods*, *43*(1), 3–31. https://doi.org/10.1016/S0167-7012(00)00201-3
[4.50] Wu, R. T., & Jahanshahi, M. R. (2019). Deep convolutional neural network for structural dynamic response estimation and system identification. *Journal of Engineering Mechanics*, *145*(1), 04018125. https://doi.org/10.1061/(asce)em.1943-7889.0001556
[4.51] Giorgi, G., & Kjeldsen, T. H. (Eds.). (2013). *Traces and Emergence of Nonlinear Programming*. Basel: Birkhäuser. https://doi.org/10.1007/978-3-0348-0439-4_11
[4.52] Kuhn, H. W., & Tucker, A. W. (1951). Nonlinear programming. In *Proceedings of the 2nd Berkeley Symposium on Mathematics, Statistics and Probability* (pp. 481–492). Berkeley: University of California Press.
[4.53] Won-Kee Hong, Thuc-Anh Le, Manh Cuong Nguyen & Tien Dat Pham (2023). ANN-based Lagrange optimization for RC circular columns having multiobjective functions. *Journal of Asian Architecture and Building Engineering*, 961–976. https://doi.org/10.1080/13467581.2022.2064864.
[4.54] Elsevier.
[4.55] Vol 2.

Chapter 5

Reverse design charts for steel-reinforced concrete (SRC) beams based on artificial neural networks

5.1 INTRODUCTION

5.1.1 Previous studies

Many studies including Vanluchene and Sun [5.1] who used ANNs to identify a location and magnitude of the maximum bending moment of a simply supported rectangular plate have focused on ANN-based analysis and design. Hajela and Berke [5.2] also optimized truss designs, where input data comprised a length and height of a truss, while output data comprised optimized bar areas and total weights of the truss. Flood and Kartam [5.3, 5.4] and Hong [5.5] applied ANNs to structural engineering. Hong [5.5] investigated an influence of a number of hidden layers and neurons on network validation and processing speed. By Wu and Jahanshahi [5.6], a convolutional neural network (CNN) applied to noisy data was used to predict a structural response more accurately than multilayer perceptron algorithm, estimating a dynamic response of a linear single-degree-of-freedom (SDOF) system, non-linear SDOF system, and full-scale three-story multi-degree-of-freedom steel frame based on a deep CNN. Lavaei and Lohrasbi [5.7] replaced sigmoid activation functions applied to neurons with wavelets to approximate a dynamic time history response of frame structures in which a backpropagation wavelet neural network is presented based on the scaled conjugate gradient algorithm. Fahmy et al. [5.8] found that ANNs proved to be a better, more cost-effective design option than codes or expert opinion for a conceptual design of orthotropic steel deck bridges. Adeli has published multiple articles on structural analysis and design problems since 1989 [5.9].

It was Lee et al. [5.10] who noted that a poor performance of ANNs and very high computation time, especially for complicated problems with multiple hidden layers, demonstrating limitations and numerical instability in structural engineering. Gupta and Sharma [5.11] also found that neural network applications in structural engineering have greatly decreased over the last decade. However, an application of ANNs has been performed

DOI: 10.1201/9781003354796-5

successfully in many studies by researchers since the early 1900s for predicting the capacity of pile foundations under axial and lateral loads (Chan et al. [5.12], Lee and Lee [5.13], Rahman et al. [5.14], Hanna et al. [5.15], Das and Basudhar [5.16]), designing and optimizing weights of steel and truss structures (Adeli and Park [5.17, 5.18], Tashakori and Adeli [5.19], Kang and Yoon [5.20]), calculating tunnels and underground opening structures (Lee and Sterling [5.21], Shi et al. 1998 [5.22], Shi [5.23], Neaupane and Achet [5.24], Yoo and Kim [5.25]), and predicting seismic responses of building structures and bridges (Oh et al. [5.26], Lagaros and Papadrakakis [5.27], Asteris et al. [5.28], Ying et al. [5.29], Huang and Huang [5.30]). However, many studies focused on verifying training accuracies rather than an practical application of networks to the holistic design of structures with unseen input parameters by the networks.

Many studies have concluded that a successful ANN's implementation depends on the quality of the training data. Holistic and robust applications of ANNs to structural design, in general, and to SRC beams, in particular, are not common design practices yet. Hong et al. [5.31, 5.32], Hong and Nguyen [5.33], and Hong and Pham [5.34] implement ANNs in designing beams and columns. Specifically, they optimized construction costs, corresponding CO_2 emissions, and structural weights using neural networks with acceptable accuracies, helping engineers in designing concrete structures imposed by design requirements.

In this chapter, ANNs are suggested to provide practical design solutions for SRC beams. Positions of network inputs and outputs are exchanged for a reverse design which is especially useful when multiple design parameters are reversed, whereas this is not possible with conventional design methods where output parameters are obtained for given input parameters. This chapter also presents ANN-based reverse design charts in which design sequences of input and output parameters are exchanged. The innovative design charts offer practical application to SRC beams encasing an H-shaped steel section. Engineers now can determine preliminary design parameters for both forward and reverse designs of SRC beams rapidly and accurately, predicting multiple design parameters in any order. ANN-based robust design of SRC beams based on design charts is possible, which is challenging to achieve with conventional design methods. The network accuracy for the SRC design with 15 input parameters and 11 output parameters was verified through conventional structural calculations. Recently, available computational powers for routine training have been greatly enhanced using multiple CPU processing units, which have let researchers train larger networks to overcome the limitations and numerical instability. This chapter was written based on the previous paper by Hong et al. [5.35].

5.1.2 Significance of Chapter 5

Design of SRC beams having over 15 input parameters and 11 output parameters can be a lengthy process, whereas understanding how these design parameters influence each other can be difficult. A reverse design of SRC beams is performed by training an artificial neural network with large structural datasets to map input parameters to output parameters rather than basing them on structural mechanics or knowledge. Two steps of a back-substitution (BS) method [5.32] are applied in which the reverse network of Step 1 calculates reverse output parameters for given reverse input parameters. A reverse network of Step 1 is followed by a forward network of Step 2 where reverse output parameters obtained in Step 1 are back-substituted into a forward network of Step 2 as input parameters to obtain design parameters. ANNs based on two steps are implemented in determining the design parameters of SRC beams for given unseen structural input parameters. Section 5.5 of this chapter provides robust design charts, offering rapid and accurate designs for SRC beams, which are challenging to achieve with conventional design methods. Design charts predict multiple design parameters in any order, which can help engineers in the preliminary design of SRC beams. The ANNs formulated in this chapter can be used to explore a behavior of SRC beams, uncovering intricate relationships among design parameters. ANN-based designs are ascertained through conventional structural calculations. Table 5.1.1 presents the nomenclature of parameters that are used in this chapter.

Table 5.1.1 Nomenclature of parameters

Nomenclature	
Forward parameters	
L (mm)	Beam length
d (mm)	Effective beam depth
b (mm)	Beam width
f_c' (MPa)	Compressive concrete strength
f_y (MPa)	Yield rebar strength
ρ_{rt}	Tensile rebar ratio
ρ_{rc}	Compressive rebar ratio
h_s (mm)	H-shaped steel height
b_s (mm)	H-shaped steel width
t_f (mm)	H-shaped steel flange thickness
t_w (mm)	H-shaped steel web thickness
f_{yS} (MPa)	Yield steel strength

(Continued)

Table 5.1.1 (Continued) Nomenclature of parameters

Y_s (mm)	Clearance between H steel section and closest rebar layer
M_D (kN·m)	Moment due to dead load
M_L (kN·m)	Moment due to live load
Reverse parameters	
ϕM_n (kN·m)	Design moment without considering the effect of self-weight at $\varepsilon_c = 0.003$
ε_{rt}	Tensile rebar strain at $\varepsilon_c = 0.003$
ε_{st}	Strain of tensile steel flange at $\varepsilon_c = 0.003$
Δ_{imme} (mm)	Immediate deflection due to live load, M_L
Δ_{long} (mm)	Sum of time-dependent deflection due to sustained loads and immediate deflection due to any additional live load
μ_ϕ	Curvature ductility
CI_b (KRW/m)	Cost index per 1m length of beam
CO_2 (t-CO_2/m)	CO_2 emission per 1m length of beam
BW (kN/m)	Beam weight per 1m length of beam
X_s (mm)	Clearance between flange of H steel section and edge of beam
SF	Safety factor

5.2 ARTIFICIAL INTELLIGENCE-BASED DESIGN SCENARIOS FOR SRC BEAMS

In Figure 5.2.1(a), the dimensions of an SRC beam encasing an H-shaped steel section include the beam height (h), width (b), rebar, and steel sections encased by concrete. The scenarios of four reverse designs (reverse scenarios 1–4) for an SRC beam are presented in Table 5.2.1. In the conventional calculation referred to a forward design, 11 design output parameters shown in Table 5.2.1 are calculated based on 15 design input parameters shown in Table 5.2.1. However, safety factor (SF) and tensile rebar strain (ε_{rt}) cannot be preassigned in forward design. As indicated in Table 5.2.1 for reverse scenario 1–4, output parameters of forward design, such as design moment strength $\phi M_{n(\text{including } Ms)}$, an SF and a tensile rebar strain (ε_{rt}), are proposed to be preassigned as input parameters on an input-side. Design moment strength $\phi M_{n(\text{including } Ms)}$, an SF, and a tensile rebar strain (ε_{rt}) control Design Scenario 1 of SRC beam, which is a significant challenge to the conventional method. As indicated in Table 5.2.1, two input parameters of forward design are placed on an output-side where b and b_s are reversely placed on an output-side for reverse scenarios 1; ρ_{rc} and t_f are reversely placed on an output-side for reverse scenario 2; ρ_{rt} and ρ_{rc}

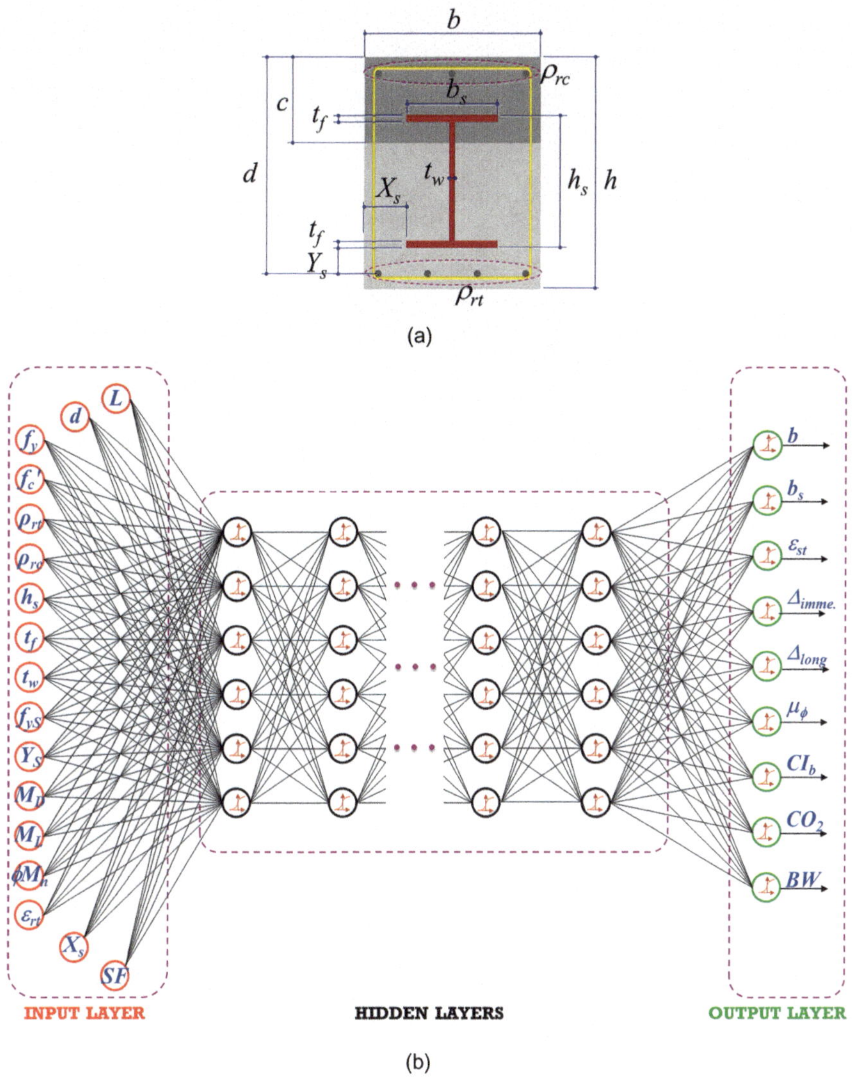

Figure 5.2.1 Geometry of SRC beam. (a) SRC beams designed based on artificial neural networks. (b) ANN topology for a reverse design.

are reversely placed on an output-side for reverse scenario 3; d and ρ_{rt} are reversely placed on an output-side for reverse scenario 4. Beam length (L), material properties [concrete strength (f_c'), rebar strength (f_y), steel strength (f_{yS})], and imposed loads [moment due to dead load (M_D), and moment due to live load (M_L)] are also preassigned on an input-side.

Table 5.2.1 Design scenarios for SRC beam

Scenarios	*Design input parameters (15 parameters)*															*Design output parameters (11 parameters)*										
	L	d	b	f_y	f_c'	ρ_{rt}	ρ_{rc}	h_s	b_s	t_f	t_w	f_{yS}	Y_s	M_D	M_L	ϕM_n	ε_{rt}	ε_{st}	$\Delta_{imme.}$	Δ_{long}	μ_ϕ	CI_b	CO_2	BW	X_s	SF
Reverse 1	i	i	o	i	i	i	i	i	o	i	i	i	i	i	i	i	i	o	o	o	o	o	o	o	i	i
Reverse 2	i	i	i	i	i	i	o	i	i	o	i	i	i	i	i	i	i	o	o	o	o	o	o	o	o	i
Reverse 3	i	i	i	i	i	o	o	i	i	i	i	i	i	i	i	i	i	o	o	o	o	o	o	o	o	i
Reverse 4	i	o	i	i	i	o	i	i	i	i	i	i	i	i	i	i	i	o	o	o	o	o	o	o	o	i

Note:

Input i

Output o

5.3 GENERATION OF LARGE DATASETS

5.3.1 Cost index (CI_b) and CO_2 emissions of concrete and metals for of SRC beams

CO_2 emissions of concrete and metals were calculated during construction [5.36] where one ton of metal emits 2.513 t-CO_2 during the construction stage, while 1 m³ of concrete emits 0.1677 t-CO_2. These data may vary slightly from those of other studies. Cost index (CI_b) includes material and manufacturing prices of all components for an SRC beam per 1 m length. Unit prices for various materials comprising SRC beams are listed in Tables 5.3.1–5.3.3 which may vary due to market volatility, but they were assumed to remain constant in this chapter. Concrete unit price depends on concrete strength as presented in Table 5.3.1.

The rebar unit price presented in Table 5.3.2 varies from 1,055 Korean won (KRW)/kgf to 1,085 KRW/kgf for rebar strengths of 500–600 MPa, indicating that rebar unit price does not depend greatly on rebar strength. Steel unit price presented in Table 5.3.3 varies from 1,800 to 1,880 KRW/kgf depending on steel strengths of 275–325 MPa. The data are subject to change.

Table 5.3.1 Concrete unit price versus concrete strength

Concrete strength (MPa)	*Concrete unit price (KRW/m³)*
30	85,000
40	94,000
50	104,000

Table 5.3.2 Rebar unit price versus rebar strength

Rebar strength (MPa)	*Rebar unit price (KRW/kgf)*
500	1,055
600	1,085

Table 5.3.3 Steel unit price versus steel strength

Steel strength (MPa)	*Steel unit price (KRW/kgf)*
275	1,800
325	1,880

5.3.2 Generation of large datasets

5.3.2.1 Big data generator

In this chapter, ANNs are trained on the big datasets generated by the SRC software AutoSRCbeam which was developed for a shallow SRC beam design. The software was verified and published in [5.38]. Figure 5.3.1(a) shows a flowchart to calculate the flexural behavior of shallow SRC beams neglecting shear deformations. However, shear behaviors must be considered for deep beams, and hence, the flowchart may be revised constantly to better predict both flexural and shear behaviors of SRC beams which will, then, be used to update big datasets. The proposed ANNs will be re-trained with an updated number of layers, neurons, epochs, validations checks when big datasets are updated by updated software.

5.3.2.2 Ranges of input parameters

An ANN-based design of an SRC beam section with fixed ends shown in Figure 5.3.1(b) is performed based on an SRC beam section with fixed ends shown in Figure 5.3.1(b) in which dimensions of the beam are a width (b)×effective depth (d)×length (L). The beam is subjected to a uniform load, which results in an immediate deflection (Δ_{imme}) and long-term deflection (Δ_{long}). Immediate deflection (Δ_{imme}) due to a live load should be less than $L/360$, and long-term deflection (Δ_{long}) due to sustained loads with additional live load should be less than $L/240$ imposed by the American Concrete Institute (ACI) design code. AutoSRCbeam developed based on the algorithm by Nguyen and Hong [5.38] is used to generate large datasets as shown in Figure 5.3.2. Input parameters are selected randomly from which output parameters are also calculated randomly. The beam length (L) and effective depth (d) are randomly selected within 6,000–12,000 and 400–1,445 mm, respectively, while the beam width (b) was randomly selected within $0.3h$–$0.8h$. Nomenclatures and ranges of random parameters for input parameters of an SRC beam are presented in Table 5.3.4 where all 15 input parameters for forward design are randomly preassigned on an input-side, whereas 11 output parameters are calculated on an output-side using AutoSRCbeam.

Tensile rebar ratios (ρ_{rt}) are randomly generated in the range $[\rho_{rt,min}; \rho_{rt,max}]$, herein, $\rho_{rt,min}$ is set by formular of ACI code $\left(\rho_{rt,min} = \max\left[1.4/f_y; \sqrt{f_c'}/4f_y\right]\right)$, whereas $\rho_{rt,max}$ is set at a value of 0.05. Compressive rebar ratios (ρ_{rc}) are randomly generated in range [$\rho_{rt}/400$; $0.5\rho_{rt}$]. Code doesn't state these limits, but these limits are implemented in this study. Tensile rebar strains are, then, calculated based on AutoSRCbeam. Tensile rebar strain (ε_{rt}) should be no less than $0.003+\varepsilon_y$ (ε_y is the yield strain of rebar) according to the ACI code 318-19. Tensile rebar ratios (ρ_{rt}) and compressive rebar ratios (ρ_{rc}) should be randomly generated to calculate tensile rebar strains (ε_{rt}) to be greater than $0.003+\varepsilon_y$ when tensile rebar strain (ε_{rt}) is smaller than $0.003+\varepsilon_y$. This is repeated until ACI 318-19 condition is satisfied ($\varepsilon_{rt} \geq \varepsilon_y + 0.003$).

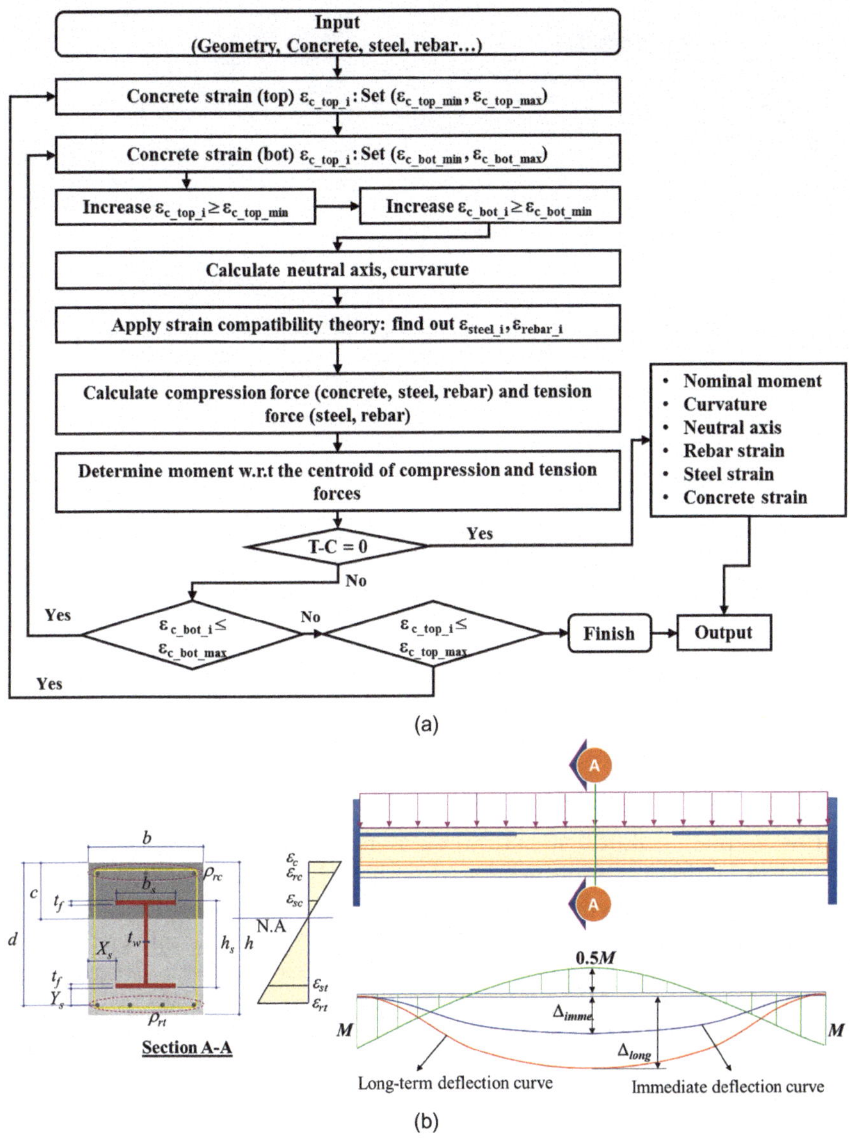

Figure 5.3.1 Structural mechanics-based software AutoSRCbeam for SRC beam. (a) Computational algorithm for shallow SRC beam [5.38]. (b) Fixed-fixed SRC beams used in ANN.

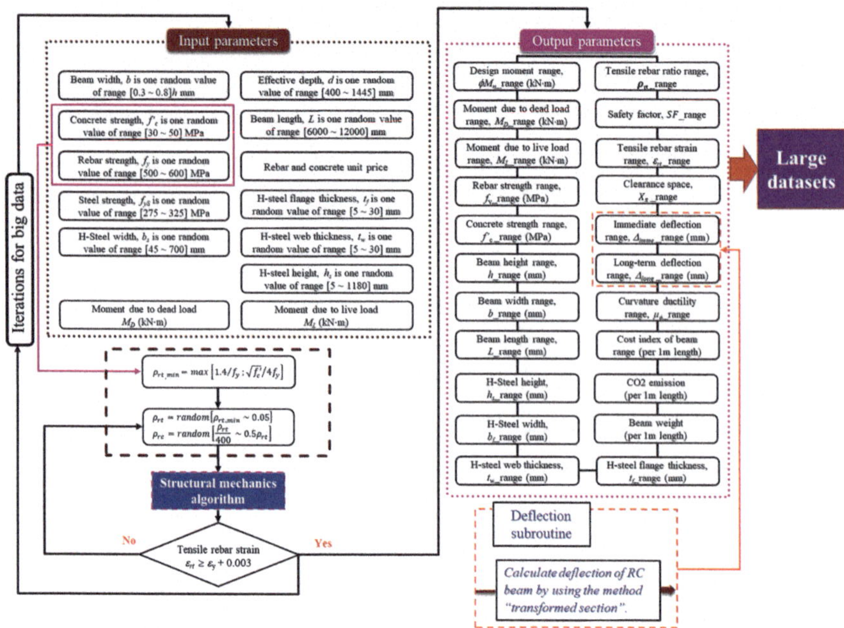

Figure 5.3.2 Flowchart for generating large structural datasets of SRC beams to train network [5.38, 5.39].

5.3.2.3 100,000 structural datasets based on a flowchart

In this chapter, 100,000 structural datasets based on a flowchart demonstrated in Figure 5.3.2 are generated with each dataset which consists of a total of 26 parameters (15 input parameters and 11 output parameters) to train ANNs. Gross beam height h is eliminated as a parameter because h is not trained as a random variable. Concrete clear cover is not random but fixed at 40 mm. The ACI code defines factored load $M_{u(\text{including } Ms)} = 1.2M_D + 1.6M_L + 1.2M_s$. The design moment strength is given by Eq. (5.3.1).

$$\phi M_{n(includingMs)} \geq M_{u(includingMs)} = 1.2M_D + 1.6M_L + 1.2M_s \tag{5.3.1}$$

$$\phi M_n = \phi M_{n(includingMs)} - 1.2M_s \geq M_u = 1.2M_D + 1.6M_L \tag{5.3.2}$$

where M_D, M_L, and M_s are moments due to dead load, live load, and self-weight, respectively. However, in Eq. (5.3.2), design loads do not include self-weight of a structure to avoid input conflict because dimensions of a beam section are not known during trainings and design stages. Input conflicts may be caused if all of M_D, M_L, M_s, and ϕM_n are used in Eq. (5.3.1) without knowing beam sections.

Table 5.3.4 Nomenclatures and ranges of parameters defining SRC beams [5.38]

	Notation		*Range*
Input parameters	L (mm)	Beam length	[6,000 ~ 12,000]
	d (mm)	Effective beam depth	[400 ~ 1,445]
	b (mm)	Beam width	[150 ~ 1,200]
	f_c' (MPa)	Concrete strength	[30 ~ 50]
	f_y (MPa)	Rebar strength	[500 ~ 600]
	ρ_{rt}	Tensile rebar ratios	$\rho_{rt,min} = max\left(\frac{0.25\sqrt{f_c'}}{f_y}; \frac{1.4}{f_y}\right)$
	ρ_{rc}	Compressive rebar ratios	[1/400 ~ 1/2] ρ_{rt}
	h_s (mm)	H-shaped steel height	[5 ~ 1,180]
	b_s (mm)	H-shaped steel width	[45 ~ 700]
	t_f (mm)	H-shaped steel flange thickness	[5 ~ 30]
	t_w(mm)	H-shaped steel web thickness	[5 ~ 30]
	f_{yS} (MPa)	Steel strength	[275 ~ 325]
	Y_s (mm)	Clearance between H steel section and closest rebar layer	[60 ~ 963]
	M_D (kN·m)	Moment due to dead load	$\left[0.2 \sim \frac{1}{1.2}\right] M_u$
	M_L (kN·m)	Moment due to service live load	$\frac{1}{1.6}(M_u - 1.2M_D)$

(Continued)

Table 5.3.4 (Continued) Nomenclatures and ranges of parameters defining SRC beams [5.38]

	Notation		*Range*
	SF	Safety factor $\boldsymbol{SF = \phi M_n / M_u}$ Where, $\boldsymbol{\phi M_n}$: Design moment excluding self-weight $\boldsymbol{M_u}$: Factored moment, $\boldsymbol{M_u = 1.2M_D + 1.6M_L}$	[1 ~ 2]
Output parameters	ϕM_n (kN·m)	Design moment without considering effect of self-weight at $\varepsilon_c = 0.003$	
	ε_{rt}	Tensile rebar strain at $\varepsilon_c = 0.003$ ε_y:Yield strain of rebar	$\varepsilon_{rt} \geq \varepsilon_y + 0.003$
	ε_{st}	Strain of tensile steel flange at $\varepsilon_c = 0.003$	
	$\Delta_{imme.}$ (mm)	Immediate deflection due to M_L service live load	($\boldsymbol{\Delta_{imme.}} \leq$ L/360, ACI 318-19)
	Δ_{long} (mm)	Sum of long-term deflection due to sustained loads and immediate deflection due to additional live load	($\boldsymbol{\Delta_{long}} \leq$ L/240, ACI 318-19)
	μ_ϕ	Curvature ductility, $\boldsymbol{\mu_\phi} = \phi_\mu / \boldsymbol{\phi_y}$, where $\boldsymbol{\phi_\mu}$: Curvature at $\varepsilon_c = 0.003$ $\boldsymbol{\phi_y}$: Curvature at tensile rebar yield	
	CI_b (KRW/m)	Cost index per 1 m length of beam	
	CO_2 (t-CO_2/m)	CO_2 emission per 1 m length of beam	
	BW (kN/m)	Beam weight per 1 m length of beam	
	X_S (mm)	Clearance between flange of H steel section and edge of beam $\boldsymbol{X_s} = 0.5(\boldsymbol{b - b_s})$	

5.3.2.4 *Means, standard deviations, variances, and histograms of randomly selected design input and output parameters*

Means, standard deviations, variances, and histograms of randomly selected design input and output parameters can be found in Table 5.3.5 based on 100,000 datasets generated for SRC beams. Figures 5.3.3(1)–(22) show the distributions of all input and output parameters. Beam length (L), beam depth (d), flange thickness of the steel section (t_f), web thickness of the steel section (t_w), concrete strength (f_c'), rebar strength (f_y), and steel strength (f_{yS}) are randomly selected with uniform distributions within minimum and maximum values, whereas others are found to distribute with a bell-shaped distribution. The range of beam length (L) is from 6,000 to 12,000 mm, whereas the distribution of beam length (L) data is uniform as shown in Figure 5.3.3(1), resulting in the mean of beam length (L) data of 8,994.3 mm which is close to an average of minimum and maximum of the data range (9,000 mm); a standard deviation and a variance are found as 1,744.6 mm and 3,043,577.9 mm^2, respectively. Figures 5.3.3(13)–(16) present histograms of the moment due to dead load (M_D), moment due to live load (M_L), design moment (ϕM_n), and tensile rebar strain (ε_{rt}) at $\varepsilon_c = 0.003$. These parameters are generated based on severe non-linearities, which led to skewed data distributions.

5.3.3 ANN-based mapping input to output parameters

Structural designs in this chapter are based on ANNs recognized as machine learning rather than being based on engineering mechanics or theory knowledge [5.5, 5.37]. ANN-based designs, thus, are performed using big datasets which are generated by structural software AutoSRCbeam. One of the significant contributions of this chapter is to find a relationships between input and output parameters of an SRC beam encasing an H-shaped steel section. Figure 5.2.1(b) shows ANN topology for a ANN-based reverse design. Weight and bias matrices are derived by training ANNs on big datasets, whereas weight and bias matrices are updated when big datasets are updated based on updated SRC algorithms. It is noted that a successful ANN-based design depends on the quality of the training data. Engineers may have to select an appropriate training method [5.31] based on a number of layers, neurons, epochs, validations, etc. when training ANNs.

Table 5.3.5 Means, standard deviations, and variances of random design inputs and outputs for SRC beams

	Parameter	Mean (μ)	Maximum	Minimum	Standard deviation (σ)	Variance (V)
Inputs	L (mm)	8,994.3	12,000.0	6,000.0	1,744.6	3,043,577.9
	d (mm)	1,059.8	1,445.5	401.4	248.8	61,904.5
	b (mm)	636.0	1,200.0	150.0	216.8	46,995.2
	f_y (MPa)	550.0	600.0	500.0	29.2	850.0
	f_c' (MPa)	40.0	50.0	30.0	6.1	36.6
	ρ_{rt}	9.89E-03	4.32E-02	2.33E-03	5.54E-03	3.07E-05
	ρ_{rc}	2.97E-03	1.94E-02	6.50E-06	2.55E-03	6.51E-06
	h_s (mm)	496.1	1,180.0	5.0	238.3	56,797.9
	b_s (mm)	280.0	700.0	45.0	114.1	130,12.2
	t_f (mm)	16.3	30.0	5.0	7.5	55.9
	t_w (mm)	15.8	30.0	5.0	7.4	54.4
	f_{yS} (MPa)	300.0	325.0	275.0	14.7	216.1
	Y_s (mm)	293.7	962.6	60.0	159.8	25,524.6
	M_D (kN·m)	1,836.2	23,107.5	9.3	1,739.6	3,026,175.2
	M_L (kN·m)	916.5	11,370.2	1.5	1,002.9	1,005,743.2
Outputs	ϕM_n (kN·m)	5,268.4	35,040.9	69.1	4,064.5	16,519,919.0
	ε_{rt}	7.83E-03	3.00E-02	5.50E-03	2.06E-03	4.24E-06
	ε_{st}	4.50E-03	2.51E-02	−3.00E-04	2.19E-03	4.80E-06
	$\Delta_{imme.}$ (mm)	1.75	16.52	0.0029	1.47	2.15
	Δ_{long} (mm)	7.93	67.06	0.43	5.35	28.66
	μ_ϕ	2.65	9.31	1.82	0.64	0.41
	CI_b (KRW/m)	407,748.4	1,579,803.8	30,450.4	202,079.7	40,836,203,417.2
	CO_2 (t-CO_2/m)	0.683	3.063	0.051	0.351	0.123
	BW (kN/m)	20.25	52.36	2.03	9.80	96.11
	X_S (mm)	178.0	420.0	30.0	66.4	4,415.5
	SF	1.50	2.00	1.00	0.30z	0.09

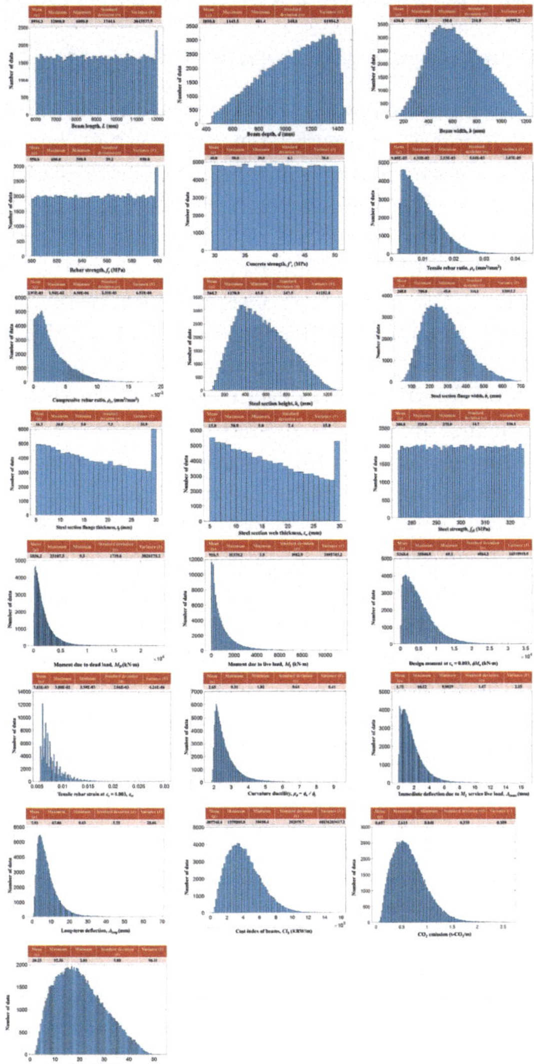

Figure 5.3.3 Distribution of ranges of input and output parameters with 100,000 datasets. (1) Beam length, L (mm). (2) Beam depth, d (mm). (3) Beam width, b (mm). (4) Rebar strength, f_y (MPa). (5) Concrete strength, $f_c^{'}$ (MPa). (6) Tensile rebar ratio, ρ_{rt} (mm^2/mm^2). (7) Compressive rebar ratio, ρ_{rc} (mm^2/mm^2). (8) Steel section height, h_s (mm). (9) Flange width of steel section, b_s (mm). (10) Flange thickness of steel section, t_f (mm). (11) Steel section web thickness, t_w (mm). (12) Steel strength, f_{yS} (MPa). (13) Moment due to dead load, M_D (MPa). (14) Moment due to live load, M_L (MPa). (15) Design moment at $\varepsilon_c = 0.003$, ϕM_n (MPa). (16) Tensile rebar strain at $\varepsilon_c = 0.003$, ε_{rt}. (17) Curvature ductility, $\mu_\phi = \phi_u/\phi_y$. (18) Immediate deflection due to M_L service live load, $\Delta_{imme.}$ (mm). (19) Long-term deflection, Δ_{long} (mm). (20) Cost index of beams, CI_b (KRW/m). (21) CO_2 emission (t-CO_2/m). (22) BW (kN).

5.4 NETWORKS BASED ON BACK SUBSTITUTION

5.4.1 Direct and back-substitution (BS) method

Useful training methods including TED, PTM, or CRS [5.31], were proposed to train big datasets of diverse structural systems based on appropriate training parameters including a number of layers, neurons, a maximum epoch, validation, etc. The authors performed ANN-based designs for RC beams [5.31, 5.32, 5.34, 5.40, 5.41], RC columns [5.33, 5.42, 5.43], PT beams [2.7, 3.9, 4.19] and SRC columns [5.44, 5.45] in their previous studies based on TED, PTM, or CRS. A design of SRC beams makes a design procedure complex, in which the strength and rigidity gained by structural steel placed in concrete contributes to a flexural and shear behavior simultaneously. Training for ANNs to design such SRC beams is not difficult as long as SRC beams are regular in shapes. ANN-wise, the complexities of structural configurations leading to complicated relationships among parameters may cause some difficulties in training.

Figure 5.4.1 shows a direct and BS method which is implemented in the design example of this chapter. The direct method requires only one step for training and design. In contrast, the BS method requires two steps for training and design. Reverse outputs (b, bs) calculated in the reverse network of Step 1 for given reverse input parameters (ϕM_n, ε_{rt}, SF, X_s) are substituted into the forward network of Step 2. Reverse outputs are used as input parameters for Step 2 to obtain design parameters. Both the direct and the reverse network of Step 1 of BS methods are performed using TED, PTM, and CRS [5.31, 5.32, 5.39, 5.46, 5.47].

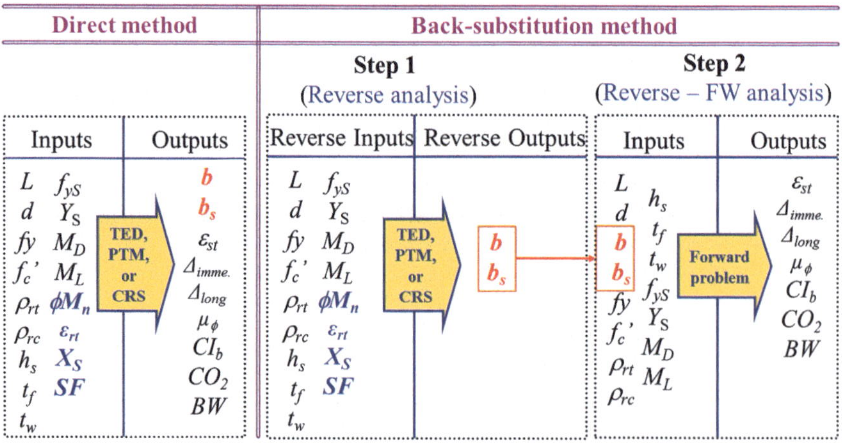

Figure 5.4.1 Network topology of direct and BS methods for Reverse Scenario 1.

5.4.2 Direct design method and enhanced accuracies of back-substitution (BS) method

Direct method [5.31] for revere scenario 1 shown in Figure 5.4.1 calculates nine output parameters (b, b_s, ε_{st}, Δ_{imme}, Δ_{long}, μ_{ϕ}, CI_b, CO_2, and beam weights (BW)) which are determined from 17 input parameters (ϕM_n, ε_{rt}, SF, X_s, L, d, f_y, f_c', ρ_{rt}, ρ_{rc}, h_s, t_f, t_w, f_{yS}, Y_s, M_D, and M_L). Design moment (ϕM_n), tensile rebar strain (ε_{rt}), and SF are the reverse input parameters of an ANN which would be calculated on an output-side the conventional forward designs.

Direct method needs only one step to finish mapping. However, the BS method needs two steps [5.32] in which 17 input parameters shown in Boxes 1–17 of Table 5.4.1(a)–(c) for Reverse Scenario 1 which are identical used in direct method are mapped to only two reverse output parameters (b and b_s) shown in Boxes 18–19 of Table 5.4.1(a)–(c) in the reverse network of Step 1, whereas a forward calculation using structural software AutoSRCbeam is followed to calculate the rest of deign parameters shown in Boxes 20–26 of Table 5.4.1(a)–(c) in the forward network of Step 2 as indicated in Figure 5.4.1. Step 2 of the BS method can be performed by any engineering software based on 15 input parameters (L, d, b, b_s, f_y, f_c', ρ_{rt}, ρ_{rc}, h_s, t_f, t_w, f_{yS}, Y_s, M_D, M_L) to calculate seven output parameters (ε_{st}, Δ_{imme}, Δ_{long}, μ_{ϕ}, CI_b, CO_2, and BW) for SRC beams as indicated in Figure 5.4.1. It is clear to see that a BS method gives more accurate results than the direct method because a BS method moves reverse output parameters (b, and b_s) from an output-side the reverse network of Step 1 into input parameters of the forward network of Step 2 when ANNs are trained.

Based on direct method for revere scenario 2, ANNs determine 10 output parameters (ρ_{rc}, t_f, ε_{st}, Δ_{imme}, Δ_{long}, μ_{ϕ}, CI_b, CO_2, BW, and X_s) from 16 input parameters (ϕM_n, ε_{rt}, SF, L, d, b, f_y, f_c', ρ_{rt}, h_s, b_s, t_w, f_{yS}, Y_s, M_D, and M_L). Based on direct method for revere scenario 3, ANNs determine 10 output parameters (ρ_{rt}, ρ_{rc}, ε_{st}, Δ_{imme}, Δ_{long}, μ_{ϕ}, CI_b, CO_2, BW, and X_s) from 16 input parameters (ϕM_n, ε_{rt}, SF, L, d, b, f_y, f_c', h_s, b_s, t_f, t_w, f_{yS}, Y_s, M_D, and M_L). Based on direct method for revere scenario 4, ANNs determine 10 output parameters (d, ρ_{rt}, ε_{st}, Δ_{imme}, Δ_{long}, μ_{ϕ}, CI_b, CO_2, BW, and X_s) from 16 input parameters (ϕM_n, ε_{rt}, SF, L, b, f_y, f_c', ρ_{rc} h_s, b_s, t_f, t_w, f_{yS}, Y_s, M_D, and M_L).

5.5 REVERSE SCENARIOS

5.5.1 Reverse Scenario 1

5.5.1.1 An ANN formulation

Scenario 1 shown in Table 5.2.1 is referred to as a reverse design, because input parameters include parameters shown in green Boxes 1–4 of Table 5.4.1(a)–(c) such as a design moment (ϕM_n), tensile rebar strain (ε_{rt}),

Table 5.4.1 Design table for reverse scenario 1 with 17 input parameters and 9 output parameters based on reverse PTM – forward AutoSRCbeam with ϕM_n = 2,000 kN·m, ϕM_n = 3,000 kNm, ϕM_n = 4,000 kN·m, and ε_{rt} = 0.008

(a) ϕM_n (kN·m) = 2,000, ε_{rt} = 0.008						
Reverse design 1 – BS (Reverse PTM – Forward AutoSRCbeam) – 100,000 dataset 17 Inputs (ϕM_n, ε_{rt}, SF, X_s, L, d, f_y, f_c', ρ_{rt}, ρ_{rc}, h_s, t_f, t_w, f_{yS}, Y_s, M_D, M_L) – 9 Outputs (b, b_s, ε_{st}, Δ_{imme}, Δ_{long}, μ_ϕ, CI_b, CO_2, BW)						
			BS		Verification by AutoSRCbeam	Error (%)
			Step 1	Step 2		
No.	Parameter	Training results	(1)		(2)	(3)
1	ϕM_n (kN·m)		2,000.0	-	1,985.2	0.74%
2	ε_{rt}		0.008	-	0.0078	2.50%
3	SF		1.43	-	1.42	0.84%
4	Xs (mm)		120.0	-	120.6	–0.48%
5	L (mm)		10,000	10,000	10,000	-
6	d (mm)		1,000	1,000	1,000	-
7	f_y (MPa)		500	500	500	-
8	f_c' (MPa)		30	30	30	-
9	ρ_{rt}		0.003	0.003	0.003	-
10	ρ_{rc}		0.002	0.002	0.002	
11	h_s (mm)		600	600	600	-
12	t_f (mm)		15	15	15	
13	t_w (mm)		6	6	6	-
14	f_{yS} (MPa)		325	325	325	-
15	Y_s (mm)		120	120	120	-
16	M_D (kN·m)		500	500	500	-
17	M_L (kN·m)		500	500	500	-

(Continued)

Table 5.4.1 (Continued) Design table for reverse scenario 1 with 17 input parameters and 9 output parameters based on reverse PTM –forward AutoSRCbeam with ϕM_n = 2,000 kN·m, ϕM_n = 3,000 kNm, ϕM_n = 4,000 kN·m, and ε_{rt} = 0.008

(a) ϕM_n (kN·m) = 2,000, ε_{rt} = 0.008						
Reverse design 1 – BS (Reverse PTM – Forward AutoSRCbeam) – 100,000 dataset *17 Inputs (ϕM_n, ε_{rt}, SF, X_s, L, d, f_y, f'_c, ρ_{rt}, ρ_{rc}, h_s, t_f, t_w, f_{yS}, Y_s, M_D, M_L) – 9 Outputs (b, b_s, ε_{st}, Δ_{imme}, Δ_{long}, μ_ϕ, CI_b, CO_2, BW)*						
			BS		*Verification by AutoSRCbeam*	*Error (%)*
			Step 1	*Step 2*		
No.	*Parameter*	*Training results*	*(1)*		*(2)*	*(3)*
18	*b* (mm)	(4 Layers – 50 Neurons) 49,991 epochs; *T.mse* = 7.94E-4; R= 1.0000	517.1	517.1	517.1	-
19	b_s (mm)	(5 Layers – 50 Neurons) 49,995 epochs; *T.mse* = 1.71E-5; R= 0.9999	275.9	275.9	275.9	-
20	ε_{st}		-	0.0068	-	-
21	Δ_{imme} (mm)		-	2.40	-	-
22	Δ_{long} (mm)		-	6.14	-	-
23	μ_ϕ		-	2.86	-	-
24	CI_b (Won/m)		-	241,887.6	-	-
25	CO_2 (t-CO_2/m)		-	0.377	-	-
26	*BW* (kN/m)		-	14.22	-	-

BS method procedure

Step 1 (*reverse analysis*): Calculating **2 reverse outputs** based on **13 ordinary inputs** and **4 reverse inputs**.

Step 2 (*forward analysis*): Calculating **7 rest outputs** using AutoSRCbeam based on **15 inputs** (substituting **2 reverse outputs** and **13 ordinary inputs**).

AutoSRCbeam check procedure

Calculating 4 outputs based on 15 inputs which are identical to 15 inputs of Step 2 of BS.

Error calculation

$(3) = ((1) - (2))/(1) \times 100\%$

Table 5.4.1 (Continued) Design table for reverse scenario 1 with 17 input parameters and 9 output parameters based on reverse PTM – forward AutoSRCbeam with ϕM_n = 2,000 kN·m, ϕM_n = 3,000 kNm, ϕM_n = 4,000 kN·m, and $\varepsilon_{rt} = 0.008$

(b) ϕM_n (kN·m) = 3,000, $\varepsilon_{rt} = 0.008$

Reverse design 1 – BS (Reverse PTM – Forward AutoSRCbeam) – 100,000 dataset
17 Inputs (ϕM_n, ε_{rt}, SF, X_s, L, d, f_y, f_c', ρ_{rt}, ρ_{rc}, h_s, t_f, t_w, f_{yS}, Y_s, M_D, M_L) – 9 Outputs (b, b_s, ε_{st}, Δ_{imme}, Δ_{long}, μ_ϕ, CI_b, CO_2, BW)

No.	Parameter	Training results	BS Step 1 (1)	BS Step 2	Verification by AutoSRCbeam (2)	Error (%) (3)
1	ϕM_n (kN·m)		3,000.0	-	2,995.3	0.16%
2	ε_{rt}		0.008	-	0.0076	5.00%
3	SF		2.14	-	2.139	0.02%
4	Xs (mm)		120.0	-	122.7	−2.23%
5	L (mm)		10,000	10,000	10,000	-
6	d (mm)		1,000	1,000	1,000	-
7	f_y (MPa)		500	500	500	-
8	f_c' (MPa)		30	30	30	-
9	ρ_{rt}		0.003	0.003	0.003	-
10	ρ_{rc}		0.002	0.002	0.002	
11	h_s (mm)		600	600	600	-
12	t_f (mm)		15	15	15	
13	t_w (mm)		6	6	6	-
14	f_{yS} (MPa)		325	325	325	-
15	Y_s (mm)		120	120	120	-
16	M_D (kN·m)		500	500	500	-

(Continued)

Table 5.4.1 (Continued) Design table for reverse scenario 1 with 17 input parameters and 9 output parameters based on reverse PTM –forward AutoSRCbeam with $\phi M_n = 2{,}000$ kN·m, $\phi M_n = 3{,}000$ kNm, $\phi M_n = 4{,}000$ kN·m, and $\varepsilon_{rt} = 0.008$

(b) ϕM_n *(kN·m) = 3,000,* ε_{rt} *= 0.008*						
Reverse design 1 – BS (Reverse PTM – Forward AutoSRCbeam) – 100,000 dataset *17 Inputs (*$\phi M_n, \varepsilon_{rt}$*, SF,* X_s*, L, d,* f_y*,* f_c'*,* ρ_{rt}*,* ρ_{rc}*,* h_s*,* t_f*,* t_w*,* f_{yS}*,* Y_s*,* M_D*,* M_L*) – 9 Outputs (b,* b_s*,* ε_{st}*,* Δ_{imme}*,* Δ_{long}*,* μ_ϕ*,* CI_b*,* CO_2*, BW)*						
			BS		*Verification by AutoSRCbeam*	*Error (%)*
			Step 1	*Step 2*		
No.	*Parameter*	*Training results*	*(1)*		*(2)*	*(3)*
17	M_L (kN·m)		500	500	500	-
18	b (mm)	(4 Layers – 50 Neurons) 49,991 epochs; *T.mse* = 7.94E-4; R= 1.0000	744.1	744.1	744.1	-
19	b_s (mm)	(5 Layers – 50 Neurons) 49,995 epochs; *T.mse* = 1.71E-5; R= 0.9999	498.7	498.7	498.7	-
20	ε_{st}		-	0.0067	-	-
21	Δ_{imme} (mm)		-	1.36	-	-
22	Δ_{long} (mm)		-	3.10	-	-
23	μ_ϕ		-	2.79	-	-
24	CI_b (Won/m)		-	370,944.8	-	-
25	CO_2 (t-CO_2/m)		-	0.572	-	-
26	*BW* (kN/m)		-	20.59	-	-

BS method procedure

Step 1 (*reverse analysis*): Calculating **2 reverse outputs** based on **13 ordinary inputs** and **4 reverse inputs**.

Step 2 (*forward analysis*): Calculating **7 rest outputs** using AutoSRCbeam based on **15 inputs** (substituting **2 reverse outputs** and **13 ordinary inputs**).

AutoSRCbeam check procedure

Calculating 4 outputs based on 15 inputs which are identical to 15 inputs of Step 2 of BS.

Error calculation

$(3) = ((1) - (2)) / (1) \times 100\%$

Table 5.4.1 (Continued) Design table for reverse scenario 1 with 17 input parameters and 9 output parameters based on reverse PTM – forward AutoSRCbeam with $\phi M_n = 2{,}000$ kN·m, $\phi M_n = 3{,}000$ kNm, $\phi M_n = 4{,}000$ kN·m, and $\varepsilon_{rt} = 0.008$

(c) ϕM_n (kN·m) = 4,000, $\varepsilon_{rt} = 0.008$

Reverse design 1 – BS (Reverse PTM – Forward AutoSRCbeam) – 100,000 dataset
17 Inputs (ϕM_n, ε_{rt}, SF, X_s, L, d, f_y, f_c', ρ_{rt}, ρ_{rc}, h_s, t_f, t_w, f_{yS}, Y_s, M_D, M_L) – 9 Outputs (b, b_s, ε_{st}, Δ_{imme}, Δ_{long}, μ_ϕ, CI_b, CO_2, BW)

No.	Parameter	Training results	BS Step 1 (1)	BS Step 2	Verification by AutoSRCbeam (2)	Error (%) (3)
1	ϕM_n (kN·m)		4,000.0	-	4,002.8	–0.07%
2	ε_{rt}		0.008	-	0.0075	6.25%
3	SF		2.86	-	2.859	0.03%
4	Xs (mm)		120.0	-	124.8	–4.03%
5	L (mm)		10,000	10,000	10,000	-
6	d (mm)		1,000	1,000	1,000	-
7	f_y (MPa)		500	500	500	-
8	f_c' (MPa)		30	30	30	-
9	ρ_{rt}		0.003	0.003	0.003	-
10	ρ_{rc}		0.002	0.002	0.002	
11	h_s (mm)		600	600	600	-
12	t_f (mm)		15	15	15	
13	t_w (mm)		6	6	6	-
14	f_{yS} (MPa)		325	325	325	-
15	Y_s (mm)		120	120	120	-
16	M_D (kN·m)		500	500	500	-

(Continued)

Table 5.4.1 (Continued) Design table for reverse scenario 1 with 17 input parameters and 9 output parameters based on reverse PTM – forward AutoSRCbeam with $\phi M_n = 2{,}000$ kN·m, $\phi M_n = 3{,}000$ kNm, $\phi M_n = 4{,}000$ kN·m, and $\varepsilon_{rt} = 0.008$

(c) ϕM_n (kN·m) = 4,000, $\varepsilon_{rt} = 0.008$						
Reverse design 1 – BS (Reverse PTM – Forward AutoSRCbeam) – 100,000 dataset *17 Inputs (ϕM_n, ε_{rt}, SF, X_s, L, d, f_y, f'_c, ρ_{rt}, ρ_{rc}, h_s, t_f, t_w, f_{yS}, Y_s, M_D, M_L) – 9 Outputs (b, b_s, ε_{st}, Δ_{imme}, Δ_{long}, μ_ϕ, CI_b, CO_2, BW)*						
			BS			
No.	*Parameter*	*Training results*	*Step 1* (1)	*Step 2*	*Verification by AutoSRCbeam* (2)	*Error (%)* (3)
17	M_L (kN·m)		500	500	500	-
18	*b* (mm)	(4 Layers – 50 Neurons) 49,991 epochs; *T.mse* = 7.94E-4; R= 1.0000	970.7	970.7	970.7	-
19	b_s (mm)	(5 Layers – 50 Neurons) 49,995 epochs; *T.mse* = 1.71E-5; R= 0.9999	721.0	721.0	721.0	-
20	ε_{st}		-	0.0066	-	-
21	Δ_{imme} (mm)		-	0.81	-	-
22	Δ_{long} (mm)		-	2.77	-	-
23	μ_ϕ		-	2.76	-	-
24	CI_b (Won/m)		-	499,693.7	-	-
25	CO_2 (t-CO_2/m)		-	0.768	-	-
26	*BW* (kN/m)		-	26.94	-	-

BS method procedure

Step 1 (*reverse analysis*): Calculating **2 reverse outputs** based on **13 ordinary inputs** and **4 reverse inputs**.

Step 2 (*forward analysis*): Calculating **7 rest outputs** using AutoSRCbeam based on **15 inputs** (substituting **2 reverse outputs** and **13 ordinary inputs**).

AutoSRCbeam check procedure

Calculating **4 outputs** based on **15 inputs** which are identical to **15 inputs** of Step 2 of BS.

Error calculation

$(3) = ((1) - (2)) / (1) \times 100\%$

Table 5.4.2 Table design for reverse scenario 2 with 16 input parameters and 10 output parameters based on reverse PTM – forward AutoSRCbeam with $\varepsilon_{rt} = 0.009$, $\varepsilon_{rt} = 0.012$, $\varepsilon_{rt} = 0.015$, and ϕM_n (kN·m) = 3,000

(a) ϕM_n (kN·m) = 3,000, $\varepsilon_{rt} = 0.009$						
Reverse design 2 – BS (Reverse PTM – Forward AutoSRCbeam) – 100,000 dataset *16 Inputs (ϕM_n, ε_{rt}, SF, L, d, b, f_y, f_c', ρ_{rt}, h_s, b_s, t_w, f_{yS}, Y_s, M_D, M_L) – 10 Outputs (ρ_{rc}, t_f, ε_{st}, Δ_{imme}, Δ_{long}, μ_ϕ, CI_b, CO_2, BW, X_s)*						
			BS		*Verification by AutoSRCbeam*	*Error (%)*
			Step 1	*Step 2*		
No.	Parameter	*Training results*	*(1)*		*(2)*	*(3)*
1	ϕM_n (kN·m)		3,000.0	-	2,990.0	0.33%
2	ε_{rt}		0.009	-	0.0089	1.11%
3	SF		1.200	-	1.196	0.33%
4	L (mm)		10,000	10,000	10,000	-
5	d (mm)		1,150	1,150	1,150	-
6	b (mm)		600	600	600	
7	f_y (MPa)		500	500	500	-
8	f_c' (MPa)		30	30	30	-
9	ρ_{rt}		0.006	0.006	0.006	-
10	h_s (mm)		400	400	400	-
11	b_s (mm)		200	200	200	
12	t_w (mm)		8	8	8	-
13	f_{yS} (MPa)		325	325	325	-
14	Y_s (mm)		80	80	80	-
15	M_D (kN·m)		750	750	750	-
16	M_L (kN·m)		1,000	1,000	1,000	-

(Continued)

Table 5.4.2 (Continued) Table design for reverse scenario 2 with 16 input parameters and 10 output parameters based on reverse PTM – forward AutoSRCbeam with $\varepsilon_{rt} = 0.009$, $\varepsilon_{rt} = 0.012$, $\varepsilon_{rt} = 0.015$, and ϕM_n (kN·m) = 3,000

(a) ϕM_n (kN·m) = 3,000, $\varepsilon_{rt} = 0.009$						
Reverse design 2 – BS (Reverse PTM – Forward AutoSRCbeam) – 100,000 dataset *16 Inputs (ϕM_n, ε_{rt}, SF, L, d, b, f_y, f_c', ρ_{rt}, h_s, b_s, t_w, f_{yS}, Y_s, M_D, M_L) – 10 Outputs (ρ_{rc}, t_f, ε_{st}, Δ_{imme}, Δ_{long}, μ_ϕ, CI_b, CO_2, BW, X_s)*						
			BS			
			Step 1	*Step 2*	*Verification by AutoSRCbeam*	*Error (%)*
No.	Parameter	*Training results*	*(1)*		*(2)*	*(3)*
17	ρ_{rc}	(5 Layers – 50 Neurons) 27,166 epochs; *T.mse* = 8.14E-4; R = 0.9952	0.00086	0.00086	0.00086	-
18	t_f (mm)	(5 Layers – 50 Neurons) 16,716 epochs; *T.mse* = 1.40E-2; R = 0.9814	5.66	5.66	5.66	-
19	ε_{st}		-	0.0082	-	-
20	Δ_{imme} (mm)		-	2.88	-	-
21	Δ_{long} (mm)		-	8.01	-	-
22	μ_ϕ		-	3.11	-	-
23	CI_b (Won/m)		-	190,205.4	-	-
24	CO_2 (t-CO_2/m)		-	0.345	-	-
25	BW (kN/m)		-	18.24	-	-
26	Xs (mm)		-	200	-	-

BS method procedure

Step 1 (*reverse analysis*): Calculating **2 reverse outputs** based on **13 ordinary inputs** and **3 reverse inputs**.

Step 2 (*forward analysis*): Calculating **8 rest outputs** using AutoSRCbeam based on **15 inputs** (substituting **2 reverse outputs** and **13 ordinary inputs**).

AutoSRCbeam check procedure

Calculating **3 outputs** based on **15 inputs** which are identical to **15 inputs** of Step 2 of BS.

Error calculation

$$(3) = ((1) - (2))/(1) \times 100\%$$

Table 5.4.2 (Continued) Table design for reverse scenario 2 with 16 input parameters and 10 output parameters based on reverse PTM – forward AutoSRCbeam with $\varepsilon_{rt} = 0.009$, $\varepsilon_{rt} = 0.012$, $\varepsilon_{rt} = 0.015$, and ϕM_n (kN·m) = 3,000

(b) ϕM_n (kN·m) = 3,000, $\varepsilon_{rt} = 0.012$

Reverse design 2 – BS (Reverse PTM – Forward AutoSRCbeam) – 100,000 dataset
16 Inputs (ϕM_n, ε_{rt}, SF, L, d, b, f_y, f_c', ρ_{rt}, h_s, b_s, t_w, f_{yS}, Y_s, M_D, M_L) – 10 Outputs (ρ_{rc}, t_f, ε_{st}, Δ_{imme}, Δ_{long}, μ_ϕ, CI_b, CO_2, BW, X_s)

No.	Parameter	Training results	BS Step 1 (1)	BS Step 2	Verification by AutoSRCbeam (2)	Error (%) (3)
1	ϕM_n (kN·m)		3,000.0	-	2,987.2	0.43%
2	ε_{rt}		0.012	-	0.0117	2.50%
3	SF		1.200	-	1.195	0.43%
4	L (mm)		10,000	10,000	10,000	-
5	d (mm)		1,150	1,150	1,150	-
6	b (mm)		600	600	600	
7	f_y (MPa)		500	500	500	-
8	f_c' (MPa)		30	30	30	-
9	ρ_{rt}		0.006	0.006	0.006	-
10	h_s (mm)		400	400	400	-
11	b_s (mm)		200	200	200	-
12	t_w (mm)		8	8	8	-
13	f_{yS} (MPa)		325	325	325	-
14	Y_s (mm)		80	80	80	-
15	M_D (kN·m)		750	750	750	-
16	M_L (kN·m)		1,000	1,000	1,000	-

(Continued)

Table 5.4.2 (Continued) Table design for reverse scenario 2 with 16 input parameters and 10 output parameters based on reverse PTM – forward AutoSRCbeam with $\varepsilon_{rt} = 0.009$, $\varepsilon_{rt} = 0.012$, $\varepsilon_{rt} = 0.015$, and ϕM_n (kN·m) = 3,000

(b) ϕM_n (kN·m) = 3,000, $\varepsilon_{rt} = 0.012$

Reverse design 2 – BS (Reverse PTM – Forward AutoSRCbeam) – 100,000 dataset
16 Inputs (ϕM_n, ε_{rt}, SF, L, d, b, f_y, f'_c, ρ_{rt}, h_s, b_s, t_w, f_{yS}, Y_s, M_D, M_L) – 10 Outputs (ρ_{rc}, t_f, ε_{st}, Δ_{imme}, Δ_{long}, μ_ϕ, CI_b, CO_2, BW, X_s)

No.	Parameter	Training results	BS Step 1 (1)	BS Step 2	Verification by AutoSRCbeam (2)	Error (%) (3)
17	ρ_{rc}	(5 Layers – 50 Neurons) 27,166 epochs; *T.mse* = 8.14E-4; R = 0.9952	0.0027	0.0027	0.0027	-
18	t_f (mm)	(5 Layers – 50 Neurons) 16,716 epochs; *T.mse* = 1.4E-2; R = 0.9814	4.81	4.81	4.81	-
19	ε_{st}		-	0.0109	-	-
20	Δ_{imme} (mm)		-	2.83	-	-
21	Δ_{long} (mm)		-	7.37	-	-
22	μ_ϕ		-	3.95	-	-
23	CI_b (Won/m)		-	192,296.3	-	-
24	CO_2 (t-CO_2/m)		-	0.355	-	-
25	BW (kN/m)		-	18.28	-	-
26	Xs (mm)		-	200	-	-

BS method procedure

Step 1 (*reverse analysis*): Calculating **2 reverse outputs** based on **13 ordinary inputs** and **3 reverse inputs**.

Step 2 (*forward analysis*): Calculating **8 rest outputs** using AutoSRCbeam based on **15 inputs** (substituting **2 reverse outputs** and **13 ordinary inputs**).

AutoSRCbeam check procedure

Calculating **3 outputs** based on **15 inputs** which are identical to **15 inputs** of Step 2 of BS.

Error calculation

$(3) = ((1) - (2))/(1) \times 100\%$

Table 5.4.2 (Continued) Table design for reverse scenario 2 with 16 input parameters and 10 output parameters based on reverse PTM – forward AutoSRCbeam with $\varepsilon_{rt} = 0.009$, $\varepsilon_{rt} = 0.012$, $\varepsilon_{rt} = 0.015$, and ϕM_n (kN·m) = 3,000

(c) ϕM_n (kN·m) = 3,000, $\varepsilon_{rt} = 0.015$

Reverse design 2 – BS (Reverse PTM – Forward AutoSRCbeam) – 100,000 dataset
16 Inputs (ϕM_n, ε_{rt}, SF, L, d, b, f_y, f_c', ρ_{rt}, h_s, b_s, t_w, f_{yS}, Y_s, M_D, M_L) – 10 Outputs (ρ_{rc}, t_f, ε_{st}, Δ_{imme}, Δ_{long}, μ_ϕ, CI_b, CO_2, BW, X_s)

		BS			
		Step 1	*Step 2*	*Verification by AutoSRCbeam*	*Error (%)*
No.	*Parameter*	*Training results* (1)		(2)	(3)
1	ϕM_n (kN·m)	3,000.0	-	3,020.1	0.82%
2	ε_{rt}	0.015	-	0.0134	4.00%
3	SF	1.200	-	1.208	0.82%
4	L (mm)	10,000	10,000	10,000	-
5	d (mm)	1,200	1,200	1,200	-
6	b (mm)	600	600	600	
7	f_y (MPa)	500	500	500	-
8	f_c' (MPa)	30	30	30	-
9	ρ_{rt}	0.006	0.006	0.006	-
10	h_s (mm)	350	350	350	-
11	b_s (mm)	150	150	150	
12	t_w (mm)	6	6	6	-
13	f_{yS} (MPa)	325	325	325	-
14	Y_s (mm)	80	80	80	-
15	M_D (kN·m)	750	750	750	-
16	M_L (kN·m)	1,000	1,000	1,000	-

(Continued)

Table 5.4.2 (Continued) Table design for reverse scenario 2 with 16 input parameters and 10 output parameters based on reverse PTM – forward AutoSRCbeam with $\varepsilon_{rt} = 0.009$, $\varepsilon_{rt} = 0.012$, $\varepsilon_{rt} = 0.015$, and ϕM_n (kN·m) = 3,000

(c) ϕM_n (kN·m) = 3,000, $\varepsilon_{rt} = 0.015$						
Reverse design 2 – BS (Reverse PTM – Forward AutoSRCbeam) – 100,000 dataset 16 Inputs (ϕM_n, ε_{rt}, SF, L, d, b, f_y, f'_c, ρ_{rt}, h_s, b_s, t_w, f_{ys}, Y_s, M_D, M_L) – 10 Outputs (ρ_{rc}, t_f, ε_{st}, Δ_{imme}, Δ_{long}, μ_ϕ, CI_b, CO_2, BW, X_s)						
			BS			
			Step 1	Step 2	Verification by AutoSRCbeam	Error (%)
No.	Parameter	Training results	(1)		(2)	(3)
17	ρ_{rc}	(5 Layers – 50 Neurons) 27,166 epochs; T.mse = 8.14E-4; R = 0.9952	0.0038	0.0038	0.0038	-
18	t_f (mm)	(5 Layers – 50 Neurons) 16,716 epochs; T.mse = 1.4E-2; R = 0.9814	6.26	6.26	6.26	-
19	ε_{st}		-	0.0135	-	-
20	Δ_{imme} (mm)		-	2.52	-	-
21	Δ_{long} (mm)		-	6.20	-	-
22	μ_ϕ		-	4.76	-	-
23	CI_b (Won/m)		-	180,456.6	-	-
24	CO_2 (t-CO_2/m)		-	0.344	-	-
25	BW (kN/m)		-	18.93	-	-
26	Xs (mm)		-	225	-	-

BS method procedure

Step 1 (*reverse analysis*): Calculating **2 reverse outputs** based on **13 ordinary inputs** and **3 reverse inputs**.

Step 2 (*forward analysis*): Calculating **8 rest outputs** using AutoSRCbeam based on **15 inputs** (substituting **2 reverse outputs** and **13 ordinary inputs**).

AutoSRCbeam check procedure

Calculating **3 outputs** based on **15 inputs** which are identical to **15 inputs** of Step 2 of BS.

Error calculation

$(3) = ((1) - (2))/(1) \times 100\%$

Table 5.4.3 Table design for reverse scenario 3 with 16 input parameters and 10 output parameters based on reverse PTM – forward AutoSRCbeam with $\varepsilon_{rt} = 0.008$, $\varepsilon_{rt} = 0.010$ and $\varepsilon_{rt} = 0.012$, and ϕM_n (kN·m) = 3,500

(a) $\phi M_n = 3500$ kN × m, $\varepsilon_{rt} = 0.008$						
Reverse design 3 – BS (Reverse PTM – Forward AutoSRCbeam) – 100,000 dataset *16 Inputs (ϕM_n, ε_{rt}, SF, L, d, b, f_y, f_c', h_s, b_s, t_f, t_w, f_{yS}, Y_s, M_D, M_L) – 10 Outputs (ρ_{rt}, ρ_{rc}, ε_{st}, Δ_{imme}, Δ_{long}, μ_ϕ, CI_b, CO_2, BW, X_s)*						
			BS		*Verification by AutoSRCbeam*	*Error (%)*
			Step 1	*Step 2*		
No.	*Parameter*	*Training results*	*(1)*		*(2)*	*(3)*
1	ϕM_n (kN·m)		3,500.0	-	3,456.5	1.24%
2	ε_{rt}		0.008	-	0.008	0.00%
3	*SF*		1.250	-	1.234	1.24%
4	L (mm)		10,000	10,000	10,000	-
5	d (mm)		1,100	1,100	1,100	-
6	b (mm)		600	600	600	
7	f_y (MPa)		500	500	500	-
8	f_c' (MPa)		30	30	30	-
9	h_s (mm)		500	500	500	-
10	b_s (mm)		200	200	200	
11	t_f (mm)		8	8	8	-
12	t_w (mm)		6	6	6	-
13	f_{yS} (MPa)		325	325	325	-
14	Y_s (mm)		150	150	150	-
15	M_D (kN·m)		1,000	1,000	1,000	-
16	M_L (kN·m)		1,000	1,000	1,000	-

(Continued)

Table 5.4.3 (Continued) Table design for reverse scenario 3 with 16 input parameters and 10 output parameters based on reverse PTM – forward AutoSRCbeam with $\varepsilon_{rt} = 0.008$, $\varepsilon_{rt} = 0.010$ and $\varepsilon_{rt} = 0.012$, and ϕM_n (kN·m) = 3,500

(a) $\phi M_n = 3500$ kN × m, $\varepsilon_{rt} = 0.008$						
Reverse design 3 – BS (Reverse PTM – Forward AutoSRCbeam) – 100,000 dataset *16 Inputs (ϕM_n, ε_{rt}, SF, L, d, b, f_y, f_c', h_s, b_s, t_f, t_w, f_{yS}, Y_s, M_D, M_L) – 10 Outputs (ρ_{rt}, ρ_{rc}, ε_{st}, Δ_{imme}, Δ_{long}, μ_ϕ, CI_b, CO_2, BW, X_s)*						
			BS		*Verification by AutoSRCbeam*	*Error (%)*
			Step 1	*Step 2*		
No.	*Parameter*	*Training results*	*(1)*		*(2)*	*(3)*
17	ρ_{rt}	(5 Layers – 50 Neurons) 27,150 epochs; *T.mse* = 6.07E-4; R = 0.9996	0.0083	0.0083	0.0083	-
18	ρ_{rc}	(5 Layers – 40 Neurons) 15,685 epochs; *T.mse* = 1.4E-3; R = 0.9917	0.0050	0.0050	0.0050	-
19	ε_{st}		-	0.0082	-	-
20	Δ_{imme} (mm)		-	2.97	-	-
21	Δ_{long} (mm)		-	9.46	-	-
22	μ_ϕ		-	3.36	-	-
23	CI_b (Won/m)		-	228,398.0	-	-
24	CO_2 (t-CO_2/m)		-	0.426	-	-
25	BW (kN/m)		-	17.86	-	-
26	Xs (mm)		-	200	-	-

BS method procedure

Step 1 (*reverse analysis*): Calculating **2 reverse outputs** based on **13 ordinary inputs** and **3 reverse inputs**.

Step 2 (*forward analysis*): Calculating **8 rest outputs** using AutoSRCbeam based on **15 inputs** (substituting **2 reverse outputs** and **13 ordinary inputs**).

AutoSRCbeam check procedure

Calculating **3 outputs** based on **15 inputs** which are identical to **15 inputs** of Step 2 of BS.

Error calculation

$$(3) = ((1) - (2))/(1) \times 100\%$$

Table 5.4.3 (Continued) Table design for reverse scenario 3 with 16 input parameters and 10 output parameters based on reverse PTM – forward AutoSRCbeam with $\varepsilon_{rt} = 0.008$, $\varepsilon_{rt} = 0.010$ and $\varepsilon_{rt} = 0.012$, and ϕM_n (kN·m) = 3,500

(b) $\phi M_n = 3{,}500$ kN·m, $\varepsilon_{rt} = 0.010$

Reverse design 3 – BS (Reverse PTM – Forward AutoSRCbeam) – 100,000 dataset
16 Inputs (ϕM_{n}, ε_{rt}, SF, L, d, b, f_y, f_c', h_s, b_s, t_f, t_w, f_{yS}, Y_s, M_D, M_L) – 10 Outputs (ρ_{rt}, ρ_{rc}, ε_{st}, Δ_{imme}, Δ_{long}, μ_{ϕ}, CI_b, CO_2, BW, X_s)

No.	Parameter	Training results	BS Step 1 (1)	BS Step 2	Verification by AutoSRCbeam (2)	Error (%) (3)
1	ϕM_n (kN·m)		3,500.0	-	3,435.6	1.84%
2	ε_{rt}		0.010	-	0.010	3.00%
3	SF		1.250	-	1.227	1.84%
4	L (mm)		10,000	10,000	10,000	-
5	d (mm)		1,100	1,100	1,100	-
6	b (mm)		600	600	600	
7	f_y (MPa)		500	500	500	-
8	f_c' (MPa)		30	30	30	-
9	h_s (mm)		500	500	500	-
10	b_s (mm)		200	200	200	
11	t_f (mm)		8	8	8	-
12	t_w (mm)		6	6	6	-
13	f_{yS} (MPa)		325	325	325	-
14	Y_s (mm)		150	150	150	-
15	M_D (kN·m)		1,000	1,000	1,000	-

(Continued)

Table 5.4.3 (Continued) Table design for reverse scenario 3 with 16 input parameters and 10 output parameters based on reverse PTM – forward AutoSRCbeam with $\varepsilon_{rt} = 0.008$, $\varepsilon_{rt} = 0.010$ and $\varepsilon_{rt} = 0.012$, and ϕM_n (kN·m) = 3,500

(b) $\phi M_n = 3{,}500$ kN·m, $\varepsilon_{rt} = 0.010$						
Reverse design 3 – BS (Reverse PTM – Forward AutoSRCbeam) – 100,000 dataset *16 Inputs (ϕM_n, ε_{rt}, SF, L, d, b, f_y, f_c', h_s, b_s, t_f, t_w, f_{ys}, Y_s, M_D, M_L) – 10 Outputs (ρ_{rt}, ρ_{rc}, ε_{st}, Δ_{imme}, Δ_{long}, μ_ϕ, CI_b, CO_2, BW, X_s)*						
			BS		*Verification by AutoSRCbeam*	*Error (%)*
			Step 1	*Step 2*		
No.	*Parameter*	*Training results*	*(1)*		*(2)*	*(3)*
16	M_L (kN·m)		1,000	1,000	1,000	-
17	ρ_{rt}	(5 Layers – 50 Neurons) 27,166 epochs; *T.mse* = 8.14E-4; R = 0.9952	0.0087	0.0087	0.0087	-
18	ρ_{rc}	(5 Layers – 50 Neurons) 16,716 epochs; *T.mse* = 1.4E-2; R = 0.9814	0.0035	0.0035	0.0035	-
19	ε_{st}		-	0.0066	-	-
20	Δ_{imme} (mm)		-	2.98	-	-
21	Δ_{long} (mm)		-	9.92	-	-
22	μ_ϕ		-	2.84	-	-
23	CI_b (Won/m)		-	225,446.1	-	-
24	CO_2 (t-CO_2/m)		-	0.419	-	-
25	BW (kN/m)		-	17.83	-	-
26	Xs (mm)		-	200	-	-

BS method procedure

Step 1 (*reverse analysis*): Calculating **2 reverse outputs** based on **13 ordinary inputs** and **3 reverse inputs**.

Step 2 (*forward analysis*): Calculating **8 rest outputs** using AutoSRCbeam based on **15 inputs** (substituting **2 reverse outputs** and **13 ordinary inputs**).

AutoSRCbeam check procedure

Calculating **3 outputs** based on **15 inputs** which are identical to **15 inputs** of Step 2 of BS.

Error calculation

$$(3) = ((1) - (2)) / (1) \times 100\%$$

Table 5.4.3 (Continued) Table design for reverse scenario 3 with 16 input parameters and 10 output parameters based on reverse PTM – forward AutoSRCbeam with $\varepsilon_{rt} = 0.008$, $\varepsilon_{rt} = 0.010$ and $\varepsilon_{rt} = 0.012$, and ϕM_n (kN·m) = 3,500

(c) $\phi M_n = 3{,}500$ kN·m, $\varepsilon_{rt} = 0.012$					
Reverse design 3 – BS (Reverse PTM – Forward AutoSRCbeam) – 100,000 dataset *16 Inputs (ϕM_n, ε_{rt}, SF, L, d, b, f_y, f_c', h_s, b_s, t_f, t_w, f_{yS}, Y_s, M_D, M_L) – 10 Outputs (ρ_{rt}, ρ_{rc}, ε_{st}, Δ_{imme}, Δ_{long}, μ_ϕ, CI_b, CO_2, BW, X_s)*					
		BS			
		Step 1	*Step 2*	*Verification by AutoSRCbeam*	*Error (%)*
No.	*Parameter* / *Training results*	*(1)*		*(2)*	*(3)*
1	ϕM_n (kN·m)	3,500.0	-	3,457.6	1.21%
2	ε_{rt}	0.012	-	0.011	5.83%
3	*SF*	1.250	-	1.235	1.21%
4	L (mm)	10,000	10,000	10,000	-
5	d (mm)	1,100	1,100	1,100	-
6	b (mm)	600	600	600	
7	f_y (MPa)	500	500	500	-
8	f_c' (MPa)	30	30	30	-
9	h_s (mm)	500	500	500	-
10	b_s (mm)	150	150	150	
11	t_f (mm)	8	8	8	-
12	t_w (mm)	6	6	6	-
13	f_{yS} (MPa)	325	325	325	-
14	Y_s (mm)	150	150	150	-
15	M_D (kN·m)	1,000	1,000	1,000	-
16	M_L (kN·m)	1,000	1,000	1,000	-

(Continued)

Table 5.4.3 (Continued) Table design for reverse scenario 3 with 16 input parameters and 10 output parameters based on reverse PTM – forward AutoSRCbeam with $\varepsilon_{rt} = 0.008$, $\varepsilon_{rt} = 0.010$ and $\varepsilon_{rt} = 0.012$, and ϕM_n (kN·m) = 3,500

No.	Parameter	Training results	BS Step 1 (1)	BS Step 2	Verification by AutoSRCbeam (2)	Error (%) (3)
(c) ϕM_n =3,500 kN·m, $\varepsilon_{rt} = 0.012$						
Reverse design 3 – BS (Reverse PTM – Forward AutoSRCbeam) – 100,000 dataset 16 Inputs (ϕM_n, ε_{rt}, SF, L, d, b, f_y, f'_c, h_s, b_s, t_f, t_w, f_{yS}, Y_s, M_D, M_L) – 10 Outputs (ρ_{rt}, ρ_{rc}, ε_{st}, Δ_{imme}, Δ_{long}, μ_ϕ, CI_b, CO_2, BW, X_s)						
17	ρ_{rt}	(5 Layers – 50 Neurons) 27,166 epochs; *T.mse* = 8.14E-4; R = 0.9952	0.0088	0.0088	0.0088	-
18	ρ_{rc}	(5 Layers – 50 Neurons) 16,716 epochs; *T.mse* = 1.4E-2; R = 0.9814	0.0060	0.0060	0.0060	-
19	ε_{st}		-	0.0095	-	-
20	Δ_{imme} (mm)		-	2.91	-	-
21	Δ_{long} (mm)		-	9.05	-	-
22	μ_ϕ		-	3.81	-	-
23	CI_b (Won/m)		-	223,088.9	-	-
24	CO_2 (t-CO_2/m)		-	0.426	-	-
25	BW (kN/m)		-	17.86	-	-
26	Xs (mm)		-	225	-	-

BS method procedure

Step 1 (*reverse analysis*): Calculating **2 reverse outputs** based on **13 ordinary inputs** and **3 reverse inputs**.

Step 2 (*forward analysis*): Calculating **8 rest outputs** using AutoSRCbeam based on **15 inputs** (substituting **2 reverse outputs** and **13 ordinary inputs**).

AutoSRCbeam check procedure

Calculating **3 outputs** based on **15 inputs** which are identical to **15 inputs** of Step 2 of BS.

Error calculation

$$(3) = \left((1) - (2)\right)/(1) \times 100\%$$

Table 5.4.4 Table design for reverse scenario 4 with 16 input parameters and 10 output parameters based on reverse PTM – forward AutoSRCbeam with $\varepsilon_{rt} = 0.008$, $\varepsilon_{rt} = 0.012$ and $\varepsilon_{rt} = 0.020$, and ϕM_n (kN·m) = 3,500

			BS			
			Step 1	Step 2	Verification by AutoSRCbeam	Error (%)
No.	Parameter	Training results	(1)		(2)	(3)
	(a) ϕM_n (kN·m) = 3,500, $\varepsilon_{rt} = 0.008$					
	Reverse design 4 – BS (Reverse PTM – Forward AutoSRCbeam) – 100,000 dataset					
	16 Inputs (ϕM_n, ε_{rt}, SF, L, b, f_y, f_c', ρ_{rc}, h_s, b_s, t_f, t_w, f_{yS}, Y_s, M_D, M_L) – 10 Outputs (d, ρ_{rt}, ε_{st}, Δ_{imme}, Δ_{long}, μ_ϕ, CI_b, CO_2, BW, X_s)					
1	ϕM_n (kN·m)		3,500.0	-	3,470.0	0.86%
2	ε_{rt}		0.008	-	0.008	0.00%
3	SF		1.250	-	1.239	0.86%
4	L (mm)		10,000	10,000	10,000	-
5	b (mm)		600	600	600	
6	f_y (MPa)		500	500	500	-
7	f_c' (MPa)		30	30	30	-
8	ρ_{rc}		0.004	0.004	0.004	-
9	h_s (mm)		500	500	500	-
10	b_s (mm)		200	200	200	
11	t_f (mm)		8	8	8	-
12	t_w (mm)		6	6	6	-
13	f_{yS} (MPa)		325	325	325	-
14	Y_s (mm)		100	100	100	-
15	M_D (kN·m)		1,000	1,000	1,000	-
16	M_L (kN·m)		1,000	1,000	1,000	-

(Continued)

Table 5.4.4 (Continued) Table design for reverse scenario 4 with 16 input parameters and 10 output parameters based on reverse PTM – forward AutoSRCbeam with $\varepsilon_{rt} = 0.008$, $\varepsilon_{rt} = 0.012$ and $\varepsilon_{rt} = 0.020$, and ϕM_n (kN·m) = 3,500

(a) ϕM_n (kN·m) = 3,500, $\varepsilon_{rt} = 0.008$						
Reverse design 4 – BS (Reverse PTM – Forward AutoSRCbeam) – 100,000 dataset *16 Inputs (ϕM_n, ε_{rt}, SF, L, b, f_y, f_c', ρ_{rc}, h_s, b_s, t_f, t_w, f_{yS}, Y_s, M_D, M_L) – 10 Outputs (d, ρ_{rt}, ε_{st}, Δ_{imme}, Δ_{long}, μ_ϕ, CI_b, CO_2, BW, X_s)*						
			BS			
			Step 1	*Step 2*	*Verification by AutoSRCbeam*	*Error (%)*
No.	*Parameter*	*Training results*	*(1)*		*(2)*	*(3)*
17	d (mm)	(4 Layers – 50 Neurons) 13,518 epochs; *T.mse* = 3.1E-3; R = 0.9941	1,073.8	1,073.8	1,073.8	-
18	ρ_{rt}	(5 Layers – 40 Neurons) 13,269 epochs; *T.mse* = 2.0E-3; R = 0.9879	0.0089	0.0089	0.0089	-
19	ε_{st}		-	0.0071	-	-
20	Δ_{imme} (mm)		-	3.03	-	-
21	Δ_{long} (mm)		-	10.26	-	-
22	μ_ϕ		-	2.81	-	-
23	CI_b (Won/m)		-	225,852.1	-	-
24	CO_2 ($t\text{-}CO_2/m$)		-	0.402	-	-
25	*BW* (kN/m)		-	17.47	-	-
26	*Xs* (mm)		-	200	-	-

BS method procedure

Step 1 (*reverse analysis*): Calculating **2 reverse outputs** based on **13 ordinary inputs** and **3 reverse inputs**.

Step 2 (*forward analysis*): Calculating **8 rest outputs** using AutoSRCbeam based on **15 inputs** (substituting **2 reverse outputs** and **13 ordinary inputs**).

AutoSRCbeam check procedure

Calculating **3 outputs** based on **15 inputs** which are identical to **15 inputs** of Step 2 of BS.

Error calculation

$$(3) = \left((1) - (2)\right)/(1) \times 100\%$$

Table 5.4.4 (Continued) Table design for reverse scenario 4 with 16 input parameters and 10 output parameters based on reverse PTM – forward AutoSRCbeam with $\varepsilon_{rt} = 0.008$, $\varepsilon_{rt} = 0.012$ and $\varepsilon_{rt} = 0.020$, and ϕM_n (kN·m) = 3,500

(b) ϕM_n (kN·m) = 3,000, $\varepsilon_{rt} = 0.012$

Reverse design 2 – BS (Reverse PTM – Forward AutoSRCbeam) – 100,000 dataset
16 Inputs (ϕM_n, ε_{rt}, SF, L, d, b, f_y, f_c', ρ_{rt}, h_s, b_s, t_w, f_{yS}, Y_s, M_D, M_L) – 10 Outputs (ρ_{rc}, t_f, ε_{st}, Δ_{imme}, Δ_{long}, μ_ϕ, CI_b, CO_2, BW, X_s)

No.	Parameter	Training results	BS Step 1 (1)	BS Step 2	Verification by AutoSRCbeam (2)	Error (%) (3)
1	ϕM_n (kN·m)		3500.0	-	3495.6	0.13%
2	ε_{rt}		0.012	-	0.012	0.00%
3	SF		1.250	-	1.248	0.13%
4	L (mm)		10000	10000	10000	-
5	d (mm)		600	600	600	-
6	b (mm)		500	500	500	
7	f_y (MPa)		30	30	30	-
8	f_c' (MPa)		0.002	0.002	0.002	-
9	ρ_{rt}		500	500	500	-
10	h_s (mm)		200	200	200	-
11	b_s (mm)		8	8	8	-
12	t_w (mm)		6	6	6	-
13	f_{yS} (MPa)		325	325	325	-
14	Y_s (mm)		100	100	100	-
15	M_D (kN·m)		1000	1000	1000	-

(Continued)

Table 5.4.4 (Continued) Table design for reverse scenario 4 with 16 input parameters and 10 output parameters based on reverse PTM – forward AutoSRCbeam with $\varepsilon_{rt} = 0.008$, $\varepsilon_{rt} = 0.012$ and $\varepsilon_{rt} = 0.020$, and ϕM_n (kN·m) = 3,500

(b) ϕM_n (kN·m) = 3,000, $\varepsilon_{rt} = 0.012$						
Reverse design 2 – BS (Reverse PTM – Forward AutoSRCbeam) – 100,000 dataset 16 Inputs (ϕM_n, ε_{rt}, SF, L, d, b, f_y, f'_c, ρ_{rt}, h_s, b_s, t_w, f_{ys}, Y_s, M_D, M_L) – 10 Outputs (ρ_{rc}, t_f, ε_{st}, Δ_{imme}, Δ_{long}, μ_ϕ, CI_b, CO_2, BW, X_s)						
			BS		Verification by AutoSRCbeam	Error (%)
			Step 1	Step 2		
No.	Parameter	Training results	(1)		(2)	(3)
16	M_L (kN·m)		1000	1000	1000	-
17	ρ_{rc}	(5 Layers – 50 Neurons) 27,166 epochs; *T.mse* = 8.14E-4; R = 0.9952	1307.8	1307.8	1307.8	-
18	t_f (mm)	(5 Layers – 50 Neurons) 16,716 epochs; *T.mse* = 1.4E-2; R = 0.9814	0.0050	0.0050	0.0050	-
19	ε_{st}		-	0.0111	-	-
20	Δ_{imme} (mm)		-	2.04	-	-
21	Δ_{long} (mm)		-	5.96	-	-
22	μ_ϕ		-	4.09	-	-
23	CI_b (Won/m)		-	212045.3	-	-
24	CO_2 (t-CO_2/m)		-	0.383	-	-
25	BW (kN/m)		-	20.60	-	-
26	Xs (mm)		-	200	-	-

BS method procedure

Step 1 (*reverse analysis*): Calculating **2 reverse outputs** based on **13 ordinary inputs** and **3 reverse inputs**.

Step 2 (*forward analysis*): Calculating **8 rest outputs** using AutoSRCbeam based on **15 inputs** (substituting **2 reverse outputs** and **13 ordinary inputs**).

AutoSRCbeam check procedure

Calculating **3 outputs** based on **15 inputs** which are identical to **15 inputs** of Step 2 of BS.

Error calculation

$(3) = ((1) - (2))/(1) \times 100\%$

Table 5.4.4 (Continued) Table design for reverse scenario 4 with 16 input parameters and 10 output parameters based on reverse PTM – forward AutoSRCbeam with $\varepsilon_{rt} = 0.008$, $\varepsilon_{rt} = 0.012$ and $\varepsilon_{rt} = 0.020$, and ϕM_n (kN·m) = 3,500

(c) ϕM_n (kN·m) = 3,500, $\varepsilon_{rt} = 0.020$						
Reverse design 4 – BS (Reverse PTM – Forward AutoSRCbeam) – 100,000 dataset 16 Inputs (ϕM_n, ε_{rt}, SF, L, b, f_y, f'_c, ρ_{rc}, h_s, b_s, t_f, t_w, f_{yS}, Y_s, M_D, M_L) – 10 Outputs (d, ρ_{rt}, ε_{st}, Δ_{imme}, Δ_{long}, μ_ϕ, CI_b, CO_2, BW, X_s)						
			BS		Verification by AutoSRCbeam	Error (%)
			Step 1	Step 2		
No.	Parameter	Training results	(1)		(2)	(3)
1	ϕM_n (kN·m)		3,500.0	-	3,510.5	−0.30%
2	ε_{rt}		0.020	-	0.019	5.00%
3	SF		1.250	-	1.254	−0.30%
4	L (mm)		10,000	10,000	10,000	-
5	b (mm)		700	700	700	
6	f_y (MPa)		500	500	500	-
7	f'_c (MPa)		30	30	30	-
8	ρ_{rc}		0.0015	0.0015	0.0015	-
9	h_s (mm)		500	500	500	-
10	b_s (mm)		150	150	150	
11	t_f (mm)		6	6	6	-
12	t_w (mm)		5	5	5	-
13	f_{yS} (MPa)		325	325	325	-
14	Y_s (mm)		100	100	100	-
15	M_D (kN·m)		1,000	1,000	1,000	-
16	M_L (kN·m)		1,000	1,000	1,000	-

(Continued)

Table 5.4.4 (Continued) Table design for reverse scenario 4 with 16 input parameters and 10 output parameters based on reverse PTM – forward AutoSRCbeam with $\varepsilon_{rt} = 0.008$, $\varepsilon_{rt} = 0.012$ and $\varepsilon_{rt} = 0.020$, and ϕM_n (kN·m) = 3,500

(c) ϕM_n (kN·m) = 3,500, $\varepsilon_{rt} = 0.020$						
Reverse design 4 – BS (Reverse PTM – Forward AutoSRCbeam) – 100,000 dataset 16 Inputs (ϕM_n, ε_{rt}, SF, L, b, f_y, f_c', ρ_{rc}, h_s, b_s, t_f, t_w, f_{yS}, Y_s, M_D, M_L) – 10 Outputs (d, ρ_{rt}, ε_{st}, Δ_{imme}, Δ_{long}, μ_ϕ, CI_b, CO_2, BW, X_s)						
			BS		Verification by AutoSRCbeam	Error (%)
			Step 1	Step 2		
No.	Parameter	Training results	(1)		(2)	(3)
17	d (mm)	(4 Layers – 50 Neurons) 13,518 epochs; T.mse = 3.1E-3; R = 0.9941	1,426.8	1,426.8	1,426.8	-
18	ρ_{rt}	(5 Layers – 40 Neurons) 13,269 epochs; T.mse = 2.0E-3; R = 0.9879	0.0041	0.0041	0.0041	-
19	ε_{st}		-	0.0177	-	-
20	Δ_{imme} (mm)		-	1.40	-	-
21	Δ_{long} (mm)		-	3.32	-	-
22	μ_ϕ		-	6.30	-	-
23	CI_b (Won/m)		-	204,965.2	-	-
24	CO_2 (t-CO_2/m)		-	0.387	-	-
25	BW (kN/m)		-	25.73	-	-
26	Xs (mm)		-	275	-	-

BS method procedure

Step 1 (*reverse analysis*): Calculating **2 reverse outputs** based on **13 ordinary inputs** and **3 reverse inputs**.

Step 2 (*forward analysis*): Calculating **8 rest outputs** using AutoSRCbeam based on **15 inputs** (substituting **2 reverse outputs** and **13 ordinary inputs**).

AutoSRCbeam check procedure

Calculating **3 outputs** based on **15 inputs** which are identical to **15 inputs** of Step 2 of BS.

Error calculation

$$(3) = ((1) - (2))/(1) \times 100\%$$

safety factor SF, clearance between flange of H steel section and edge of beam X_s *(mm)* which would be calculated on an output-side the conventional forward designs. It is noted that *SF* and X_s (mm) are reversely preassigned as 1.43 and 120, respectively, as shown in Boxes 3 and 4 of Table 5.4.1(a)–(c). Reverse output parameters (widths of the concrete and steel sections (b and b_s)) shown in purple Boxes 18–19 of Table 5.4.1(a)–(c) are obtained on an output-side the reverse network of Step 1. The material properties of the concrete, rebar, and steel (f_c', f_y, and f_{yS}); concrete geometry (L and d); steel dimensions (h_s, t_f, and t_w); rebar ratios (ρ_{rt} and ρ_{rc}); design moment (ϕM_n); and lower clearance between a rebar layer and H-shaped steel section (Y_s) are ordinary input parameters shown in red parameters of Table 5.4.1(a)–(c) which are assigned on an input-side when they are predetermined by engineers.

5.5.1.2 Training in two steps for back-substitution method

In the first step of the BS method using the PTM, a shallow neural network (SNN) is used to map the 17 input parameters including three reverse input parameters shown in green (M_n, ε_{rt}, SF, X_s in Box 1–4 of Table 5.4.1(a)–(c)) and 13 ordinary input parameters shown in red parameters (L, d, f_y, f_c', ρ_{rt}, ρ_{rc}, h_s, t_f, t_w, f_{yS}, Y_s, M_D, M_L, in Box 5–17 of Table 5.4.1(a)–(c)) to the two reverse output parameters shown in purple parameters (b and b_s), in Box 18–19 of Table 5.4.1(a)–(c).

In Box 18 of Table 5.4.1(a)–(c) of the first step of the reverse network, one output parameter (beam width (b) in purple) is calculated by mapping17 input parameters shown in green and red Boxes 1–17 of Table 5.4.1(a)–(c) to b using PTM where four hidden layers and 50 neurons are used with 49,991 epochs, resulting in mean squared error (MSE) = 7.94 × 10^{-4} and Regression (R) index = 1.0 as shown in Table 5.4.1(a)–(c). In Box 19 of Table 5.4.1(a)–(c), another parameter (width of H-shaped steel section (b_s) in purple) is also calculated by mapping 17 input parameters shown in green and red Boxes 1–17 of Table 5.4.1(a)–(c) to b_s using PTM where five hidden layers and 50 neurons are used with 49,995 epochs, resulting in MSE = 1.71×10^{-5} and Regression (R) index = 0.9999 as shown in Table 5.4.1(a)–(c). In the second step of the forward network, structural mechanics (AutoSRCbeam software) calculates seven forward output parameters (ε_{st}, Δ_{imme}, Δ_{long}, μ_{ϕ}, CI_b, CO_2, BW, in brown) shown in Boxes 20–26 of Table 5.4.1(a)–(c) for given 15 input parameters (L, b, d, f_y, f_c', ρ_{rt}, ρ_{rc}, h_s, b_s, t_f, t_w, f_{yS}, Y_s, M_D, and M_L, in pink) shown in Boxes 5–19 of Table 5.4.1(a)–(c). Input parameters of Step 2 shown in the pink Boxes 5–19 of the Step 2 of Table 5.4.1(a)–(c) are identical to 13 ordinary inputs shown in red Boxes 5–17 of Table 5.4.1(a)–(c) and two reverse output parameters (b and b_s) obtained in the Step 1 shown in purple Boxes 18–19 of Table 5.4.1(a)–(c).

5.5.1.3 Verifications

To verify the BS network results, four forward output parameters (design moment (ϕM_n), tensile rebar strain (ε_{rt}), *SF*, and X_s in sky blue Boxes 1–4 of Table 5.4.1(a)–(c)) which were also preassigned as reverse input parameters shown in green Boxes 1–4 of Table 5.4.1(a)–(c) of Step 1 are re-calculated by structural mechanics (AutoSRCbeam software) using 15 parameters (L, d, b, f_y, f_c', ρ_{rt}, ρ_{rc}, h_s, b_s, t_f, t_w, f_{yS}, Y_s, M_D, and M_L) shown in blue Boxes 5–19 of Table 5.4.1(a)–(c) as input parameters. Three SRC beams are designed for preassigned design moments (ϕM_n) of 2,000, 3,000, and 4,000 kN·m based on a preassigned tensile rebar strain (ε_{rt}) fixed at 0.008 as presented in Table 5.4.1(a)–(c). As can be seen in Table 5.4.1(a)–(c) obtained for preassigned design moments (ϕM_n) of 2,000, 3,000, and 4,000 kN·m based on a preassigned tensile rebar strain (ε_{rt}) fixed at 0.008, insignificant errors of 0.74%, 0.16%, and −0.07% for design moments (ϕM_n) are observed, respectively, whereas maximum errors of 2.50%, 5.00%, and 6.25% for tensile rebar strain (ε_{rt}) are obtained, respectively.

Three SRC beams share the same material properties and geometry (L = 10,000 mm, d = 1,000 mm, f_y = 500 MPa, f_c' = 30 MPa, ρ_{rt} = 0.003, ρ_{rc} = 0.002, h_s = 600 mm, t_f = 15 mm, t_w = 6 mm, f_{yS} = 325 MPa, Y_s = 230 mm, M_D = 500 kN·m, and M_L = 500 kN·m) except for beam widths (b). In Table 5.4.1(a)–(c), the design moments are assigned as 2,000, 3,000, and 4,000 kN·m, calculating beam widths of 517.1, 744.1, and 970.7 mm, respectively. It is noted that some input parameters are adjusted for the three designs to avoid input conflicts among the reverse input parameters. The stronger SRC beams with increased beam widths lead to the reduction of deflections, immediate ($\Delta_{imme.}$), and long-term deflection (Δ_{long}). The beam widths of 517.1, 744.1, and 970.7 mm result in 2.40, 1.36, and 0.81 mm of immediate deflections, and 6.14, 3.10, and 2.77 mm of long-term deflections, respectively. Designs shown in Table 5.4.1(a)–(c) based on ANNs present tensile rebar strains (ε_{rt}) of 0.0078(2.50%), 0.0076(5.00%), and 0.0075(6.25%), respectively, leading to the curvature ductilities of μ_ϕ = 2.86, 2.79, and 2.76 for three SRC beams, respectively.

5.5.2 Reverse Scenario 2

5.5.2.1 An ANN formulation

In reverse scenario 2 shown in Table 5.2.1, reverse input parameters among the input parameters which would be calculated on an output-side the conventional forward designs are a design moment (ϕM_n), tensile rebar strain (ε_{rt}), and *SF* shown in green Boxes 1–3 of Table 5.4.2(a)–(c). It is noted that *SF* and X_s (mm) are reversely preassigned as 2.14 and 120, respectively, as shown in Boxes 3 and 4 of Table 5.4.2(a)–(c). Reverse output parameters, compressive rebar ratio (ρ_{rc}), and flange thickness of the steel section and (t_f), shown in purple Boxes 17–18 of Table 5.4.2(a)–(c) are obtained on an output-side the reverse network of Step 1. Material properties of the concrete, rebar, and steel (f_c', f_y, and f_{yS}); concrete geometry (L, d, and b); steel dimensions (h_s, b_s, and t_w); rebar ratio (ρ_{rt}); design moment (ϕM_n); and lower clearance between a rebar layer and H-shaped steel section (Y_s) are also assigned on an input-side as they are predetermined by engineers.

5.5.2.2 Training in two steps for back-substitution method

In the reverse networks of Step 1 for the BS method, an SNN is used to map 16 input parameters (ϕM_n, ε_{rt}, SF, L, d, b, f_y, f_c', ρ_{rt}, h_s, b_s, t_w, f_{yS}, Y_s, M_D, M_L) shown in green and red Boxes 1–16 of Table 5.4.2(a)–(c) to two reverse output parameters (ρ_{rc}, t_f) shown in purple Boxes 17–18 of Table 5.4.2(a)–(c) using PTM.

In Box 17 of Table 5.4.2(a)–(c) of the reverse networks of Step 1 for the BS method, one output parameter (ρ_{rc} in purple) is calculated by mapping 16 input parameters shown in green and red Boxes 1–16 of Table 5.4.2(a)–(c) to ρ_{rc} using PTM where five hidden layers and 50 neurons are implemented with 27,166 epochs, resulting in $MSE = 8.14\times10^{-4}$ and $R = 0.9952$. In Box 18 of Table 5.4.2(a)–(c), another parameter (flange thickness of H-shaped steel section (t_f) in purple) is also calculated by mapping 16 inputs shown in green and red Boxes 1–16 of Table 5.4.2(a)–(c) to t_f using PTM where five hidden layers and 50 neurons are implemented with 16,716 epochs, resulting in $MSE = 1.40\times10^{-2}$ and $R = 0.9814$.

In the forward networks of Step 2 for the BS method, structural mechanics (AutoSRCbeam software) calculate eight forward output parameters (ε_{st}, Δ_{imme}, Δ_{long}, μ_ϕ, CI_b, CO_2, BW, X_S) shown in brown Boxes 19–26 of Table 5.4.2(a)–(c) from 15 input parameters (L, b, d, f_y, f_c', ρ_{rt}, ρ_{rc}, h_s, b_s, t_f, t_w, f_{yS}, Y_s, M_D, and M_L) shown in the pink Boxes 4–18 of the Step 2 of Table 5.4.2(a)–(c) that are identical to 13 ordinary inputs shown in red parameters and two reverse output parameters (ρ_{rc} and t_f) shown in purple parameters from Step 1.

5.5.2.3 Verifications

To verify the BS network results, three forward output parameters [design moment (ϕM_n), tensile rebar strain (ε_{rt}), and SF] shown in sky blue Boxes 1–3 of Table 5.4.2(a)–(c) which were also preassigned as reverse input parameters shown in green Boxes 1–3 of Table 5.4.2(a)–(c) of Step 1 are calculated by structural mechanics (AutoSRCbeam software) using 15 input parameters (L, d, b, f_y, f_c', ρ_{rt}, ρ_{rc}, h_s, b_s, t_f, t_w, f_{yS}, Y_s, M_D, and M_L) shown in blue Boxes 4–18 of Table 5.4.2(a)–(c) as input parameters using PTM.

Three SRC beams are designed based on tensile rebar strains (ε_{rt}) of 0.009, 0.012, and 0.015 and a preassigned design moment (ϕM_n) fixed at 3,000 kN·m as shown in Table 5.4.2(a)–(c). Some input parameters are adjusted to avoid input conflicts among the reverse input parameters. Insignificant errors for the tensile rebar strains (ε_{rt}) of 1.11%, 2.50%, and 4.00%, respectively, are observed based on ANNs as indicated in Table 5.4.2(a)–(c) for preassigned tensile rebar strains (ε_{rt}) of 0.009, 0.012, and 0.015, respectively based on a preassigned design moment (ϕM_n) fixed at 3,000 kN·m.

The compressive rebar ratios (ρ_{rc}) are calculated as 0.00086, 0.0027, and 0.0038 as presented in Table 5.4.2(a)–(c), respectively, at a fixed tensile rebar ratio (ρ_{rt}) of 0.006. This demonstrates that ANNs tend to increase the compressive rebar ratio (ρ_{rc}) when the tensile rebar strain (ε_{rt}) is preassigned as 0.009, 0.012, and 0.015 as stated in Table 5.4.2(a)–(c), while the tensile rebar ratio (ρ_{rt}) is fixed. Note that the compressive rebar ratio (ρ_{rc}) is extrapolated in the ranges of large datasets where compressive rebar ratios (ρ_{rc} = 0.0038) for preassigned ε_{rt} of 0.015 are calculated greater than they are limited as half of the tensile rebar ratio (ρ_{rt} = 0.006). Observation also shows that tensile rebar strains (ε_{rt}) are 0.009, 0.012, and 0.015 with the curvature ductility (μ_ϕ) of 3.11, 3.95, and 4.76, respectively, resulting in beam depths (1,150, 1,150, and 1,200 mm) similar among three SRC beams. The beam deflections are also 2.88, 2.83, and 2.52 mm for immediate deflections and 8.01, 7.37, and 6.20 mm for long-term deflections corresponding to tensile rebar strains (ε_{rt}) of 0.009, 0.012, and 0.015, respectively. The compressive rebar ratios increase from 0.00086 to 0.0038 to meet design moments (ϕM_n) of 3,000 kN·m and *SF* of 1.2 when tensile rebar strains (ε_{rt}) increase from 0.009 to 0.015 as shown in three SRC beams of Table 5.4.2(a) and (c). It is noted that ANNs well predict the structural behavior of the steel-concrete composite beams having 26 parameters.

5.5.3 Reverse Scenario 3

5.5.3.1 An ANN formulation

In reverse scenario 3 shown in Table 5.2.1, reverse input parameters among the input parameters which would be calculated on an output-side the conventional forward designs are a design moment (ϕM_n), tensile rebar strain (ε_{rt}), and *SF* shown in green Boxes 1–3 of Table 5.4.3(a)–(c). It is noted that *SF* and X_s (mm) are reversely preassigned as 2.86 and 120, respectively, as shown in Boxes 3 and 4 of Table 5.4.3(a)–(c). Reverse output parameters, tensile and compressive rebar ratios (ρ_{rt} and ρ_{rc}), shown in purple Boxes 17–18 of Table 5.4.3(a)–(c) are obtained on an output-side the reverse network of Step 1. Material properties of the concrete, rebar, and steel (f_c', f_y, and f_{yS}); concrete geometry (L, d, and b); steel dimensions (h_s, b_s, t_f, and t_w); design moments (ϕM_n); and lower clearance between a rebar layer and H-shaped steel section (Y_s) are assigned as inputs as they are predetermined by engineers.

5.5.3.2 Training in two steps for back-substitution method

In the reverse networks of Step 1 for the BS method, an SNN is used to map 16 input parameters (ϕM_n, ε_{rt}, SF, L, d, b, f_y, f_c' h_s, b_s, t_f, t_w, f_{yS}, Y_s, M_D, and M_L) shown in green and red Boxes 1–16 of Table 5.4.3(a)–(c) to two reverse output parameters (ρ_{rt} and ρ_{rc}) shown in purple Boxes 17–18 of Table 5.4.3(a)–(c) using PTM. In Box 17 of Table 5.4.3(a)–(c) of the reverse networks of Step 1 for the BS method, one output parameter (ρ_{rt} in purple) is calculated by mapping16 input parameters shown in green and red Boxes 1–16 of Table 5.4.3(a)–(c) to ρ_{rt} using PTM where five hidden layers and 50 neurons were used with 27,150 epochs, resulting in $MSE = 6.07\times10^{-4}$ and $R = 0.9996$. In Box 18 of Table 5.4.3(a)–(c), another parameter (ρ_{rc} in purple) is also calculated by mapping16 inputs shown in green and red Boxes 1–16 of Table 5.4.3(a)–(c) to ρ_{rc} using PTM where five hidden layers and 50 neurons were implemented with 15,685 epochs, resulting in $MSE = 1.40\times10^{-3}$ and $R = 0.9917$. In the forward networks of Step 2 for the BS method, structural mechanics (AutoSRCbeam software) calculate eight forward output parameters (ε_{st}, Δ_{imme}, Δ_{long}, μ_{ϕ}, CI_b, CO_2, BW, and X_s) shown in brown Boxes 19–26 of Table 5.4.3(a)–(c) from 15 input parameters (L, b, d, f_y, f_c', ρ_{rt}, ρ_{rc}, h_s, b_s, t_f, t_w, f_{yS}, Y_s, M_D, and M_L) shown in the pink Boxes 4–18 of of the Step 2 of Table 5.4.3(a)–(c) that are identical to 13 ordinary inputs shown in red parameters and two reverse output parameters (ρ_{rc} and t_f) shown in purple parameters from Step 1.

5.5.3.3 Verifications

To verify the BS network results, three forward output parameters (design moment (ϕM_n), tensile rebar strain (ε_{rt}), and SF) shown in sky blue Boxes 1–3 of Table 5.4.3(a)–(c)) which were also preassigned as reverse input parameters shown in green Boxes 1–3 of Table 5.4.3(a)–(c) of Step 1 are calculated by structural mechanics (AutoSRCbeam software) using 16 input parameters (L, d, b, f_y, f_c', ρ_{rt}, ρ_{rc}, h_s, b_s, t_f, t_w, f_{yS}, Y_s, M_D, and M_L) shown in blue Boxes 4–18 of Table 5.4.3(a)–(c) as input parameters using PTM.

Three SRC beams are designed for tensile rebar strains (ε_{rt}) of 0.008, 0.010, and 0.012 and a preassigned design moment (ϕM_n) fixed at 3,500 kN·m as shown in Table 5.4.3(a)–(c). Some input parameters are adjusted to avoid input conflicts among the reverse input parameters. Insignificant errors of 0.00%, 3.00%, and 5.83% for tensile rebar strains, ε_{rt}, are observed based on ANNs corresponding to tensile rebar strain (ε_{rt}) of 0.008, 0.010, and 0.012, respectively, as indicated in Table 5.4.3(a)–(c).

Figure 5.5.1 shows the distribution of tensile rebar strains (ε_{rt}) corresponding to a compressive concrete strain (ε_c) of 0.003. No considerable errors occur in the sparse data zone where the tensile rebar strain ε_{rt} is greater than 0.01 because ANNs can perform some limited extrapolation if reverse input parameters are selected appropriately even when inputs and outputs lie outside the datasets. However, input conflicts are often caused by weak capability with extrapolations, resulting in significant errors when reverse input parameters are inappropriately selected.

Observation shows that tensile rebar ratios (ρ_{rt}) increase slightly 0.0083, 0.0087, and 0.0088 in three SRC beams, whereas compressive rebar ratios (ρ_{rc})

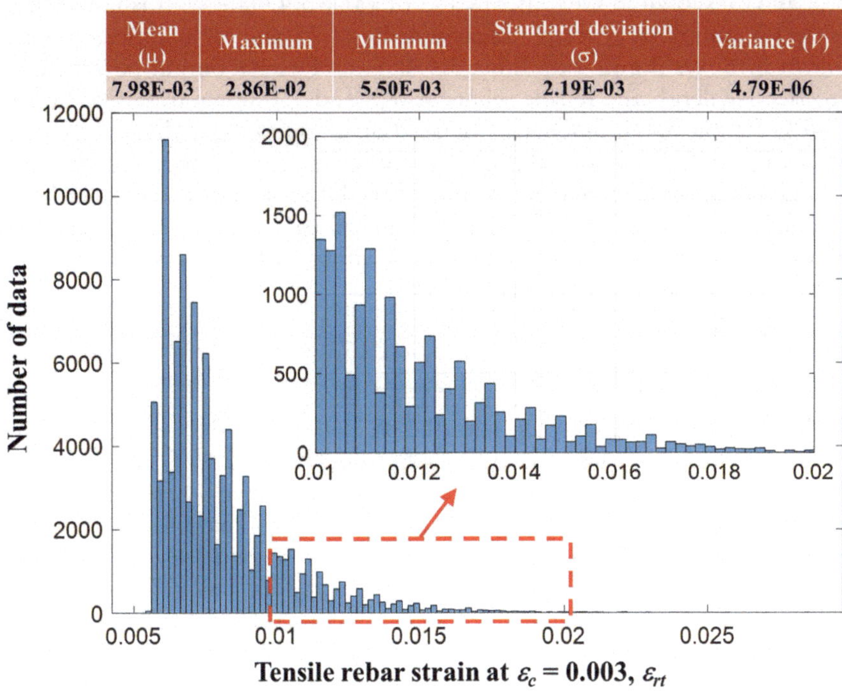

Mean (μ)	Maximum	Minimum	Standard deviation (σ)	Variance (V)
7.98E-03	2.86E-02	5.50E-03	2.19E-03	4.79E-06

Figure 5.5.1 Distribution of tensile rebar strain (ε_{rt}).

are obtained as 0.005, 0.0035, and 0.006 when tensile rebar strains increase to 0.008, 0.01, and 0.012, respectively, when the target design moment is set at $\phi M_n = 3{,}500$ kN·m. It is worth noting that deflections of immediate and long term of the SRC beam shown in Table 5.4.3(c) corresponding to a tensile rebar strain $\varepsilon_{rt} = 0.012$ are the smallest (Δ_{imme}= 2.91 mm, and Δ_{long}= 9.01 mm due to the largest compressive rebar ratio of $\rho_{rc} = 0.006$ even if the smallest H-shaped steel section ($h_s \times b_f \times t_f \times t_w = 500 \times 150 \times 8 \times 6$ mm) is used. The immediate and long-term deflections of the SRC beam shown in Table 5.4.3(b) corresponding to tensile rebar strain $\varepsilon_{rt} = 0.010$ are the biggest (Δ_{imme}= 2.98 mm, and Δ_{long}= 9.92 mm) due to the least of compressive rebar ratio ($\rho_{rc} = 0.0035$). Curvature ductilities of $\mu_\phi = 3.36$, 2.84, and 3.81 are obtained in three SRC beams for compressive rebar ratios of 0.005, 0.0035, and 0.006 as shown in Table 5.4.3(a)–(c).

5.5.4 Reverse Scenario 4

5.5.4.1 An ANN formulation

In reverse scenario 4 shown in Table 5.2.1, reverse input parameters among the input parameters which would be calculated on an output-side the conventional forward designs are a design moment (ϕM_n), tensile rebar strain

(ε_{rt}), and SF shown in green Boxes 1–3 of Table 5.4.4(a)–(c). It is noted that SF and X_s (mm) are reversely preassigned as 1.25 and 120, respectively, as shown in Boxes 3 and 4 of Table 5.4.4(a)–(c). Reverse output parameters, an effective depth and tensile rebar ratio (d and ρ_{rt}) shown in purple Boxes 17–18 of Table 5.4.4(a)–(c), are obtained on an output-side the reverse network of Step 1. The material properties of the concrete, rebar, and steel (f_c', f_y, and f_{yS}); concrete geometry (L and b); steel dimensions (h_s, b_s, t_f, and t_w); compressive rebar ratio (ρ_{rc}); design moments (ϕM_n); and lower clearance between a rebar layer and H-shaped steel section (Y_s) are assigned as inputs as they are predetermined by engineers.

5.5.4.2 Training in two steps for back-substitution method

In the reverse networks of Step 1 for the BS method, an SNN is used to map 16 input parameters (ϕM_n, ε_{rt}, SF, L, b, f_y, f_c', ρ_{rc}, h_s, b_s, t_f, t_w, f_{yS}, Y_s, M_D, M_L) shown in green and red Boxes 1–16 of Table 5.4.4(a)–(c) to two reverse output parameters (d and ρ_{rt}) shown in purple Boxes 17–18 of Table 5.4.4(a)–(c) using PTM. In Box 17 of Table 5.4.4(a)–(c) of the reverse networks of Step 1 for the BS method, one output parameter (d in purple) is calculated by mapping 16 input parameters shown in green and red Boxes 1–16 of Table 5.4.4(a)–(c) to d using PTM where four hidden layers and 50 neurons were implemented with 13,518 epochs with $MSE = 3.10\times10^{-3}$ and $R = 0.9941$. In Box 18 of Tables 5.4.4(a)–(c), another parameter (ρ_{rt} in purple) is also calculated by mapping 16 inputs shown in green and red Boxes 1–16 of Table 5.4.4(a)–(c) to ρ_{rt} using PTM where five hidden layers and 40 neurons were used with 13,269 epochs, yielding $MSE = 2.00\times10^{-3}$ and $R = 0.9879$. In the forward networks of Step 2 for the BS method, structural mechanics (AutoSRCbeam software) calculate eight forward output parameters (ε_{st}, Δ_{imme}, Δ_{long}, μ_ϕ, CI_b, CO_2, BW, and X_s) shown in brown Boxes 19–26 of Tables 5.4.4(a)–(c) from 15 input parameters (L, b, d, f_y, f_c', ρ_{rt}, ρ_{rc}, h_s, b_s, t_f, t_w, f_{yS}, Y_s, M_D, and M_L) shown in the pink Boxes 4–18 of the Step 1 of Table 5.4.4(a)–(c) that are identical to 13 ordinary inputs shown in red parameters and two reverse output parameters (d and ρ_{rt}) shown in purple parameters from Step 1.

5.5.4.3 Verifications

To verify the BS network results, three forward output parameters (design moment (ϕM_n), tensile rebar strain (ε_{rt}), and SF) shown in sky blue Boxes 1–3 of Table 5.4.4(a)–(c) which were also preassigned as reverse input parameters shown in green Boxes 1–3 of Table 5.4.4(a)–(c) of Step 1 are calculated by structural mechanics (AutoSRCbeam software) using 15 input parameters (L, d, b, f_y, f_c', ρ_{rt}, ρ_{rc}, h_s, b_s, t_f, t_w, f_{yS}, Y_s, M_D, and M_L) shown in blue Boxes 4–18 of Table 5.4.4(a)–(c) as input parameters. Three SRC beams are designed for tensile rebar strains (ε_{rt}) of 0.008, 0.012, and 0.02, and a preassigned design moment (ϕM_n) fixed at 3,500 kN·m as shown in Table 5.4.4(a)–(c). Some input parameters are adjusted to avoid input conflicts among the reverse input parameters. As shown in Table 5.4.4(a) and (b), tensile rebar strains (ε_{rt}) identical to the preassigned tensile rebar strains (ε_{rt}) of 0.008 and 0.012 are obtained for the first two designs, respectively. An insignificant error of 5.00% is also observed for the preassigned tensile rebar strain (ε_{rt}) of 0.020 as obtained in Table 5.4.4(c).

The calculated tensile rebar ratios (ρ_{rt}) are 0.0089, 0.0050, and 0.0041 corresponding to tensile rebar strains (ε_{rt}) of 0.008, 0.012, and 0.020 as indicated in the Step 1 of Table 5.4.4(a)–(c). These are more than twice the preassigned compressive rebar ratios (ρ_{rc}) of 0.004, 0.002, and 0.0015 as shown in Table 5.4.4(a)–(c), respectively. The compressive rebar ratios (ρ_{rc}) are limited to half the tensile rebar ratios (ρ_{rt}) when large datasets are generated. It is worth noting that increasing tensile rebar strains (ε_{rt}) in three SRC beam designs (0.008, 0.012, and 0.02) decreases tensile rebar ratios (ρ_{rt} = 0.0089, 0.0050, and 0.0041).

The least tensile rebar strain (ε_{rt} = 0.008) of the SRC beam shown in Table 5.4.4(a) results in the least beam depth (d = 1073.8 mm), whereas the largest rebar strain (ε_{rt} = 0.019) of the SRC beam shown in Table 5.4.4(c) results in the largest beam depth (d = 1,426.8 mm). Beam widths are 600, 600, and 700, whereas depths are 1,073.8, 1,307.8, and 1,426.8 based on ANNs corresponding to tensile rebar strains (ε_{rt}) of 0.008, 0.012, and 0.020 as shown in Table 5.4.4(a)–(c), respectively, resulting in decreasing beam deflections, Δ_{imme}= 3.03, 2.04, and 1.40 mm; Δ_{long}= 10.26, 5.96, and 3.32 mm for Table 5.4.4(a)–(c), respectively. Increasing tensile rebar strains (ε_{rt}) in three SRC designs (0.008, 0.012, and 0.02) also increases the curvature ductility (μ_{ϕ} = 2.81, 4.09, and 6.30).

Errors shown in Table 5.4.4(a)–(c) are insignificant for the preassigned input parameters such as the design moment (ϕM_n) of 3,500 kN·m. This is because preassigned input parameters are well selected with tensile rebar strains (ε_{rt}) of 0.008, 0.012, and 0.020, which eliminates input conflicts. The ANN can perform some limited extrapolation when reverse input parameters are selected appropriately even when design parameters are preassigned outside the datasets. The ANN shows some extrapolation capability when input conflicts are avoided. However, network accuracies become weak in general when design parameters preassigned too far outside the training datasets are used, such as a tensile rebar strain (ε_{rt}) of 0.020. This is because input conflicts are often caused by weak extrapolation capability, which results in significant errors when reverse input parameters are inappropriately selected.

5.5.5 Development of design charts

5.5.5.1 Design charts for reverse scenario 4

Design charts for reverse scenario 4 are developed using Table 5.4.4 obtained for three design moments (ϕM_n = 3,000, 3,500, and 4,000 kN·m) and tensile rebar strains (ε_{rt}) which vary between 0.0055 and 0.012. Design charts shown in Figures 5.5.2–10 are constructed for SRC beams having design parameters: $M_D = 1{,}000$ kN·m, $M_L = 1{,}000$ kN·m, $L = 10{,}000$ mm, $b = 600$ mm, $f_y = 500$ MPa, $f_c' = 30$ MPa, $b_s = 150$ mm, $h_s = 600$ mm, $t_f = 10$ mm, $t_w = 8$ mm, $f_{yS} = 325$ MPa, $X_s = 225$ mm, and $Y_s = 100$ mm. Table 5.4.4 is repeatedly obtained by varying tensile rebar strains (ε_{rt}) from 0.0055 to 0.012 for each of three design moments ($\phi M_n = 3{,}000$, 3,500, and 4,000 kN·m), respectively, resulting in design parameters including design moment (ϕM_n), tensile rebar strain (ε_{rt}), and SF, d, ρ_{rt}, ε_{st}, μ_ϕ, Δ_{imme}, Δ_{long}, CI_b, CO_2, and BW. Design charts shown in Figures 5.5.2–10 are then plotted using the design parameters. Sequence of determining design parameters can be in any order, and hence, engineers can determine them according to their design needs.

5.5.5.2 A design moment (ϕM_n)

In Figure 5.5.2, a design moment (ϕM_n) obtained by AutoSRCbeam using parameters that ANNs identified is shown in sky blue Box 1 of Table 5.4.4(a)–(c) and given in the left ordinate: AutoSRCbeam, while the error between the AutoSRCbeam-based design moment (ϕM_n) using parameters that ANNs identified, shown in sky blue Box 1 of Table 5.4.4 and preassigned ϕM_n shown in green Box 1 of Table 5.4.4 is given in the right ordinate, demonstrating that a good accuracy of the design moment (ϕM_n) was obtained by the reverse network. All errors of design moment (ϕM_n) are less than 4% with respect to the preassigned tensile rebar strains (ε_{rt}) of 0.0055–0.012.

5.5.5.3 A tensile rebar strain (ε_{rt})

Figure 5.5.3 shows an accuracy of the tensile rebar strains (ε_{rt}) corresponding to a concrete strain of 0.003. The tensile rebar strain (ε_{rt}) of the network prediction which is verified by AutoSRCbeam is shown in sky blue Box 1 of Table 5.4.4(a)–(c) and given in the left ordinate: AutoSRCbeam is used to calculate a design moment (ϕM_n) based on parameters that ANNs identified, while errors between the AutoSRCbeam-based design moment (ϕM_n) using parameters that ANNs identified shown in sky blue Box 1 of Table 5.4.4 and preassigned ε_{rt} shown in green Box 1 of Table 5.4.4 is given in the right ordinate. The errors of the tensile rebar strains (ε_{rt}) with respect to all design moments ($\phi M_n = 3{,}000$, 3,500, and 4,000 kN·m) are in a range between −6% and 3% for $\varepsilon_{rt} = 0.0055$–$0.012$.

5.5.5.4 A safety factor (SF)

Figure 5.5.4 shows an accuracy of an SF corresponding to a concrete strain of 0.003 similarly to a design moment (ϕM_n) and tensile rebar strain (ε_{rt}).

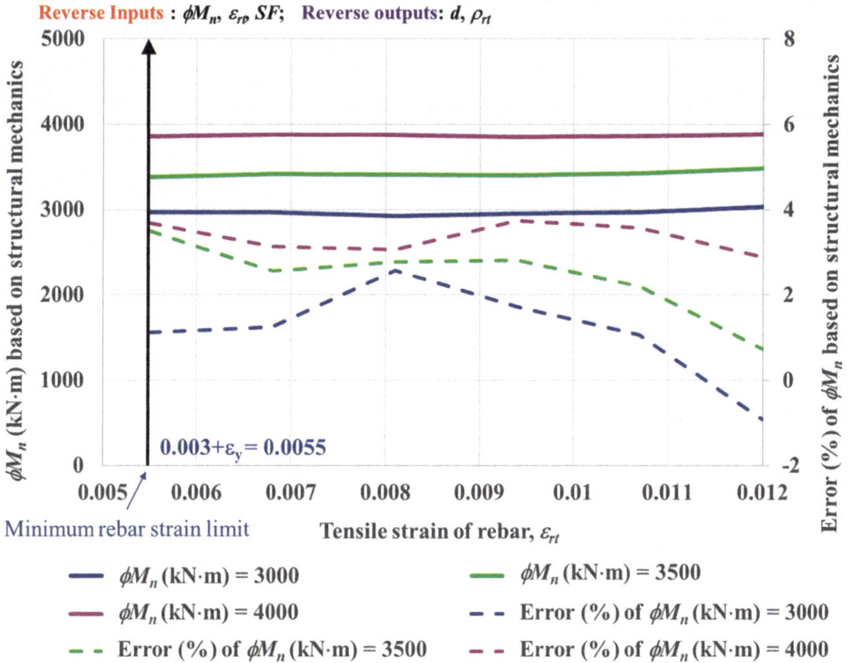

Figure 5.5.2 Verification of design moments (ϕM_n).

The maximum error of an SF is less than 4% with respect to all design moments (ϕM_n = 3,000, 3,500, and 4,000 kN·m) for ε_{rt} = 0.0055–0.012.

5.5.5.5 *Effective depths (d) and tensile rebar ratios (ρ_{rt}) vs. tensile rebar strains (ε_{rt})*

Figure 5.5.5 shows how effective depth (d) calculated in Box 17 of Table 5.4.4 and tensile rebar ratios (ρ_{rt}) calculated in Box 18 of Table 5.4.4 vary against a tensile rebar strain (ε_{rt}) corresponding to a concrete strain of 0.003. Effective depth (d) is plotted in the direction opposite to a tensile rebar ratio (ρ_{rt}) to keep an equilibrium of beam sections with respect to design moments (ϕM_n = 3,000, 3,500, and 4,000 kN·m) as shown in

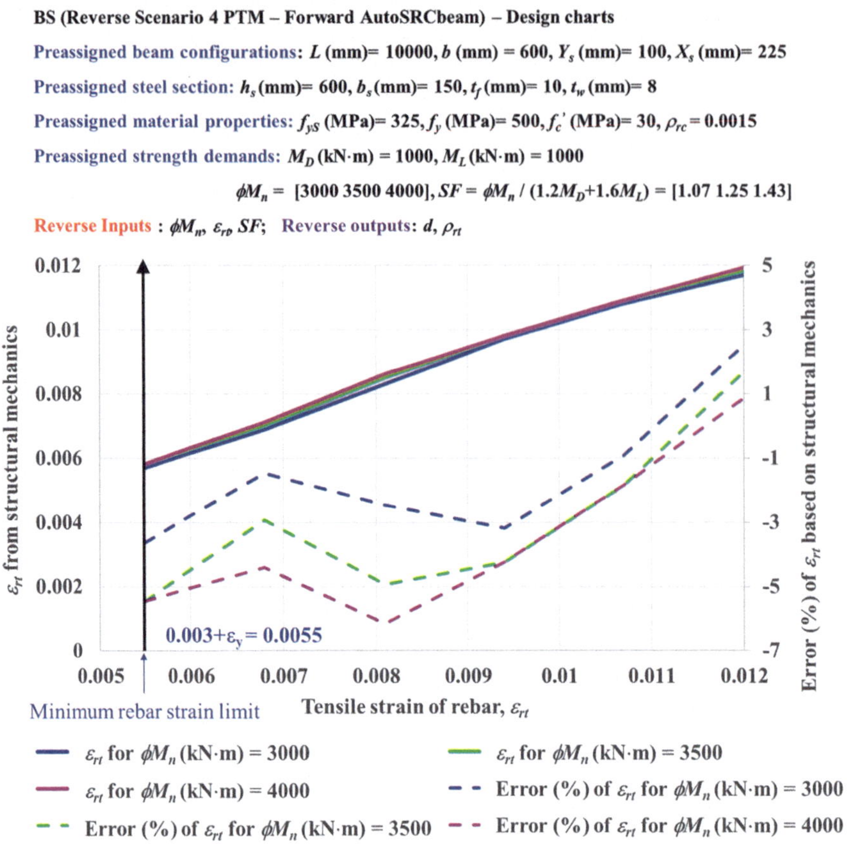

BS (Reverse Scenario 4 PTM – Forward AutoSRCbeam) – Design charts

Preassigned beam configurations: L (mm)= 10000, b (mm) = 600, Y_s (mm)= 100, X_s (mm)= 225

Preassigned steel section: h_s(mm)= 600, b_s(mm)= 150, t_f(mm)= 10, t_w(mm)= 8

Preassigned material properties: f_{yS} (MPa)= 325, f_y (MPa)= 500, f_c' (MPa)= 30, ρ_{rc} = 0.0015

Preassigned strength demands: M_D (kN·m) = 1000, M_L (kN·m) = 1000

ϕM_n = [3000 3500 4000], $SF = \phi M_n / (1.2M_D+1.6M_L)$ = [1.07 1.25 1.43]

Reverse Inputs : ϕM_n, ε_{rt}, SF; **Reverse outputs:** d, ρ_{rt}

0.003+ε_y = 0.0055

Minimum rebar strain limit

Tensile strain of rebar, ε_{rt}

ε_{rt} from structural mechanics

Error (%) of ε_{rt} based on structural mechanics

— ε_{rt} for ϕM_n (kN·m) = 3000
— ε_{rt} for ϕM_n (kN·m) = 3500
— ε_{rt} for ϕM_n (kN·m) = 4000
- - Error (%) of ε_{rt} for ϕM_n (kN·m) = 3000
- - Error (%) of ε_{rt} for ϕM_n (kN·m) = 3500
- - Error (%) of ε_{rt} for ϕM_n (kN·m) = 4000

Figure 5.5.3 Verification of tensile rebar strain (ε_{rt}) corresponding to ε_c = 0.003.

Figure 5.5.5. Effective depths (d) increase proportional to tensile rebar strains (ε_{rt}), whereas tensile rebar ratios (ρ_{rt}) decrease as tensile rebar strains (ε_{rt}) increase.

5.5.5.6 *Tensile steel strains (ε_{st}) and curvature ductilities (μ_ϕ) vs. tensile rebar strains (ε_{rt})*

Figure 5.5.6 shows how tensile steel strains (ε_{st}) calculated in Box 19 of Table 5.4.4 and curvature ductilities (μ_ϕ) calculated in Box 22 of Table 5.4.4 vary against tensile rebar strains (ε_{rt}) corresponding to a concrete strain of 0.003. Tensile steel strains (ε_{st}) increase almost linearly from 0.0045 to 0.011 with tensile rebar strains between ε_{rt} = 0.0055 and 0.012 based on strain compatibility. Besides, curvature ductilities (μ_ϕ) increase from 2.15 to 4.08 for ε_{rt} = 0.0055 to 0.012 and they had clustered as shown in Figure 5.5.6.

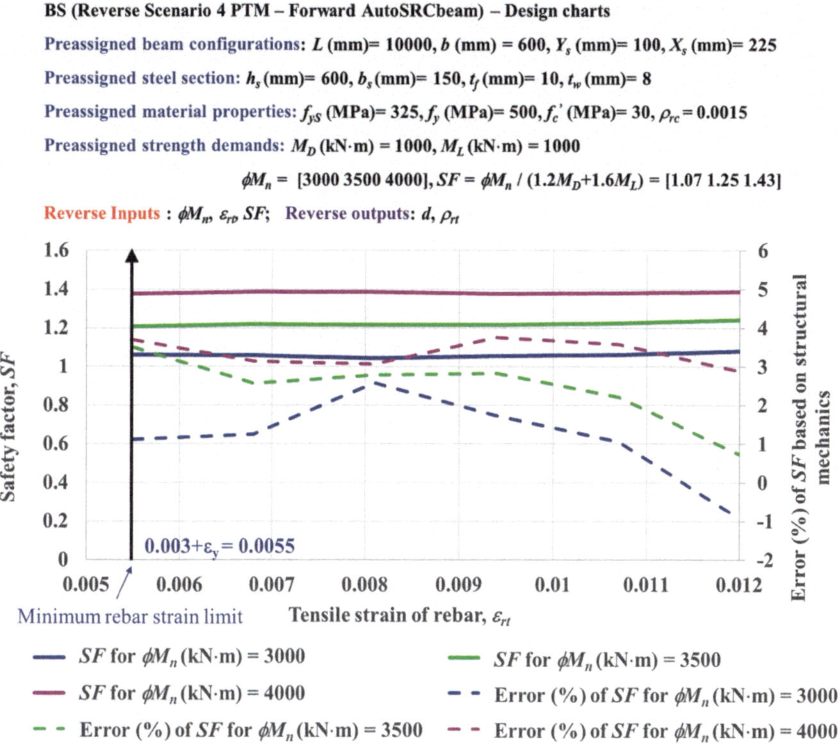

Figure 5.5.4 Verification of SF corresponding to $\varepsilon_c = 0.003$.

5.5.5.7 *Immediate and long-term deflections (Δ_{imme} and Δ_{long}) vs. tensile rebar strains (ε_{rt})*

Figure 5.5.7 shows the immediate and long-term deflections (Δ_{imme} calculated in Box 20 of Table 5.4.4 and Δ_{long} calculated in Box 21 of Table 5.4.4) as functions of the tensile rebar strain (ε_{rt}) at a concrete strain of 0.003. Immediate deflections (Δ_{imme}) decrease slightly between ε_{rt} = 0.0055 and 0.012 as shown in Figure 5.5.7 because effective depths (d) shown in Figure 5.5.5 increase, when tensile rebar strains (ε_{rt}) increase from 0.0055 to 0.012. Long-term deflections (Δ_{long}) decrease more rapidly than immediate deflections (Δ_{imme}) when effective depths (d) shown in Figure 5.5.5 increase because the effect of increasing effective depths (d) shown in Figure 5.5.5 on long-term deflections (Δ_{long}) is more influential than on immediate deflections, indicating that tensile rebar ratios (ρ_{rt}) shown in Figure 5.5.5 reduce long-term deflections (Δ_{long}) more than immediate deflections. Long-term deflections (Δ_{long}) also decrease when compressive rebar ratios (ρ_{rc}) increase. Long-term effect factor of 1.86 ($\lambda_D = 2/(1 + 50 \times \rho_{rc}) = 1.86$) is calculated based on compressive rebar ratios (ρ_{rc}) of 0.0015 shown in Figure 5.5.7.

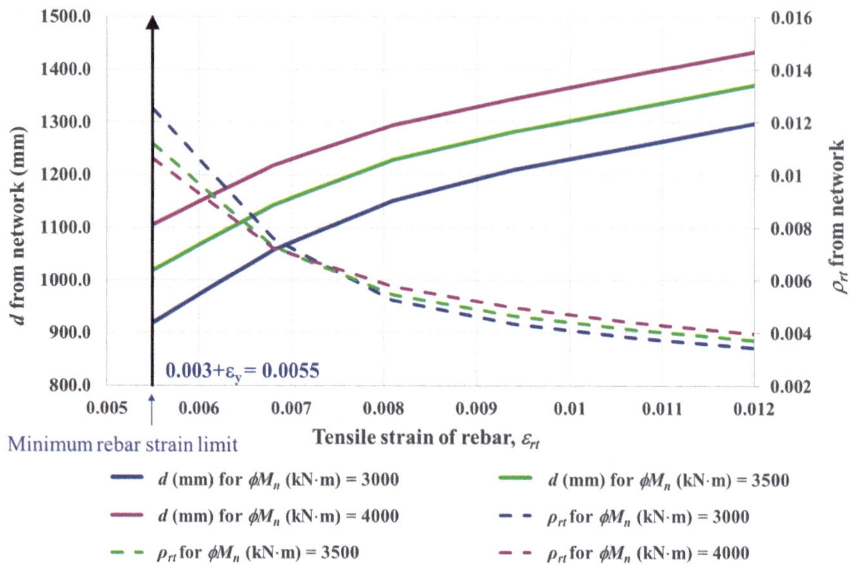

Figure 5.5.5 Design chart for determining d and ρ_{rt} as a function of tensile rebar strains (ε_{rt}) corresponding to a concrete strain of 0.003.

5.5.5.8 A cost index of an SRC beam (CI_b), and CO_2 emissions, and beam weight (BW) vs. tensile rebar strains (ε_{rt})

As shown in Figure 5.5.8, a cost index of an SRC beam (CI_b) calculated in Box 23 of Table 5.4.4 and CO_2 emissions calculated in Box 24 of Table 5.4.4 decrease as tensile rebar strains (ε_{rt}) increase from 0.0055 to 0.012. This is because that effective depths (d) increase and tensile rebar ratios (ρ_{rt}) decrease as tensile rebar strains (ε_{rt}) increase as shown in Figure 5.5.5. A cost index of an SRC beam (CI_b) and CO_2 emissions decrease rapidly when ε_{rt} = 0.0055–0.008, whereas they decrease slightly when ε_{rt} = 0.008–0.012. This is due to tensile rebar ratios (ρ_{rt}) which decrease rapidly when ε_{rt} = 0.0055–0.008 and slightly decrease when ε_{rt} = 0.008–0.012. A decrease in tensile rebar ratios (ρ_{rt}) had a larger influence than an increase in effective depth (d) on a decrease in the amount of materials to reduce a cost

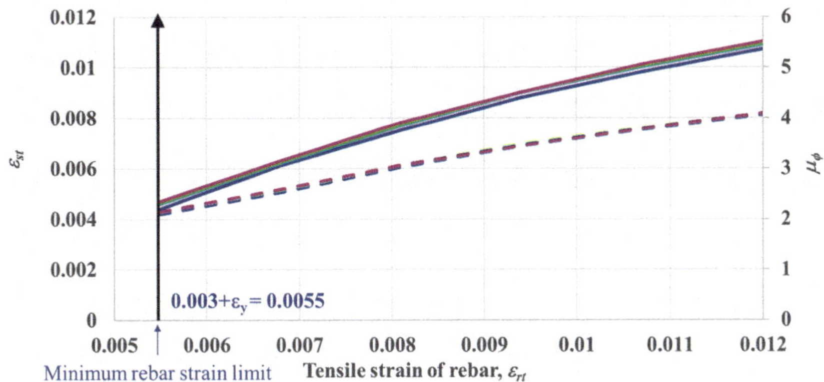

Figure 5.5.6 Design chart for determining ε_{st} and μ_ϕ as a function of tensile rebar strains (ε_{rt}) corresponding to a concrete strain of 0.003.

index (CI_b) and CO_2 emissions as indicated in Figure 5.5.8. In Figure 5.5.9, *BW* calculated in Box 25 of Table 5.4.4 increase rapidly when tensile rebar strains (ε_{rt}) increase from 0.0055 to 0.008 because effective depths (d) increase rapidly in this rebar strain range. However, *BW* increase gradually for ε_{rt} = 0.008 to 0.012 due to effective depths (*d*) which increase gradually in this rebar strain range as shown in Figure 5.5.5. An increase in effective depths (*d*) requiring more concrete volume to increase *BW* had a greater influence than a decrease in tensile rebar ratios (ρ_{rt}) on raising a concrete volume to increase *BW*. Preassigning design parameters on an input-side the design charts can cause input conflicts, and hence, preassigning design parameters appropriately on an input-side the design charts is important to reduce errors when using design chart.

5.5.6 Application of design charts

Figures 5.5.2–9 present graphical design charts developed based on reverse designs shown in Table 5.4.4 comprising ten output parameters (d, ρ_{rt}, ε_{st},

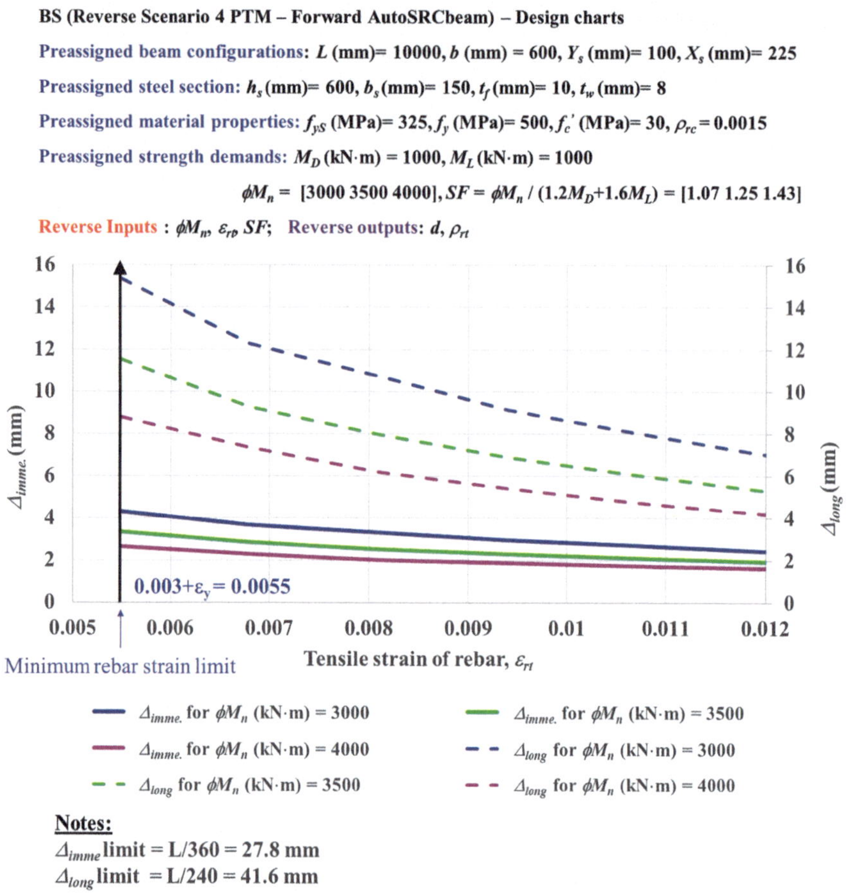

Figure 5.5.7 Design chart for determining Δ_{imme} and Δ_{long} as a function of tensile rebar strains (ε_{rt}) corresponding to a concrete strain of 0.003.

Δ_{imme}, Δ_{long}, μ_ϕ, CI_b, CO_2, BW, and X_s) and 16 input parameters (ϕM_n, ε_{rt}, SF, L, b, f_y, f_c', ρ_{rc}, h_s, b_s, t_f, t_w, f_{yS}, Y_s, M_D, and M_L). Design charts shown in Figures 5.5.2–9 assist engineers to design SRC beams. SRC beams can be designed when ϕM_n, μ_ϕ, and SF are preassigned as input parameters on an input-side. This type of design is challenging using conventional methods.

Designing SRC beams meeting design moment (ϕM_n) of 3,200 kN·m is shown in Figure 5.5.10. As shown in Steps 1–4 in Figure 5.5.10(a), tensile rebar strain (ε_{rt}) and tensile steel strain (ε_{st}) are determined to be 0.0096 and 0.0091 for design moment (ϕM_n) of 3,200 kN·m, respectively, when a curvature ductility (μ_ϕ) is specified as 3.5. The SF is selected as 1.143 for tensile rebar strain (ε_{rt}) of 0.0096 and design moment (ϕM_n) of 3,200

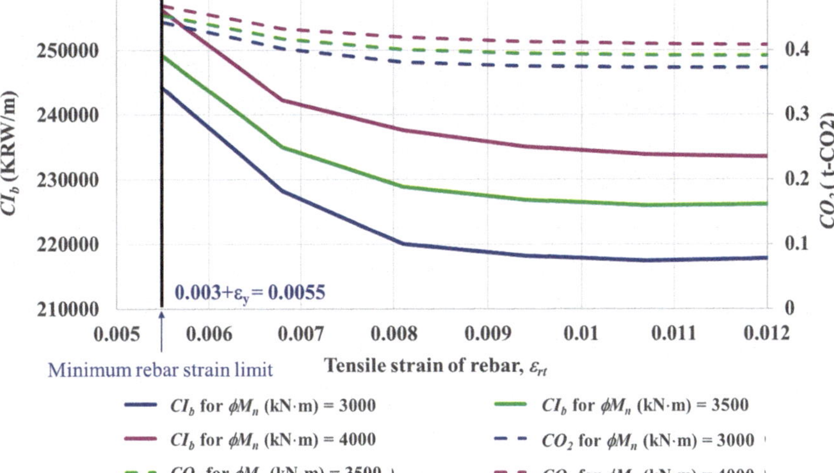

Figure 5.5.8 Design chart for determining CI_b and CO_2 as a function of tensile rebar strains (ε_{rt}) corresponding to a concrete strain of 0.003.

kN·m as shown in Figure 5.5.4, and hence, an *SF*, curvature ductility (μ_ϕ) and design moment (ϕM_n) of Figure 5.5.10 are selected as 1.143, 3.5, and 3,200 kN·m, respectively. As shown in Steps 5–8 in Figure 5.5.10(b), the effective depth (d) and tensile rebar ratio (ρ_{rt}) are determined as 1,247 mm and 0.0042, respectively, based on $\varepsilon_{rt} = 0.0096$ and design moment (ϕM_n) of 3,200 kN·m.

The immediate deflection (Δ_{imme}) of 2.60 mm and long-term deflection (Δ_{long}) of 8.00 mm are then obtained as shown in Steps 9–12 in Figure 5.5.10(c). The cost index of the SRC beam (CI_b) of 222,000 KRW/m, CO_2 emissions of 0.380 t-CO_2/m, and *BW* of 19.80 kN/m are also obtained based on Steps 13–16 in Figure 5.5.10(d) and based on Steps 17–18 in Figure 5.5.10(e).

Two steps of Figure 5.5.11 summarize the design of an SRC beam with preassigned $\phi M_n = 3{,}200$ kN·m, $\mu_\phi = 3.5$, and $SF = 1.143$. Section A–A determined based on the ANN-based reverse design charts shows an immediate

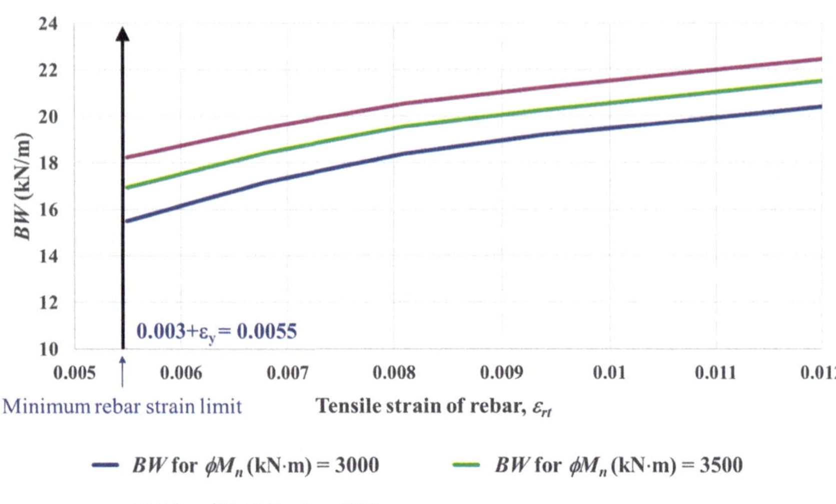

Figure 5.5.9 Design chart for determining BW as a function of tensile rebar strains (ε_{rt}) corresponding to a concrete strain of 0.003.

deflection of 2.60 mm and a long-term deflection of 8.00 mm. These values are less than the design limits of 27.8 ($L/360$) and 41.6 mm ($L/240$), respectively. CI_b, CO_2 emissions, and BW are also obtained, which can help engineers with making the final design decisions. Table 5.4.5 validates ANN-based graphical design results for an SRC beam section corresponding to ϕM_n = 3,200 kN·m, μ_ϕ = 3.5, and SF = 1.143, which is challenging to perform with conventional methods. A maximum error of –2.08% is obtained for a tensile strain of rebar (ε_{rt}) compared with structural mechanics. An acceptable accuracy is obtained for most of the other design parameters. An SRC beam has been designed based on the reverse design charts shown in Figures 5.5.2–10 which is summarized in Table 5.4.5. The shear strength capacity of an SRC beam is also calculated as shown in the following Eq. (5.5.1).

$$V_{SRC} = V_S + V_R + V_C \tag{5.5.1}$$

where

V_{SRC} : Shear design capacity of SRC beam,
V_S : Shear design capacity provided by H steel-shaped,
V_R : Shear design capacity provided by stirrups,
V_C : Shear design capacity provided by concrete

As shown in Table 5.4.5, $M_D = 1{,}000$ kN·m, $M_L = 1{,}000$ kN·m, and M_s (self-weight) = 15.4 kN·m, and hence, the factored $M_u = 1.2(M_D + M_S) + 1.6M_L$ = 2,818.5 kN·m, resulting in the following factored shear load for fixed-fixed end: $V_u = 6M_u/l = 1{,}691.1$ kN

The design shear capacity needs to satisfy; $\phi V_{SRC} \geq V_u$

Design capacity is provided by H steel-shaped as shown in Eq. (5.5.2).

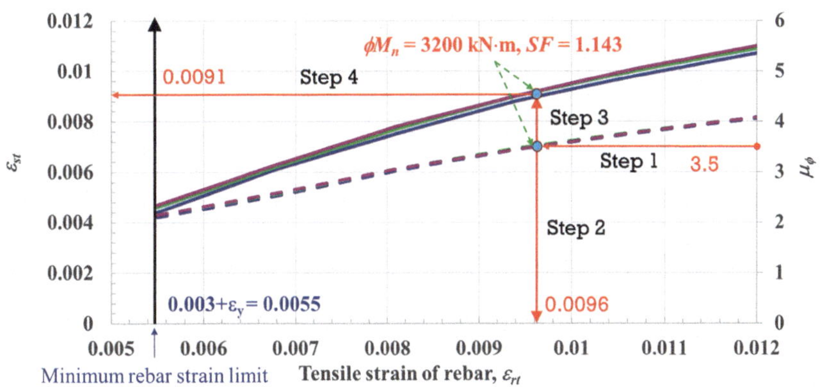

Figure 5.5.10 Design of an SRC beam using design charts. (a) ε_{rt} and μ_ϕ corresponding to ϕM_n = 3,200 kN·m, μ_ϕ = 3.5, and SF = 1.143. (b) d and ρ_{rt} corresponding to ϕM_n = 3,200 kN·m, μ_ϕ = 3.5, and SF = 1.143. (c) Δ_{imme} and Δ_{long} corresponding to ϕM_n = 3,200 kN·m, μ_ϕ= 3.5, and SF = 1.143. (d) CI_b and CO_2 emission corresponding to ϕM_n = 3,200 kN·m, μ_ϕ = 3.5, and SF = 1.143. (e) BW corresponding to ϕM_n = 3,200 kN·m, μ_ϕ= 3.5, and SF = 1.143.

(Continued)

$$\phi V_S = \phi \times 0.6 f_{yS} A_w = 0.9 \times 0.6 \times 325 \times (600 \times 8) = 842.4 \times 10^3 N = 842.4 \text{ kN} \tag{5.5.2}$$

Design shear design capacity is provided by concrete in ACI 318-19 as shown in Eq. (5.5.3).

$$\phi V_c = \phi \times \left[0.17 \lambda \sqrt{f_c^{'}} + \frac{N_u}{6A_g} \right] bd = 0.75 \times \left[0.17 \times 1 \times \sqrt{30} \right] \times 600 \times 1247 = 522.4 \text{ kN} \tag{5.5.3}$$

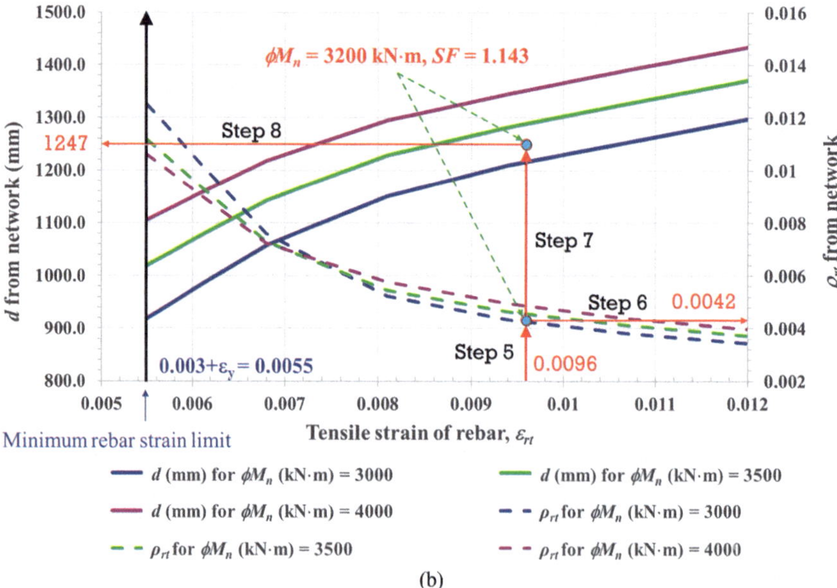

(b)

Figure 5.5.10 (Continued) Design of an SRC beam using design charts. (a) ε_{rt} and μ_ϕ corresponding to ϕM_n = 3,200 kN·m, μ_ϕ= 3.5, and SF = 1.143. (b) d and ρ_{rt} corresponding to ϕM_n = 3,200 kN·m, μ_ϕ = 3.5, and SF = 1.143. (c) Δ_{imme} and Δ_{long} corresponding to ϕM_n = 3,200 kN·m, μ_ϕ = 3.5, and SF = 1.143. (d) CI_b and CO_2 emission corresponding to ϕM_n = 3,200 kN·m, μ_ϕ = 3.5, and SF = 1.143. (e) BW corresponding to ϕM_n = 3,200 kN·m, μ_ϕ = 3.5, and SF = 1.143.

(Continued)

Design shear design capacity is provided by stirrups in ACI 318-19 as shown in Eq. (5.5.4).

$$\phi V_R = \phi \times \frac{A_v f_{yt} d}{s} = 0.75 \times \frac{157 \times 400 \times 1247}{100} = 587.3 \text{ kN} \tag{5.5.4}$$

where

A_v : area of stirrup $\phi 10$ with two legs,

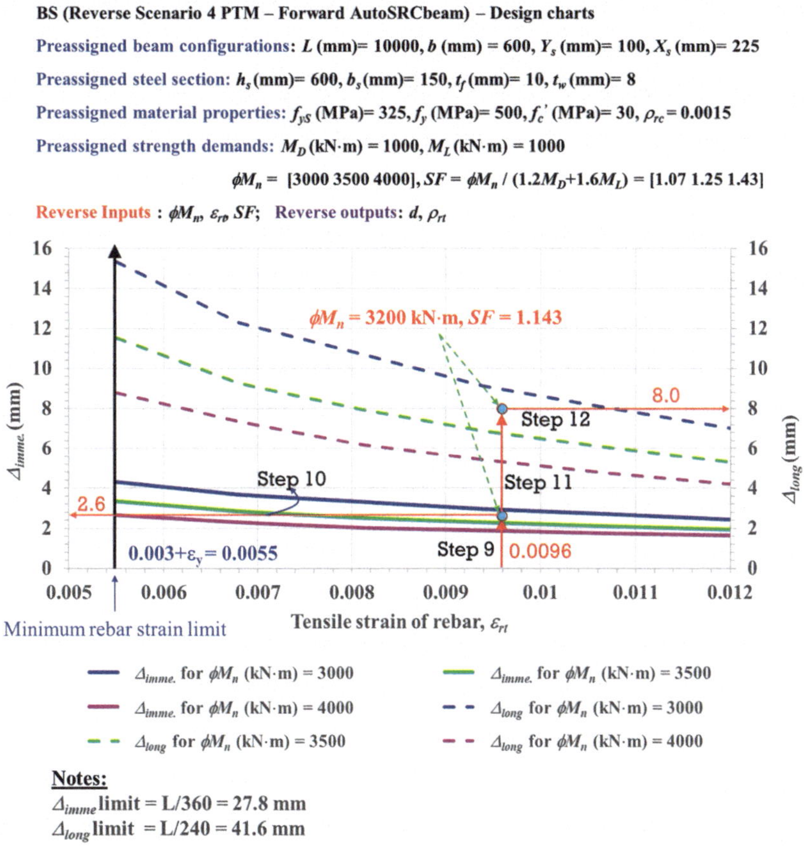

(c)

Figure 5.5.10 (Continued) Design of an SRC beam using design charts. (a) ε_{rt} and μ_ϕ corresponding to ϕM_n = 3,200 kN·m, μ_ϕ = 3.5, and SF = 1.143. (b) d and ρ_{rt} corresponding to ϕM_n = 3,200 kN·m, μ_ϕ = 3.5, and SF = 1.143. (c) Δ_{imme} and Δ_{long} corresponding to ϕM_n = 3,200 kN·m, μ_ϕ= 3.5, and SF = 1.143. (d) CI_b and CO_2 emission corresponding to ϕM_n = 3,200 kN·m, μ_ϕ = 3.5, and SF = 1.143. (e) BW corresponding to ϕM_n = 3,200 kN·m, μ_ϕ = 3.5, and SF = 1.143.

(Continued)

f_{yt}: yield strength of stirrup ($f_{yt} = 400\,\text{MPa}$),
s: spacing of stirrups (s = 100 mm at fixed end)

$$\phi V_{SRC} = \phi V_S + \phi V_R + \phi V_C = 842.4 + 522.4 + 587.3 = 1952.1$$

$$kN > V_u = 1691.1 \text{ kN}$$

Therefore, design shear capacity of SRC satisfies the factored shear force.

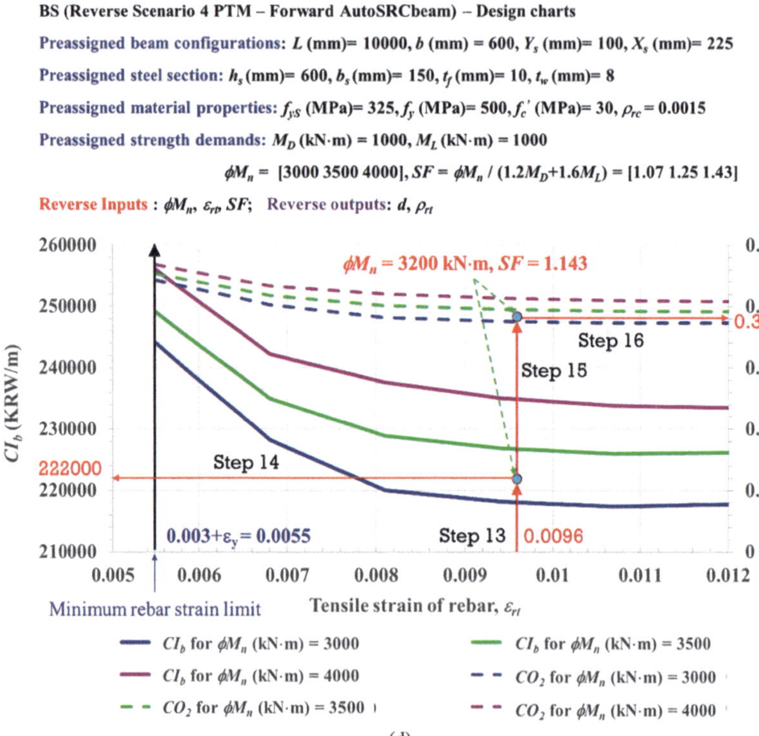

Figure 5.5.10 (Continued) Design of an SRC beam using design charts. (a) ε_{rt} and μ_ϕ corresponding to ϕM_n = 3,200 kN·m, μ_ϕ = 3.5, and SF = 1.143. (b) d and ρ_{rt} corresponding to ϕM_n = 3,200 kN·m, μ_ϕ = 3.5, and SF = 1.143. (c) Δ_{imme} and Δ_{long} corresponding to ϕM_n = 3,200 kN·m, μ_ϕ = 3.5, and SF = 1.143. (d) CI_b and CO_2 emission corresponding to ϕM_n = 3,200 kN·m, μ_ϕ = 3.5, and SF = 1.143. (e) BW corresponding to ϕM_n = 3,200 kN·m, μ_ϕ = 3.5, and SF = 1.143.

(Continued)

BS (Reverse Scenario 4 PTM – Forward AutoSRCbeam) – Design charts

Preassigned beam configurations: L (mm)= 10000, b (mm) = 600, Y_s (mm)= 100, X_s (mm)= 225

Preassigned steel section: h_s (mm)= 600, b_s (mm)= 150, t_f (mm)= 10, t_w (mm)= 8

Preassigned material properties: f_{yS} (MPa)= 325, f_y (MPa)= 500, f_c' (MPa)= 30, ρ_{rc} = 0.0015

Preassigned strength demands: M_D (kN·m) = 1000, M_L (kN·m) = 1000

ϕM_n = [3000 3500 4000], $SF = \phi M_n$ / (1.2M_D+1.6M_L) = [1.07 1.25 1.43]

Reverse Inputs : ϕM_n, ε_{rt}, SF; Reverse outputs: d, ρ_{rt}

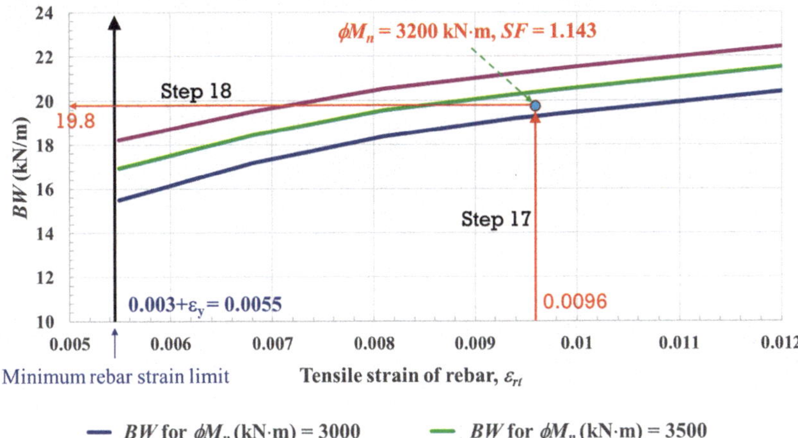

(e)

Figure 5.5.10 (Continued) Design of an SRC beam using design charts. (a) ε_{rt} and μ_ϕ corresponding to ϕM_n = 3,200 kN·m, μ_ϕ = 3.5, and SF = 1.143. (b) d and ρ_{rt} corresponding to ϕM_n = 3,200 kN·m, μ_ϕ = 3.5, and SF = 1.143. (c) Δ_{imme} and Δ_{long} corresponding to ϕM_n = 3,200 kN·m, μ_ϕ = 3.5, and SF = 1.143. (d) CI_b and CO_2 emission corresponding to ϕM_n = 3,200 kN·m, μ_ϕ = 3.5, and SF = 1.143. (e) BW corresponding to ϕM_n = 3,200 kN·m, μ_ϕ = 3.5, and SF = 1.143.

Step 1

Preassigning beam, steel section configurations, strength demand, and expected curvature ductility

Preassigned beam geometries				Preassigned steel section dimensions				Preassigned material properties			
L (mm)	b (mm)	X_s (mm)	Y_s (mm)	h_s (mm)	b_s (mm)	t_f (mm)	t_w (mm)	ρ_{rc}	f_y (MPa)	f_{yS} (MPa)	f_c' (MPa)
10000	600	225	100	600	150	10	8	0.0015	500	325	30

Preassigned design strength and expected curvature ductility – Reverse inputs				
M_D (kN·m)	M_L (kN·m)	ϕM_n (kN·m)	μ_ϕ (ε_{rt})	SF
1000	1000	3200	3.5 (0.0096)	1.143

Finding

d	ρ_{rt}	ε_{st}
Δ_{imme} (mm)	Δ_{long} (mm)	CI_b (KRW/m)
CO_2 (t-CO_2/m)	BW (kN/m)	

Section A-A (with ordinary input parameters)

Legend: - Black text indicates to input parameters

- Red text indicates to output parameters

Reverse inputs

Step 2

Design results based on design charts

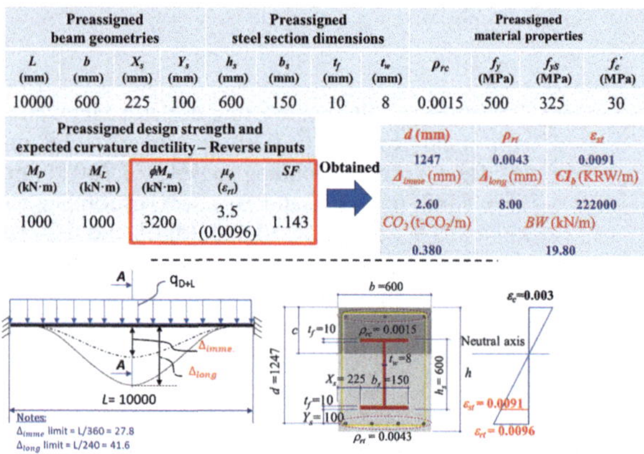

Preassigned beam geometries				Preassigned steel section dimensions				Preassigned material properties			
L (mm)	b (mm)	X_s (mm)	Y_s (mm)	h_s (mm)	b_s (mm)	t_f (mm)	t_w (mm)	ρ_{rc}	f_y (MPa)	f_{yS} (MPa)	f_c' (MPa)
10000	600	225	100	600	150	10	8	0.0015	500	325	30

Preassigned design strength and expected curvature ductility – Reverse inputs				
M_D (kN·m)	M_L (kN·m)	ϕM_n (kN·m)	μ_ϕ (ε_{rt})	SF
1000	1000	3200	3.5 (0.0096)	1.143

Obtained

d (mm)	ρ_{rt}	ε_{st}
1247	0.0043	0.0091
Δ_{imme} (mm)	Δ_{long} (mm)	CI_b (KRW/m)
2.60	8.00	222000
CO_2 (t-CO_2/m)	BW (kN/m)	
0.380	19.80	

Section A-A (with ordinary input parameters)

Legend: - Black text indicates to input parameters

- Red text indicates to output parameters

- Blue bold text indicates to calculated output parameters

Reverse input

Figure 5.5.11 Summary of design SRC beams with preassigned parameters. [ϕM_n = 3,200 kN·m, μ_ϕ = 3.5, SF = 1.143, M_D = 1,000 kN·m, M_L = 1,000 kN·m, L = 10,000 mm, b = 600 mm, f_y = 500 MPa, f_c' = 30 MPa, b_s = 150 mm, h_s = 600 mm, t_f = 10 mm, t_w = 8 mm, f_{yS} = 325 MPa, X_s = 225 mm, Y_s = 100 mm.]

Table 5.4.5 Verification of reverse design for SRC beam sections corresponding to $\phi M_n = 3{,}200$ kN·m, $\mu_\phi = 3.5$

Design charts – BS – 100,000 dataset
17 Inputs (ϕM_n, μ_ϕ, SF, L, b, f_y, f'_c, ρ_{rc}, h_s, b_s, t_f, t_w, f_{yS}, Y_s, M_D, M_L, X_s) – 9 Outputs (d, ρ_{rt}, ε_{rt}, ε_{st}, Δ_{imme}, Δ_{long}, CI_b, CO_2, BW)

No.	Parameter	Training results	Design chart results	AutoSRCbeam check	Error (%)
1	ϕM_n (kN·m)		3,200.0	3,150.0	1.56%
2	ε_{rt}		0.0096	0.0098	−2.08%
3	SF		1.143	1.125	1.26%
4	L (mm)		10,000	10,000	0.00%
5	b (mm)		600	600	0.00%
6	f_y (MPa)		500	500	0.00%
7	f'_c (MPa)		30	30	0.00%
8	ρ_{rc}		0.0015	0.0015	0.00%
9	h_s (mm)		600	600	0.00%
10	b_s (mm)		150	150	0.00%
11	t_f (mm)		10	10	0.00%
12	t_w (mm)		8	8	0.00%
13	f_{yS} (MPa)		325	325	0.00%
14	Y_s (mm)		100	100	0.00%
15	M_D (kN·m)		1,000	1,000	0.00%
16	M_L (kN·m)		1,000	1,000	0.00%
17	d (mm)		1,247	1,247	0.00%
18	ρ_{rt}		0.0043	0.0043	0.00%
19	ε_{st}		0.0091	0.0090	1.10%
20	Δ_{imme} (mm)		2.60	2.58	0.77%
21	Δ_{long} (mm)		8.00	7.92	1.00%
22	μ_ϕ		3.50	3.48	0.57%
23	CI_b (Won/m)		222,000.0	221,890.4	0.05%
24	CO_2 (t-CO_2/m)		0.380	0.384	−1.05%
25	BW (kN/m)		19.80	19.73	0.35%
26	X_s (mm)		225	225	0.00%

Reverse inputs
Reverse outputs
Preassigned inputs
Outputs of design chart
Inputs of AutoSRCbeam
Outputs of AutoSRCbeam

5.6 CONCLUSIONS

In this chapter, ANN-based reverse design charts are generated for rapid and accurate designs of SRC beams. The calculation sequences of the input and output parameters can be exchanged. An ANN is applied to the design of SRC beams, which can be a lengthy process with a conventional method because understanding how parameters influence one another for SRC beams is complex. The objective of this chapter is to develop an ANN-based design method for SRC beams, which can have over 26 design parameters. Reverse designs are obtained by training ANNs with large datasets to map the input to output parameters rather than through Fast and accurate reverse design of SRC beams is now possible. Design sequences of input and output parameters are arbitrarily selected when using ANN-based reverse design charts. The accuracy has been verified through structural calculations using conventional methods. The following recommendations are suggested for the robust and sustainable design of SRC beams based on ANNs:

1. Direct and BS methods are introduced. A training takes two steps in the BS method. Reverse output parameters obtained in a reverse network of Step 1 are back-substituted into a forward network of Step 2. Any standard software can be used for a forward network of Step 2.
2. Rapid and accurate designing of SRC beams can be performed based on ANN-based reverse design charts in which design sequences of input and output parameters can be exchanged, which is challenging to achieve with conventional design methods. ANNs are trained on multiple non-linear inputs and output parameters to obtain weight and bias matrices. Developed neural network and design charts are capable of designing SRC beams with various scenarios while accurately determining design parameters for given unseen structural input parameters.
3. Design accuracies of the SRC design are verified through structural calculations with the conventional method. Limited extrapolations are demonstrated with acceptable accuracies when input conflicts are avoided, even if inputs and outputs are preassigned insignificantly outside the training datasets. However, the network accuracies become weak in general with input conflicts when design parameters are preassigned outside the training datasets.
4. Design charts predicting multiple design parameters in any order to assist engineers to obtain preliminary designs for SRC beams are provided. Robust design of SRC beams based on design charts is also possible in which practical design application to SRC beams enables a fast and accurate determination of design parameters for both forward and reverse designs. Many areas of design can be benefited from the reverse techniques provided in this chapter.
5. ANN-based design charts for SRC beams are developed based on the big datasets generated according to well-known latest American

standards (ACI 318-19) for real-world applications. Contributions of steel, concrete, rebars, and stirrups to stiffnesses and strengths are well described in many design codes such as ACI, EC, and Korean Design Standard (KDS), and hence, data generations for SRC beams are not complex once code-based manuals are followed. ANN-based flowchart described in Sections 5.4 and 5.5 of this chapter can be reconstructed for different design codes such as ACI, EC, and KDS. Training can be repeated whenever big datasets are updated using any type of code.

6. Reverse design can pose an input conflict when preassigned input parameters are not consistent one another, preventing ANN-based design from being practical for engineers. Chapters 6 and 7 will identify optimized parameters in a forward manner to avoid input conflicts using ANN-based Lagrange optimization with equality and inequality conditions.

REFERENCES

[5.1] Vanluchene, R. D., & Sun, R. (1990). Neural networks in structural engineering. *Computer-Aided Civil and Infrastructure Engineering*, *5*(3): 207–215. https://doi.org/10.1111/j.1467-8667.1990.tb00377.x

[5.2] Hajela, P., & Berke, L. (1991). Neurobiological computational models in structural analysis and design. *Computers & Structures*, *41*(4), 657–667. https://doi.org/10.1016/0045-7949(91)90178-O

[5.3] Flood, I., & Kartam, N. (1994). Neural networks in civil engineering. I: Principles and understanding. *Journal of Computing in Civil Engineering*, *8*(2), 131–148. https://doi.org/10.1061/(ASCE)0887-3801(1994)8:2(131)

[5.4] Flood, I., & Kartam, N. (1994). Neural networks in civil engineering. II: Systems and application. *Journal of Computing in Civil Engineering*, *8*(2), 149–162. https://doi.org/10.1061/(ASCE)0887-3801(1994)8:2(149)

[5.5] Hong, W. K. (2019). *Hybrid Composite Precast Systems: Numerical Investigation to Construction*. Woodhead Publishing, Elsevier, Cambridge, MA.

[5.6] Wu, R. T., & Jahanshahi, M. R. (2019). Deep convolutional neural network for structural dynamic response estimation and system identification. *Journal of Engineering Mechanics*, *145*(1), 0401812. https://doi.org/10.1061/(ASCE)EM.1943-7889.0001556

[5.7] Lavaei, A., & Lohrasbi, A. (2012). Dynamic analysis of structures using neural networks. In *Proceedings of 15th World Conference on Earthquake Engineering*, Lisbon, Portugal.

[5.8] Fahmy, A. S., El-Madawy, M. E. T., & Gobran, Y. A. (2016). Using artificial neural networks in the design of orthotropic bridge decks. *Alexandria Engineering Journal*, *55*(4), 3195–3203. https://doi.org/10.1016/j.aej.2016.06.034

[5.9] Adeli, H. (2001). Neural networks in civil engineering: 1989–2000. *Computer-Aided Civil and Infrastructure Engineering*, *16*(2), 126–142. https://doi.org/10.1111/0885-9507.00219

[5.10] Lee, S., Ha, J., Zokhirova, M., Moon, H., & Lee, J. (2018). Background information of deep learning for structural engineering. *Archives of Computational Methods in Engineering*, *25*(1), 121–129. https://doi.org/10.1007/s11831-017-9237-0

[5.11] Gupta, T., & Sharma, R. K. (2011). Structural analysis and design of buildings using neural network: A review. *International Journal of Engineering and Management Sciences*, 2(4), 216–220.

[5.12] Chan, W. T., Chow, Y. K., & Liu, L. F. (1995). Neural network: An alternative to pile driving formulas. *Computers and Geotechnics*, *17*(2), 135–156.

[5.13] Lee, I. M., & Lee, J. H. (1996). Prediction of pile bearing capacity using artificial neural networks. *Computers and Geotechnics*, *18*(3), 189–200.

[5.14] Rahman, M. S., Wang, J., Deng, W., & Carter, J. P. (2001). A neural network model for the uplift capacity of suction caissons. *Computers and Geotechnics*, *28*(4), 269–287. https://doi.org/10.1016/S0266-352X(00)00033-1

[5.15] Hanna, A. M., Morcous, G., & Helmy, M. (2004). Efficiency of pile groups installed in cohesionless soil using artificial neural networks. *Canadian Geotechnical Journal*, *41*(6), 1241–1249. https://doi.org/10.1139/t04-050

[5.16] Das, S. K., & Basudhar, P. K. (2006). Undrained lateral load capacity of piles in clay using artificial neural network. *Computers and Geotechnics*, *33*(8), 454–459. https://doi.org/10.1016/j.compgeo.2006.08.006

[5.17] Adeli, H., & Park, H. S. (1995). A neural dynamics model for structural optimization—theory. *Computers & Structures*, *57*(3), 383–390. https://doi.org/10.1016/0045-7949(95)00048-L

[5.18] Adeli, H., & Park, H. S. (1995). Optimization of space structures by neural dynamics. *Neural Networks*, *8*(5), 769–781. https://doi.org/10.1016/0893-6080(95)00026-V

[5.19] Tashakori, A., & Adeli, H. (2002). Optimum design of cold-formed steel space structures using neural dynamics model. *Journal of Constructional Steel Research*, *58*(12), 1545–1566. https://doi.org/10.1016/S0143-974X(01)00105-5

[5.20] Kang, H. T., & Yoon, C. J. (1994). Neural network approaches to aid simple truss design problems. *Computer-Aided Civil and Infrastructure Engineering*, *9*(3), 211–218. https://doi.org/10.1111/j.1467-8667.1994.tb00374.x

[5.21] Lee, C., & Sterling, R. (1992, January). Identifying probable failure modes for underground openings using a neural network. In *International Journal of Rock Mechanics and Mining Sciences & Geomechanics Abstracts* (Vol. 29, No. 1, pp. 49–67). Pergamon. https://doi.org/10.1016/0148-9062(92)91044-6

[5.22] Shi, J., Ortigao, J. A. R., & Bai, J. (1998). Modular neural networks for predicting settlements during tunneling. *Journal of Geotechnical and Geoenvironmental Engineering*, *124*(5), 389–39. https://doi.org/10.1061/(ASCE)1090-0241(1998)124:5(389)

[5.23] Shi, J. J. (2000). Reducing prediction error by transforming input data for neural networks. *Journal of Computing in Civil Engineering*, *14*(2), 109–116. https://doi.org/10.1061/(ASCE)0887-3801(2000)14:2(109)

[5.24] Neaupane, K., & Achet, S. (2004). Some applications of a backpropagation neural network in geo-engineering. *Environmental Geology*, *45*(4), 567–57 https://doi.org/10.1007/s00254-003-0912-0

[5.25] Yoo, C., & Kim, J. M. (2007). Tunneling performance prediction using an integrated GIS and neural network. *Computers and Geotechnics*, *34*(1), 19–30. https://doi.org/10.1016/j.compgeo.2006.08.007

[5.26] Oh, B. K., Glisic, B., Park, S. W., & Park, H. S. (2020). Neural network-based seismic response prediction model for building structures using artificial earthquakes. *Journal of Sound and Vibration*, *468*, 115109. https://doi.org/10.1016/j.jsv.2019.115109

[5.27] Lagaros, N. D., & Papadrakakis, M. (2012). Neural network based prediction schemes of the non-linear seismic response of 3D buildings. *Advances in Engineering Software*, *44*(1), 92–11. https://doi.org/10.1016/j.advengsoft.2011.05.033

[5.28] Asteris, P. G., Nozhati, S., Nikoo, M., Cavaleri, L., & Nikoo, M. (2019). Krill herd algorithm-based neural network in structural seismic reliability evaluation. *Mechanics of Advanced Materials and Structures*, *26*(13), 1146–1153. https://doi.org/10.1080/15376494.2018.1430874

[5.29] Ying, W., Chong, W., Hui, L., & Renda, Z. (2009, November). Artificial neural network prediction for seismic response of bridge structure. In 2009 International Conference on Artificial Intelligence and Computational Intelligence (Vol. 2, pp. 503–506). IEEE. https://doi.org/10.1109/AICI.2009.303

[5.30] Huang, C., & Huang, S. (2020, October). Predicting capacity model and seismic fragility estimation for RC bridge based on artificial neural network. In *Structures* (Vol. 27, pp. 1930–1939). Elsevier. https://doi.org/10.1016/j.istruc.2020.07.063

[5.31] Hong, W. K., Pham, T. D., & Nguyen, V. T. (2022). Feature selection based reverse design of doubly reinforced concrete beams. *Journal of Asian Architecture and Building Engineering*, *21*(4), 1472–1496. https://doi.org/10.1080/13467581.2021.1928510

[5.32] Hong, W. K., Nguyen, V. T., & Nguyen, M. C. (2022). Artificial intelligence-based novel design charts for doubly reinforced concrete beams. *Journal of Asian Architecture and Building Engineering*, *21*(4), 1497–1519. https://doi.org/10.1080/13467581.2021.1928511

[5.33] Hong, W. K., & Nguyen, M. C. (2022). AI-based Lagrange optimization for designing reinforced concrete columns. *Journal of Asian Architecture and Building Engineering*, *21*(6), 2330–2344. https://doi.org/10.1080/13467581.2021.1971998

[5.34] Hong, W. K., & Pham, T. D. (2022). Reverse designs of doubly reinforced concrete beams using Gaussian process regression models enhanced by sequence training/designing technique based on feature selection algorithms. *Journal of Asian Architecture and Building Engineering*, *21*(6), 2345–2370. https://doi.org/10.1080/13467581.2021.1971999

[5.35] Hong, W. K., Nguyen, D. H., & Nguyen, V. T. (2022). Reverse design charts for flexural strength of steel-reinforced concrete beams based on artificial neural networks. *Journal of Asian Architecture and Building Engineering*, 1–39. https://doi.org/10.1080/13467581.2022.2097238

[5.36] Hong, W. K., Kim, J. M., Park, S. C., Lee, S. G., Kim, S. I., Yoon, K. J.,... & Kim, J. T. (2010). A new apartment construction technology with effective CO_2 emission reduction capabilities. *Energy*, *35*(6), 2639–2646. https://doi.org/10.1016/j.energy.2009.0 036

[5.37] Berrais, A. (1999). Artificial neural networks in structural engineering: Concept and applications. *Engineering Sciences*, *12*(1), 53–67.

[5.38] Nguyen, D. H., & Hong, W. K. (2019). Part I: The analytical model predicting post-yield behavior of concrete-encased steel beams considering various confinement effects by transverse reinforcements and steels. *Materials*, *12*(14), 2302. https://doi.org/10.3390/ma12142302

[5.39] Nguyen, D.H. Artificial Intelligence (AI)-based design methodology for steel-reinforced concrete (SRC) members. Ph.D. dissertation, Kyung Hee University 2021.

[5.40] Hong, W. K., Nguyen, V. T., & Nguyen, M. C. (2022). Optimizing reinforced concrete beams cost based on AI-based Lagrange functions. *Journal of Asian Architecture and Building Engineering*, 21(6), 2426–2443. https://doi.org/10.1080/13467581.2021.200710

[5.41] Pham, T. D., & Hong, W. K. (2022). Genetic algorithm using probabilistic-based natural selections and dynamic mutation ranges in optimizing precast beams. *Computers & Structures*, *258*, 106681. https://doi.org/10.1016/j.compstruc.2021.106681

[5.42] Hong, W. K., Nguyen, M. C., & Pham, T. D. (2023). Optimized interaction PM diagram for rectangular reinforced concrete column based on artificial neural networks. *Journal of Asian Architecture and Building Engineering*, 22(1), 201–225. https://doi.org/0.1080/13467581.2021.2018697

[5.43] Hong, W. K., Le, T. A., Nguyen, M. C., & Pham, T. D. (2022). ANN-based Lagrange optimization for RC circular columns having multi-objective functions. *Journal of Asian Architecture and Building Engineering*. https://doi.org/10.1080/13467581.2022.2064864

[5.44] Hong, W. K., Nguyen, V. T., Nguyen, D. H., & Nguyen, M. C. (2022). An AI-based Lagrange optimization for a design for concrete columns encasing H-shaped steel sections under a biaxial bending. *Journal of Asian Architecture and Building Engineering*. https://doi.org/10.1080/13467581.2022.2060985

[5.45] Hong, W. K., Nguyen, V. T., Nguyen, D. H., & Nguyen, M. C. (2022). Reverse design-based optimizations for reinforced concrete columns encasing H-shaped steel section using ANNs. *Journal of Asian Architecture and Building Engineering*, 660–674. https://doi.org/10.1080/13467581.2022.2047985

[5.46] Hong, W. K. (2021). *Artificial Intelligence-Based Design of Reinforced Concrete Structures*. Daega, Seoul.

[5.47] Hong, W. K. (2023). *Artificial Neural Network-based Optimized Design of Reinforced Concrete Structures*. CRC Press, Boca Raton, FL.

Chapter 6

An optimization of steel-reinforced concrete beams using an ANN-based Lagrange algorithm

6.1 INTRODUCTION AND SIGNIFICANCE OF CHAPTER 6

6.1.1 Previous studies

SRC members include concrete, steel section, and rebar. H-shaped steel section and rebar are embedded inside concrete, and thus, they can be isolated from high temperature and corrosive agents of the surrounding environment, leading to an increase in the sustainability of the structures. SRC members are being widely used for various building structures and transportation facilities in recent decades to enhance strength, stiffness, and ductility of structures. Behaviors of SRC members under external loads have been investigated by many studies, including Li and Matsui [6.1], El-Tawil and Deerling [6.2], Bridge and Roderick [6.3], Mirza et al. [6.4], Ricles and Paboojian [6.5], Furlong [6.6], Mirza and Skrabek [6.7], Chen and Lin [6.8], Dundar et al. [6.9], Munoz and Hsu [6.10], Rong and Shi [6.11], Virdi and Dowling [6.12], Roik and Bergmann [6.13], and Brettle [6.14]. A capacity of SRC members can be calculated by structural mechanics-based traditional method. This chapter introduces a new method based on ANNs. ANNs have been implemented successfully in structural analysis by many researchers (Abambres and Lantsoght [6.15], Sharifi, Lotfi, and Moghbeli [6.16], Asteris et al. [6.17], and Armaghani et al. [6.18]). An optimization of reinforced concrete members was performed using the ANN-based Hong–Lagrange algorithm in previous studies (Hong and Nguyen [6.19], Hong, Nguyen, and Nguyen [6.20], Hong, Nguyen, and Pham [6.21], Hong and Le [6.22], Hong et al. [6.23], Hong 2021 [6.24]).

In the present chapter, the objective functions including cost index (CI_b), CO_2 emissions, and member weight (W) of SRC beams are optimized based on ANN-based Lagrange algorithm. Finding studies for minimizing cost index (CI_b), CO_2 emissions, and beam weights (W) of structural frames is difficult, whereas this chapter presents an ANN-based design of SRC beams to seek a reduction of greenhouse effects to mitigate climate change and to reduce construction cost and time. Optimal designs of SRC beams can

DOI: 10.1201/9781003354796-6

also be performed by the trial-and-error method based on a conventional method, however, requiring much effort and time than those when using ANNs. This study introduces ANN-based optimal design for SRC beams, replacing traditional approaches. Large datasets to train ANNs have been generated by a structural mechanics-based AutoSRCbeam (Nguyen and Hong [6.25] and Hong 2019 [6.26]). A design accuracy of SRC beams using ANNs is validated both through structural analysis and a large structural dataset. This chapter was written based on the previous paper by Hong et al. [6.27].

6.1.2 Innovations and significances

Explicit objective functions and constraints are difficult to derive for an application of a conventional Lagrange optimization, and hence, these functions are replaced by ANN-based generalized functions. Any design target can be selected as an objective function to be minimized for a design of SRC beams. It is noted that cost (CI_b) is an interest of structural engineers, whereas CO_2 emissions, and weight (W) are the interests of governments and contractors, respectively. Objective functions of CI_b, CO_2 emissions, and W of SRC beams and their constraints are derived based on ANNs trained by a large dataset of 200,000 as functions of input parameters. Solutions under KKT conditions should be sought when inequality constraints are used to impose design requirements. Newton–Raphson iterations are, then, used to converge initial input parameters to stationary points, minimizing Lagrange functions leading to minimized design targets of CI_b, CO_2 emissions, and W, whereas design parameters including beam dimensions and rebar ratios are also obtained on an output-side. A large dataset and structural mechanics-based software (AutoSRCbeam) developed by Nguyen and Hong [6.25] are used to verify the accuracy of the minimized objective functions of CI_b, CO_2 emissions, and W and corresponding design parameters. Innovative, robust, and practical methods are presented to optimize SRC beam designs.

6.2 ANN-BASED DESIGN SCENARIOS FOR SRC BEAMS

Table 6.2.1 presents a forward scenario for a design SRC beam encasing H-shaped steel section, in which an optimization is performed based on 26 parameters including 15 forward inputs and 11 forward outputs. Dimensions of an SRC beam with beam height (h) and width (b) shown in Figure 6.2.1 are selected as input parameters for forward scenario. Concrete section encases steel section with height (h_s), flange (b_s), web thickness (t_f), and flange thickness (t_w). Material properties of concrete (f_c'), rebar (f_y),

Table 6.2.1 Design scenarios for SRC beam

	Forward inputs (15 parameters)															*Forward outputs (11 parameters)*										
Scenarios	L	d	b	f_y	f_c'	ρ_{rt}	ρ_{rc}	h_s	b_s	t_f	t_w	f_{yS}	Y_s	M_D	M_L	ϕM_n	ε_{rt}	ε_{st}	$\Delta_{imme.}$	Δ_{long}	μ_ϕ	CI_b	CO_2	W	X_s	SF
Forward	i	i	i	i	i	i	i	i	i	i	i	i	i	i	i	o	o	o	o	o	o	o	o	o	o	o
Note:																										
Input	i																									
Output	o																									

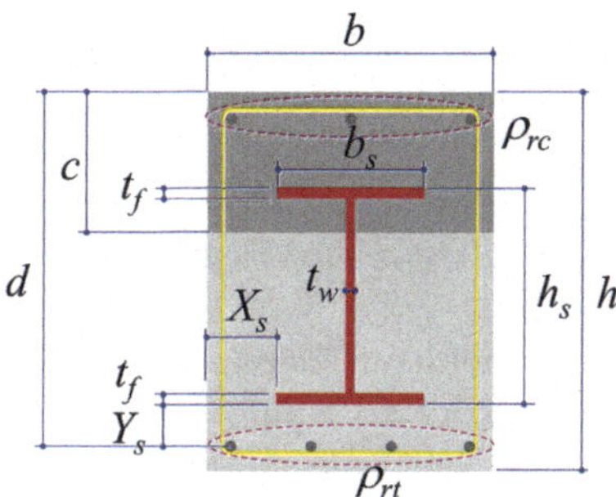

Figure 6.2.1 Geometry of SRC beam. (Hong, Nguyen, and Nguyen [6.28].)

steel (f_{ys}), and rebar ratios for compressive rebars (ρ_c), and tensile rebars (ρ_t) are preassigned on an input-side. External loads including moment due to dead load (M_D) and live load (M_L) are also imposed on an input-side. Eleven output parameters include design moment (ϕM_n), tensile strain of rebar (ε_{rt}) and steel (ε_{st}), immediate deflection ($\Delta_{imme.}$), long-term deflection (Δ_{long}), curvature ductility (μ_ϕ), cost index per 1 m length (CI_b), CO_2 emissions per 1 m length, beam weight per 1 m length (W), horizontal clearance (X_s), and safety factor (SF). Cost index (CI_b), CO_2 emissions, and member weight (W) are selected as objective functions to optimize an SRC beam.

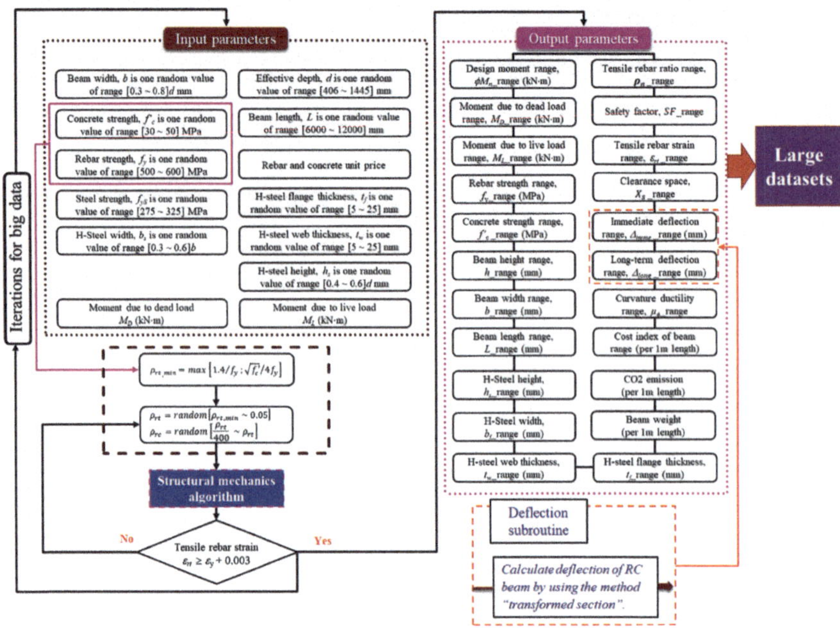

Figure 6.3.1 Flowchart for big data generation of SRC beams used to train ANNs. (Nguyen and Hong [6.25].)

6.3 GENERATION OF LARGE DATASETS

Large datasets are generated with output parameters which are randomly calculated based on randomly selected input parameters using AutoSRCbeam [6.25] which are developed based on an algorithm shown in flowchart of Figure 6.3.1. Nomenclatures and ranges of input parameters defining an SRC beam are shown in Table 6.3.1. SRC beams with fixed-fixed end conditions shown in Figure 6.3.2 are considered. SRC beams subjected to uniform loads that consist of dead load and live load producing M_D and M_L, respectively. A self-weight of member is not included in the design moment capacity of a beam (ϕM_n) because beam sections are not known in the preliminary design stage. A design must satisfy safety factor $SF=\phi M_n/M_u \geq 1.0$, where M_u is a factored moment that is calculated by a load combination of M_D and M_L with the load factors ($M_u=1.2M_D+1.6M_L$).

The length of beam is selected within the range of [6,000 ~ 12,000] mm, which is the most typical range for building frames. An effective depth of a beam depth (d) is also randomly selected from 406.1 to 1,445.5 mm, whereas a beam width (b) is selected in a range from 0.3d to 0.8d when heights of beam h are selected in a range of [500–1,500] mm. Table 6.3.1 presents dimensions of embedded H-shaped steel sections which

Table 6.3.1 Nomenclatures and ranges of parameters defining SRC beams

		Notation	Range
Input parameters	L (mm)	Beam length	[6,000 ~ 12,000]
	d (mm)	Effective beam depth	[406.1 ~ 1,445.5]
	b (mm)	Beam width	$[0.3 \sim 0.8]d$
	h_s (mm)	H-shaped steel height	$[0.4 \sim 0.6]d$
	b_s (mm)	H-shaped steel width	$[0.3 \sim 0.6]b$
	t_f (mm)	H-shaped steel flange thickness	[5 ~ 25]
	t_w(mm)	H-shaped steel web thickness	[5 ~ 25]
	f_c' (MPa)	Concrete strength	[30 ~ 50]
	f_y (MPa)	Rebar strength	[500 ~ 600]
	f_{yS} (MPa)	Steel strength	[275 ~ 325]
	ρ_{rt}	Tensile rebar ratios	$\rho_{rt,\min} = \max\left(\frac{0.25\sqrt{f_c'}}{f_y}; \frac{1.4}{f_y}\right)$
	ρ_{rc}	Compressive rebar ratios	$[1/400 \sim 1.5]\ \rho_{rt}$
	Ys	Vertical clearance	
	M_D (kN·m)	Moment due to dead load	$\left[0.2 \sim \frac{1}{1.2}\right]M_u$ ($M_u = 1.2M_D + 1.6M_L$)
	M_L (kN·m)	Moment due to service live load	$\frac{1}{1.6}(M_u - 1.2M_D)$ ($M_u = 1.2M_D + 1.6M_L$)
Output parameters	ϕM_n (kN·m)	Design moment without considering effect of self-weight at ε_c=0.003	
	μ_ϕ	Curvature ductility, $\mu_\phi = \phi_u/\phi_y$ Where, ϕ_u: Curvature at ε^c=0.003 ϕ_y: Curvature at tensile rebar yield	
	ε_{rt}	Tensile rebar strain at ε^c=0.003 ε_y :Yield strain of rebar	$\varepsilon_{rt} \geq 0.004$
	ε_{rc}	Compressive rebar strain at $\varepsilon^c = 0.003$	
	$\Delta_{imme.}$	Immediate deflection due to M_L service live load	($\Delta_{imme.} \leq$ L/360, ACI 318-19)
	Δ_{long}	Sum of long-term deflection due to sustained loads and immediate deflection due to additional live load	($\Delta_{long} \leq$ L/240, ACI 318-19)
	CI_b (KRW/m)	Cost index per 1 m length of beam	
	CO_2 (t-CO_2/m)	CO_2 emissions per 1 m length of beam	
	W (kN/m)	Beam weight per 1 m length of beam	
	X_s	Horizontal clearance	
	SF	Safety factor ($\phi M_n/M_u$)	

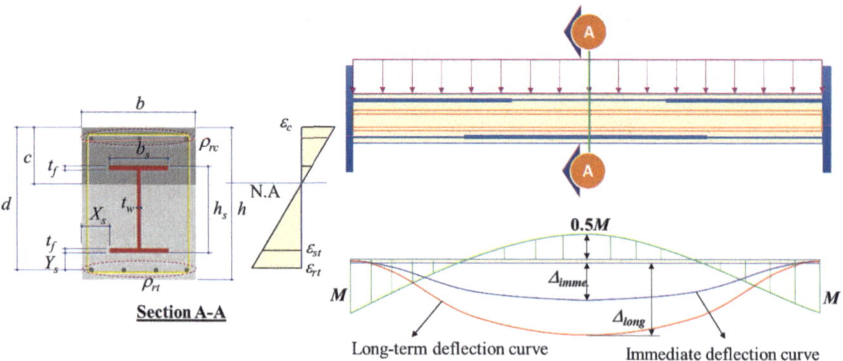

Figure 6.3.2 Fixed-fixed SRC beams used for ANN. (Hong, Nguyen, and Nguyen [6.28].)

are randomly selected. Material strengths of concrete (f_c'), rebar (f_y), and steel (f_{yS}) are selected in ranges from 30 to 50 MPa, 500 to 600 MPa, and 275 to 325 MPa, respectively. Minimum tensile rebar ratios are selected as $\rho_{rt,min} = max\left(0.25\sqrt{f_c'}/f_y;1.4/f_y\right)$ following American Concrete Institute (ACI) code, whereas compressive rebar ratios are randomly selected in a range of $[1/400 \sim 1.5]\rho_{rt}$. Tensile rebar strain ($\varepsilon_{rt}$) should be larger than $0.003+\varepsilon_y$ according to ACI code (ACI Standard [6.29]). Immediate deflection ($\Delta_{imme.}$) and long-term deflection (Δ_{long}) should not be greater than the deflection limits $L/360$ and $L/240$ according to ACI 318–19 (ACI Standard [6.29]), respectively. Table 6.3.2 provides a statistical summary of the input and output parameters, including maxima, means, minima, standard deviations, and variances.

6.4 FORMULATIONS OF OBJECTIVE FUNCTIONS USING ANN-BASED LAGRANGE ALGORITHM FOR SRC BEAMS

6.4.1 Training ANN on each output parameter based on a parallel training method (PTM)

In this chapter, PTM (Hong, Pham, and Nguyen [6.30], and Hong [6.24]) maps 15 input parameters (L, d, b, f_y, f_c', ρ_{sc}, ρ_{st}, h_s, b_s, t_f, t_w, f_{yS}, Y_s, M_D, M_L) to each of 11 output parameters (ϕM_n, ε_{rt}, ε_{st}, $\Delta_{\text{imme.}}$, Δ_{long}, μ_ϕ, CI_b, CO_2, W, X_s, SF) based on 11 independent ANNs as demonstrated in Figure 6.4.1. Eleven trainings are performed by MATLAB Deep Learning Toolbox™ platform (MATLAB Deep Learning Toolbox [6.31]) based on 200,000 datasets which are generated by AutoSRCbeam. Large datasets of 200,000 are divided into three separate parts covering 70% (140,000 datasets) for

Table 6.3.2 Maxima, means, minima, standard deviations, and variances of random design inputs and outputs for SRC beams

	Parameter	Mean (μ)	Maximum	Minimum	Standard deviation (σ)	Variance (V)
Inputs	L (mm)	8,996.7	12,000.0	6,000.0	1,748.5	3,057,200.7
	d (mm)	1,039.9	1,445.5	406.1	246.6	60,803.2
	b (mm)	632.0	1,200	150	215.6	46,470.9
	h_s (mm)	427.7	1,180.0	5.0	221.6	49,085.0
	b_s (mm)	277.0	710.0	45.0	113.6	12,893.6
	t_f (mm)	16.1	30.0	5.0	7.5	55.7
	t_w (mm)	16.3	30.0	5.0	7.4	54.7
	f_c' (MPa)	40.0	50	30	6.1	36.7
	f_y (MPa)	550.0	600	500	29.1	849.6
	f_{yS} (MPa)	300.0	325.0	275.0	14.7	216.4
	ρ_{rt}	0.0166	0.05897	0.00243	0.01025	1.05×10^{-04}
	ρ_{rc}	0.0167	0.08130	7.00E-06	0.01382	1.91×10^{-04}
	Ys	245.4	943.0	60.0	139.6	19,495.1
	M_D (kN·m)	2,959.1	36,961.7	12.2	3,108.4	9,662,085.2
	M_L (kN·m)	1,480.4	24,343.4	1.3	1,772.1	3,140,272.3
Outputs	ϕM_n (kN·m)	7,539.5	44,085.5	84.8	6341.4	40,213,677.0
	ε_{rt}	0.00832	0.042	0.0055	0.0026	6.78E-06
	ε_{st}	0.00504	0.031	−0.0004	0.0025	6.04E-06
	μ_ϕ	2.738	13.749	1.784	0.788	0.621
	$\Delta_{imme.}$(mm)	2.06	27.98	0.10	1.85	3.41
	Δ_{long} (mm)	9.40	84.17	0.47	6.73	45.29
	CI_b (KRW/m)	493,406.1	2,121,983.0	26,970.3	271,908.2	73,934,071,181.9
	CO_2 (t-CO_2/m)	0.899	4.292	0.044	0.548	0.300
	W (kN/m)	20.93	57.3	1.97	10.39	107.87
	Xs (mm)	177.5	420.0	30.0	66.0	4,352.9
	SF	1.37	2.0	0.75	0.363	0.132

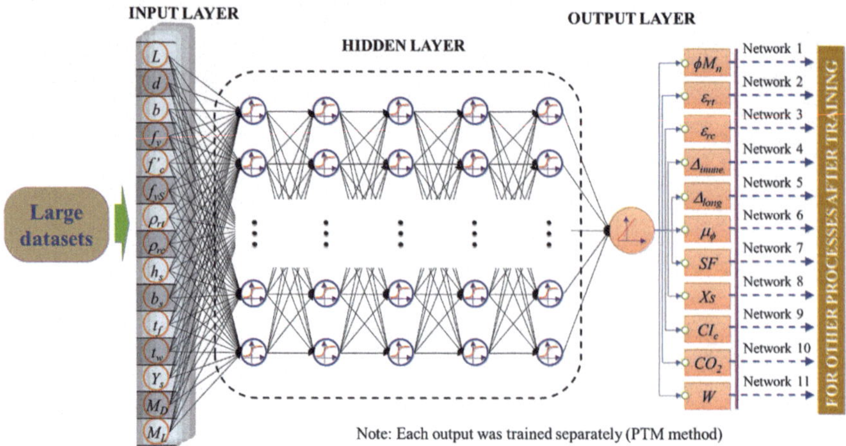

Figure 6.4.1 Topology of ANN using PTM.

training, 15% (30,000 datasets) for validation, and 15% (30,000 datasets) for testing data, resulting in training accuracies shown in Table 6.4.1 where test MSE and regression (R) ranging from 2.77E-04 to 1.69E-07 and 0.9987 to 1.0000 are obtained based on suggested and validation epochs preassigned as 50,000 and 1,000, respectively.

6.4.2 Optimized objective functions using ANN-based Lagrange algorithm

6.4.2.1 *Formulation of objective functions based on ANNs*

Objective functions of cost (CI_b), CO_2 emissions, and weight (W) which are the main interests of engineers, governments, and contractors, respectively, are to be minimized. The governments need to keep a clean environment by reducing CO_2 emissions as much as possible. Objective functions CI_b, CO_2 emissions, and W are derived based on ANNs with multilayer perceptron using four layers and 64 neurons as shown in Table 6.4.1. Forward networks with neural links based on weights and biases are shown in Eqs. (6.4.1)–(6.4.3) (Krenker, Bešter, and Kos [6.32]) which are trained using PTM. It is noted that an activation function is implemented in yielding nonlinear behaviors of the objective functions.

$$\underbrace{CI_b}_{[1\times1]} = g_{CI_b}^{D}\left(f_{lin}^{4}\left(\underbrace{\mathbf{W}_{CI_b}^{4}}_{[1\times64]} f_t^{3}\left(\underbrace{\mathbf{W}_{CI_b}^{3}}_{[64\times64]} \ldots f_t^{1}\left(\underbrace{\mathbf{W}_{CI_b}^{1}}_{[64\times15]} \underbrace{g_{CI_b}^{N}(\mathbf{X})}_{[15\times1]} + \underbrace{\mathbf{b}_{CI_b}^{1}}_{[64\times1]}\right) \ldots + \underbrace{\mathbf{b}_{CI_b}^{3}}_{[64\times1]}\right) + \underbrace{b_{CI_b}^{4}}_{[1\times1]}\right)\right) \quad (6.4.1)$$

Table 6.4.1 Training accuracies based on PTM

No.	*Training with an output*	*Data*	*Layers*	Neurons	*Suggested Epoch*	*Best Epoch*	*Stopped Epoch*	*Validation Epoch*	*Test MSE*	*R at Best Epoch*
1	ϕM_n	200,000	4	64	50,000	48,865	49,865	1,000	8.25E-06	1.0000
2	ε_{rt}	200,000	4	64	50,000	34,638	35,638	1,000	5.30E-05	0.9987
3	ε_{st}	200,000	4	64	50,000	49,997	50,000	1,000	4.17E-05	0.9992
4	$\Delta_{imme.}$	200,000	4	64	50,000	49,997	50,000	1,000	9.96E-06	0.9997
5	Δ_{long}	200,000	4	64	50,000	49,748	50,000	1,000	1.80E-05	0.9997
6	μ_ϕ	200,000	4	64	50,000	35,898	36,898	1,000	4.22E-05	0.9989
7	SF	200,000	4	64	50,000	27,140	27,141	1,000	1.69E-07	1.0000
8	X_s	200,000	4	64	50,000	15,079	16,079	1,000	2.08E-06	1.0000
9	CI_b	200,000	4	64	50,000	32,795	33,795	1,000	1.47E-06	1.0000
10	CO_2	200,000	4	64	50,000	21,661	21,661	1,000	2.15E-08	1.0000
11	W	200,000	4	64	50,000	49,944	50,000	1,000	2.77E-04	0.9996

$$\underbrace{CO_2}_{[1\times1]} = g_{CO_2}^{D}\left(f_{lin}^{4}\left(\underbrace{\mathbf{W}_{CO_2}^{4}}_{[1\times64]} f_t^{3}\left(\underbrace{\mathbf{W}_{CO_2}^{3}}_{[64\times64]} \cdots f_t^{1}\left(\underbrace{\mathbf{W}_{CO_2}^{1}}_{[64\times15]} \underbrace{g_{CO_2}^{N}(\mathbf{X})}_{[15\times1]} + \underbrace{\mathbf{b}_{CO_2}^{1}}_{[64\times1]}\right) \cdots + \underbrace{\mathbf{b}_{CO_2}^{3}}_{[64\times1]}\right) + \underbrace{b_{CO_2}^{4}}_{[1\times1]}\right)\right) \tag{6.4.2}$$

$$\underbrace{W}_{[1\times1]} = g_{W}^{N}\left(f_{lin}^{L}\left(\underbrace{\mathbf{W}_{W}^{4}}_{[1\times64]} f_t^{3}\left(\underbrace{\mathbf{W}_{W}^{3}}_{[64\times64]} \cdots f_t^{1}\left(\underbrace{\mathbf{W}_{W}^{1}}_{[64\times15]} \underbrace{g_{W}^{N}(\mathbf{X})}_{[15\times1]} + \underbrace{\mathbf{b}_{W}^{1}}_{[64\times1]}\right) \cdots + \underbrace{\mathbf{b}_{W}^{3}}_{[64\times1]}\right) + \underbrace{b_{W}^{4}}_{[1\times1]}\right)\right) \tag{6.4.3}$$

where

$\mathbf{X}$; input parameters, $\mathbf{X}=[L, d, b, f_y, f_c', \rho_{sc}, \rho_{st}, h_s, b_s, t_f, t_w, f_{yS}, Y_s, M_D, M_L]^T$

g^N, g^D; normalizing and de-normalizing functions

$\mathbf{W}_{CI_b}^{i}, \mathbf{W}_{CO_2}^{i}, \mathbf{W}_{W}^{i}$; weight matrices $(i = \overline{1-4})$ obtained by training based on PTM.

Weight matrices connect two hidden layers for objective functions. $\mathbf{b}_{CI_b}^{i}, \mathbf{b}_{CO_2}^{i}, \mathbf{b}_{W}^{i}$; bias matrices $(i = \overline{1-4})$ obtained by training based on PTM, helping networks best fit the large datasets when deriving ANN-based objective functions.

Forward networks are linked by weight and bias matrices, resulting in non-linear behaviors of the objective functions through activation functions. An activation function of ANNs used in this chapter is ***tansig*** as shown in Eq. (6.4.4).

$$f_t(x) = tansig(x) = \frac{2}{1+e^{-2x}} - 1 \tag{6.4.4}$$

The dimension of first weight matrix connecting input layer to first hidden layer is 64×15, where 64 represents a number of neurons and 15 is a number of input parameters for ANNs for objective functions CI_b, CO_2, and W as shown in Tables 6.4.2a-(1)–6.4.2c-(1), respectively, where bias matrices of the first hidden layer are also presented. In Tables 6.4.2a-(2)–6.4.2c-(2), the dimension of the fourth weight matrix connecting the fourth hidden layer to output layer for the objective functions CI_b, CO_2, and W, respectively, is 1×64, in which one output parameter and 64 neurons represent ANNs. Dimensions of the first and the fourth bias matrices of the objective functions including CI_b, CO_2, and W are 1×64, and 1×1, respectively.

Table 6.4.2 Weight and bias matrices used to derive objective functions CI_b, CO_2 emissions, and W

(a) For an objective function CI_b

1. Connecting input layer to the first hidden layer

$\mathbf{W}^1_{CI_b[64\times15]}$	$\boldsymbol{b}^1_{CI_b[64\times1]}$
$\begin{bmatrix} -0.342 & -0.481 & 0.297 & -0.222 & . & 0.087 & -0.153 & 0.092 & -0.444 \\ 0.072 & 0.312 & -0.348 & -0.249 & . & -0.297 & 0.029 & 0.692 & 0.608 \\ 0.201 & 0.632 & -0.276 & 0.116 & . & 0.075 & 0.144 & 0.625 & 0.746 \\ -0.037 & 0.385 & -0.337 & 0.092 & . & -0.079 & -0.035 & -0.513 & -0.782 \\ . & . & . & . & . & . & . & . & . \\ . & . & . & . & . & . & . & . & . \\ . & . & . & . & . & . & . & . & . \\ -0.017 & 0.528 & -0.049 & 0.023 & . & -0.048 & 0.047 & 0.030 & -0.218 \\ -0.463 & -0.199 & -0.628 & 0.129 & . & 0.604 & -0.247 & 0.661 & 0.897 \\ -0.037 & -0.199 & 0.043 & 0.059 & . & -0.049 & -0.153 & -0.100 & -0.130 \\ -0.086 & 0.568 & 0.037 & 0.061 & . & -0.187 & 0.176 & 0.295 & -0.656 \end{bmatrix}$	$\begin{bmatrix} 2.113 \\ -1.887 \\ -1.869 \\ -1.670 \\ . \\ . \\ . \\ -1.582 \\ -1.999 \\ -1.529 \\ -1.925 \end{bmatrix}$

2. Connecting the fourth hidden layer to output layer

$\mathbf{W}^4_{CI_b[1\times64]}$	$\boldsymbol{b}^4_{CI_b[1\times1]}$
[0.1299 −0.8455 0.4698 −0.2894 .... 0.4341 0.9981 −0.1363 −0.1348]	[0.0462]

(Continued)

Table 6.4.2 (Continued) Weight and bias matrices used to derive objective functions CI_b, CO_2 emissions, and W

(b) For an objective function CO_2 emissions

1. Connecting input layer to the first hidden layer

$\mathbf{W}^{1}_{CO_2[64\times15]}$									$b^{1}_{CO_2[64\times1]}$
0.068	0.854	0.848	−0.050	.	−0.067	0.085	−0.254	−0.091	−1.811
0.144	0.026	0.302	0.612	.	0.491	0.160	−0.015	0.263	−2.023
0.372	0.377	−0.276	−0.156	.	0.113	0.449	−0.462	0.185	−1.841
0.336	−0.034	−0.607	0.034	.	0.355	0.399	0.171	−0.575	−1.767
.	.	.	.	.	.	.	.	.	.
.	.	.	.	.	.	.	.	.	.
.	.	.	.	.	.	.	.	.	.
−0.193	−0.184	−0.495	0.374	.	0.290	−0.578	−0.504	−0.067	−1.728
0.013	0.607	0.703	0.552	.	0.049	−0.171	−0.703	−0.475	1.709
0.300	0.270	−0.658	−0.413	.	0.521	−0.142	−0.162	−0.538	1.985
−0.087	0.041	0.716	−0.282	.	0.461	−0.625	0.530	0.137	−2.005

2. Connecting the fourth hidden layer to output layer

$\mathbf{W}^{4}_{CO_2[1\times64]}$									$b^{4}_{CO_2[1\times1]}$
[−0.441	−0.462	0.234	−0.814	····	−0.090	−0.212	0.664	0.127]	[−0.8767]

(Continued)

Table 6.4.2 (Continued) Weight and bias matrices used to derive objective functions CI_b, CO_2 emissions, and W

(c) For an objective function W

1. Connecting input layer to the first hidden layer

$\mathbf{W}^1_{W[64\times15]}$	$b^1_{w[64\times1]}$
$\begin{bmatrix} -0.036 & -0.822 & 0.442 & 0.010 & . & -0.085 & 0.108 & -0.340 & 0.611 \\ -0.339 & -0.271 & -0.605 & 0.334 & . & -0.396 & -0.242 & -0.448 & 0.404 \\ 0.088 & -0.527 & 0.630 & -0.089 & . & 0.235 & -0.283 & -0.302 & -0.714 \\ -0.007 & 0.021 & -0.152 & -0.347 & . & -0.482 & 0.652 & -0.442 & -0.708 \\ . & . & . & . & . & . & . & . & . \\ . & . & . & . & . & . & . & . & . \\ . & . & . & . & . & . & . & . & . \\ 0.289 & 0.805 & -0.547 & -0.046 & . & 0.111 & -0.802 & -0.517 & 0.165 \\ -0.060 & -0.423 & -0.279 & 0.036 & . & -0.011 & -0.247 & -0.638 & 0.008 \\ 0.042 & -0.850 & -0.599 & 0.026 & . & -0.103 & -0.034 & -0.104 & 0.299 \\ -0.281 & -0.410 & 0.309 & 0.110 & . & 0.060 & -0.466 & 0.725 & -0.515 \end{bmatrix}$	$\begin{bmatrix} 1.874 \\ 1.995 \\ -1.680 \\ 1.927 \\ . \\ . \\ . \\ 1.742 \\ -1.393 \\ 1.838 \\ -1.939 \end{bmatrix}$

2. Connecting the fourth hidden layer to output layer

$\mathbf{W}^4_{CO_2[1\times64]}$	$b^4_{CO_2[1\times1]}$
$[-0.455 \quad -0.368 \quad 0.646 \quad 0.574 \quad \cdots \quad -0.510 \quad 0.204 \quad -0.138 \quad 0.830]$	$[-0.331]$

Table 6.4.3 Equality and inequality constraints imposed by ACI 318-19 for optimization designs

No.	*Equality conditions*			*Inequality conditions*		
1	L	=	10,000 mm	$\rho_{rt,\min} = \max\left(\frac{0.25\sqrt{f_c'}}{f_y};\frac{1.4}{f_y}\right)$	≤	ρ_{rt}
2	d	=	950 mm	ρ_{rt}	≤	0.05
3	f_y	=	500 MPa	0	≤	ρ_{rc}
4	f_c'	=	30 MPa	ρ_{rc}	≤	ρ_{rt}
5	t_f	=	13 mm	0.3*b*	≤	b_s
6	t_w	=	8 mm	b_s	≤	0.6*b*
7	f_{yS}	=	325 MPa	0.4*d*	≤	h_s
8	Y_s	=	70 mm	h_s	≤	0.6*d*
9	M_D	=	500 kN·m	0.3*d*	≤	b
10	M_L	=	1,500 kN·m	b	≤	0.8*d*
11				$0.003+f_y/200{,}000$	≤	ε_{rt}
12				140 mm	≤	b_s
13				$\Delta_{imme.}$	≤	L/360
14				Δ_{long}	≤	L/240
15				50 mm	≤	X_s
16				1.0	≤	SF

6.4.2.2 Ten equality and 16 inequality constraints

In this chapter, ANN-based Lagrange Algorithm is implemented in designing an SRC beam to minimize objective functions of CI_b, CO_2, and W individually. Constraining conditions of 10 equalities and 16 inequalities are presented in Table 6.4.3. Ten equality conditions are selected with beam length L=100,000 mm, beam depth d=950 mm, strength of rebar f_y=500 MPa, strength of concrete f_c'=30 MPa, flange thickness of steel H-shaped t_f =13 mm, web thickness of steel H-shaped t_w=8 mm, vertical clearance Y_s=70 mm, moment due to dead load M_D=500 kN·m, and moment due to live load M_L=1,500 kN·m. In addition to 10 equality constraints, 16 inequality constraints are also imposed according to ACI 318-19 (ACI Standard [6.29]). Ratios of tensile rebar (ρ_{rt}) should be larger than the minimum requirement based on $\rho_{rt,\min} = \max\left(\frac{0.25\sqrt{f_c'}}{f_y};\frac{1.4}{f_y}\right) = 0.0028$, whereas tensile rebar strains should not be less than $0.003+f_y/200{,}000$= 0.0055 according to ACI 318–19 (ACI Standard [6.29]), resulting in beams designed in the tension-controlled region ensuring that rebars yield well before concrete crushes in the compressive region, providing ductility enough for an SRC beam to prevent brittle failures of an SRC beam. ACI

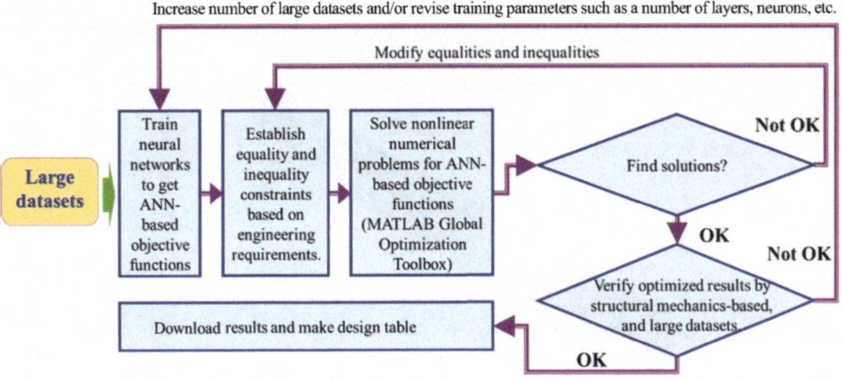

Figure 6.4.2 Flowchart of ANN-based Lagrange algorithm.

318-19 also imposes immediate deflection ($\Delta_{imme.}$) and long-term deflection (Δ_{long}) of the beam to be limited by $L/360$ and $L/240$, respectively, as shown in Table 6.4.3. Horizontal clearance (X_s) is greater than 50mm to secure spaces for an installation of stirrups for beams. Finally, SF must be greater than 1.0. An ANN-based Lagrange algorithm is shown in Figure 6.4.2.

ANNs are trained on large datasets of 200,000 using PTM to obtain weights and biases for deriving objective functions. MATLAB Global Optimization Toolbox (MATLAB Global Optimization Toolbox [6.33], MATLAB Optimization Toolbox [6.34], MATLAB Parallel Computing Toolbox [6.35], MATLAB Statistics and Machine Learning Toolbox [6.36], MATLAB [6.37]) are used to find stationary points of the Lagrange functions with constraints including 10 equality and 16 inequality conditions shown in Table 6.4.3, leading to optimizing objective functions of CI_b, CO_2 emissions, and W which are, then, are verified with those of structural mechanics-based calculation and large datasets as illustrated in Figures 6.4.3–6.4.5.

6.4.3 Optimization designs of SRC beams based on ANN-based Lagrange forward network

6.4.3.1 An optimized design for costs of SRC beams (CI_b)

1. Minimizing costs of SRC beams (CI_b)

   A minimized cost of SRC beams is calculated as 213,793.7 KRW/m based on ANN-based Lagrange algorithm which is close to 214,343.8 KRW/m calculated when design parameters that ANNs identify during an optimization of costs are used for AutoSRCbeam, demonstrating an insignificant error of –0.26% as shown in Table 6.4.4. A minimized objective function of CI_b provided by ANNs is also verified by 133,711 observations, which are filtered from 500,000 datasets

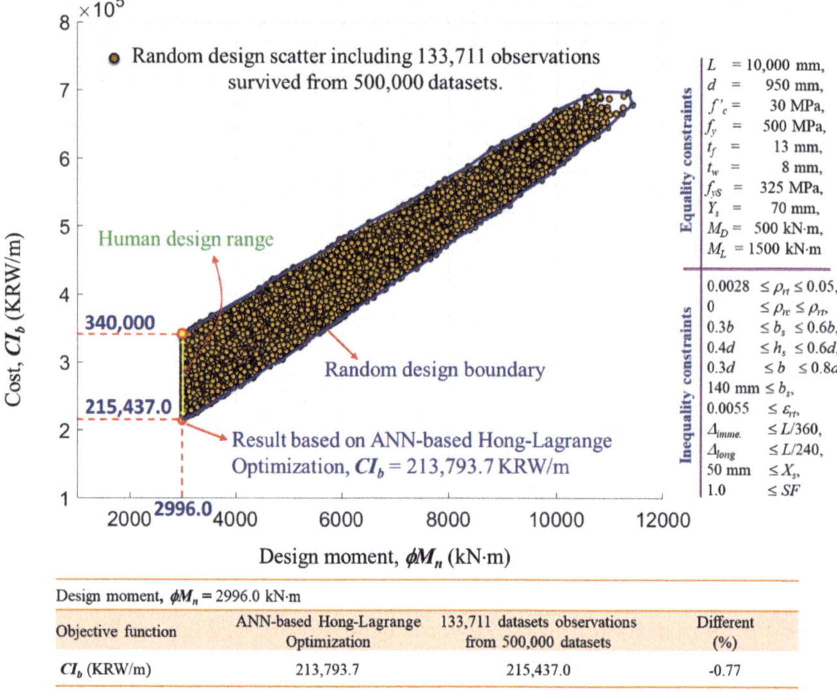

Design moment, ϕM_n = 2996.0 kN·m

Objective function	ANN-based Hong-Lagrange Optimization	133,711 datasets observations from 500,000 datasets	Different (%)
CI_b (KRW/m)	213,793.7	215,437.0	-0.77

Figure 6.4.3 Beam cost CI_b and ϕM_n relationship based on large datasets.

through design parameters (L=10,000 mm, d=950 mm, f_y=500 MPa, f'_c=30 MPa, t_f= 13 mm, t_w=8 mm, f_{yS}=325 MPa, Y_s=70 mm, M_D=500 kN·m, and M_L=1,500 kN·m). Figure 6.4.3 plots CI_b as a function of design moment (ϕM_n = 2996.0 kN·m) based on 133,711 datasets, showing that CI_b varies from 215,437.0 to 340,000 KRW/m which can be regarded as a human design range. The minimum CI_b of 215,437.0 KRW/m identified by 133,711 datasets is sufficiently close to 213,793.7 KRW/m optimized by ANN-based Lagrange algorithm, indicating an insignificant error of −0.77% as shown in Figure 6.4.3. ANN-based Lagrange algorithm is an efficient tool for engineering applications to optimize a cost of a SRC beam that is challenging to obtain by a conventional structural method.

2. Design parameters minimizing costs of SRC beams (CI_b)

   Eleven output parameters in blue Cells 16-26 of Table 6.4.4 are obtained using AutoSRCbeam, based on preassigned 15 input parameters in red Cells 1-15 of Table 6.4.4 which are obtained during minimizing an objective function of CI_b by ANN-based Lagrange forward network. Design moment (ϕM_n) of 2996.0 kN·m obtained by ANNs

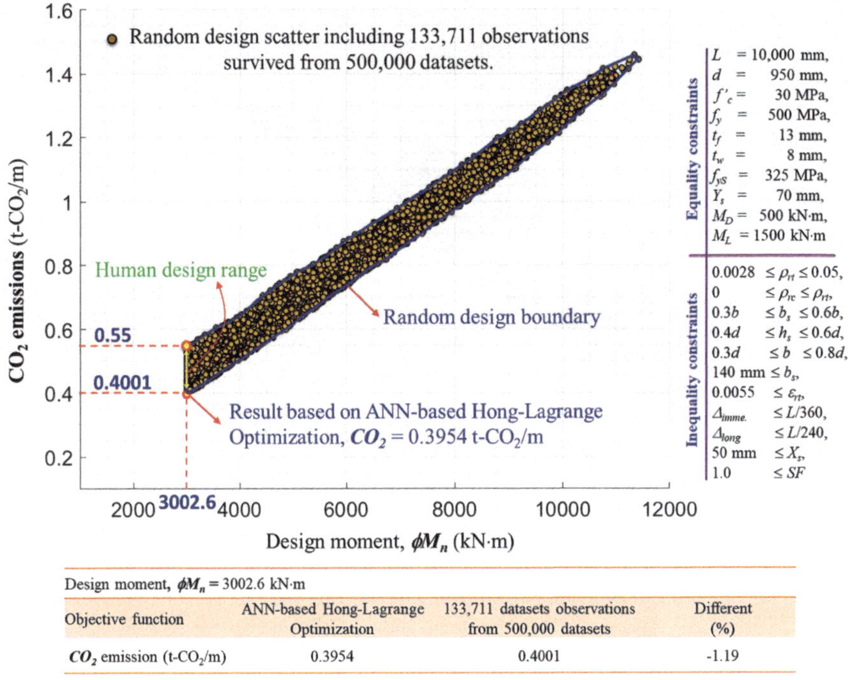

Design moment, ϕM_n = 3002.6 kN·m

Objective function	ANN-based Hong-Lagrange Optimization	133,711 datasets observations from 500,000 datasets	Different (%)
CO_2 emission (t-CO_2/m)	0.3954	0.4001	-1.19

Figure 6.4.4 CO_2 emissions and ϕM_n relationship based on large datasets.

vs. 3,010.1 kN·m based on AutoSRCbeam is verified with an insignificant error of −0.47% as shown in Table 6.4.4. Tensile rebar strain (ε_{rt}) of 0.0055 calculated by ANNs identical to tensile rebar strain (ε_{rt}) of 0.0055 based on AutoSRCbeam is also obtained. Immediate deflections ($\Delta_{imme.}$) of 6.24 mm and 5.95 mm are obtained by ANNs and AutoSRCbeam, respectively, yielding a relative error of 4.65%, demonstrating only a difference of 0.29 mm. Long-term deflection (Δ_{long}) of 13.94 and 12.98 mm are also obtained by ANNs and AutoSRCbeam, respectively, presenting an insignificant difference of 0.96 mm with an error of 6.89%. It is noted that immediate deflections ($\Delta_{imme.}$) and long-term deflection (Δ_{long}) calculated based on ANN-based Lagrange algorithm are smaller than 27.8 mm (L/360) and 41.7 mm (L/240) required by ACI 318-19 code for immediate and long-term deflections, respectively. Table 6.4.4 demonstrates a robust possibility in designing of SRC beams using ANNs, showing accuracies of ANN-based Lagrange algorithm in designing SRC beams with a maximum error of 2.17% for all parameters except for errors of immediate and long-term deflection reaching 4.65% and 6.89%, respectively.

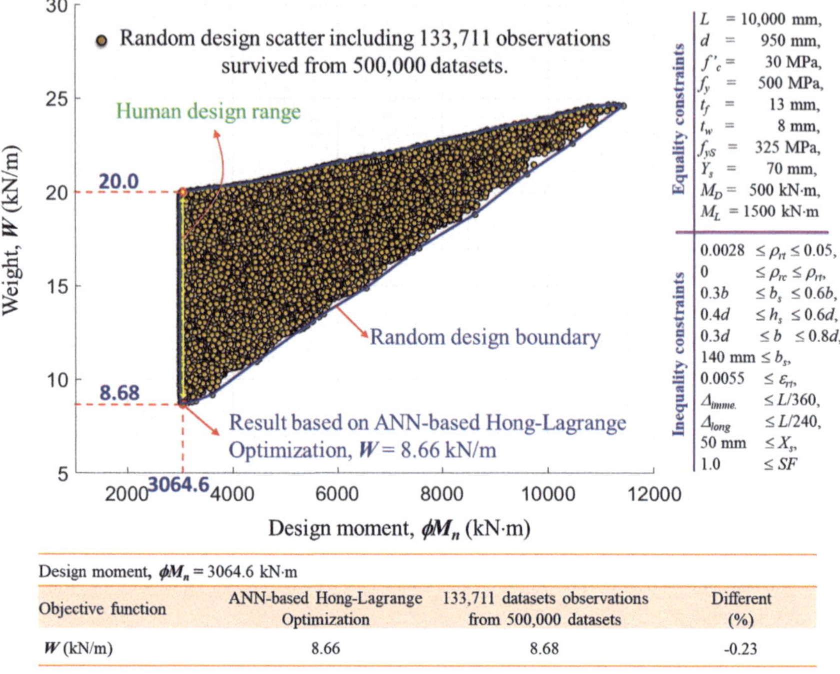

Design moment, ϕM_n = 3064.6 kN·m

Objective function	ANN-based Hong-Lagrange Optimization	133,711 datasets observations from 500,000 datasets	Different (%)
W (kN/m)	8.66	8.68	-0.23

Figure 6.4.5 Beam weight W and ϕM_n relationship based on large datasets.

6.4.3.2 An optimized design for CO_2 emissions of SRC beams

1. Minimizing CO_2 emissions of SRC beams

   CO_2 emissions of SRC beams are optimized as 0.3954 t-CO_2/m based on ANN-based Lagrange algorithm which is close to 0.3966 t-CO_2/m calculated when design parameters that ANNs identify during an optimization of CO_2 emissions are used for AutoSRCbeam, demonstrating an insignificant error of -0.30% as shown in Table 6.4.5. A minimized objective function of CO_2 emissions provided by ANNs is also verified by 133,711 observations, which are filtered from 500,000 datasets through design parameters (L=10,000 mm, d=950 mm, f_y=500 MPa, f_c'=30 MPa, t_f = 13 mm, t_w=8 mm, f_{yS}=325 MPa, Y_s=70 mm, M_D=500 kN·m, and M_L=1500 kN·m). Figure 6.4.4 presents CO_2 emissions as a function of design moment (ϕM_n = 3002.6 kN·m) based on 133,711 datasets, showing that CO_2 emissions vary from 0.4001 t-CO_2/m to 0.5500 t-CO_2/m which can be regarded as a human design range. The minimum CO_2 emission of 0.4001 t-CO_2/m identified by 133,711 datasets is sufficiently close to 0.3954 t-CO_2/m optimized by ANN-based Lagrange method,

Table 6.4.4 Verification of ANN-based Lagrange algorithm for beam cost CI_b by structural mechanics-based calculation

ANN-based Lagrange optimization
Training PTM based on 200,000 datasets
15 Inputs ($L, b, d, f_y, f_c', \rho_{rc}, \rho_{rt}, h_s, b_s, t_f, t_w, f_{yS}, Y_s, M_D, M_L$) – 11 Outputs ($\phi M_n, \varepsilon_{rt}, \varepsilon_{st}, \Delta_{imme}, \Delta_{long}, m_{\phi}, CI_b, CO_2, W, X_s, SF$)

Parameters		Training network and accuracies	ANN-based Lagrange optimization	Structural mechanics-based calculations (AutoSRCbeam)	Error (%)
1	L (mm)		10,000	10,000	-
2	d (mm)		950	950	-
3	b (mm)		388.6	388.6	-
4	f_y (MPa)		500	500	-
5	f_c' (MPa)		30	30	-
6	ρ_{rt}		0.0151	0.0151	-
7	ρ_{rc}		0.0113	0.0113	-
8	h_s (mm)		380.0	380.0	-
9	b_s (mm)		140	140	-
10	t_f (mm)		13	13	-
11	t_w (mm)		8	8	-
12	f_{yS} (MPa)		325	325	-
13	Y_s (mm)		70	70	-
14	M_D (kN·m)		500	500	-
15	M_L (kN·m)		1,500	1,500	–

(Continued)

Table 6.4.4 (Continued) Verification of ANN-based Lagrange algorithm for beam cost CI_b by structural mechanics-based calculation

ANN-based Lagrange optimization Training PTM based on 200,000 datasets 15 Inputs ($L, b, d, f_y, f_c', \rho_{rc}, \rho_{rt}, h_s, b_s, t_f, t_w, f_{yS}, Y_s, M_D, M_L$) – 11 Outputs ($\phi M_n, \varepsilon_{rt}, \varepsilon_{st}, \Delta_{imme.}, \Delta_{long}, m_\phi, CI_b, CO_2, W, X_s, SF$)					
Parameters		Training network and accuracies	ANN-based Lagrange optimization	Structural mechanics-based calculations (AutoSRCbeam)	Error (%)
16	ϕM_n (kN·m)	4 layers-64 neurons; 48,865 epochs; T.MSE=8.25E-06; R=1.0000	2,996.0	3,010.1	−0.47
17	ε_{rt}	4 layers-64 neurons; 34,638 epochs; T.MSE=5.30E-05; R=0.9987	0.0055	0.0055	0.00
18	ε_{st}	4 layers-64 neurons; 49,997 epochs; T.MSE=4.17E-05; R=0.9992	0.0046	0.0045	2.17
19	$\Delta_{imme.}$ (mm)	4 layers-64 neurons; 49,997 epochs; T.MSE=9.96E-06; R=0.9997	6.24	5.95	4.65
20	Δ_{long} (mm)	4 layers-64 neurons; 35,898 epochs; T.MSE=1.80E-05; R=0.9997	13.94	12.98	6.89
21	m_ϕ	4 layers-64 neurons; 48,865 epochs; T.MSE=4.22E-05; R=0.9989	2.01	2.01	0.00
22	CI_b (KRW/m)	4 layers-64 neurons; 32,795 epochs; T.MSE=1.47E-06; R=1.0000	213,793.7	214,343.8	**−0.26**
23	CO_2 (t-CO_2/m)	4 layers-64 neurons; 21,661 epochs; T.MSE=2.15E-08; R=1.0000	0.3961	0.3969	−0.20
24	W (kN/m)	4layers-64 neurons; 49,944 epochs; T.MSE=2.77E-04; R=0.9996	10.85	10.89	−0.37
25	Xs (mm)	4 layers-64 neurons; 15,079 epochs; T.MSE=2.08E-06; R=1.0000	124.33	124.28	0.04
26	SF	4 layers-64 neurons; 27,140 epochs; T.MSE=1.69E-07; R=1.0000	1.00	1.00	0.00

Note: Error=(ANN−AutoSRCbeam)/ANN.

indicating an insignificant error of –1.19% as shown in Figure 6.4.4. ANN-based Lagrange algorithm method is an efficient tool for engineering applications to optimize CO_2 emissions of an SRC beam that is challenging to obtain by a conventional structural method.

2. Design parameters minimizing CO_2 emissions of SRC beams

   Eleven output parameters in blue Cells 16-26 of Table 6.4.5 are obtained using AutoSRCbeam based on preassigned 15 input parameters in red Cells 1-15 of Table 6.4.5 which are obtained during minimizing an objective function of CO_2 emissions by ANN-based Lagrange forward network. Design moment (ϕM_n) of 3,002.6 kN·m obtained by ANNs compared with 3,026.6 kN·m based on AutoSRCbeam is verified with an insignificant error of –0.80% as shown in Table 6.4.5. Tensile rebar strain (ε_{rt}) of 0.0055 calculated by ANNs similar to tensile rebar strain (ε_{rt}) of 0.0056 based on AutoSRCbeam is obtained, demonstrating an error of –1.82%. Immediate deflections ($\Delta_{imme.}$) of 6.22 mm and 6.05 mm obtained using ANNs and AutoSRCbeam, respectively, yield a relative error of 2.73% which demonstrates only a difference of 0.17 mm. An insignificant difference of 0.70 mm with an error of 4.98% is also demonstrated for long-term deflections (Δ_{long}) of 14.06 and 13.36 mm obtained by ANNs and AutoSRCbeam, respectively. Immediate deflections ($\Delta_{imme.}$) and long-term deflections (Δ_{long}) calculated based on ANN-based Lagrange algorithm are smaller than 27.8 mm (L/360) and 41.7 mm (L/240) which are required by ACI 318–19 code. Table 6.4.5 demonstrates a robust possibility in designing of SRC beams using ANNs, resulting in accuracies of ANN-based Lagrange algorithm in designing SRC beams with a maximum error of 2.73% for all parameters except for an error of long-term deflection reaching 4.98%.

6.4.3.3 An optimized design for beam weights (W) of SRC beams

1. Minimizing structural weights (W) of SRC beams

   Structural weight (W) of SRC beams minimized as 8.66 kN/m based on ANN-based Lagrange algorithm is close to 8.64 kN/m calculated when design parameters that ANNs identify during an optimization of structural weights (W) are used for AutoSRCbeam, demonstrating an insignificant error of 0.23% as shown in Table 6.4.6. A minimized objective function of structural weights (W) provided by ANNs is also verified by 133,711 observations, which are filtered from 500,000 datasets through design parameters (L=10,000 mm, d=950 mm, f_y=500 MPa, f_c'=30 MPa, t_f=13 mm, t_w=8 mm, f_{yS}=325 MPa, Y_s=70 mm, M_D=500 kN·m, and M_L=1,500 kN·m). As shown in Figure 6.4.5, structural weights (W) as a function of design moment (ϕM_n)=3064.6 kN·m based on 133,711 datasets show that structural

Table 6.4.5 Verification of ANN-based Lagrange algorithm for CO_2 emissions by structural mechanics-based calculation

			ANN-based Lagrange optimization Training PTM based on 200,000 datasets 15 Inputs ($L, b, d, f_y, f_c', \rho_{rc}, \rho_{rt}, h_s, b_s, t_f, t_w, f_{yS}, Y_s, M_D, M_L$) – 11 Outputs ($\phi M_n, \varepsilon_{rt}, \varepsilon_{st}, \Delta_{imme}, \Delta_{long}, m_{\phi}, CI_b, CO_2, W, X_s, SF$)		
Parameters		Training network and accuracies	ANN-based Lagrange optimization	Structural mechanics-based calculations (AutoSRCbeam)	Error (%)
1	L (mm)		10,000	10,000	-
2	d (mm)		950	950	-
3	b (mm)		409.3947	409.3947	-
4	f_y (MPa)		500	500	-
5	f_c' (MPa)		30	30	-
6	ρ_{rt}		0.0144	0.0144	-
7	ρ_{rc}		0.0099	0.0099	-
8	h_s (mm)		380.0	380.0	-
9	b_s (mm)		140	140	-
10	t_f (mm)		13	13	-
11	t_w (mm)		8	8	-
12	f_{yS} (MPa)		325	325	-
13	Y_s (mm)		70	70	-
14	M_D (kN·m)		500	500	-
15	M_L (kN·m)		1,500	1,500	-

(*Continued*)

Table 6.4.5 (Continued) Verification of ANN-based Lagrange algorithm for CO_2 emissions by structural mechanics-based calculation

ANN-based Lagrange optimization Training PTM based on 200,000 datasets 15 Inputs ($L, b, d, f_y, f_c^{'}, \rho_{rc}, \rho_{rt}, h_s, b_s, t_f, t_w, f_{yS}, Y_s, M_D, M_L$) – 11 Outputs ($\phi M_n, \varepsilon_{rt}, \varepsilon_{st}, \Delta_{imme}, \Delta_{long}, m_\phi, CI_b, CO_2, W, X_s, SF$)					
Parameters		*Training network and accuracies*	*ANN-based Lagrange optimization*	*Structural mechanics-based calculations (AutoSRCbeam)*	*Error (%)*
16	ϕM_n (kN·m)	4 layers-64 neurons; 48,865 epochs; T.MSE = 8.25E-06; R = 1.0000	3,002.6	3,026.6	–0.80
17	ε_{rt}	4 layers-64 neurons; 34,638 epochs; T.MSE = 5.30E-05; R = 0.9987	0.0055	0.0056	–1.82
18	ε_{st}	4 layers-64 neurons; 49,997 epochs; T.MSE = 4.17E-05; R = 0.9992	0.0046	0.0046	0.00
19	$\Delta_{imme.}$ (mm)	4 layers-64 neurons; 49,997 epochs; T.MSE = 9.96E-06; R = 0.9997	6.22	6.05	2.73
20	Δ_{long} (mm)	4 layers-64 neurons; 35,898 epochs; T.MSE = 1.80E-05; R = 0.9997	14.06	13.36	4.98
21	m_ϕ	4 layers-64 neurons; 48,865 epochs; T.MSE = 4.22E-05; R = 0.9989	2.02	2.04	–0.99
22	CI_b (KRW/m)	4 layers-64 neurons; 32,795 epochs; T.MSE = 1.47E-06; R = 1.0000	214141.8	214529.5	–0.18
23	CO_2 (t-CO_2/m)	4 layers-64 neurons; 21,661 epochs; T.MSE = 2.15E-08; R = 1.0000	0.3954	0.3966	**–0.30**
24	W (kN/m)	4 layers-64 neurons; 49,944 epochs; T.MSE = 2.77E-04; R = 0.9996	11.32	11.33	–0.09
25	Xs (mm)	4 layers-64 neurons; 15,079 epochs; T.MSE = 2.08E-06; R = 1.0000	134.74	134.70	0.03
26	SF	4layers-64 neurons; 27,140 epochs; T.MSE = 1.69E-07; R = 1.0000	1.00	1.01	–1.00

Note: Error = (ANN – AutoSRCbeam)/ANN.

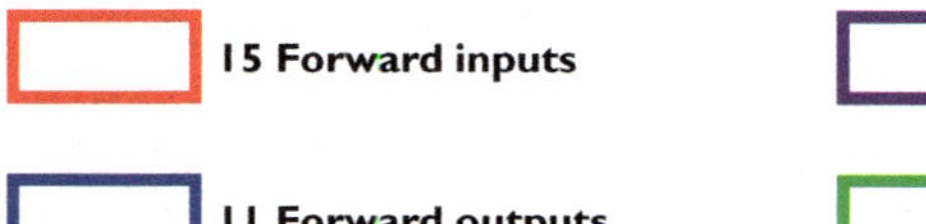

weights (W) vary from 8.68 to 20.00 kN/m, which can be regarded as a human design range. The minimum structural weight (W) of 8.68 kN/m identified by 133,711 datasets is sufficiently close to 8.66 kN/m minimized by the ANN-based Lagrange algorithm, indicating an insignificant error of −0.23% as shown in Figure 6.4.5. ANN-based Hong–Lagrange algorithm is an efficient tool for engineering applications to optimize structural weights (W) of an SRC beam that is challenging to obtain by a conventional structural method.

2. Design parameters minimizing structural weights (W) of SRC beams

   Eleven output parameters in blue Cells 16-26 of Table 6.4.6 are obtained using AutoSRCbeam, based on preassigned 15 input parameters in red Cells 1-15 of Table 6.4.6 which are obtained during minimizing an objective function of structural weights (W) by ANN-based Lagrange forward network. In Table 6.4.6, design moment (ϕM_n) of ϕM_n=3,064.6 kN·m obtained by ANNs is compared with 3,058.0 kN·m calculated by AutoSRCbeam, being verified with an insignificant error of 0.22%. An error of 1.82% is demonstrated when tensile rebar strain (ε_{rt}) of 0.0055 calculated by ANNs is compared with the tensile rebar strain (ε_{rt}) of 0.0054 obtained based on AutoSRCbeam. Immediate deflections ($\Delta_{imme.}$) of 5.79 and 5.49 mm calculated by ANNs and AutoSRCbeam, respectively, result in a relative error of 5.22% with only a difference of 0.30 mm. An insignificant difference of 0.63 mm with an error of 5.70% is also demonstrated by long-term deflection (Δ_{long}) of 11.03 and 10.40 mm which are obtained by ANNs and AutoSRCbeam, respectively. It is noted that immediate deflections ($\Delta_{imme.}$) and long-term deflection (Δ_{long}) calculated based on ANN-based Lagrange algorithm are smaller than 27.8 mm (L/360) and 41.7 mm (L/240) required by ACI 318-19 code, respectively. In Table 6.4.6, a robust design of SRC beams using ANNs is demonstrated, showing an accuracy of ANN-based Lagrange algorithm with a maximum error of −2.97% for all parameters except for errors of immediate and long-term deflection reaching 5.22% and 5.70%, respectively.

6.4.3.4 Overall designs efficiencies optimizing objective functions of CI_b, CO_2 emission, and W based on ANN-based Lagrange algorithm

The overall designs and their accuracies of objective functions, CI_b, CO_2 emission, and W, are performed using ANN-based Lagrange algorithm as shown in Table 6.4.7 where beam width (b), tensile rebar ratio (ρ_{rt}), compressive rebar ratio (ρ_{rc}), height of steel H-shaped (h_s), and width of steel H-shaped (b_s) shown in blue are adjusted while optimizing objective functions based on inequality constraints shown in Table 6.4.3. Observation shows that flange widths (b_s) of

Table 6.4.6 Verification of ANN-based Lagrange algorithm for beam weight W by structural mechanics-based calculation

Parameters		Training network and accuracies	ANN-based Lagrange optimization	Structural mechanics-based calculations (AutoSRCbeam)	Error (%)
ANN-based Lagrange optimization Training PTM based on 200,000 datasets 15 Inputs ($L, b, d, f_y, f_c^{\prime}, \rho_{rc}, \rho_{rt}, h_s, b_s, t_f, t_w, f_{yS}, Y_s, M_D, M_L$) – 11 Outputs ($\phi M_n, \varepsilon_{rt}, \varepsilon_{st}, \Delta_{imme}, \Delta_{long}, m_{dp}, CI_b, CO_2, W, X_s, SF$)					
1	L (mm)		10,000	10,000	-
2	d (mm)		950	950	-
3	b (mm)		285	285	-
4	f_y (MPa)		500	500	-
5	$f_c^{\prime}$ (MPa)		30	30	-
6	ρ_{rt}		0.0229	0.0229	-
7	ρ_{rc}		0.0229	0.0229	-
8	h_s (mm)		416.0	416.0	-
9	b_s (mm)		140	140	-
10	t_f (mm)		13	13	-
11	t_w (mm)		8	8	-
12	f_{yS} (MPa)		325	325	-
13	Y_s (mm)		70	70	-
14	M_D (kN·m)		500	500	-
15	M_L (kN·m)		1,500	1,500	-

(Continued)

Table 6.4.6 (Continued) Verification of ANN-based Lagrange algorithm for beam weight *W* by structural mechanics-based calculation

ANN-based Lagrange optimization Training PTM based on 200,000 datasets 15 Inputs ($L, b, d, f_y, f_c', \rho_{rc}, \rho_{rt}, h_s, b_s, t_f, t_w, f_{yS}, Y_s, M_D, M_L$) – 11 Outputs ($\phi M_n, \varepsilon_{rt}, \varepsilon_{st}, \Delta_{imme.}, \Delta_{long}, m_\phi, CI_b, CO_2, W, X_s, SF$)					
Parameters		*Training network and accuracies*	*ANN-based Lagrange optimization*	*Structural mechanics-based calculations (AutoSRCbeam)*	*Error (%)*
16	ϕM_n (kN·m)	4 layers-64 neurons; 48,865 epochs; T.MSE=8.25E-06; R=1.0000	3,064.6	3,058.0	0.22
17	ε_{rt}	4 layers-64 neurons; 34,638 epochs; T.MSE=5.30E-05; R=0.9987	0.0055	0.0054	1.82
18	ε_{st}	4 layers-64 neurons; 49,997 epochs; T.MSE=4.17E-05; R=0.9992	0.0039	0.0040	−2.97
19	$\Delta_{imme.}$ (mm)	4 layers-64 neurons; 49,997 epochs; T.MSE=9.96E-06; R=0.9997	5.79	5.49	5.22
20	Δ_{long} (mm)	4layers-64 neurons; 35,898 epochs; T.MSE=1.80E-05; R=0.9997	11.03	10.40	5.70
21	μ_ϕ	4layers-64 neurons; 48,865 epochs; T.MSE=4.22E-05; R=0.9989	1.98	1.96	1.05
22	CI_b (KRW/m)	4 layers-64 neurons; 32,795 epochs; T.MSE=1.47E-06; R=1.0000	227560.7	227729.6	−0.07
23	CO_2 (t-CO_2/m)	4 layers-64 neurons; 21,661 epochs; T.MSE=2.15E-08; R=1.0000	0.4325	0.4278	1.08
24	W (kN/m)	4 layers-64 neurons; 49,944 epochs; T.MSE=2.77E-04; R=0.9996	**8.66**	**8.64**	**0.23**
25	Xs (mm)	4 layers-64 neurons; 15,079 epochs; T.MSE=2.08E-06; R=1.0000	72.51	72.50	0.01
26	SF	4 layers-64 neurons; 27,140 epochs; T.MSE=1.69E-07; R=1.0000	1.01	1.02	−0.99

Note: Error=(ANN−AutoSRCbeam)/ANN.

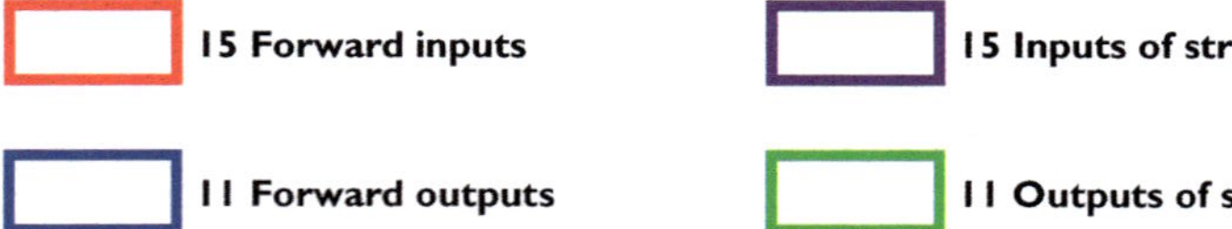

Table 6.4.7 Overall designs and accuracies of objective functions of CI_b, CO_2 emission, and W based on ANN-based Lagrange algorithm

ANN-based Hong–Lagrange optimization Training PTM based on 200,000 datasets 15 Inputs ($L, b, d, f_y, f_c', \rho_{rc}, \rho_{rt}, h_s, b_s, t_f, t_w, f_{yS}, Y_s, M_D, M_L$) – 11 Outputs ($\phi M_n, \varepsilon_{rt}, \varepsilon_{st}, \Delta_{imme}, \Delta_{long}, m_\phi, CI_b, CO_2, W, X_s, SF$)					
Parameters		Training network and accuracies	Objective function cost, CI_b	Objective function CO_2 emissions	Objective function weight, W
1	L (mm)		10,000	10,000	10,000
2	d (mm)		950	950	950
3	b (mm)		388.6	409.3947	285
4	f_y (MPa)		500	500	500
5	f_c' (MPa)		30	30	30
6	ρ_{rt}		0.0151	0.0144	0.0229
7	ρ_{rc}		0.0113	0.0099	0.0229
8	h_s (mm)		380.0	380.0	416.0
9	b_s (mm)		140	140	140
10	t_f (mm)		13	13	13
11	t_w (mm)		8	8	8
12	f_{yS} (MPa)		325	325	325
13	Y_s (mm)		70	70	70
14	M_D (kN·m)		500	500	500
15	M_L (kN·m)		1,500	1,500	1,500

(Continued)

Table 6.4.7 (Continued) Overall designs and accuracies of objective functions of CI_b, CO_2 emission, and *W* based on ANN-based Lagrange algorithm

		ANN-based Hong–Lagrange optimization *Training PTM based on 200,000 datasets* *15 Inputs* ($L, b, d, f_y, f_c', \rho_{rc}, \rho_{rt}, h_s, b_s, t_f, t_w, f_{yS}, Y_s, M_D, M_L$) *– 11 Outputs* ($\phi M_n, \varepsilon_{rt}, \varepsilon_{st}, \Delta_{imme.}, \Delta_{long}, m_\phi, CI_b, CO_2, W, X_s, SF$)			
Parameters		*Training network and accuracies*	*Objective function cost,* CI_b	*Objective function* CO_2 *emissions*	*Objective function weight, W*
16	ϕM_n (kN·m)	4 layers-64 neurons; 48,865 epochs; T.MSE=8.25E-06; R=1.0000	2,996.0	3,002.6	3,064.6
17	ε_{rt}	4 layers-64 neurons; 34,638 epochs; T.MSE=5.30E-05; R=0.9987	0.0055	0.0055	0.0055
18	ε_{st}	4 layers-64 neurons; 49,997 epochs; T.MSE=4.17E-05; R=0.9992	0.0046	0.0046	0.0039
19	$\Delta_{imme.}$ (mm)	4 layers-64 neurons; 49,997 epochs; T.MSE=9.96E-06; R=0.9997	6.24	6.22	5.79
20	Δ_{long} (mm)	4 layers-64 neurons; 35,898 epochs; T.MSE=1.80E-05; R=0.9997	13.94	14.06	11.03
21	μ_ϕ	4 layers-64 neurons; 48,865 epochs; T.MSE=4.22E-05; R=0.9989	2.01	2.02	1.98
22	CI_b (KRW/m)	4 layers-64 neurons; 32,795 epochs; T.MSE=1.47E-06; R=1.0000	**213,793.7**	214,141.8	227,560.7
23	CO_2 (t-CO_2/m)	4 layers-64 neurons; 21,661 epochs; T.MSE=2.15E-08; R=1.0000	0.3961	**0.3954**	0.4325
24	W (kN/m)	4layers-64 neurons; 49,944 epochs; T.MSE=2.77E-04; R=0.9996	10.85	11.32	**8.66**
25	Xs (mm)	4 layers-64 neurons; 15,079 epochs; T.MSE=2.08E-06; R=1.0000	124.33	134.74	72.51
26	*SF*	4 layers-64 neurons; 27,140 epochs; T.MSE=1.69E-07; R=1.0000	1.00	1.00	1.01

Blue **color font indicates the parameters which are adjusted**
Yellow means design tagets which are optimized.

the H-shaped steel for the three objective functions are all 140 mm as shown in Box 9 of Table 6.4.7. ANN-based Lagrange algorithm attempts to reduce beam width (b) to a minimum level ($0.3d$=285 mm) based on an inequality constraint as shown in Box 3 of Table 6.4.7 and Table 6.4.3, leading to minimum beam weight (W=8.66 kN/m). The objective function of W is optimized when tensile rebar ratio (ρ_{rt}), compressive rebar ratio (ρ_{rc}), and height of steel H-shaped (h_s) are 0.0229, 0.229, and 416.0 mm, respectively, which are greater than 0.0151, 0.0113, and 380.0 mm obtained when minimizing the objective function of CI_b while meeting the requirements $SF \geq 1$. However, concrete volume (dxb = 950x388.6mm) decreases to (dxb = 950x285mm) to decrease W. The proposed method minimizes objective functions of CI_b, CO_2 emissions, and W at 213,793.7 KRW/m, 0.3954 t-CO_2/m, and 8.66 kN/m, respectively, as shown in Table 6.4.7. It is noted that costs (CI_b) are obtained at 213,793.7 and 214,141.8 KRW/m when CI_b and CO_2 emissions are minimized, respectively, showing that costs (CI_b) increase by 0.16% when CO_2 emissions are minimized. CO_2 emissions are also obtained at 0.3961 and 0.3954 t-CO_2/m when CI_b and CO_2 emissions are minimized, respectively, showing that CO_2 emissions increase by 0.18% when CI_b are minimized. It is also noted that CI_b and CO2 emissions increase to 227,560.7 from 213,793.7 (6.4% increase) and 0.4325 from 0.3954 (9.4% increase),respectively, when W is optimized, indicating efficiencies of CI_b and CO_2 emissions are neutralized to minimize W. It is also shown that the safety factors are 1.00, 1.00, and 1.01 when CI_b, CO_2 emissions, and W, respectively, as preassigned on an input-side which is difficult to perform with conventional designs.

6.5 CONCLUSION

In this chapter, cost index (CI_b), CO_2 emissions, and weight (W) of SRC beams are optimized based on an ANN-based Lagrange algorithm. The three objective functions are design targets that are optimized for a design of SRC beams. ANN-based objective functions for CI_b, CO_2 emissions, and W of SRC beams, and their constraints are derived as functions of input parameters based on as large as 200,000 datasets. The following conclusions are drawn from this chapter:

1. A robust and sustainable tool to design SRC beams based on ANN-based Lagrange algorithm is introduced, enabling fast and accurate structural design optimizations. Conventional design methods can be replaced by ANN-based design, eliminating human errors, saving engineers' effort and time, improving design accuracy, and reducing construction costs, CO_2 emissions in designing an SRC beam.
2. ANN-based objective functions of cost index CI_b, CO_2 emissions, and beam weights (W) are formulated as a function of 15 forward design

inputs to minimize cost (CI_b) of materials and manufacture, CO_2 emissions, and beam weights (W) constrained by equality and inequality conditions based on design codes and engineer's requirements. Large datasets of 200,000 are used to train ANN using PTM to derive ANN-based objective functions of cost CI_b, CO_2 emissions, and weights. A robust ANN-based Lagrange algorithm is, then, presented to optimize objective functions of CI_b, CO_2 emissions, and W for a design of an SRC beam.

3. Newton–Raphson iterations are used to converge initial input parameters to stationary points, at which Lagrange functions are minimized. Solutions under KKT conditions should be sought when 10 equality and 16 inequality conditions following design codes and engineer's requirements are imposed to govern minimization of cost index, CO_2 emissions, and beam weights.
4. Cost index of 213,793.7 KRW/m, CO_2 emissions of 0.3954 t-CO_2/m, and beam weights of 8.66 kN/m obtained by ANN-based Lagrange algorithm are lower than those identified from 133,711 observations extracted from 500,000 datasets which are 215,437.0–340,000.0 KRW/m for cost, 0.4001–0.5500 t-CO_2/m for CO_2 emissions, and from 8.68 – 20.00 kN/m for beam weights W, respectively. These ranges are regarded as a human design range, demonstrating a sufficient accuracy of the proposed ANN-based design.
5. It is noted that objective functions of CO_2 emissions and W are optimized in an opposite way, whereas objective functions of CO_2 emissions and CI_b are optimized in a similar way.
6. The entire design parameters that are identified by an ANN-based Lagrange algorithm during optimizing objective functions of CI_b, CO_2 emissions, and W are accurately verified by the structural mechanics-based method. The proposed method is verified as a robust tool for designing SRC beams.

REFERENCES

[6.1] Li, L. I., & Matsui, C. (2000). Effects of axial force on deformation capacity of steel encased reinforced concrete beam–columns. In *Proceedings of 12th World Conference on Earthquake Engineering*, Auckland, New Zealand.

[6.2] El-Tawil, S., & Deierlein, G. G. (1999). Strength and ductility of concrete encased composite columns. *Journal of Structural engineering*, *125*(9), 1009–1019. https://doi.org/10.1061/(ASCE)0733-9445(1999)125:9(1009)

[6.3] Bridge, R. Q., & Roderick, J. W. (1978). Behavior of built-up composite columns. *Journal of the Structural Division*, *104*(7), 1141–1155. https://doi.org/10.1061/JSDEAG.0004956

[6.4] Mirza, S. A., Hyttinen, V., & Hyttinen, E. (1996). Physical tests and analyses of composite steel-concrete beam-columns. *Journal of Structural Engineering*, *122*(11), 1317–1326. https://doi.org/10.1061/(ASCE)0733-9445(1996)122:11(1317)

[6.5] Ricles, J. M., & Paboojian, S. D. (1994). Seismic performance of steel-encased composite columns. *Journal of Structural Engineering*, *120*(8), 2474–2494. https://doi.org/10.1061/(ASCE)0733-9445(1994)120:8(2474)

[6.6] Furlong, R. W. (1974). Concrete encased steel columns—design tables. *Journal of the Structural Division*, *100*(9), 1865–1882. https://doi.org/10.1061/JSDEAG.0003878

[6.7] Mirza, S. A., & Skrabek, B. W. (1992). Statistical analysis of slender composite beam-column strength. *Journal of Structural Engineering*, *118*(5), 1312–1332. https://doi.org/10.1061/(ASCE)0733-9445(1992)118:5(1312)

[6.8] Chen, C. C., & Lin, N. J. (2006). Analytical model for predicting axial capacity and behavior of concrete encased steel composite stub columns. *Journal of Constructional Steel Research*, *62*(5), 424–433. https://doi.org/10.1016/j.jcsr.2005.04.021

[6.9] Dundar, C., Tokgoz, S., Tanrikulu, A. K., & Baran, T. (2008). Behaviour of reinforced and concrete-encased composite columns subjected to biaxial bending and axial load. *Building and environment*, *43*(6), 1109–1120. https://doi.org/10.1016/j.buildenv.2007.02.010

[6.10] Munoz, P. R., & Hsu, C. T. T. (1997). Behavior of biaxially loaded concrete-encased composite columns. *Journal of structural Engineering*, *123*(9), 1163–1171. https://doi.org/10.1061/(ASCE)0733-9445(1997)123:9(1163)

[6.11] Rong, C., & Shi, Q. (2021). Analysis constitutive models for actively and passively confined concrete. *Composite Structures*, *256*, 113009. https://doi.org/10.1016/j.compstruct.2020.113009

[6.12] Virdi, K. S., Dowling, P. J., BS 449, & BS 153. (1973). The ultimate strength of composite columns in biaxial bending. *Proceedings of the Institution of Civil Engineers*, *55*(1), 251–272. https://doi.org/10.1680/iicep.1973.4958

[6.13] Roik, K., & Bergmann, R. (1990). Design method for composite columns with unsymmetrical cross-sections. *Journal of Constructional Steel Research*, *15*(1–2), 153–168. https://doi.org/10.1016/0143-974X(90)90046-J

[6.14] Brettle, H. J. (1973). Ultimate strength design of composite columns. *Journal of the Structural Division*, *99*(9), 1931–1951. https://doi.org/10.1061/JSDEAG.0003606

[6.15] Abambres, M., & Lantsoght, E. O. (2020). Neural network-based formula for shear capacity prediction of one-way slabs under concentrated loads. *Engineering Structures*, *211*, 110501. https://doi.org/10.1016/j.engstruct.2020.110501

[6.16] Sharifi, Y., Lotfi, F., & Moghbeli, A. (2019). Compressive strength prediction using the ANN method for FRP confined rectangular concrete columns. *Journal of Rehabilitation in Civil Engineering*, *7*(4), 134–153. https://doi.org/10.22075/JRCE.2018.14362.1260

[6.17] Asteris, P. G., Armaghani, D. J., Hatzigeorgiou, G. D., Karayannis, C. G., & Pilakoutas, K. (2019). Predicting the shear strength of reinforced concrete beams using Artificial Neural Networks. *Computers and Concrete, An International Journal*, *24*(5), 469–488. https://doi.org/10.12989/CAC.2019.24.5.469

[6.18] Armaghani, D. J., Hatzigeorgiou, G. D., Karamani, C., Skentou, A., Zoumpoulaki, I., & Asteris, P. G. (2019). Soft computing-based techniques for concrete beams shear strength. *Procedia Structural Integrity*, *17*, 924–933. https://doi.org/10.1016/j.prostr.2019.08.123

[6.19] Hong, W. K., & Nguyen, M. C. (2022). AI-based Lagrange optimization for designing reinforced concrete columns. *Journal of Asian Architecture and Building Engineering*, *21*(6), 2330–2344. https://doi.org/10.1080/13467581.2021.1971998

[6.20] Hong, W. K., Nguyen, V. T., & Nguyen, M. C. (2022). Optimizing reinforced concrete beams cost based on AI-based Lagrange functions. *Journal of Asian Architecture and Building Engineering*, *21*(6), 2426–2443. https://doi.org/10.1080/13467581.2021.2007105

[6.21] Hong, W. K., Nguyen, M. C., & Pham, T. D. (2023). Optimized interaction PM diagram for rectangular reinforced concrete column based on artificial neural networks abstract. *Journal of Asian Architecture and Building Engineering*, *22*(1), 201–225. https://doi.org/10.1080/13467581.2021.2018697

[6.22] Hong, W. K., & Le, T. A. (2022). ANN-based optimized design of doubly reinforced rectangular concrete beams based on multi-objective functions. *Journal of Asian Architecture and Building Engineering*, 1–17. https://doi.org/10.1080/13467581.2022.2085720

[6.23] Hong, W. K., Le, T. A., Nguyen, M. C., & Pham, T. D. (2022). ANN-based Lagrange optimization for RC circular columns having multi-objective functions. *Journal of Asian Architecture and Building Engineering*, 1–16. https://doi.org/10.1080/13467581.2022.2064864

[6.24] Hong, W. K. 2021. *Artificial intelligence-based Design of Reinforced Concrete Structures*. Daega, Seoul.

[6.25] Nguyen, D. H., & Hong, W. K. (2019). Part I: The analytical model predicting post-yield behavior of concrete-encased steel beams considering various confinement effects by transverse reinforcements and steels. *Materials*, *12*(14), 2302. https://doi.org/10.3390/ma12142302

[6.26] Hong, W. K. 2019. *Hybrid Composite Precast Systems: Numerical Investigation to Construction*. Elsevier: Woodhead Publishing.

[6.27] Hong, W. K., Nguyen, D. H. (2022). Optimization of steel-reinforced concrete beams using artificial neural network-based Hong-Lagrange optimization. *Journal of Asian Architecture and Building Engineering* (under review).

[6.28] Hong, W. K., Nguyen, D. H., & Nguyen, V. T. (2022). Reverse design charts for flexural strength of steel-reinforced concrete beams based on artificial neural networks. *Journal of Asian Architecture and Building Engineering*, 1–39. https://doi.org/10.1080/13467581.2022.2097238

[6.29] An ACI Standard. (2019). *Building Code Requirements for Structural Concrete (ACI 318-19)*. American Concrete Institute, Farmington Hills, MI.

[6.30] Hong, W. K., Pham, T. D., & Nguyen, V. T. (2022). Feature selection based reverse design of doubly reinforced concrete beams. *Journal of Asian Architecture and Building Engineering*, *21*(4), 1472–1496. https://doi.org/10.1080/13467581.2021.1928510

[6.31] MathWorks. (2022a). Deep Learning Toolbox: User's Guide (R2022a). Retrieved July 26, 20122 from: https://www.mathworks.com/help/pdf_doc/deeplearning/nnet_ug.pdf

[6.32] Krenker, A., Bešter, J., & Kos, A. (2011). Introduction to the artificial neural networks. In *Artificial Neural Networks: Methodological Advances and Biomedical Applications* (pp. 1–18). InTech. https://doi.org/10.5772/15751

[6.33] MathWorks. (2022a). Global Optimization: User's Guide (R2022a). Retrieved July 26, 20122 from: https://www.mathworks.com/help/pdf_doc/gads/gads.pdf

[6.34] MathWorks. (2022a). Optimization Toolbox: Documentation (R2022a). Retrieved July 26, 2022, from: https://www.mathworks.com/help/pdf_doc/optim/optim.pdf

[6.35] MathWorks. (2022a). Parallel Computing Toolbox: Documentation (R2022a). Retrieved July 26, 2022, from: https://www.mathworks.com/help/pdf_doc/parallel-computing/parallel-computing.pdf

[6.36] MathWorks. (2022a). Statistics and Machine Learning Toolbox: Documentation (R2022a). Retrieved July 26, 2022, from: https://www.mathworks.com/help/pdf_doc/stats/stats.pdf

[6.37] MathWorks. (2022). *MATLAB R2022a, Version 9.9.0*. Natick, MA: MathWorks.

Chapter 7

Multi-objective optimization (MOO) for steel-reinforced concrete beam developed based on ANN-based Lagrange algorithm

7.1 INTRODUCTION AND SIGNIFICANCE OF CHAPTER 7

7.1.1 Introduction

In the area of structural engineering, an optimization of design has always been a challenging issue over the last few decades. Four major categories of structural optimization can be classified, such as cost reduction, environmental impact reduction, structural performance improvement, and MOO according to a study by Mei and Wang [7.1]. A study of an optimization of rectangular RC beams is presented by Shariat et al. [7.2] using a computational LMM. However, calculations of SRC members are much more complex than common RC structures due to the contribution of an H-shaped steel section embedded inside concrete material, thus, being difficult for calculations of strengths, deflections, and flexibility.

SRC members have been used in various types of infrastructure facilities, such as buildings, parking, and transportation. A sustainability of SRC structures increases significantly due to the protection of concrete outside for H-shaped steel sections inside, preventing a structure from high temperature and corrosive agents of the surrounding environment. When SRC members are subjected to imposed load, thus, steel, rebar, and concrete work simultaneously to prevent damage to structures. Many studies have been performed to investigate behaviors of SRC members under applied loads [7.3–7.16]. SRC members' capacity can be calculated by conventional structural mechanics-based calculations, whereas an ANN-based method has been developed to design structural members that do not require complex structural mechanics theory. Many researchers successfully applied ANNs in structural analysis, including notable studies of [7.17–7.20]. The author and his students perform several studies for an optimization based on either single objective functions or multi-objective functions for RC members in previous studies [7.21–7.28]. Multi-objective functions for SRC beams including cost (CI_b), CO_2 emissions, and weight (W) are introduced in this chapter, applying an ANN-based Lagrange algorithm to a structural design for engineers.

DOI: 10.1201/9781003354796-7

This chapter is a steppingstone for a design of the next generation, not based on structural mechanics but based on ANNs. An ANN-based Lagrange algorithm introduced in his chapter simultaneously optimizes multi-objective functions for engineers and decision-makers. Large datasets used to train ANNs are obtained by structural mechanics-based calculations, called *AutoSRCbeam*, which was developed by Hong et al. in previous studies [7.29,7.30]. The design accuracy of SRC beams obtained using an ANN-based Lagrange algorithm is verified both through structural analysis and large structural datasets as shown in the previous study by Hong et al. [6.27]. This chapter was written based on the previous paper by Hong et al. [7.32].

7.1.2 Research significance

A forward design includes 15 input parameters (L, d, b, f_y, f_c', ρ_{sc}, ρ_{st}, h_s, b_s, t_f, t_w, f_{yS}, Y_s, M_D, M_L), and 11 outputs (ϕM_n, ε_{rt}, ε_{st}, Δ_{imme}, Δ_{long}, μ_ϕ, CI_b, CO_2 emission, W, X_s, SF) as shown in Table 7.1.1. No studies were presented based on ANNs to simultaneously optimize the three objective functions including CI_b, CO_2 emission, and W for an SRC beam. In this chapter, an ANN-based Lagrange algorithm is implemented in optimizing three objective functions (CI_b, CO_2 emission, and W) simultaneously.

This chapter presents a hybrid network using an ANN and Lagrange algorithm to design an SRC beam, capable of simultaneously optimizing three objective functions (CI_b, CO_2 emission, and W) with significant accuracy. A unified objective function (called *UFO*) is formulated based on three objective functions via weighted fractions, integrating them into one objective function to simultaneously optimize all objective functions (CI_b, CO_2 emission, and W) [7.27]. A Pareto frontier, also called Pareto front for an SRC beam, is constructed based on a combination of MOO results. A contribution of the individual objective function is represented by the weight fractions, selected based on an interest of engineers and decision-makers.

Table 7.1.1 Forward design scenario for SRC beam

Forward design scenario							
Input parameters				*Output parameters*			
1	L	9	b_s	1	ϕM_n	9	W
2	d	10	t_f	2	ε_{rt}	10	X_s
3	b	11	t_w	3	ε_{st}	11	SF
4	f_y	12	f_{yS}	4	$\Delta_{imme.}$		
5	f_c'	13	Y_s	5	Δ_{long}		
6	ρ_{rt}	14	M_D	6	μ_ϕ		
7	ρ_{rc}	15	M_L	7	CI_b		
8	h_s			8	CO_2		

A Pareto frontier derived based on an ANN-based Lagrange algorithm to simultaneously optimize designs for three objective functions (CI_b, CO_2, W) is well compared with the lower boundary of large datasets randomly generated by structural mechanics-based calculations (*AutoSRCbeam*). This chapter introduces a powerful tool for the design of next generation in optimizing multiple objective functions for SRC beams that meet imposed constraints and design standards. Both equalities and inequalities are taken into consideration based on engineers' requirements and regional design standards.

7.2 ANN-BASED DESIGN SCENARIOS FOR SRC BEAMS

Figure 7.2.1 demonstrates a section of an SRC beam, including beam section (h, b), steel H-shaped section (h_s, b_s, $t_{f,}$ and t_w) which is encased in concrete material. A forward design scenario for an SRC beam is presented, including 15 inputs and 11 outputs as shown in Table 7.1.1. Fifteen input parameters contain length of beam (L), beam dimensions (b, d) material strengths of concrete (f_c'), rebar (f_y), and steel (f_{yS}), compressive and tensile rebar ratio (ρ_c, ρ_t), steel height (h_s), steel flange (b_s), steel web thickness (t_f), and steel flange thickness (t_w), moment due to dead load (M_D), and moment due to live load (M_L). Eleven output parameters include design moment capacity (ϕM_n) excluding beam weight, tensile strains of steel and rebar (ε_{st}, ε_{rt}), immediate and long-term deflections (Δ_{imme}, Δ_{long}), curvature ductility (μ_ϕ), materials and manufacture cost (CI_b) per 1 m length, CO_2 emission per 1 m length, beam weight (W) per 1 m length, horizontal clearance (X_s), and safety factor (SF). A cost (CI_b) for materials and manufacture, CO_2 emissions, and beam weights (W) are selected as multiple objective functions for an optimization of SRC beam in the present design example.

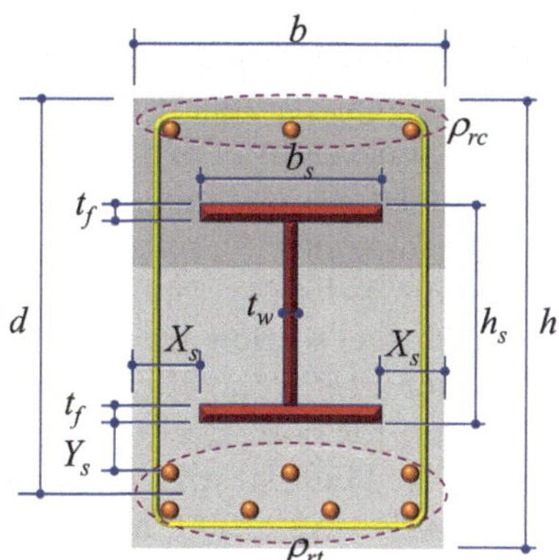

Figure 7.2.1 Geometry of SRC beams [7.31].

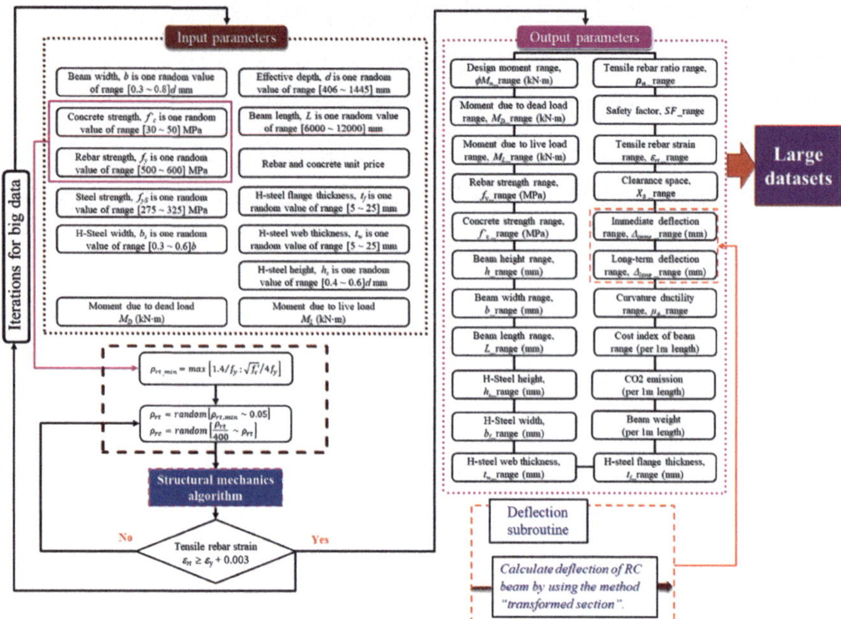

Figure 7.3.1 Flowchart for generating big data of SRC beams used to train network [7.29].

7.3 GENERATION OF LARGE DATASETS

A structural mechanics-based calculation called AutoSRCbeam is used to generate 200,000 datasets. AutoSRCbeam is established based on an algorithm as indicated in Figure 7.3.1. This program was developed by Hong et al. in the previous study [7.29]. Fifteen input parameters for AutoSRCbeam are randomly selected in designated ranges, randomly providing 11 output parameters. Ranges for dimensions of SRC beams are designated from 6,000 to 12,000 mm, 500 to 1,500 mm, and 0.3d to 0.8d for beam length (L), beam height (h), and beam width (b), respectively, where d is effective beam depth in a range of [406.1–1,444.5] mm, as referred to Table 7.3.1. The dimensions of H-shaped steel section are randomly selected in appropriate ranges of [0.4–0.6]d for steel section height (h_s), [0.3–0.6]b for steel section width (b_s), [5–25] mm for both steel web and flange thickness (t_w and t_f), as shown in Table 7.3.1. Material strengths of beam components are chosen in ranges of [30–50] MPa, [500–600] MPa, and [275–325] MPa for concrete (f'_c), rebar (f_y), and steel (f_{yS}), respectively. Compressive rebar ratio is randomly chosen in a range of [1/400 ~ 1.5]ρ_{rt}, where ρ_{rt} is tensile rebar

Table 7.3.1 Nomenclatures and ranges of parameters defining SRC beams [7.31]

		Notation	*Range*
Input parameters	L (mm)	Beam length	[6,000 ~ 12,000]
	d (mm)	Effective beam depth	[406.1 ~ 1,445.5]
	b (mm)	Beam width	$[0.3 \sim 0.8]d$
	h_s (mm)	H-shaped steel height	$[0.4 \sim 0.6]d$
	b_s (mm)	H-shaped steel width	$[0.3 \sim 0.6]b$
	t_f (mm)	H-shaped steel flange thickness	[5 ~ 25]
	t_w (mm)	H-shaped steel web thickness	[5 ~ 25]
	f_c' (MPa)	Concrete strength	[30 ~ 50]
	f_y (MPa)	Rebar strength	[500 ~ 600]
	f_{yS} (MPa)	Steel strength	[275 ~ 325]
	ρ_{rt}	Tensile rebar ratios	$\rho_{rt,\min} = \max\left(\frac{0.25\sqrt{f_c'}}{f_y}; \frac{1.4}{f_y}\right)$
	ρ_{rc}	Compressive rebar ratios	$[1/400 \sim 1.5]\ \rho_{rt}$
	Ys	Vertical clearance	
	M_D (kN·m)	Moment due to dead load	$\left[0.2 \sim \frac{1}{1.2}\right] M_u (M_u = 1.2\ M_D + 1.6\ M_L)$
	M_L (kN·m)	Moment due to service live load	$\frac{1}{1.6}(M_u - 1.2M_D)$ $(M_u = 1.2M_D + 1.6M_L)$
Output parameters	ϕM_n (kN·m)	Design moment without considering effect of self-weight at $\varepsilon_c = 0.003$	
	μ_ϕ	Curvature ductility, $\mu_\phi = \phi_u / \phi_y$ Where, ϕ_u : Curvature at $\varepsilon^c = 0.003$ ϕ_y : Curvature at tensile rebar yield	
	ε_{rt}	Tensile rebar strain at $\varepsilon^c = 0.003$	$\varepsilon_{rt} \geq 0.003 + \varepsilon_{ty}$
	ε_{rc}	Compressive rebar strain at $\varepsilon^c = 0.003$	
	$\Delta_{imme.}$	Immediate deflection due to M_L service live load	($\Delta_{imme.} \leq L/360$, ACI 318-19)
	Δ_{long}	Sum of long-term deflection due to sustained loads and immediate deflection due to additional live load	($\Delta_{long} \leq L/240$, ACI 318-19)
	CI_b (KRW/m)	Cost index per 1 m length of beam	
	CO_2 (t-CO_2/m)	CO_2 emission per 1 m length of beam	
	W (kN/m)	Beam weight per 1 m length of beam	
	X_s	Horizontal clearance	
	SF	Safety factor ($\phi M_n / M_u$)	

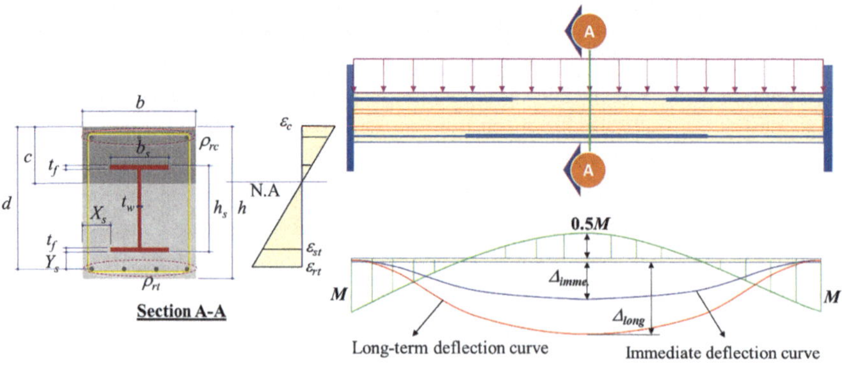

Figure 7.3.2 Fixed-fixed SRC beams used in ANN [7.33].

ratio with a minimum $\rho_{rt,min} = max\left(0.25\sqrt{f_c'}/f_y\ ; 1.4/f_y\right)$, following ACI standard [7.34]. Notations and ranges of 15 input parameters defining an SRC beam are indicated in Table 7.3.1. ACI 319-18 code requests tensile rebar strain (ε_{rt}) greater than 0.003 + ε_{ty} (ε_{ty} is yield strain of rebar) to ensure enough ductility of beam.

This study focuses on SRC beams with fixed-fixed end conditions, as illustrated in Figure 7.3.2. SRC beams are subjected to uniform loads, including dead and live loads, yielding M_D and M_L, respectively. According to ACI 318-19 [7.34], deflections are limited to $L/360$ for immediate deflection ($\Delta_{imme.}$) and $L/240$ for long-term deflection (Δ_{long}). In the preliminary design stage, beam sections are unknown, thus, the design moment capacity of a beam (ϕM_n) is formulated by excluding a self-weight of a beam when generating large datasets. Factored moment (M_u) represents a magnitude of external forces, calculated by load combination of M_D and M_L with load factors (M_u=1.2M_D+1.6M_L). A safety factor represents how safe beam is against applied loads, which is calculated as a ratio between a design moment and factored moment (SF=$\phi M_n/M_u$), and the safety factor must be greater than 1.0. Original and normalized large datasets of 200,000 are shown in Tables 7.3.2(a) and (b), respectively, generated by AutoSRCbeam. Inputs of structural mechanics-based calculation including 15 parameters (L, d, b, f_y, f_c', ρ_{sc}, ρ_{st}, h_s, b_s, t_f, t_w, f_{yS}, Y_s, M_D, M_L) are randomly selected, producing 11 corresponding output parameters (ϕM_n, ε_{rt}, ε_{st}, $\Delta_{imme.}$, Δ_{long}, μ_ϕ, CI_b, CO_2, W, X_s, SF) for design SRC beams. Mean, maxima, and minima of overall 26 parameters based on 200,000 datasets are indicated in Table 7.3.2(a). All input and output parameters are normalized in a range from –1 to 1 by using MAPMINMAX function of MATLAB [7.35] as shown in Table 7.3.2(b).

Table 7.3.2 Large datasets of **SRC** beams generated by AutoSRCbeam

			(a) Non-normalized data								
			200,000 datasets generated by AutoSRCbeam								
		Parameter	*Data 1*	*Data 2*	*Data 3*	*Data 4*	…	*Data 200,000*	*Mean (μ)*	*Maxima*	*Minima*
Fifteen input parameters for structural mechanics-based (AutoSRCbeam)	1	L (mm)	8,400	11,900	9,900	11,050	…	9,850	8,996.7	12,000.0	6,000.0
	2	d (mm)	786	1,097.8	972.1	680.5	…	910.5	1,039.9	1,445.5	406.1
	3	b (mm)	470	850	430	290	…	765	632.0	1,200	150
	4	h_s (mm)	513	548	423	412	…	518	427.7	1,180.0	5.0
	5	b_s (mm)	41	22	29	43	…	23	277.0	710.0	45.0
	6	t_f (mm)	0.0196	0.0186	0.0435	0.0132	…	0.0034	16.1	30.0	5.0
	7	t_w (mm)	0.0229	0.0271	0.0465	0.0140	…	0.0010	16.3	30.0	5.0
	8	f_c' (MPa)	160	645	270	155	…	450	40.0	50	30
	9	f_y (MPa)	170	395	160	130	…	230	550.0	600	500
	10	f_{yS} (MPa)	6	7	28	9	…	11	300.0	325.0	275.0
	11	ρ_{rt}	15	27	16	9	…	5	0.0166	0.05897	0.00243
	12	ρ_{rc}	296	358	267	289	…	379	0.0167	0.08130	7.00E-06
	13	Ys	110.8	137.4	242.4	267.9	…	231.6	245.4	943.0	60.0
	14	M_D (kN·m)	290.9	4,956.5	1,809.1	426.3	…	630.3	2,959.1	36,961.7	12.2
	15	M_L (kN·m)	691.0	3,166.6	1,469.9	272.3	…	133.3	1,480.4	24,343.4	1.3

(Continued)

Table 7.3.2 (Continued) Large datasets of **SRC** beams generated by AutoSRCbeam

		(a) Non-normalized data									
		200,000 datasets generated by AutoSRCbeam									
		Parameter	*Data 1*	*Data 2*	*Data 3*	*Data 4*	…	*Data 200,000*	*Mean* (μ)	*Maxima*	*Minima*
Eleven Output parameters for structural mechanics-based (AutoSRCbeam)	16	ϕM_n (kN·m)	2,865.8	10,794.1	6,286.4	795.7	…	1,512.6	7,539.5	44,085.5	84.8
	17	ε_{rt}	0.0107	0.0074	0.0056	0.0114	…	0.0094	0.00832	0.042	0.0055
	18	ε_{st}	0.0081	0.0052	0.0020	0.0060	…	0.0064	0.00504	0.031	−0.0004
	19	μ_ϕ	2.15	4.57	1.94	7.00	…	0.87	2.738	13.749	1.784
	20	$\Delta_{imme.}$(mm)	3.73	18.63	5.85	28.90	…	6.35	2.06	27.98	0.10
	21	Δ_{long} (mm)	3.45	2.34	2.39	4.76	…	3.32	9.40	84.17	0.47
	22	CI_b (KRW/m)	226,654.0	754,831.5	515,783.1	114,789.7	…	199,231.1	493,406.1	2,121,983.0	2,6970.3
	23	CO_2 (t-CO_2/m)	0.453	1.399	1.059	0.211	…	0.337	0.899	4.292	0.044
	24	*W* (kN/m)	11.23	28.97	15.50	5.81	…	18.57	20.93	57.3	1.97
	25	*Xs* (mm)	150.0	227.5	135.0	80.0	…	267.5	177.5	420.0	30.0
	26	*SF*	1.97	0.98	1.39	0.84	…	1.56	1.37	2.0	0.75

(Continued)

TABLE 7.3.2 *(Continued)* Large datasets of **SRC** beams generated by AutoSRCbeam

		Parameter	Data 1	Data 2	Data 3	Data 4	...	Data 200,000	Mean (μ)	Maxima	Minima
		(b) Normalized data									
		200,000 datasets generated by AutoSRCbeam									
Fifteen input parameters for structural mechanics-based (AutoSRCbeam)	1	L (mm)	−0.800	0.950	−0.050	0.525	...	−0.075	−0.002	1.000	−1.000
	2	d (mm)	−0.255	0.338	0.099	−0.456	...	−0.018	0.184	1.000	−1.000
	3	b (mm)	−0.390	0.333	−0.467	−0.733	...	0.171	−0.104	1.000	−1.000
	4	h_s (mm)	0.130	0.480	−0.770	−0.880	...	0.180	0.001	1.000	−1.000
	5	b_s (mm)	0.050	−0.900	−0.550	0.150	...	−0.850	0.000	1.000	−1.000
	6	t_f (mm)	−0.440	−0.473	0.342	−0.647	...	−0.969	−0.434	1.000	−1.000
	7	t_w (mm)	−0.487	−0.393	0.040	−0.687	...	−0.979	−0.565	1.000	−1.000
	8	f_c' (MPa)	−0.729	0.093	−0.542	−0.737	...	−0.237	−0.322	1.000	−1.000
	9	f_y (MPa)	−0.630	0.037	−0.659	−0.748	...	−0.452	−0.327	1.000	−1.000
	10	f_{yS} (MPa)	−0.600	−0.533	0.867	−0.400	...	−0.267	0.072	1.000	−1.000
	11	ρ_{rt}	−0.200	0.760	−0.120	−0.680	...	−1.000	−0.083	1.000	−1.000
	12	ρ_{rc}	−0.040	0.580	−0.330	−0.110	...	0.790	0.001	1.000	−1.000
	13	Ys	−0.885	−0.825	−0.587	−0.530	...	−0.612	−0.600	1.000	−1.000
	14	M_D (kN·m)	−0.985	−0.730	−0.902	−0.977	...	−0.966	−0.844	1.000	−1.000
	15	M_L (kN·m)	−0.941	−0.730	−0.875	−0.977	...	−0.989	−0.878	1.000	−1.000

(Continued)

Table 7.3.2 (Continued) Large datasets of SRC beams generated by AutoSRCbeam

		(b) Normalized data								
		200,000 datasets generated by AutoSRCbeam								
		Parameter	*Data 1*	*Data 2*	*Data 3*	*Data 4*	...	*Data 200,000*	*Mean (μ)*	*Maxima*	*Minima*
Eleven output parameters for structural mechanics-based (AutoSRCbeam)	16	ϕM_n (kN·m)	−0.880	−0.540	−0.733	−0.968	...	−0.938	−0.690	1.000	−1.000
	17	ε_{rt}	−0.723	−0.883	−0.971	−0.689	...	−0.786	−0.845	1.000	−1.000
	18	ε_{st}	−0.513	−0.677	−0.858	−0.632	...	−0.609	−0.697	1.000	−1.000
	19	μ_ϕ	−0.835	−0.650	−0.852	−0.465	...	−0.933	−0.820	1.000	−1.000
	20	$\Delta_{imme.}$ (mm)	−0.934	−0.593	−0.885	−0.358	...	−0.874	−0.772	1.000	−1.000
	21	Δ_{long} (mm)	−0.799	−0.930	−0.924	−0.643	...	−0.814	−0.857	1.000	−1.000
	22	CI_b (KRW/m)	−0.824	−0.358	−0.569	−0.922	...	−0.848	−0.581	1.000	−1.000
	23	CO_2 (t-CO_2/m)	−0.827	−0.428	−0.571	−0.929	...	−0.876	−0.623	1.000	−1.000
	24	W (kN/m)	−0.686	−0.082	−0.540	−0.870	...	−0.436	−0.367	1.000	−1.000
	25	*Xs* (mm)	−0.377	0.026	−0.455	−0.740	...	0.234	−0.251	1.000	−1.000
	26	*SF*	0.952	−0.632	0.024	−0.856	...	0.296	0.001	1.000	−1.000

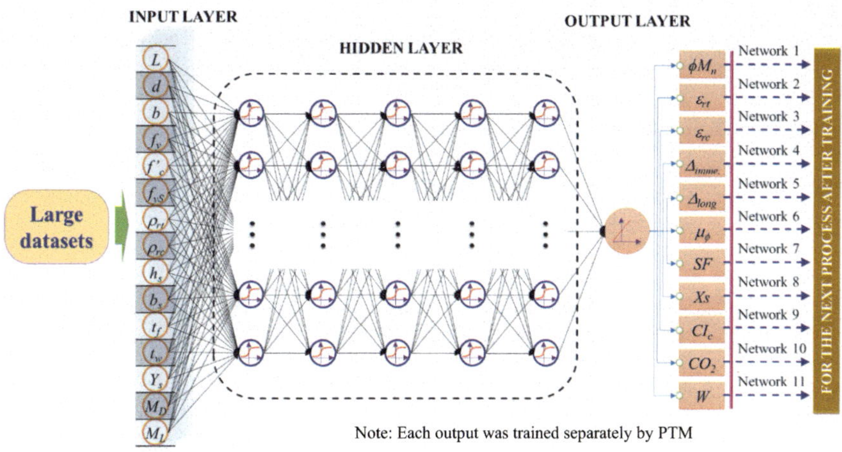

Figure 7.4.1 Topology of ANN for PTM method [7.31].

7.4 MOO USING ANN-BASED LAGRANGE ALGORITHM FOR SRC BEAMS

7.4.1 Training ANNs based on PTM

In this study, ANNs are trained using a PTM [7.26,7.36]. As shown in Figure 7.4.1, 11 training networks are formulated independently when 15 input parameters (L, d, b, f_y, f_c', ρ_{sc}, ρ_{st}, h_s, b_s, t_f, t_w, f_{yS}, Y_s, M_D, M_L) are mapped to each of 11 output parameters (ϕM_n, ε_{rt}, ε_{st}, $\Delta_{imme.}$, Δ_{long}, μ_ϕ, CI_b, CO_2, W, X_s, SF) as demonstrated in Figure 7.4.1. Table 7.4.1 presents the training results of 11 networks using PTM, based on 200,000 datasets that are divided into three distinct portions, covering 70% (140,000 datasets) for training, 15% (30,000 datasets) for validation, and 15% (30,000 datasets) for testing data. The lowest and highest test MSE are 2.77E-04 and 1.69E-07 for training weight (W) and SF, respectively. The lowest and highest regression (R) are 0.9987 and 1.0000 for training strain of tensile rebar (ε_{rt}) and several parameters (ϕM_n, SF, X_s, CI_b, CO_2), respectively, as indicated in Table 7.4.1. Neural networks are trained based on 50,000 and 1,000 for suggested and validation epochs, respectively. Trainings are performed by MATLAB Deep Learning Toolbox™ platform [7.37].

Table 7.4.1 Training results based on PTM method [7.31]

No.	*Training with an output*	*Data*	*Layers*	*Neurons*	*Validation epoch*	*Suggested epoch*	*Best epoch*	*Stopped epoch*	*Test MSE*	*R at best epoch*
1	ϕM_n	200,000	4	64	1,000	50,000	48,865	49,865	8.25E-06	1.0000
2	ε_{rt}	200,000	4	64	1,000	50,000	34,638	35,638	5.30E-05	0.9987
3	ε_{st}	200,000	4	64	1,000	50,000	49,997	50,000	4.17E-05	0.9992
4	$\Delta_{imme.}$	200,000	4	64	1,000	50,000	49,997	50,000	9.96E-06	0.9997
5	Δ_{long}	200,000	4	64	1,000	50,000	49,748	50,000	1.80E-05	0.9997
6	μ_ϕ	200,000	4	64	1,000	50,000	35,898	36,898	4.22E-05	0.9989
7	*SF*	200,000	4	64	1,000	50,000	27,140	27,141	1.69E-07	1.0000
8	X_s	200,000	4	64	1,000	50,000	15,079	16,079	2.08E-06	1.0000
9	CI_b	200,000	4	64	1,000	50,000	32,795	33,795	1.47E-06	1.0000
10	CO_2	200,000	4	64	1,000	50,000	21,661	21,661	2.15E-08	1.0000
11	W	200,000	4	64	1,000	50,000	49,944	50,000	2.77E-04	0.9996

7.4.2 Optimized objective functions using ANN-based Lagrange algorithm

7.4.2.1 *Derivation of objective functions based on forward neural networks*

Derivation of objective functions including cost (CI_b), CO_2 emissions, and weight (W) based on forward ANNs is described in this section, where the relationships among 15 input parameters (L, d, b, f_y, f_c', ρ_{sc}, ρ_{st}, h_s, b_s, t_f, t_w, f_{yS}, Y_s, M_D, M_L) and 11 output parameters (ϕM_n, ε_{rt}, ε_{st}, $\Delta_{imme.}$, Δ_{long}, μ_ϕ, CI_b, CO_2, W, X_s, SF) are linked weight and bias matrices obtained by mapping entire input parameters to each output parameter using PTM. An activation function is also implemented to yield non-linear behaviors of objective functions as indicated in Eqs. (7.4.1)–(7.4.3) [7.31,7.38,7.39]. An activation function *tansig* is used in this paper, as expressed in Eq. (7.4.4). ANNs with multilayer perceptron trained using PTM are based on four layers and 64 neurons as shown in Table 7.4.1.

$$\underbrace{CI_b}_{[1\times1]} = g_{CI_b}^{D}\left(f_{lin}^{4}\left(\underbrace{\mathbf{W}_{CI_b}^{4}}_{[1\times64]} f_t^{3}\left(\underbrace{\mathbf{W}_{CI_b}^{3}}_{[64\times64]} \cdots f_t^{1}\left(\underbrace{\mathbf{W}_{CI_b}^{1}}_{[64\times15]} \underbrace{g_{CI_b}^{N}(\mathbf{X})}_{[15\times1]} + \underbrace{\mathbf{b}_{CI_b}^{1}}_{[64\times1]} \right) \cdots + \underbrace{\mathbf{b}_{CI_b}^{3}}_{[64\times1]} \right) + \underbrace{b_{CI_b}^{4}}_{[1\times1]} \right)\right) \quad (7.4.1)$$

$$\underbrace{CO_2}_{[1\times1]} = g_{CO_2}^{D}\left(f_{lin}^{4}\left(\underbrace{\mathbf{W}_{CO_2}^{4}}_{[1\times64]} f_t^{3}\left(\underbrace{\mathbf{W}_{CO_2}^{3}}_{[64\times64]} \cdots f_t^{1}\left(\underbrace{\mathbf{W}_{CO_2}^{1}}_{[64\times15]} \underbrace{g_{CO_2}^{N}(\mathbf{X})}_{[15\times1]} + \underbrace{\mathbf{b}_{CO_2}^{1}}_{[64\times1]} \right) \cdots + \underbrace{\mathbf{b}_{CO_2}^{3}}_{[64\times1]} \right) + \underbrace{b_{CO_2}^{4}}_{[1\times1]} \right)\right) \quad (7.4.2)$$

$$\underbrace{W}_{[1\times1]} = g_{W}^{D}\left(f_{lin}^{L}\left(\underbrace{\mathbf{W}_{W}^{4}}_{[1\times64]} f_t^{3}\left(\underbrace{\mathbf{W}_{W}^{3}}_{[64\times64]} \cdots f_t^{1}\left(\underbrace{\mathbf{W}_{W}^{1}}_{[64\times15]} \underbrace{g_{W}^{N}(\mathbf{X})}_{[15\times1]} + \underbrace{\mathbf{b}_{W}^{1}}_{[64\times1]} \right) \cdots + \underbrace{\mathbf{b}_{W}^{3}}_{[64\times1]} \right) + \underbrace{b_{W}^{4}}_{[1\times1]} \right)\right) \quad (7.4.3)$$

$$f_t(x) = tansig(x) = \frac{2}{1+e^{-2x}} - 1 \quad (7.4.4)$$

where

$\mathbf{X}$; input parameters, $\mathbf{X} = \left[L, d, b, f_y, f_c', \rho_{sc}, \rho_{st}, h_s, b_s, t_f, t_w, f_{yS}, Y_s, M_D, M_L\right]^T$.
g^N, g^D; normalizing and de-normalizing functions.
$\mathbf{W}^i_{CI_b}, \mathbf{W}^i_{CO_2}, \mathbf{W}^i_W$; weight matrices $(i = \overline{1-4})$ obtained using training PTM.
$\mathbf{b}^i_{CI_b}, \mathbf{b}^i_{CO_2}, \mathbf{b}^i_W$; bias matrices $(i = \overline{1-4})$ obtained using training PTM.

7.4.2.2 Ten equality and 16 inequality constraints

Table 7.4.2 presents 10 equality and 16 inequality constraints to optimize three objective functions (CI_b, CO_2, and W) simultaneously. Ten equality constraints including eight input parameters (L = 100,000 mm, d=950 mm, f_y=500 MPa, f_c'=30 MPa, t_f = 12 mm, t_w = 8 mm, f_{yS} = 500 MPa Y_s=70 mm) and two applied loads (M_D = 500 kN·m, M_L=1,500 kN·m). Besides 10 equalities selected, 16 inequality conditions are also selected according to ACI 318-19 standard [7.34]. A minimum requirement for rebar ratio

TABLE 7.4.2 Equality and inequality constraints imposed by **ACI 318-19** for optimization designs **[7.31]**

	Equality conditions				*Inequality conditions*		
EC_1	L	=	10,000 mm	IC_1	$\rho_{rt,\min} = \max\left(\frac{0.25\sqrt{f_c'}}{f_y}; \frac{1.4}{f_y}\right) = 0.0028$	≤	ρ_{rt}
EC_2	d	=	950 mm	IC_2	ρ_{rt}	≤	0.05
EC_3	f_y	=	500 MPa	IC_3	$\rho_{rt}/400$	≤	ρ_{rc}
EC_4	f_c'	=	30 MPa	IC_4	ρ_{rc}	≤	ρ_{rt}
EC_5	t_f	=	12 mm	IC_5	$0.3b$	≤	b_s
EC_6	t_w	=	8 mm	IC_6	b_s	≤	$0.6b$
EC_7	f_{yS}	=	325 MPa	IC_7	$0.4d$	≤	h_s
EC_8	Y_s	=	70 mm	IC_8	h_s	≤	$0.6d$
EC_9	M_D	=	500 kN·m	IC_9	$0.3d$	≤	b
EC_{10}	M_L	=	1,500 kN·m	IC_{10}	b	≤	$0.8d$
				IC_{11}	$0.003+f_y/200{,}000=0.0055$	≤	ε_{rt}
				IC_{12}	140 mm	≤	b_s
				IC_{13}	$\Delta_{imme.}$	≤	$L/360$
				IC_{14}	Δ_{long}	≤	$L/240$
				IC_{15}	50 mm	≤	X_s
				IC_{16}	1.0	≤	SF

is expressed by equation $\rho_{rt,\min} = \max\left(\frac{0.25\sqrt{f_c'}}{f_y};\frac{1.4}{f_y}\right) = 0.0028$ according to ACI 318-19 code, and maximum rebar ratio is selected at 0.05, which is indicated by inequality IC_1 and IC_2 in Table 7.4.2. A maximum rebar ratio of 0.05 is established arbitrarily when generating big data. ACI 318-19 also recommends tensile rebar strain should not be less than $0.003+f_y/200{,}000=0.0055$ to ensure enough ductility of beams, indicated by inequality IC_{11} in Table 7.4.2. Immediate and long-term deflections ($\Delta_{imme.}$ and Δ_{long}) are limited by $L/360$ and $L/240$ according to ACI 318-19, respectively, indicated by inequalities IC_{13} and IC_{14}. Other equality constraints shown in Table 7.4.2 are also selected. Ten equalities and 16 inequalities are used to optimize multi-objective functions simultaneously based on MATLAB Global Optimization Toolbox [7.40–7.43].

7.4.2.3 Derivation of a unified function of objective (UFO) for a SRC beam

A UFO for SRC beams is defined using algorithms based on the *weighted sum technique* [7.44], which is created by integrating three objective functions (CI_b, CO_2, and W) with their respective weight fractions w_{CI_b}, w_{CO_2}, w_W, respectively, as indicated in Eqs. (7.4.5) and (7.4.6). These fraction weights are in a range from 0 to 1, whose sum is 1 as shown in Eq. (7.4.6). A specific tradeoff among objectives is used to holistically evaluate a design project for engineers. Individual objective function of CI_b, CO_2, and W is a specific case of UFO when establishing $w_{CI_b} : w_{CO_2} : w_W = 1{:}0{:}0;\ 0{:}1{:}0;$ or $0{:}0{:}1$, respectively. A Lagrange function of a *UFO* shown in Eqs. (7.4.7)–(7.4.9) which are based on three objective functions is simultaneously optimized. The optimization is constrained by equality and inequality constraints shown in Table 7.4.2.

The Lagrange function is substituted into the built-in optimization toolbox of MATLAB [7.40] to obtain optimized design parameters. Figure 7.4.2 demonstrates a flowchart for five steps to solve MOO problems, whereas detailed descriptions of algorithms for RC columns are shown in Hong et al. [7.27].

$$\begin{aligned} UFO &= \sum_{1}^{3} w_i F_i^{ANN}(\mathbf{x}) = w_1 F_1^{ANN}(\mathbf{x}) + w_2 F_2^{ANN}(\mathbf{x}) + w_3 F_3^{ANN}(\mathbf{x}) \\ &= w_{CI_b} CI_b^{ANN}(\mathbf{x}) + w_{CO_2} CO_2^{ANN}(\mathbf{x}) + w_W W^{ANN}(\mathbf{x}) \end{aligned} \quad (7.4.5)$$

where

$$\begin{aligned} & w_{CI_b} + w_{CO_2} + w_W = 1 \\ & 0 \le w_{CI_b}, w_{CO_2}, w_W \le 1 \ (\text{dimensionless}) \end{aligned} \quad (7.4.6)$$

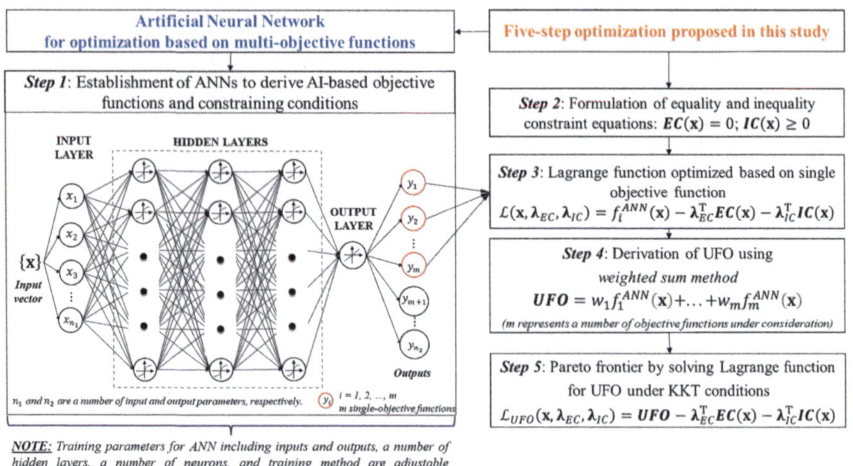

Figure 7.4.2 ANN-based Lagrange optimization algorithm of five steps based on *UFO* [7.28].

Lagrange function utilizing *UFO* function:

$$\mathcal{L}_{\text{UFO}}(\text{x}, \lambda_c, \lambda_v) = UFO - \lambda_c^{\text{T}} EC(\text{x}) - \lambda_v^{\text{T}} IC(\text{x}) \tag{7.4.7}$$

$$EC(\text{x}) = \left[EC_1(\text{x}),\ EC_2(\text{x}), \ldots,\ EC_{m_1}(\text{x})\right]^T; \quad m_1 = 10:\ \text{number of equality constraints} \tag{7.4.8}$$

$$IC(\text{x}) = \left[IC_1(\text{x}),\ IC_2(\text{x}), \ldots,\ IC_{m_2}(\text{x})\right]^T; \quad m_2 = 16:\ \text{number of inequality constraints} \tag{7.4.9}$$

7.4.3 Results of optimization MOO of SRC beams

7.4.3.1 *Definition of four specific cases on a Pareto frontier based on a three-dimensional Pareto frontier*

A Pareto frontier is obtained based on a combination of multiple optimized designs for three objective functions (CI_b, CO_2, and W) with their weight fractions ($w_{CI_b} : w_{CO_2} : w_W$) for SRC beams. These weight fractions ($w_{CI_b} : w_{CO_2} : w_W$) represent tradeoff ratios contributed by each of the three

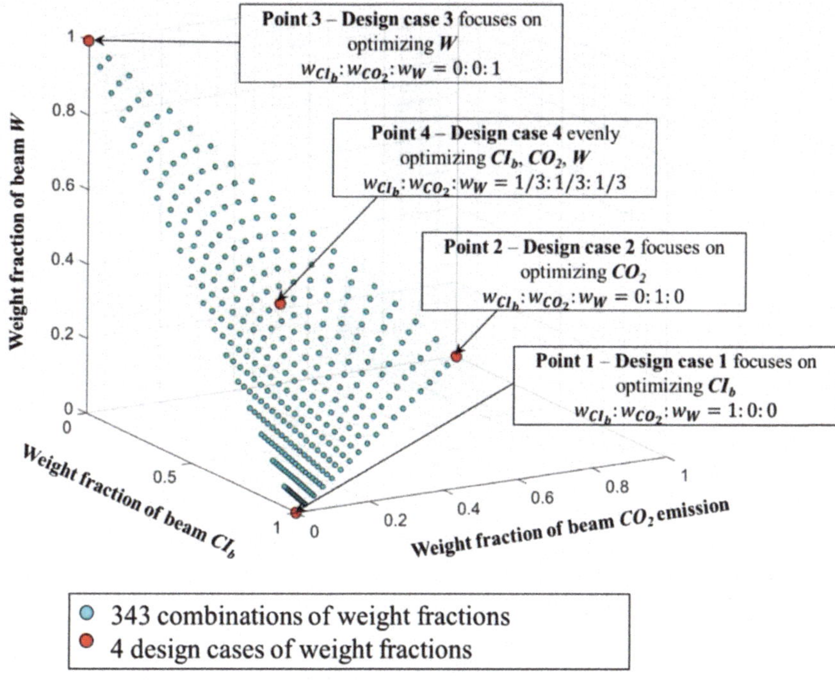

Figure 7.4.3 Weight fractions generated for Pareto frontier [7.35].

objective functions to real-life optimizations for engineers and decision-makers. The 343 combinations of weight fractions are generated for constructing a Pareto frontier, including four specific cases as indicated in Figure 7.4.3. It is noted that the sum of three weight fractions is always equivalent to 1. Four specific cases on a Pareto frontier include an individual optimization of objective function, CI_b, CO_2, and W represented by Points 1, 2, and 3, respectively, and Point 4 representing an evenly optimized design for CI_b, CO_2, and W with their weight fractions $w_{CI_b} : w_{CO_2} : w_W = 1/3:1/3:1/3$. Design parameters ($b$, h_s, b_s, ρ_{rc}, ρ_{rt}) are then obtained by solving MOO using an ANN-based Lagrange algorithm with 10 equality and 16 inequality conditions shown in Table 7.4.2.

A Pareto frontier includes 343 optimized designs for SRC beams, indicated by red dots as shown in Figure 7.4.4(a)–(c). Design parameters (b, h_s, b_s, ρ_{rc}, ρ_{rt}) are calculated for four specific combinations while all 10 equalities (L=10,000 mm, d=950 mm, f_y=500 MPa, f_c'=30 MPa, f_{yS}=325 MPa, t_f = 12 mm, t_w=8 mm, Y_s=70 mm, M_D=500 kN·m, and M_L=1500 kN·m) and 16 inequalities given in Table 7.4.2 are satisfied.

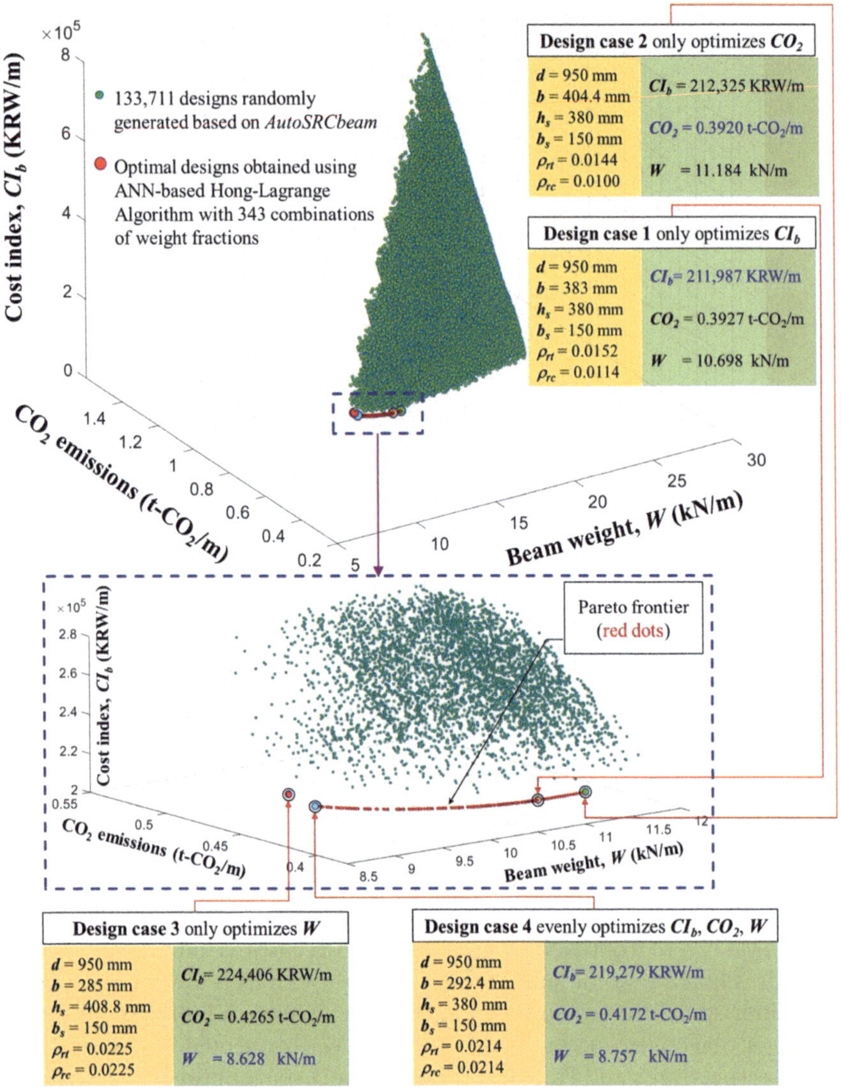

(a)

Figure 7.4.4 Pareto-efficient designs verified by structural datasets at EC_9: M_D=500 kN·m; EC_{10}: M_L=1,500 kN·m. (a) A three-dimensional Pareto frontier for the three objective functions (CI_b, CO_2, and W). (b) Design points projected to a $CI_b - W$ plane. (c) Design points projected to a $CO_2 - W$ plane. (d) Design points projected to a $CI_b - CO_2$ plane.

(Continued)

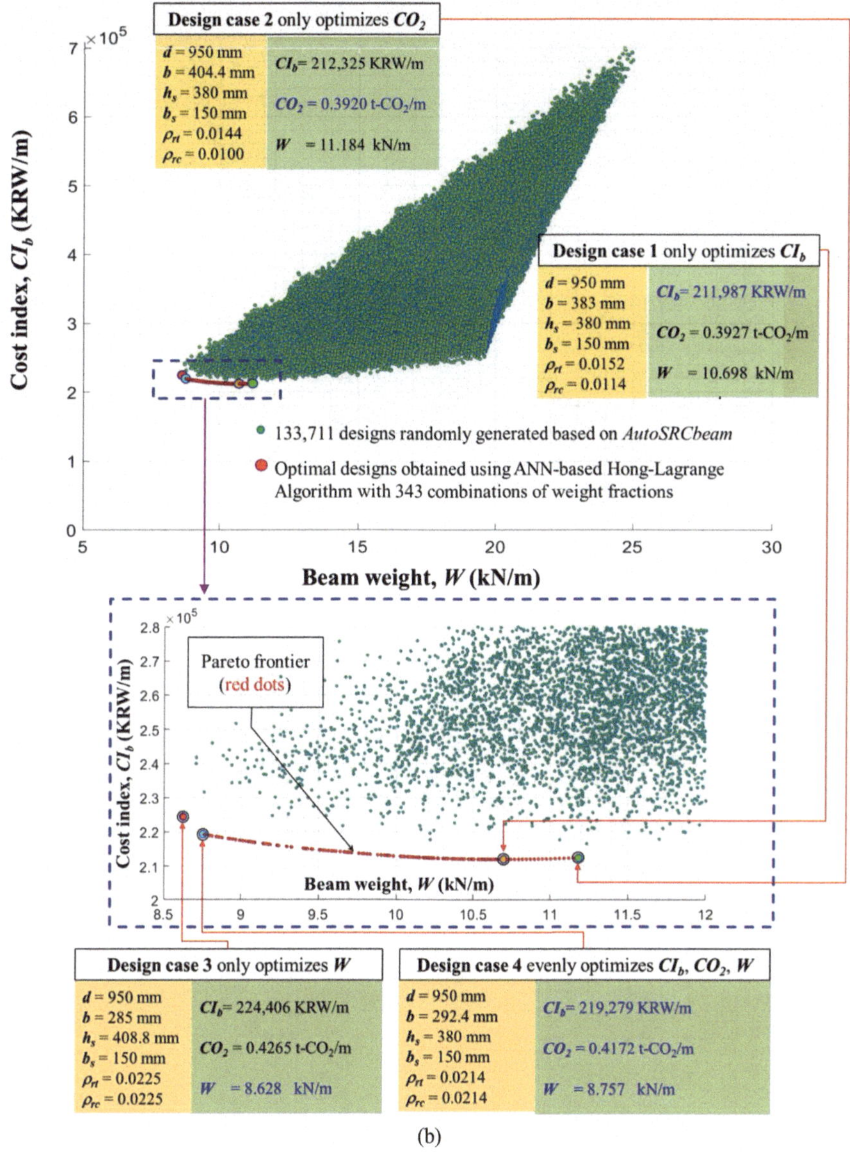

Figure 7.4.4 (Continued) Pareto-efficient designs verified by structural datasets at EC_9: M_D=500 kN·m; EC_{10}: M_L=1,500 kN·m. (a) A three-dimensional Pareto frontier for the three objective functions (CI_b, CO_2, and W). (b) Design points projected to a $CI_b - W$ plane. (c) Design points projected to a $CO_2 - W$ plane. (d) Design points projected to a $CI_b - CO_2$ plane.

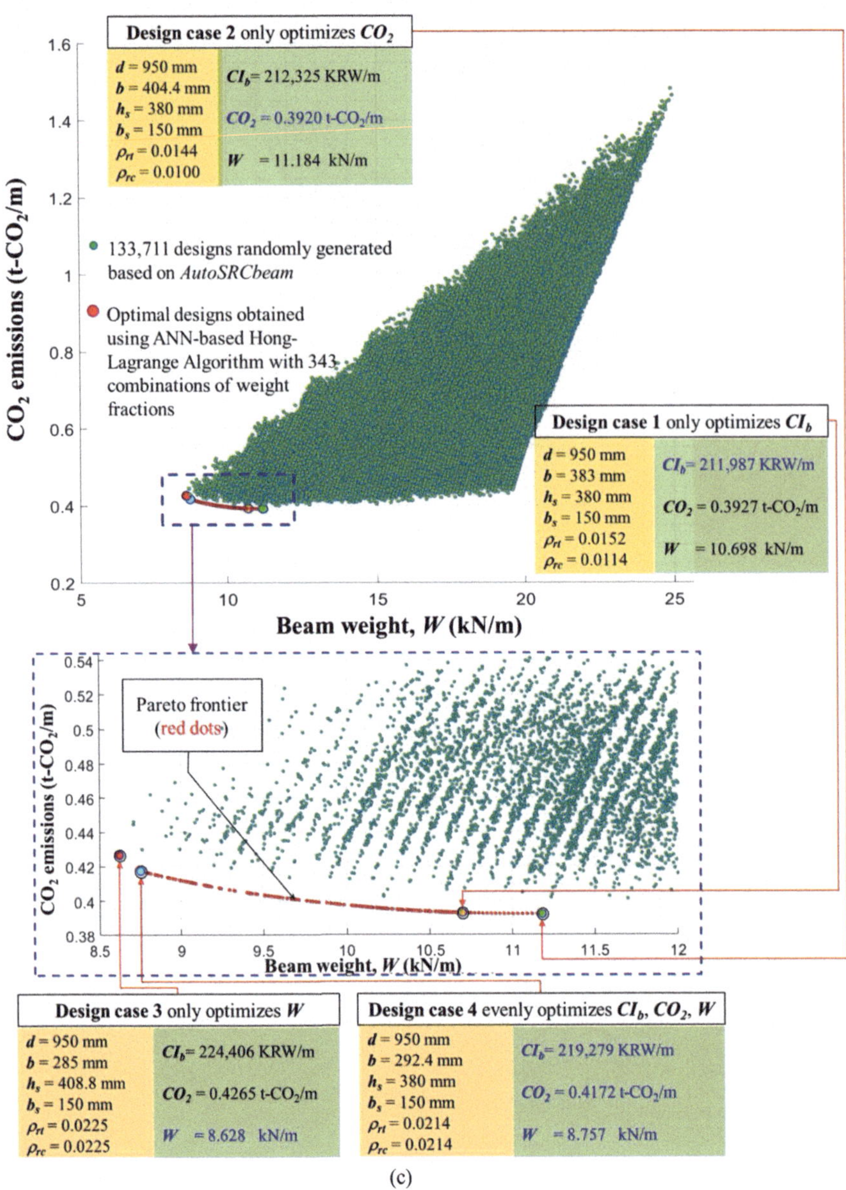

Figure 7.4.4 (Continued) Pareto-efficient designs verified by structural datasets at EC_9: M_D=500 kN·m; EC_{10}: M_L=1,500 kN·m. (a) A three-dimensional Pareto frontier for the three objective functions (CI_b, CO_2, and W). (b) Design points projected to a $CI_b - W$ plane. (c) Design points projected to a $CO_2 - W$ plane. (d) Design points projected to a $CI_b - CO_2$ plane.

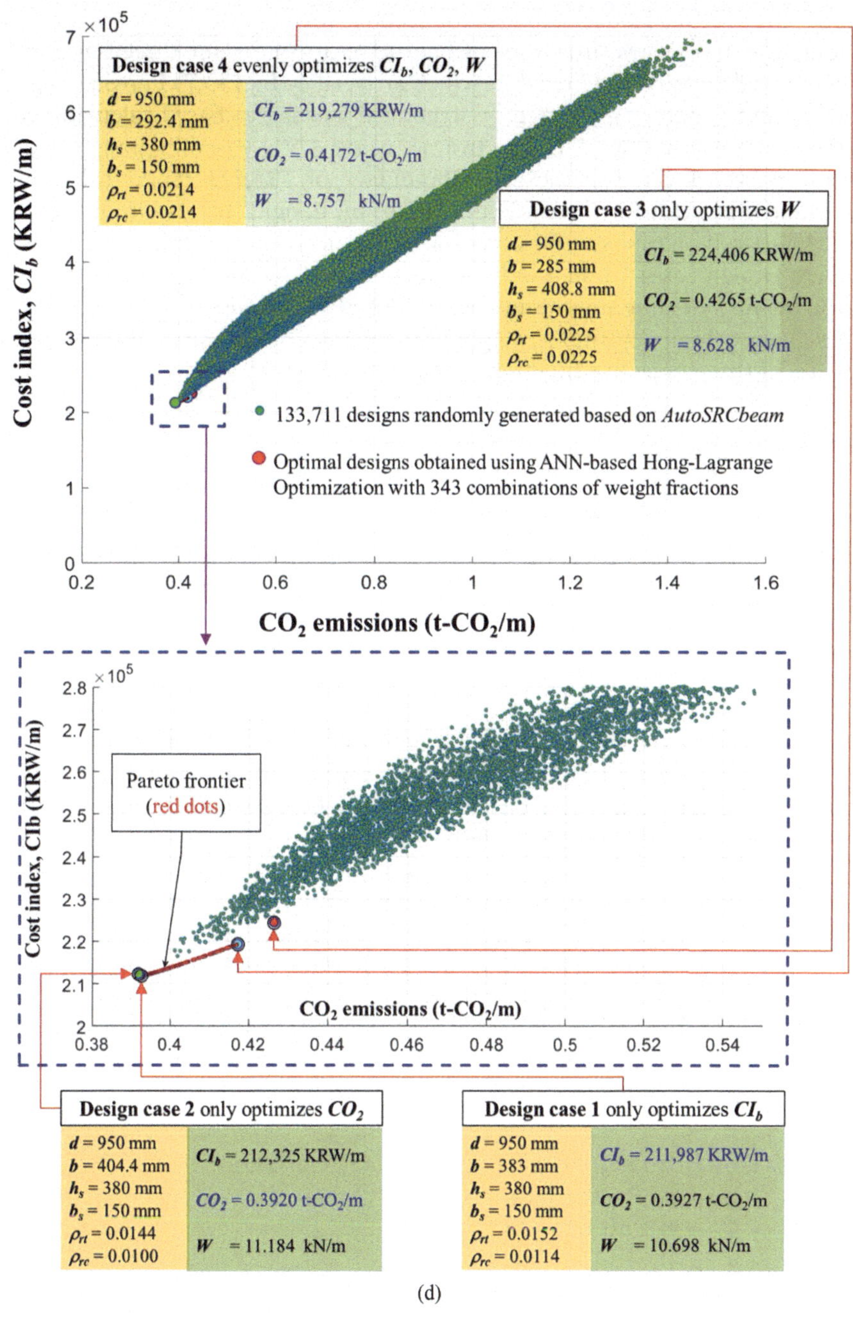

(d)

Figure 7.4.4 (Continued) Pareto-efficient designs verified by structural datasets at EC_9: M_D=500 kN·m; EC_{10}: M_L=1,500 kN·m. (a) A three-dimensional Pareto frontier for the three objective functions (CI_b, CO_2, and W). (b) Design points projected to a $CI_b - W$ plane. (c) Design points projected to a $CO_2 - W$ plane. (d) Design points projected to a $CI_b - CO_2$ plane.

7.4.3.2 Four design cases verified by big datasets

Figure 7.4.4(a) shows three-dimensional Pareto frontier for three objective functions denoted by 343 red dots which are verified by 133,711 green design points (green points) generated by structural mechanics-based calculations. Observation shows that a Pareto frontier using ANN-based Lagrange algorithm is well located at the lower boundary of 133,711 random designs calculated by *AutoSRCbeam*. Four SRC beam designs are shown in Figure 7.4.4(a) based on four specific cases for optimizing three objective functions, where Design cases 1 and 2 only optimize cost (CI_b) and CO_2 emissions of SRC beams, yielding minimum cost (CI_b) and CO_2 emissions, respectively. Design case 3 only optimizes weight (W), whereas design case 4 evenly optimizes all three objective functions (CI_b, CO_2, and W).

7.4.3.3 Design case 1 only optimizing costs CI_b based on Pareto frontier projected on CI_b –W and CI_b –CO_2 planes

The 343 Pareto points are generated with different tradeoff ratios, also optimizing three objective functions simultaneously. In Figure 7.4.4(b)–(d), Pareto frontier are projected on CI_b –W, CO_2 –W, and CI_b – CO_2 planes, respectively, where the tradeoffs shown on Pareto frontier which are used for optimizing three objective functions. Design case 1 only optimizes costs CI_b or individual objective function CI_b, with corresponding weight factors assigned to the three objective functions (CI_b, CO_2, and W) based on $w_{CI_b} : w_{CO_2} : w_W = 1:0:0$, as indicated in Figure 7.4.4(b) and (d). Design case 1 considers a tradeoff only contributed by costs (CI_b), ignoring an interest in CO_2 emissions and weights (W). Design parameters ($d = 950$ mm, b = 383 mm, h_s = 380 mm, b_s = 150 mm, ρ_{rt} = 0.0152, ρ_{rc} = 0.0114) are obtained while all 10 equalities and 16 inequalities given in Table 7.4.2 are satisfied. Optimized cost (CI_b) is obtained at 211,987 KRW/m, which is the minimum cost of all Pareto points, as clearly shown in Figure 7.4.4(b), and (d). CO_2 emissions and weights (W) for design case 1 correspond to 0.3927 t-CO_2/m and 10.698 kN/m, respectively, which are larger than 0.3920 t-CO_2/m and 8.628 kN/m obtained when CO_2 emissions and weights (W) are only optimized based on Design cases 2 and 3, respectively.

7.4.3.4 Design case 2 only optimizing CO_2 emissions based on Pareto frontier projected on CO_2 – W and CI_b – CO_2 planes

Design case 2 only optimizes CO_2 emissions or individual objective function CO_2, with corresponding weight factors for the three objective functions (CI_b, CO_2, and W) based on $w_{CI_b} : w_{CO_2} : w_W = 0:1:0$, as illustrated in Figure 7.4.4(c) and (d). Design case 2 considers a tradeoff only contributed by CO_2 emissions, ignoring an interest in costs CI_b and weights (W). Design parameters (d=950 mm, b=404.4 mm, h_s=380 mm, b_s=150 mm,

ρ_{rt}=0.0144, ρ_{rc}=0.0100) are obtained while all 10 equalities and 16 inequalities given in Table 7.4.2 are satisfied. Optimized CO_2 emission is obtained at 0.3920 t-CO_2/m, which is the minimum CO_2 emissions of all Pareto points as shown in Figure 7.4.4(c) and (d). Costs (CI_b) and weights (W) for design case 2 correspond to 212,325 KRW/m and 11.184 kN/m, respectively, which are larger than 211,987 KRW/m and 8.628 kN/m obtained when Costs (CI_b) and weights (W) are only optimized based on Design cases 1 and 3, respectively. The SRC beam designed with design cases 1 and 2 is the largest in volume to use more concrete, while reducing costs (CI_b) and CO_2 emissions compared with design cases 3 and 4.

7.4.3.5 Design case 3 only optimizing weights (W) based on Pareto frontier projected on $CI_b - W$ and $CO_2 - W$ planes

The minimum beam width (b = 0.3d = 0.3 × 950 = 285 mm) is recommended based on an inequality constraint (IC_9) shown in Table 7.4.2 for Design case 3 which optimizes the individual objective function of weights (W), resulting in the lightest beam weights W=8.628 kN/m as shown in Figure 7.4.4(b) and (c). However, tensile and compressive rebar ratios (ρ_{rt}, ρ_{rc}), and height of steel H-shaped (h_s) are 0.0225, 0.0225, and 408.8 mm, respectively, larger than those obtained when optimizing the other objective functions of costs CI_b, CO_2 emissions, and multiple objective functions based on $w_b : w_{CI_b} : w_{Weight} = 1/3 : 1/3 : 1/3$ in order to decrease beam weights W. Costs and CO_2 emissions are obtained at 224,406 KRW/m and 0.4265 t-CO_2/m in design case 3, respectively, when objective function W is optimized as indicated in Figure 7.4.4(b) and (c), which are larger than 211,987 KRW/m and 0.3920 t-CO_2/m obtained when Costs (CI_b) and CO_2 emissions are only optimized based on Design cases 1 and 2, respectively.

7.4.3.6 Design case 4 with evenly weight fractions $w_h : w_{CI_b} : w_{Weight} = 1/3 : 1/3 : 1/3$ based on Pareto frontier projected on $CI_b - W$, $CO_2 - W$, and $CI_b - CO_2$ planes

Evenly weight fractions $w_b : w_{CI_b} : w_{Weight} = 1/3 : 1/3 : 1/3$ are implemented in Design case 4, which optimizes all three objective functions at the same time with an equal contribution. As shown in Figure 7.4.4, observation shows that minimized objective functions (costs CI_b, CO_2 emissions, and beam weights W) by Design case 4 is obtained in the middle, indicating that median values of CI_b=219,279 KRW/m, CO_2 emissions = 0.4172 t-CO_2/m and W=8.757 kN/m by Design case 4 are obtained between Design case 1 (or Design case 2) and Design case 3, as shown in Figure 7.4.4. ANN-based Lagrange algorithm calculates b=292.4 mm for Design case 4 with three equal weight fractions ($w_b : w_{CI_b} : w_{Weight} = 1/3 : 1/3 : 1/3$), which is the most favorable value of beam widths that a proposed method finds when three

Table 7.4.3 Multiple objective optimization using ANN-based Hong–Lagrange algorithm based on $w_{CI_b} : w_{CO_2} : w_W = 1/3 : 1/3 : 1/3$

	ANN-based Hong–Lagrange optimization Training PTM based on 200,000 datasets 15 Inputs ($L, b, d, f_y, f_c', \rho_{rc}, \rho_{rt}, h_s, b_s, t_f, t_w, f_{yS}, Y_s, M_D, M_L$) - 11 Outputs ($\phi M_n, \varepsilon_{rt}, \varepsilon_{st}, \Delta_{imme}, \Delta_{long}, m_\phi, CI_b, CO_2, W, X_s, SF$)				
	Parameters	*Training network and accuracies*	*ANN-based Hong–Lagrange algorithm*	*Structural mechanics-based calculations (AutoSRCbeam)*	*Error (%)*
1	L (mm)		10,000	10,000	-
2	d (mm)		950	950	-
3	b (mm)		292.4	292.4	-
4	f_y (MPa)		500	500	-
5	f_c' (MPa)		30	30	-
6	ρ_{rt}		0.0214	0.0214	-
7	ρ_{rc}		0.0214	0.0214	-
8	h_s (mm)		380	380	-
9	b_s (mm)		150	150	-
10	t_f (mm)		12	12	-
11	t_w (mm)		8	8	-
12	f_{yS} (MPa)		325	325	-
13	Y_s (mm)		70.0	70.0	-
14	M_D (kN . m)		500	500	-
15	M_L (kN . m)		1,500	1,500	-

(*Continued*)

Table 7.4.3 (Continued) Multiple objective optimization using ANN-based Hong–Lagrange algorithm based on $w_{CI_b} : w_{CO_2} : w_W = 1/3 : 1/3 : 1/3$

	ANN-based Hong–Lagrange optimization Training PTM based on 200,000 datasets 15 Inputs ($L, b, d, f_y, f'_c, \rho_{rc}, \rho_{rt}, h_s, b_s, t_f, t_w, f_{yS}, Y_s, M_D, M_L$) - 11 Outputs ($\phi M_n, \varepsilon_{rt}, \varepsilon_{st}, \Delta_{imme}, \Delta_{long}, m_\phi, CI_b, CO_2, W, X_s, SF$)				
	Parameters	*Training network and accuracies*	*ANN-based Hong–Lagrange algorithm*	*Structural mechanics-based calculations (AutoSRCbeam)*	*Error (%)*
16	ϕM_n (kN·m)	4 layers-64 neurons; 48,865 epochs; T.MSE = 8.25E-06; R = 1.0000	3,031.9	3,005.8	0.86%
17	ε_{rt}	4 layers-64 neurons; 34,638 epochs; T.MSE = 5.30E-05; R = 0.9987	0.0055	0.0054	1.82%
18	ε_{st}	4 layers-64 neurons; 49,997 epochs; T.MSE = 4.17E-05; R = 0.9992	0.0039	0.0040	−2.56%
19	$\Delta_{imme.}$ (mm)	4 layers-64 neurons; 49,997 epochs; T.MSE = 9.96E-06; R = 0.9997	5.96	5.67	4.87%
20	Δ_{long} (mm)	4 layers-64 neurons; 35,898 epochs; T.MSE = 1.80E-05; R = 0.9997	12.07	11.44	5.22%
21	μ_ϕ	4 layers-64 neurons; 48,865 epochs; T.MSE = 4.22E-05; R = 0.9989	1.98	1.98	0.00%
22	CI_b (KRW/m)	4 layers-64 neurons; 32,795 epochs; T.MSE = 1.47E-06; R = 1.0000	219,279.1	219,583.6	**−0.14%**
23	CO_2 (t-CO_2/m)	4 layers-64 neurons; 21,661 epochs; T.MSE = 2.15E-08; R = 1.0000	0.42	0.41	2.38%
24	W (kN/m)	4 layers-64 neurons; 49,944 epochs; T.MSE = 2.77E-04; R = 0.9996	8.76	8.74	0.23%
25	Xs (mm)	4 layers-64 neurons; 15,079 epochs; T.MSE = 2.08E-06; R = 1.0000	71.23	71.21	0.03%
26	SF	4 layers-64 neurons; 27,140 epochs; T.MSE = 1.69E-07; R = 1.0000	1.00	1.00	0.00%

Note: Error = (ANN – AutoSRCbeam)/ANN.

15 inputs for ANN forward network

15 Inputs of structural mechanics

11 outputs for ANN forward network

11 Outputs of structural mechanics

Yellow means design targets

objective functions are optimized evenly. Beam width of b=292.4 mm for Design case 4 is close to a minimum of b=285 mm for Design case 3.

7.4.3.7 Verifications

Table 7.4.3 summarizes designs when an ANN-based Lagrange algorithm evenly minimizes three objective functions CI_b, CO_2, and W (w_{CIb}: w_{CO2}: w_W = 1/3:1/3:1/3). Green cells (Rows 16–26) in Table 7.4.3 show 11 output parameters obtained using structural mechanics-based calculations (*AutoSRCbeam*) using 15 input parameters in red cells (Rows 1–15) which are obtained by an ANN-based Lagrange algorithm while all 10 equalities and 16 inequalities given in Table 7.4.2 are met. A cost of an SRC beam is obtained as 219,279.1 KRW/m by an ANN-based Lagrange algorithm which is close to 219,583.6 KRW/m calculated by structural mechanics-based calculation, indicating an insignificant error of −0.14%. Insignificant errors in CO_2 emissions and beam weight are also obtained at 2.38% and 0.23%, respectively, between an ANN-based Lagrange algorithm and structural mechanics-based calculations using design parameters obtained by ANNs.

An insignificant error of 0.86% is also observed when comparing design moment (ϕM_n) of 3,031.9 kN·m obtained by ANNs and 3,005.8 kN·m from structural mechanics-based calculations using design parameters obtained by ANNs, as indicated in Table 7.4.3. Tensile rebar strain (ε_{rt}) is calculated at 0.0055 and 0.0054, obtained by ANNs and structural mechanics-based calculations, respectively, using design parameters obtained by ANNs leading to an insignificant error of 1.82%. ANNs and structural mechanics-based calculation provide immediate deflections ($\Delta_{imme.}$) of 5.96 mm and 5.67 mm, respectively, yielding a relative error of 4.87%, indicating only a difference of 0.29 mm. Long-term deflection (Δ_{long}) of 12.07 and 11.44 mm obtained by ANNs, and structural mechanics-based calculation also indicate a negligible difference of 0.63 mm with an error of 5.22%. It is noted that limited deflections following ACI 318–19 code are 27.8 mm (*L*/360) and 41.7 mm (*L*/240) for immediate deflection ($\Delta_{imme.}$) and long-term deflection (Δ_{long}), respectively. Observation shows that immediate deflection ($\Delta_{imme.}$) and long-term deflection (Δ_{long}) are obtained by ANN-based Lagrange algorithm which meets ACI 318-19 standard. Table 7.4.3 demonstrates a practical methodology for the ANN-based Lagrange algorithm in optimizing SRC beams with accuracies presenting a maximum error of -2.56% of ε_{st} except for errors of immediate ($\Delta_{imme.}$) and long-term deflection (Δ_{long}) reaching 4.87% and 5.22%, respectively. It is noteworthy that inequality *IC16* of Table 7.4.2 ensures that safety factor ($\phi Mn/Mu$) is obtained as 1.0 in #26 of Table 7.4.3 based on ANNs which is verified by structural mechanics-based calculations using using design parameters obtained by ANNs. Safety factor of 1.0 cannot be obtained based on conventional design method.

7.5 CONCLUSIONS

An ANN-based Lagrange algorithm is introduced for simultaneously optimizing multi-objective functions of an SRC beam that can be widely used by engineers. Engineers and decision-makers can use the method presented in this chapter as a guideline for optimizing multi-objective functions in the practical designs for any type of structures. A *UFO* of SRC beams with a fixed-fixed condition is optimized, satisfying various standard restrictions simultaneously, showing the significance of the proposed method. Equality and inequality constraints can be established by an interest of engineers and codes to reflect the design requirements. A Pareto frontier consists of 343 optimized designs in this design example, minimizing three objective functions (CI_b, CO_2, and W) simultaneously with tradeoff weight fractions, which is verified by the lower boundary of large datasets including 133,711 designs randomly generated based on structural mechanics-based calculations. The proposed ANN-based Lagrange algorithm can be extended to various fields in real-world designs. The following conclusions are further drawn from this chapter:

1. The aim of this chapter focuses on the technique to optimize multi-objective functions simultaneously, which is interesting to engineers and decision-makers for SRC beam designs. The proposed method can be extended to various fields in real-life designs.
2. A *UFO* is established by integrating the three objective functions consisting of costs (CI_b), CO_2 emissions, and weights (W) into one with their tradeoffs of weight factions. MOO based on *UFO* is implemented under 10 equality and 16 inequality constraints, which are used to impose design requirements using MATLAB Global Optimization Toolbox.
3. A Pareto frontier constructed based on 343 optimized designs for three objective functions (CI_b, CO_2, W) is verified by large datasets, yielding a good comparison with a lower boundary of 133,711 designs randomly generated based on structural mechanics-based calculations (*AutoSRCbeam*) for preassigned external loads, M_D=500 kN·m, M_L=1,500 kN·m.
4. A robust and sustainable tool for engineers is proposed in this chapter to design SRC beams using ANN-based Lagrange algorithm, which optimizes multi-objective functions (CI_b, CO_2 emissions, and W) of an SRC beam simultaneously under external loads.

REFERENCES

[7.1] Mei, L., & Wang, Q. (2021). Structural optimization in civil engineering: A literature review. *Buildings*, *11*(2), 66. https://doi.org/10.3390/buildings11020066

[7.2] Shariat, M., Shariati, M., Madadi, A., & Wakil, K. (2018). Computational Lagrangian Multiplier Method by using for optimization and sensitivity analysis of rectangular reinforced concrete beams. *Steel and Composite Structures*, *29*(2), 243–256. https://doi.org/10.12989/scs.2018.29.2.243

[7.3] Li, L. I., & Matsui, C. (2000). Effects of axial force on deformation capacity of steel encased reinforced concrete beam–columns. In *Proceedings of 12th World Conference on Earthquake Engineering*.

[7.4] El-Tawil, S., & Deierlein, G. G. (1999). Strength and ductility of concrete encased composite columns. *Journal of Structural Engineering*, *125*(9), 1009–1019. https://doi.org/10.1061/(ASCE)0733-9445(1999)125:9(1009)

[7.5] Bridge, R. Q., & Roderick, J. W. (1978). Behavior of built-up composite columns. *Journal of the Structural Division*, *104*(7), 1141–1155. https://doi.org/10.1061/JSDEAG.0004956

[7.6] Mirza, S. A., Hyttinen, V., & Hyttinen, E. (1996). Physical tests and analyses of composite steel-concrete beam-columns. *Journal of Structural Engineering*, *122*(11), 1317–1326. https://doi.org/10.1061/(ASCE)0733-9445(1996)122:11(1317)

[7.7] Ricles, J. M., & Paboojian, S. D. (1994). Seismic performance of steel-encased composite columns. *Journal of Structural Engineering*, *120*(8), 2474–2494. https://doi.org/10.1061/(ASCE)0733-9445(1994)120:8(2474)

[7.8] Furlong, R. W. (1974). Concrete encased steel columns—design tables. *Journal of the Structural Division*, *100*(9), 1865–1882. https://doi.org/10.1061/JSDEAG.0003878

[7.9] Mirza, S. A., & Skrabek, B. W. (1992). Statistical analysis of slender composite beam-column strength. *Journal of Structural Engineering*, *118*(5), 1312–1332. https://doi.org/10.1061/(ASCE)0733-9445(1992)118:5(1312)

[7.10] Chen, C. C., & Lin, N. J. (2006). Analytical model for predicting axial capacity and behavior of concrete encased steel composite stub columns. *Journal of Constructional Steel Research*, *62*(5), 424–433. https://doi.org/10.1016/j.jcsr.2005.04.021

[7.11] Dundar, C., Tokgoz, S., Tanrikulu, A. K., & Baran, T. (2008). Behaviour of reinforced and concrete-encased composite columns subjected to biaxial bending and axial load. *Building and Environment*, *43*(6), 1109–1120. https://doi.org/10.1016/j.buildenv.2007.02.010

[7.12] Munoz, P. R., & Hsu, C. T. T. (1997). Behavior of biaxially loaded concrete-encased composite columns. *Journal of structural Engineering*, *123*(9), 1163–1171. https://doi.org/10.1061/(ASCE)0733-9445(1997)123:9(1163)

[7.13] Rong, C., & Shi, Q. (2021). Analysis constitutive models for actively and passively confined concrete. *Composite Structures*, *256*, 113009. https://doi.org/10.1016/j.compstruct.2020.113009.

[7.14] Virdi, K. S., Dowling, P. J., BS 449, & BS 153. (1973). The ultimate strength of composite columns in biaxial bending. *Proceedings of the Institution of Civil Engineers*, *55*(1), 251–272. https://doi.org/10.1680/iicep.1973.4958

[7.15] Roik, K., & Bergmann, R. (1990). Design method for composite columns with unsymmetrical cross-sections. *Journal of Constructional Steel Research*, *15*(1–2), 153–168. https://doi.org/10.1016/0143-974X(90)90046-J

[7.16] Brettle, H. J. (1973). Ultimate strength design of composite columns. *Journal of the Structural Division*, *99*(9), 1931–1951. https://doi.org/10.1061/JSDEAG.0003606

[7.17] Abambres, M., & Lantsoght, E. O. (2020). Neural network-based formula for shear capacity prediction of one-way slabs under concentrated loads. *Engineering Structures*, *211*, 110501. https://doi.org/10.1016/j.engstruct.2020.110501

[7.18] Sharifi, Y., Lotfi, F., & Moghbeli, A. (2019). Compressive strength prediction using the ANN method for FRP confined rectangular concrete columns. *Journal of Rehabilitation in Civil Engineering*, 7(4), 134–153. https://doi.org/10.22075/JRCE.2018.14362.1260

[7.19] Asteris, P. G., Armaghani, D. J., Hatzigeorgiou, G. D., Karayannis, C. G., & Pilakoutas, K. (2019). Predicting the shear strength of reinforced concrete beams using Artificial Neural Networks. *Computers and Concrete, an International Journal*, *24*(5), 469–488. https://doi.org/10.12989/CAC.2019.24.5.469

[7.20] Armaghani, D. J., Hatzigeorgiou, G. D., Karamani, C., Skentou, A., Zoumpoulaki, I., & Asteris, P. G. (2019). Soft computing-based techniques for concrete beams shear strength. *Procedia Structural Integrity*, *17*, 924–933. https://doi.org/10.1016/j.prostr.2019.08.123

[7.21] Hong, W. K., & Nguyen, M. C. (2022). AI-based Lagrange optimization for designing reinforced concrete columns. *Journal of Asian Architecture and Building Engineering*, *21*(6), 2330–2344. https://doi.org/10.1080/13467581.2021.1971998

[7.22] Hong, W. K., Nguyen, V. T., & Nguyen, M. C. (2022). Optimizing reinforced concrete beams cost based on AI-based Lagrange functions. *Journal of Asian Architecture and Building Engineering*, *21*(6), 2426–2443. https://doi.org/10.1080/13467581.2021.2007105

[7.23] Hong, W. K., Nguyen, M. C., & Pham, T. D. (2023). Optimized Interaction PM diagram for Rectangular Reinforced Concrete Column based on Artificial Neural Networks Abstract. *Journal of Asian Architecture and Building Engineering*, 22(1), 201–225. https://doi.org/10.1080/13467581.2021.2018697

[7.24] Hong, W. K., & Le, T. A. (2022). ANN-based optimized design of doubly reinforced rectangular concrete beams based on multi-objective functions. *Journal of Asian Architecture and Building Engineering*. https://doi.org/10.1080/13467581.2022.2085720

[7.25] Hong, W. K., Le, T. A., Nguyen, M. C., & Pham, T. D. (2022). ANN-based Lagrange optimization for RC circular columns having multi-objective functions. *Journal of Asian Architecture and Building Engineering*, 1–16. https://doi.org/10.1080/13467581.2022.2064864

[7.26] Hong, W. K. (2021). *Artificial intelligence-based Design of Reinforced Concrete Structures*. Daega, Seoul.

[7.27] Hong, W. K., Le, T. A., Nguyen, M. C., & Pham, T. D. (2022). ANN-based Lagrange optimization for RC circular columns having multi-objective functions. *Journal of Asian Architecture and Building Engineering*. https://doi.org/10.1080/13467581.2022.2064864

[7.28] Hong, W. K., & Le, T. A. (2022). ANN-based optimized design of doubly reinforced rectangular concrete beams based on multi-objective functions. *Journal of Asian Architecture and Building Engineering*. https://doi.org/10.1080/13467581.2022.2085720

[7.29] Nguyen, D. H., & Hong, W. K. (2019). Part I: The analytical model predicting post-yield behavior of concrete-encased steel beams considering various confinement effects by transverse reinforcements and steels. *Materials*, *12*(14), 2302. https://doi.org/10.3390/ma12142302

[7.30] Hong, W. K. (2019). *Hybrid Composite Precast Systems: Numerical Investigation to Construction*. Elsevier: Woodhead Publishing.

[7.31] Hong, W. K., Nguyen, D. H. (2022). Pareto frontier for steel-reinforced concrete beam developed based on ANN-based Hong-Lagrange algorithm. *Journal of Asian Architecture and Building Engineering* (under review).

[7.32] Hong, W. K., Nguyen, D. H., & Nguyen, V. T. (2022). Reverse design charts for flexural strength of steel-reinforced concrete beams based on artificial neural networks. *Journal of Asian Architecture and Building Engineering*, 1–39. https://doi.org/doi.org/10.1080/13467581.2022.2097238

[7.33] An ACI Standard. (2019). *Building Code Requirements for Structural Concrete (ACI 318-19)*. American Concrete Institute, Farmington Hills, MI.

[7.34] MathWorks. (2022). *MATLAB R2022a, Version 9.9.0*. Natick, MA: MathWorks.

[7.35] Hong, W. K., Pham, T. D., & Nguyen, V. T. (2022). Feature selection based reverse design of doubly reinforced concrete beams. *Journal of Asian Architecture and Building Engineering*, *21*(4), 1472–1496. https://doi.org/10.1080/13467581.2021.1928510

[7.36] MathWorks, (2022a). Deep Learning Toolbox: User's Guide (R2022a). Retrieved July 26, 20122 from: https://www.mathworks.com/help/pdf_doc/deeplearning/nnet_ug.pdf

[7.37] Krenker, A., Bešter, J., & Kos, A. (2011). Introduction to the artificial neural networks. In *Artificial Neural Networks: Methodological Advances and Biomedical Applications* (pp. 1–18). InTech. https://doi.org/10.5772/15751

[7.38] Villarrubia, G., De Paz, J. F., Chamoso, P., & De la Prieta, F. (2018). Artificial neural networks used in optimization problems. *Neurocomputing*, *272*, 10–16.https://doi.org/10.1016/j.neucom.2017.04.075

[7.39] MathWorks, (2022a). Global Optimization: User's Guide (R2022a). Retrieved July 26, 2022, from: https://www.mathworks.com/help/pdf_doc/gads/gads.pdf

[7.40] MathWorks, (2022a). Optimization Toolbox: Documentation (R2022a). Retrieved July 26, 2022, from: https://www.mathworks.com/help/pdf_doc/optim/optim.pdf

[7.41] MathWorks, (2022a). Parallel Computing Toolbox: Documentation (R2022a). Retrieved July 26, 2022, from: https://www.mathworks.com/help/pdf_doc/parallel-computing/parallel-computing.pdf

[7.42] MathWorks, (2022a). Statistics and Machine Learning Toolbox: Documentation (R2022a). Retrieved July 26, 2022, from: https://www.mathworks.com/help/pdf_doc/stats/stats.pdf

[7.43] Afshari, H., Hare, W., and Tesfamariam, S. (2019). Constrained multi-objective optimization algorithms: Review and comparison with application in reinforced concrete structures. *Applied Soft Computing Journal*, *83*, 105631. https://doi.org/10.1016/j.asoc.2019.105631

Chapter 8

An ANN-based reverse design of reinforced concrete columns encasing H-shaped steel section

8.1 INTRODUCTION

8.1.1 Previous studies

Reinforced concrete columns encasing steel sections (SRC columns) have been popular in building infrastructures, bridge piers, and earthquake-resistant constructions. The most common type of steel section used in SRC columns is H-shaped steel sections. SRC columns are highly efficient and economical structures due to a compressive strength of concrete and both tensile and compressive strength with ductility of steel sections being used together. Another advantage of SRC columns is the durability and cost-effectiveness of concrete and its high resistance to fire. Several studies have investigated an experimental behavior of SRC columns under an axial load (Chicoine et al., [8.1]; Wang, [8.2]), under uniaxial bending and axial compressive load (Mirza et al., [8.3]; Ricles and Paboojian, [8.4]; El-Tawil and Deierlein, [8.5]; Li and Matsui, [8.6]), and under biaxial bending and axial compressive load (Munoz and Hsu, [8.7]; Morino et al., [8.8]). Some analytical studies have also been conducted by Chen and Lin [8.9], Nguyen and Hong [8.10], Virdi and Dowling [8.11], Roik and Bergmann [8.12], Dundar et al. [8.13], Munoz and Hsu [8.14] to investigate the performance of SRC columns. Foraboschi [8.15–8.17] investigated theoretical designs of reinforced concrete columns and beams which can be used to generate large datasets for ANN trainings. Some studies have applied ANNs successfully to predict capacities of complex structures such as one-way slabs by Abambres and Lantsoght [8.18], FRP confined rectangular concrete columns by Sharifi et al. [8.19], and reinforced concrete beams by Asteris et al. [8.20], Armaghani et al. [8.21], and Hong et al. [8.22,8.23].

8.1.2 Motivations of the optimizations

Despite these recent analytical and experimental studies on SRC columns, applications of SRC columns are still limited due to complexities in structural analysis and designs. This chapter introduces ANN-based reverse designs

DOI: 10.1201/9781003354796-8

Table 8.1.1 Nomenclatures of parameters for SRC columns

No.	*Nomenclature*	
	Forward input parameters	
1	h (mm)	Column height
2	b (mm)	Column width
3	h_s (mm)	Height of a H-shaped steel section
4	b_s (mm)	Width of a H-shaped steel section
5	t_f (mm)	Flange thickness of a H-shaped steel section
6	t_w (mm)	Web thickness of a H-shaped steel section
7	f_c' (MPa)	Compressive concrete strength
8	f_{yr} (MPa)	Yield rebar strength
9	f_{ys} (MPa)	Yield steel strength
10	ρ_{rX}	Rebar ratio along horizontal axis (X-axis)
11	ρ_{rY}	Rebar ratio along vertical axis (Y-axis)
12	P_u(kN)	Factored axial load
13	M_{uX}(kN·m)	Factored moment about the horizontal axis (X-axis)
14	M_{uY}(kN·m)	Factored moment about the vertical axis (Y-axis)
	Forward output parameters	
15	ϕM_n (kN·m)	Design moment strength
16	ϕP_n (kN)	Design axial strength
17	M_u (kN·m)	Combined factored bending moment ($\sqrt{M_{uX}^2 + M_{uY}^2}$)
18	SF	Safety factor ($\phi M_n / M_u = \phi P_n / P_u$)
19	Y_s (mm)	Vertical clearance between column and steel flange
20	X_s (mm)	Horizontal clearance from column and steel flange
21	ρ_s	Steel ratio
22	ε_s	Steel strain
23	ε_r	Rebar strain
24	b/h	Column aspect ratio
25	$D_{r,max}$	A maximum rebar diameter
26	CI_c (KRW/m)	Cost index of column
27	CO_2 (t-CO_2/m)	CO_2 emission
28	W_c (kN/m)	Column weight

to predict capacities of SRC columns including design axial strength (ϕP_n), design moment strengths (ϕM_{nX} and ϕM_{nY}) about X- and Y- axes. A capacity of ANNs to learn a trend of large datasets is a basis for predicting capacities of complex concrete-steel composite SRC structures. A theory of strain compatibility proposed by Nguyen and Hong [8.10] is used to generate large datasets of SRC columns under a biaxial load based on parameters used to define SRC columns listed in Table 8.1.1. A reverse design is performed while minimizing cost (CI_c), CO_2 emission, and weight (W_c), yielding corresponding

Table 8.1.2 Design scenarios for SRC columns

(a) Forward design scenario

Input		*Output*	
h	f_y	SF	$D_{r,max}$
b	f_{ys}	ε_r	CI_c
h_s	ρ_{rX}	ε_s	CO_2
b_s	ρ_{rY}	Y_s	W_c
t_w	P_u	X_s	
t_f	M_{uX}	b/h	
f_c'	M_{uY}	ρ_s	

(b) Reverse design scenario

Input		*Output*	
SF*	f_{ys}	h**	ε_s
ρ_s*	ρ_{rX}	b**	CI_c
b/h*	ρ_{rY}	h_s**	CO_2
t_f	P_u	b_s**	W_c
t_w	M_{uX}	Y_s	$D_{r,max}$
f_c'	M_{uY}	X_s	
f_{yr}		ε_r	

*: Reverse input parameters.
**: Reverse output parameters.

design parameters including column and steel dimensions of SRC columns on output-side ANNs. Constraints required by ANSI/AISC 360-16 [8.24] and ACI 318-19 [8.26] are satisfied when reverse design parameters including safety factor, rebar and steel ratios shown in Table 8.1.2(b) are preassigned on an input-side. Preassigning safety factor (*SF*) equivalent to 1.0 for design optimizations of SRC columns means that factored loads are equal to design strengths. Forward design parameters are also shown in Table 8.1.2(a). The proposed design replaces conventional software. This chapter was written based on the previous paper by Hong et al. [8.25].

8.2 CONTRIBUTIONS OF THE PROPOSED METHOD TO SRC COLUMN DESIGNS

8.2.1 Research significance based on novelty and innovation

Many analytical and experimental studies recently have investigated a behavior of SRC columns under a biaxial load (P_u, M_{uX}, and M_{uY}), as shown in Figure 8.2.1(a), whereas the strain compatibility of SRC columns under a biaxial load is illustrated in Figure 8.2.1(b). Most designs of concrete-steel composite SRC columns are based on conventional design method. Finding an optimized size of H-shaped steel sections is a goal of all engineers. However, the only way to optimize composite SRC columns is based on a trial-and-error technique that is time-consuming and laborious. It is also challenging for engineers to determine the proper sizes of concrete and steel sections in a preliminary design stage while minimizing cost index (CI_c), CO_2 emission, and column weights (W_c).

In this chapter, a reverse design technique with sufficient accuracies optimizes designs of SRC columns without repeating based on trial-and-error technique that the conventional design method must be based on when optimizing members. Optimizations of SRC columns that is important from the practical point of view are performed by using artificial intelligence-based design. Figures 8.2.1 and 8.2.2 show theoretical mechanics on which the large datasets have been generated. ANN-based forward and reverse scenario shown in Table 8.1.2 are introduced to help engineers design concrete and steel sections with preassigned *SF*s on an input layer of the ANN. Accuracies of the proposed method are found adequate for use in practical designs as designs optimized by ANNs based on reverse design are verified using structural mechanics. Cost index (CI_c), CO_2 emissions, and W_c, are now minimized, yielding various rebar (ρ_{rX} and ρ_{rY}) and steel ratios (ρ_s) accordingly, whereas column height (h) and width (b) as well as height (h_s) and width (b_s) of H-shaped steel sections are also obtained on an output-side the ANNs when minimizing cost index (CI_c), CO_2 emissions, and W_c. This chapter uses a BS method with a CTS and CRS proposed by Hong et al. [8.19,8.20] to train ANNs. Novel and innovative ANN-based designs of SRC columns under a biaxial load are presented, enabling engineers to optimize column sizes and steel sections in a preliminary design stage while minimizing cost index (CI_c), CO_2 emission, and column weights (W_c). A four-step algorithm for optimizing designs of SRC columns using ANN-based reverse designs is illustrated in Figure 8.2.3.

8.2.2 Innovations readers will learn in this chapter

1. ANNs are proposed as an alternative tool for a design optimization of SRC columns under a biaxial load when *SF* and steel ratio (ρ_s) are prescribed as reverse inputs in a reverse scenario which meets code requirements from ANSI/AISC 360-16 [8.24].
2. Based on the forward Step 1 of BS network trained by large datasets using CRS, four reverse output parameters which are column height (h), width (b), dimensions of steel section (h_s and b_s), are calculated for given reverse input parameters such as safety factor (*SF*) for a factored biaxial load, steel ratio (ρ_s), and aspect ratio of columns (b/h).

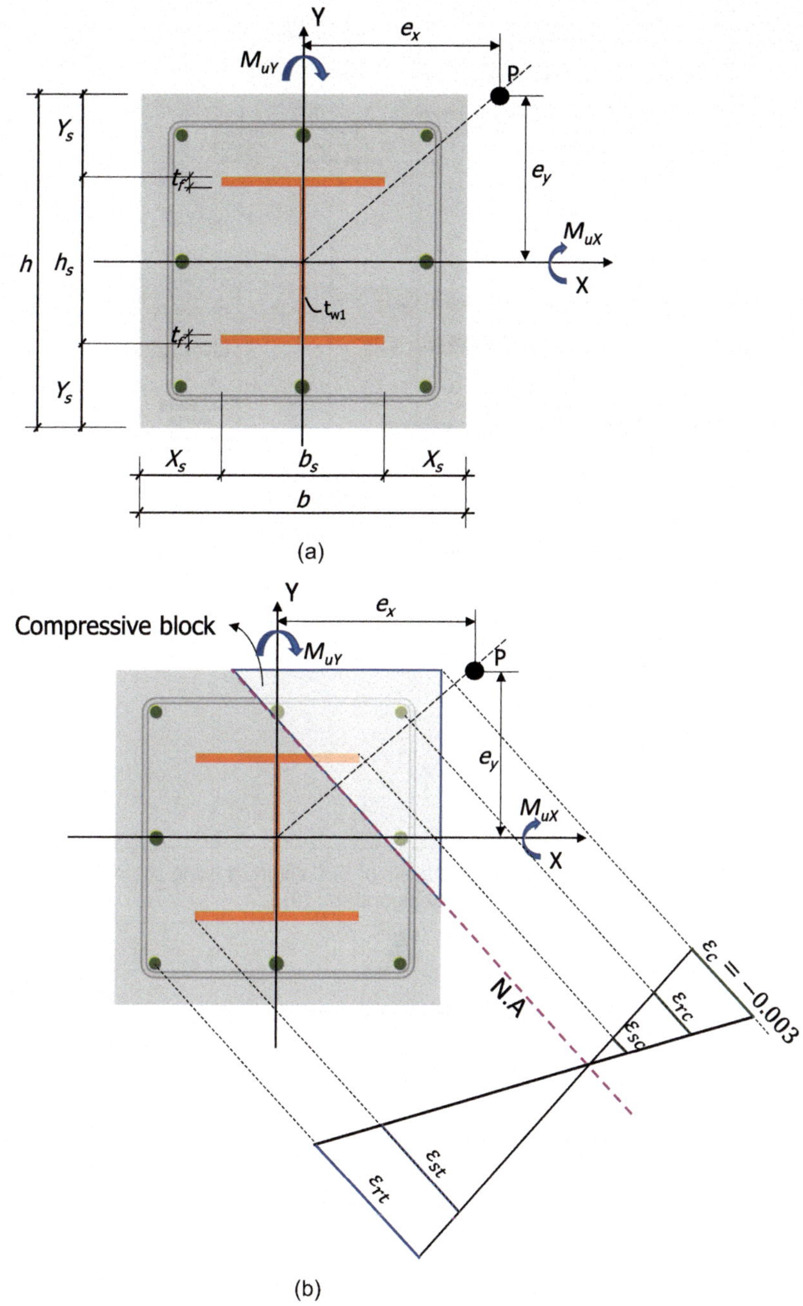

Figure 8.2.1 Cross section and strain compatibility of SRC columns. (a) Cross section of SRC columns under a biaxial load. (b) Strain compatibility of SRC columns under a biaxial load.

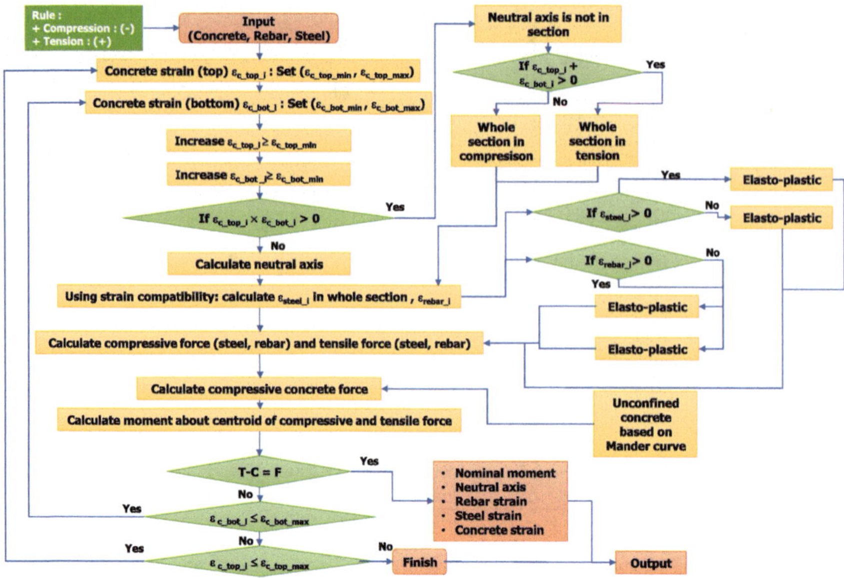

Figure 8.2.2 Computational algorithm for SRC columns [10].

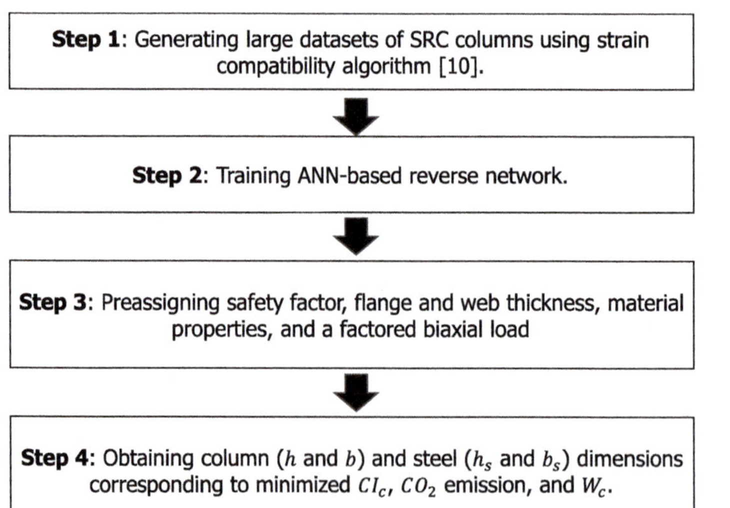

Figure 8.2.3 An algorithm for minimizing designs of SRC columns.

Four reverse output parameters are, then, back-substituted into the reverse Step 2 of BS to calculate the other eight output parameters $(\varepsilon_r, \varepsilon_s, D_{r,\max}, Y_s, X_s, CI_c, CO_2, W_c)$.

3. The effects of rebar and steel ratios on CI_c, CO_2 emissions, and W_c are uncovered in which rebar and steel ratios decrease to minimize CI_c and CO_2 emissions, whereas rebar and steel ratios increase to minimize column weights (W_c).
4. CI_c and CO_2 emissions are minimized similarly when rebar and steel ratios decrease to their minima of 0.004 and 0.01 for rebar and steel ratios, respectively, required by ANSI/AISC 360-16 [8.24]. Designs with minimized column weight (W_c) are achieved in an opposite way when rebar and steel ratios increase.
5. Optimized designs of SRC columns based on the proposed ANNs are verified by those calculated from structural mechanics. Observed errors are insignificantly smaller than 2% for *SF* when using the BS (reverse CRS – Forward AutoSRCHCol) method.
6. Three-dimensional interaction diagrams corresponding to minimized CI_c, CO_2 emissions, and W_c are presented when a safety factor ($SF = 1.0$) for a factored biaxial load is preassigned.
7. Robust, expeditious, and facile but sufficiently accurate design methods of SRC columns using ANNs are now available to engineers, enabling them to explore a behavior of SRC columns for practical designs.
8. Next chapters introduce derivative optimization methods such as a Lagrane Multiplier method (LMM) under multiple loads. Simultaneous optimizations of multi-objective design targets including CI_c, CO_2 emission, and W_c will be also investigated in the next chapters.

8.3 REVERSE DESIGN-BASED OPTIMIZATIONS

8.3.1 Generation of large structural datasets

Computational algorithm based on strain compatibility for SRC columns was developed by Nguyen and Hong [8.10] to study a behavior of SRC columns as shown in Figure 8.2.2. Mander curve [8.27] is used to model concrete behaviors, whereas an elasto-plastic model is used for describing rebar and steel materials. Figure 8.3.1 shows flow chart for generating large structural datasets of SRC columns. ANNs are trained on a large dataset of 100,000 generated based on strain compatibility proposed by Nguyen and Hong [8.10]. Fourteen input parameters $(h, b, h_s, b_s, t_w, t_f, f_c', f_{yr}, f_{ys}, \rho_{rX}, \rho_{rY}, P_u, M_{uX}, M_{uY})$ and 11 output parameters $(SF, \varepsilon_r, \varepsilon_s, \rho_s, D_{r,\max}, Y_s, X_s, b/h, CI_c, CO_2, W_c)$ are presented in Table 8.1.2(a) for a forward design scenario, whereas Table 8.1.2(b) demonstrates a reverse scenario where *SF*, ρ_s, and b/h are placed in an input layer. Table 8.3.1 summarizes statistics of 100,000 datasets generated for

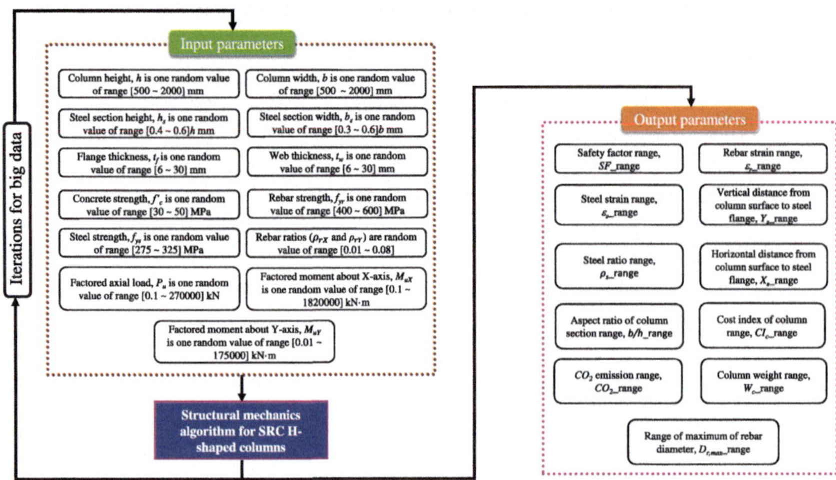

Figure 8.3.1 Flow chart for generating large structural datasets of SRC columns.

Table 8.3.1 Statistics of 100,000 datasets generated for SRC columns

Number of datasets	*100,000*						
Parameters	*h (mm)*	*b (mm)*	*h_s (mm)*	*b_s (mm)*	*t_w (mm)*	*t_f (mm)*	*f'_c (MPa)*
Maximum	2,000	2,000	1,200	1,190	30	30	50
Mean	1,412.1	1,375.7	694.8	607.5	18.8	19.0	40.0
Minimum	500	500	200	150	6	6	30
Standard deviation	365.02	375.87	208.31	220.52	7.06	6.85	6.04
Parameters	*f_{yr} (MPa)*	*f_{ys} (MPa)*	*ρ_{rX}*	*ρ_{rY}*	*P_u (kN)*	*M_{uX} (kN·m)*	*M_{uY} (kN·m)*
Maximum	600	325	0.079	0.079	269,332	181,046	172,629
Mean	499.9	300.0	0.022	0.023	22,935.8	10,403.3	10,185.1
Minimum	400	275	0.001	0.001	0.021	0.026	0.165
Standard deviation	58.03	14.69	0.02	0.02	27,560.32	11,960.87	11,517.30

SRC columns including ranges of the large datasets, maxima, and minima with standard deviations. Column dimensions (height, h and width, b) are generated randomly in a range of 500–2,000 mm, whereas height (h_s) and width (b_s) of an H-shaped steel section are also randomly selected in a range of 200–1,200 mm and 150–1,200 mm, respectively. The thicknesses of flange (t_f) and web (t_w) are generated from 6 to 30 mm. Figure 8.3.2 shows distribution of 100,000 rebar ratios. Concrete strengths (f'_c) are generated

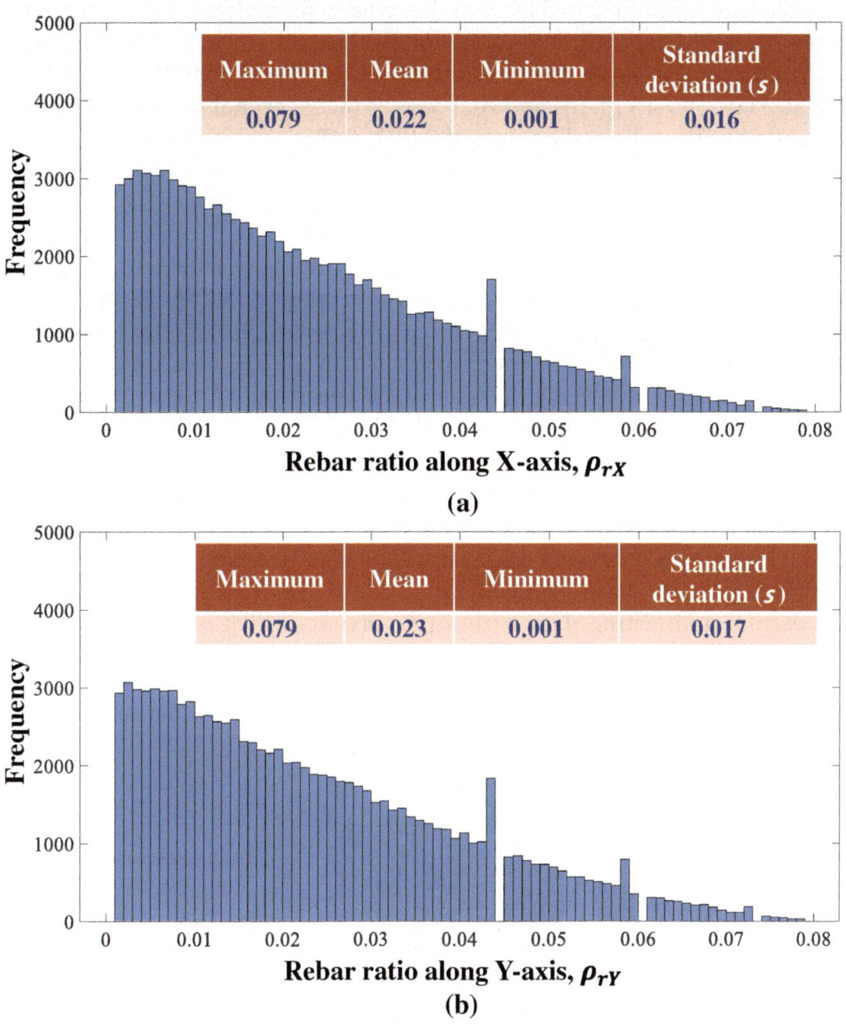

Figure 8.3.2 Distribution of rebar ratios based on 100,000 datasets. (a) Rebar ratio along X-axis, ρ_{rX}. (b) Rebar ratio along Y-axis, ρ_{rY}.

between 30 and 50 MPa, whereas material properties of rebar (f_{yr}) and steel (f_{ys}) are randomly selected in a range of 400–600 MPa and 275–325 MPa, respectively, based on uniform distribution. For SRC columns, a rebar ratio is limited not to be lower than 0.004 by Section I2.1a, ANSI/AISC 360-16 [8.24] to prevent rebars from yielding under sustained service loads, whereas a maximum rebar ratio is restricted to 0.08 according to Section 10.6.1, ACI 318-19 [8.26] to make sure that concrete is consolidated sufficiently around rebars. Distributions of rebar ratios observed from 100,000 datasets are exhibited in Figure 8.3.2 where skew distributions of rebar ratios along X- and Y-directions are found. This is because total rebar ratios are constrained to be greater than 0.004 and smaller than 0.08 according to ANSI/AISC 360-16 [8.24] and ACI 318-19 [8.26] even if each rebar ratio is generated uniformly in the range of 0.001–0.08.

8.3.2 BS procedure to perform reverse design

No specific technique for finding optimal training parameters such as a number of hidden layers and neurons is available [8.28]. Figures 8.3.3 and 8.3.4 show an ANN in Step 1 of BS with CRS method in which Deep Learning Toolbox, MathWorks 2021a [8.29] was used to train four output parameters (h, b, h_s, and b_s) based on 13 input parameters $\left(SF, \rho_s, b/h, t_w, t_f, f_c', f_{yr}, f_{ys}, \rho_{rX}, \rho_{rY}, P_u, M_{uX}, M_{uY}\right)$. In Table 8.3.2, best training accuracies found for each reverse output parameter (h, b, h_s, and b_s) in Step 1 of BS-CRS method based on nine trainings consisting of 3, 4, and 5 hidden layers with 50, 60, and 70 neurons. The training is performed on 100,000 datasets generated using structural algorithm suggested by Nguyen and Hong [8.10]. Large datasets of 100,000 are separated into training, validation, and testing data with a proportion of 70% (70,000 datasets),

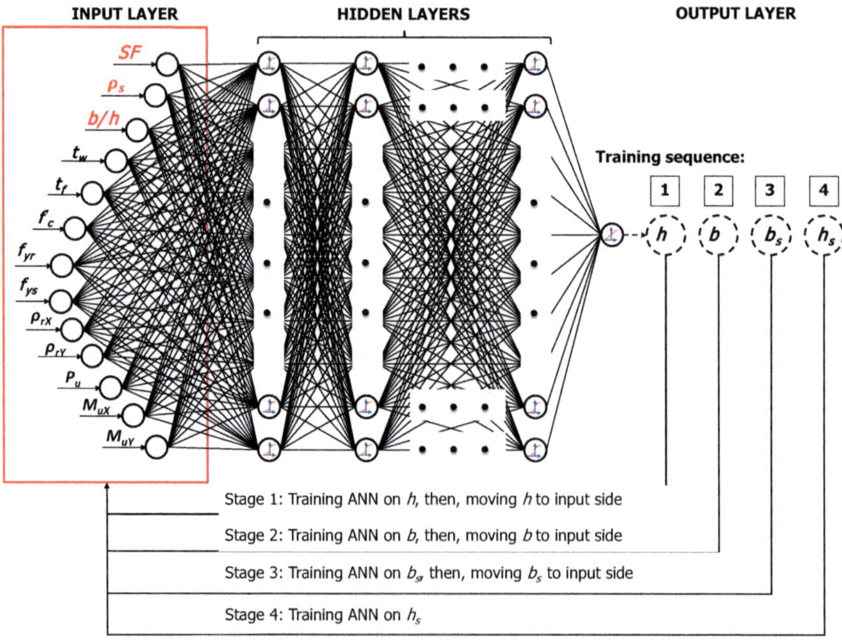

Figure 8.3.3 A reverse artificial neural network based on Step 1 of BS method for reverse scenario shown in Table 8.1.2(b).

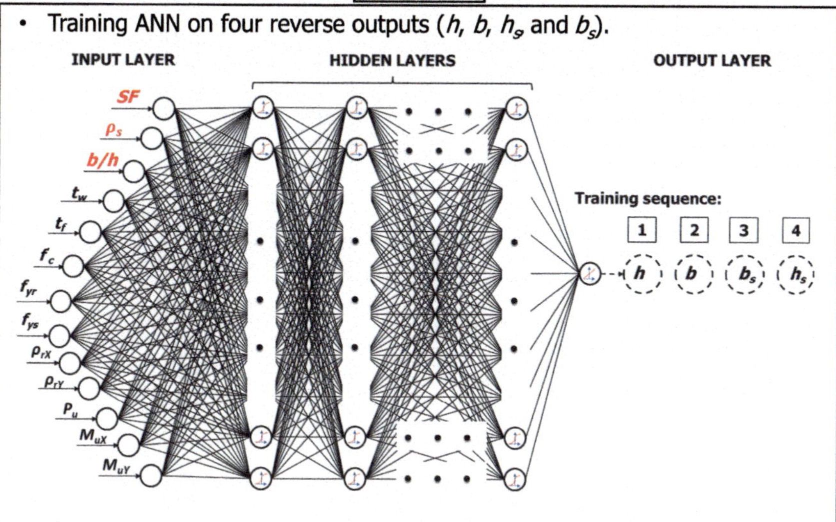

STEP 2

- Back-substituting four reverse outputs (h, b, h_s and b_s) with ten forward inputs ($t_w, t_f, f_c', f_{yr}, f_{ys}, \rho_{rX}, \rho_{rY}, P_u, M_{uX}, M_{uY}$) to calculate eight remaining forward outputs ($\varepsilon_r, \varepsilon_s, D_{r,max}, Y_s, X_s, CI_c, CO_2, W_c$) using AutoSRC HCol.
- It is noted that any structural software can be implemented in Step 2 of BS.

Figure 8.3.4 Algorithm of BS method.

15% (15,000 dataset), and 15% (15,000 datasets). A number of epoch and validation checks are preassigned as 50,000 and 500, respectively.

The training sequence of $h \rightarrow b \rightarrow b_s \rightarrow h_s$ is determined for a Step 1 of BS-CRS using three layers with 30 neurons as shown in Figure 8.3.3 which demonstrates four stages of training reverse output parameters of $h \rightarrow b \rightarrow b_s \rightarrow h_s$ [8.23], [8.31]. Using reverse networks of Step 1 of BS, training sequence $h \rightarrow b \rightarrow b_s \rightarrow h_s$ shown in Table 8.3.2 is used to obtain design Table 8.3.3 which is based on minimized CI_c and CO_2 emission. A BS method was proposed by Hong et al. [8.22] to solve a reverse design scenario of doubly reinforced concrete beams. A BS method consists of reverse CRS (BS-CRS method) (Hong, Pham, and Nguyen [8.23]) networks in the first step and forward calculations based on structural mechanics in the second step. Figures 8.3.3 and 8.3.4 present BS procedure to perform reverse design for SRC columns. In Step 1 of BS method of Figures 8.3.3 and 8.3.4, a reverse ANN for reverse scenario shown in Table 8.1.2(b) is derived to calculate four reverse outputs (h, b, h_s, and b_s) on an output-side. Input layer consists

Table 8.3.2 Best training results of Step 1 of BS-CRS method based on one, four, and five hidden layers with 50, 60, and 70 neurons

(1) 13 Inputs $\left(SF, \rho_s, b/h, t_w, t_f, f'_c, f_{yr}, f_{ys}, \rho_{rX}, \rho_{rY}, P_u, M_{uX}, M_{uY}\right)$ - 1 Output (h)							
No.	*Data*	*Hidden layers*	*Neurons*	*Required epoch*	*Best epoch for training*	*Test T.MSE*	*R at best epoch*
1	100,000	5	60	50,000	48,955	4.53E-5	0.999
(2) 14 Inputs $\left(SF, \rho_s, b/h, t_w, t_f, f'_c, f_{yr}, f_{ys}, \rho_{rX}, \rho_{rY}, P_u, M_{uX}, M_{uY}, h^{[a]}\right)$ - 1 Output (b)							
No.	*Data*	*Hidden layers*	*Neurons*	*Required epoch*	*Best epoch for training*	*Test T.MSE*	*R at best epoch*
2	100,000	5	50	50,000	49,997	5.2E-9	1.00
(3) 15 Inputs $\left(SF, \rho_s, b/h, t_w, t_f, f'_c, f_{yr}, f_{ys}, \rho_{rX}, \rho_{rY}, P_u, M_{uX}, M_{uY}, h, b^{[a]}\right)$ - 1 Output (b_s)							
No.	*Data*	*Hidden layers*	*Neurons*	*Required epoch*	*Best epoch for training*	*Test T.MSE*	*R at best epoch*
3	100,000	5	50	50,000	3,168	7.0E-3	0.981
(4) 16 Inputs $\left(SF, \rho_s, b/h, t_w, t_f, f'_c, f_{yr}, f_{ys}, \rho_{rX}, \rho_{rY}, P_u, M_{uX}, M_{uY}, h, b, b_s^{[a]}\right)$ - 1 Output (h_s)							
No.	*Data*	*Hidden layers*	*Neurons*	*Required epoch*	*Best epoch for training*	*Test T.MSE*	*R at best epoch*
4	100,000	5	60	50,000	50,000	1.2E-7	1.00

[a] Feature indexes gained from sequence

Table 8.3.3 Design table based on minimized CI_c and CO_2 emissions

No.	Parameters	BS - CRS		Structural mechanics	Error* (%)
		Step 1	Step 2		
1	SF	1.00	1.00	1.01	-0.73%
2	ρ_s	0.01	0.01	0.01	-0.03%
3	b/h	1.00	1.00	1.00	0.00%
4	t_f (mm)	8	8	8	-
5	t_w (mm)	8	8	8	-
6	f_c' (MPa)	30	30	30	-
7	f_{yr} (MPa)	500	500	500	-
8	f_{ys} (MPa)	325	325	325	-
9	ρ_{rX}	0.002	0.002	0.002	-
10	ρ_{rY}	0.002	0.002	0.002	-
11	P_u (kN)	10000	10000	10000	-
12	M_{uX} (kN·m)	5000	5000	5000	-
13	M_{uY} (kN·m)	7000	7000	7000	-
14	b (mm)	1503.17	1503.17	1503.17	-
15	h (mm)	1503.16	1503.16	1503.16	-
16	h_s (mm)	992.08	992.08	992.08	-
17	b_s (mm)	924.58	924.58	924.58	-
18	ε_r	-	0.0037	0.0037	-
19	ε_s	-	0.0027	0.0027	-
20	Y_s (mm)	-	255.54	255.54	-
21	X_s (mm)	-	289.30	289.30	-
22	CI_c (KRW/m)	-	600,466	600,466	-
23	CO_2 (t-CO_2/m)	-	1.00	1.00	-
24	W_c (kN/m)	-	54.84	54.84	-
25	$D_{r,max}$	-	16	16	-

CRS sequence: $h \rightarrow b \rightarrow b_s \rightarrow h_s$

BS method procedure

Step 1 (*reverse network*): Calculating **4 reverse outputs** based on **10 ordinary inputs** and **3 reverse inputs**.

Step 2 (*forward analysis*): Calculating **8 outputs** using AutoSRC HCol based on **14 inputs** (**4 reverse outputs** calculated from Step 1 and **10 ordinary inputs**).

AutoSRC HCol check procedure

Calculating **3 outputs** based on **14 inputs** which are identical to **14 inputs** of Step 2 of BS.

*Note: Errors $= \frac{(ANN-AutoSRCHCol)}{ANN} \times 100\%$

of 13 parameters including three reverse input parameters (SF, ρ_s, and b/h) and ten ordinary input parameters ($t_w, t_f, f_c', f_{yr}, f_{ys}, \rho_{rX}, \rho_{rY}, P_u, M_{uX}, M_{uY}$). In a forward ANN based on Step 2 of BS method of Figure 8.3.4, 11 parameters including three reverse input parameters (SF, ρ_s, and $\frac{b}{h}$) and eight ordinary output parameters $\left(\varepsilon_r, \varepsilon_s, D_{r,\max}, Y_s, X_s, CI_c, CO_2, W_c\right)$ are calculated in an output layer for given 14 output parameters including ten ordinary input parameters ($t_w, t_f, f_c', f_{yr}, f_{ys}, \rho_{rX}, \rho_{rY}, P_u, M_{uX}, M_{uY}$) and four reverse output-based input parameters (reverse outputs of column dimensions (h, b) and steel sections (h_s, b_s) obtained in Step 1) using conventional structural software AutoSRCHCol. Three reverse input parameters (SF, ρ_s, and b/h) were preassigned on an input-side Step 1 and calculated on an output-side Step 2, and hence, they are compared to verify a reverse design performed by BS procedure. Table 8.3.3 shows −0.73%, −0.03%, and 0.00% errors for SF, ρ_s, and b/h, respectively. It is noted that any type of structural software can be used in Step 2 shown in Figure 8.3.4. Reverse scenario is performed to design SRC columns, where SF is optimized to equal 1.0, resulting in design strengths equivalent to a factored biaxial load. It is difficult to obtain a safety factor of 1.0 using conventional design method.

8.3.3 Network training for four reverse outputs (h, b, h_s, and b_s) based on BS-CRS

No specific technique for finding optimal training parameters such as a number of hidden layers and neurons is available [8.28]. Figures 8.3.3 and 8.3.4 show an ANN in Step 1 of BS with CRS in which Deep Learning Toolbox, MathWorks 2021a [8.29] was used to train four output parameters (h, b, h_s, and b_s) based on 13 input parameters $\left(SF, \rho_s, b/h, t_w, t_f, f_c', f_{yr}, f_{ys}, \rho_{rX}, \rho_{rY}, P_u, M_{uX}, M_{uY}\right)$. Training accuracies of the reverse network used for Step 1 of the BS method are summarized in Table 8.3.2 where the best training accuracies of Step 1 of BS-CRS method are shown for each output parameter (h, b, h_s, and b_s) based on nine trainings consisting of 3, 4, and 5 hidden layers with 50, 60, and 70 neurons.

The training is performed on 100,000 datasets generated using structural algorithm suggested by Nguyen and Hong [8.10]. Sequence $h \rightarrow b \rightarrow b_s \rightarrow h_s$ shown in Table 8.3.2 is used to obtain design Table 8.3.3 which minimizes CIc and CO_2 emission.

8.4 REVERSE DESIGNS TO MINIMIZE CI_C, CO_2, W_C

8.4.1 Reverse design using BS method

In reverse scenarios, the positions of the input and output parameters of ANN are exchanged. An ANN memorizes trends of large datasets which can be used to solve reverse designs. In the reverse scenario as shown in

Table 8.1.2(b), three reverse inputs which are *SF*, steel ratio (ρ_s), and aspect ratio of columns (b/h) are preassigned with ten ordinary input parameters $\left(t_w, t_f, f_c', f_{yr}, f_{ys}, \rho_{rX}, \rho_{rY}, P_u, M_{uX}, M_{uY}\right)$ on an input-side an ANN, whereas five output parameters such as vertical (Y_s) and horizontal (X_s) clearances, rebar (ε_r) and steel (ε_s) strain, column diameter ($D_{r,\max}$) and three design targets such as cost CI_c, CO_2 emissions, weight W_c are calculated on an output-side an ANN. As shown in Table 8.3.3, three reverse inputs which are safety factor (*SF*) of 1, aspect ratio of columns (b/h) of 1, and steel ratio (ρ_s) preassigned on an input-side are also calculated for verification on an output-side an ANN.

8.4.2 Reverse design-based optimization using back-substitution (BS) method

In the reverse scenario in Step 1 of the BS method, four reverse outputs h, b, h_s, and b_s are calculated when 13 design parameters $\left(SF, \rho_s, b/h, t_w, t_f, f_c', f_{yr}, f_{ys}, \rho_{rX}, \rho_{rY}, P_u, M_{uX}, M_{uY}\right)$, including three reverse inputs $\left(SF, \rho_s, \text{and } b/h\right)$ and ten ordinary inputs ($t_w, t_f, f_c', f_{yr}, f_{ys}, \rho_{rX}, \rho_{rY}, P_u, M_{uX}, M_{uY}$), are preassigned.

In the reverse scenario, safety factors of 1.0 constrain factored loads equal to design strengths. Minimum and maximum steel ratio (ρ_s) are limited to 0.01 and 0.08 according to Section I2.1a, ANSI/AISC 360-16 [8.24] and Section 10.6.1, ACI 318-19 [8.26], respectively, whereas rebar ratios for SRC columns are also constrained not to be smaller than 0.004 to prevent rebars from yielding under sustained service loads according to Section I2.1a, ANSI/AISC 360-16 [8.24], whereas maximum rebar ratios for SRC columns are restricted to 0.08 according to Section 10.6.1, ACI 318-19 [8.26].

Figure 8.4.1 can be constructed by solving for rebar diameters $D_{r.\max}$ from reverse design using BS method when varying rebar ratios (ρ_{rX} and ρ_{rY}) and steel ratio (ρ_s) in a range of 0.003–0.09 and 0.009–0.015, respectively. The effects of rebar (ρ_{rX} and ρ_{rY}) and steel ratio (ρ_s) on a maximum of rebar diameters $D_{r.\max}$ and column dimensions (h and b) are identified in Figure 8.4.1(a) and (b), respectively, where two reverse inputs (*SF* and b/h) are each preassigned as 1.0, whereas both thicknesses of the flange and web of the H-shaped steel section are defined as 8 mm. Compressive concrete strength (f_c'), rebar strength (f_{yr}), and steel strength (f_{ys}) are predetermined to be 30, 500, and 325 MPa, respectively, whereas a factored axial load (P_u), moment about the X-axis (M_{uX}), and moment about the Y-axis (M_{uY}) are defined as P_u = 10,000 kN, M_{uX} = 5,000 kN·m, and M_{uY} = 7,000 kN·m.

The valid range of rebar is 0.004–0.08 and steel ratios should be greater than 0.01 as presented in Figure 8.4.1(a) and (b) even if Figure 8.4.1(a) and (b) are plotted over the rebar ratios (ρ_{rX} and ρ_{rY}) and steel ratios (ρ_s) which vary in a range of 0.003–0.09 and 0.009–0.015, respectively. A bundle of rebars is considered in 100,000 datasets. Section 25.6.1, ACI 318-19 [8.26] limits a bundle of rebars to a maximum four rebars. The maximum diameter

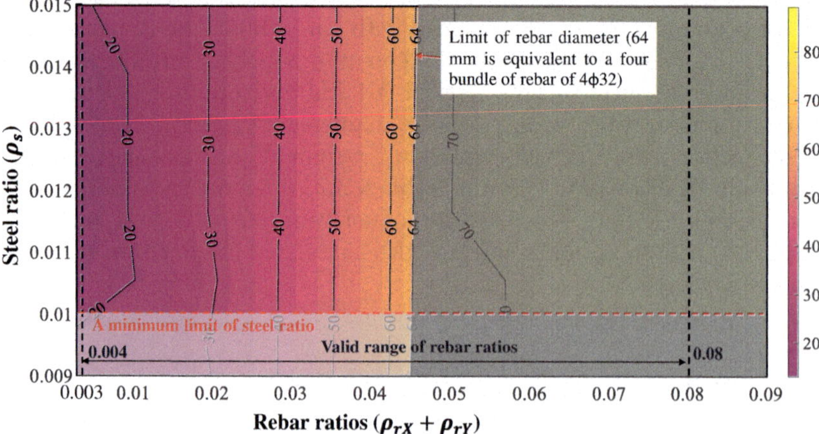

: A maximum of diameter (=16 mm) for a bundle of rebar with preassigned parameters:
+ Reverse inputs: $SF = 1$, $b/h = 1$, ρ_{rX} and ρ_{rY}
+ An H-shaped steel section: $t_f = 8$, $t_w = 8$
+ Material properties: $f_c' = 30\ MPa$, $f_{yr} = 500\ MPa$, $f_{ys} = 325\ MPa$
+ Factored loads: $P_u = 10000\ kN$, $M_{uX} = 5000\ kN \cdot m$, $M_{uY} = 7000\ kN \cdot m$

(a)

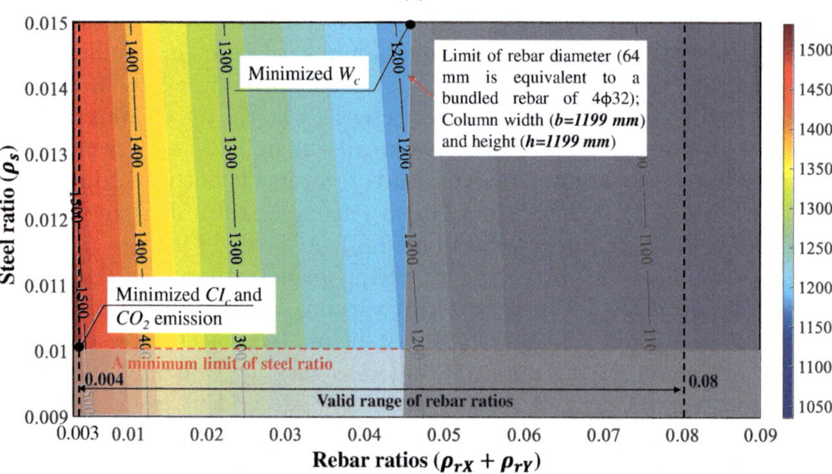

: Column width (*b=1503 mm*) and height (*h=1503 mm*) with preassigned parameters:
+ Reverse inputs: $SF = 1$, $b/h = 1$, ρ_{rX} and ρ_{rY}
+ An H-shaped steel section: $t_f = 8$, $t_w = 8$
+ Material properties: $f_c' = 30\ MPa$, $f_{yr} = 500\ MPa$, $f_{ys} = 325\ MPa$
+ Factored loads: $P_u = 10000\ kN$, $M_{uX} = 5000\ kN \cdot m$, $M_{uY} = 7000\ kN \cdot m$

(b)

Figure 8.4.1 Effects of rebar ($\rho_{rX} + \rho_{rY}$) and steel (ρ_s) ratios on $D_{r,max}$ and column dimensions (h and b). (a) Effects of rebar and steel ratios on $D_{r,max}$. (b) Effects of rebar and steel ratios on column height (h) and width (b).

of single rebar considered in this chapter is 32 mm, and hence, a maximum diameter of bundle of rebar using four rebars is 64 mm. The shadow blocks in Figures 8.4.1(a) and (b) show infeasible regions for the reverse design, illustrating that a maximum rebar diameter, rebar ratios, and steel ratios for the reverse design are limited when minimizing CI_c, CO_2 emissions, and column weight W.

The influence of rebar and steel ratios on diameter of rebar ($D_{r,max}$) is exhibited in Figure 8.4.1(a). Diameter of a rebar ($D_{r,max}$) of 16 mm at lower left corner of Figure 8.4.1(a) is found where CI_c, CO_2 emissions are minimized. The influence of rebar and steel ratios on column dimensions (h and b) is also exhibited in Figure 8.4.1(b). Column dimensions h=b =1,503 mm at lower left corner of Figure 8.4.1(b) are found where CI_c, CO_2 emissions are minimized. Diameter of rebar ($D_{r,max}$) of 63.5 mm at upper right corner of Figure 8.4.1(a) is found where W is minimized, whereas column dimensions h=b=1,199 mm at upper right corner of Figure 8.4.1(b) are found where W emissions are minimized. Table 8.3.3 shows lower left point on the Figure 8.4.1(a) and (b) where rebar ratios (ρ_{rX} and ρ_{rY}) and steel ratios (ρ_s) are preassigned as 0.002, 0.002, and 0.01, respectively, are identified while minimizing CI_c and CO_2 emissions and calculating diameter of single rebar ($D_{r,max}$) of 16 mm. Table 8.4.1 also shows upper right point on Figure 8.4.1(a) and (b) where rebar ratios (ρ_{rX} and ρ_{rY}) and steel ratios (ρ_s) are preassigned as 0.022, 0.022, and 0.015, respectively, where design based on minimized W_c are identified, calculating diameter of single rebar ($D_{r,max}$) of 63.5 mm. Figure 8.4.1(a) and Table 8.3.3 show the minimized CI_c, CO_2, corresponding to rebar ratios (ρ_{rX} and ρ_{rY}) and steel ratios (ρ_s) which are preassigned as 0.002, 0.002, and 0.01, whereas Figure 8.4.1(b) and Table 8.4.1 show the minimized column weight (W_c) corresponding to rebar ratios (ρ_{rX} and ρ_{rY}) and steel ratios (ρ_s) which are preassigned as 0.022, 0.022, 0.015. It is noted, as shown in Table 8.3.3, that ANNs select small rebars and steel sections while selecting large concrete sections to minimize CI_c, CO_2 emissions. However, as shown in Table 8.4.1, ANNs select large rebars and steel sections while selecting small concrete sections to minimize Wc. Designs shown in Tables 8.3.3 and 8.4.1 are obtained with safety factor of 1.02 which cannot be obtained using conventional design method.

8.4.3 Optimized designs of the CI_c and CO_2 emission

Figure 8.4.2(a)–(c) can also be constructed by solving for cost index CI_c, CO_2 emissions, and column weight W, respectively, from reverse design using BS method when varying rebar ratios (ρ_{rX} and ρ_{rY}) and steel ratio (ρ_s) in a range of 0.003–0.09 and 0.009–0.015, respectively. Influence of rebar ratios (ρ_{rX} and ρ_{rY}) and steel ratios (ρ_s) on design targets such as cost index (CI_c) and CO_2 emissions is shown in Figure 8.4.2(a) and (b). Rebar ratios (ρ_{rX} and ρ_{rY}) and steel ratios vary in a range of 0.003–0.09 and

Table 8.4.1 *Design table based on minimized W_c*

No.	Parameters	BS - CRS		Structural mechanics	Error* (%)
		Step 1	Step 2		
1	SF	1.00	1.02	1.02	-1.58%
2	ρ_s	0.015	0.015	0.015	-0.01%
3	b/h	1.00	1.00	1.00	0.00%
4	t_f (mm)	8	8	8	-
5	t_w (mm)	8	8	8	-
6	f_c' (MPa)	30	30	30	-
7	f_{yr} (MPa)	500	500	500	-
8	f_{ys} (MPa)	325	325	325	-
9	ρ_{rX}	0.022	0.022	0.022	-
10	ρ_{rY}	0.022	0.022	0.022	-
11	P_u (kN)	10000	10000	10000	-
12	M_{uX} (kN·m)	5000	5000	5000	-
13	M_{uY} (kN·m)	7000	7000	7000	-
14	b (mm)	1199.03	1199.03	1199.03	-
15	h (mm)	1199.02	1199.02	1199.02	-
16	h_s (mm)	1040.43	1040.43	1040.43	-
17	b_s (mm)	835.73	835.73	835.73	-
18	ε_r	-	0.0023	0.0023	-
19	ε_s	-	0.0021	0.0021	-
20	Y_s (mm)	-	79.30	79.30	-
21	X_s (mm)	-	181.65	181.65	-
22	CI_c (KRW/m)	-	964,366	964,366	-
23	CO_2 (t-CO_2/m)	-	1.91	1.91	-
24	W_c (kN/m)	-	38.45	38.45	-
25	$D_{r,max}$	-	63.5	63.5	-

CRS sequence: $h \rightarrow b \rightarrow b_s \rightarrow h_s$

BS method procedure

Step 1 (*reverse network*): Calculating **4 reverse outputs** based on **10 ordinary inputs** and **3 reverse inputs**.

Step 2 (*forward analysis*): Calculating **8 outputs** using AutoSRCHCol based on **14 inputs** (**4 reverse outputs** calculated from Step 1 and **10 ordinary inputs**).

AutoSRCHCol procedure

Calculating **3 outputs** based on **14 inputs** which are identical to **14 inputs** of Step 2 of BS.

*Note: $Errors = \frac{(ANN - AutoSRCHCol)}{ANN} \times 100\%$

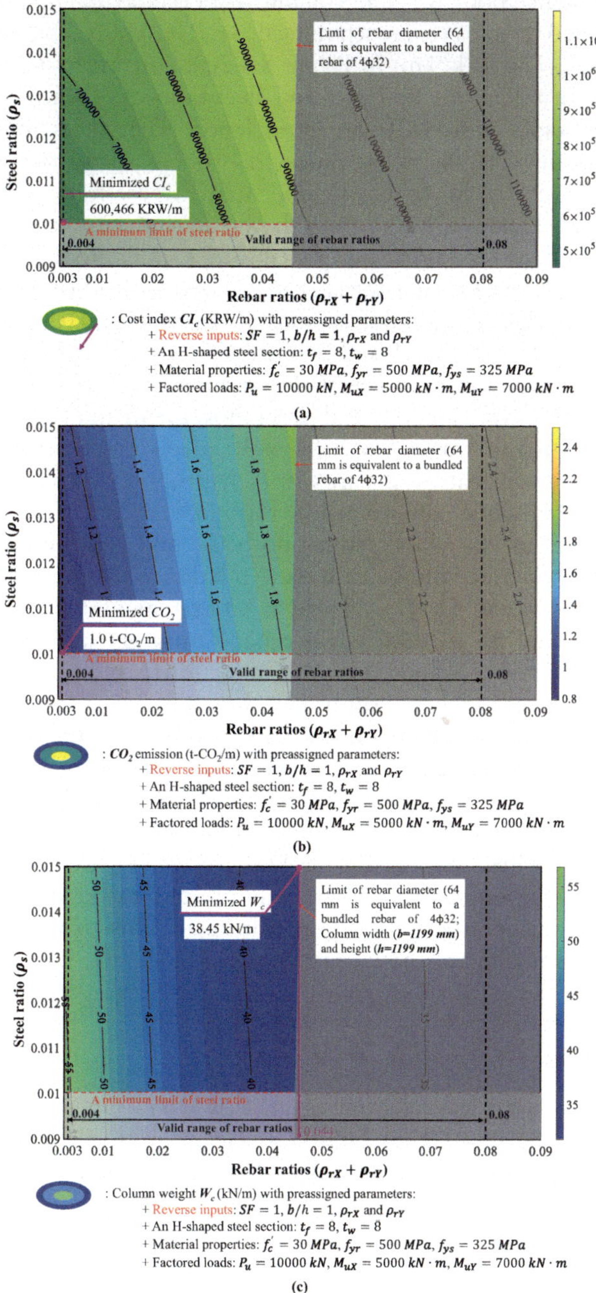

Figure 8.4.2 Effects of rebar ($\rho_{rX} + \rho_{rY}$) and steel (ρ_s) ratios on objective parameters (CI_c, CO_2 emissions, and W_c). (a) Effects of rebar and steel ratios on CI_c. (b) Effects of rebar and steel ratios on CO_2 emissions. (c) Effects of rebar and steel ratios on weight (W_c)

0.009–0.015 while both *SF* and column aspect ratio (b / h) are assigned as 1.0. Factored loads are predetermined as P_u=10,000 kN, M_{uX}=5,000 kN·m, and M_{uY}=7,000 kN·m. In Figure 8.4.2(a), cost index (CI_c) is derived as a function of rebar (ρ_{rX} and ρ_{rY}) and steel ratios (ρ_s) when f_c'=30 MPa, f_{yr}=500 MPa, and f_{ys}=325 MPa are preassigned. The CI_c is minimized as 600,466 KRW/m when the rebar ratios (ρ_{rX} and ρ_{rY}) and steel ratios (ρ_s) are minimized at 0.004 ($\rho_{rX}+\rho_{rY}$) and 0.01, respectively, as shown in Figure 8.4.2(a). Figure 8.4.2(b) also shows an influence of rebar ratios (ρ_{rX} and ρ_{rY}) and steel ratios (ρ_s) on CO_2 emissions similar to that obtained for cost index (CI_c). The CO_2 emissions are minimized as 1.00 t-CO_2/m at minimized rebar ratios ($\rho_{rX}+\rho_{rY}$) and steel ratios (ρ_s) of 0.004 and 0.01, respectively. Figure 8.4.2(a) and (b) shows CI_c and CO_2 emissions which are minimized when column width (b) and column height (h) are maximized at h=b=1,503 mm, whereas rebar ratios (ρ_{rX} and ρ_{rY}) and steel ratios (ρ_s) are minimized.

Figure 8.4.1(b) and Tables 8.3.3 and 8.4.1 also show that rebar ratios (ρ_{rX} and ρ_{rY}) and steel ratios (ρ_s) are inversely proportion to column width (b) and height (h) when column aspect ratio (b/h) of a 1.0 is preassigned. Rebar and steel ratios which reduce both CI_c and CO_2 emission, while increasing the concrete volume to achieve a balanced section. The h, b, h_s, and b_s shown in Boxes 14–17 of Table 8.3.3 are calculated on an output-side Step 1 of the BS method when rebar ratios ρ_{rX} and ρ_{rY} shown in Boxes 9 and 10 are assigned as 0.002 and 0.002 as input parameters, respectively, whereas and steel ratios (ρ_s) shown in Box 2 are also assigned as reach 0.01. Accuracies of minimized CI_c and CO_2 emissions are sufficient for use in practical designs, indicating that *SF* of −0.73% error is observed in Table 8.3.3. The *SF* calculated from AutoSRCHCol using design parameters obtained by BS method is compared with the preassigned SF in the first step of BS method. As shown in Figure 8.4.3, a three-dimensional interaction diagram is plotted based on column and steel dimensions corresponding to the minimized design of CI_c and CO_2. All design parameters are shown in Figure 8.4.3.

8.4.4 Optimized designs of W_c

Figure 8.4.2(c) shows how column weights (W_c) vary when rebar and steel ratios vary from 0.003–0.09 and 0.009–0.015, respectively, whereas the *SF* and column aspect ratio (b/h) are assigned as 1.0. Factored loads are predetermined as P_u=10,000 kN, M_{uX}=5,000 kN·m, and M_{uY}=7,000 kN·m while material properties are also preassigned as f_c'=30 MPa, f_{yr}=500 MPa, and f_{ys}=325 MPa. W_c is minimized as 38.45 kN/m when the rebar (ρ_{rX} and ρ_{rY}) and steel (ρ_s) ratios are maximized as shown in Figure 8.4.2(c). The rebars

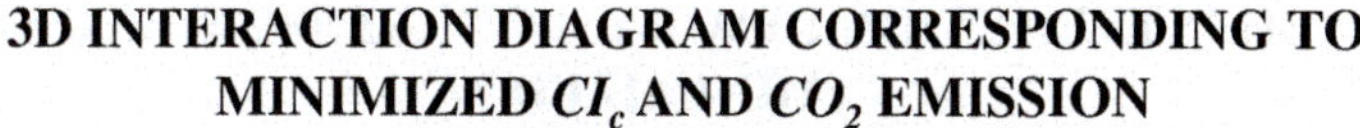

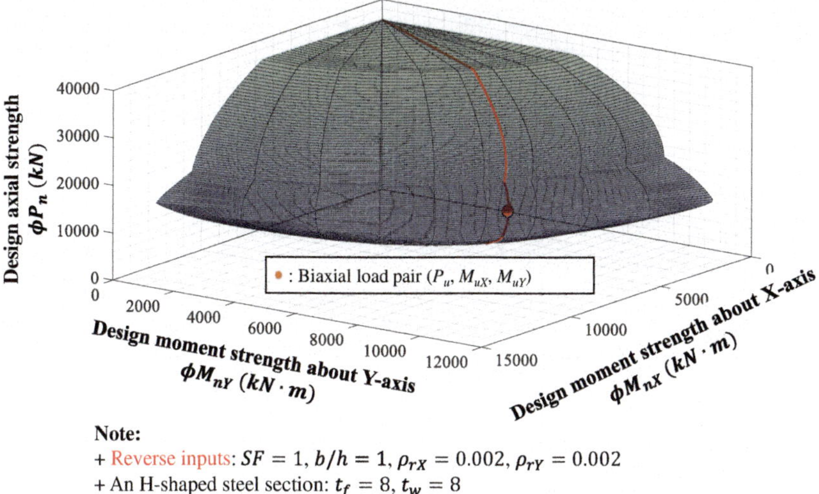

Note:

+ Reverse inputs: $SF = 1$, $b/h = 1$, $\rho_{rX} = 0.002$, $\rho_{rY} = 0.002$
+ An H-shaped steel section: $t_f = 8$, $t_w = 8$
+ Material properties: $f_c' = 30\ MPa$, $f_{yr} = 500\ MPa$, $f_{ys} = 325\ MPa$
+ Factored loads: $P_u = 10000\ kN$, $M_{uX} = 5000\ kN \cdot m$, $M_{uY} = 7000\ kN \cdot m$

Figure 8.4.3 Three-dimensional interaction diagram based on minimized CI_c and CO_2 emission.

$(\rho_{rX} + \rho_{rY})$ are maximized to 0.044 when a maximum rebar diameter of 64 mm is obtained as shown in Figs. 8.4.1(c) where the steel (ρ_s) ratio maximized to 0.015 also minimizes W_c. Reverse outputs (h, b, h_s, and b_s) are minimized on an output-side Step 1 the BS method when rebar (0.004) and steel (0.015) ratios are maximized as shown in Table 8.4.1 which presents a minimized W_c. The SF error of –1.58% is demonstrated in which the safety factor preassigned on an input-side as a reverse parameter is well compared with that calculated by AutoSRCHCol software using design parameters obtained by BS method Figure 8.4.4 shows a three-dimensional interaction diagram plotted based on column and steel dimensions corresponding to the minimized design of W. The SF was preassigned as 1.0 when P_u=10,000 kN, M_{uX}=5,000 kN·m, and M_{uY}=7,000 kN·m are preassigned. All design parameters are shown in Figure 8.4.4. It is worth noting that W_c are minimized when column dimensions decrease, whereas rebar and steel ratios increase. However, CI_c and CO_2 emissions are minimized with increased column dimensions and decreased rebar and steel ratios.

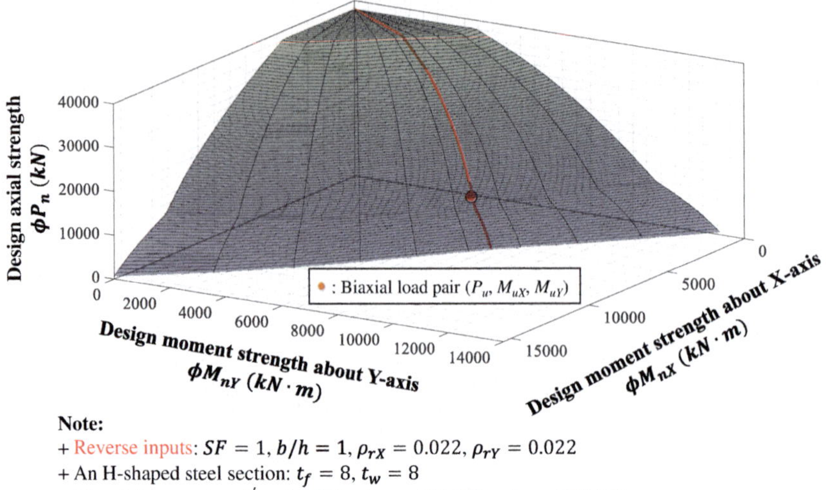

Figure 8.4.4 Three-dimensional interaction diagram based on minimized W_c.

8.5 CONCLUSION

In this chapter, non-Lagrange-based design optimizations for reinforced concrete columns encasing H-shaped steel sections are introduced. ANNs-based reverse design is performed to optimize a design of SRC columns with *SF* preassigned on an input-side for given factored biaxial loads, steel ratios (ρ_s), and aspect ratio of columns (b/h) while satisfying code requirements imposed by in ANSI/AISC 360-16 and ACI 318-19.

In Step 1 of BS shown in Figures 8.3.3 and 8.3.4, three reverse inputs (*SF*, ρ_s, and b/h) are the parameters which were preassigned on an input-side, whereas four reverse outputs (h, b, h_s, and b_s) are calculated on an output-side. Three reverse input parameters (*SF*, ρ_s, and b/h) preassigned on an input-side are used to predict four reverse outputs (h, b, h_s, and b_s) which will be used as input parameters of forward networks of Step 2 of Tables 8.3.3 and 8.4.1. Effects of rebar and steel ratios on cost index (CI_c), CO_2 emission, and column weight (W_c) are also identified, resulting in three-dimensional interaction diagrams which are optimized with respect to CI_c, CO_2 emission, and W_c based on preassigned *SF* equivalent to 1.0. Predictions by ANNs are verified using structural mechanics using design parameters obtained by BS method, demonstrating a significant accuracy of the proposed ANN-based design. In this chapter, a useful and practical designs for SRC columns are offered to lessen engineer's effort while securing design accuracies.

The following conclusions can be further drawn for readers:

1. SRC columns under a biaxial load are optimized by a reverse design, implementing a BS method (reverse CRS – Forward AutoSRCHCol) (Hong [8.30]). Based on a network trained by large datasets using CRS training technique for the Step 1 of BS, four reverse outputs, which are column height (h), width (b), and dimensions of steel section (h_s and b_s), are calculated for given the three reverse input parameters based on a factored biaxial load, steel ratio (ρ_s), and aspect ratio of columns (b/h). Four reverse outputs are, then, back-substituted in Step 2 of BS shown in Tables 8.3.3 and 8.4.1 to calculate the other eight output parameters $\left(\varepsilon_r, \varepsilon_s, D_{r,\max}, Y_s, X_s, CI_c, CO_2, W_c\right)$ and the three reverse input parameters, SF, ρ_s, and b/h which were already preassigned on an input-side. In the reverse network of the Step 2, s structural mechanics -based AutoSRCHCol proposed by Nguyen and Hong [8.10] is used when h, b, h_s, and b_s obtained from Step 1 of BS are back-substituted with the other ten forward inputs $\left(t_w, t_f, f_c^{'}, f_{yr}, f_{ys}, \rho_{rX}, \rho_{rY}, P_u, M_{uX}, M_{uY}\right)$. Any structural software can be implemented in Step 2 of a BS to find forward outputs based on the reverse outputs obtained in Step 1 of the BS.
2. Rebar and steel ratios influence CI_c, CO_2 emission, and W_c. Rebar and steel ratios decrease to minimize CI_c and CO_2 emission, whereas rebar and steel ratios increase to minimize column weights (W_c). Cost CI_c, CO_2 emission, and weight W_c are minimized in the ranges of 0.003–0.09 and 0.009–0.015 for steel (ρ_s) and rebar ratios (ρ_{rX} and ρ_{rY}), respectively. In Step 1, column (h and b) and steel dimensions (h_s and b_s) are determined corresponding to minimized cost CI_c, CO_2 emissions, and weight W_c. ANNs ensure to provide optimized designs by preassigning safety factor equal to or greater than 1.0, whereas optimizing designs using conventional method based on a trial-and-error technique is challenging for engineers to determine proper column sizes and steel sections in a preliminary design stage, minimizing cost index (CI_c), CO_2 emission, and column weights (W_c).
3. CI_c and CO_2 emissions are minimized similarly, when rebar and steel ratios are reduced to their minima of 0.004 and 0.01, respectively, as imposed by ANSI/AISC 360-16 [8.24]. Designs with minimized column weight (W_c) are achieved in an opposite way when rebar and steel ratios should be maximized.
4. Optimized design of SRC columns based on the proposed ANNs are verified by those calculated from structural mechanics using design parameters obtained by BS method, demonstrating observed errors

which are as small as −0.73% for SF when minimizing CI_c and CO_2 emissions. The error of SF is also as small as −1.58% when minimizing W_c for *SF*.

5. Three-dimensional interaction diagrams corresponding to minimized CI_c, CO_2 emission, and W_c are presented when safety factors ($SF = 1.0$) for a factored biaxial load are preassigned based on for a factored biaxial load.
6. A behavior of concrete-steel composite columns for practical design applications now can be performed using the ANN-based robust, expeditious, and dependable but sufficiently accurate design tools.
7. In Chapter 9, design optimizations of SRC columns under multiple loads will be performed based on derivative optimization methods, such as a Lagrange Mulplier method(LMM), whereas in Chapter 10, multi-objective functions, including CI_c, CO_2 emission, and W_c will be also optimized simultaneously.

REFERENCES

[8.1] Chicoine, T., Tremblay, R., Massicotte, B., Ricles, J. M., & Lu, L. W. (2002). Behavior and strength of partially encased composite columns with built-up shapes. *Journal of Structural Engineering*, 128(3), 279–288. https://doi.org/10.1061/(ASCE)0733-9445(2002)128:3(279)

[8.2] Wang, Y. C. (1999). Tests on slender composite columns. *Journal of Constructional Steel Research*, *49*(1), 25–41. https://doi.org/10.1016/S0143-974X(98)00202-8

[8.3] Mirza, S. A., Hyttinen, V., & Hyttinen, E. (1996). Physical tests and analyses of composite steel-concrete beam-columns. *Journal of Structural Engineering*, *122*(11), 1317–1326. https://doi.org/10.1061/(ASCE)0733-9445(1996)122:11(1317)

[8.4] Ricles, J. M., & Paboojian, S. D. (1994). Seismic performance of steel-encased composite columns. *Journal of Structural Engineering*, *120*(8), 2474–2494. https://doi.org/10.1061/(ASCE)0733-9445(1994)120:8(2474)

[8.5] El-Tawil, S., & Deierlein, G. G. (1999). Strength and ductility of concrete encased composite columns. *Journal of Structural Engineering*, *125*(9), 1009–1019. https://doi.org/10.1061/(ASCE)0733-9445(1999)125:9(1009)

[8.6] Li, L. I., & Matsui, C. (2000). Effects of axial force on deformation capacity of steel encased reinforced concrete beam–columns. In *Proceedings of 12th World Conference on Earthquake Engineering.*

[8.7] Munoz, P. R., & Hsu, C. T. T. (1997). Behavior of biaxially loaded concrete-encased composite columns. *Journal of Structural Engineering*, *123*(9), 1163–1171. https://doi.org/10.1061/(ASCE)0733-9445(1997)123:9(1163)

[8.8] Morino, S., Matsui, C., & Watanabe, H. (1984, July). Strength of biaxially loaded SRC columns. In: Frangopol, D. M. (ed) *Composite and Mixed Construction* (pp. 185–194). ASCE, Atlantic City, NJ.

[8.9] Chen, C. C., & Lin, N. J. (2006). Analytical model for predicting axial capacity and behavior of concrete encased steel composite stub columns. *Journal of Constructional Steel Research*, *62*(5), 424–433. https://doi.org/10.1016/j.jcsr.2005.04.021

[8.10] Nguyen, D. H., & Hong, W. K. (2020). An analytical model computing the flexural strength and performance of the concrete columns confined by both transverse reinforcements and steel sections. *Journal of Asian Architecture and Building Engineering*, *19*(6), 647–669. https://doi.org/10.1080/13467581.2020.1775603

[8.11] Virdi, K. S., Dowling, P. J., BS 449, & BS 153. (1973). The ultimate strength of composite columns in biaxial bending. *Proceedings of the Institution of Civil Engineers*, *55*(1), 251–272. https://doi.org/10.1680/iicep.1973.4958

[8.12] Roik, K., & Bergmann, R. (1990). Design method for composite columns with unsymmetrical cross-sections. *Journal of Constructional Steel Research*, *15*(1–2), 153–168. https://doi.org/10.1016/0143-974X(90)90046-J

[8.13] Dundar, C., Tokgoz, S., Tanrikulu, A. K., & Baran, T. (2008). Behaviour of reinforced and concrete-encased composite columns subjected to biaxial bending and axial load. *Building and Environment*, *43*(6), 1109–1120. https://doi.org/10.1016/j.buildenv.2007.02.010

[8.14] Munoz, P. R., & Hsu, C. T. T. (1997). Biaxially loaded concrete-encased composite columns: design equation. *Journal of Structural Engineering*, *123*(12), 1576–1585. https://doi.org/10.1061/(ASCE)0733-9445(1997)123:12(1576)

[8.15] Foraboschi, P. (2019). Bending load-carrying capacity of reinforced concrete beams subjected to premature failure. *Materials*, *12*(19), 3085. https://doi.org/10.3390/ma12193085

[8.16] Foraboschi, P. (2020). Optimal design of seismic resistant RC columns. *Materials*, *13*(8), 1919. https://doi.org/10.3390/ma13081919

[8.17] Foraboschi, P. (2016). Versatility of steel in correcting construction deficiencies and in seismic retrofitting of RC buildings. *Journal of Building Engineering*, *8*, 107–122. https://doi.org/10.1016/j.jobe.2016.10.003

[8.18] Abambres, M., & Lantsoght, E. O. (2020). Neural network-based formula for shear capacity prediction of one-way slabs under concentrated loads. *Engineering Structures*, *211*, 110501. https://doi.org/10.1016/j.engstruct.2020.110501

[8.19] Sharifi, Y., Lotfi, F., & Moghbeli, A. (2019). Compressive strength prediction using the ANN method for FRP confined rectangular concrete columns. *Journal of Rehabilitation in Civil Engineering*, 7(4), 134–153. https://doi.org/10.22075/JRCE.2018.14362.1260

[8.20] Asteris, P. G., Armaghani, D. J., Hatzigeorgiou, G. D., Karayannis, C. G., & Pilakoutas, K. (2019). Predicting the shear strength of reinforced concrete beams using Artificial Neural Networks. *Computers and Concrete*, 24(5), 469–488.https://doi.org/10.12989/CAC.2019.24.5.469

[8.21] Armaghani, D. J., Hatzigeorgiou, G. D., Karamani, C., Skentou, A., Zoumpoulaki, I., & Asteris, P. G. (2019). Soft computing-based techniques for concrete beams shear strength. *Procedia Structural Integrity*, 17, 924–933. https://doi.org/10.1016/j.prostr.2019.08.123

[8.22] Hong, W. K., Nguyen, V. T., & Nguyen, M. C. (2022). Artificial intelligence-based novel design charts for doubly reinforced concrete beams. *Journal of Asian Architecture and Building Engineering*, *21*(4), 1497–1519. https://doi.org/10.1080/13467581.2021.1928511

[8.23] Hong, W. K., Pham, T. D., & Nguyen, V. T. (2022). Feature selection based reverse design of doubly reinforced concrete beams. *Journal of Asian Architecture and Building Engineering*, *21*(4), 1472–1496. https://doi.org/10.1080/13467581.2021.1928510

[8.24] ANSI/AISC 360-16 (2016). *Specification for Structural Steel Buildings*. American Institute of Steel Construction, Chicago, IL.

[8.25] Hong, W. K., Nguyen, V. T., Nguyen, D. H., & Nguyen, M. C. (2022). Reverse design-based optimizations for reinforced concrete columns encasing H-shaped steel section using ANNs. *Journal of Asian Architecture and Building Engineering*, 1–15. https://doi.org/10.1080/13467581.2022.2047985

[8.26] ACI Committee. (2019). *Building Code Requirements for Structural Concrete (ACI 318-19) and Commentary*. American Concrete Institute, Farmington Hills, MI.

[8.27] Mander, J. B., Priestley, M. J., & Park, R. (1988). Theoretical stress-strain model for confined concrete. *Journal of Structural Engineering*, *114*(8), 1804–1826. https://doi.org/10.1061/(ASCE)0733-9445(1988)114:8(1804)

[8.28] Arafa, M., Alqedra, M., & An-Najjar, H. (2011). Neural network models for predicting shear strength of reinforced normal and high strength concrete deep beams. *Journal of Applied Sciences*, *11*(2). https://doi.org/10.3923/jas.2011.266.274

[8.29] MathWorks. (2021a). *Deep Learning Toolbox™ Getting Started Guide*. The MathWorks, Inc, Natick, MA.

[8.30] Hong, W. K (2023). *Artificial Neural Network-based Optimized Design of Reinforced Concrete Structures*. Taylor & Francis, Boca Raton, FL.

[8.31] Won-kee Hong, (2023). Artificial Intelligence-Based Design of Reinforced Concrete Structures. Elsevier.

Chapter 9

Design optimizations of concrete columns encasing H-shaped steel sections under a biaxial bending using an ANN-based Lagrange algorithm

9.1 INTRODUCTION AND SIGNIFICANCE OF THIS STUDY

9.1.1 Research background

ANNs have been used successfully by Abambres and Lantsoght [9.1], Sharifi et al. [9.2], Asteris et al. [9.3], and Armaghani et al. [9.4] for structural analysis. Hong et al. [9.5,9.6] have performed optimization studies of reinforced concrete beams based on ANN. Hong and Nguyen [9.7] also successfully applied the Lagrange Multiplier method (LMM) to obtain optimal designs of reinforced rectangular concrete columns based on ANN.

Steel-encased reinforced concrete columns (SRC) are being used extensively to enhance a strength and ductility of steel sections with an increased rigidity and strength of concrete. Steel sections can be protected from fire and corrosion by encasing them in concrete columns. SRC members have been recently used in high-rise buildings and transportation facilities. Mirza et al. [9.8], Ricles and Paboojian [9.9], El-Tawil and Deierlein [9.10], Li and Matsui [9.11], Brettle [9.12], Bridge and Roderick [9.13], Furlong [9.14], and Mirza and Skrabek [9.15] have experimentally investigated a behavior of SRC columns when subjected to uniaxial bending and axial compressive loads, whereas Munoz and Hsu [9.16] and Morino et al. [9.17] investigated a performance of SRC columns under biaxial bending and axial compressive loads. Chen and Lin [9.18], Dundar et al. [9.19], Munoz and Hsu [9.20], Furlong [9.21], Kato [9.22], Virdi and Dowling [9.23], and Roik and Bergmann [9.24] also investigated behaviors of SRC columns. Effects of confined concrete on performance of SRC columns have been investigated by Rong et al. [9.25] and Rong and Shi [9.26]. However, in the conventional method, optimizing designs of SRC columns can be obtained only by trial-and-error, which is challenging for engineers to perform, whereas ANNs-based Lagrange multipliers perform, with an acceptable accuracy, design optimization of reinforced concrete columns encasing H-shaped steel sections subjected to biaxial bending and concentric axial loads. Nguyen and Hong [9.27] and Hong [9.28] investigated a numerical and experimental

DOI: 10.1201/9781003354796-9

behavior of such SRC columns based on a strain compatibility to generate large datasets for use in the training of ANNs.

9.1.2 Research innovations and significances

In this chapter, cost index (CI_c), CO_2 emissions, and weight (W_c) of SRC columns under a biaxial load are minimized. Explicit objective functions and constraint constraints which are difficult to derive for an application of a conventional Lagrange optimization are replaced by ANN-based generalized functions for objective and constraining functions. Solutions under KKT conditions should be sought when inequality constraints are used to impose design requirements. Newton–Raphson iterations are used to converge initial input parameters to stationary points, minimizing Lagrange functions. ANN-based objective functions for CI_c, CO_2 emissions, and W_c of SRC columns and their constraints as functions of input parameters are derived based on a large dataset of 100,000. Cost index (CI_c), CO_2 emissions, and weight (W_c) are objective targets which need to be optimized for a design of SRC columns in this chapter. Design parameters including column dimensions and rebar ratios are also obtained on an output-side while minimizing CI_c, CO_2 emissions, and W_c. Any design target can be adopted as an objective function. Cost (CI_c) is an interest of structural engineers, whereas CO_2 emissions, and weight (W_c) are the interests of governments and contractors, respectively. Errors are insignificant, indicating that the proposed network could be implemented in practical designs to help engineers optimize SRC column designs as the accuracy of the minimized objective functions (CI_c,CO_2 emissions, and W_c) and corresponding design parameters are verified using a structural mechanics-based software (AutoSRCHCol) developed by Nguyen and Hong [9.27]. This chapter was written based on the previous paper by Hong et al. [9.29].

9.1.3 Tasks readers can perform after this chapter

1. Readers can optimize SRC columns under a biaxial load while optimizing ANN-based Lagrange functions with respect to cost index (CI_c), CO_2 emissions, and weight (W_c) of SRC columns. Cost index (CI_c), CO_2 emissions, and weight (W_c) of SRC columns are minimized while yielding steel height (h_s), flange width (b_s), flange (t_f) thickness, and web (t_w) thickness under factored biaxial loads P_u, M_{uX}, M_{uY}. Both cost index (CI_c) and CO_2 are minimized when rebar ratios and a steel section are minimized while dimensions of the concrete section are maximized. On the other hand, column dimensions decrease to lower column weight, resulting in increased rebar and steel quantity to compensate concrete volume reduction. CI_c and CO_2 emissions are minimized in opposite way to minimizing W_c.

2. Material properties such as concrete (f_c'), rebar (f_{yr}), and steel (f_{ys}) are predetermined on an input-side. Factored biaxial loads are factored axial load (P_u) and factored bending moments along X- and Y-axes (M_{uX} and M_{uY}, respectively) preassigned on an input-side a forward design network to design SRC columns. Factored biaxial loads (P_u, M_{uX}, M_{uY}) are preassigned on an input-side. ANNs will design a column section to obtain design strengths (ϕP_n, ϕM_{nX}, ϕM_{nY}) greater than preassigned factored biaxial loads (P_u, M_{uX}, M_{uY}).
3. Optimizations for concrete columns encasing H-shaped steel sections under a biaxial bending are performed based on 9 equality functions $c(x)$ and 17 inequality constraints $v(x)$. Inequality constraint matrix (S) is developed to designate activated and inactivated inequality constraints. A diagonal matrix S comprises an inequality term of the Lagrange function, which takes a complementary slack condition into $Sv(x)=0$ when formulating KKT equations. An activated inequality v_j is demonstrated by setting the $s_j=1$, whereas s_j is set as 0 when the inequality v_j is inactive, designating s_1, s_2, ..., s_l with a status of inequalities $v_1(x)$, $v_2(x)$, ..., $v_l(x)$, respectively.
4. Conventional Lagrange multipliers are difficult to apply to optimizations of SRC column designs when explicit objective functions and constraint constraints are difficult to derive for a conventional Lagrange optimization. In this chapter, weight and bias matrices are obtained based on large datasets to formulate functions based on ANNs. ANN-based differentiable objective and constraint functions are derived to obtain Lagrange functions. ANN-based generalized functions for CI_c, CO_2 emissions, W_c and design constraints replace explicit functions when formulating Lagrange functions.
5. Newton–Raphson iterations are used to solve for KKT conditions based on inequality constraints imposed by design requirements. Design parameters including column dimensions and rebar ratios are obtained on an output-side while minimizing CI_c, CO_2 emissions, and W_c.
6. Structural weight of precast module frames can be minimized for lifting efficiency using the proposed ANN-based Lagrange optimization.
7. Lagrange-based training and Newton–Raphson-based iterations were performed based on the latest MATLAB toolboxes (MATLAB Deep Learning Toolbox [9.30], MATLAB Global Optimization Toolbox [9.31], MATLAB Optimization Toolbox [9.32], MATLAB Parallel Computing Toolbox [9.33], MATLAB Statistics and Machine Learning Toolbox [9.34]). The author expects accuracies and applicability can be significantly extended in near future when the computation speed does not hinder the optimizations.

9.2 SRC COLUMN DESIGNS USING ANNs AND LAGRANGE MULTIPLIERS

9.2.1 Design scenario for SRC columns

In this chapter, an ANN-based forward network for a single biaxial load is formulated. Column dimensions including height (h), width (b), steel height (h_s), flange width (b_s), flange thickness (t_f), and web thickness (t_w) are preassigned on an input-side for forward design scenario, when optimizing SRC columns. Material properties of concrete (f_c'), rebar (f_{yr}), and steel (f_{ys}) are also predetermined on an input-side. Factored biaxial loads selected as factored axial load (P_u) and factored bending moments with respect to X- and Y- axes (M_{uX} and M_{uY}, respectively) are input parameters to optimize SRC columns as shown in Figure 9.2.1(a) and Table 9.2.1.

Figure 9.2.1(a) and (b) illustrates cross section and strain compatibility of SRC columns imposed by factored biaxial loads consisting of P_u, M_{uX}, M_{uY}, respectively, which are preassigned on an input-side. Column sections with design strengths (ϕP_n, ϕM_{nX}, ϕM_{nY}) greater than preassigned factored biaxial loads (P_u, M_{uX}, M_{uY}) are obtained using ANNs. Parameters calculated on an output-side the ANN include safety factor (SF), rebar (ε_r) and steel (ε_s) strain, a maximum rebar diameter ($D_{r,max}$), vertical (Y_s) and horizontal (X_s) clearances, a column aspect ratio ($b\,/\,h$), and three objective functions: cost index (CI_c), CO_2 emissions, and weight (W_c). Forward design scenario is summarized in Table 9.2.2.

Rebar ratios (ρ_{rX} and ρ_{rY}) are preassigned on an input-side a forward design network, and hence, a maximum rebar diameter ($D_{r,\max}$) is calculated accordingly on an output-side based on column dimensions and rebar ratios. A location of steel sections is also determined for SRC columns, and hence, vertical (Y_s) and horizontal (X_s) clearances between column and steel flange shown in Figure 9.2.1(a) are determined most appropriately on an output-side based on a steel section and column dimensions as shown in Table 9.2.2. The rest of the corresponding design parameters are then calculated to minimize the three design targets, cost index (CI_c), CO_2 emissions, and weight (W_c) which are minimized on an output-side based on column dimensions, rebar ratios, and steel section. Cost index (CI_c), CO_2 emission, and weight (W_c) are good design targets for optimizing a design of SRC columns. Cost (CI_c) is an interest of structural engineers, whereas CO_2 emission, and weight (W_c) are the interests of governments and contractors, respectively. Any design target can be adopted as an objective function. Both cost index (CI_c) and CO_2 are minimized as shown in Tables 9.4.1 and 9.4.3 when rebar ratios and a steel section are minimized while dimensions of the concrete section are maximized. On the other hand, column dimensions decrease to reduce column weight, resulting in increased rebar and steel quantity to compensate concrete volume reduction as shown in Table 9.4.5. CI_c and CO_2 emissions are minimized in opposite way to minimizing W_c.

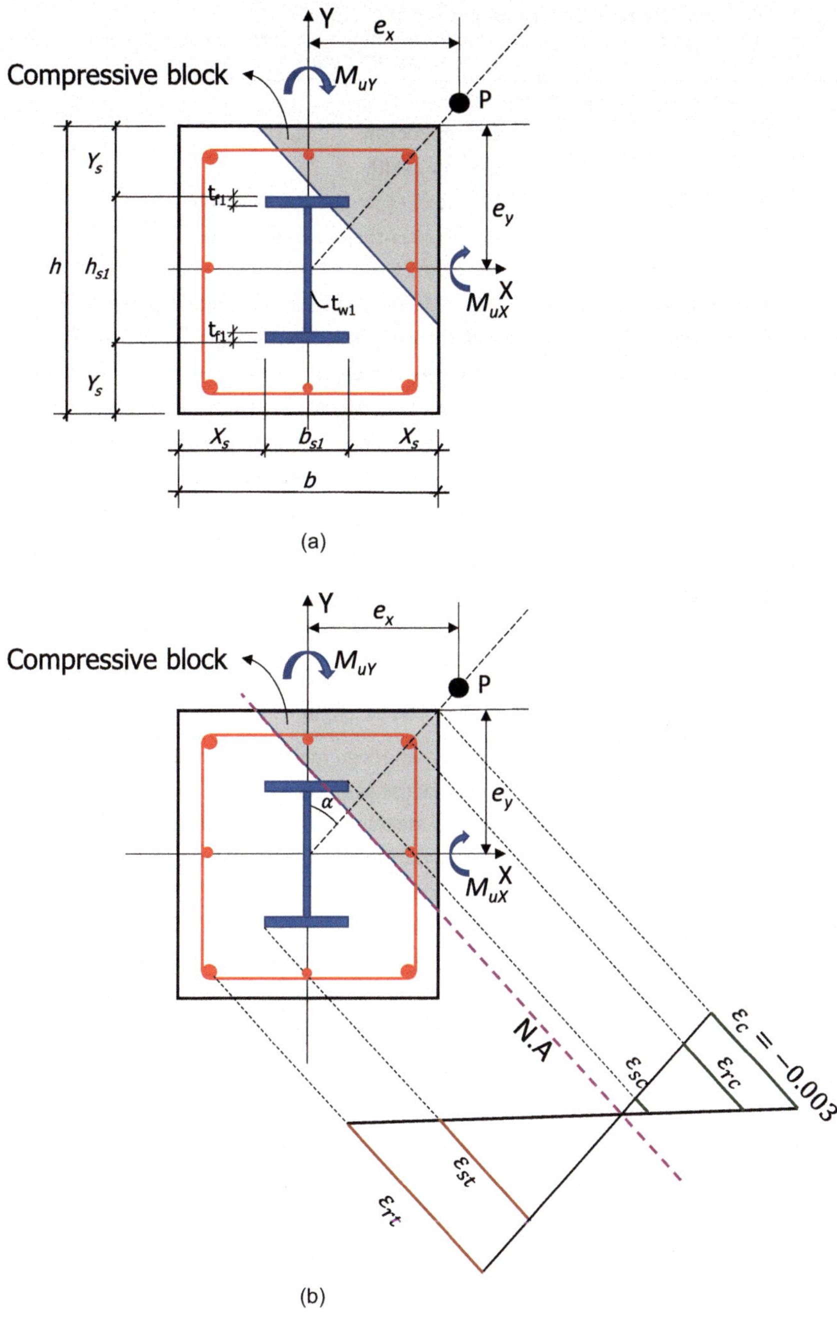

Figure 9.2.1 Cross section and strain compatibility of an SRC column. (a) Cross section of an SRC column under biaxial loads. (b) Strain compatibility of an SRC column under biaxial loads.

Table 9.2.1 Nomenclatures of parameters for SRC columns

No.	*Nomenclature*	
	Forward input parameters	
1	h (mm)	Column height
2	b (mm)	Column width
3	h_s (mm)	Height of H-shaped steel section
4	b_s (mm)	Width of H-shaped steel section
5	t_f (mm)	Flange thickness of H-shaped steel section
6	t_w (mm)	Web thickness of H-shaped steel section
7	f_c' (MPa)	Compressive concrete strength
8	f_{yr} (MPa)	Yield rebar strength
9	f_{ys} (MPa)	Yield steel strength
10	ρ_{rX}	Rebar ratio along to horizontal axis (X-axis)
11	ρ_{rY}	Rebar ratio along to vertical axis (Y-axis)
12	P_u(kN)	Axial force
13	M_{uX}(kN·m)	Factored moment about the horizontal axis (X-axis)
14	M_{uY}(kN·m)	Factored moment about the vertical axis (Y-axis)
	Forward output parameters	
15	ϕM_n(kN·m)	Design moment strength
16	ϕP_n(kN)	Design axial strength
17	M_u(kN·m)	Combined design moment strength ($\sqrt{M_{uX}^2 + M_{uY}^2}$)
18	SF	Safety factor ($\phi M_n / M_u = \phi P_n / P_u$)
19	Y_s(mm)	Vertical clearance between column and steel flange
20	X_s (mm)	Horizontal clearance from column and steel flange
21	ρ_s	Steel ratio
22	ε_s	Steel strain
23	ε_r	Rebar strain
24	b/h	Column aspect ratio
25	$D_{r,max}$	A maximum rebar diameter
26	CI_c (KRW/m)	Cost index of column
27	CO_2 (t-CO_2/m)	CO_2 emission
28	W_c (kN/m)	Column weight

9.2.2 Objective and constraining functions for Lagrange optimizations

Cost index (CI_c), CO_2 emissions, and weight (W_c) of an SRC column are considered as objective functions when implementing ANN-based Lagrange method based on the following design conditions. The material properties of concrete (f_c'), rebars (f_{yr}), and steel sections (f_{ys}) and the thickness of the flanges (t_f) and webs (t_w) of the steel sections are considered

Table 9.2.2 Forward design scenarios for SRC columns

Forward design scenario							
		Input				*Output*	
1	h	8	f_y	1	SF	8	$D_{r,max}$
2	b	9	f_{ys}	2	ε_r	9	CI_c
3	h_s	10	ρ_{rX}	3	ε_s	10	CO_2
4	b_s	11	ρ_{rY}	4	Y_s	11	W_c
5	t_w	12	P_u	5	X_s		
6	t_f	13	M_{uX}	6	b/h		
7	f_c'	14	M_{uY}	7	ρ_s		

by equality constraints, whereas inequality constraints are implemented in controlling an optimization, meeting code requirements of reinforced concrete columns encasing steel sections. The ANN-based Lagrange method is implemented in optimizing designs of SRC columns by optimizing CI_c, CO_2 emissions, and W_c based on 14 forward input parameters $\left(h, b, h_s, b_s, t_w, t_f, f_c', f_{yr}, f_{ys}, \rho_{rX}, \rho_{rY}, P_u, M_{uX}, M_{uY}\right)$ and 11 forward output parameters $\left(SF, \varepsilon_r, \varepsilon_s, \rho_s, D_{r,\max}, Y_s, X_s, {}^{b}\!/_{h}, CI_c, CO_2, W_c\right)$ presented in Table 9.2.2.

For SRC columns, a rebar ratio along X- and Y-axes $\left(\rho_{rX}\text{ and }\rho_{rY}\right)$ is limited not to be lower than 0.004 by Section I2.1a, ANSI/AISC 360-16 [9.35]. A minimum limit is required for rebar ratios to reduce effects of concrete creep and shrinkage, and to prevent rebars from yielding under sustained service loads. A maximum rebar ratio along X- and Y-axes $\left(\rho_{rX}\text{ and }\rho_{rY}\right)$ is restricted to 0.08 according to Section 10.6.1, ACI 318-19 [9.36]. A maximum limit of 0.08 is also suggested for the rebar ratios to ensure that concrete is consolidated sufficiently around the rebars. Bundled rebars are also considered when generating large datasets. The maximum diameter of bundled rebars using four rebars (each rebar diameter of 32 mm) has to be less than 64 mm. Steel ratios encased in reinforced concrete columns have to be less than 0.01 according to Section I2.1a of ANSI/AISC 360-16 [9.31]. A design strength $(\phi P_n, \phi M_n)$ of columns has to be high enough to withstand factored biaxial loads $\left(P_u, M_{uX}, M_{uY}\right)$. *SF* of a column, a ratio of its design strength to factored loads, has to be greater than or equal to 1.0. Cost index (CI_C), CO_2 emissions, and weight (W_c) of columns against biaxial loads are optimized individually using AI-based Lagrange method, while, on an output-side, determining dimensions of columns and steel sections, and rebar ratios corresponding to optimized CI_C, CO_2 emissions, and W_c . Three-dimensional interaction diagrams with respect to bending moment and axial load are also presented based on optimized CI_C, CO_2 emissions, and W_c.

9.2.3 Generation of large structural datasets

Table 9.2.2 presents 14 input and 11 output parameters which are used to generate large datasets for training ANNs to design SRC columns using ANNs. The 14 input parameters include dimensions of columns (height h and width b) and steel sections (height h_s, width b_s, and flange and web thicknesses t_f and t_w, respectively); material properties of concrete (f_c'), rebars (f_{yr}), and steel sections (f_{ys}); rebar ratios (ρ_{rX} and ρ_{rY}); and biaxial loads (P_u, M_{uX}, M_{uY}). The 11 output parameters include SF; strains of rebar (ε_r) and steel (ε_s); aspect ratio of a column (b / h); steel ratio ρ_s(#21; steel ratio) shown in Table 9.2.1.; maximum rebar diameter ($D_{r,\max}$); vertical (Y_s) and horizontal (X_s) clearances; and cost index (CI_c), CO_2 emissions, and weight (W_c) of a column. Figures 9.2.1 and 9.2.2 illustrate a cross section and strain compatibility of an SRC column. Structural design software AutoSRCHCol developed based on a strain compatibility algorithm, shown in Figure 9.2.2, proposed by Nguyen and Hong [9.27] is used to generate datasets for the forward training of an SRC column. Nguyen and Hong [9.27] used a constitutive model for concrete proposed by Mander, whereas an elasto-plastic model is used for simulating rebar and steel materials when developing a structural design software (AutoSRCHCol) to investigate performance of SRC columns. Accuracies of an AutoSRCHCol software are verified using Abaqus and experimental data by Nguyen and Hong [9.27]. Accuracies of ANN-based Lagrange optimization are, then, verified by AutoSRCHCol for use in practical designs. A flow chart for a generation of large structural datasets of an SRC column for training ANNs is shown in Figure 9.2.3. Column height (h) and width (b) are randomly generated in the 500–2,000 mm range, whereas compressive concrete strength (f_c') is in the 30–50 MPa range. Strengths of rebar (f_{yr}) and (f_{ys}) are also generated randomly within the 400–600 MPa and 275–325 MPa ranges, respectively. A large dataset consisting of 100,000 datasets is generated to train ANNs for designing an SRC column. Table 9.2.3 summarizes the statistics of the input parameters, including maxima, means, minima, standard deviations, and variances. The histograms of randomly generated input parameters are presented in Figure 9.2.4(a)–(n), describing their distribution. The biaxial load pair (P_u, M_{uX}, M_{uY}) is first investigated.

9.2.4 Network training based on PTM

As shown in Figure 9.2.5, the 14 input parameters (h, b, h_s, b_s, t_w, t_f, f_c', f_{yr}, f_{ys}, ρ_{rX}, ρ_{rY}, P_u, M_{uX}, M_{uY}) are mapped to each of the 11 output parameters $\left(SF, \varepsilon_r, \varepsilon_s, \rho_s, D_{r,\max}, Y_s, X_s, {}^{b}\!/_{h}, CI_c, CO_2, W_c\right)$ using PTM [9.5]. In Table 9.2.4, the best training parameters including a number of hidden

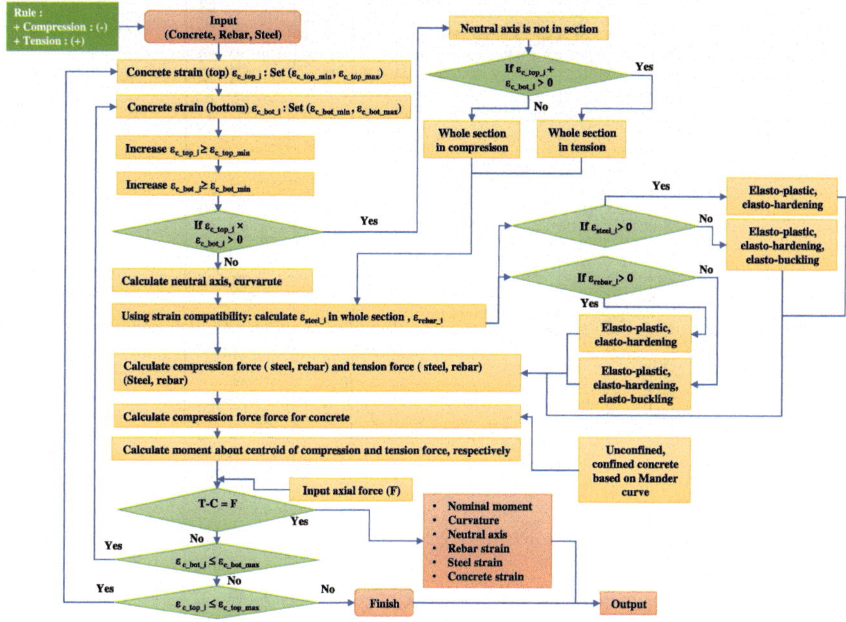

Figure 9.2.2 Computational algorithm for an SRC column [9.27].

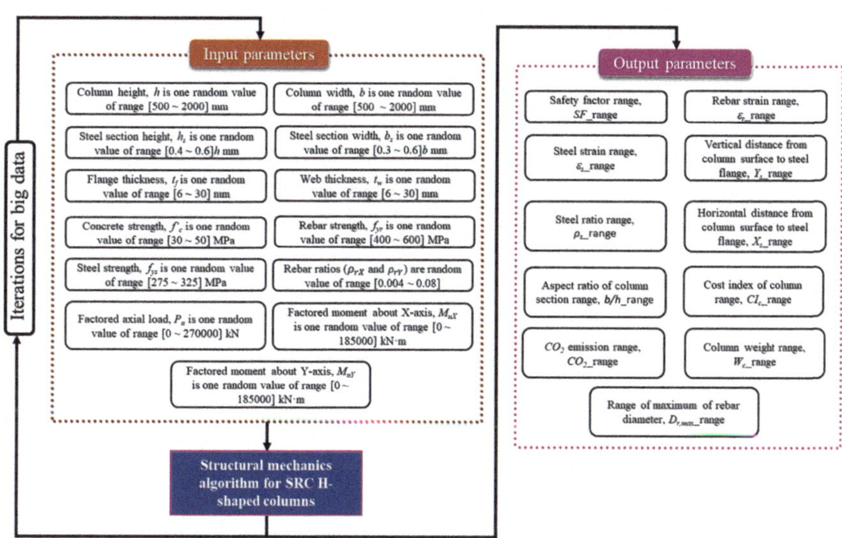

Figure 9.2.3 Flow chart for a generation of large structural datasets of an SRC column for training neural networks.

Table 9.2.3 Statistics including means, standard deviations, and variances of input parameters of SRC columns to generate 100,000 datasets

Number of datasets	*100,000*						
Parameters	h *(mm)*	b *(mm)*	h_s *(mm)*	b_s *(mm)*	t_w *(mm)*	t_f *(mm)*	f'_c *(MPa)*
Maximum	2,000	2,000	1,200	1,190	30	30	50
Mean	1,412.1	1,375.7	694.8	607.5	18.8	19.0	40.0
Minimum	500	500	200	150	6	6	30
Variance (V)	133,237	141,277	43,394	48,630	50	47	37
Standard deviation	365.02	375.87	208.31	220.52	7.06	6.85	6.04
Parameters	f_{yr} *(MPa)*	f_{ys} *(MPa)*	ρ_{rX}	ρ_{rY}	P_u *(kN)*	M_{uX} *(kN·m)*	M_{uY} *(kN·m)*
Maximum	600	325	0.079	0.079	269,332	181,046	172,629
Mean	499.9	300.0	0.022	0.023	22,935.8	10,403.3	10,185.1
Minimum	400	275	0.001	0.001	0.021	0.026	0.165
Variance (V)	3367	216	2.67E-04	2.77E-04	759,571,154	143,062,412	132,648,133
Standard deviation	58.03	14.69	0.02	0.02	27,560.32	11,960.87	11,517.30

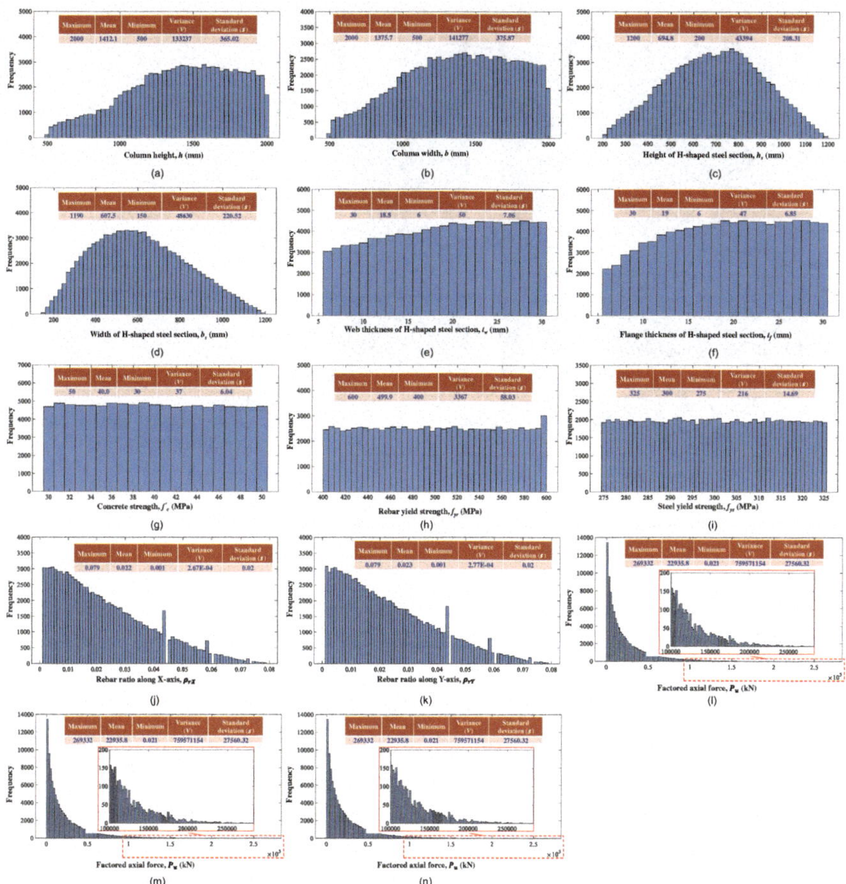

Figure 9.2.4 Distribution of input and output parameters with 100,000 datasets. (a) Column height, h (mm). (b) Column width, b (mm). (c) Height of H-shaped steel section, h_s (mm). (d) Width of H-shaped steel section, b_s (mm). (e) Web thickness of H-shaped steel section, t_w (mm). (f) Flange thickness of H-shaped steel section, t_f (mm). (g) Concrete strength, f_c' (MPa). (h) Rebar yield strength, f_{yr} (MPa). (i) Steel yield strength, f_{ys} (MPa). (j) Rebar ratio along X-axis, ρ_{rX}. (k) Rebar ratio along Y-axis, ρ_{rY}. (l) Factored axial force, P_u (kN). (m) Factored moment with respect to X-axis, M_{uX} (kN·m). (n) Factored moment with respect to Y-axis, M_{uY} (kN·m).

layers and a number of neurons for hidden layers are found based on nine training networks (8, 9, and 10 hidden layers with 40, 50, and 60 neurons). Best forward training results for each output parameter selected by implementing 8, 9, and 10 hidden layers with 40, 50, and 60 neurons are summarized in Table 9.2.4. Accuracy of testing datasets is evaluated based on mean square error (MSE) as shown in Table 9.2.4.

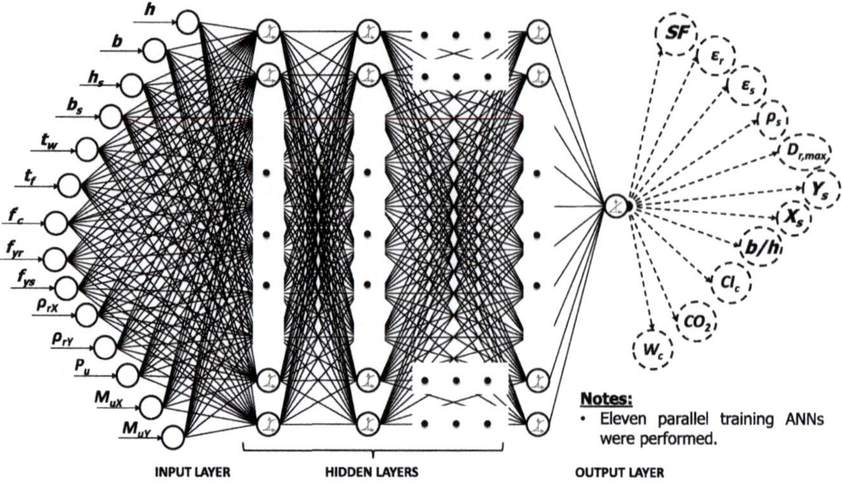

Figure 9.2.5 ANN forward network considering one load pair.

9.3 DESIGN OPTIMIZATION USING THE LAGRANGE MULTIPLIER METHOD

9.3.1 Formulation of the Lagrange function

Equations (9.3.1-1) and (9.3.1-2) show functions which are generalized based on ANNs. Red part of Eq. (9.3.1-1) represents an ANN connecting the input layer to the first hidden layer, whereas blue part of Eq. (9.3.1-2) represents an ANN connecting the first hidden layer to the second hidden layer.

$$y=\sigma^N\left(\mathbf{W}^N\sigma^{N-1}\left(\mathbf{W}^{N-1}\ldots\sigma^2\left(\mathbf{W}^2\sigma^1\left(\mathbf{W}^1\mathbf{x}+\mathbf{b}^1\right)+\boldsymbol{b}^2\right)\ldots+\mathbf{b}^{N-1}\right)+b^N\right) \tag{9.3.1-1}$$

$$y=\sigma^N\left(\mathbf{W}^N\sigma^{N-1}\left(\mathbf{W}^{N-1}\ldots\sigma^2(\mathbf{W}^2\sigma^1\left(\mathbf{W}^1\mathbf{x}+\mathbf{b}^1\right)+\boldsymbol{b}^2)\ldots+\mathbf{b}^{N-1}\right)+b^N\right) \tag{9.3.1-2}$$

Figure 9.2.5 presents a feed-forward network where each node in hidden layers or an output layer is fully connected with all nodes of the previous layer; layers by weighted combination of the previous nodes and bias of the previous layer. Weights are adjusted during a training of ANNs via an iterative process referred to as a backpropagation. The collected information of connecting weights is captured and stored when iterative

Table 9.2.4 Optimal training results implementing 8, 9, and 10 hidden layers with 40, 50, and 60 neurons

Forward PTM training 14 Forward Inputs ($h, b, h_s, b_s, t_w, t_f, f'_c, f_y, f_{ys}, \rho_{rX}, \rho_{rY}, P_u, M_{uX}, M_{uY}$) 11 Forward Outputs ($SF, \varepsilon_r, \varepsilon_s, \rho_s, D_{r,max}, Y_s, X_s, b/h, CI_c, CO_2, W_c$)							
No.	*Data*	*Hidden layers*	*Neurons*	*Suggested epoch*	*Best epoch for training*	*Test MSE*	*R at best epoch*
(1) Training PTM on SF							
1	100,000	10	50	50,000	48,942	3.14E-04	0.999
(2) Training PTM on ε_r							
2	100,000	8	50	50,000	38,404	4.21E-05	0.999
(3) Training PTM on ε_s							
3	100,000	10	60	50,000	38,546	3.99E-05	0.999
(4) Training PTM on ρ_s							
4	100,000	8	40	50,000	49,866	3.65E-08	1.000
(5) Training PTM on $D_{r,max}$							
5	100,000	8	60	50,000	18,507	8.05E-04	0.998
(6) Training PTM on Y_s							
6	100,000	10	50	50,000	48,616	1.46E-08	1.000
(7) Training PTM on X_s							
7	100,000	8	50	50,000	43,365	3.81E-09	1.000
(8) Training PTM on b/h							
8	100,000	8	40	50,000	16,317	1.73E-07	1.000
(9) Training PTM on CI_c							
9	100,000	10	40	50,000	18,080	3.74E-07	1.000
(10) Training PTM on CO_2							
10	100,000	10	50	50,000	46,398	6.54E-08	1.00
(11) Training PTM on W_c							
11	100,000	10	40	50,000	33,965	1.47E-07	1.00

processes have converged (Roman [9.37]). Input parameters are mapped into output parameters in numerous ways; for example, Figure 9.2.5 presents a network for an SRC column, in which 14 input parameters $\left(h, b, h_s, b_s, t_w, t_f, f'_c, f_{yr}, f_{ys}, \rho_{rX}, \rho_{rY}, P_u, M_{uX}, M_{uY}\right)$ are mapped into 11 output parameters $\left(SF, \varepsilon_r, \varepsilon_s, \rho_s, D_{r,\max}, Y_s, X_s, b/h, CI_c, CO_2, W_c\right)$ based on weighted linear combination with an activation function.

Generalized objective functions (CI_c, CO_2 emissions, and W_c), and equality and inequality constraints are derived using ANNs as shown in Eq. (9.3.1–3) [9.7].

$$f(\mathbf{x}) = g^D\left(f_{lin}^N\left(\mathbf{W}^N f_t^{N-1}\left(\mathbf{W}^{N-1} \ldots f_t^1\left(\mathbf{W}^1 g^N(\mathbf{x}) + \mathbf{b}^1\right) \ldots + \mathbf{b}^{N-1}\right) + b^N\right)\right) \tag{9.3.1-3}$$

where $\mathbf{x}$ is the input parameters; N is a number of layers, including hidden layers and output layer; $\mathbf{W}^n$ is a weight matrix between layer $n-1$ and layer n; $\mathbf{b}^n$ is a bias matrix added to each neuron of layer n; and g^N and g^D are normalization and de-normalization functions, respectively. Activation functions f_t^n at layer n are implemented in formulating non-linear relationships of networks; a linear activation function f_{lin}^n is selected for an output layer when output values are unbounded. An ANN-based objective function as shown in Eq. (9.3.1-3) is used for Lagrange function.

As shown in Eq. (9.3.2), Hong and Nguyen [9.7] presented the Lagrange function $\mathcal{L}$ as a function of input variables $\mathbf{x} = [x_1, x_2, \ldots, x_n]^T$, equality $[\mathbf{c}(\mathbf{x})]$ and inequality $[\mathbf{v}(\mathbf{x})]$ constraints, Lagrange multipliers of equality $\boldsymbol{\lambda}_c = [\lambda_1, \lambda_2, \ldots, \lambda_m]^T$, and inequality constraints $\boldsymbol{\lambda}_v = [\lambda_1, \lambda_2, \ldots, \lambda_l]^T$, respectively. Eqs. (9.3.3) and (9.3.4) are equality $[c(x)]$ and inequality $[v(x)]$ constraints, respectively. As shown in Eqs. (9.3.5)–(9.3.7) which are based on the Jacobian and Hessian matrix, Hong and Nguyen [9.7] use Newton–Raphson method to solve non-linear equations and find stationary points of Lagrange functions. The Newton–Raphson approximation is repeated until convergence is achieved [9.7].

$$\mathcal{L}\left(\mathbf{x}, \boldsymbol{\lambda}_c^T, \boldsymbol{\lambda}_v^T\right) = f(\mathbf{x}) - \boldsymbol{\lambda}_c^T \mathbf{c}(\mathbf{x}) - \boldsymbol{\lambda}_v^T \mathbf{S}\mathbf{v}(\mathbf{x}) = CI_b - \boldsymbol{\lambda}_c^T \mathbf{c}(\mathbf{x}) - \boldsymbol{\lambda}_v^T \mathbf{S}\mathbf{v}(\mathbf{x}) \tag{9.3.2}$$

$$\boldsymbol{c}(\boldsymbol{x}) = \left[c_1(\boldsymbol{x}),\, c_2(\boldsymbol{x}),\, \ldots,\, c_6(\boldsymbol{x})\right]^T \tag{9.3.3}$$

$$\boldsymbol{v}(\boldsymbol{x}) = \left[v_1(\boldsymbol{x}), v_2(\boldsymbol{x}), \ldots, v_{11}(\boldsymbol{x})\right]^T \tag{9.3.4}$$

$$\begin{bmatrix} \mathbf{x}^{(k+1)} \\ \boldsymbol{\lambda}_c^{(k+1)} \\ \boldsymbol{\lambda}_v^{(k+1)} \end{bmatrix} = \begin{bmatrix} \mathbf{x}^{(k)} \\ \boldsymbol{\lambda}_c^{(k)} \\ \boldsymbol{\lambda}_v^{(k)} \end{bmatrix} - \left[\boldsymbol{H}_L\left(\mathbf{x}^{(k)}, \boldsymbol{\lambda}_c^{(k)}, \boldsymbol{\lambda}_v^{(k)}\right)\right]^{-1} \nabla L\left(\mathbf{x}^{(k)}, \boldsymbol{\lambda}_c^{(k)}, \boldsymbol{\lambda}_v^{(k)}\right) \tag{9.3.5}$$

where $\left[\boldsymbol{H}_L\left(\mathbf{x}^{(k)}, \boldsymbol{\lambda}_c^{(k)}, \boldsymbol{\lambda}_v^{(k)}\right)\right]$ is the Hessian matrix of the Lagrange function, whereas $\nabla\mathcal{L}\left(\mathbf{x}^{(k)}, \boldsymbol{\lambda}_c^{(k)}, \boldsymbol{\lambda}_v^{(k)}\right)$ is the first derivation of the Lagrange function based on KTT conditions [9.38, 9.39]. The derivations of ANN based Hessian matrix were developed by the MathWorks Technical Support Department [9.7, 9.40].

$$\nabla \mathcal{L}(\mathbf{x}, \boldsymbol{\lambda}_c, \boldsymbol{\lambda}_v) = \begin{bmatrix} \nabla f(\mathbf{x}) - \boldsymbol{J}_c(\mathbf{x})^T \boldsymbol{\lambda}_c - \boldsymbol{J}_v(\mathbf{x})^T S\boldsymbol{\lambda}_v \\ -\mathbf{c}(\mathbf{x}) \\ -Sv(\mathbf{x}) \end{bmatrix} \tag{9.3.6}$$

$$\boldsymbol{J}_c(\mathbf{x}) = \begin{bmatrix} \nabla c_1(\mathbf{x}) \\ \nabla c_2(\mathbf{x}) \\ \vdots \\ \nabla c_m(\mathbf{x}) \end{bmatrix} \text{ and } \boldsymbol{J}_v(\mathbf{x}) = \begin{bmatrix} \nabla v_1(\mathbf{x}) \\ \nabla v_2(\mathbf{x}) \\ \vdots \\ \nabla v_l(\mathbf{x}) \end{bmatrix} \tag{9.3.7}$$

In Table 9.3.1, (40×14) weight and (40×1) bias matrices for cost (CI_c) and column weight (W_c) are obtained by training ANNs, whereas (50×14) weight and (50×1) bias matrices for CO_2 emissions are obtained by training ANNs shown in Table 9.2.4, where 40 neurons are used to predict cost (CI_c) and column weight (W_c) with the best training accuracies, whereas 50 neurons are used to predict CO_2 emissions with the best training accuracies. Objective functions for cost (CI_c), CO_2 emissions, and weight (W_c) of an SRC column are generalized based on weight and bias matrices obtained by training ANNs (Hong [9.41]). Lagrange optimization derived in terms of ANN-based objective functions such as cost (CI_c),CO_2 emissions, and weight (W_c) of an SRC column is minimized, meeting design requirements based on equality and inequality constraints shown in Eqs. (9.3.3) and (9.3.4). The ANN-based functions replace explicit functions removing difficulties of both deriving and differentiating analytical functions.

9.3.2 Derivation of AI-based objective functions for CI_c, CO_2 emissions, and weight (W_c) and their Lagrange functions to optimize

Lagrange function shown in Eq. (9.3.2) is obtained in terms of objective functions such as cost index (CI_c), CO_2 emissions, and weight (W_c) of an SRC column which are generalized by ANNs as shown in Table 9.3.2(a). Fourteen input parameters $(h, b, h_s, b_s, t_w, t_f, f_c', f_{yr}, f_{ys}, \rho_{rX}, \rho_{rY}, P_u, M_{uX}, M_{uY})$ are used to derive objective functions of CI_c, CO_2 emissions, and W_c of SRC columns when being subjected to biaxial loads (P_u, M_{uX}, M_{uY}). In Table 9.2.4, best training accuracies of 11 output predictions including objective functions, cost index (CI_c), CO_2 emissions, and weight (W_c) of SRC columns are obtained based on PTM using 8, 9, and 10 hidden layers with 40, 50, and 60 neurons. As shown in Table 9.3.2(b), equality and inequality constraints

Table 9.3.1 Weight and bias matrices of a forward network for objective functions (CI_c, CO_2, and W_c) considering one biaxial load

(a) For cost index of columns (CI_c)

$\mathbf{W}_{[40\times14]}$									$\mathbf{b}_{[40\times1]}$
−0.314	−0.308	−0.066	0.129	⋯	0.102	0.362	0.457	−0.156	1.841
0.009	−0.653	0.021	0.187	⋯	0.041	−0.049	0.577	−0.073	−1.754
−0.148	0.172	0.143	0.605	⋯	0.030	−0.453	−0.659	−0.890	1.793
0.562	0.010	0.193	−0.299	⋯	−0.031	0.561	−0.196	0.535	−1.753
⋮	⋮	⋮	⋮	⋱	⋮	⋮	⋮	⋮	⋮
−0.079	0616	−0.110	−0.150	⋯	0.004	0.167	0.414	0.544	1.427
−0.306	−0.462	−0.385	−0.157	⋯	−0.124	0.359	−0.309	0.859	−1.888
0.316	0.579	0.056	0.286	⋯	−0.019	0.082	0.503	−0.380	1.707
0.266	−0.020	0.099	0.056	⋯	−0.027	0.268	0.502	0.555 $_{[40\times14]}$	1.864

(Continued)

Table 9.3.1 (Continued) Weight and bias matrices of a forward network for objective functions (CI_c, CO_2, and W_c) considering one biaxial load

(b) For CO_2 emissions

$\mathbf{W}_{[50\times14]}$	$\mathbf{b}_{[50\times1]}$
$\begin{bmatrix} 0.153 & -0.592 & 0.698 & 0.216 & \cdots & 0.428 & 0.551 & 0.544 & 0.756 \\ -0528 & 0.832 & 0.143 & 0.221 & \cdots & -0.006 & 0.192 & -0.545 & -0.44 \\ -0.744 & 0.554 & 0.082 & 0.143 & \cdots & 0.049 & 0.168 & -0.377 & -0.03 \\ -0.295 & -0.154 & -0.693 & 0.405 & \cdots & -0.136 & 0.010 & -0.515 & -0.29 \\ \vdots & \vdots & \vdots & \vdots & \ddots & \vdots & \vdots & \vdots & \vdots \\ 0.401 & -0.544 & -0.390 & -0.517 & \cdots & -0.281 & -0.524 & -0.821 & 0.17 \\ 0.343 & -0.408 & -0.139 & 0.096 & \cdots & 0.437 & -0.917 & -0.340 & -0.40 \\ 0.419 & -0.637 & -0.138 & -0.136 & \cdots & -0.004 & 0.225 & -0.410 & 0.353 \\ 0.547 & 0.043 & 0.092 & 0.043 & \cdots & 0.002 & 0.394 & -0.208 & 0.063 \end{bmatrix}_{[50\times14]}$	$\begin{bmatrix} -1.937 \\ 1.715 \\ 1.832 \\ 1.934 \\ \vdots \\ 1.881 \\ 1.999 \\ 1.873 \\ 1.757 \end{bmatrix}$

(Continued)

Table 9.3.1 (Continued) Weight and bias matrices of a forward network for objective functions (Cl_c, CO_2, and W_c) considering one biaxial load

(c) For column weight (W_c)

$\mathbf{W}_{[40\times14]}$

$$\begin{bmatrix} -0.848 & 1.079 & -0.136 & -0.211 & \cdots & 0.059 & 0.419 & 0.156 & -0.16 \\ -0.031 & -0.395 & -0.072 & 0.143 & \cdots & -0.654 & 0.809 & -0.068 & 0.091 \\ 0.015 & 0.973 & 0.097 & 0.262 & \cdots & 0.038 & -0.630 & -0.083 & 0.395 \\ -0.381 & -0.193 & 0.265 & 0.567 & \cdots & -0.183 & -0.267 & -1.084 & -0.02 \\ \vdots & \vdots & \vdots & \vdots & \ddots & \vdots & \vdots & \vdots & \vdots \\ 0.367 & 0.389 & 0.153 & 0.165 & \cdots & 0.014 & 0.234 & 0.323 & 0.316 \\ -0.118 & 0.692 & 0.693 & 0.540 & \cdots & 0.308 & -0.078 & 0.481 & 0.207 \\ 0.861 & 0.769 & -0.173 & -0.195 & \cdots & -0.078 & 0.310 & 0.397 & -0.82 \\ 0.336 & -0.705 & -0.190 & -0.112 & \cdots & 0.039 & 0.000 & -0.061 & -0.66 \end{bmatrix}_{[40\times14]}$$

$\mathbf{b}_{[40\times1]}$

$$\begin{bmatrix} 1.783 \\ -1.949 \\ 1.783 \\ 1.833 \\ \vdots \\ 1.630 \\ -1.952 \\ 1.809 \\ -1.861 \end{bmatrix}$$

Table 9.3.2 Design optimization for CI_c, CO_2 emissions, and column weight (W_c) considering one biaxial load pair

(a) ANN-based CI_c, CO_2 emissions, and column weight (W_c) based on forward formulation

Minimize

$$\underbrace{CI_c}_{[1\times1]} = g_{CI_c}^{D}\left(f_l^{11}\left(\underbrace{\mathbf{W}_{CI_c}^{11}}_{[1\times40]} f_t^{10}\left(\underbrace{\mathbf{W}_{CI_c}^{10}}_{[40\times40]}\cdots f_t^{1}\left(\underbrace{\mathbf{W}_{CI_c}^{1}}_{[40\times14]}\underbrace{g_{CI_c}^{N}(\mathbf{x})}_{[14\times1]}+\underbrace{\mathbf{b}_{CI_c}^{1}}_{[40\times1]}\right)\cdots+\underbrace{\mathbf{b}_{CI_c}^{10}}_{[40\times1]}\right)+\underbrace{b_{CI_c}^{11}}_{[1\times1]}\right)\right)$$

$$\underbrace{CO_2}_{[1\times1]} = g_{CO_2}^{D}\left(f_l^{11}\left(\underbrace{\mathbf{W}_{CO_2}^{11}}_{[1\times50]} f_t^{10}\left(\underbrace{\mathbf{W}_{CO_2}^{10}}_{[50\times50]}\cdots f_t^{1}\left(\underbrace{\mathbf{W}_{CO_2}^{1}}_{[50\times14]}\underbrace{g_{CO_2}^{N}(\mathbf{x})}_{[14\times1]}+\underbrace{\mathbf{b}_{CO_2}^{1}}_{[50\times1]}\right)\cdots+\underbrace{\mathbf{b}_{CO_2}^{10}}_{[50\times1]}\right)+\underbrace{b_{CO_2}^{11}}_{[1\times1]}\right)\right)$$

$$\underbrace{W_c}_{[1\times1]} = g_{W_c}^{D}\left(f_l^{11}\left(\underbrace{\mathbf{W}_{W_c}^{11}}_{[1\times40]} f_t^{10}\left(\underbrace{\mathbf{W}_{W_c}^{10}}_{[40\times40]}\cdots f_t^{1}\left(\underbrace{\mathbf{W}_{W_c}^{1}}_{[40\times14]}\underbrace{g_{W_c}^{N}(\mathbf{x})}_{[14\times1]}+\underbrace{\mathbf{b}_{W_c}^{1}}_{[40\times1]}\right)\cdots+\underbrace{\mathbf{b}_{W_c}^{10}}_{[40\times1]}\right)+\underbrace{b_{W_c}^{11}}_{[1\times1]}\right)\right)$$

(b) Equality and inequality conditions for optimizations

EQUALITY CONSTRAINTS	*INEQUALITY CONSTRAINTS*
$c_1(\mathbf{x}): b/h = 1$	$v_1(\mathbf{x}): SF - 1 \geq 0$
$c_2(\mathbf{x}): f_c' = 30$ MPa	$v_2(\mathbf{x}): \rho_s - 0.01 \geq 0$
$c_3(\mathbf{x}): f_{yr} = 500$ MPa	$v_3(\mathbf{x}): \rho_{rX} + \rho_{rY} - 0.004 \geq 0$
$c_4(\mathbf{x}): f_{ys} = 325$ MPa	$v_4(\mathbf{x}): -\rho_{rX} - \rho_{rY} + 0.08 \geq 0$
$c_5(\mathbf{x}): P_u = 12,000$ kN	$v_5(\mathbf{x}): \rho_{rX} \geq 0$
$c_6(\mathbf{x}): M_{uX} = 6,500$ kN·m	$v_6(\mathbf{x}): \rho_{rY} \geq 0$
$c_7(\mathbf{x}): M_{uY} = 5,000$ kN·m	$v_7(\mathbf{x}): Y_s - 100 \geq 0$
$c_8(\mathbf{x}): t_w = 8$ mm	$v_8(\mathbf{x}): X_s - 100 \geq 0$
$c_9(\mathbf{x}): t_f = 8$ mm	$v_9(\mathbf{x}): h - 500 \geq 0$
	$v_{10}(\mathbf{x}): -h + 2,000 \geq 0$
	$v_{11}(\mathbf{x}): b - 500 \geq 0$
	$v_{12}(\mathbf{x}): -b + 2,000 \geq 0$
	$v_{13}(\mathbf{x}): h_s - 200 \geq 0$
	$v_{14}(\mathbf{x}): -h_s + 1,000 \geq 0$
	$v_{15}(\mathbf{x}): b_s - 200 \geq 0$
	$v_{16}(\mathbf{x}): -b_s + 1,000 \geq 0$
	$v_{17}(\mathbf{x}): -D_{r.\max} + 64 \geq 0$

are imposed to govern the Lagrange optimization. Material properties of concrete, rebars, and steel sections are set as equality constraints ($c_2(\mathbf{x}) = f_c' = 30$ MPa, $c_3(\mathbf{x}) = f_{yr} = 500$ MPa, and $c_4(\mathbf{x}) = f_{ys} = 325$ MPa, respectively), whereas the flange ($c_8(\mathbf{x}) = t_f$) and web ($c_9(\mathbf{x}) = t_w$) thicknesses of the steel sections are taken as 8 mm as shown in Table 9.3.2(b). *SF* denoted as v_1 in Table 9.3.2(b) is considered as an inequality constraint greater than 1.0. Steel ratio (ρ_s) presented as v_2 in Table 9.3.2(b) is set at 0.01 or above in accordance with Section I2.1a of ANSI/AISC 360-16 [9.35]. Rebar ratios for an SRC column are constrained by v_3 and v_4 as shown in Table 9.3.2(b) because they are limited to the range 0.004–0.08 in accordance with ANSI/AISC 360-16 [9.35] and ACI 318-19 [9.36], respectively. The maximum, minimum allowable strains in rebar (ε_r) and steel (ε_s) could be of any value, not constrained by both equalities and inequalities because they are not specified in ACI 318-19 [9.36] and ANSI/AISC 360-16 [9.35]. Dimensions of an SRC column including a location of steel sections, Y_s and X_s, are constrained by inequality conditions v_7 and v_8, respectively, whereas rebar diameters are constrained by v_{17}.

9.3.3 Formulation of active and inactive conditions

Nine equality constraints shown in Table 9.3.2(b) are imposed in the Lagrange function to make optimization simple. In Table 9.3.2(b), 17 inequality constraints (v_i) are imposed by design codes and architects when optimizing an SRC column. Active inequality constraints are designated by an inequality constraint (S) matrix shown in Eq. (9.3.8) where inequality constraints are activated when $s_i = 1$, whereas inequality constraints are inactivated when $s_i = 0$.

$$\mathbf{S} = \begin{bmatrix} s_1 & 0 & \cdots & 0 \\ 0 & s_2 & \cdots & 0 \\ \vdots & \vdots & \ddots & \vdots \\ 0 & 0 & \cdots & s_{17} \end{bmatrix}_{[17\times17]} \tag{9.3.8}$$

For example shown in Table 9.3.2(b), inequality constraint v_2 is imposed because Section I2.1a of ANSI/AISC 360-16 [9.35] requires a minimum steel ratio of 0.01. Inequality v_2 is active when its corresponding inequality factor (s_2) is equal to 1.0, indicating that steel ratio is bound to its allowable minimum value (0.01), and leading to Eq. (9.3.9) where active inequality constraint v_2 is treated as equality constraint during optimization.

$$\boldsymbol{S} = \begin{array}{c} \\ \end{array}\begin{array}{cccccc} 1^{st} & 2^{nd} & 3^{rd} & & 16^{th} & 17^{th} \\ \left[\begin{array}{c} s_1=0 \\ 0 \\ 0 \\ \vdots \\ 0 \\ 0 \end{array}\right. & \begin{array}{c} 0 \\ s_2=1 \\ 0 \\ \vdots \\ 0 \\ 0 \end{array} & \begin{array}{c} 0 \\ 0 \\ 0 \\ \vdots \\ 0 \\ 0 \end{array} & \begin{array}{c} \cdots \\ \cdots \\ \cdots \\ \ddots \\ \cdots \\ \cdots \end{array} & \begin{array}{c} 0 \\ 0 \\ 0 \\ \vdots \\ s_{16}=0 \\ 0 \end{array} & \left.\begin{array}{c} 0 \\ 0 \\ 0 \\ \vdots \\ 0 \\ s_{17}=0 \end{array}\right]_{17\times 17} \end{array} \begin{array}{c} 1^{st} \\ 2^{nd} \\ 3^{rd} \\ \mathrm{M} \\ 16^{th} \\ 17^{th} \end{array} \quad (9.3.9)$$

In Eq. (9.3.10), Lagrange function is derived with an active inequality constraint v_2 which can be obtained based on Eqs. (9.3.2) and (9.3.9).

$$\mathcal{L}(\mathbf{x}, \boldsymbol{\lambda}_c, \boldsymbol{\lambda}_v)$$

$$= f(\mathbf{x}) - \boldsymbol{\lambda}_c^T \mathbf{c}(\mathbf{x}) - \boldsymbol{\lambda}_v^T \begin{bmatrix} s_1=0 & 0 & \cdots & 0 \\ 0 & s_2=1 & \cdots & 0 \\ \vdots & \vdots & \ddots & \vdots \\ 0 & 0 & \cdots & s_{17}=0 \end{bmatrix}_{[17\times 17]} \mathbf{v}(\mathbf{x}) \quad (9.3.10)$$

where $f(\mathbf{x})$ are the ANN-based generalized functions of CI_c, CO_2 emissions, and W_c.

9.4 OPTIMIZATION VERIFICATIONS

9.4.1 Optimized cost (CI_c) of an SRC column

9.4.1.1 Optimized cost (CI_c) verified by structural design software (AutoSRCHCol)

In Table 9.4.1, optimized cost based on a forward design implementing ANN-based Lagrange optimization is compared with those obtained using structural software (AutoSRCHCol) using design parameters obtained by ANNs [9.27] when a biaxial bending (M_{uX} = 6,500 kN·m, and M_{uY} = 5,000 kN·m) with an axial load (P_u = 12,000 kN) are acting. The cost CI_c of the SRC column is optimized when biaxial loads are set equivalent to a design strength of the SRC columns indicated by $SF = 1$. Cost (CI_c) of the SRC column is minimized as 569,126 KRW/m based on a forward design implementing ANN-based Lagrange optimization, whereas cost (CI_c) of 562,496 KRW/m is calculated by structural design using design parameters obtained by ANNs, showing 1.16% error. In Table 9.4.1, 14 input parameters minimizing cost (CI_c) of an SRC column are used in AutoSRCHCol to calculate

Table 9.4.1 Minimized cost of columns (CI_c) based on training results in Table 9.2.4

MINIMIZED COST OF COLUMN (CI_c)
Training results: 10 hidden layers-40 neurons; T.MSE=3.74E-7

No.	*Inputs for generating big data and training*		No.	*Outputs for generating big data and training*			
	Notation	*Value*		*Notation*	*ANN*	*Structural calculations*	*Errors** *(%)*
				(1)	*(2)*	*(3)*	*(4)*
1	$b^{1,3}$ (mm)	1,457.10	1	SF^3	1.00	1.00	–0.43%
2	$h^{1,3}$ (mm)	1,456.44	2	ε_r	0.0029	0.0030	–3.41%
3	$\rho_{rX}^{1,3}$	0.003	3	ε_s	0.0021	0.0021	–0.57%
4	$\rho_{rY}^{1,3}$	0.001	4	ρ_s^3	0.010	0.010	0.44%
5	$h_s^{1,3}$ (mm)	1,000.00	5	$D_{r,max}^3$	18.5	19.0	–2.75%
6	$b_s^{1,3}$ (mm)	828.52	6	Y_s^3 (mm)	228.21	228.22	0.00%
7	t_f^2 (mm)	8	7	X_s^3 (mm)	314.28	314.29	0.00%
8	t_w^2 (mm)	8	8	b/h^2	1.00	1.00	–0.05%
9	$f_c'^2$ (MPa)	30	9	CI_c (KRW/m)	569,126	562,496	1.16%
10	f_{yr}^2 (MPa)	500	10	CO_2 (t-CO_2/m)	0.946	0.940	0.61%
11	f_{ys}^2 (MPa)	325	11	W_c (kN/m)	51.50	51.50	0.00%
12	P_u^2 (kN)	12,000					
13	M_{uX}^2 (kN·m)	6,500					
14	M_{uY}^2 (kN·m)	5,000					

[1] Calculated on an output-side when optimizing CI_c.
[2] Equality constraints.
[3] Inequality constraints.
* $Error\ (4) = \frac{(2)-(3)}{(2)} \times 100\%$

11 output parameters, verifying the biggest error of –3.41% for the rebar strain (ε_r). The errors in SF (–0.43%) and optimized CI_c (1.16%) are insignificant, and hence, the accuracy of ANNs which minimize Lagrange function is acceptable, indicating that ANNs can be used in practical designs.

9.4.1.2 Column interaction diagram corresponding to an optimized CI_c

In Figure 9.4.1, a three-dimensional P-M interaction diagram which are obtained using parameters obtained based on ANNs shown in Table

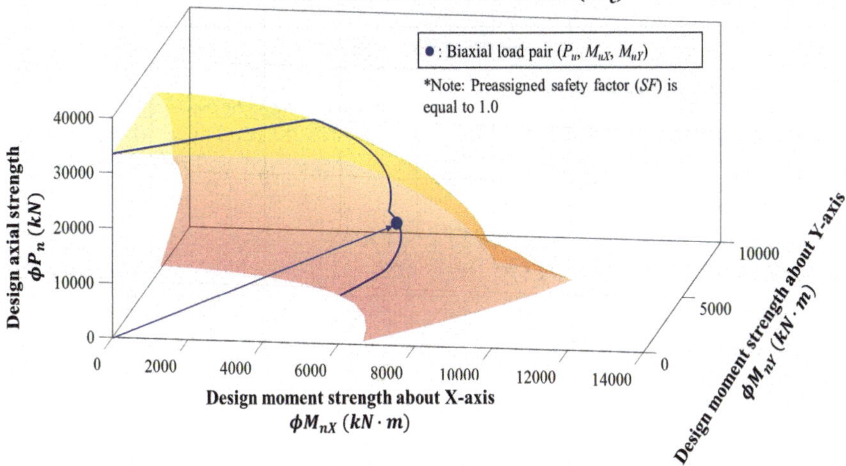

Figure 9.4.1 3D P-M diagram corresponding to optimized CI_c considering one biaxial load.

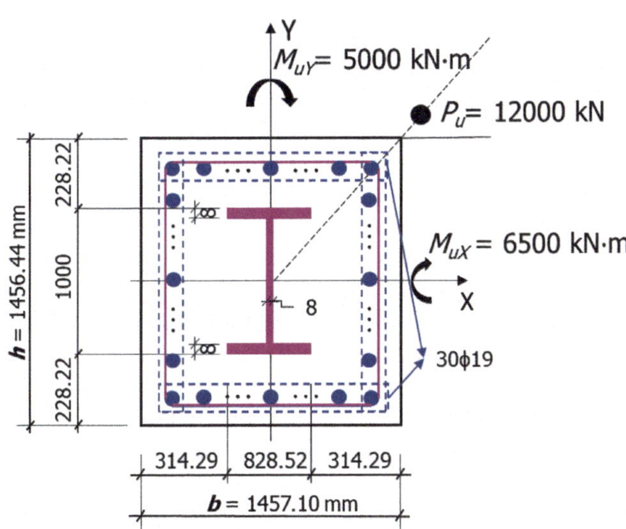

Figure 9.4.2 Section of an SRC column corresponding to minimized CI_c.

9.4.1 corresponding to the minimized cost (CI_c) is plotted for the column illustrated in Figure 9.4.2 where dimensions of column sections, steel sections, and the rebar ratios corresponding to a minimized cost (CI_c)

Table 9.4.2 Verification of optimized CI_c based on ANN by large datasets based on preassigned factored load P_u = 12,000 kN, M_{uX}=6,500 kNm, M_{uY}=5,000 kNm

Cost CI_c of SRC columns			
Min. cost based on big data		*Proposed method based on ANN (KRW/m)*	*Difference (%)*
A number of big data	*Min. CI_c (KRW/m)*		
650,000 data	621,669	569,126	−9.2%
1,750,000 data	621,669		−9.2%
3,750,000 data	572,670		−0.6%
5,750,000 data	572,670		−0.6%

are obtained by a forward design implementing ANN-based Lagrange optimization.

9.4.1.3 Verification of optimized CI_c based on ANN by large datasets

The maximum rebar diameter ($D_{r,\max}$) is set at 64 mm, which is equivalent to a bundle of four single rebars with 32 mm-diameter each. *SF* and $D_{r,\max}$ are the output parameters, and hence, they are not generated when generating large datasets. To verify the minimized cost index (CI_c) obtained by the proposed method, a large dataset containing 5,750,000 data points is generated, which meet preassigned equality constraints shown in Table 9.3.2(b). A dataset of 96,902 in which *SF*s are greater than 1.0SF are extracted from a large dataset of 5,750,000, ensuring that extracted SRC columns are strong enough to resist factored loads. The minimum CI_c obtained from 96,902 observations which are obtained from a 5,750,000 dataset is 572,670 KRW/m as shown in Table 9.4.2.

In Figure 9.4.3, the moments of the two different axes (M_{uX} and M_{uY}) are set at 6,500 kN·m and 5,000 kN·m as shown in equality constraints c_6 and c_7 of Table 9.3.2(b), respectively, whereas axial load (P_u) is set to 12,000 kN as shown in equality constraint c_5 of Table 9.3.2(b). The minimized CI_c obtained from the proposed method is compared with those based on four different big datasets, 650,000 datasets shown in Figure 9.4.3(a), 1,750,000 datasets shown in Figure 9.4.3(b), 3,750,000 datasets shown in Figure 9.4.3(c). The relationship between the axial load (P_u) and cost index of a column (CI_c) is illustrated in Figure 9.4.3(d) based on 96,902 extracted observations from 5,750,000 datasets. The minimum cost CI_c of 621,669 KRW/m obtained from 650,000 and 1,750,000 datasets are identical, demonstrating a difference of −9.2% compared with the minimized cost of 569,126 KRW/m based on the proposed method as shown in Table 9.4.2, Figure 9.4.3(a), and (b). The difference between the minimum CI_c obtained using large datasets and the proposed method is

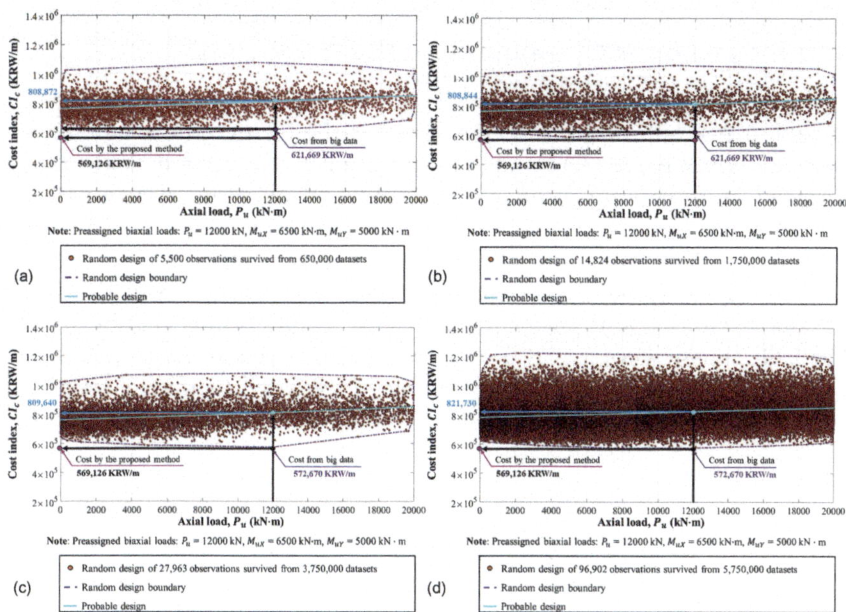

Figure 9.4.3 Verification of optimized CI_c based on large datasets. (a) Random design survived from 650,000 datasets. (b) Random design survived from 1,750,000 datasets. (c) Random design survived from 3,750,000 datasets. (d) Random design survived from 5,750,000 datasets (30.7% reduction).

reduced to −0.6% when a number of the datasets increases to 3,750,000 and 5,750,000 as shown in Table 9.4.2, Figure 9.4.3(c), and (d), respectively. It is noted that the minimum CI_c obtained from 96,902 observations which are extracted from a 5,750,000 dataset is 572,670 KRW/m, whereas the proposed method based on ANN offers a column design with a minimized CI_c of 569,126 KRW/m, which is less than 572,670 KRW/m obtained from a 5,750,000 dataset by −0.6% as presented in Figure 9.4.3(d). The minimized cost of the SRC column based on ANN is also less than that of a probable column design of 821,730 KRW/m by 30.7% determined based on trendline functions ("polyfit" and "polyval" commands) provided in MATLAB [9.40].

As shown in Table 9.4.1, the minimized CI_c obtained based on an ANN is 569,126 KRW/m showing only a 1.16% difference when being compared with structural calculations of 562,496 KRW/m calculated using design parameters obtained by ANNs at a preassigned factored moments (M_{uX} and M_{uY}) of 6,500 kN·m, 5,000 kN·m and axial load (P_u) of 12,000 kN. Minimized cost of columns (CI_c) is calculated based on training results shown in Table 9.2.4.

9.4.2 Optimized CO_2 emissions of an SRC column

9.4.2.1 Optimized CO_2 emissions verified by structural design software (AutoSRCHCol)

In Table 9.4.3, optimized CO_2 emissions based on a forward design implementing ANN-based Lagrange optimization is compared with those obtained using structural software (AutoSRCHCol) using design parameters obtained by ANNs [9.27] when a biaxial bending ($M_{uX} = 6,500$ kN·m, and $M_{uY} = 5,000$ kN·m) with an axial load ($P_u = 12,000$ kN) are acting. The CO_2 emissions of the SRC column are optimized when biaxial loads are set equivalent to a design strength of the SRC columns indicated by $SF = 1$. The CO_2 emissions of an SRC column are minimized as 0.946 (t-CO_2/m) based on a forward design implementing ANN-based Lagrange optimization whereas CO_2 emissions of 0.940 (t-CO_2/m) are calculated by structural

Table 9.4.3 Minimized CO_2 emissions based on training results in Table 9.2.4

MINIMIZED CO_2 EMISSIONS *Training results: 10 hidden layers-50 neurons; T.MSE=6.54E-8*							
No.	*Input parameters*		*No.*	*Output parameters*			
	Notation	*Value*		*Notation*	*ANN*	*Structural calculations*	*Errors* (%)*
				(1)	*(2)*	*(3)*	*(4)*
1	$b^{1,3}$ (mm)	1,457.10	1	SF^3	1.00	1.00	−0.43%
2	$h^{1,3}$ (mm)	1,456.44	2	ε_r	0.0029	0.0030	−3.41%
3	$\rho_{rX}^{1,3}$	0.003	3	ε_s	0.0021	0.0021	−0.57%
4	$\rho_{rY}^{1,3}$	0.001	4	ρ_s^3	0.010	0.010	0.44%
5	$h_s^{1,3}$ (mm)	1,000.00	5	$D_{r,max}^3$	18.5	19.0	−2.75%
6	$b_s^{1,3}$ (mm)	828.52	6	Y_s^3 (mm)	228.21	228.22	0.00%
7	t_f^2 (mm)	8	7	X_s^3 (mm)	314.28	314.29	0.00%
8	t_w^2 (mm)	8	8	b/h^2	1.00	1.00	−0.05%
9	$f_c'^2$ (MPa)	30	9	CI_c (KRW/m)	569,126	562,496	1.16%
10	f_{yr}^2 (MPa)	500	10	CO_2 (t-CO_2/m)	0.946	0.940	0.61%
11	f_{ys}^2 (MPa)	325	11	W_c (kN/m)	51.50	51.50	0.00%
12	P_u^2 (kN)	12,000					
13	M_{uX}^2 (kN·m)	6,500					
14	M_{uY}^2 (kN·m)	5,000					

1 Calculated on an output-side when optimizing CO_2.
2 Equality constraints.
3 Inequality constraints.

* $\text{Error }(4) = \frac{(2)-(3)}{(2)} \times 100\%$

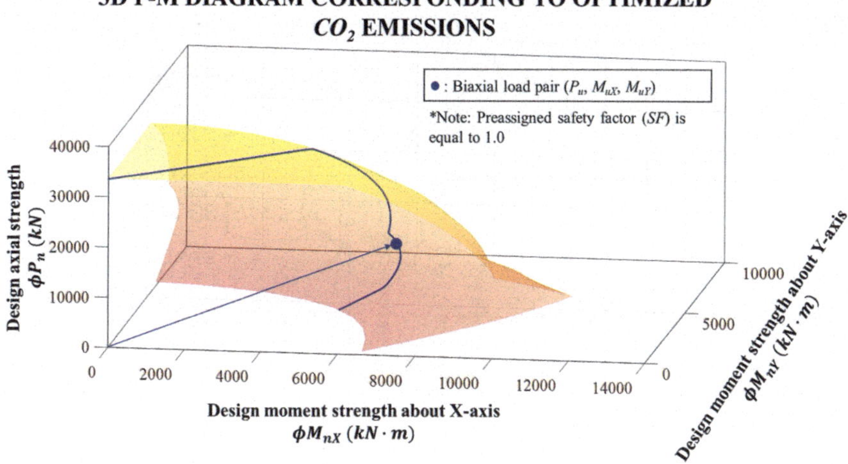

Figure 9.4.4 3D P-M diagram corresponding to optimized CO_2 emissions considering one biaxial load.

design, using design parameters obtained by ANNs, which shows 0.61% error. In Table 9.4.3, 14 input parameters minimizing CO_2 emissions of an SRC column are used in AutoSRCHCol to calculate 11 output parameters, verifying the biggest error of –3.41% for the rebar strain (ε_r). The errors in *SF* (–0.43%) and optimized CO_2 emissions (0.61%) are insignificant, and hence, the accuracy of ANNs which minimize Lagrange function is acceptable, indicating that ANNs can be used in practical designs.

9.4.2.2 Column interaction diagram corresponding to an optimized CO_2 emissions

In Figure 9.4.4, a three-dimensional P-M interaction diagram obtained using parameters obtained in Table 9.4.3 corresponding to the minimized CO_2 emissions is plotted for the SRC column illustrated in Figure 9.4.5 in which dimensions of column sections, steel sections, and the rebar ratios corresponding to a minimized CO_2 emissions are obtained by a forward design implementing ANN-based Lagrange optimization.

9.4.2.3 Verification of an optimized CO_2 emissions based on ANN by large datasets

To verify the minimized CO_2 emissions obtained by the proposed method, a large dataset containing 5,750,000 data points is generated, which meet preassigned equality constraints shown in Table 9.3.2(b). A dataset of 96,902 in which *SF*s are greater than 1.0SF are extracted from a large dataset of

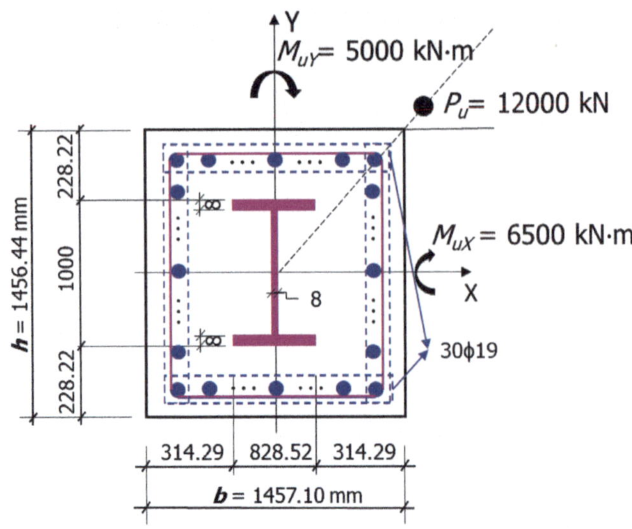

Figure 9.4.5 Section of an SRC column corresponding to minimized CO_2 emissions.

Table 9.4.4 Verification of optimized CO_2 from Lagrange method by large datasets with preassigned factored load P_u= 12,000 kN, M_{uX}=6,500 kNm, M_{uY}=5,000 kNm

CO_2 emission of SRC columns			
Min. CO_2 emission based on big data		*Proposed method based on ANN (t-CO_2/m)*	*Difference (%)*
A number of big data	*Min. CO_2 emission (t-CO_2/m)*		
650,000 data	1.122	0.946	−18.6%
1,750,000 data	1.122		−18.6%
3,750,000 data	0.955		−1.0%
5,750,000 data	0.955		−1.0%

5,750,000, ensuring that the SRC columns are strong enough to resist factored loads. The minimum CO_2 emissions obtained from 96,902 observations which are obtained from a 5,750,000 dataset is 0.955 (t-CO_2/m) as shown in Table 9.4.4.

The minimum CO_2 emissions of 1.122 (t-CO_2/m) obtained from 650,000 and 1,750,000 datasets are identical, demonstrating a difference of −18.6% compared with the minimized CO_2 emissions of 0.946 (t-CO_2/m) based on the proposed method as shown in Table 9.4.4, Figure 9.4.6(a), and (b). The difference between the minimum CO_2 emissions obtained using large

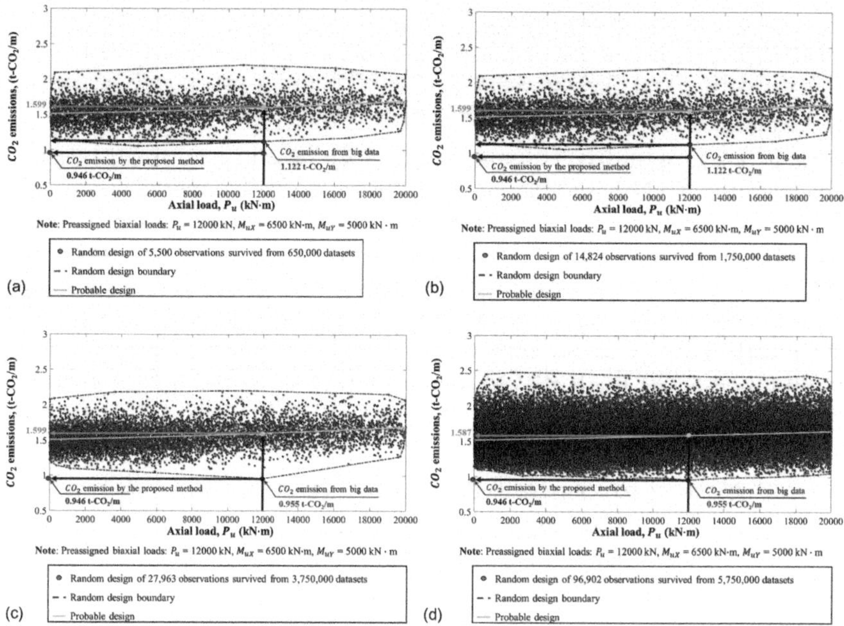

Figure 9.4.6 Verification of optimized CO_2 emissions from Lagrange method by large datasets. (a) Random design survived from 650,000 datasets. (b) Random design survived from 1,750,000 datasets. (c) Random design survived from 3,750,000 datasets. (d) Random design survived from 5,750,000 datasets (40.4% reduction).

datasets and the proposed method is reduced to –1.0% when the number of the datasets increases to 3,750,000 and 5,750,000 as shown in Table 9.4.4, Figure 9.4.6(c), and (d), respectively. It is noted that the minimum CO_2 emissions obtained from 96,902 observations which are extracted from a 5,750,000 dataset is 0.955 (t-CO_2/m), whereas the proposed method based on ANN offers an SRC column design with a minimized CO_2 emissions of 0.946 (t-CO_2/m), which is less than 0.955 (t-CO_2/m) identified from a 5,750,000 dataset by –1.0% as presented in Fig. 9.4.6(d). It is noted that the minimized CO_2 emissions based on ANN is less than that of a probable column design of 1.587 (t-CO_2/m) by 40.4% determined based on trend-line functions ("polyfit" and "polyval" commands) provided in MATLAB [9.40]. The minimized CO_2 emissions obtained based on an ANN is 0.946 (t-CO_2/m) showing only 0.61% difference when being compared with structural calculations of 0.940 (t-CO_2/m) obtained using design parameters obtained by ANNs at a preassigned factored moments (M_{uX} and M_{uY}) of 6,500 kN·m, 5,000 kN·m and axial load (P_u) of 12,000 kN as shown in Table 9.4.3. Minimized CO_2 emissions of 0.946 (t-CO_2/m) are obtained based on training results shown in Table 9.2.4.

9.4.3 Optimized column weight (W_c) of an SRC column

9.4.3.1 Optimized column weight (W_c) verified by structural design software (AutoSRCHCol)

In Table 9.4.5, optimized column weights (W_c) based on a forward design implementing ANN-based Lagrange optimization are compared with those obtained using structural software (AutoSRCHCol) using design parameters obtained by ANNs [9.27] when a biaxial bending ($M_{uX} = 6{,}500$ kN·m, and $M_{uY} = 5{,}000$ kN·m) with an axial load ($P_u = 12{,}000$ kN) are acting. The column weight (W_c) of an SRC column is optimized when biaxial loads are set equivalent to a design strength of an SRC column indicated by $SF = 1$. The column weight (W_c) of an SRC column is minimized as 36.08 (kN/m) based

Table 9.4.5 Minimized column weights (W_c) based on training results in Table 9.2.4

MINIMIZED COLUMN WEIGHT (W_c)
Training results: 10 hidden layers-40 neurons; T.MSE = 1.47E-7

No.	Input parameters		No.	Output parameters			
	Notation	*Value*		*Notation*	*ANN*	*Structural calculations*	*Errors** *(%)*
				(1)	*(2)*	*(3)*	*(4)*
1	$b^{1,3}$ (mm)	1,159.14	1	SF^3	1.00	0.98	1.67%
2	$h^{1,3}$ (mm)	1,160.24	2	ε_r	0.0021	0.0020	5.78%
3	$\rho_{rX}^{1,3}$	0.024	3	ε_s	0.00170	0.00174	−2.98%
4	$\rho_{rY}^{1,3}$	0.023	4	ρ_s^3	0.014	0.014	1.01%
5	$h_s^{1,3}$ (mm)	960.61	5	$D_{r,max}^3$	64.0	63.6	0.59%
6	$b_s^{1,3}$ (mm)	702.93	6	Y_s^3 (mm)	100.00	99.81	0.19%
7	t_f^2 (mm)	8	7	X_s^3 (mm)	228.07	228.10	−0.02%
8	t_w^2 (mm)	8	8	b/h^2	1.00	1.00	0.10%
9	$f_c'^2$ (MPa)	30	9	CI_c (KRW/m)	907,208.31	911,788.88	−0.50%
10	f_{yr}^2 (MPa)	500	10	CO_2 (t-CO_2/m)	1.842	1.835	0.40%
11	f_{ys}^2 (MPa)	325	11	W_c (kN/m)	36.08	36.09	−0.04%
12	P_u^2 (kN)	12,000					
13	M_{uX}^2 (kN·m)	6,500					
14	M_{uY}^2 (kN·m)	5,000					

[1] Calculated on an output-side when optimizing W_c.
[2] Equality constraints.
[3] Inequality constraints.
* Error $(4) = \frac{(2)-(3)}{(2)} \times 100\%$.

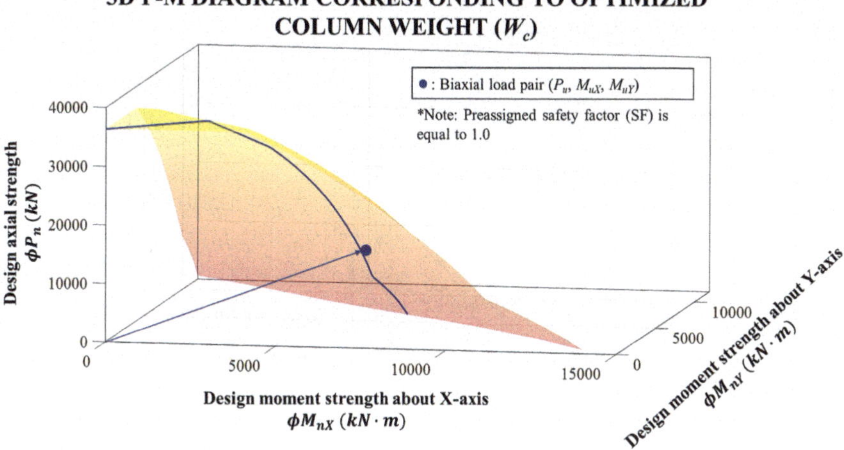

Figure 9.4.7 3D P-M diagram corresponding to optimized column weights (W_c) considering one biaxial load.

on a forward design implementing ANN-based Lagrange optimization, whereas the column weight (W_c) of 36.09 (kN/m) is calculated by structural design using design parameters obtained by ANNs, showing −0.04% error. In Table 9.4.5, 14 input parameters minimizing column weights (W_c) of an SRC column are used in AutoSRCHCol to calculate 11 output parameters, verifying the biggest error of 5.78% for the rebar strain (ε_r). The errors in *SF* (1.67%) and optimized column weight (W_c) (−0.04%) are insignificant, and hence, the accuracy of ANNs which minimize Lagrange function is acceptable, indicating that ANNs can be used in practical designs.

9.4.3.2 Column interaction diagram corresponding to an optimized column weight (W_c)

In Figure 9.4.7, a three-dimensional P-M interaction diagram obtained using parameters obtained in Table 9.4.5 corresponding to the minimized column weights (W_c) is plotted for the column illustrated in Figure 9.4.8 where dimensions of column sections, steel sections, and the rebar ratios corresponding to a minimized column weights (W_c) are obtained by a forward design implementing ANN-based Lagrange optimization.

9.4.3.3 Verification of an optimized column weight (W_c) based on ANN by large datasets

To verify the minimized column weight (W_c) obtained by the proposed method, a large dataset containing 5,750,000 data points is generated, which meet preassigned equality constraints shown in Table 9.3.2(b).

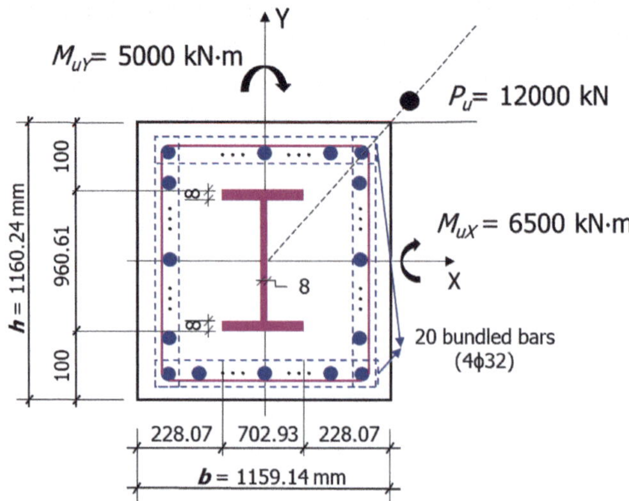

Figure 9.4.8 Section of an SRC column corresponding to minimized W_c.

Table 9.4.6 Verification of optimized W_c from Lagrange method by large datasets with preassigned factored load P_u = 12,000 kN, M_{uX}=6,500 kNm, M_{uY}=5,000 kNm

Column weight, W_c of SRC columns			
Min. W_c based on big data		*Proposed method based on ANN (kN/m)*	*Difference (%)*
A number of big data	*Min. W_c (kN/m)*		
650,000 data	38.52	36.08	−6.8%
1,750,000 data	37.98		−5.3%
3,750,000 data	37.83		−4.9%
5,750,000 data	37.83		−4.9%

Datasets of 96,902 in which *SF*s are greater than 1.0SF are extracted from a large dataset of 5,750,000, ensuring that the SRC columns are strong enough to resist factored loads. The minimum column weight (W_c) obtained from 96,902 observations which are obtained from a 5,750,000 dataset is 37.83 (kN/m) as shown in Table 9.4.6.

The minimum column weight (W_c) of 38.52 (kN/m) and 37.98 (kN/m) are obtained from 650,000 and 1,750,000 datasets, respectively, demonstrating a difference of −6.8% and −5.3%, respectively, compared with the minimized column weight (W_c) of 36.08 (kN/m) based on the proposed method as shown in Table 9.4.6, Figure 9.4.9(a), and (b). The difference between the minimum column weight (W_c) identified using large datasets

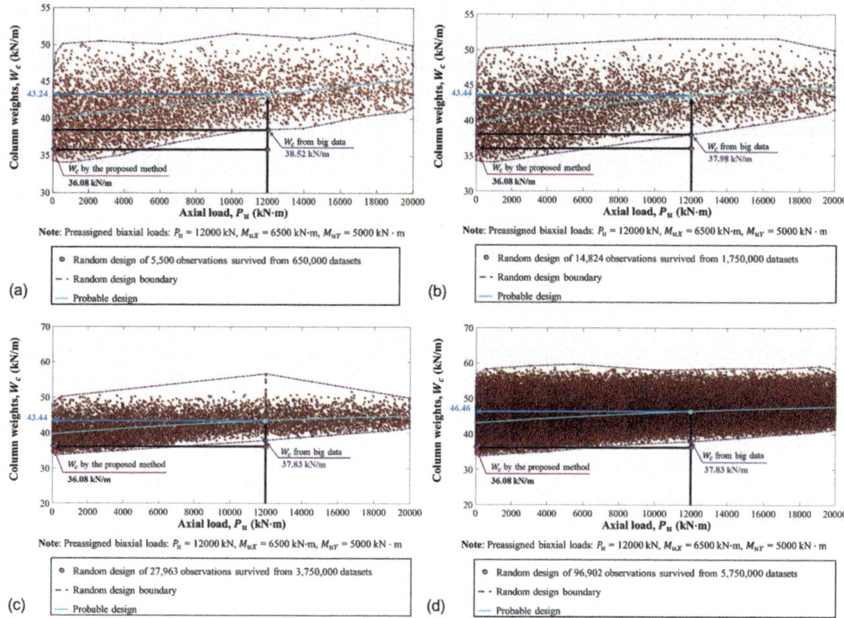

Figure 9.4.9 Verification of optimized W_c from Lagrange method by large datasets. (a) Random design survived from 650,000 datasets. (b) Random design survived from 1,750,000 datasets. (c) Random design survived from 3,750,000 datasets. (d) Random design survived from 5,750,000 datasets (22.3% reduction).

and the proposed method is reduced to –4.9% when a number of the datasets increases to 3,750,000 and 5,750,000 as shown in Table 9.4.6, Figure 9.4.9(c), and (d), respectively. It is noted that the minimum column weight (W_c) obtained from 96,902 observations which are extracted from a 5,750,000 dataset is 37.83 (kN/m), whereas the proposed method based on an ANN offers a column design with a minimized column weight (W_c) of 36.08 (kN/m), which is less than 37.83 (kN/m) identified from a 5,750,000 dataset by –4.9% as presented in Figure 9.4.9(d). As shown in Figure 9.4.9(d) obtained from 96,902 datasets, the minimized column weight (W_c) based on an ANN is also less than that of a probable column design of 46.46 (kN/m) by 22.3% determined based on trendline functions ("polyfit" and "polyval" commands) provided in MATLAB [9.40]. As shown in Table 9.4.5, the minimized column weight (W_c) obtained based on an ANN is 36.08 (kN/m) shows only –0.04% difference when being compared with structural calculations of 36.09 (kN/m) calculated using design parameters obtained by ANNs at a preassigned factored moments (M_{uX} and M_{uY}) of 6,500 kN·m, 5,000 kN·m and axial load (P_u) of 12,000 kN. Minimized column weight (W_c) is calculated based on training results shown in Table 9.2.4.

9.5 CONCLUSION

In this chapter, ANNs are formulated to optimize designs of reinforced concrete columns encasing H-shaped steel sections subjected to biaxial loads. The design parameters were identified which reduces cost index, CO_2 emissions and weight of the columns when compared to conventionally designed beams. Objective functions such cost index (CI_c), CO_2 emissions, and weight (W_c) of reinforced concrete columns encasing H-shaped steel sections subjected to biaxial loads are optimized based on ANN-based Lagrange optimization in which objective parameters CI_c, CO_2 emissions, and W_c are generalized based on ANNs. ANN-based generalized functions can replace complex explicit functions that are difficult to derive when objective functions are mathematically complex for the formulation of Lagrange functions. Dimensions of columns and steel sections are calculated as output parameters corresponding to minimized CI_c, CO_2 emissions, and W_c. Three-dimensional interaction diagrams of SRC columns subjected to biaxial bending and concentric axial loads are also formulated based on minimized CI_c, CO_2 emissions, and W_c. An ANN-based LMM identifies design parameters which reduce CI_c, CO_2 emissions, and W_c of a column by 30.7%, 40.4%, and 22.3%, respectively, when being compared with those of a conventionally designed SRC column. An accuracy of the ANN-based optimized designs is demonstrated using structural mechanics using design parameters obtained by ANNs.

Some of the findings of this chapter are highlighted as follows:

1. Optimizing designs of an SRC column based on conventional Lagrange multipliers are complex because explicit objective functions and constraint constraints are difficult to derive for an application of a conventional Lagrange optimization. In this chapter, conventional Lagrange multipliers which require explicit objective functions are replaced by differentiable objective and constraint functions which are generalized by ANNs. ANN-based generalized functions are used to formulate Lagrange functions. Differentiable objective and constraint functions are derived to implement an ANN-based Lagrange optimization. ANN-based design targets for CI_c, CO_2 emissions, W_c and design constraints imposed by design codes are derived as functions of input parameters based on large datasets of 100,000.
2. ANNs minimizing Lagrange functions are proposed to optimize designs of an SRC column under a biaxial load. Objective functions (three design targets), cost index (CI_c), CO_2 emissions, and weight (W_c) of an SRC column, are minimized while yielding steel height (h_s), flange width (b_s), flange (t_f), and web (t_w) thickness under factored biaxial loads P_u, M_{uX}, M_{uY}. Both cost index (CI_c) and CO_2 are minimized when rebar ratios and a steel section are minimized while dimensions of the

concrete section are maximized. On the other hand, reduced column dimensions decrease column weight (W_c), resulting in increased rebar and steel quantity to compensate concrete volume reduction. CI_c and CO_2 emissions are minimized in opposite way to minimizing W_c.

3. Material properties of concrete (f_c'), rebar (f_{yr}), and steel (f_{ys}) are predetermined on an input-side. Factored biaxial loads including factored axial load (P_u), bending moments (M_{uX} and M_{uY}) about X- and Y- axes are selected as input parameters to design an SRC column. Factored biaxial loads (P_u, M_{uX}, M_{uY}) are preassigned on an input-side a forward design network with safety factor equal to or greater than 1.0, to design column section to obtain design strengths ($\phi P_n, \phi M_{nX}, \phi M_{nY}$) greater than preassigned factored biaxial loads (P_u, M_{uX}, M_{uY}).
4. Optimizations are performed based on 9 equality functions $c(x)$ and 17 inequality constraints $v(x)$. Lagrange functions are formulated, then, implemented in deriving KKT equations. Inequality constraint matrix (S) is proposed to designate activated and inactivated inequality constraints. A diagonal matrix S comprises inequality terms of the Lagrange function. Each value s_1, s_2, ..., s_l indicates a status of inequalities $v_1(x)$, $v_2(x)$, ..., $v_l(x)$, respectively, whereas a complementary slack condition is represented by $Sv(x)=0$. An activated inequality v_j is demonstrated by setting the s_j value=1, whereas s_j is set as 0 when the inequality v_j is inactive.
5. Solutions under KKT conditions based on inequality constraints defined by design requirements are, then, solved using Newton–Raphson iterations. Design parameters including SRC column dimensions and rebar ratios are obtained on an output-side while minimizing CIc, CO_2 emissions, and W_c. The accuracy of the minimized objective functions (CIc, CO_2 emissions, and W_c) and corresponding design parameters is verified by using a structural-mechanics-based software (AutoSRCHCol) using design parameters obtained by ANNs. Errors are insignificant, indicating that the proposed network can be implemented in practical designs to help engineers optimize SRC column designs. They were minimized less than those of probable design of SRC columns by 30.7%, 40.4%, and 22.3%, respectively.
6. In Chapter 10, the three objective targets (CIc, CO_2 emissions, and W_c) of SRC columns are optimized using ANN-based Lagrange optimization in a simultaneous way. An optimization curve called Pareto frontier will provide contributions made by each of the three objective functions, uncovering the inter-relationships of the optimization among the three design targets for decision-makers.
7. Most of research focused on a SRC columns through experimental and theoretical analysis. This chapter has a significance and highly innovative aspects by presenting structural designs based on the combination between artificial intelligence and traditional designs.

The ANN-based designs suggested this chapter are implemented in designing an SRC column with H-shaped steel sections shown in Figure 9.2.1, which is a typical SRC column gaining a broad industrial application. Large datasets should be also generated to perform ANN-based design for an SRC column having shapes of steel section such as cross-shaped steel or T sections.

8. Weight (W_c) of precast module frames can also be minimized for lifting efficiency of a modular construction using the proposed optimization.
9. The proposed method was performed based on the latest MATLAB toolboxes ([9.30–9.34]). The proposed ANN-based method significantly depends on fast computing facilities for generating large datasets, Lagrange-based training, and Newton–Raphson-based iterations. Accuracies and applicability can be significantly extended in near future when the computation speed does not hinder the optimizations.

REFERENCES

[9.1] Abambres, M., & Lantsoght, E. O. (2020). Neural network-based formula for shear capacity prediction of one-way slabs under concentrated loads. *Engineering Structures*, *211*, 110501. https://doi.org/10.1016/j.engstruct.2020.110501

[9.2] Sharifi, Y., Lotfi, F., & Moghbeli, A. (2019). Compressive strength prediction using the ANN method for FRP confined rectangular concrete columns. *Journal of Rehabilitation in Civil Engineering*, 7(4), 134–153. https://doi.org/10.22075/JRCE.2018.14362.1260

[9.3] Asteris, P. G., Armaghani, D. J., Hatzigeorgiou, G. D., Karayannis, C. G., & Pilakoutas, K. (2019). Predicting the shear strength of reinforced concrete beams using artificial neural networks. *Computers and Concrete*, *24*(5), 469–488. https://doi.org/10.12989/CAC.2019.24.5.469

[9.4] Armaghani, D. J., Hatzigeorgiou, G. D., Karamani, C., Skentou, A., Zoumpoulaki, I., & Asteris, P. G. (2019). Soft computing-based techniques for concrete beams shear strength. *Procedia Structural Integrity*, *17*, 924–933. https://doi.org/10.1016/j.prostr.2019.08.123

[9.5] Hong, W. K., Pham, T. D., & Nguyen, V. T. (2022). Feature selection based reverse design of doubly reinforced concrete beams. *Journal of Asian Architecture and Building Engineering*, *21* (4), 1472–1496. https://doi.org/10.1080/13467581.2021.1928510

[9.6] Hong, W. K., Nguyen, V. T., & Nguyen, M. C. (2022). Artificial intelligence-based novel design charts for doubly reinforced concrete beams. *Journal of Asian Architecture and Building Engineering*, *21*(4), 1497–1519. https://doi.org/10.1080/13467581.2021.1928511

[9.7] Hong, W. K., & Nguyen, M. C. (2022). AI-based Lagrange optimization for designing reinforced concrete columns. *Journal of Asian Architecture and Building Engineering*, *21*(6), 2330–2344. http://dx.doi.org/10.1080/13467581.2021.1971998

[9.8] Mirza, S. A., Hyttinen, V., & Hyttinen, E. (1996). Physical tests and analyses of composite steel-concrete beam-columns. *Journal of Structural Engineering*, *122*(11), 1317–1326. https://doi.org/10.1061/(ASCE)0733-9445(1996)122:11(1317)

[9.9] Ricles, J. M., & Paboojian, S. D. (1994). Seismic performance of steel-encased composite columns. *Journal of Structural Engineering*, *120*(8), 2474–2494. https://doi.org/10.1061/(ASCE)0733-9445(1994)120:8(2474)

[9.10] El-Tawil, S., & Deierlein, G. G. (1999). Strength and ductility of concrete encased composite columns. *Journal of Structural Engineering*, *125*(9), 1009–1019.

[9.11] Li, L. I., & Matsui, C. (2000). Effects of axial force on deformation capacity of steel encased reinforced concrete beam–columns. In *Proceedings of 12th World Conference on Earthquake Engineering*.

[9.12] Brettle, H. J. (1973). Ultimate strength design of composite columns. *Journal of the Structural Division*, *99*(9), 1931–1951. https://doi.org/10.1061/JSDEAG.0003606

[9.13] Bridge, R. Q., & Roderick, J. W. (1978). Behavior of built-up composite columns. *Journal of the Structural Division*, *104*(7), 1141–1155. https://doi.org/10.1061/JSDEAG.0004956.

[9.14] Furlong, R. W. (1974). Concrete encased steel columns—design tables. *Journal of the Structural Division*, *100*(9), 1865–1882. https://doi.org/10.1061/JSDEAG.0003878.

[9.15] Mirza, S. A., & Skrabek, B. W. (1992). Statistical analysis of slender composite beam-column strength. *Journal of Structural Engineering*, *118*(5), 1312–1332. https://doi.org/10.1061/(ASCE)0733-9445(1992)118:5(1312).

[9.16] Munoz, P. R., & Hsu, C. T. T. (1997). Behavior of biaxially loaded concrete-encased composite columns. *Journal of Structural Engineering*, *123*(9), 1163–1171. https://doi.org/10.1061/(ASCE)0733-9445(1997)123:9(1163)

[9.17] Morino, S., Matsui, C., & Watanabe, H. (1984, July). Strength of biaxially loaded SRC columns. In: Frangopol, D. M. (ed), *Composite and Mixed Construction* (pp. 185–194). ASCE, Atlantic City, NJ.

[9.18] Chen, C. C., & Lin, N. J. (2006). Analytical model for predicting axial capacity and behavior of concrete encased steel composite stub columns. *Journal of Constructional Steel Research*, *62*(5), 424–433. https://doi.org/10.1016/j.jcsr.2005.04.021

[9.19] Dundar, C., Tokgoz, S., Tanrikulu, A. K., & Baran, T. (2008). Behaviour of reinforced and concrete-encased composite columns subjected to biaxial bending and axial load. *Building and Environment*, *43*(6), 1109–1120. https://doi.org/10.1016/j.buildenv.2007.02.010

[9.20] Munoz, P. R., & Hsu, C. T. T. (1997). Biaxially loaded concrete-encased composite columns: Design equation. *Journal of Structural Engineering*, *123*(12), 1576–1585. https://doi.org/10.1061/(ASCE)0733-9445(1997)123:12(1576)

[9.21] Furlong, R. W. (1968). Design of steel-encased concrete beam-columns. *Journal of the Structural Division*, *94*(1), 267–282. https://doi.org/10.1061/JSDEAG.0001854

[9.22] Kato, B. (1996). Column curves of steel-concrete composite members. *Journal of Constructional Steel Research*, *39*(2), 121–135. https://doi.org/10.1016/S0143-974X(96)00030-2

[9.23] Virdi, K. S., Dowling, P. J., BS 449, & BS 153. (1973). The ultimate strength of composite columns in biaxial bending. *Proceedings of the Institution of Civil Engineers*, *55*(1), 251–272. https://doi.org/10.1680/iicep.1973.4958

[9.24] Roik, K., & Bergmann, R. (1990). Design method for composite columns with unsymmetrical cross-sections. *Journal of Constructional Steel Research*, *15*(1–2), 153–168. https://doi.org/10.1016/0143-974X(90)90046-J

[9.25] Rong, C., Shi, Q., & Wang, B. (2021). Seismic performance of angle steel frame confined concrete columns: Experiments and FEA model. *Engineering Structures*, *235*, 111983. https://doi.org/10.1016/j.engstruct.2021.111983

[9.26] Rong, C., & Shi, Q. (2021). Analysis constitutive models for actively and passively confined concrete. *Composite Structures*, *256*, 113009. https://doi.org/10.1016/j.compstruct.2020.113009

[9.27] Nguyen, D. H., & Hong, W. K. (2020). An analytical model computing the flexural strength and performance of the concrete columns confined by both transverse reinforcements and steel sections. *Journal of Asian Architecture and Building Engineering*, *19*(6), 647–669. https://doi.org/10.1080/13467581.2020.1775603

[9.28] Hong, W. K. (2019). *Hybrid Composite Precast Systems: Numerical Investigation to Construction*. Woodhead Publishing, Elsevier, Cambridge, MA.

[9.29] Hong, W. K., Nguyen, V. T., Nguyen, D. H., & Nguyen, M. C. (2022). An AI-based Lagrange optimization for a design for concrete columns encasing H-shaped steel sections under a biaxial bending. *Journal of Asian Architecture and Building Engineering*, *22*, 821–841.

[9.30] MathWorks. (2022a). Deep Learning Toolbox: User's Guide (R2022a). Retrieved July 26, 2022, from: https://www.mathworks.com/help/pdf_doc/deeplearning/nnet_ug.pdf

[9.31] MathWorks. (2022a). Global Optimization: User's Guide (R2022a). Retrieved July 26, 20122 from: https://www.mathworks.com/help/pdf_doc/gads/gads.pdf

[9.32] MathWorks. (2022a). Optimization Toolbox: Documentation (R2022a). Retrieved July 26, 2022, from: https://www.mathworks.com/help/pdf_doc/optim/optim.pdf

[9.33] MathWorks. (2022a). Parallel Computing Toolbox: Documentation (R2022a). Retrieved July 26, 2022, from: https://www.mathworks.com/help/pdf_doc/parallel-computing/parallel-computing.pdf

[9.34] MathWorks. (2022a). Statistics and Machine Learning Toolbox: Documentation (R2022a). Retrieved July 26, 2022, from: https://www.mathworks.com/help/pdf_doc/stats/stats.pdf

[9.35] American Institute of Steel Construction. (2016). *ANSI/AISC 360-16 Specification for Structural Steel Buildings*. AISC, Chicago, IL.

[9.36] An ACI Standard. (2019). *Building Code Requirements for Structural Concrete (ACI 318–19)*. American Concrete Institute, Farmington Hills, MI.

[9.37] Roman. (2020). Neural Networks: From Zero to Hero. Retrieved from: https://towardsdatascience.com/neural-networks-from-hero-to-zero-afc30205df05

[9.38] Kuhn, H. W., & Tucker, A. W. (1951). Nonlinear programming. In *Proceedings of 2nd Berkeley Symposium* (pp. 481–492). Berkeley: University of California Press.

[9.39] Karush, W. (1939). *Minima of Functions of Several Variables with Inequalities as Side Constraints*. M.Sc. thesis. Department of Mathematics, University of Chicago, Chicago, IL.

[9.40] MathWorks. (2021). *MATLAB R2021a, Version 9.9.0*. Natick, MA: The MathWorks Inc.

[9.41] Hong, W.K. (2023). *Artificial Neural Network- Based Optimized Design of Reinforced Concrete Structures*, Taylor & Francis, Boca Raton, FL.

Chapter 10

A Pareto frontier using an ANN-based multi-objective optimization (MOO) for concrete columns encasing H-shaped steel sustaining multi-biaxial loads

10.1 INTRODUCTION AND SIGNIFICANCE OF THE CHAPTER 10

10.1.1 Research background

Steel-encased reinforced concrete columns (SRC) encasing H-shaped steel sections are used popularly in high-rise buildings and heavy storage facilities to utilize the rigidity and strength of concrete as well as the ductility and strength of steel sections. Some studies have been conducted successfully to investigate experimental behaviors of SRC columns sustaining uniaxial bending moments and axial compressions by Mirza et al. [10.1], Ricles and Paboojian [10.2], El-Tawil and Deierlein [10.3], Li and Matsui [10.4], Brettle [10.5], Bridge and Roderick [10.6], Furlong [10.7], and Mirza and Skrabek [10.8]. Analytical analyses have been performed by Chen and Lin [10.9], Dundar et al. [10.10], Munoz and Hsu [10.11], Furlong [10.12], Kato [10.13], Virdi and Dowling [10.14], Roik and Bergmann [10.15], and Nguyen and Hong [10.16]. In this chapter, a theory of strain compatibility proposed by Nguyen and Hong [10.16] is used to generate large datasets of reinforced concrete columns encasing H-shaped steel sections (SRC columns) which are implemented in training ANNs. ANNs have been applied successfully in the field of structural analysis by Abambres and Lantsoght [10.17], Sharifi et al. [10.18], Asteris et al. [10.19], Armaghani et al. [10.20], and Hong et al. [10.21,10.22].

Effects of multiple biaxial loads on a performance of SRC columns are also explored in this chapter in which a novel optimization design combining ANNs and Lagrange optimizations is proposed. Hong and Nguyen et al. [10.23] implemented ANNs and Lagrange optimizations successfully in obtaining optimized designs of rectangular reinforced concrete columns with respect to a single-objective function such as a column cost index (CI_c), CO_2 emissions (CO_2), and column weight (W_c), separately, as introduced in Chapter 9. Many studies focused on optimizing RC structures [10.24–10.28]. MOO designs have been performed by Yoon et al. [10.29] in designs of reinforced concrete columns investigating a relation of cost index of columns (CI_c)

DOI: 10.1201/9781003354796-10

and CO_2 emissions. Gholami et al. [10.30] proposed optimization designs of a composite sandwich panel considering panel weight and cost index as objective functions. Chapter 10 proposes a Pareto frontier with respect to multi-objective functions which are cost index of columns (CI_c), CO_2 emissions, and column weights (W_c) based on ANN-based Lagrange MOO design of SRC columns sustaining multiple biaxial loads.

10.1.2 Research objectives and innovations

Structural engineers encounter several design objectives in the design of SRC columns encasing H-shaped steel sections subject to multiple biaxial load pairs. However, MOO applications in structural engineering practice are uncommon despite their high demand and efficiency. This chapter aims to combine ANNs with a Lagrange method for optimizing multi-objective functions (a column cost index (CI_c), CO_2 emissions (CO_2), and column weight (W_c)) at the same time, meeting contractors' interests and producing a robust and sustainable design. ANNs based on Lagrange optimization are proposed to design SRC columns sustaining multiple biaxial loads while optimizing multi-objective functions simultaneously. Components of weight matrices derived by training ANNs subject to one biaxial load pair are reused to derive weight matrices for ANNs which are subject to multiple biaxial load pairs. *ANN-1LP* is firstly trained on large datasets, and hence, weight matrices of *ANN-1LP* associated with one biaxial load pair are reused to derive weight matrix of ANNs to design SRC columns sustaining multiple load pairs (denoted as *ANN-nLP*).

ANN-based generalized objective functions and constraints are challenging to be derived analytically in some complex design problems. Hong et al. [10.31] derived ANN-based objective functions including a column cost index (CI_c), CO_2 emissions (CO_2), and column weight (W_c) as an alternative to complex explicit functions. They formulated ANN-based functions as a function of forward input variables. For MOO designs, A *UFO* is established by integrating the three objective parameters including cost index (CI_c), CO_2 emissions, and column weights (W_c). A set of ANN-based Lagrange MOO designs known as a Pareto frontier is, then, established in which *UFO* is optimized simultaneously with respect to their fractions indicating contributions made by each of the objective functions (CI_c, CO_2 emissions, and W_c). A Pareto frontier based on ANN-based MOO uncovers the fractions contributed by each objective function to the overall optimizations, being able to inform engineers and decision-makers of contributions made by each objective function. A Pareto frontier is used to assess design projects which is beneficial, such as a three-dimensional P-M interaction diagram based on simultaneously optimized objective functions (CI_c, CO_2, and W_c). P-M interaction diagrams are developed with respect to combined biaxial bending and axial loads. The optimized objective functions are verified by structural calculations using design parameters obtained by ANNs; the error was <2%, which demonstrates that the proposed ANN-based MOO offers accuracy sufficient for holistically practical design applications. This chapter was written based on the previous paper by Hong et al. [10.31].

10.1.3 Research significance

SRC columns encasing H-shaped steel sections shown in Figure 10.2.1 sustain multiple biaxial loads in which multi-objective functions, cost index (CI_c), CO_2 emissions, and column weights (W_c) are optimized simultaneously. ANNs are used to establish generalized objective functions for a column cost index (CI_c), CO_2 emissions (CO_2), and column weight (W_c), which are then optimized based on Lagrange multipliers, yielding design parameters including column dimensions, rebar ratios, and steel dimensions. ANN-based functions replace analytical functions which are sometimes challenging to derive with conventional Lagrange multipliers for complex design problems. In this chapter, each ANN-based objective function is integrated into a *UFO* which is used to derive a global Lagrange function. Based on separate single-objective function $\left\{f_{\mathrm{CI}_c}(\boldsymbol{x}),\, f_{\mathrm{CO}_2}(\boldsymbol{x}),\, f_{\mathrm{W}_c}(\boldsymbol{x})\right\}^T$, UFO shown in Eq. (10.1.1) is derived using tradeoff fractions, $\boldsymbol{w} = \left\{w_{\mathrm{CI}_c}, w_{\mathrm{CO}_2}, w_{\mathrm{W}_c}\right\}^T$ to simultaneously optimize multi-objective functions.

$$f_{\mathrm{UFO}} = w_{\mathrm{CI}c} f_{\mathrm{CI}_c}^N + w_{\mathrm{CO}_2} f_{\mathrm{CO}_2}^N + w_{\mathrm{W}_c} f_{\mathrm{W}_c}^N \tag{10.1.1}$$

Using conventional design methods, an optimization of a behavior of SRC columns encasing H-shaped steels subjected to multiple biaxial bending moment pairs and compressions needs repeated iterations, requiring considerable effort and time. All requirements for strength and serviceability should be met, and a balanced section should be achieved, and hence, the trial-and-error method is not a daily design practice.

In this chapter, a modularized weight matrix for SRC columns subject to multiple biaxial load pairs is derived. Robust and rapid but practically accurate ANN-based Lagrange method is introduced to optimize multi-objective functions for designs of SRC columns sustaining multiple biaxial loads, in particular, for structural components such as foundations and walls in general. The proposed method offers a straightforward, simplified, and fast approach for optimizing an UFO of SRC columns under multiple loads with adequate accuracy. A network can help engineers with optimizing columns and steel sections in preliminary design stages.

10.1.4 Tasks readers can perform after this chapter

1. The readers can derive weight matrices for ANNs which are subject to multiple biaxial load pair
    1. Weight matrices for ANNs which are subject to multiple biaxial load pairs can be derived from components of weight matrices derived based on one biaxial load pair. Weight matrices of an ANN subject to multiple biaxial load pairs (*ANN-nLP*) are derived by

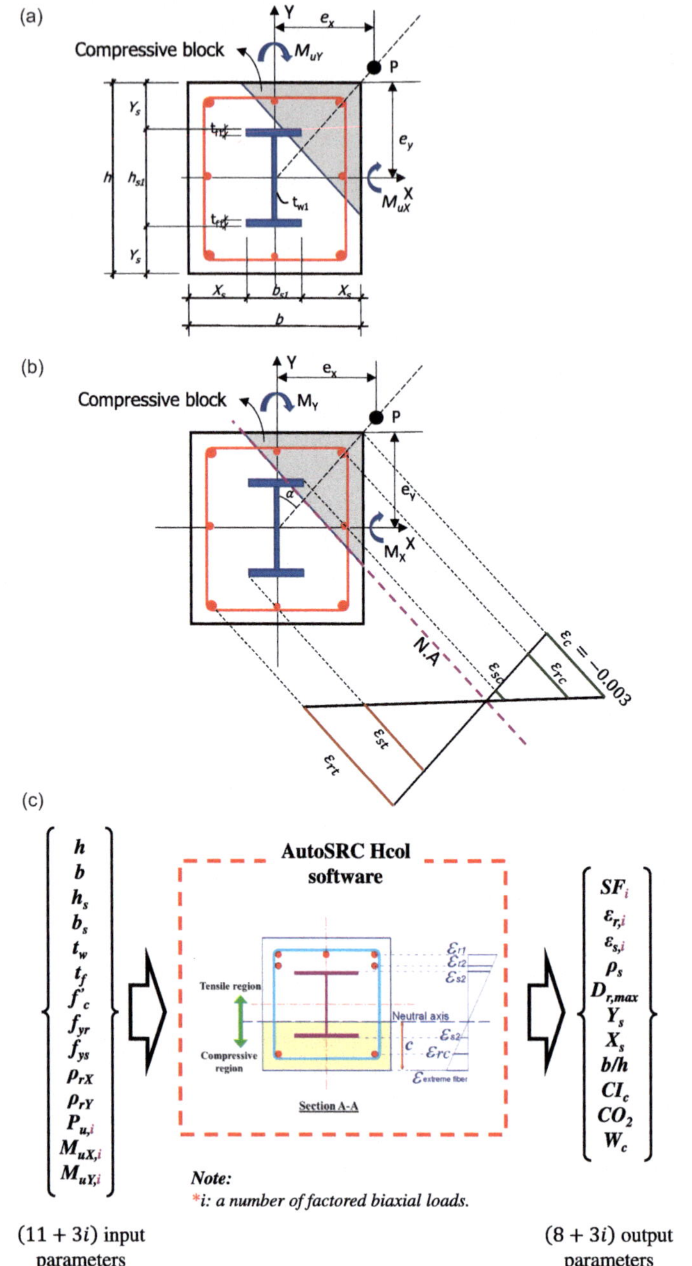

Figure 10.2.1 Cross section for forward design scenarios of SRC columns sustaining multiple biaxial loads. (a) Cross section of SRC with H-shaped columns under biaxial loads. (b) Strain compatibility of SRC with H-shaped columns under biaxial loads. (c) Forward design scenarios of SRC columns sustaining multiple biaxial loads.

reusing components of weight matrices of an ANN subject to one biaxial load pair (*ANN-1LP*). An ANN with weight matrices subject to one biaxial load pair (*ANN-1LP*) needs to be trained only once, whereas an ANN with weight matrices subject to multiple biaxial load pairs (*ANN-nLP*) does not need to be trained. This method modularizes a weight matrix to reduce a computation and training time required.

2. A network combining an ANN and Lagrange multipliers is derived to minimize CI_c, CO_2, and W_c of SRC columns encasing H-shaped steels.
3. Three-dimensional P–M diagrams subjected to multiple biaxial bending and axial load pairs can be plotted when CI_c, CO_2, and W_c are minimized.
4. The proposed method is applied to minimize a design of SRC columns subjected to multiple biaxial load combinations. ANN-based optimization results are verified by those based on structural calculations using design parameters obtained by ANNs.

2. The readers can minimize a *UFO* subjected to multiple biaxial load pairs
   1. An ANN-based *UFO* integrating cost index (CI_c), CO_2 emissions, and column weights (W_c) is derived based on contribution fractions which can assist engineers in deciding how much tradeoffs between objective functions are to be used. Contribution to the three objective functions (CI_c, CO_2 emissions, and W_c) are determined by tradeoff fractions w_{CI_c}, w_{CO_2}, and w_{W_c}, which are implemented in deriving *UFO*. ANN-based *UFO*, then, are implemented in Lagrange optimization to design SRC columns sustaining multiple biaxial loads.
   2. A design example optimizing UFO with preassigned tradeoff fractions is performed. Structural mechanics-based software uses design parameters obtained by ANNs to verify, that ANN-based MOO algorithm offers practical design tools optimizing multi-objective targets with sufficient accuracies.
   3. A Pareto frontier which is a set of multi-optimal design results is verified by using 970,000 datasets, showing a lower boundary of 970,000 datasets coincides with a Pareto frontier.
   4. Proportional trends between optimal designs of CI_c and CO_2 emissions are observed, whereas reversely proportional trends between optimal W_c and each CI_c or CO_2 emissions are found.

10.2 DESIGN OF SRC COLUMNS SUSTAINING MULTIPLE BIAXIAL LOADS OPTIMIZING UFO

10.2.1 SRC column section for ANN-based design scenario

Figure 10.2.1 and Table 10.2.1 show design parameters of an SRC column encasing H-shaped steel subjected to a biaxial load pair. In Figure 10.2.1(c),

Table 10.2.1 Nomenclatures of parameters for defining SRC with H-shaped columns

No.	Nomenclature	
	Forward input parameters	
1	h (mm)	Column height
2	b (mm)	Column width
3	h_s (mm)	Height of H-shaped steel section
4	b_s (mm)	Width of H-shaped steel section
5	t_f (mm)	Flange thickness of H-shaped steel section
6	t_w (mm)	Web thickness of H-shaped steel section
7	f_c' (MPa)	Compressive concrete strength
8	f_{yr} (MPa)	Yield rebar strength
9	f_{ys} (MPa)	Yield steel strength
10	ρ_{rX}	Rebar ratio along to horizontal axis (X-axis)
11	ρ_{rY}	Rebar ratio along to vertical axis (Y-axis)
12	P_u (kN)	Axial force
13	M_{uX} (kN·m)	Factored moment about horizontal axis (X-axis)
14	M_{uY} (kN·m)	Factored moment about vertical axis (Y-axis)
	Forward output parameters	
15	ϕM_n (kN·m)	Design moment strength
16	ϕP_n (kN)	Design axial strength
17	M_u (kN·m)	Combined design moment strength ($\sqrt{M_{uX}^2 + M_{uY}^2}$)
18	SF	Safety factor ($\varphi M_n / M_u = \varphi P_n / P_u$)
19	Y_s (mm)	Vertical distance from column surface to surface of steel flange
20	X_s (mm)	Horizontal distance from column surface to surface of steel flange
21	ρ_s	Steel ratio
22	ε_s	Steel strain
23	ε_r	Rebar strain
24	b/h	Aspect ratio of column section
25	$D_{r,max}$	A maximum diameter of rebar
26	CI_c (KRW/m)	Cost index of column
27	CO_2 (t-CO_2/m)	CO_2 emission
28	W_c (kN/m)	Column weight

a forward design scenario of SRC columns sustaining three biaxial loads is presented with 20 input and 17 output parameters when $n=3$. A section of the SRC column is illustrated in Figure 10.2.1(a)–(c) where 20 input parameters include section dimensions of a rectangular column (height, h and width, b) and H-shaped steel section having height (h_s), width (b_s), and thickness of flange (t_f), web (t_w). Input parameters also include material properties of concrete (f_c'), rebar (f_{yr}), steel (f_{ys}), and ratios of reinforcing bars, ρ_{rY} and ρ_{rX}, along Y- and X-directions, respectively. The three pairs of factored biaxial loads ($P_{u,i}$, $M_{uX,i}$, $M_{uY,i}$) with $i=1,2,3$ are also shown in Figure 10.2.1(c), whereas safety factor (SF), rebar strain (ε_r), and steel strain (ε_s) are considered as output parameters. Eight other output parameters which are steel ratio (ρ_s), a maximum rebar diameter ($D_{r,\max}$), vertical (Y_s) and horizontal (X_s) clearances between steel sections and concrete columns, aspect ratio of columns (b/h), cost index (CI_c), CO_2 emissions, and column weights (W_c) are calculated on an output-side. Relationships among three objective parameters (CI_c, CO_2 emissions, and W_c) in UFO are also investigated when multiple biaxial loads are acting on SRC columns.

10.2.2 Generation of large datasets

A strain compatibility algorithm proposed by Nguyen and Hong et al. [10.16] as shown in Figure 10.2.2 is implemented in developing an analytical software (AutoSRCHCol), which investigates behaviors of SRC columns. Large datasets of 100,000, then, are generated by using AutoSRCHCol for SRC columns under biaxial loads. The data distribution of input and output parameters are illustrated in Figure 10.2.3. Table 10.2.2 presents unit price, unit CO_2 emission, and unit weight of structural components such as concretes, rebars, and steels. Statistics of input parameters including minimum, maximum, mean, variance, and standard deviation used for generating 100,000 datasets are summarized in Table 10.2.3. The 100,000 datasets generated based on AutoSRCHCol are implemented in training ANNs.

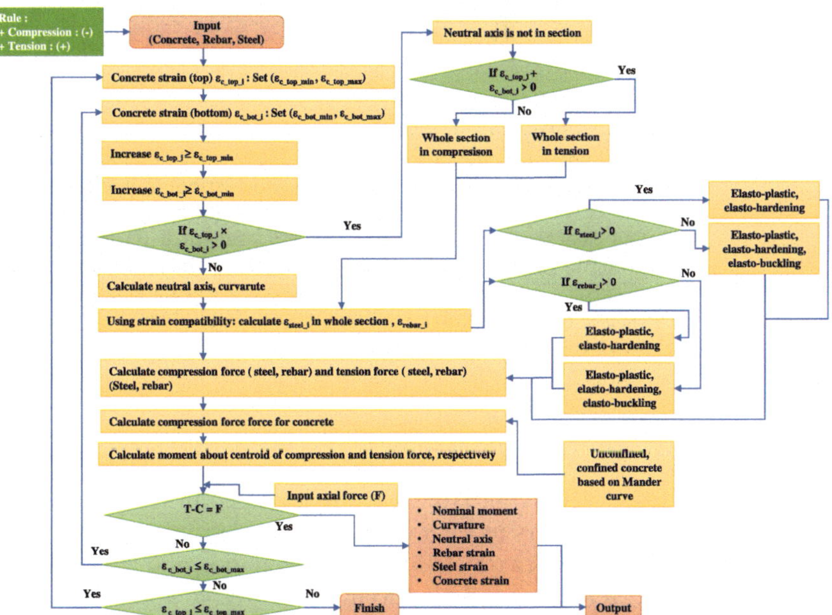

Figure 10.2.2 Computation algorithm used for generating large datasets [10.16].

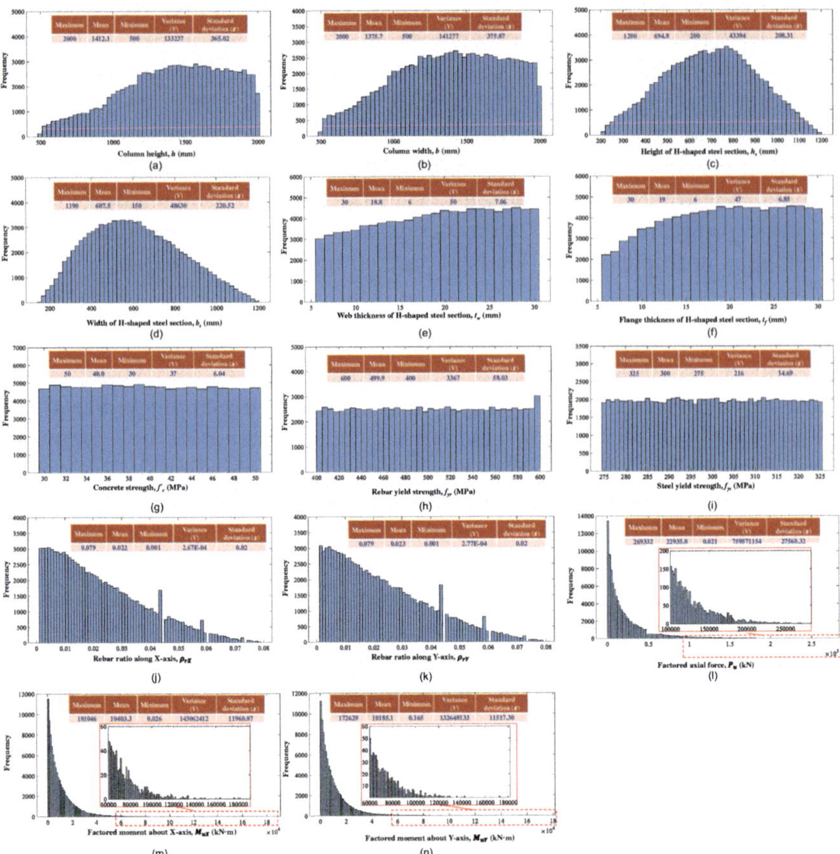

Figure 10.2.3 Distribution of large datasets generated using AutoSRCHCol. (a) Column height, h (mm). (b) Column width, b (mm). (c) Height of H-shaped steel section, h_s (mm). (d) Width of H-shaped steel section, b_s (mm). (e) Web thickness of H-shaped steel section, t_w (mm). (f) Flange thickness of H-shaped steel section, t_f (mm). (g) Concrete strength, $f_c^{'}$ (MPa). (h) Rebar yield strength, f_{yr} (MPa). (i) Steel yield strength, f_{ys} (MPa). (j) Rebar ratio along X-axis, ρ_{rX}. (k) Rebar ratio along Y-axis, ρ_{rY}. (l) Factored axial force, P_u (kN). (m) Factored moment about X-axis, M_{uX} (kN·m). (n) Factored moment about Y-axis, M_{uY} (kN·m).

10.2.3 Training ANNs on each output parameter based on PTM

PTM proposed by Hong et al. [10.21] is implemented in training ANNs to map all input parameters to each output parameter, individually. PTM trains output parameters separately for less computation and less time. An *ANN-3LP* [10.32] for forward design scenario of SRC columns sustaining three pairs of biaxial loads is illustrated in Figure 10.2.4 which depicts

Table 10.2.2 Unit price, CO_2 emissions, and weights of materials [10.39]

Material	*Strength*	*Unit price*	*Unit CO_2 emission*	*Unit weight*
Concrete	30 MPa	85,000 (KRW/m^3)	0.1677 (t-CO_2/m^3)	23.56 kN/m^3
	40 MPa	94,000 (KRW/m^3)		
	50 MPa	104,000 (KRW/m^3)		
Rebar	500 MPa	1,055 (KRW/kg)	2.512 (t-CO_2/ton)	78.5 kN/m^3
	600 MPa	1,085 (KRW/kg)		
Steel	275 MPa	1,800 (KRW/kg)		
	325 MPa	1,880 (KRW/kg)		

Table 10.2.3 Statistics of input parameters used for generating 100,000 datasets

Number of datasets	*100,000*						
Parameters	h *(mm)*	b *(mm)*	h_s *(mm)*	b_s *(mm)*	t_w *(mm)*	t_f *(mm)*	f'_c *(MPa)*
Maximum	2,000	2,000	1,200	1,190	30	30	50
Mean	1,412.1	1,375.7	694.8	607.5	18.8	19.0	40.0
Minimum	500	500	200	150	6	6	30
Variance (*V*)	133,237	141,277	43,394	48,630	50	47	37
Standard deviation	365.02	375.87	208.31	220.52	7.06	6.85	6.04
Parameters	f_{yr} (MPa)	f_{ys} (MPa)	ρ_{rX}	ρ_{rY}	P_u (kN)	M_{uX} (kN·m)	M_{uY} (kN·m)
Maximum	600	325	0.079	0.079	269,332	181,046	172,629
Mean	499.9	300.0	0.022	0.023	22,935.8	10,403.3	10,185.1
Minimum	400	275	0.001	0.001	0.021	0.026	0.165
Variance (*V*)	3367	216	2.67E-04	2.77E-04	759,571,154	143,062,412	132,648,133
Standard deviation	58.03	14.69	0.02	0.02	27,560.32	11,960.87	11,517.30

input, hidden, and output layers with neurons. Input and output layers consist of $(11 + 3i)$ and $(8 + 3i)$ $(i = 1,2,3)$ parameters where i is a number of biaxial load pairs as shown in Figures 10.2.1(c) and 10.2.4. Each hidden layer is activated by *tansig* functions [10.33]. Training is performed based on 100,000 datasets generated based on AutoSRCHCol which are separated into training, validation, and testing data with a proportion of 70% (70,000 datasets), 15% (15,000 datasets), and 15% (15,000 datasets), respectively [10.34]. Best training for each output parameter is sought among nine training attempts based on a combination of eight, nine, and ten hidden layers with 40, 50, and 60 neurons as shown in Table 10.2.4. Suggested and validation epochs are preassigned as 50,000 and 500, respectively.

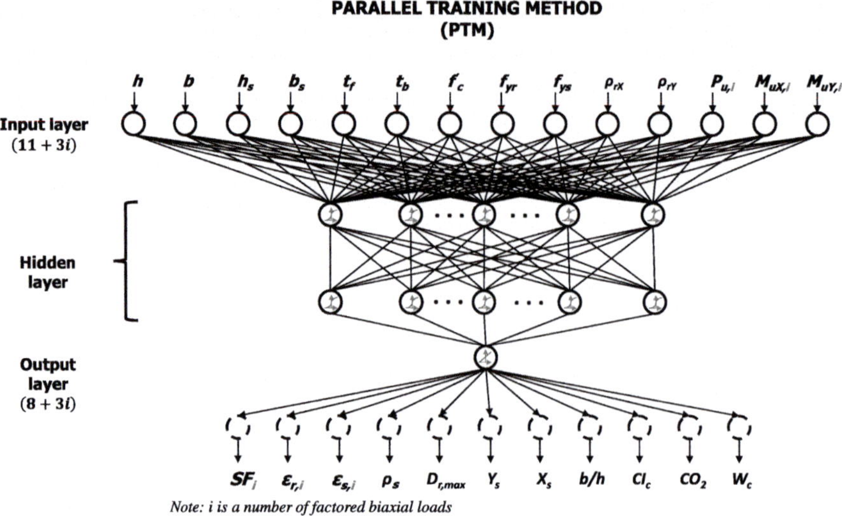

Figure 10.2.4 PTM training of *ANN-iLP* for SRC columns sustaining multiple biaxial loads.

10.3 ANN-BASED COLUMN DESIGN SCENARIO

10.3.1 Fourteen forward input parameters and 11 forward output parameters for a design

Table 10.3.1(a) shows parameters used to formulate the ANN-based forward network (*ANN-1LP*) considering one biaxial load pair (P_u, M_{uX}, M_{uY}). There are 14 forward input parameters $(h, b, h_s, b_s, t_w, t_f, f_c', f_{yr}, f_{ys}, \rho_{rX}, \rho_{rY}, P_u, M_{uX}, M_{uY})$ and 11 forward output parameters $(SF, \varepsilon_r, \varepsilon_s, \rho_s, D_{r,max}, Y_s, X_s, b/h, CI_c, CO_2, W_c)$ when one biaxial load pair (P_u, M_{uX}, M_{uY}) is applied. Material properties such as f_c', f_{yr}, f_{ys} are predetermined, which are denoted as equality constraints for Lagrange multipliers. For SRC columns, rebar ratios $(\rho_{rX}$ and $\rho_{rY})$ along the X- and Y-axes are limited to no smaller than 0.004 as per Section I2.1a in ANSI/AISC 360-16 [10.35] and no larger than 0.08 as per Section 10.6.1 in ACI 318-19 [10.36]. The minimum rebar ratio is set to reduce the effects of creep and shrinkage of concrete and prevent the rebar from yielding under sustained service loads. The maximum rebar ratio is set to ensure that concrete can effectively consolidate around rebars. Bundled rebar is considered, and therefore, a maximum rebar diameter was set to 64 mm, which is equivalent to four rebars bundles. SRC columns should have a minimum steel section of 0.01 as per Section I2.1a, ANSI/AISC 360-16 [10.35]; however, a maximum steel section is not specified in ANSI/AISC 360-16. The column strengths $\left(\phi P_n, \phi M_n = \left(\phi M_{n,i} = \sqrt{M_{nX,i}^2 + M_{nY,i}^2}, (i = 1, .., n)\right)\right)$ should be designed to resist the factored biaxial load pairs (P_u, M_{uX}, M_{uY}). The safety factor is

Table 10.2.4 Selecting best training for each output parameter (PTM) based on a combination of 8, 9, and 10 hidden layers with 40, 50, and 60 neurons

Forward PTM training 14 Forward Inputs ($h, b, h_s, b_s, t_w, t_f, f'_c, f_y, f_{ys}, \rho_{rX}, \rho_{rY}, P_u, M_{uX}, M_{uY}$) 11 Forward Outputs ($SF, \varepsilon_r, \varepsilon_s, \rho_s, D_{r,max}, Y_s, X_s, b/h, CI_c, CO_2, W_c$)							
No.	*Data*	*Hidden layers*	*Neurons*	*Suggested epoch*	*Best epoch for training*	*Test MSE*	*R at best epoch*
(1) Training PTM on *SF*							
1	100,000	10	50	50,000	48,942	3.14E-04	0.999
(2) Training PTM on ε_r							
2	100,000	8	50	50,000	38,404	4.21E-05	0.999
(3) Training PTM on ε_s							
3	100,000	10	60	50,000	38,546	3.99E-05	0.999
(4) Training PTM on ρ_s							
4	100,000	8	40	50,000	49,866	3.65E-08	1.000
(5) Training PTM on $D_{r,max}$							
5	100,000	8	60	50,000	18,507	8.05E-04	0.998
(6) Training PTM on Y_s							
6	100,000	10	50	50,000	48,616	1.46E-08	1.000
(7) Training PTM on X_s							
7	100,000	8	50	50,000	43,365	3.81E-09	1.000
(8) Training PTM on *b/h*							
8	100,000	8	40	50,000	16,317	1.73E-07	1.000
(9) Training PTM on CI_c							
9	100,000	10	40	50,000	18,080	3.74E-07	1.000
(10) Training PTM on CO_2							
10	100,000	10	50	50,000	46,398	6.54E-08	1.00
(11) Training PTM on W_c							
11	100,000	10	40	50,000	33,965	1.47E-07	1.00

then calculated as the ratio of the design strength to the factored load ($SF = \phi P_n / P_u = \phi M_n / M_u$). Table 10.3.1(b) summarizes the input and output parameters for *ANN-nLP* to resist three biaxial load pairs.

Lagrange multipliers are used to optimize a column cost index (CI_C), CO_2 emissions (CO_2), and column weight (W_c) for SRC columns subjected to multiple biaxial load pairs. The optimized CI_C, CO_2, and W_c are then used to determine the corresponding column dimensions (column height h and column width b), rebar ratios (ρ_{rX} and ρ_{rY}), and steel dimensions (section height h_s, section width b_s, and thicknesses of web and flange t_w and t_f, respectively). Table 10.3.1(c) shows load pairs used for design scenarios of SRC columns with H-shaped steel core.

Table 10.3.1 Scenarios for SRC columns with H-shaped steel core

(a) ANN-1LP considering one biaxial load pair

Forward design scenarios

Input		Output	
h	f_y	SF	$D_{r,max}$
b	f_{ys}	ε_r	CI_c
h_s	ρ_{rX}	ε_s	CO_2
b_s	ρ_{rY}	Y_s	W_c
t_w	P_u	X_s	
t_f	M_{uX}	b/h	
f_c'	M_{uY}	ρ_s	

(b) ANN-3LPs considering three biaxial load pairs

Forward design scenarios with three biaxial load pairs

Input		Output	
h	f_y	SF_1, SF_2, SF_3	$D_{r,max}$
b	f_{ys}	$\varepsilon_{r,1}, \varepsilon_{r,2}, \varepsilon_{r,3}$	CI_c
h_s	ρ_{rX}	$\varepsilon_{s,1}, \varepsilon_{s,2}, \varepsilon_{s,3}$	CO_2
b_s	ρ_{rY}	Y_s	W_c
t_w	$P_{u,1}, P_{u,2}, P_{u,3}$	X_s	
t_f	$M_{uX,1}, M_{uX,2}, M_{uX,3}$	b/h	
f_c'	$M_{uY,1}, M_{uY,2}, M_{uY,3}$	ρ_s	

(c) Load pairs for scenarios for SRC columns with H-shaped steel core

$P_{u,i}$ (kN)	$M_{uX,i}$ (kN)	$M_{uY,i}$ (kN)
5,000	4,000	5,000
10,000	6,700	5,000
4,500	3,500	2,000

10.3.2 ANN-based Lagrange multi-objective optimization (MOO)

ANN-based generalized objective and constraining functions are derived to replace explicit functions which are sometimes difficult to obtain for conventional Lagrange optimizations. ANN-based functions are derived in Eq. (10.3.1) as introduced by Hong and Nguyen et al. [10.23], and Krenker et al. [10.37].

$$\boldsymbol{f}(\mathbf{x}) = g^D\left(f_{lin}^N\left(\mathbf{W}^N f_t^{N-1}\left(\mathbf{W}^{N-1}\ldots f_t^1\left(\mathbf{W}^1 g^N(\mathbf{x}) + \mathbf{b}^1\right)\ldots + \mathbf{b}^{N-1}\right) + b^N\right)\right) \tag{10.3.1}$$

in which $\mathbf{x}$ is a set of input parameters; N is a number of layers including hidden layers and output layer; $\mathbf{W}^n$ is a weight matrix between layer $n-1$ and layer n; $\mathbf{b}^n$ is a bias matrix assigned neurons of layer n; and g^N, g^D

are normalization and de-normalization function, respectively. Activation functions f_t^n at layer n are used to formulate non-linear relationships of ANNs, whereas a linear activation function, f_{lin}^n, is selected for an output layer when output values are unbounded, allowing linear behavior. ANN-based objective functions including cost index (CI_c), CO_2 emissions, and W_c are derived in Eqs. (10.3.2)–(10.3.4), whereas a UFO is, then, established by integrating objective functions as shown in Eq. (10.3.5) [10.23,10.37].

$$\underbrace{f_{CI_c}^{ANN}}_{[1\times1]} = g_{CI_c}^{D}\left(f_l^{11}\left(\underbrace{\mathbf{W}_{CI_c}^{11}}_{[1\times40]} f_t^{10}\left(\underbrace{\mathbf{W}_{CI_c}^{10}}_{[40\times40]} \cdots f_t^{1}\left(\underbrace{\mathbf{W}_{CI_c}^{1}}_{[40\times20]}\underbrace{g_{CI_c}^{N}(\mathbf{x})}_{[20\times1]} + \underbrace{\mathbf{b}_{CI_c}^{1}}_{[40\times1]}\right)\cdots + \underbrace{\mathbf{b}_{CI_c}^{10}}_{[40\times1]}\right) + \underbrace{b_{CI_c}^{11}}_{[1\times1]}\right)\right) \tag{10.3.2}$$

$$\underbrace{f_{CO_2}^{ANN}}_{[1\times1]} = g_{CO_2}^{D}\left(f_l^{11}\left(\underbrace{\mathbf{W}_{CO_2}^{11}}_{[1\times50]} f_t^{10}\left(\underbrace{\mathbf{W}_{CO_2}^{10}}_{[50\times50]} \cdots f_t^{1}\left(\underbrace{\mathbf{W}_{CO_2}^{1}}_{[50\times20]}\underbrace{g_{CO_2}^{N}(\mathbf{x})}_{[20\times1]} + \underbrace{\mathbf{b}_{CO_2}^{1}}_{[50\times1]}\right)\cdots + \underbrace{\mathbf{b}_{CO_2}^{10}}_{[50\times1]}\right) + \underbrace{b_{CO_2}^{11}}_{[1\times1]}\right)\right) \tag{10.3.3}$$

$$\underbrace{f_{W_c}^{ANN}}_{[1\times1]} = g_{W_C}^{D}\left(f_l^{11}\left(\underbrace{\mathbf{W}_{W_c}^{11}}_{[1\times40]} f_t^{10}\left(\underbrace{\mathbf{W}_{W_c}^{10}}_{[40\times40]} \cdots f_t^{1}\left(\underbrace{\mathbf{W}_{W_c}^{1}}_{[40\times20]}\underbrace{g_{W_c}^{N}(\mathbf{x})}_{[20\times1]} + \underbrace{\mathbf{b}_{W_c}^{1}}_{[40\times1]}\right)\cdots + \underbrace{\mathbf{b}_{W_c}^{10}}_{[40\times1]}\right) + \underbrace{b_{W_c}^{11}}_{[1\times1]}\right)\right) \tag{10.3.4}$$

$$f_{UFO} = w_{CIc}\frac{f_{CI_c}^{ANN} - CI_c^{\min}}{CI_c^{\max} - CI_c^{\min}} + w_{CO_2}\frac{f_{CO_2}^{ANN} - CO_2^{\min}}{CO_2^{\max} - CO_2^{\min}} + w_{W_c}\frac{f_{W_c}^{ANN} - W_c^{\min}}{W_c^{\max} - W_c^{\min}} \tag{10.3.5}$$

Hong and Nguyen [10.23] calculated a Lagrange function $\mathcal{L}$ as a function of input variable $\mathbf{x} = [x_1, x_2, \ldots, x_n]^T$, equality $\mathbf{c}(\mathbf{x})$ and inequality $\mathbf{v}(\mathbf{x})$ constraints, and Lagrange multipliers corresponding to equality constraints $\lambda_c = [\lambda_1, \lambda_2, \ldots, \lambda_m]^T$ and inequality constrains $\lambda_v = [\lambda_1, \lambda_2, \ldots, \lambda_l]^T$, respectively, as shown in Eq. (10.3.6). The 20 forward input variables $\mathrm{x} = [x_1, x_2, \ldots, x_n]^T$ are given as $[h,\ b,\ h_s,\ b_s,\ t_w,\ t_f,\ f_c',\ f_{yr},\ f_{ys},\ \rho_{rX},\ \rho_{rY},\ P_{u,1},\ M_{uX,1}, M_{uY,1},\ P_{u,2},\ M_{uX,2},\ M_{uY,2},\ P_{u,3},\ M_{uX,3},\ M_{uY,3}]^T$ when three biaxial load pairs are considered as explained in Figures 10.2.1(c), 10.2.4, and Table 10.3.1 where a number of input and output parameters are $(11+3i)$ and $(8+3i)$, respectively. Eleven input parameters and eight output parameters with red color shown in Table 10.3.1(a) and (b) are the parameters which do not change whereas a number of load pairs are subject to changes. Newton–Rapson iteration is used to solve partial differential equations $\nabla\mathcal{L}(\mathrm{x}, \lambda_c, \lambda_v)$ in order to find stationary points of Lagrange function, $\mathcal{L}(\mathrm{x}, \lambda_c^T, \lambda_v^T)$ with respect to 20 input variables x, Lagrange multiplier of equality constraints λ_c, and

Lagrange multiplier of inequality constraints, λ_v. MATLAB [10.38] and Hong and Nguyen et al. [10.23] derive stationary points of a Lagrange function using Newton–Raphson technique as summarized in Eqs. (10.3.7)–(10.3.9). Jacobian matrix of constrain vectors $\mathbf{c}(\mathbf{x})$ and $\mathbf{v}(\mathbf{x})$ are represented by $\boldsymbol{J}_c(\mathbf{x})$ and $\boldsymbol{J}_v(\mathbf{x})$, respectively. Both Lagrange multipliers (λ_c and λ_v) for equality and inequality constraints are applied to convert constrained optimization problems into unconstrained problems. As shown in Eq. (10.3.7), $\left[\boldsymbol{H}_{\mathcal{L}}\left(\mathbf{x}^{(k)},\lambda_c^{(k)},\lambda_v^{(k)}\right)\right]$ is a Hessian matrix of Lagrange function with respect to $\mathbf{x}^{(k)},\lambda_c^{(k)},\lambda_v^{(k)}$; $\nabla\mathcal{L}\left(\mathbf{x}^{(k)},\lambda_c^{(k)},\lambda_v^{(k)}\right)$ is a Jacobian matrix of Lagrange function which is the first derivation of Lagrange function [10.23].

$$\mathcal{L}\left(\mathbf{x},\lambda_c^T,\lambda_v^T\right)=f(\mathbf{x})-\lambda_c^T\mathbf{c}(\mathbf{x})-\lambda_v^T\mathbf{S}\mathbf{v}(\mathbf{x}) \quad (10.3.6)$$

$$\begin{bmatrix} \mathbf{x}^{(k+1)} \\ \lambda_c^{(k+1)} \\ \lambda_v^{(k+1)} \end{bmatrix}=\begin{bmatrix} \mathbf{x}^{(k)} \\ \lambda_c^{(k)} \\ \lambda_v^{(k)} \end{bmatrix}-\left[\boldsymbol{H}_{\mathrm{L}}\left(\mathbf{x}^{(k)},\lambda_c^{(k)},\lambda_v^{(k)}\right)\right]^{-1}\nabla\mathrm{L}\left(\mathbf{x}^{(k)},\lambda_c^{(k)},\lambda_v^{(k)}\right) \quad (10.3.7)$$

$$\nabla\mathcal{L}(\mathbf{x},\lambda_c,\lambda_v)=\begin{bmatrix} \nabla f(\mathbf{x})-\boldsymbol{J}_c(\mathbf{x})^T\lambda_c-\boldsymbol{J}_v(\mathbf{x})^T\mathbf{S}\lambda_v \\ -\mathbf{c}(\mathbf{x}) \\ -Sv(\mathbf{x}) \end{bmatrix} \quad (10.3.8)$$

$$\boldsymbol{J}_c(\mathbf{x})=\begin{bmatrix} \nabla c_1(\mathbf{x}) \\ \nabla c_2(\mathbf{x}) \\ \vdots \\ \nabla c_m(\mathbf{x}) \end{bmatrix} \text{ and } \boldsymbol{J}_v(\mathbf{x})=\begin{bmatrix} \nabla v_1(\mathbf{x}) \\ \nabla v_2(\mathbf{x}) \\ \vdots \\ \nabla v_l(\mathbf{x}) \end{bmatrix} \quad (10.3.9)$$

10.3.3 Formulation of weigh matrices for concrete columns encasing H-shaped steel sustaining multi-biaxial loads

Table 10.2.1 lists the design parameters used in this chapter. Figure 10.2.1 shows an SRC column subjected to a biaxial load pair. SRC columns should be designed to resist all combinations of flexural moments and axial forces, whereas beam designs meet flexural moment requirements only. ANNs are used to establish objective functions for a column cost index (CI_C), CO_2 emissions (CO_2), and column weight (W_c) when being subject to multiple biaxial load pairs. Design parameters including column and steel sections are determined while optimizing objective functions. Two types of ANNs are considered in which the first ANN subject to only one biaxial load pair (P_u, M_{uX}, M_{uY}) is referred to as *ANN-1LP*. Large datasets with one biaxial load pair are generated based on a strain compatibility proposed by Nguyen and Hong et al. [10.16] shown in Figure 10.2.2. *ANN-1LP* comprises an input layer with 14 input parameters $(h, b, h_s, b_s, t_w, t_f, f_c', f_{yr}, f_{ys}, \rho_{rX}, \rho_{rY}, P_u, M_{uX}, M_{uY})$ including three parameters defining one biaxial load pair and an output layer with 11 output parameters $\left(SF, \varepsilon_r, \varepsilon_s, \rho_s, D_{r,\max}, Y_s, X_s, {}^{b}\!/_{h}, CI_c, CO_2, W_c\right)$ including three parameters defining safety factors, steel strains and rebar strains corresponding to the biaxial load pair as shown in Figure 10.2.1(c) and Table 10.3.1. The ANN forward network subject to one load pair (*ANN-1LP)* is presented in Figure 10.3.1(a), whereas the ANN forward network subject to multiple *n* biaxial load pairs are referred to as *ANN-nLP* which is obtained by reusing a weight matrix obtained by training *ANN-1LP*, which is presented in Figure 10.3.1(a) [10.32].

When three multiple biaxial load pairs $\left(P_{u,i}, M_{uX,i}, M_{uY,i}; i = 1,2,3\right)$ are considered, the dimensions of output parameters including safety factors and strains $\left(SF_i, \varepsilon_{r,i}, \varepsilon_{s,i}; i = 1,2,3\right)$ are accordingly affected. It is noted that other eight output parameters $\left(\rho_s, D_{r,\max}, Y_s, X_s, {}^{b}\!/_{h}, CI_c, CO_2, W_c\right)$ of the weight matrix shown in Figure 10.3.1(b) are not affected at the first hidden layer. For example, for *ANN-1LP* being subject to one biaxial load pair, weight matrix of CI_c at Hidden Layer 1 obtained with the best training shown in Table 10.2.4 is 40×14 because there are 14 input parameters based on 40 neurons.

Figure 10.3.1(a) shows an *ANN-1LP*, which is subject to one biaxial load pair (P_u, M_{uX}, M_{uY}). The input layer includes 14 input parameters $(h, b, h_s, b_s, t_w, t_f, f_c', f_{yr}, f_{ys}, \rho_{rX}, \rho_{rY}, P_u, M_{uX}, M_{uY}; (11 + i = 11 + 3 = 14))$, and the output layer includes 11 output parameters $\left(SF, \varepsilon_r, \varepsilon_s, \rho_s, D_{r,\max}, Y_s, X_s, {}^{b}\!/_{h}, CI_c, CO_2, W_c; (8 + i = 8 + 3 = 11)\right)$. Table 10.2.4 shows that the optimal training of *ANN-1LP* among nine training runs achieved with 8, 9, and 10 hidden layers with 40, 50, and 60 neurons. Table 10.3.2 presents matrices of weight and bias matrices connecting the input layer and the first hidden layer of a forward network for CI_c, CO_2, and W_c. For example, for *ANN-1LP* being subject to one biaxial load pair, weight matrix of CI_c at Hidden Layer 1 obtained with the best training shown in Table 10.2.4 is 40×14 because there are 14 input parameters based on 40 neurons. Three load pairs considered in *ANN-3LP* lead to 20 input parameters, resulting in the dimension of weight matrix of CI_c at Hidden Layer 1 is 40×20. Thus, reusing a part of weight matrix from *ANN-1LP* associated with one biaxial load pair shown in Figure 10.3.1(a) to obtain *ANN-nLP* can help reduce the computations and time needed to train the ANN with three biaxial load pairs shown in Figure 10.3.1(b). The reusing of a part of weight matrix subject to single factored load (*ANN-1LP*) facilitates a process for obtaining ANNs subject to multiple factored loads (*ANN-nLP*), which facilitates the computational process. *ANN-nLP*

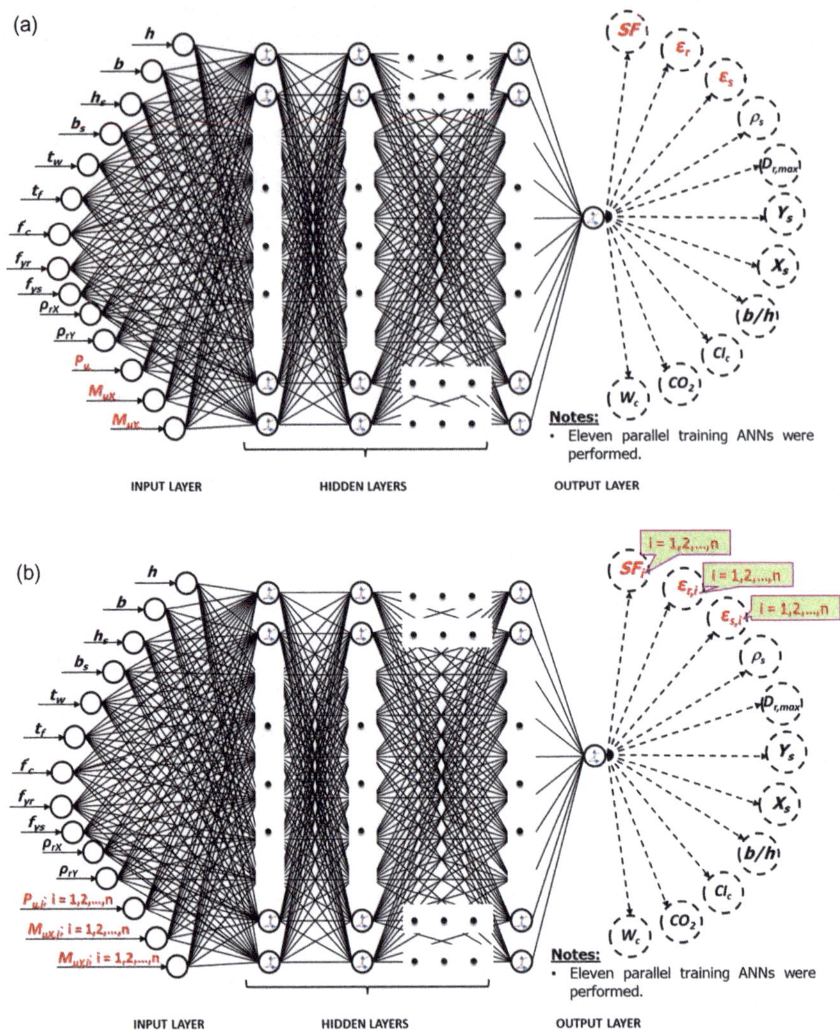

Figure 10.3.1 ANN forward network. (a) ANN forward network considering one load pair. (b) ANN forward network considering multiple load pairs.

can, then, be used to generalize objective functions for CI_C, CO_2, and W_c. Lagrange multipliers are applied for optimizing the ANN-based objective functions (optimized CI_C, CO_2, and W_c) to yield design parameters for columns. ANN forward network sustaining n multiple load pairs is presented in Figure 10.3.1(b), whereas an ANN sustaining three load pairs is presented in Figure 10.3.2 which illustrates an *ANN-3LP* subject to three biaxial load pairs $\left(P_{u,i},\ M_{uX,i}, M_{uY,i}; i=1,\ 2,\ 3\right)$ where $n=3$.

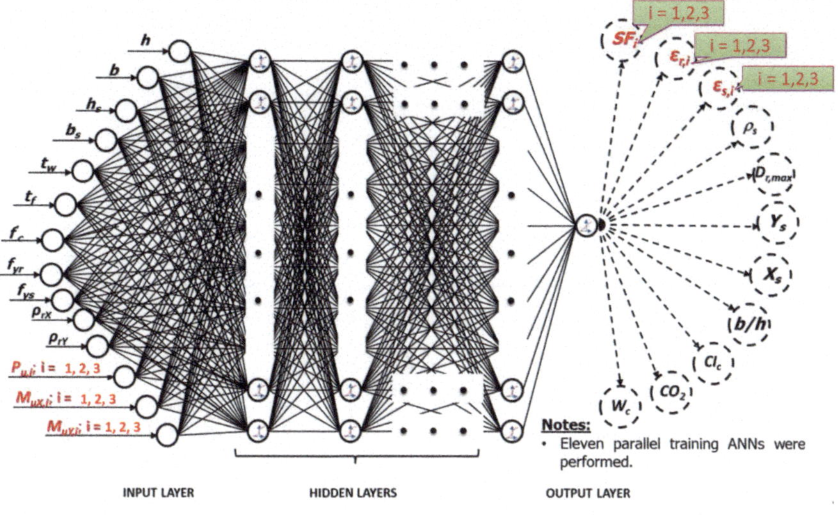

Figure 10.3.2 ANN forward network considering three biaxial load pairs.

10.3.4 Generalized *ANN-nLP* subject to n multiple biaxial load pairs based on network module

10.3.4.1 Generalized ANN-nLP *from* ANN-1LP

As shown in red input parameters of Table 10.3.1(b), the input parameters of *ANN-nLP* sustaining multiple n biaxial load pairs $\left(P_{u,i},\ M_{uX,i}, M_{uY,i};\ i = 1, 2,\ldots,n\right)$ are based on 11 ordinary input parameters $\left(h,\ b,\ h_s,\ b_s,\ t_w,\ t_f,\ f_c',\ f_{yr},\ f_{ys},\ \rho_{rX},\ \rho_{rY}\right)$. The output layer of *ANN-nLP* includes the parameters $\left(SF_i,\ \varepsilon_{r,i},\ \varepsilon_{s,i};\ i = 1,2,\ldots,n\right)$ corresponding to each biaxial load pair over the eight ordinary red output parameters $\left(\rho_s,\ D_{r,\max},\ Y_s,\ X_s,\ {}^{b}\!/_{h},\ CI_c,\ CO_2,\ W_c\right)$ which is shown in Table 10.3.1(b) and Figure 10.3.1(b). PTM is then used to extract the following ANN-based function shown in Eq. (10.3.10) [10.23].

$$f(\mathbf{x}) = g_{\mathbf{x}}^{D}\left(f_l^{N}\left(\mathbf{W}_{\mathbf{x}}^{N} f_t^{N-1}\left(\mathbf{W}_{\mathbf{x}}^{N-1} \ldots f_t^{1}\left(\mathbf{W}_{\mathbf{x}}^{1} g_{\mathbf{x}}^{N}(\mathbf{x}) + \mathbf{b}^{1}\right) \ldots + \mathbf{b}^{N-1}\right) + b^{N}\right)\right) \quad (10.3.10)$$

where $\mathbf{x}$ is an input vector; N is a number of layers including hidden layers and output layer; $\mathbf{W}_{\mathbf{x}}^{n}$ is a weight matrix between layer $n-1$ and layer n; $\mathbf{b}^{n}$ is a bias matrix added to neurons of layer n; and $g_{\mathbf{x}}^{N}$ and $g_{\mathbf{x}}^{D}$ are a normalization and de-normalization function, respectively. Activation functions f_t^{n} are implemented at layer n to formulate non-linear relationships of networks. A linear activation function f_l^{n} is then selected for an output layer because

Table 10.3.2 Weight and bias matrices of a forward network for objective functions (CI_c, CO_2, and W_c) considering three biaxial load pairs

(a) For cost index of columns (CI_c)

$\mathbf{W}_{[40\times20]}$									$\mathbf{b}_{[40\times1]}$
−0.314	−0.308	−0.066	0.129	⋯	0.000	−0.156	0.000	0.000	1.841
0.009	−0.653	0.021	0.187	⋯	0.000	−0.073	0.000	0.000	−1.754
−0.148	0.172	0.143	0.605	⋯	0.000	−0.890	0.000	0.000	1.793
0.562	0.010	0.193	−0.299	⋯	0.000	0.535	0.000	0.000	−1.753
⋮	⋮	⋮	⋮	⋱	⋮	⋮	⋮	⋮	⋮
−0.079	0.616	−0.110	−0.150	⋯	0.000	0.544	0.000	0.000	1.427
−0.306	−0.462	−0.385	−0.157	⋯	0.000	0.859	0.000	0.000	−1.888
0.316	0.579	0.056	0.286	⋯	0.000	−0.380	0.000	0.000	1.707
0.266	−0.020	0.099	0.056	⋯	0.000	0.555	0.000	0.000	1.864
								$[40\times20]$	

(b) For CO_2 emissions

$\mathbf{W}_{[50\times20]}$									$\mathbf{b}_{[50\times1]}$
0.153	−0.592	0.698	0.216	⋯	0.000	0.756	0.000	0.000	−1.937
−0.528	0.832	0.143	0.221	⋯	0.000	−0.448	0.000	0.000	1.715
−0.744	0.554	0.082	0.143	⋯	0.000	−0.036	0.000	0.000	1.832
−0.295	−0.154	−0.693	0.405	⋯	0.000	−0.292	0.000	0.000	1.934
⋮	⋮	⋮	⋮	⋱	⋮	⋮	⋮	⋮	⋮
0.401	−0.554	−0.390	−0.517	⋯	0.000	−0.172	0.000	0.000	1.881
0.343	−0.408	−0.139	0.096	⋯	0.000	−0.400	0.000	0.000	1.999
0.419	−0.637	−0.138	−0.136	⋯	0.000	0.353	0.000	0.000	1.873
0.547	0.043	0.092	0.046	⋯	0.000	0.063	0.000	0.000	1.757
								$[50\times20]$	

(Continued)

Table 10.3.2 (Continued) Weight and bias matrices of a forward network for objective functions (CI_c, CO_2, and W_c) considering three biaxial load pairs

(c) For column weight (W_c)									
$\mathbf{W}_{[40\times20]}$									$\mathbf{b}_{[40\times1]}$
−0.848	1.079	−0.136	−0.211	⋯	0.000	−0.161	0.000	0.000	1.783
−0.031	−0.395	−0.072	0.143	⋯	0.000	0.091	0.000	0.000	−1.949
0.015	0.973	0.097	0.262	⋯	0.000	0.395	0.000	0.000	1.783
−0.381	−0.193	0.265	0.567	⋯	0.000	−0.020	0.000	0.000	1.833
⋮	⋮	⋮	⋮	⋱	⋮	⋮		⋮ ⋮	⋮
0.367	0.389	0.153	0.165	⋯	0.000	0.316	0.000	0.000	1.630
−0.118	0.692	−0.693	0.540	⋯	0.000	0.207	0.000	0.000	−1.952
0.861	0.769	−0.173	−0.195	⋯	0.000	−0.820	0.000	0.000	1.809
−0.336	−0.705	0.190	−0.112	⋯	0.000	−0.662	0.000	0.000	−1.861
$[40\times20]$									

outputs are unbounded. By PTM, the input layer of *ANN-1LP* is trained by mapping input parameters including one biaxial load pair (P_u, M_{uX}, M_{uY}) to three associated output parameters [SF, rebar strain (ε_r), and steel strain (ε_s)] and eight fixed outputs $\left(\rho_s, D_{r,\max}, Y_s, X_s, {}^{b}\!/_{h}, CI_c, CO_2, W_c\right)$. Note that only SF, ε_r, and ε_s are affected when a number of load pairs changes. ANN-based functions of SF , ε_r, and ε_s are derived in Eqs. (10.3.11)–(10.3.13), respectively. The training results of SF, ε_r, and ε_s are shown in Table 10.2.4.

$$\underbrace{SF}_{[1\times1]} = g_{SF}^{D}\left(f_l^{11}\left(\underbrace{\mathbf{W}_{SF}^{11}}_{[1\times50]} f_t^{10}\left(\underbrace{\mathbf{W}_{SF}^{10}}_{[50\times50]} \ldots f_t^{1}\left(\underbrace{\mathbf{W}_{SF}^{1}}_{[50\times14]}\underbrace{g_{SF}^{N}(\mathbf{x})}_{[14\times1]} + \underbrace{\mathbf{b}_{SF}^{1}}_{[50\times1]}\right)\ldots + \underbrace{\mathbf{b}_{SF}^{10}}_{[50\times1]}\right) + \underbrace{b_{SF}^{11}}_{[1\times1]}\right)\right) \tag{10.3.11}$$

$$\underbrace{\varepsilon_r}_{[1\times1]} = g_{\varepsilon_r}^{D}\left(f_l^{9}\left(\underbrace{\mathbf{W}_{\varepsilon_r}^{9}}_{[1\times50]} f_t^{8}\left(\underbrace{\mathbf{W}_{\varepsilon_r}^{8}}_{[50\times50]} \ldots f_t^{1}\left(\underbrace{\mathbf{W}_{\varepsilon_r}^{1}}_{[50\times14]}\underbrace{g_{\varepsilon_r}^{N}(\mathbf{x})}_{[14\times1]} + \underbrace{\mathbf{b}_{\varepsilon_r}^{1}}_{[50\times1]}\right)\ldots + \underbrace{\mathbf{b}_{\varepsilon_r}^{8}}_{[50\times1]}\right) + \underbrace{b_{\varepsilon_r}^{9}}_{[1\times1]}\right)\right) \tag{10.3.12}$$

$$\underbrace{\varepsilon_s}_{[1\times1]} = g_{\varepsilon_s}^{D}\left(f_l^{11}\left(\underbrace{\mathbf{W}_{\varepsilon_s}^{11}}_{[1\times60]} f_t^{10}\left(\underbrace{\mathbf{W}_{\varepsilon_s}^{10}}_{[60\times60]} \ldots f_t^{1}\left(\underbrace{\mathbf{W}_{\varepsilon_s}^{1}}_{[60\times14]}\underbrace{g_{\varepsilon_s}^{N}(\mathbf{x})}_{[14\times1]} + \underbrace{\mathbf{b}_{\varepsilon_s}^{1}}_{[60\times1]}\right)\ldots + \underbrace{\mathbf{b}_{\varepsilon_s}^{10}}_{[60\times1]}\right) + \underbrace{b_{\varepsilon_s}^{11}}_{[1\times1]}\right)\right) \tag{10.3.13}$$

10.3.4.2 Formulation of generalized ANN-nLP considering two biaxial load pairs by network duplication

A weight matrix of ANN subject to n multiple biaxial load pairs $(P_{u,i},\ M_{uX,i},\ M_{uY,i}; i=1,2,\ldots,n)$ at for each output parameter $(SF_i,\ \varepsilon_{r,i},\ \varepsilon_{s,i};\ i=1,2,\ldots,\ n)$ can be obtained by reusing the components of weight matrix corresponding to one biaxial load pair (P_u, M_{uX}, M_{uY}), avoiding training ANNs to derive weight matrices with multiple biaxial load pairs. The weight matrix of the first hidden layer $\left(W_{SF}^{1}\right)$ subjected to one biaxial load pair to predict the safety factor (SF_1) of a network (*ANN-1LP*) is given in Eq. (10.3.14). The weight matrix W_{SF}^{1} has a dimension of 50 × 14, which corresponds to 50 neurons used for mapped from 14 $(11+i=11+3=14)$ input parameters to SF using PTM. It is noted that, to predict SF_1, 12th component corresponding to P_u, 13th component corresponding to M_{uX}, and 14th component corresponding to M_{uY} for weight matrix W_{SF}^{1} at the first hidden layer are derived by training ANNs when an SRC column is subject to one biaxial load pair (P_u, M_{uX}, M_{uY}).

$$W_{SF}^{1} = \begin{pmatrix} w_{1,1}^{ANN-1LP} & w_{1,2}^{ANN-1LP} & w_{1,3}^{ANN-1LP} & \cdots & w_{1,12}^{ANN-1LP} & w_{1,13}^{ANN-1LP} & w_{1,14}^{ANN-1LP} \\ w_{2,1}^{ANN-1LP} & w_{2,2}^{ANN-1LP} & w_{2,3}^{ANN-1LP} & \cdots & w_{2,12}^{ANN-1LP} & w_{2,13}^{ANN-1LP} & w_{2,14}^{ANN-1LP} \\ \vdots & \vdots & \vdots & \cdots & \vdots & \vdots & \vdots \\ w_{49,1}^{ANN-1LP} & w_{49,2}^{ANN-1LP} & w_{49,3}^{ANN-1LP} & \cdots & w_{49,12}^{ANN-1LP} & w_{49,13}^{ANN-1LP} & w_{49,14}^{ANN-1LP} \\ w_{50,1}^{ANN-1LP} & w_{50,2}^{ANN-1LP} & w_{50,3}^{ANN-1LP} & \cdots & w_{50,12}^{ANN-1LP} & w_{50,13}^{ANN-1LP} & w_{50,14}^{ANN-1LP} \end{pmatrix}_{[50\times14]} \quad (10.3.14)$$

$$W_{SF1}^{1} = \begin{pmatrix} w_{1,1}^{ANN-1LP} & w_{1,2}^{ANN-1LP} & w_{1,3}^{ANN-1LP} & \cdots & w_{1,12}^{ANN-1LP} & 0 & w_{1,13}^{ANN-1LP} & 0 & w_{1,14}^{ANN-1LP} & 0 \\ w_{2,1}^{ANN-1LP} & w_{2,2}^{ANN-1LP} & w_{2,3}^{ANN-1LP} & \cdots & w_{2,12}^{ANN-1LP} & 0 & w_{2,13}^{ANN-1LP} & 0 & w_{2,14}^{ANN-1LP} & 0 \\ \vdots & \vdots & \vdots & \cdots & \vdots & \vdots & \vdots & \vdots & \vdots & \vdots \\ w_{49,1}^{ANN-1LP} & w_{49,2}^{ANN-1LP} & w_{49,3}^{ANN-1LP} & \cdots & w_{49,12}^{ANN-1LP} & 0 & w_{49,13}^{ANN-1LP} & 0 & w_{49,14}^{ANN-1LP} & 0 \\ w_{50,1}^{ANN-1LP} & w_{50,2}^{ANN-1LP} & w_{50,3}^{ANN-1LP} & \cdots & w_{50,12}^{ANN-1LP} & 0 & w_{50,13}^{ANN-1LP} & 0 & w_{50,14}^{ANN-1LP} & 0 \end{pmatrix}_{[50\times17]^{*}} \quad (10.3.15)$$

$$W_{SF2}^{1} = \begin{bmatrix} w_{1,1}^{ANN-1LP} & w_{1,2}^{ANN-1LP} & w_{1,3}^{ANN-1LP} & \cdots & 0 & w_{1,12}^{ANN-1LP} & 0 & w_{1,13}^{ANN-1LP} & 0 & w_{1,14}^{ANN-1LP} \\ w_{2,1}^{ANN-1LP} & w_{2,2}^{ANN-1LP} & w_{2,3}^{ANN-1LP} & \cdots & 0 & w_{2,12}^{ANN-1LP} & 0 & w_{2,13}^{ANN-1LP} & 0 & w_{2,14}^{ANN-1LP} \\ \vdots & \vdots & \vdots & \cdots & \vdots & \vdots & \vdots & \vdots & \vdots & \vdots \\ w_{49,1}^{ANN-1LP} & w_{49,2}^{ANN-1LP} & w_{49,3}^{ANN-1LP} & \cdots & 0 & w_{49,12}^{ANN-1LP} & 0 & w_{49,13}^{ANN-1LP} & 0 & w_{49,14}^{ANN-1LP} \\ w_{50,1}^{ANN-1LP} & w_{50,2}^{ANN-1LP} & w_{50,3}^{ANN-1LP} & \cdots & 0 & w_{50,12}^{ANN-1LP} & 0 & w_{50,13}^{ANN-1LP} & 0 & w_{50,14}^{ANN-1LP} \end{bmatrix}_{[50\times17]^{*}} \quad (10.3.16)$$

$$W^1_{\varepsilon_r} = \begin{bmatrix} w^{ANN-1LP}_{1,1} & w^{ANN-1LP}_{1,2} & w^{ANN-1LP}_{1,3} & \cdots & w^{ANN-1LP}_{1,12} & w^{ANN-1LP}_{1,13} & w^{ANN-1LP}_{1,14} \\ w^{ANN-1LP}_{2,1} & w^{ANN-1LP}_{2,2} & w^{ANN-1LP}_{2,3} & \cdots & w^{ANN-1LP}_{2,12} & w^{ANN-1LP}_{2,13} & w^{ANN-1LP}_{2,14} \\ \vdots & \vdots & \vdots & \cdots & \vdots & \vdots & \vdots \\ w^{ANN-1LP}_{49,1} & w^{ANN-1LP}_{49,2} & w^{ANN-1LP}_{49,3} & \cdots & w^{ANN-1LP}_{49,12} & w^{ANN-1LP}_{49,13} & w^{ANN-1LP}_{49,14} \\ w^{ANN-1LP}_{50,1} & w^{ANN-1LP}_{50,2} & w^{ANN-1LP}_{50,3} & \cdots & w^{ANN-1LP}_{50,12} & w^{ANN-1LP}_{50,13} & w^{ANN-1LP}_{50,14} \end{bmatrix}_{[50\times14]} \tag{10.3.17}$$

$$W^1_{\varepsilon_{r,1}} = \begin{bmatrix} w^{ANN-1LP}_{1,1} & w^{ANN-1LP}_{1,2} & w^{ANN-1LP}_{1,3} & \cdots & w^{ANN-1LP}_{1,12} & 0 & w^{ANN-1LP}_{1,13} & 0 & w^{ANN-1LP}_{1,14} & 0 \\ w^{ANN-1LP}_{2,1} & w^{ANN-1LP}_{2,2} & w^{ANN-1LP}_{2,3} & \cdots & w^{ANN-1LP}_{2,12} & 0 & w^{ANN-1LP}_{2,13} & 0 & w^{ANN-1LP}_{2,14} & 0 \\ \vdots & \vdots & \vdots & \cdots & \vdots & \vdots & \vdots & \vdots & \vdots & \vdots \\ w^{ANN-1LP}_{49,1} & w^{ANN-1LP}_{49,2} & w^{ANN-1LP}_{49,3} & \cdots & w^{ANN-1LP}_{49,12} & 0 & w^{ANN-1LP}_{49,13} & 0 & w^{ANN-1LP}_{49,14} & 0 \\ w^{ANN-1LP}_{50,1} & w^{ANN-1LP}_{50,2} & w^{ANN-1LP}_{50,3} & \cdots & w^{ANN-1LP}_{50,12} & 0 & w^{ANN-1LP}_{50,13} & 0 & w^{ANN-1LP}_{50,14} & 0 \end{bmatrix}_{[50\times17]^*} \tag{10.3.18}$$

$$W^1_{\varepsilon_{r,2}} = \begin{bmatrix} w^{ANN-1LP}_{1,1} & w^{ANN-1LP}_{1,2} & w^{ANN-1LP}_{1,3} & \cdots & 0 & w^{ANN-1LP}_{1,12} & 0 & w^{ANN-1LP}_{1,13} & 0 & w^{ANN-1LP}_{1,14} \\ w^{ANN-1LP}_{2,1} & w^{ANN-1LP}_{2,2} & w^{ANN-1LP}_{2,3} & \cdots & 0 & w^{ANN-1LP}_{2,12} & 0 & w^{ANN-1LP}_{2,13} & 0 & w^{ANN-1LP}_{2,14} \\ \vdots & \vdots & \vdots & \cdots & \vdots & \vdots & \vdots & \vdots & \vdots & \vdots \\ w^{ANN-1LP}_{49,1} & w^{ANN-1LP}_{49,2} & w^{ANN-1LP}_{49,3} & \cdots & 0 & w^{ANN-1LP}_{49,12} & 0 & w^{ANN-1LP}_{49,13} & 0 & w^{ANN-1LP}_{49,14} \\ w^{ANN-1LP}_{50,1} & w^{ANN-1LP}_{50,2} & w^{ANN-1LP}_{50,3} & \cdots & 0 & w^{ANN-1LP}_{50,12} & 0 & w^{ANN-1LP}_{50,13} & 0 & w^{ANN-1LP}_{50,14} \end{bmatrix}_{[50\times17]^*} \tag{10.3.19}$$

$$W_{\varepsilon_s}^{1} = \begin{bmatrix} w_{1,1}^{ANN-1LP} & w_{1,2}^{ANN-1LP} & w_{1,3}^{ANN-1LP} & \cdots & w_{1,12}^{ANN-1LP} & w_{1,13}^{ANN-1LP} & w_{1,14}^{ANN-1LP} \\ w_{2,1}^{ANN-1LP} & w_{2,2}^{ANN-1LP} & w_{2,3}^{ANN-1LP} & \cdots & w_{2,12}^{ANN-1LP} & w_{2,13}^{ANN-1LP} & w_{2,14}^{ANN-1LP} \\ \vdots & \vdots & \vdots & \cdots & \vdots & \vdots & \vdots \\ w_{59,1}^{ANN-1LP} & w_{59,2}^{ANN-1LP} & w_{59,3}^{ANN-1LP} & \cdots & w_{59,12}^{ANN-1LP} & w_{59,13}^{ANN-1LP} & w_{59,14}^{ANN-1LP} \\ w_{60,1}^{ANN-1LP} & w_{60,2}^{ANN-1LP} & w_{60,3}^{ANN-1LP} & \cdots & w_{60,12}^{ANN-1LP} & w_{60,13}^{ANN-1LP} & w_{60,14}^{ANN-1LP} \end{bmatrix}_{[60\times14]} \quad (10.3.20)$$

$$W_{\varepsilon_{s,1}}^{1} = \begin{bmatrix} w_{1,1}^{ANN-1LP} & w_{1,2}^{ANN-1LP} & w_{1,3}^{ANN-1LP} & \cdots & w_{1,12}^{ANN-1LP} & 0 & w_{1,13}^{ANN-1LP} & 0 & w_{1,14}^{ANN-1LP} & 0 \\ w_{2,1}^{ANN-1LP} & w_{2,2}^{ANN-1LP} & w_{2,3}^{ANN-1LP} & \cdots & w_{2,12}^{ANN-1LP} & 0 & w_{2,13}^{ANN-1LP} & 0 & w_{2,14}^{ANN-1LP} & 0 \\ \vdots & \vdots & \vdots & \cdots & \vdots & \vdots & \vdots & \vdots & \vdots & \vdots \\ w_{59,1}^{ANN-1LP} & w_{59,2}^{ANN-1LP} & w_{59,3}^{ANN-1LP} & \cdots & w_{59,12}^{ANN-1LP} & 0 & w_{59,13}^{ANN-1LP} & 0 & w_{59,14}^{ANN-1LP} & 0 \\ w_{60,1}^{ANN-1LP} & w_{60,2}^{ANN-1LP} & w_{60,3}^{ANN-1LP} & \cdots & w_{50,12}^{ANN-1LP} & 0 & w_{60,13}^{ANN-1LP} & 0 & w_{60,14}^{ANN-1LP} & 0 \end{bmatrix}_{[60\times17]^*} \quad (10.3.21)$$

$$W_{\varepsilon_{s,2}}^{1} = \begin{bmatrix} w_{1,1}^{ANN-1LP} & w_{1,2}^{ANN-1LP} & w_{1,3}^{ANN-1LP} & \cdots & 0 & w_{1,12}^{ANN-1LP} & 0 & w_{1,13}^{ANN-1LP} & 0 & w_{1,14}^{ANN-1LP} \\ w_{2,1}^{ANN-1LP} & w_{2,2}^{ANN-1LP} & w_{2,3}^{ANN-1LP} & \cdots & 0 & w_{2,12}^{ANN-1LP} & 0 & w_{2,13}^{ANN-1LP} & 0 & w_{2,14}^{ANN-1LP} \\ \vdots & \vdots & \vdots & \cdots & \vdots & \vdots & \vdots & \vdots & \vdots & \vdots \\ w_{59,1}^{ANN-1LP} & w_{59,2}^{ANN-1LP} & w_{59,3}^{ANN-1LP} & \cdots & 0 & w_{59,12}^{ANN-1LP} & 0 & w_{59,13}^{ANN-1LP} & 0 & w_{59,14}^{ANN-1LP} \\ w_{60,1}^{ANN-1LP} & w_{60,2}^{ANN-1LP} & w_{60,3}^{ANN-1LP} & \cdots & 0 & w_{60,12}^{ANN-1LP} & 0 & w_{60,13}^{ANN-1LP} & 0 & w_{60,14}^{ANN-1LP} \end{bmatrix}_{[60\times17]^*} \quad (10.3.22)$$

Let's now derive a weight matrix with two biaxial load pairs (*ANN-2LPs*) when an SRC column is subject to two biaxial load pairs ($P_{u,1}$, $P_{u,2}$, $M_{uX,1}$, $M_{uX,2}$, $M_{uY,1}$, $M_{uY,2}$). The weight matrices $\left(W^1_{SF1}\text{ and }W^1_{SF2}\right)$ of SF_1 and SF_2 shown in Eqs. (10.3.15) and (10.3.16), respectively, reused the 12th, 13th, and 14th columns of W^1_{SF} from *ANN-1LP* shown in Eq. (10.3.14) to calculate the safety factors (SF_1 and SF_2) corresponding to each of the two biaxial load pairs. Note that SF_1 is controlled only by $P_{u,1}$, $M_{uX,1}$, and $M_{uY,1}$, and hence, matrix components in the columns of W^1_{SF1} corresponding to $P_{u,2}$, $M_{uX,2}$, and $M_{uY,2}$ should be 0 because these matrix components do not affect the calculation of SF_1. Modularized weight matrices for the rebar (ε_r) and steel strains (ε_s) shown in Eqs. (10.3.17)–(10.3.22) can be also obtained by reusing the 12th, 13th, and 14th columns corresponding to each of the two biaxial load pairs of the weight matrices for rebar and steel strains (*ANN-1LP*). Similarly to modularized weight matrices of the safety factor.

Note that W^1_{SF}, $W^1_{\varepsilon_r}$, and $W^1_{\varepsilon_s}$ have dimensions of $\left[50\times(11+3n)\right]$, $\left[50\times(11+3n)\right]$, and $\left[60\times(11+3n)\right]$ where n is a number of biaxial load pairs, whereas $n=2$ in equations above. It is noted that 50 neurons are implemented in training an ANN for $\boldsymbol{SF}$ and $\boldsymbol{\varepsilon_r}$, whereas 60 neurons are implemented in training an ANN for $\boldsymbol{\varepsilon_s}$ as shown in Table 10.2.4.

10.4 DESIGN OF AN SRC COLUMN ENCASING H-SHAPED STEEL SECTION SUSTAINING THREE BIAXIAL LOAD PAIRS

10.4.1 Design scenario

10.4.1.1 Selection of design parameters and ranges based on structural analysis

Figure 10.2.1(c) illustrates a design of SRC columns with H-shaped steels. Design parameters presented in this ANN-based example include parameters shown in Table 10.2.1 and Figure 10.2.1(c) which can also be determined by conventional designs. ANN-based cost index, CO_2 emission, and column weight denoted as CI_c, CO_2, and W_c, respectively, are determined for SRC columns with H-shaped steels. The material cost based on the Korean the unit prices and the weight of SRC columns with H-shaped steels are presented in Table 10.2.2, whereas the determination of CO_2 emissions followed Hong et al. [10.39]. Table 10.2.1 presents notations and nomenclatures of the design parameters.

10.4.1.2 Establishing equality and inequality constraints imposed by design codes

MOO designs of SRC columns sustaining three pairs of the three biaxial loads ($M_{uX,1}$, $M_{uY,1}$, $P_{u,2}$, $M_{uX,2}$, $M_{uY,2}$, $P_{u,3}$, $M_{uX,3}$, $M_{uY,3}$) are investigated in which cost index (CI_c), CO_2 emissions, and column weights (W_c) are

Table 10.4.1 Equality and inequality conditions for optimization designs

(a) Equality constraints

No.	*Equality constraints*	*Description*
1	$c_1(\mathbf{x}): b/h = 1$	Square columns are considered.
2	$c_2(\mathbf{x}): f_c' = 30$ MPa	Material properties are preassigned.
3	$c_3(\mathbf{x}): f_{yr} = 500$ MPa	
4	$c_4(\mathbf{x}): f_{ys} = 325$ MPa	
5	$c_5(\mathbf{x}): P_{u,1} = 6,200$ kN	Three factored biaxial loads are considered to design SRC columns.
6	$c_6(\mathbf{x}): P_{u,2} = 12,000$ kN	
7	$c_7(\mathbf{x}): P_{u,3} = 4,500$ kN	
8	$c_8(\mathbf{x}): M_{uX,1} = 4,000$ kN·m	
9	$c_9(\mathbf{x}): M_{uX,2} = 6,500$ kN·m	
10	$c_{10}(\mathbf{x}): M_{uX,3} = 3,500$ kN·m	
11	$c_{11}(\mathbf{x}): M_{uY,1} = 5,000$ kN·m	
12	$c_{12}(\mathbf{x}): M_{uY,2} = 5,000$ kN·m	
13	$c_{13}(\mathbf{x}): M_{uY,3} = 2,000$ kN·m	
14	$c_{14}(\mathbf{x}): t_w = 8$ mm	Web and flange thicknesses are preassigned as 8 mm.
15	$c_{15}(\mathbf{x}): t_f = 8$ mm	

(b) Inequality constraints

No.	*Inequality constraints*	*Description*
1	$v_1(\mathbf{x}): SF_1 - 1 \geq 0$	Safety factors are constrained to be greater than 1.0 to meet strength requirements.
2	$v_2(\mathbf{x}): SF_2 - 1 \geq 0$	
3	$v_3(\mathbf{x}): SF_3 - 1 \geq 0$	
4	$v_4(\mathbf{x}): \rho_s - 0.01 \geq 0$	Steel ratio is restrained to be bigger than 0.01 according to ANSI/AISC 360-16 [10.38].
5	$v_5(\mathbf{x}): \rho_{rX} + \rho_{rY} - 0.004 \geq 0$	Rebar ratios are limited to be greater than 0.004 and smaller than 0.08 according to ANSI/AISC 360-16 and ACI 318-19 [10.35].
6	$v_6(\mathbf{x}): -\rho_{rX} - \rho_{rY} + 0.08 \geq 0$	
7	$v_7(\mathbf{x}): \rho_{rX} \geq 0$	Rebar ratios along to X- and Y-directions must be positive numbers.
8	$v_8(\mathbf{x}): \rho_{rY} \geq 0$	
9	$v_9(\mathbf{x}): Y_s - D_{r.max} - 75 \geq 0$	Vertical and horizontal clearances are constrained to be greater than a summation of $D_{r.max}$, clear cover (50 mm), and a distance from steel flange to rebar (25 mm).
10	$v_{10}(\mathbf{x}): X_s - D_{r.max} - 75 \geq 0$	
11	$v_{11}(\mathbf{x}): h - 500 \geq 0$	Column height is limited in a range of 500–2,000 mm.
12	$v_{12}(\mathbf{x}): -h + 2,000 \geq 0$	

(*Continued*)

Table 10.4.1 (Continued) Equality and inequality conditions for optimization designs

(b) *Inequality constraints*		
No.	*Inequality constraints*	*Description*
13	$v_{13}(\mathbf{x}): b - 500 \geq 0$	Column width is limited in a range of 500–2,000 mm.
14	$v_{14}(\mathbf{x}): -b + 2,000 \geq 0$	
15	$v_{15}(\mathbf{x}): h_s - 200 \geq 0$	Steel height is limited in a range of 200–1,000 mm.
16	$v_{16}(\mathbf{x}): -h_s + 1,000 \geq 0$	
17	$v_{17}(\mathbf{x}): b_s - 200 \geq 0$	Steel width is limited in a range of 200–1,000 mm.
18	$v_{18}(\mathbf{x}): -b_s + 1,000 \geq 0$	
19	$v_{19}(\mathbf{x}): -D_{r.max} + 64 \geq 0$	A maximum bundled rebar diameter is restrained to be smaller than 64 mm.

considered as objective functions. ANN-based objective and constraint functions are derived using weight and bias matrixes obtained by training ANNs. There are 20 forward input parameters $\left(h,\ b,\ h_s,\ b_s,\ t_w,\ t_f,\ f_c',\ f_{yr},\ f_{ys},\ \rho_{rX},\ \rho_{rY},\ P_{u,1},\ M_{uX,1},\ M_{uY,1},\ P_{u,2},\ M_{uX,2}\ M_{uY,2},\ P_{u,3},\ M_{uX,3},\ M_{uY,3}\right)$ and 17 forward output parameters $\left(SF_1,\ SF_2,\ SF_3,\ \varepsilon_{r,1},\ \varepsilon_{r,2},\ \varepsilon_{r,3},\ \varepsilon_{s,1},\ \varepsilon_{s,2},\ \varepsilon_{s,3},\ \rho_s,\ D_{r,\max},\ Y_s,\ X_s,\ {}^{b}\!/_{h},\ CI_c,\ CO_2, W_c\right)$ which are demonstrated in Figures 10.2.1(c), 10.2.4, and 10.3.2 for optimizing *UFO* when the three pairs of the biaxial loads are applied. Material properties of concrete (f_c'), rebar (f_{yr}), steel (f_{ys}) are preassigned as equality constraints when formulating Lagrange function as shown in equality constraints c_2, c_3, and c_4, respectively, shown in Table 10.4.1(a). Three pairs of the three factored biaxial loads are predetermined as equality constraint to design SRC columns. Equality constraints $c_5 - c_{34}$ shown in Table 10.4.1(a) presents three pairs of the three factored biaxial loads.

Inequality constraints shown in Table 10.4.1(b) are used to impose all requirements for SRC column designs following ACI 318-19 [10.36] and AISC 360-16 [10.35]. Three SFs shown in inequality $v_1 - v_3$ corresponding to three biaxial load pairs are controlled not to be smaller than 1.0 to ensure that columns' capacities resist factored loads sufficiently. A steel ratio of SRC columns is also constrained to be not smaller than 0.01 in accordance with Section I2.1a, ANSI/AISC 360-16 [10.35] which is presented by inequality v_4. A minimum of rebar ratios (ρ_{rX} and ρ_{rY}) is limited as 0.004 as shown in inequality v_5 in accordance with Section I2.1a, ANSI/AISC 360-16 [10.35] to avoid rebar from yielding under sustained service loads. In according to Section 10.6.1, ACI 318-19 [10.36], rebar ratios (ρ_{rX} and ρ_{rY}) are limited to be not greater than 0.08 to ensure that rebars can be consolidated sufficiently by concrete as demonstrated inequality v_6 in Table 10.4.1(b). Bundled rebars are taken into consideration when generating large datasets for SRC columns. A maximum rebar diameter is constrained as 64 mm which is a bundle of four rebar each with a diameter of 32 mm as shown inequality v_{19} in Table 10.4.1(b). *UFO* including CI_c, CO_2

Table 10.4.2 Normalized functions of objectives

Objectives	*Minimum value*	*Maximum value*	*Normalized function*
CI_c (KRW/m)	570,352	1,269,606	$f^N_{CI_c} = \dfrac{CI_c - 570{,}352}{1{,}269{,}606 - 570{,}352}$
CO_2 (t-CO_2/m)	0.947	2.563	$f^N_{CO_2} = \dfrac{CO_2 - 0.947}{2.563 - 0.947}$
W_c (kN/m)	36.55	62.15	$f^N_{W_c} = \dfrac{W_c - 36.55}{62.15 - 36.55}$

emissions, and W_c is established using tradeoff weight fractions as shown in Eq. (10.3.5). Normalized *UFO* is obtained by using max–min normalization as shown from Eqs. (10.3.2) to (10.3.4). Maxima and minima of objective functions are obtained based on Lagrange optimization as shown in Table 10.4.2 where maxima and minima of each objective function are found independently. Each objective function (CI_c, CO_2 emissions, or W_c) is accompanied with its weight fraction W_{CI_c}, W_{CO_2}, and W_{W_c}, respectively, in which a summation of these weight fraction is equal to 1.0. Weight fractions by which a set of multi-objective functions are optimized vary from 0 to 1.0 in *UFO* known as a Pareto frontier. The stationary points of *UFO* based on equality and inequality constraints shown in Table 10.4.1 are calculated using Newton–Raphson iteration, resulting in optimal designs defined as a Pareto frontier.

10.4.2 Formulation of ANN forward network subjected to three biaxial load pairs

Lagrange multipliers have been used to optimize a design of an SRC column subject to three biaxial load pairs which are ($P_{u,1}, P_{u,2}, P_{u,3}, P_{u,1}$, $P_{u,2}, P_{u,3}, M_{uX,1}, M_{uX,2}, M_{uX,3}, M_{uY,1}, M_{uY,2}$, and $M_{uY,3}$). Three output parameters [safety factors (SF_1, SF_2, and SF_3), rebar strains ($\varepsilon_{r,1}, \varepsilon_{r,2}$, and $\varepsilon_{r,3}$), and steel strains ($\varepsilon_{s,1}$, $\varepsilon_{s,2}$, and $\varepsilon_{s,3}$)] are then influenced by three biaxial load pairs. Figure 10.3.2 shows an *ANN-3LPs*. For forward design scenarios for SRC columns subject to three biaxial load pairs, Table 10.3.1(b) lists 20 forward input parameters $\left(h, b, h_s, b_s, t_w, t_f, f_c', f_y, f_{ys}, \rho_{rX}, \rho_{rY}, P_{u,1}, P_{u,2}, P_{u,3}, M_{uX,1}, M_{uX,2}, M_{uX,3}, M_{uY,1}, M_{uY,2}, M_{uY,3}\right)$ and 17 forward output parameters $\left(SF_1, SF_2, SF_3, \varepsilon_{r,1}, \varepsilon_{r,2}, \varepsilon_{r,3}, \varepsilon_{s,1}, \varepsilon_{s,2}, \varepsilon_{s,3}, \rho_s, D_{r,\max}, Y_s, X_s, b/h, CI_c, CO_2, W_c\right)$ for given three biaxial load pairs. The parameters of the SRC column sections are presented in the Nomenclature shown in Table 10.2.1. Table 10.2.3 presents the parameter statistics (maxima, means, minima, variances, and standard deviations) of the randomly generated design inputs. The column cost index (CI_C), CO_2 emissions (CO_2), and column weight (W_c) of the SRC columns are optimized based on the ANN-based objective functions which are presented in Eqs. (10.3.2)–(10.3.4), whereas equality and inequality constraints with descriptions are summarized in Table 10.4.1. Hong and Nguyen et al. [10.23] reported how stationary points of Lagrange function $\mathcal{L}$ shown in Eq. (10.4.1) are identified as a function of input variables $\mathbf{x} = [x_1, x_2, \ldots, x_n]^T$ which is derived with Lagrange multipliers $\boldsymbol{\lambda}_c = [\lambda_1, \lambda_2, \ldots, \lambda_m]^T$ for equality constraints $\mathbf{c}(\mathbf{x})$ and $\boldsymbol{\lambda}_v = [\lambda_1, \lambda_2, \ldots, \lambda_l]^T$ for inequality $\mathbf{v}(\mathbf{x})$ constraints. Equality $\mathbf{c}(\mathbf{x})$ and inequality $\mathbf{v}(\mathbf{x})$ constraints are given by Eqs. (10.4.2) and (10.4.3), respectively. Table 10.4.1 shows equality and inequality constraints used to optimize CI_c, CO_2, and W_c. Fifteen simple equalities, including b/h, f'_c, f_{yr}, f_{ys}, $P_{u,1}$, $P_{u,2}$, $P_{u,3}$, $M_{uX,1}$, $M_{uX,2}$, $M_{uX,3}$, $M_{uY,1}$, $M_{uY,2}$, $M_{uY,3}$, t_w, and t_f, can be directly substituted in ANNs. Nineteen inequality constraints are introduced to meet mostly code requirements, ANSI/AISC 360-16 [10.35] and ACI 318-19 [10.36], limiting the range of the stationary points [10.23].

$$\mathrm{L}(\mathbf{x}, \boldsymbol{\lambda}_c, \boldsymbol{\lambda}_v) = f(\mathbf{x}) - \boldsymbol{\lambda}_c^T \mathbf{c}(\mathbf{x}) - \boldsymbol{\lambda}_v^T \mathbf{v}(\mathbf{x}) \tag{10.4.1}$$

$$\mathbf{c}(\mathbf{x}) = [c_1(\mathbf{x}), c_2(\mathbf{x}), \ldots, c_m(\mathbf{x})]^T \tag{10.4.2}$$

$$\mathbf{v}(\mathbf{x}) = [v_1(\mathbf{x}), v_2(\mathbf{x}), \ldots, v_l(\mathbf{x})]^T \tag{10.4.3}$$

10.4.3 Normalized UFO capturing three objective functions

All tradeoff ratios shown in Eqs. (10.4.4) and (10.4.5) are non-negative and they vary in a range of 0–1 to generate a Pareto frontier that contains non-dominated design points. In *UFO*, each objective parameter (CI_c, CO_2 emissions, and W_c) has its own tradeoff ratio ranging from 0.0 to 1.0. Units of objective functions (CI_c, CO_2 emissions, and W_c) are KRW/m, t-CO_2/m, and kN/m, respectively, and hence, they are calculated in different units. All objective functions must be normalized to be unit-free functions when deriving *UFO* as indicated in Eqs. (10.4.6)–(10.4.8) [10.40]. This can be done by normalizing each objective function based on its maximum and minimum as shown in Table 10.4.2. The maximum and minimum of an individual objective function (CI_c, CO_2 emissions, and W_c) are obtained by an ANN-based Lagrange optimization based on equality and inequality constraints. An objective function (CI_c, CO_2 emissions, and W_c) is independently optimized when a tradeoff ratio corresponding to one objective function is set as 1.0, whereas the other tradeoff ratios are set as 0.

$$f_{\mathrm{UFO}} = w_{\mathrm{CI}_c} f_{\mathrm{CI}_c}^N + w_{\mathrm{CO_2}} f_{\mathrm{CO_2}}^N + w_{\mathrm{W}_c} f_{\mathrm{W}_c}^N \tag{10.4.4}$$

$$w_{\mathrm{CI}c} + w_{\mathrm{CO_2}} + w_{\mathrm{W}_c} = 1.0 \text{ and } 0 \le w_{\mathrm{CI}c}, w_{\mathrm{CO_2}}, w_{\mathrm{W}_c} \le 1.0 \tag{10.4.5}$$

in which,

f_{UFO}: Normalized unified functions of objectives
$f_{CI_c}^N$, $f_{CO_2}^N$, $f_{W_c}^N$: A normalized function of CI_c, CO_2 emissions, and W_c, respectively

$$f_{CI_c}^N = \frac{CI_c - CI_c^{min}}{CI_c^{max} - CI_c^{min}} \tag{10.4.6}$$

$$f_{CO_2}^N = \frac{CO_2 - CO_2^{min}}{CO_2^{max} - CO_2^{min}} \tag{10.4.7}$$

$$f_{W_c}^N = \frac{W_c - W_c^{min}}{W_c^{max} - W_c^{min}} \tag{10.4.8}$$

10.4.4 ANN-based Lagrange optimization based on UFO

10.4.4.1 Five steps for UFO optimization

An ANN is developed to map 20 input parameters to 17 output parameters which are selected to perform a forward design of SRC columns under the three pairs of axial loads and flexural moments. Input parameters, $\mathbf{x} = \left(h,\ b,\ h_s,\ b_s,\ t_w,\ t_f,\ f_c',\ f_{yr},\ f_{ys},\ \rho_{rX},\ \rho_{rY},\ P_{u,1},\ M_{uX,1},\ M_{uY,1},\ P_{u,2},\ M_{uX,2},\ M_{uY,2},\ P_{u,3},\ M_{uX,3},\ M_{uY,3}\right)^T$, denote section dimensions of a rectangular column (height, h and width, b) and H-shaped steel section (height (h_s), width (b_s), and thickness of flange (t_f), web (t_w)), material properties of concrete (f_c'), rebar (f_{yr}), steel (f_{ys}), ratios of reinforcing bars, ρ_{rY} and ρ_{rX}, along Y- and X-directions, respectively, and three pairs of factored biaxial loads ($P_{u,i}$, $M_{uX,i}$, $M_{uY,i}$) with $i = 1, 2, 3$. Output parameters $\mathbf{y} = \left(SF_1,\ SF_2,\ SF_3,\ \varepsilon_{r,1},\ \varepsilon_{r,2},\ \varepsilon_{r,3},\ \varepsilon_{s,1},\ \varepsilon_{s,2},\ \varepsilon_{s,3},\ \rho_s,\ D_{r,max},\ Y_s,\ X_s,\ {}^{b}\!/_{h},\ CI_c, CO_2, W_c\right)^T$ are selected in which SF_1, SF_2, and SF_3 represent three safety factors (SF_i), three rebar strain ($\varepsilon_{r,i}$), three steel strain ($\varepsilon_{s,i}$) corresponding to the three biaxial load pairs ($P_{u,i}$, $M_{uX,i}$, $M_{uY,i}$) with $i = 1, 2, 3$ in which the safety factor, $SF_i = P_{n,i}/P_{u,i} = M_{n,i}/M_{u,i} \geq 1.0 \left(M_{u,i} = \sqrt{M_{uX,i}^2 + M_{uY,i}^2}\right)$, must reflect the design strength requirement of the SRC column. The 100,000 datasets are generated with respect to input parameters shown in Table 10.2.3. The PTM is then used to train ANN using 8, 9, and 10 hidden layers with 40, 50, and 60 neurons. Figure 10.3.2 illustrates the topology of neural networks used in this training. Table 10.2.4 presents training results and training accuracies. Design optimizing UFO of SRC columns is a multiply constrained problem, implying that solving for stationary points optimizing multiple objective functions is difficult. The Lagrange multipliers

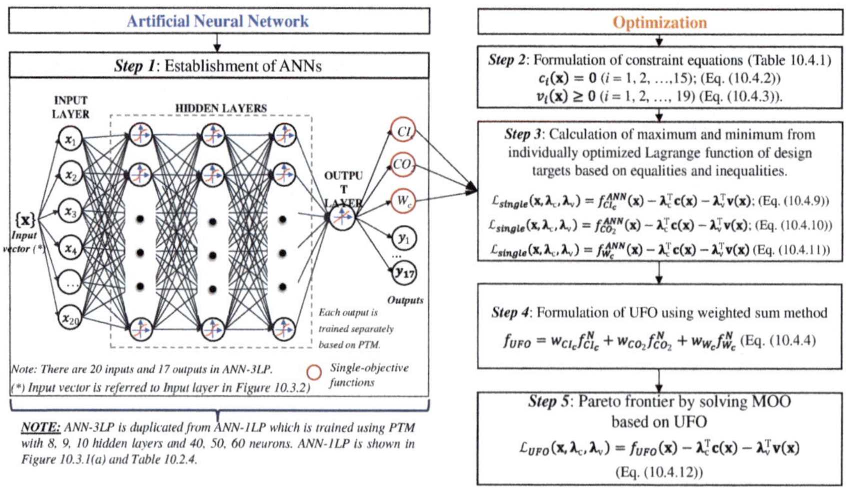

Figure 10.4.1 ANN-based Lagrange optimization algorithm for MOO [10.40].

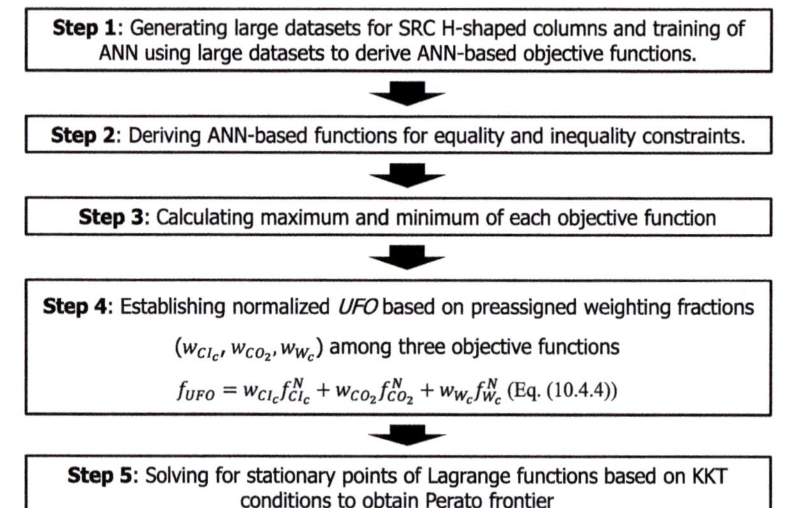

Figure 10.4.2 Algorithm to obtain ANN-based Lagrange MOO design of SRC columns.

are used to transfer constrained problems into non-boundary problems to optimize multiple objective functions. Figures 10.4.1 and 10.4.2 demonstrate an algorithm to solve a MOO based on *UFO* for SRC columns, which is described in five steps as follows.

Step 1: Single-objective function $\{f_{CI_c}(x), f_{CO_2}(x),$ and $f_{W_c}(x)\}$ is derived by ANNs. ANNs are established based on three sub-steps, which are described as follows.

Step 1.1: Selection of inputs $\left(h,\ b,\ h_s,\ b_s,\ t_w,\ t_f,\ f_c',\ f_{yr},\ f_{ys},\ \rho_{rX},\ \rho_{rY},\ P_{u,1},\ M_{uX,1},h,\ b,\ h_s,\ b_s,\ t_w,\ t_f,\ f_c',\ f_{yr},\ f_{ys},\ \rho_{rX},\ \rho_{rY},\ P_{u,1},\ M_{uX,1},\right)$ and outputs $\left(SF_1,\ SF_2,\ SF_3,\ \varepsilon_{r,1},\ \varepsilon_{r,2},\ \varepsilon_{r,3},\ \varepsilon_{s,1},\ \varepsilon_{s,2},\ \varepsilon_{s,3},\ \rho_s,\ D_{r,\max},\ Y_s,\ X_s,\ b/h,\ CI_c,\ CO_2,\ W_c\right)$ for ANNs.

Step 1.2: Generation of large structural datasets.

Step 1.3: Derivation of single-objective functions $\{f_{CI_c}(x),\ f_{CO_2}(x),$ and $f_{W_c}(x)\}$ by training ANNs on large structural datasets.

Step 2: Constrained conditions for optimizations are developed based on 15 equality equations $\{\ c_i(x)=0\ \}(i=1, 2, ..., 15)$ and 19 inequality equations $\{v_i(x)\ge 0\ \}\ (i=1, 2, ..., 19)$ shown in Table 10.4.1.

Step 3: This step calculates maximum and minimum from individually optimized Lagrange function of design targets based on equalities and inequalities to define their boundaries for the normalizations discussed in Step 4. The Lagrange functions of each objective function (CI_c, CO_2 and W_c) of SRC columns derived in Eqs. (10.4.9)–(10.4.11) are individually minimized and maximized as presented in Table 10.4.2, satisfying equalities and inequalities shown in Table 10.4.1. Lagrange functions are solved following first-order necessary conditions (KKT conditions) based on the Newton–Raphson iteration. Equations (10.4.9)–(10.4.11) are optimized to yield $CI_c^{\max}$, $CI_c^{\min}$, $CO_2^{\max}$, $CO_2^{\min}$, $W_c^{\max}$, and $W_c^{\min}$ for each objective function shown in Table 10.4.2 [10.23].

$$\mathcal{L}_{single}\left(\mathbf{x},\boldsymbol{\lambda}_c,\boldsymbol{\lambda}_v\right)=f_{CI_c}^{ANN}\left(\mathbf{x}\right)-\boldsymbol{\lambda}_c^{T}\mathbf{c}\left(\mathbf{x}\right)-\boldsymbol{\lambda}_v^{T}\mathbf{v}\left(\mathbf{x}\right) \tag{10.4.9}$$

$$\mathcal{L}_{single}\left(\mathbf{x},\boldsymbol{\lambda}_c,\boldsymbol{\lambda}_v\right)=f_{CO_2}^{ANN}\left(\mathbf{x}\right)-\boldsymbol{\lambda}_c^{T}\mathbf{c}\left(\mathbf{x}\right)-\boldsymbol{\lambda}_v^{T}\mathbf{v}\left(\mathbf{x}\right) \tag{10.4.10}$$

$$\mathcal{L}_{single}\left(\mathbf{x},\boldsymbol{\lambda}_c,\boldsymbol{\lambda}_v\right)=f_{W_c}^{ANN}\left(\mathbf{x}\right)-\boldsymbol{\lambda}_c^{T}\mathbf{c}\left(\mathbf{x}\right)-\boldsymbol{\lambda}_v^{T}\mathbf{v}\left(\mathbf{x}\right) \tag{10.4.11}$$

Step 4: Normalized *UFO* function shown in Eq. (10.4.4) in terms of individual objective function is obtained by substituting maxima and minima of single-objective function, $CI_c^{\max}$, $CI_c^{\min}$, $CO_2^{\max}$, $CO_2^{\min}$, $W_c^{\max}$, and $W_c^{\min}$, into Eqs. (10.4.6)–(10.4.8).

Step 5: Objective *UFO* function which is calculated in ***Step 4*** is used to calculate Lagrange function shown in Eq. (10.4.12) and its KKT conditions

An ANN-based *UFO* formulated to optimize CI_c, CO_2 emissions, and W_c, simultaneously is implemented in a Lagrange function shown in Eq. (10.4.12) which is constrained by 15 equality functions $\mathbf{c}(\mathbf{x})$ and 19 inequality functions $\mathbf{v}(\mathbf{x})$ shown in Eqs. (10.4.13) and (10.4.14), and Table 10.4.1. Each equality and inequality constraint is accompanied with Lagrange multipliers as shown in Eqs. (10.4.15) and (10.4.16). Inequality matrix S designating inequality constraints is presented in Eq. (10.4.17) in order to indicate which inequality constraints are activated when applying KKT conditions. Newton–Rapson iteration is used to solve for stationary points of Lagrange function, $\mathcal{L}(\mathbf{x}, \boldsymbol{\lambda}_c^T, \boldsymbol{\lambda}_v^T)$ shown in Eq. (10.4.12) with respect to stationary points which include 20 input variables $\mathbf{x}$, Lagrange multiplier of equality constraints $\boldsymbol{\lambda}_c$, and Lagrange multiplier of inequality constraints, $\boldsymbol{\lambda}_v$. A maximum number of 100 iterations are used for 100 combinations of tradeoff ratios of three objectives $(w_{CI_c}, w_{CO_2}, w_{W_c})$ to generate a Pareto frontier. The tradeoff ratios of each objective w_{CI_c}, w_{CO_2}, and w_{W_c} shown in Eqs. (10.4.4) and (10.4.5) are equally spaced in a 0–1 range [10.23].

$$\mathcal{L}_{\mathrm{UFO}}(\boldsymbol{x}, \boldsymbol{\lambda}_c, \boldsymbol{\lambda}_v) = f_{\mathrm{UFO}}(\boldsymbol{x}) - \boldsymbol{\lambda}_c^T \boldsymbol{c}(\boldsymbol{x}) - \boldsymbol{\lambda}_v^T \boldsymbol{v}(\boldsymbol{x}) \tag{10.4.12}$$

$$\boldsymbol{c}(\boldsymbol{x}) = \left[c_1(\boldsymbol{x}),\ c_2(\boldsymbol{x}),\ \ldots,\ c_{15}(\boldsymbol{x})\right]^T \tag{10.4.13}$$

$$\boldsymbol{v}(\boldsymbol{x}) = \left[v_1(\boldsymbol{x}), v_2(\boldsymbol{x}), \ldots,\ v_{19}(\boldsymbol{x})\right]^T \tag{10.4.14}$$

$$\boldsymbol{\lambda}_c = \left[\lambda_{c1}, \lambda_{c2}, \ldots, \lambda_{c15}\ \right]^T \tag{10.4.15}$$

$$\boldsymbol{\lambda}_v = \left[\lambda_{v1}, \lambda_{v2}, \ldots, \lambda_{v19}\ \right]^T \tag{10.4.16}$$

$$\mathrm{S} = \begin{bmatrix} s_1 & 0 & \cdots & 0 \\ 0 & s_2 & \cdots & 0 \\ \vdots & \vdots & \ddots & \vdots \\ 0 & 0 & \cdots & s_{19} \end{bmatrix}_{[19\times19]} \tag{10.4.17}$$

10.4.4.2 Two gradient vectors $\nabla f(x)$, $\nabla c(x)$ defining a contact of functions based on KKT conditions

The stationary points can be found by proving an gradient vector of objective function $\nabla f(x)$ and a constraint function $\nabla c(x)$ are tangent at constraining variables (x_i, y_i) [10.41]. However, tangents on functions are not uniquely determined at one point as shown in Figure 10.4.3. This is why gradients need to be introduced to prove the two gradient vectors of an objective function $\nabla f(x)$ and a constraint function $\nabla c(x)$ pointing in the same (or opposite direction) with a scalar multiple of the other, resulting in an objective function $\nabla f(x)$ and a constraint function $\nabla c(x)$ which are tangent as shown in Figure 10.4.4.

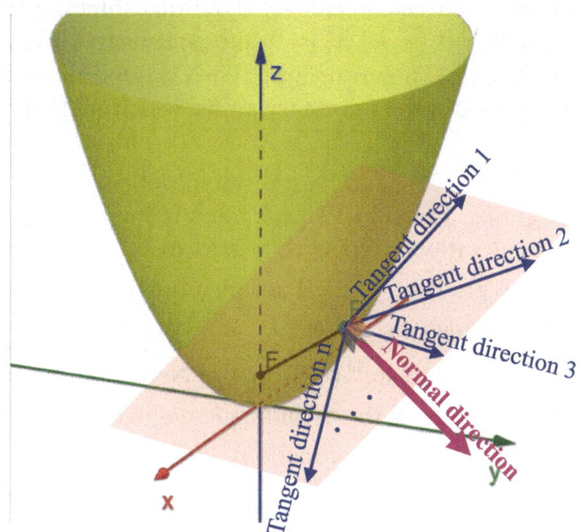

Figure 10.4.3 Tangents on surface and a gradient vector.

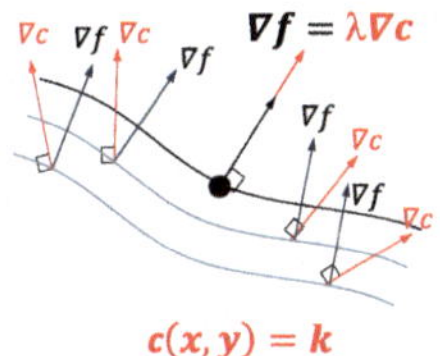

Figure 10.4.4 Two gradient vectors $\nabla f(x), \nabla c(x)$ defining a contact of functions $(f(x), c(x))$ ([10.41]; two gradient vectors $\nabla f(x), \nabla c(x)$ pointing in the same (or opposite direction) with a scalar multiple of the other.

10.4.4.3 Newton–Rapson iteration to solve for constraining stationary points optimizing UFO based on KKT conditions

Hong and Nguyen et al. [10.23] used Newton–Raphson iteration to solve non-linear equations derived based on KKT conditions [10.42,10.43]. Newton–Raphson iteration is applied to find stationary points $\left(\mathbf{x}^*, \lambda_c^*, \text{and } \lambda_v^*\right)$ which are the roots of the non-linear equations shown in Eqs. (10.4.9)–(10.4.11), as explained in **Step 3** of Section 10.4.4.1. Stationary points of Lagrange functions are identified using Jacobian and Hessian matrices shown in Eqs. (10.3.7)–(10.3.9), where $\left[\boldsymbol{H}_{\mathrm{L}}\left(\mathbf{x}^{(k)}, \boldsymbol{\lambda}_c^{(k)}, \boldsymbol{\lambda}_v^{(k)}\right)\right]$ is the Hessian matrix of the Lagrange function, and $\nabla \mathrm{L}\left(\mathbf{x}^{(k)}, \boldsymbol{\lambda}_c^{(k)}, \boldsymbol{\lambda}_v^{(k)}\right)$ is Jacobian of the Lagrange function. The ANN-based Hessian matrix was derived using MathWorks

Technical Support Department [10.44]. Lagrange function is optimized when its first derivative $\nabla L(\mathbf{x}, \boldsymbol{\lambda}_c, \boldsymbol{\lambda}_v) = 0$ at the stationary points of Lagrange function, $(\mathbf{x}^*, \boldsymbol{\lambda}_c^*, \text{and } \boldsymbol{\lambda}_v^*)$ shown in Eq. (10.4.12). Optimized design of an SRC column is obtained at the stationary points which are input parameters, $\mathbf{x} = (h,\ b,\ h_s,\ b_s,\ t_w,\ t_f,\ f_c',\ f_{yr},\ f_{ys},\ \rho_{rX},\ \rho_{rY},\ P_{u,1},\ M_{uX,1},\ M_{uY,1},\ P_{u,2},\ M_{uX,2},\ M_{uY,2},\ P_{u,3},\ M_{uX,3},\ M_{uY,3})^T$, and Lagrange multipliers, $\boldsymbol{\lambda}_c$ and $\boldsymbol{\lambda}_v$. The first derivative of Lagrange function $\nabla L(\mathbf{x}, \boldsymbol{\lambda}_c, \boldsymbol{\lambda}_v)$ is linearized [10.45] before Newton–Raphson iteration starts with respect to an assumed initial vector of $(\mathbf{x}^0, \boldsymbol{\lambda}_c^0, \boldsymbol{\lambda}_v^0)$. The gradient of the Lagrange function at the initial input parameters is determined as $\nabla \mathcal{L}(\mathbf{x}^0, \boldsymbol{\lambda}_c^0, \boldsymbol{\lambda}_v^0)$. A convergence criterion is defined as $\nabla L(\mathbf{x}^0, \boldsymbol{\lambda}_c^0, \boldsymbol{\lambda}_v^0) \le 1e^{-15} (\approx 0)$ in the Newton–Raphson iteration. A location $(\Delta\mathbf{x}^0, \Delta\boldsymbol{\lambda}_c^0, \Delta\boldsymbol{\lambda}_v^0)$ in the vicinity of the assumed initial vector of $(\mathbf{x}^0, \boldsymbol{\lambda}_c^0, \boldsymbol{\lambda}_v^0)$ on the multi-variate functions shown in Eq. (10.4.18) with k being equal to 0 is calculated to update first derivative of Lagrange function $\nabla L(\mathbf{x}, \boldsymbol{\lambda}_c, \boldsymbol{\lambda}_v)$ where $\mathbf{g}^{(k=0)}$ and $\mathbf{H}^{(k=0)}$ are the gradient and Hessian of Lagrange function $\mathcal{L}$, respectively, at an initial vector $(\mathbf{x}^0, \boldsymbol{\lambda}_c^0, \boldsymbol{\lambda}_v^0)$ [10.23,10.46].

The variation $(\Delta\mathbf{x}^0, \Delta\boldsymbol{\lambda}_c^0, \Delta\boldsymbol{\lambda}_v^0)$ calculated in Eq. (10.4.18) is added to update next input parameters $(\mathbf{x}^1, \boldsymbol{\lambda}_c^1, \boldsymbol{\lambda}_v^1)$ for next iteration. Updated input parameters of $(\mathbf{x}^1 = \mathbf{x}^0 + \Delta\mathbf{x}^0, \boldsymbol{\lambda}_c^1 = \boldsymbol{\lambda}_c^0 + \Delta\boldsymbol{\lambda}_c^0, \boldsymbol{\lambda}_v^1 = \boldsymbol{\lambda}_v^0 + \Delta\boldsymbol{\lambda}_v^0)$ are implemented if the criterion for the first iteration is not satisfied. The convergence criterion is now checked with $(\mathbf{x}^1, \boldsymbol{\lambda}_c^1, \boldsymbol{\lambda}_v^1)$. The same calculation is repeated until convergence is met ($\mathcal{L}(\mathbf{x}, \boldsymbol{\lambda}_c, \boldsymbol{\lambda}_v) \le 1e^{-15}$), followed by Eqs. (10.4.18) and (10.4.19) for the kth iteration. Another terminating criterion is defined by a maximum number of iterations (as 50 iterations in this design example) to save computation effort. Notably, the Newton–Raphson method depends on initial input parameters, and hence, the proposed ANN-based Lagrange optimization adopts multiple initial input parameters to ensure stationary points are rapidly captured. A number of initial trial input parameters is determined as 2^n, where n is a number of variables to be sought in designs. In SRC column designs sustaining three pairs of the three biaxial loads, input parameters contain 20 parameters $x = (h,\ b,\ h_s,\ b_s,\ t_w,\ t_f,\ f_c',\ f_{yr},\ f_{ys},\ \rho_{rX},\ \rho_{rY},\ P_{u,1},\ M_{uX,1},\ M_{uY,1},\ P_{u,2},\ M_{uX,2},\ M_{uY,2},\ P_{u,3},\ M_{uX,3},\ M_{uY,3})^T$, in which $t_w,\ t_f,\ f_c',\ f_{yr},\ f_{ys},\ f_{ys},\ P_{u,1},\ M_{uX,1},\ M_{uY,1},\ P_{u,2},\ M_{uX,2},\ M_{uY,2},\ P_{u,3},\ M_{uX,3},\ M_{uY,3}$ are typically predefined in designs, and hence, $h,\ b,\ h_s,\ b_s,\ \rho_{rX},\ \rho_{rY}$ vary, resulting in $n = 6$ and $2^6 = 64$ initial trial input parameters. The Newton–Raphson algorithm to solve KKT conditions of Lagrange functions shown in **Step 5** of Section 10.4.4.1 of optimization process is summarized in Figure 10.4.5 [10.40].

$$\begin{Bmatrix} \Delta \boldsymbol{x}^k \\ \Delta \boldsymbol{\lambda}_c^k \\ \Delta \boldsymbol{\lambda}_v^k \end{Bmatrix} = -\left(\mathbf{H}^k\right)^{-1} \mathbf{g}^k \qquad (10.4.18)$$

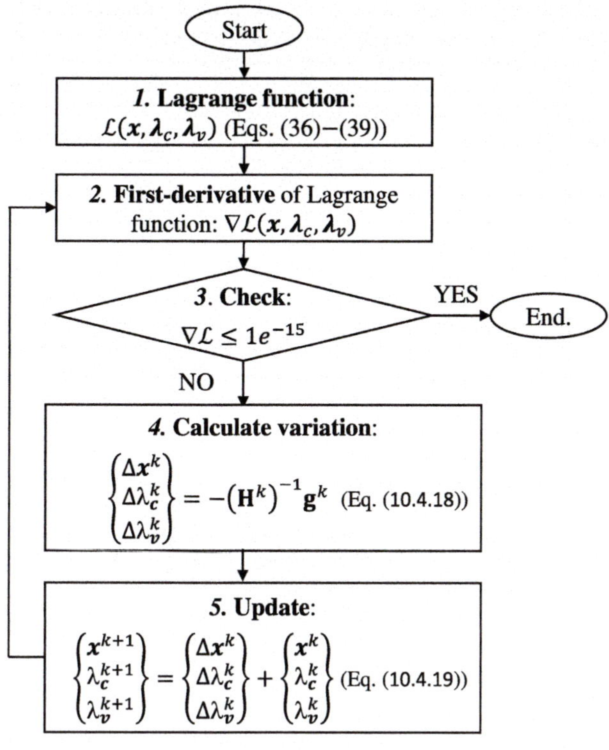

Figure 10.4.5 Newton–Raphson algorithm to solve KKT conditions of Lagrange functions utilized in Step 3 and Step 5 of optimization process [10.40]

$$\begin{Bmatrix} x^{k+1} \\ \lambda_c^{k+1} \\ \lambda_v^{k+1} \end{Bmatrix} = \begin{Bmatrix} \Delta x^{k} \\ \Delta\lambda_c^{k} \\ \Delta\lambda_v^{k} \end{Bmatrix} + \begin{Bmatrix} x^{k} \\ \lambda_c^{k} \\ \lambda_v^{k} \end{Bmatrix} \tag{10.4.19}$$

10.4.5 ANN-based Pareto frontier based on Lagrange optimizations

10.4.5.1 Verification of Pareto frontier

Cost index (CI_c), CO_2 emissions, and column weights (W_c) are treated as objective functions for a design of SRC columns sustaining three pairs of the three biaxial loads. Pareto frontier is known as a set of MOO. In this design. weight fractions ($w_{CI_c}; w_{CO_2}; w_{W_c}$) of each objective function is equally spaced from 0 to 1.0 using "*linspace*" code in MATLAB and their summation is equal to 1.0. A Pareto frontier of SRC columns is obtained when a number of 100 tradeoff ratios $\left[w_{CI_c}; w_{CO_2}; w_{W_c}\right]$ is considered as shown in Table 10.4.3. Figure 10.4.6 shows Pareto frontier, demonstrating relationships among cost index (CI_c), CO_2 emissions, and column weights (W_c) in a

Table 10.4.3 100 weight fractions (tradeoffs) based on "*linspace*" code in MATLAB

No.	w_{CI_c}	w_{CO_2}	w_{W_c}
1	0	0	1
2	0	0.11	0.89
3	0	0.22	0.78
4	0	0.33	0.67
5	0	0.44	0.56
⋮	⋮	⋮	⋮
100	1	0	0

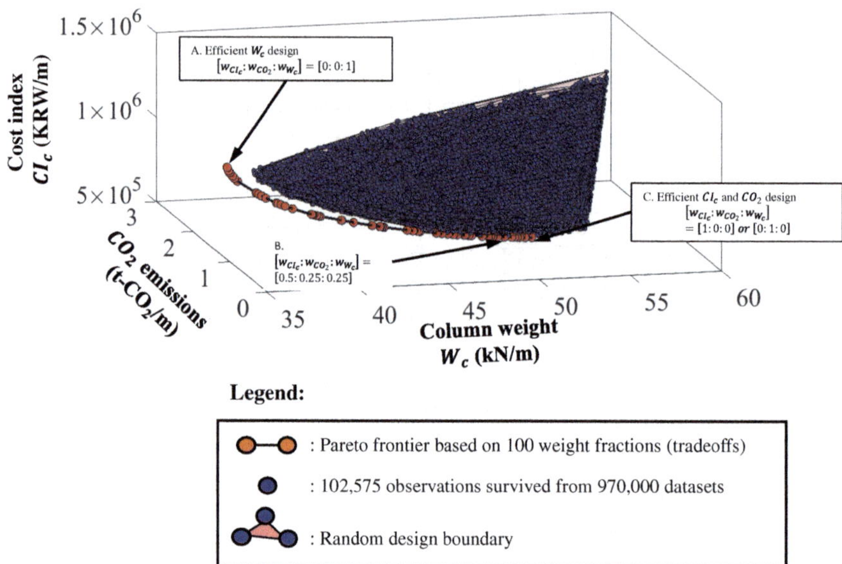

Figure 10.4.6 Verification of Pareto frontier by 970,000 datasets.

three-dimensional view which shows 100 combinations of tradeoff ratios for an SRC column design that can withstand three pairs of factored biaxial load ($P_{u,1}$ = 6,200 kN, $M_{uX,1}$ = 4,000 kN.m, $M_{uY,1}$ = 5,000 kN.m), ($P_{u,2}$ = 12,000 kN, $M_{uX,2}$ = 4,000 kN.m, $M_{uY,2}$ = 5,000 kN.m), ($P_{u,3}$ = 4,500 kN, $M_{uX,3}$ = 3,500 kN.m, $M_{uY,3}$ = 2,000 kN.m). Large datasets of 970,000 are generated to ascertain a Pareto frontier with preassigned equality constraints shown in Table 10.4.1. SF and a maximum rebar diameter are output parameters when generating large datasets, and hence, SF smaller than 1.0 and a maximum rebar diameter greater than 64 mm are removed, having 102,575 observations available from large datasets of 970,000. Factored loads and rebar ratios are randomly selected inputs when generating large datasets, therefore, SF and a maximum rebar diameter are calculated on an output-side which cannot be controlled initially to be greater than 1.0 and smaller than 64mm, respectively. Observations of 102,575 from 970,000 large datasets are used to verify Pareto frontier as shown in Figure 10.4.6, in which a Pareto frontier coincides with a lower boundary of 102,575 observations. It is noted that negligible differences found from Pareto frontier shown in Figure 10.4.6 can be further narrowed by repeating with increased number of big datasets or verifying using more big datasets. A trend in optimization proportional between CI_c and CO_2 emission is observed in Figure 10.4.7(a). It is noted that optimal designs for CI_c and CO_2 emission are obtained when concrete sections increase, whereas rebar and steel ratios decrease to lower CI_c and CO_2 emission. However, W_c and CI_c are reversely proportional as illustrated in Figure 10.4.7(b) because rebar and steel ratios should increase while decreasing concrete volumes to lower the weight of the columns, showing a trend in optimization opposite between W_c and CI_c emission. An opposite trend between CO_2 emissions and W_c are also illustrated in Figure 10.4.7(c).

10.4.5.2 A design results optimizing UFO based on tradeoff ratios

UFO shown in as shown in Eq. (10.4.20) is implemented in designing an SRC column. A tradeoff ratio $\left[w_{CI_c}; w_{CO_2}; w_{W_c}\right] = [0.5;\ 0.25; 0.25]$ shown in Eq. (10.4.20) is assumed to optimize multi-objective functions of the design example. A tradeoff ratio $\left[w_{CI_c}; w_{CO_2}; w_{W_c}\right] = [0.5;\ 0.25; 0.25]$ represents 50%, 25%, and 25% contributions to *UFO* by each objective function when optimizing *UFO*. Table 10.4.4 prepared with a tradeoff ratio $\left[w_{CI_c}; w_{CO_2}; w_{W_c}\right] = [0.5;\ 0.25; 0.25]$ verifies the design optimized based on a *UFO* by AutoSRCHCol software using design parameters obtained by ANNs [10.16]. As shown in Table 10.4.4, the largest error is found as –3.53% for rebar strain ($\varepsilon_{r,3}$) which is calculated when being subject to the third biaxial load, however, the rebar strain ($\varepsilon_{r,3}$) obtained by ANN-based design (0.0031) differs from that (0.0032) calculated based on AutoSRCHCol using design parameters obtained by ANNs. The difference is only 0.0001. The largest error for the SFs is only 0.24%, indicating that ANN-based design optimizing UFO yields sufficiently accurate designs for a use in practical designs. Errors for objective functions of CI_c, CO_2 emissions, and W_c are also negligible, showing errors 0.32%, 0.39%, and 0.007%. The proposed ANN-based design optimizing *UFO* can assist engineers to design an SRC column optimizing multiple objective functions based on their tradeoff weight fraction. A three-dimensional P-M interaction diagram plotted in Figure 10.4.8 is optimized based on *UFO* when objective functions (CI_c, CO_2 emissions, and W_c) are optimized based on a tradeoff weight fraction of three objectives $\left[w_{CI_c}; w_{CO_2}; w_{W_c}\right] = [0.5;\ 0.25; 0.25]$. Load pairs 1 and 3 are located inside the P-M interaction diagram, whereas Load pair

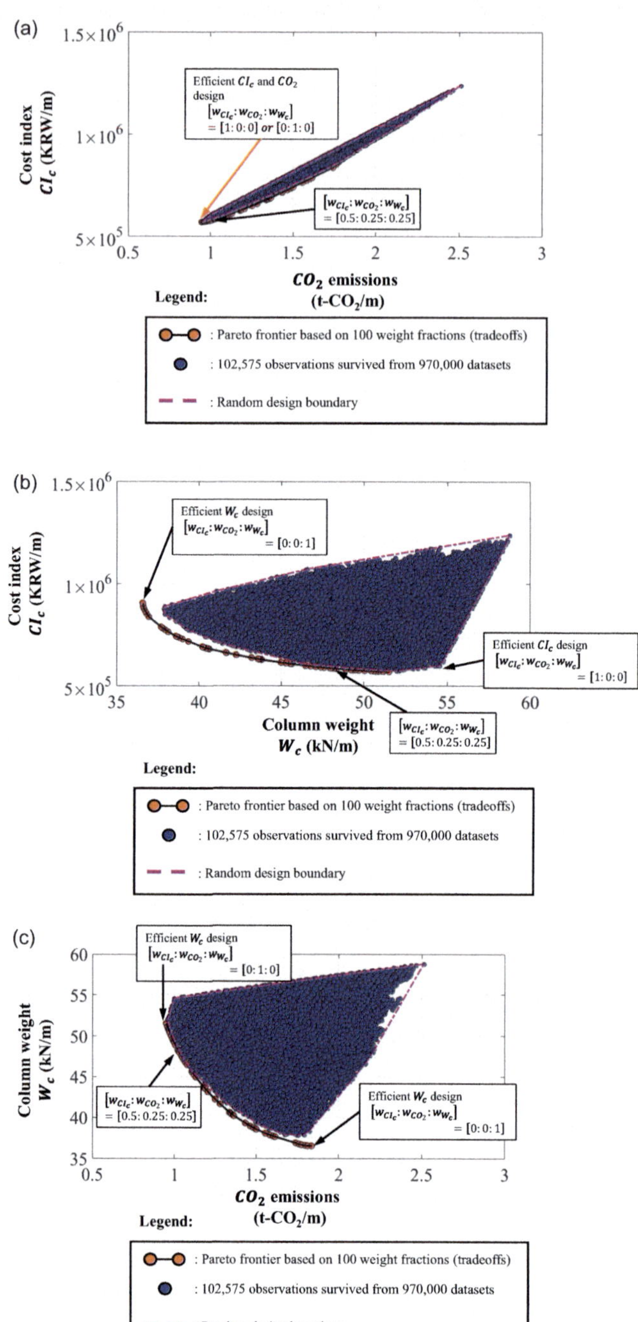

Figure 10.4.7 Relationships among objective functions (CI_c, CO_2 emissions, and W_c). (a) CI_c vs. CO_2 emissions. (b) CI_c vs. W_c. (c) CO_2 emission vs. W_c.

Table 10.4.4 ANN-based Lagrange MOO designs with $[w_{CI_c}; w_{CO_2}; w_{W_c}] = [0.5; 0.25; 0.25]$

MINIMIZED COLUMN WEIGHT								
Training results: 10 hidden layers-40 neurons; T.MSE = 1.47E-7								
No.	*Input for generating big data and training*		No.	*Output for generating big data and training*				
	Notation	*Value*		*Notation* (1)	*ANN* (2)	*Structural calculations* (3)	*Errors (%)* (4)
1	$b^{1,3}$ (mm)	1396.34	1	$SF_1^{\,3}$	1.24	1.25	−0.22
2	$h^{1,3}$ (mm)	1395.61	2	$SF_2^{\,3}$	1.00	1.00	−0.11
3	$\rho_{rX}^{\ 1,3}$	0.007	3	$SF_3^{\,3}$	2.24	2.23	0.24
4	$\rho_{rY}^{\ 1,3}$	0.001	4	$\varepsilon_{r,1}$	0.0035	0.0036	−2.84
5	$h_s^{\,1,3}$ (mm)	1,000	5	$\varepsilon_{r,2}$	0.0026	0.0027	−1.37
6	$b_s^{\,1,3}$ (mm)	716.11	6	$\varepsilon_{r,3}$	0.0031	0.0032	−3.53
7	$t_f^{\,1,3}$ (mm)	8	7	$\varepsilon_{s,1}$	0.0026	0.0026	0.31
8	$t_w^{\,1,3}$ (mm)	8	8	$\varepsilon_{s,2}$	0.0019	0.00188	1.23
9	$f_c'^{\,2}$ (MPa)	30	9	$\varepsilon_{s,3}$	0.0023	0.0024	−0.26
10	$f_{yr}^{\ 2}$ (MPa)	500	10	$\rho_s^{\,3}$	0.01	0.01	0.80
11	$f_{ys}^{\ 2}$ (MPa)	325	11	$D_{r,max}^{\ \ 3}$	27.4	25	8.84
12	$P_{u,1}^{\ \ 2}$ (kN)	6,200	12	Y_s (mm)	197.80	197.80	−0.002
13	$M_{uX,1}^{\ \ 2}$ (kNm)	12,000	13	X_s (mm)	340.10	340.11	−0.002
14	$M_{uY,1}^{\ \ 2}$ (kNm)	4,500	14	b/h^2	1.00	1.00	−0.05
15	$P_{u,2}^{\ \ 2}$ (kN)	4,000	15	CI_c (KRW/m)	588,457.82	586,538.96	0.32
16	$M_{uX,2}^{\ \ 2}$ (kNm)	6,500	16	CO_2 (t-CO_2/m)	1.03	1.03	0.39
17	$M_{uY,2}^{\ \ 2}$ (kNm)	3,500	17	W_c (kN/m)	47.76	47.75	0.007
18	$P_{u,3}^{\ \ v}$ (kN)	5,000					
19	$M_{uX,3}^{\ \ 2}$ (kNm)	5,000					
20	$M_{uY,3}^{\ \ 2}$ (kNm)	2,000					

[1] Calculated on an output-side according to optimized designs.
[2] Equality constraints.
[3] Inequality constraints.

2 is placed on the P-M interaction diagram when being optimized based on *UFO*.

$$f_{UFO} = 0.5\frac{CI_c - 570{,}352}{1{,}269{,}606 - 570{,}352} + 0.25\frac{CO_2 - 0.947}{2.563 - 0.947} + 0.25\frac{W_c - 36.55}{62.15 - 36.55} \tag{10.4.20}$$

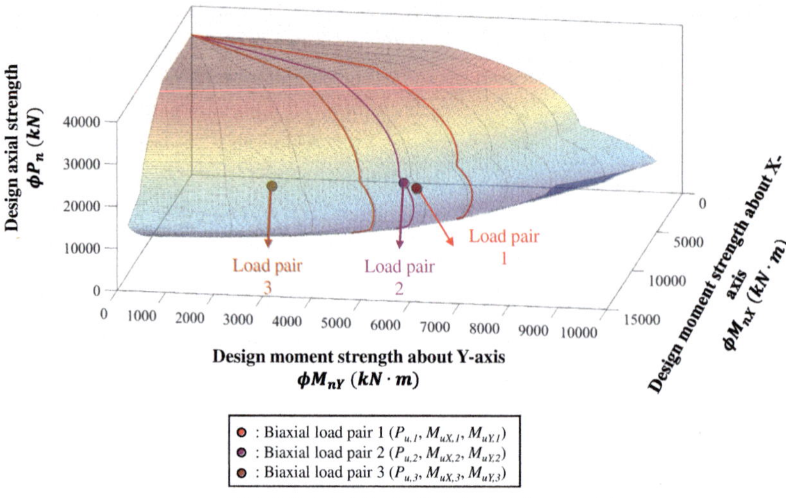

Figure 10.4.8 P-M diagram optimized based on multi-objective functions using weight fractions; tradeoff$=\left[w_{CI_c}; w_{CO_2}; w_{W_c}\right]=\left[0.5; 0.25; 0.25\right]$.

10.4.6 Discussion of the diversity of the proposed ANN-based Lagrange algorithm

The gradient-based ANN-based Lagrange MOO algorithm optimizes multiple objective functions (CI_c, CO_2 emissions, and W_c) at the same time based on any given tradeoff weight fractions of the three objectives $\left[w_{CI_c}; w_{CO_2}; w_{W_c}\right]$. ANN-based designs optimizing UFO are solutions of KKT conditions calculated using the Newton–Raphson iteration, which depends on its initial input parameters. The design example uses 64 sets of initial trial input parameters, as presented in Section 10.4.4.3, to perform a wide search. The three tradeoff weight fractions of the three objective functions ranging from 0 to 1 used by MOO designs are presented in Table 10.4.3 which demonstrates a searching region of Pareto. Figure 10.4.6 shows that three point A (0.0, 0.0, 0.1), point B (0.5, 0.25, 0.25), and point C (1.0, 0.0, 0.0) represent objective functions (CI_c, CO_2 emissions, and W_c) optimized based on tradeoff weight ratios of three objectives $w_{CI_c} : w_{CO_2} : w_{W_c}$ as 0.0:0.0:0.1, 0.5:0.25:0.25, and 1.0:0.0:0.0, respectively.

10.5 CONCLUSIONS

An ANN-based Lagrange MOO algorithm known as a Pareto frontier is introduced to design SRC columns with H-shaped steel sections sustaining multiple biaxial loads, optimizing multi-objective functions simultaneously.

A set of multi-objective-based optimized designs known as a Pareto frontier is performed by varying tradeoff weight fractions among objective functions (CI_c, CO_2 emissions, and W_c). The design example is shown to derive ANN-based generalized objective functions and constraints which are challenging to be derived analytically in some complex design problems. A UFO is established with respect to three objective functions which are cost index (CI_c), CO_2 emissions, and column weights (W_c). The following conclusions can be further drawn for practical design applications.

1. An ANN-based *UFO* is an integrated function of multi-objective functions, CI_c, CO_2 emissions, and W_c, based on tradeoffs ratios. Contribution to the three objective functions (CI_c, CO_2 emissions, and W_c) are determined by tradeoff weight fractions $w_{CI_c}, w_{CO_2}, \text{and}\, w_{W_c}$, which are implemented in deriving *UFO*. ANN-based *UFO* is implemented in Lagrange optimization to design SRC columns sustaining multiple biaxial loads. MOO design can assist engineers in deciding how much tradeoffs among objective functions are contributed.
2. A Pareto frontier based on ANN-based MOO uncover the weight fractions ($w_{CI_c}, w_{CO_2}, \text{and}\, w_{W_c}$) of each objective and target, presenting contributions made by each of the three objective functions (CI_c, CO_2 emissions, and W_c). A Pareto frontier informs engineers and decision-makers of particular tradeoff fractions contributed by the objective functions to assess design projects holistically. An optimization curve will identify how the three objective functions are optimized based on tradeoff fractions.
3. A design example optimizing *UFO* with preassigned tradeoff weight fractions is performed and verified by structural mechanics-based software using design parameters obtained by ANNs, showing that ANN-based MOO algorithm offers practical design tools optimizing multi-objective targets with sufficient accuracies. A Pareto frontier which is a set of multi-optimal design results is also verified by using 970,000 datasets, showing a lower boundary of 970,000 datasets coincides with a Pareto frontier.
4. Proportional trends between optimal designs of CI_c and CO_2 emissions are observed, whereas reverse proportional trends between optimal W_c and each CI_c or CO_2 emissions are observed.
5. Large datasets are generated to confirm the minimized CI_c, CO_2 emissions, and W_c. In the design based on a tradeoff weight fraction of three objectives $[w_{CI_c}; w_{CO_2}; w_{W_c}] = [0.5; 0.25; 0.25]$, the cost CI_c (569,126 KRW/m) is minimized by 22.5% compared with the probable cost (734,478 KRW/m) obtained by 970,000 datasets. Furthermore, the proposed method reduced CO_2 emissions and W_c by 37.3% and 15.2%, respectively.
6. Robust, simple, rapid, and accurate ANN-based optimizations are proposed as an alternative way to perform multi-objective optimal designs of SRC columns sustaining multiple biaxial loads in

conventional manner. ANN-based optimization method outperforms over the conventional optimizations when objective functions are difficult to obtain explicitly. ANN-based optimizations can be applied to various structural components such as foundations and walls.

REFERENCES

[10.1] Mirza, S. A., Hyttinen, V., & Hyttinen, E. (1996). Physical tests and analyses of composite steel-concrete beam-columns. *Journal of Structural Engineering*, *122*(11), 1317–1326. https://doi.org/10.1061/(ASCE)0733-9445(1996)122:11(1317)

[10.2] Ricles, J. M., & Paboojian, S. D. (1994). Seismic performance of steel-encased composite columns. *Journal of Structural Engineering*, *120*(8), 2474–2494. https://doi.org/10.1061/(ASCE)0733-9445(1994)120:8(2474)

[10.3] El-Tawil, S., & Deierlein, G. G. (1999). Strength and ductility of concrete encased composite columns. *Journal of Structural engineering*, *125*(9), 1009–1019. https://doi.org/10.1061/(ASCE)0733-9445(1999)125:9(1009)

[10.4] Li, L. I., & Matsui, C. (2000). Effects of axial force on deformation capacity of steel encased reinforced concrete beam–columns. In *Proceedings of 12th World Conference on Earthquake Engineering*.

[10.5] Brettle, H. J. (1973). Ultimate strength design of composite columns. *Journal of the Structural Division*, *99*(9), 1931–1951. https://doi.org/10.1061/JSDEAG.0003606

[10.6] Bridge, R. Q., & Roderick, J. W. (1978). Behaviour of built-up composite columns. *Journal of the Structural Division*, *104*(7), 1141–1155. https://doi.org/10.1061/JSDEAG.0004956

[10.7] Furlong, R. W. (1974). Concrete encased steel columns design tables. *Journal of the Structural Division*, *100*(9), 1865–1882. https://doi.org/10.1061/JSDEAG.0003878

[10.8] Mirza, S. A., & Skrabek, B. W. (1992). Statistical analysis of slender composite beam-column strength. *Journal of Structural Engineering*, *118*(5), 1312–1332. https://doi.org/10.1061/(ASCE)0733-9445(1992)118:5(1312)

[10.9] Chen, C. C., & Lin, N. J. (2006). Analytical model for predicting axial capacity and behavior of concrete encased steel composite stub columns. *Journal of Constructional Steel Research*, *62*(5), 424–433. https://doi.org/10.1016/j.jcsr.2005.04.021

[10.10] Dundar, C., Tokgoz, S., Tanrikulu, A. K., & Baran, T. (2008). Behaviour of reinforced and concrete-encased composite columns subjected to biaxial bending and axial load. *Building and Environment*, *43*(6), 1109–1120. https://doi.org/10.1016/j.buildenv.2007.02.010

[10.11] Munoz, P. R., & Hsu, C. T. T. (1997). Biaxially loaded concrete-encased composite columns: Design equation. *Journal of Structural Engineering*, *123*(12), 1576–1585. https://doi.org/10.1061/(ASCE)0733-9445(1997)123:12(1576)

[10.12] Furlong, R. W. (1968). Design of steel-encased concrete beam-columns. *Journal of the Structural Division*, *94*(1), 267–282. https://doi.org/10.1061/JSDEAG.0001854

[10.13] Kato, B. (1996). Column curves of steel-concrete composite members. *Journal of Constructional Steel Research*, *39*(2), 121–135. https://doi.org/10.1016/S0143-974X(96)00030-2

[10.14] Virdi, K. S., Dowling, P. J., BS 449, & BS 153. (1973). The ultimate strength of composite columns in biaxial bending. *Proceedings of the Institution of Civil Engineers*, *55*(1), 251–272.

[10.15] Roik, K., & Bergmann, R. (1990). Design method for composite columns with unsymmetrical cross-sections. *Journal of Constructional Steel Research*, *15*(1–2), 153–168. https://doi.org/10.1016/0143-974X(90)90046-J

[10.16] Nguyen, D. H., & Hong, W. K. (2020). An analytical model computing the flexural strength and performance of the concrete columns confined by both transverse reinforcements and steel sections. *Journal of Asian Architecture and Building Engineering*, *19*(6), 647–669. https://doi.org/10.1080/13467581.2020.1775603

[10.17] Abambres, M., & Lantsoght, E. O. (2020). Neural network-based formula for shear capacity prediction of one-way slabs under concentrated loads. *Engineering Structures*, *211*, 110501. https://doi.org/10.1016/j.engstruct.2020.110501

[10.18] Sharifi, Y., Lotfi, F., & Moghbeli, A. (2019). Compressive strength prediction using the ANN method for FRP confined rectangular concrete columns. *Journal of Rehabilitation in Civil Engineering*, 7(4), 134–153. https://doi.org/10.22075/JRCE.2018.14362.1260

[10.19] Asteris, P. G., Armaghani, D. J., Hatzigeorgiou, G. D., Karayannis, C. G., & Pilakoutas, K. (2019). Predicting the shear strength of reinforced concrete beams using artificial neural networks. *Computers and Concrete*, *24*(5), 469–488. https://doi.org/10.12989/CAC.2019.24.5.469

[10.20] Armaghani, D. J., Hatzigeorgiou, G. D., Karamani, C., Skentou, A., Zoumpoulaki, I., & Asteris, P. G. (2019). Soft computing-based techniques for concrete beams shear strength. *Procedia Structural Integrity*, *17*, 924–933. https://doi.org/10.1016/j.prostr.2019.08.123

[10.21] Hong, W. K., Pham, T. D., & Nguyen, V. T. (2021). Feature selection based reverse design of doubly reinforced concrete beams. *Journal of Asian Architecture and Building Engineering*. https://doi.org/10.1080/13467581.2021.1928510

[10.22] Hong, W. K., Nguyen, V. T., & Nguyen, M. C. (2021). Artificial intelligence-based novel design charts for doubly reinforced concrete beams. *Journal of Asian Architecture and Building Engineering*. https://doi.org/10.1080/13467581.2021.1928511

[10.23] Hong, W. K., & Nguyen, M. C. (2021). AI-based Lagrange optimization for designing reinforced concrete columns. *Journal of Asian Architecture and Building Engineering*, *21*, 2330–2344.

[10.24] Aghaee, K., Yazdi, M. A., & Tsavdaridis, K. D. (2015). Investigation into the mechanical properties of structural lightweight concrete reinforced with waste steel wires. *Magazine of Concrete Research*, *67*(4), 197–205. https://doi.org/10.1680/macr.14.00232

[10.25] Fanaie, N., Aghajani, S., & Dizaj, E. A. (2016). Theoretical assessment of the behavior of cable bracing system with central steel cylinder. *Advances in Structural Engineering*, *19*(3), 463–472. https://doi.org/10.1177/1369433216630052

[10.26] Madadi, A., Eskandari-Naddaf, H., Shadnia, R., & Zhang, L. (2018). Characterization of ferrocement slab panels containing lightweight expanded clay aggregate using digital image correlation technique. *Construction and Building Materials, 180*, 464–476. https://doi.org/10.1016/j.conbuildmat.2018.06.024

[10.27] Nasrollahi, S., Maleki, S., Shariati, M., Marto, A., & Khorami, M. (2018). Investigation of pipe shear connectors using push out test. *Steel and Composite Structures, 27*(5), 537–543. http://dx.doi.org/10.12989/scs.2018.27.5.537

[10.28] Paknahad, M., Shariati, M., Sedghi, Y., Bazzaz, M., & Khorami, M. (2018). Shear capacity equation for channel shear connectors in steel-concrete composite beams. *Steel and Composite Structures, 28*(4), 483–494. https://doi.org/10.12989/scs.2018.28.4.483

[10.29] Yoon, Y. C., Kim, K. H., Lee, S. H., & Yeo, D. (2018). Sustainable design for reinforced concrete columns through embodied energy and CO_2 emission optimization. *Energy and Buildings*, 174, 44–53. https://doi.org/10.1016/j.enbuild.2018.06.013.

[10.30] Gholami, M., Fathi, A., & Baghestani, A. M. (2021). Multi–objective optimal structural design of composite superstructure using a novel MONMPSO algorithm. *International Journal of Mechanical Sciences, 193*, 106149. https://doi.org/10.1016/j.ijmecsci.2020.106149.

[10.31] Hong W. K, Nguyen, V. T., Le, T. A., Pham, T. D., & Nguyen, D. H. (2022). A multiobjective optimization method for steel-reinforced concrete (SRC) columns encasing H-shaped steel sections. *Journal of Asian Architecture and Building Engineering* (under review).

[10.32] Hong, W. K., Le, T.A., & Nguyen, M.C. (2022). Capturing multiple load combinations for optimizing circular column behavior by an AI-based modularized weight matrix. *Journal of Asian Architecture and Building Engineering* (under review).

[10.33] Hong, W.K. (2021). *Artificial Intelligence-Based Design of Reinforced Concrete Structures*. Daega, Seoul .

[10.34] MathWorks (2022). Deep Learning Toolbox: User's Guide (R2022a). Retrieved July 26, 20122 from: https://www.mathworks.com/help/pdf_doc/deeplearning/nnet_ug.pdf

[10.35] AISC (American Institute of Steel Construction). (2016). *ANSI/AISC 360-16: Specification for Structural Steel Buildings*, AISC, Chicago, IL.

[10.36] An ACI Standard. (2019). *Building Code Requirements for Structural Concrete (ACI 318-19)*. American Concrete Institute, Farmington Hills, IL.

[10.37] Krenker, A., Bešter, J., & Kos, A. (2011). Introduction to the artificial neural networks. In *Artificial Neural Networks: Methodological Advances and Biomedical Applications* (pp. 1–18). InTech. https://doi.org/ 10.5772/15751.

[10.38] MathWorks (2020). *MATLAB R2020b, Version 9.9.0*. Natick, MA: MathWorks.

[10.39] Hong, W. K., Kim, J. M., Park, S. C., Lee, S. G., Kim, S. I., Yoon, K. J., ... & Kim, J. T. (2010). A new apartment construction technology with effective CO_2 emission reduction capabilities. *Energy, 35*(6), 2639–2646. https://doi.org/10.1016/j.energy.2009.05.036

[10.40] Hong, W. K., Le, T. A., & Nguyen, M. C. (2022). ANN-based Lagrange optimization for RC circular columns having multi-objective functions. *Journal of Asian Architecture and Building Engineering* (under review), *22*, 961–976.

[10.41] Osborne, J., Hicks, G., & Fuentes, R. (2008). Global analysis of the double-gimbal mechanism. *IEEE Control Systems Magazine*, *28*(4), 44–64. https://doi.org/10.1109/MCS.2008.924794

[10.42] Kuhn, H. W., & Tucker, A. W. (1951). Nonlinear programming. In *Proceedings of 2nd Berkeley Symposium* (pp. 481–492). Berkeley: University of California Press.

[10.43] Karush, W. (1939). *Minima of Functions of Several Variables with Inequalities as Side Constraints*. M.Sc. Dissertation. Department of Mathematics, the University of Chicago.

[10.44] MathWorks (2021). *MATLAB R2021a, Version 9.9.0*. Natick, MA: The MathWorks Inc.

[10.45] Hong, W.K. (2023). *Artificial Neural Network- based Optimized Design of Reinforced Concrete Structures*. Taylor & Francis, Boca Raton, FL.

[10.46] Hong, W. K., Pham, T. D., & Nguyen, V. T. (2022). Feature selection based reverse design of doubly reinforced concrete beams. *Journal of Asian Architecture and Building Engineering*, *21*(4), 1472–1496. https://doi.org/10.1080/13467581.2021.1928510

Index

Note: Bold page numbers refer to tables; *italic* page numbers refer to figures.

absolute error 208, 261
ACI 318-19 7
Adapt 19, 22
ANN-based design, cost minimization
  beam cost 207–208, *208*
  cracking moment 217
  crack widths 208–209, *209*
  deflections *213*, 216
  large datasets 218–219
  mechanics-based design 218–219
  moment demand *214*, 217
  moment resistance 217
  neutral axis depth *214*, 217–218
  reinforcement strains *214*, 217–218
  shear stirrup ratios *215*, 218
  steel-reinforced concrete beams 280–282
  tendon and concrete stresses *210*, 210–216, *211*
American Concrete Institute (ACI) design code 284, 286
anchorage draw-in 30–31
anchorage-seating loss 27
ANN-1LP 478
AutoPTbeam 46, 48, 50, 51, 70, 105, 113, 120, 124, 144, 187, 219, 253, 256, 269
AutoSRCbeam 284, 318, 322, 384
AutoSRCHCol 483
  CO_2 emissions **462**, 462–463
  column weight **466**, 466–467
  cost index 457–458, **458**
axial compression 22
axial tensile strength 187
back-substitution (BS) method 279, 420, **422**, 424–425, *426*, 427
  chained training scheme with revised sequence 136–159
  parallel training method 86–136
  steel-reinforced concrete beams 292–293, 318, 320–322, 324
beam depth minimization, ANN-based design charts
  concrete stresses 230–232, *231*
  cost and beam weight 235–238, *237*
  cracking moment 233–234, *234*
  crack widths 232–233, *233*
  deflections 228–230, *230*
  moment demand 233–234, *234*
  moment resistance 233–234, *234*
  neutral axis depth 234–235, *235*
  reinforcement strains 234–235, *235*
  shear stirrup ratios 235, *236*
beam weight 330–331, *331*
beam weight minimization, ANN-based design charts
  beam weights 227–228
  concrete and tendon stresses *221*, 221–222
  cracking moment 224, *224*
  crack widths 222, *223*
  deflections 219, *220*
  moment demand 224, *224*
  moment resistance 224, *224*
  neutral axis depth 224–225, *225*
  reinforcement strains 224–225, *225*
  shear stirrup ratios 225–226, *226*

biaxial load pairs
design codes 500, **501–502**, 502, **503**
design parameters 500
formulation of forward network 503–504
Lagrange MOO algorithm 516
Lagrange optimization 505–516
normalized UFO capturing 504–505
structural analysis 50
bonded tendons 5
BS method *see* Back-substitution (BS) method

casting beds 3
chained training scheme with revised sequence (CRS) 45
Class 2 low-relaxation steels 33, *33*
coefficient of determination (R^2) 51
CO_2 emissions 278, 283, 330–331, *331*
Pareto frontier 402
reinforced concrete columns 427, **428**, *429*, 430
steel-reinforced concrete beams 362, *363*, 364, **365–366**, 367, **371–374**
by structural design software **462**, 462–463
three-dimensional P-M interaction diagram 463, *463*
verification 463–465, **464**, *464*, *465*
column weight (W_c)
pareto frontier 403
reinforced concrete columns 430–431, *431*
steel-reinforced concrete beams 367, **368–369**, 370, **371–374**, 375
by structural design software **466**, 466–467
three-dimensional P-M interaction diagram 467, *467*
verification 467–469, *468*, **468**, *469*
compressive stress 2
computer-aided design (CAD) 46
concrete
creep 32
elasticity 8
posttensioning 1, 4–5
prestressing 1
pretensioning 1, *3*, 3–4
shrinkage strain 32
strength 7
stress control 35–37
convolutional neural network (CNN) 277
corrosions, by chlorides 36
cost index (CI_c) 184, 235, *236*, 253, 265, 283, 330–331, *331*, 438
derivation 451, **455**, 456
pareto frontier 402
reinforced concrete columns 427, **428**, *429*, 430
steel-reinforced concrete beams 361, *362*, **371–374**
by structural design software 457–458, **458**
three-dimensional P-M interaction diagram 458, *459*, **460**
verification 460–461, *461*
cost minimization, AI-based design charts
beam cost 207–208, **208**
cracking moment 217
crack widths 208–209, *209*
deflections *213*, 216
large datasets 218–219
mechanics-based design 218–219
moment demand *214*, 217
moment resistance 217
neutral axis depth *214*, 217–218
reinforcement strains *214*, 217–218
shear stirrup ratios *215*, 218
tendon and concrete stresses *210*, 210–216, *211*
crack control 37–40
cracked section analysis 35
cracking moment 268
crack propagation 22, *23*
crack widths 37, **39**, 40, 50, 196–197, **253**
creep 10
of concrete 32
curvature ductility 328, 333
curvature friction 28

damage assessment 45
data generation 183–185
Deep artificial Neural Networks (DNNs) 47
deflection 197
deflection control 40–43

deflection ductility ratio 46, 62, 114, 116, 117, 120–122, 124, 125, 127, 132–135, 139, 140, 142, 144, 145, 146, 148, 150, 151, 153, 155, 156, 167, 169, 178
deformation 10

effective depths 327–328, *328*
elastic deflection 41
elastic shortening 31–32
equality 360–361, *361*
equality constraints **394**, 394–395
equivalent load, hyperstatic moments 14–16
Eurocode 194
Eurocode 0 (EC0) 252
Eurocode 2 (EC2) 7, 8, 10, 30, 33, 42, 250
excessive cracks 191

flexural design
  section analysis 34–35
  at service limit stage 35–43
    crack control 37–40
    deflection control 40–43
    stress control 35–37
  at ultimate stage 43
flexural strength 197
force loss 30, 31
forward design, design charts 70–86, **71–84**, *85*
forward networks 89, 103, 114, 393–394
friction loss 28–30, *30*

gradient vectors 508, *509*
Grey Wolf optimization 245–246

Hessian matrix 202
histograms *55–58*
Hong–Lagrange algorithm 391, **392**, 393–396, **394**, *396*
horizontal load 190
hybrid glowworm swarm algorithm 183
hyperstatic forces 19
  equivalent load 14–16
  on flexural capacity of posttensioned frame 21–25
  general 13–14
  prestressing sequence 18–20
  secondary prestressing effects 16–25

inequality constraints 203–206, 360–361, *361*, **394**, 394–395
input parameters *486*, 486–487, **487**
Inward movement 30

jacking force 15, 27, 33
jacking length, of tendon 6, **6**
Jacobi of Lagrange function 202, 203

Karush–Kuhn–Tucker (KKT) 183, 198, 203, 206

Lagrange function 184, 198, 200, 206
Lagrange multiplier method (LMM) 204
  active and inactive conditions 456–457
  AI-based objective functions 451, 456
  CO_2 emissions, and column weight 451, **455**
  cost index 451, **455**
  Lagrange function 448–451, **449**, **452–454**
Lagrange multipliers 203, 205, 205, 258, 269, 487
layer and neurons 112–113, 130, 141, 147
loads
  classification 190
  combination **190**, 252, **252**
longitudinal cracks 35
longitudinal load 190

MATLAB 45, 258, 269
mean absolute error (MAE) 51
mean absolute percentage error 51
mean squared error (MSE) 318
modulus of elasticity 7, 8, 187
multi-objective optimization (MOO)
  bigdata generation 253–257
  Hong–Lagrange algorithm 391, **392**, 393–396, **394**, *396*
  input parameters **247**, 250–253
  Lagrange algorithm 516
  Lagrange optimizations 477, 488–490
  motivations of 246
  multiple biaxial loads 477
  objectives and innovations 478
  output parameters **247**, 253
  PT beam

multi-objective optimization (*cont.*)
influence of objective functions on concrete sections 267–269
negligible errors 261
objective functions derivation based on forward ANNs 257–258
optimization 258
optimized objective functions 262–266
Pareto frontiers 261–262
practical applications 269
tradeoff ratios, generation of 260–261
UFO, derivation of 258–259
UFO, formulation of 259–26
verification of optimized design parameters 269
research significance 246–248
for steel-reinforced concrete beam 381–391
multiple biaxial loads
biaxial load pairs 496–500
design parameters 481–483, **482**
generalized ANN-nLP from ANN-1LP 493–496, *493*, **494–495**
generation of large datasets 483, *483*, *484*, **485**
training ANNs 484, 485, *486*

neighborhood component analysis (NCA) 136, 139
neural dynamic-based iteration 183
Newton–Raphson approximation 203
Newton–Raphson iteration 207, 438, 439, 509–510, *511*
non-linear relationships 200
normalized stress 193

output parameters *486*, 486–487, **487**

parabola–rectangle stress–strain model 8
parallel training method (PTM) 45, 70–86, **71–84**, 352, **353**, 354, *354*
Pareto frontiers 245, 246, 249, 478
by big datasets 402
CO_2 emissions 402–403
cost index 402
three-dimensional 396–397, *398–401*
for three objective functions 402
tradeoff ratios 261–262
two and three-dimensional 262–265
verification of 266, 406, **404–405**, 511–513, *512*
weight fractions 262, *262*
weights 403
posttensioning
bonded *4*
concrete 1, 4–5
unbonded *4*
prestressing 1–2
assessment 27–28, *28*
long-term 27, *27*
principle *2*, 2–3
short-term 25, *27*, 28, *29*
pretensioned concrete beams
AI-based design charts minimizing cost beam cost 207–208, *208*
cracking moment 217
crack widths 208–209, *209*
deflections *213*, 216
large datasets 218–219
mechanics-based design 218–219
moment demand *214*, 217
moment resistance 217
neutral axis depth *214*, 217–218
reinforcement strains *214*, 217–218
shear stirrup ratios *215*, 218
tendon and concrete stresses *210*, 210–216, *211*
ANN-based design charts minimizing beam depths
concrete stresses 230–232, *231*
cost and beam weight 235–238, *237*
cracking moment 233–234, *234*
crack widths 232–233, *233*
deflections 228–230, *230*
moment demand 233–234, *234*
moment resistance 233–234, *234*
neutral axis depth 234–235, *235*
reinforcement strains 234–235, *235*
shear stirrup ratios 235, *236*
ANN-based design charts minimizing weight
beam weights 227–228
concrete and tendon stresses *221*, 221–222
cracking moment 224, *224*

crack widths 222, *223*
deflections 219, *220*
moment demand 224, *224*
moment resistance 224, *224*
neutral axis depth 224–225, *225*
reinforcement strains 224–225, *225*
shear stirrup ratios 225–226, *226*
back-substitution
chained training scheme with revised sequence 136–159
parallel training method 86–136
design scenarios 60–62
design using ANN-based Lagrange algorithm
input parameters **186–187,** 186–191
large dataset generation 194–198
objective functions derivation based on forward ANNS 198–207
output parameters **186–187,** 191–194
input parameters **48,** 48–49
multi-objective optimizations
influence of objective functions on concrete sections 267–269
negligible errors 261
objective functions derivation based on forward ANNs 257–258
optimization 258
optimized objective functions 262–266
Pareto frontiers 261–262
practical applications 269
PTM method 257, **257**
tradeoff ratios, generation of 260–261
UFO 258–260
verification of optimized design parameters 269
network training 49–60
output parameters 49–51, **50**
parallel training method 70–86, **71–84**
reverse design
based on CRS using DNN 136–159
based on CRS using SNN 159–169
based on PTM using DNN 86–136
structural datasets, large 51–60, **52–54,** *55–60*
training entire data 62, **63–69**
training methods 62–86
pretensioning *3,* 3–4
PTM *see* parallel training method (PTM)

rebars 12, 34
elasticity of modulus 187–188
stress control 37
stress–strain relationship 187, *189*
regression model 51
reinforced concrete (RC) beams 248
reinforced concrete (RC) columns 411–413, **412, 413**
back-substitution method 420–421, **422, 423,** 424–425, *426,* 427
CO_2 emission 427, **428,** *429,* 430
cost index 427, **428,** *429,* 430
large structural datasets 417–420, **418,** *418–420*
network training *421,* **423**, 424
weights 430–431, *431*
reinforcement
slab 32
transverse 36
relaxation, tendon 33, **33**
reverse artificial neural network *420*
reverse design, PT concrete beams
based on CRS using DNN 136–159
based on CRS using SNN 159–169
based on PTM using DNN 86–136
reverse network 89, 103, 114
root mean squared error 51

safety factor (SF) 280, 326–327
secondary moments *see* hyperstatic moments
secondary prestressing effects 16–25
7-wire strand 11, *11*
Shallow artificial Neural Networks (SNNs) 47
shallow neural network (SNN) 318
shear strength 198
shrinkage 10
shrinkage strain, of concrete 32
single-degree-of-freedom (SDOF) system 277

steel-encased reinforced concrete (SRC) columns 249, 437, 477
analytical and experimental studies 413–414, *415*
ANN-based design 348–349, *349*, *350*, 383, *383*, *384*
biaxial load pairs
design codes 500, **501–502**, 502–503, **503**
design parameters 500
formulation of forward network 503–504
Lagrange MOO algorithm 516
Lagrange optimization 505–516
normalized UFO capturing 504–505
structural analysis 500
CO_2 emissions 362, *363*, 364, **365–366**, 367, **371–374**, **462**, 462–465, *463*, **464**, *464*, *465*
column weight **466**, 466–467, *467*, *468*, **468**, 469, *469*
computational algorithm *416*
cost index 361, *362*, **371–374**, 457–461, **458**, **459**, **460**, *461*
design 440, *441*, **442–443**
equality and inequality 360–361, *361*
forward design 382
H-shaped steel 490–492, *492*
Lagrange multiplier method
active and inactive conditions 456–457
ANN-based objective functions 451, 456
CO_2 emissions, and column weight 451, **455**
cost index 451, **455**
Lagrange function 448–451, **449**, **452–454**
Lagrange optimizations 422–443
large datasets 350, **351**, 352, *352*, 384, **385**, 386, *386*, **387–390**, *391*
large structural datasets 444, *445*, **446**, *447*
multiple biaxial loads
biaxial load pairs 496–500
design parameters 481–483, **482**
generalized ANN-nLP from ANN-1LP 493–496, *493*, **494–495**
generation of large datasets 483, *483*, *484*, **485**
training ANNs 484, 485, *486*
objective functions 354, **355**, 356, **357–360**
parallel training method 352, **353**, 354, *354*
PTM 444, 447, *448*
weights 367, **368–369**, 370, **371–374**, 375
steel-reinforced concrete (SRC) beams 274
ANN-based mapping input to output parameters 289
big data generator 284
cost index 283
design charts, development of
application 331–338
cost index, CO_2 emissions and beam weight emissions *vs.* tensile rebar strains 330–331, *331*
design moment 326
effective depths and tensile rebar ratios *vs.* tensile rebar strains 327–328, *328*
immediate and long-term deflections *vs.* tensile rebar strains 329–330, *330*
reverse scenario 325–326
safety factor 326–327
tensile steel strains and curvature ductilities *vs.* tensile rebar strains 328
direct and back-substitution method 292–293
input parameters 284–286, **287**, 289
material and manufacturing prices 283, **283**
means 289
output parameters **288**, 289
reverse design
ANN formulation 293–318, 319, 321, 323–324
back-substitution method 318, 320–322, 324
verifications 319–325
standard deviations 289
structural datasets 286
variances 289

stress control 35–37
stress–strain relationship 8–9, *9*
  bi-linear 11, *12*
  prestressing strand 11, *12*
  of rebars 187, *189*

TCVN 5574-2018 7
TED *see* training entire data (TED)
tendon 10–12
  bonded system 5
  jacking length of 6, **6**
  relaxation 33, **33**
  stress control 37
  unbonded system 6
tendon ratio 87, 95, 110
tensile rebar ratios 284, 327–328, *328*
tensile rebar strains 280, 284, 323, 324, 326, 327–328, 328, *327*, 330–331, *331*
tensile steel strains 328
tensile stress 1, 3, 22, 35
tension-stiffening effect 37
three-dimensional Pareto frontier 396–397, *398–401*
time-dependent loss 32
  evaluation 33
tradeoff ratio 260–261, 513, *514*, **515**, *516*
training entire data (TED) 45, 62, **63–69**
trial-and-error method 34

UFO *see* unified objective function (UFO)
ultimate limit state (ULL) 62, 95
unbonded tendons 6
unified objective function (UFO) 246, 248
  derivation 258–259, 395–396, *396*
  formulation 259–260
  gradient vectors 508, *509*
  optimization 505, *506*, 507–508
  for PT beam 258–259
  three objective functions 504–505

vertical load 190

weighted sum method 258
weight fractions 262, *262*
wobble friction 28

Young's modulus 8